CAMBRIDGE LIBRARY COLLECTION

Books of enduring scholarly value

Technology

The focus of this series is engineering, broadly construed. It covers technological innovation from a range of periods and cultures, but centres on the technological achievements of the industrial era in the West, particularly in the nineteenth century, as understood by their contemporaries. Infrastructure is one major focus, covering the building of railways and canals, bridges and tunnels, land drainage, the laying of submarine cables, and the construction of docks and lighthouses. Other key topics include developments in industrial and manufacturing fields such as mining technology, the production of iron and steel, the use of steam power, and chemical processes such as photography and textile dyes.

Electrical Papers

A self-taught authority on electromagnetic theory, telegraphy and telephony, Oliver Heaviside (1850–1925) dedicated his adult life to the improvement of electrical technologies. Inspired by James Clerk Maxwell's field theory, he spent the 1880s presenting his ideas as a regular contributor to the weekly journal, *The Electrician*. The publication of *Electrical Papers*, a year after his election to the Royal Society in 1891, established his fame beyond the scientific community. An eccentric figure with an impish sense of humour, Heaviside's accessible style enabled him to educate an entire generation in the importance and application of electricity. In so doing he helped to establish that very British phenomenon, the garden-shed inventor. Combining articles on the electromagnetic wave surface and electromagnetic induction with notes on nomenclature and the self-induction of wires, Volume 2 serves as an excellent source for both electrical engineers and historians of science.

Electrical Papers

VOLUME 2

OLIVER HEAVISIDE

CAMBRIDGE
UNIVERSITY PRESS

CAMBRIDGE UNIVERSITY PRESS

Cambridge, New York, Melbourne, Madrid, Cape Town,
Singapore, São Paolo, Delhi, Tokyo, Mexico City

Published in the United States of America by Cambridge University Press, New York

www.cambridge.org
Information on this title: www.cambridge.org/9781108028578

This edition first published 1892
This digitally printed version 2011

ISBN 978-1-108-02857-8 Paperback

ELECTRICAL PAPERS.

VOL. II.

ELECTRICAL PAPERS

BY

OLIVER HEAVISIDE

IN TWO VOLUMES

VOL. II.

London

MACMILLAN AND CO.

AND NEW YORK

1892

CONTENTS OF VOL. II.

ELECTRICAL PAPERS.

XXXI. ON THE ELECTROMAGNETIC WAVE-SURFACE.

[*Phil. Mag.*, June, 1885, p. 397, S. 5, vol. 19.]

MAXWELL showed (*Electricity and Magnetism*, vol. ii., art. 794) that his equations of electromagnetic disturbances, on the assumption that the electric capacity varies in different directions in a crystal, lead to the Fresnel form of wave-surface. There is no obscurity arising from the ignored wave of normal disturbance, because the very existence of a plane wave requires that there be none. In fact, the electric displacement and the magnetic induction are both in the wave-front, and are perpendicular to one another. The magnetic force and induction are parallel, on account of the constant permeability; whilst the electric force, though not parallel to the displacement, is yet perpendicular to the magnetic induction (and force); the normal to the wave-front, the electric force, and the displacement being in one plane. The ray is also in this plane, perpendicular to the electric force. There are of course two rays for (in general) every direction of wave-normal, each with separate electromagnetic variables to which the above remarks apply.

It is easily proved, and it may be legitimately inferred without a formal demonstration, from a consideration of the equations of induction, that if we consider the dielectric to be isotropic as regards capacity, but eolotropic as regards permeability, the same general results will follow, if we translate capacity to permeability, electric to magnetic force, and electric displacement to magnetic induction. The three principal velocities will be $(c\mu_1)^{-\frac{1}{2}}$, $(c\mu_2)^{-\frac{1}{2}}$, and $(c\mu_3)^{-\frac{1}{2}}$, if c is the constant value of the capacity, and μ_1, μ_2, μ_3 are the three principal permeabilities. The wave-surface will be of the same character, only differing in the constants.

But a dielectric may be eolotropic both as regards capacity and permeability. The electric displacement is then a linear function of the electric force, and the magnetic induction another linear function of the magnetic force. The principal axes of capacity, or lines of parallelism of electric force and displacement, cannot, in the general case, be assumed to have any necessary relation to the principal axes of permeability, or lines of parallelism of magnetic force and induction.

Disconnecting the matter altogether from the hypothesis that light consists of electromagnetic vibrations, we shall inquire into the conditions of propagation of plane electromagnetic waves in a dielectric which is eolotropic as regards both capacity and permeability, and determine the equation to the wave-surface.

For any direction of the normal (to the wave-front, understood) there are in general two normal velocities, *i.e.*, there are two rays differently inclined to the normal whose ray-velocities and normal wave-velocities are different. And for any direction of ray there are in general two ray-velocities, *i.e.*, two parallel rays having different velocities and wave-fronts.

In any wave (plane) the electric displacement and the magnetic induction must be always in the wave-front, *i.e.*, perpendicular to the normal. But they are only exceptionally perpendicular to one another.

In any ray the electric force and the magnetic force are both perpendicular to the direction of the ray. But they are only exceptionally perpendicular to one another.

The magnetic force is always perpendicular to the electric displacement, and the electric force perpendicular to the magnetic induction. This of course applies to either wave. If we have to rotate the plane through the normal and the magnetic force through an angle θ to bring it to coincide with the magnetic induction, we must rotate the plane through the normal and the electric displacement through the same angle θ in the same direction to bring it to coincide with the electric force, the axis of rotation being the normal itself.

In the two waves having a common wave-normal, the displacement of either is parallel to the induction of the other. And in the two rays having a common direction, the magnetic force of either is parallel to the electric force of the other.

Nearly all our equations are symmetrical with respect to capacity and permeability; so that for every equation containing some electric variables there is a corresponding one to be got by exchanging electric force and magnetic force, etc. And when the forces, inductions, etc., are eliminated, leaving only capacities and permeabilities, these may be exchanged in any formula without altering its meaning, although its immediate Cartesian expansion after the exchange may be entirely different, and only convertible to the former expression by long processes.

If either μ or c be constant, we have the Fresnel wave-surface. Perhaps the most important case besides these is that in which the principal axes of permeability are parallel to those of capacity. There are then six principal velocities instead of only three, for the velocity of a wave depends upon the capacity in the direction of displacement as well as upon the permeability in the direction of induction. For instance, if μ_1, μ_2, μ_3, and c_1, c_2, c_3, are the principal permeabilities and capacities, and the wave-normal be parallel to the common axis of μ_1 and c_1, the other principal axes are the directions of induction and displacement, and the two normal velocities are $(c_2\mu_3)^{-\frac{1}{2}}$ and $(c_3\mu_2)^{-\frac{1}{2}}$.

The principal sections of the wave-surface in this case are all ellipses (instead of ellipses and circles, as in the one-sided Fresnel-wave); and two of these ellipses always cross, giving two axes of single-ray velocity. But should the ratio of the capacity to the permeability be the same for all the axes ($\mu_1/c_1 = \mu_2/c_2 = \mu_3/c_3$), the wave-surface reduces to a single ellipsoid, and any line is an optic axis. There is but one velocity, and no particular polarization. If the ratio is the same for two of the axes, the third is an optic axis.

Owing to the extraordinary complexity of the investigation when written out in Cartesian form (which I began doing, but gave up aghast), some abbreviated method of expression becomes desirable. I may also add, nearly indispensable, owing to the great difficulty in making out the meaning and mutual connections of very complex formulæ. In fact the transition from the velocity-equation to the wave-surface by proper elimination would, I think, baffle any ordinary algebraist, unassisted by some higher method, or at any rate by some kind of shorthand algebra. I therefore adopt, with some simplification, the method of vectors, which seems indeed the only proper method. But some of the principal results will be fully expanded in Cartesian form, which is easily done. And since all our equations will be either wholly scalar or wholly vector, the investigation is made independent of quaternions by simply defining a scalar product to be so and so, and a vector product so and so. The investigation is thus a Cartesian one modified by certain simple abbreviated modes of expression.

I have long been of opinion that the sooner the much needed introduction of quaternion methods into practical mathematical investigations in Physics takes place the better. In fact every analyst to a certain extent adopts them: first, by writing only one of the three Cartesian scalar equations corresponding to the single vector equation, leaving the others to be inferred; and next, by writing the first only of the three products which occur in the scalar product of two vectors. This, systematized, is I think the proper and natural way in which quaternion methods should be gradually brought in. If to this we further add the use of the vector product of two vectors, immensely increased power is given, and we have just what is wanted in the tridimensional analytical investigations of electromagnetism, with its numerous vector magnitudes.

It is a matter of great practical importance that the notation should be such as to harmonize with Cartesian formulæ, so that we can pass from one to the other readily, as is often required in mixed investigations, without changing notation. This condition does not appear to me to be attained by Professor Tait's notation, with its numerous letter prefixes, and especially by the $-S$ before every scalar product, the negative sign being the cause of the greatest inconvenience in transitions. I further think that Quaternions, as applied to Physics, should be established more by definition than at present; that scalar and vector products should be defined to mean such or such combinations, thus avoiding some extremely obscure and quasi-metaphysical reasoning, which is quite unnecessary.

The first three sections of the following preliminary contain all we want as regards definitions; most of the rest of the preliminary consists of developments and reference-formulæ, which, were they given later, in the electromagnetic problem, would inconveniently interrupt the argument, and much lengthen the work.

Scalars and Vectors.—In a scalar equation every term is a scalar, or algebraic quantity, a mere magnitude; and + and − have the ordinary signification. But in a vector equation every term stands for a vector, or directed magnitude, and + and − are to be understood as compounding like velocities, forces, etc. Putting all vectors upon one side, we have the general form

$$\mathbf{A}+\mathbf{B}+\mathbf{C}+\mathbf{D}+\ldots=0;$$

where **A**, **B**, ..., are any vectors, which, if n in number, may be represented, since their sum is zero, by the n sides of a polygon. Let A_1, A_2, A_3 be the three ordinary scalar components of **A** referred to any set of three rectangular axes, and similarly for the other vectors. This notation saves multiplication of letters. Then the above equation stands for the three scalar equations

$$\left.\begin{aligned} A_1+B_1+C_1+D_1+\ldots&=0,\\ A_2+B_2+C_2+D_2+\ldots&=0,\\ A_3+B_3+C_3+D_3+\ldots&=0.\end{aligned}\right\}$$

The − sign before a vector simply reverses its direction—that is, negatives its three components.

According to the above, if **i**, **j**, **k**, be rectangular vectors of unit length, we have

$$\mathbf{A}=\mathbf{i}A_1+\mathbf{j}A_2+\mathbf{k}A_3, \quad \ldots\ldots\ldots\ldots\ldots\ldots\ldots(1)$$

etc.; if A_1, A_2, A_3 be the components of **A** referred to the axes of **i**, **j**, **k**. That is, **A** is the sum of the three vectors $\mathbf{i}A_1$, $\mathbf{j}A_2$, $\mathbf{k}A_3$, of lengths A_1, A_2, A_3, parallel to **i**, **j**, **k** respectively.

Scalars Product.—We define **AB** thus,

$$\mathbf{AB}=A_1B_1+A_2B_2+A_3B_3, \quad \ldots\ldots\ldots\ldots\ldots\ldots(2)$$

and call it the scalar product of the vectors **A** and **B**. Its magnitude is that of **A** × that of **B** × the cosine of the angle between them. Thus, by (1) and (2),

$$A_1=\mathbf{Ai}, \qquad A_2=\mathbf{Aj}, \qquad A_3=\mathbf{Ak};$$

and in general, **N** being any unit vector, **AN** is the scalar component of **A** parallel to **N**, or, briefly, the **N** component of **A**. Similarly,

$$\mathbf{i}^2=1, \qquad \mathbf{j}^2=1, \qquad \mathbf{k}^2=1,$$

because **i** and **i** are parallel and of length unity, etc. And

$$\mathbf{ij}=0, \qquad \mathbf{jk}=0, \qquad \mathbf{ki}=0,$$

because **i** and **j**, for instance, are perpendicular. Notice that **AB** = **BA**.

We have also

$$\mathbf{A}=\frac{\mathbf{A}^2}{\mathbf{A}}=\frac{\mathbf{A}^3}{\mathbf{A}^2}=\text{etc.},$$

and

$$\frac{1}{\mathbf{A}} \text{ or } \mathbf{A}^{-1}=\frac{\mathbf{A}}{\mathbf{A}^2}=\frac{\mathbf{A}^2}{\mathbf{A}^3}=\text{etc.}$$

Thus $\mathbf{A}^{-1}$ has the same direction as $\mathbf{A}$; its length is the reciprocal of that of $\mathbf{A}$.

Vector Product.—We define $\mathbf{VAB}$ thus,

$$\mathbf{VAB} = \mathbf{i}(A_2B_3 - A_3B_2) + \mathbf{j}(A_3B_1 - A_1B_3) + \mathbf{k}(A_1B_2 - A_2B_1), \quad \text{.....(3)}$$

and call $\mathbf{VAB}$ the vector product of $\mathbf{A}$ and $\mathbf{B}$. Its magnitude is that of $\mathbf{A}$ × that of $\mathbf{B}$ × the sine of the angle between them. Its direction is perpendicular to $\mathbf{A}$ and to $\mathbf{B}$ with the usual conventional relation between positive directions of translation and of rotation (the vine system). Thus,

$$\mathbf{Vij} = \mathbf{k}, \qquad \mathbf{Vjk} = \mathbf{i}, \qquad \mathbf{Vki} = \mathbf{j}.$$

Notice that $\mathbf{VAB} = -\mathbf{VBA}$, the direction being reversed by reversing the order of the letters; for, by exchanging $\mathbf{A}$ and $\mathbf{B}$ in (3), we negative each term.

Hamilton's ∇.—The operator

$$\nabla = \mathbf{i}\frac{d}{dx} + \mathbf{j}\frac{d}{dy} + \mathbf{k}\frac{d}{dz} \quad \text{.............................(4)}$$

may, since the differentiations are scalar, be treated as a vector, of course with either a scalar or a vector to follow it. If it operate on a scalar P we have the vector

$$\nabla P = \mathbf{i}\frac{dP}{dx} + \mathbf{j}\frac{dP}{dy} + \mathbf{k}\frac{dP}{dz}, \quad \text{.....................(5)}$$

whose three components are dP/dx, etc. If it operate on a vector $\mathbf{A}$, we have, by (2), the scalar product

$$\nabla\mathbf{A} = \frac{dA_1}{dx} + \frac{dA_2}{dy} + \frac{dA_3}{dz}, \quad \text{........................(6)}$$

and, by (3), the vector product

$$\mathbf{V}\nabla\mathbf{A} = \mathbf{i}\left(\frac{dA_3}{dy} - \frac{dA_2}{dz}\right) + \mathbf{j}\left(\frac{dA_1}{dz} - \frac{dA_3}{dx}\right) + \mathbf{k}\left(\frac{dA_2}{dx} - \frac{dA_1}{dy}\right). \quad \text{........(7)}$$

The scalar product $\nabla\mathbf{A}$ is the divergence of the vector $\mathbf{A}$, the amount leaving the unit volume, if it be a flux. The vector product (7) is the curl of $\mathbf{A}$, which will occur below. There are three remarkable theorems relating to ∇, viz.,

$$P_2 - P_1 = \int_1^2 \nabla P d\mathbf{s}, \quad \text{............................(8)}$$

$$\int \mathbf{A}d\mathbf{s} = \iint \mathbf{B}d\mathbf{S}, \quad \text{............................(9)}$$

$$\iint \mathbf{C}d\mathbf{S} = \iiint \nabla\mathbf{C}dv. \quad \text{........................(10)}$$

Starting with P, a single-valued scalar function of position, the rise in its value from any point to another is expressed in (8) as the line-integral, along any line joining the points, of $\nabla Pd\mathbf{s}$, the scalar product of ∇P and $d\mathbf{s}$, the vector element of the curve.

Then passing from an unclosed to a closed curve, let $\mathbf{A}$ be any vector function of position (single-valued, of course). Its line-integral round the closed curve is expressed in (9) as the surface-integral over any surface bounded by the curve of another vector **B**, which $=\mathrm{V}\nabla\mathbf{A}$. $\mathbf{B}d\mathbf{S}$ is the scalar product of **B** and the vector element of surface $d\mathbf{S}$, whose direction is defined by its unit normal.

Finally, passing from an unclosed to a closed surface, (10) expresses the surface-integral of any vector **C** over the closed surface (normal positive outward), as the volume-integral of its divergence within the included space.

Linear Vector Operators.—If **H** be the magnetic force at a point, **B** the induction, **E** the electric force, and **D** the displacement, all vectors, then

$$\mathbf{B}=\mu\mathbf{H}, \qquad \text{and} \qquad \mathbf{D}=c\mathbf{E}/4\pi \quad \text{.................... (11)}$$

express the relation of **B** to **H** and of **D** to **E** in a dielectric medium. If it be isotropic as regards displacement, c is the electric capacity; and if it be isotropic as regards induction, μ is the magnetic permeability; c and μ are then constants, if the medium be homogeneous, or scalar functions of position if it be heterogeneous.

We shall not alter the form of the above equations in the case of eolotropy, when c and μ become linear operators. For instance, the induction will always be $\mu\mathbf{H}$, to be understood as a definite vector, got from **H** another vector, in a manner fully defined by (in case we want the developments) the following equations (not otherwise needed). Let H_1, ..., and B_1, ..., be the components of **H** and **B** referred to any rectangular axes. Then

$$\left.\begin{aligned} B_1&=\mu_{11}H_1+\mu_{12}H_2+\mu_{13}H_3,\\ B_2&=\mu_{21}H_1+\mu_{22}H_2+\mu_{23}H_3,\\ B_3&=\mu_{31}H_1+\mu_{32}H_2+\mu_{33}H_3,\end{aligned}\right\} \quad \text{........................ (12)}$$

where μ_{11}, etc., are constants, which may have any values not making **HB** negative; with the identities $\mu_{12}=\mu_{21}$, etc. Or,

$$B_1=\mu_1H_1, \qquad B_2=\mu_2H_2, \qquad B_3=\mu_3H_3, \quad \text{........... (13)}$$

when the components are those referred to the principal axes of permeability, μ_1, μ_2, μ_3 being the principal permeabilities, all positive.

Inverse Operators.—Since $\mathbf{B}=\mu\mathbf{H}$, we have $\mathbf{H}=\mu^{-1}\mathbf{B}$, where μ^{-1} is the operator inverse to μ. When referred to the principal axes, we have

$$\mu_1'=\frac{1}{\mu_1}, \qquad \mu_2'=\frac{1}{\mu_2}, \qquad \mu_3'=\frac{1}{\mu_3}. \quad \text{................ (14)}$$

But when referred to any rectangular axes, we have

$$\mu_{11}'=\frac{\mu_{11}\mu_{22}-\mu_{12}^2}{\mu_1\mu_2\mu_3}, \qquad \mu_{12}'=\frac{\mu_{13}\mu_{23}-\mu_{12}\mu_{33}}{\mu_1\mu_2\mu_3}, \quad \text{etc.,} \quad \text{......... (15)}$$

by solution of (12). The accents belong to the inverse coefficients. The rest may be written down symmetrically, by cyclical changes of the figures. In the index-surface the operators are inverse to those in the wave-surface.

Conjugate Property.—The following property will occur frequently. **A** and **B** being any vectors,

$$\mathbf{A}\mu\mathbf{B}=\mathbf{B}\mu\mathbf{A}, \quad \text{.............................. (16)}$$

or the scalar product of A and μB equals that of B and μA. It only requires writing out the full scalar products to see its truth, which results from the identities $\mu_{12}=\mu_{21}$, etc. Similarly,

$$\mathbf{A}\mu c\mathbf{B}=\mu\mathbf{A}c\mathbf{B}=c\mu\mathbf{AB},\quad \text{etc.},$$

$$\mathbf{AB}=\mathbf{A}\mu\mu^{-1}\mathbf{B}=\mu\mathbf{A}\mu^{-1}\mathbf{B},\quad \text{etc.},$$

where in the first line c is another self-conjugate operator.

D is expressed in terms of E similarly to (12) by coefficients c_{11}, c_{12}, etc.; or, as in (13), by the principal capacities c_1, c_2, c_3.

Theorem.—The following important theorem will be required. A and B being any vectors,

$$\mu_1\mu_2\mu_3\mathbf{VAB}=\mu\mathbf{V}\mu\mathbf{A}\mu\mathbf{B}. \qquad \text{.........................(17)}$$

For completeness a proof is now inserted, adapted from that given by Tait. Since VAB is perpendicular to A and B, by definition of a vector product, therefore

$$\mathbf{AVAB}=0, \qquad \text{and} \qquad \mathbf{BVAB}=0,$$

by definition of a scalar product. Therefore

$$\mathbf{A}\mu\mu^{-1}\mathbf{VAB}=0, \qquad \text{and} \qquad \mathbf{B}\mu\mu^{-1}\mathbf{VAB}=0,$$

by introducing $\mu\mu^{-1}=1$. Hence

$$\mu\mathbf{A}\mu^{-1}\mathbf{VAB}=0, \qquad \text{and} \qquad \mu\mathbf{B}\mu^{-1}\mathbf{VAB}=0,$$

by the conjugate property; that is, μ^{-1}VAB is perpendicular to μA and to μB. Or

$$h\mu^{-1}\mathbf{VAB}=\mathbf{V}\mu\mathbf{A}\mu\mathbf{B},$$

where h is a scalar. Or

$$h\mathbf{VAB}=\mu\mathbf{V}\mu\mathbf{A}\mu\mathbf{B},$$

by operating by μ. To find h, multiply by any third vector C (not to be in the same plane as A and B), giving

$$h\mathbf{CVAB}=\mathbf{C}\mu\mathbf{V}\mu\mathbf{A}\mu\mathbf{B};$$

therefore

$$h=\frac{\mu\mathbf{CV}\mu\mathbf{A}\mu\mathbf{B}}{\mathbf{CVAB}},$$

by the conjugate property. Now expand this quotient of two scalar products, and it will be found to be independent of what vectors A, B, C may be. Choose them then to be i, j, k, three unit vectors parallel to the principal axes of μ. Then

$$h=\frac{\mu_3\mathbf{kV}\mu_1\mathbf{i}\mu_2\mathbf{j}}{\mathbf{kVij}}=\mu_1\mu_2\mu_3,$$

by the i, j, k properties before mentioned. This proves (17).

Transformation-Formula.—The following is very useful. A, B, C being any vectors,

$$\mathbf{VAVBC}=\mathbf{B}(\mathbf{CA})-\mathbf{C}(\mathbf{AB}). \qquad \text{......................(18)}$$

Here CA and AB are scalar products, merely set in brackets to separate distinctly from the vectors B and C they multiply. This formula is evident on expansion.

The Equations of Induction.—**E** and **H** being the electric and magnetic forces at a point in a dielectric, the two equations of induction are [vol. I., p. 449, equations (22), (23)]

$$\text{curl}\,\mathbf{H} = c\dot{\mathbf{E}}, \quad \text{.............................(19)}$$

$$-\text{curl}\,\mathbf{E} = \mu\dot{\mathbf{H}}\,; \quad \text{.............................(20)}$$

c and μ being the capacity and permeability operators, and curl standing for $V\nabla$ as defined in equation (7). Let Γ and **G** be the electric and the magnetic current, then

$$\Gamma = c\dot{\mathbf{E}}/4\pi, \qquad \mathbf{G} = \mu\dot{\mathbf{H}}/4\pi. \quad \text{.....................(21)}$$

The dot, as usual, signifies differentiation to the time. The electric energy is $\mathbf{E}c\mathbf{E}/8\pi$ per unit volume, and the magnetic energy $\mathbf{H}\mu\mathbf{H}/8\pi$ per unit volume. If **A** is Maxwell's vector potential of the electric current, we have also

$$\text{curl}\,\mathbf{A} = \mu\mathbf{H}, \qquad \mathbf{E} = -\dot{\mathbf{A}}. \quad \text{....................(21}a\text{)}$$

Similarly, we may make a vector **Z** the vector potential of the magnetic current, such that [vol. I., p. 467]

$$-\text{curl}\,\mathbf{Z} = c\mathbf{E}, \qquad \mathbf{H} = -\dot{\mathbf{Z}}. \quad \text{....................(22)}$$

The complete magnetic energy of any current system may, by a well-known transformation, be expressed in the two ways

$$T = \Sigma\,\mathbf{H}\mu\mathbf{H}/8\pi = \Sigma\,\tfrac{1}{2}\mathbf{A}\Gamma,$$

the Σ indicating summation through all space. Similarly, the electric energy, if there be no electrification, may be written in the two ways

$$U = \Sigma\,\mathbf{E}c\mathbf{E}/8\pi = \Sigma\,\tfrac{1}{2}\mathbf{Z}\mathbf{G}.$$

If there be electrification, we have also another term to add, the real electro*static* energy, in terms of the scalar potential and electrification. And if there be impressed electric force in the dielectric, part of **G** will be imaginary magnetic current, analogous to the imaginary electric current which may replace a system of intrinsic magnetization.

Plane Wave.—Let there be a plane wave in the medium. Its direction is defined by its normal. Let then **N** be the vector normal of unit length, and z be distance measured along the normal. If v be the velocity of the wave-front, the rate the disturbance travels along the normal, or the component parallel to the normal of the actual velocity of propagation of the disturbance, we have

$$\mathbf{H} = f(z - vt),$$

if the wave be a positive one, as we shall suppose, giving

$$-v\frac{d}{dz} = \frac{d}{dt}, \quad \text{..............................(23)}$$

applied to **H** or **E**.

Next, examine what the operator $V\nabla$ or curl becomes when, as at present, the disturbance is assumed not to change direction, but only magnitude, as we pass along the normal. Apply the theorem of Version (9) to the elementary rectangular area bounded by two sides parallel to **E** of length a, and two sides of length b perpendicular to **E** and in the same plane as **E** and the normal **N**. Since its area is ab, and $b = dz \sin\theta$,

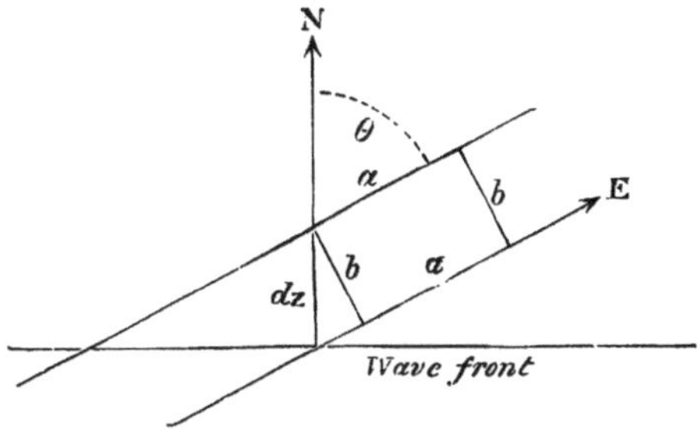

and the two sides b contribute nothing to the line-integral, we find that

$$\text{curl} = \mathrm{V}\mathbf{N}\frac{d}{dz}, \qquad \text{.............................(24)}$$

applied to **E** or **H** or other vectors, in the case of a plane wave. Using this, and (23), in the equations of induction (19), (20), they become

$$\mathrm{V}\mathbf{N}\frac{d\mathbf{H}}{dz} = -vc\frac{d\mathbf{E}}{dz},$$

$$\mathrm{V}\mathbf{N}\frac{d\mathbf{E}}{dz} = v\mu\frac{d\mathbf{H}}{dz}.$$

Here, since the z-differentiation is scalar, and occurs on both sides, it may be dropped, giving us

$$\mathrm{V}\mathbf{N}\mathbf{H} = -vc\mathbf{E}, \qquad \text{.............................(25)}$$

$$\mathrm{V}\mathbf{N}\mathbf{E} = v\mu\mathbf{H}. \qquad \text{.............................(26)}$$

The induction and the displacement are therefore necessarily in the wave-front, by the definition of a vector product, being perpendicular to **N**. Also the displacement is perpendicular to the magnetic force, and the induction is perpendicular to the electric force.

Index-Surface.—Let *

$$\mathbf{s} = \frac{\mathbf{N}}{v} \qquad \text{.................................(27)}$$

be a vector parallel to the normal, whose length is the reciprocal of the normal velocity v. It is the vector of the index-surface. By (25) and (26) we have

$$c\mathbf{E} = -\mathrm{V}\mathbf{s}\mathbf{H}, \quad \text{therefore} \quad -\mathbf{E} = c^{-1}\mathrm{V}\mathbf{s}\mathbf{H}; \qquad \text{.........(28)}$$

and $$\mu\mathbf{H} = \mathrm{V}\mathbf{s}\mathbf{E}, \quad \text{therefore} \quad \mathbf{H} = \mu^{-1}\mathrm{V}\mathbf{s}\mathbf{E}. \qquad \text{.........(29)}$$

Now use the theorem (17). Then, if

$$m = \mu_1\mu_2\mu_3, \qquad n = c_1c_2c_3 \qquad \text{....................(30)}$$

* [In order to secure the advantage of black letters for vectors, I have changed the notation thus :—The original σ is now **s** ; ρ is **r** ; β is **b** ; γ is **g** ; and α is **a**.]

be the products of the principal permeabilities and capacities, the theorem gives, applied to (28) and (29),

$$-n\mathbf{E} = \mathrm{V}c\mathbf{s}c\mathbf{H}, \qquad (31)$$
$$m\mathbf{H} = \mathrm{V}\mu\mathbf{s}\mu\mathbf{E}. \qquad (32)$$

Putting the value of H given by (32) in (28) first, and then the value of E given by (31) in (29), we have

$$-m\mathbf{E} = c^{-1}\mathrm{V}\mathbf{s}\mathrm{V}\mu\mathbf{s}\mu\mathbf{E}, \qquad (33)$$
$$-n\mathbf{H} = \mu^{-1}\mathrm{V}\mathbf{s}\mathrm{V}c\mathbf{s}c\mathbf{H}. \qquad (34)$$

To these apply the transformation-formula (18), giving

$$-mc\mathbf{E} = \mu\mathbf{s}(\mathbf{s}\mu\mathbf{E}) - \mu\mathbf{E}(\mathbf{s}\mu\mathbf{s}), \qquad (33a)$$

and

$$-n\mu\mathbf{H} = c\mathbf{s}(\mathbf{s}c\mathbf{H}) - c\mathbf{H}(\mathbf{s}c\mathbf{s}), \qquad (34a)$$

where the bracketed quantities are scalar products. Put in this form,

$$\{(\mathbf{s}\mu\mathbf{s})\mu - mc\}\mathbf{E} = \mu\mathbf{s}(\mathbf{s}\mu\mathbf{E}), \qquad (35)$$
$$\{(\mathbf{s}c\mathbf{s})c - n\mu\}\mathbf{H} = c\mathbf{s}(\mathbf{s}c\mathbf{H}), \qquad (36)$$

and perform on them the inverse operations to those contained in the { }'s, dividing also by the scalar products on the right sides. Then

$$\frac{\mathbf{E}}{\mathbf{s}\mu\mathbf{E}} = \frac{\mu\mathbf{s}}{(\mathbf{s}\mu\mathbf{s})\mu - mc}, \qquad (37)$$

$$\frac{\mathbf{H}}{\mathbf{s}c\mathbf{H}} = \frac{c\mathbf{s}}{(\mathbf{s}c\mathbf{s})c - n\mu}. \qquad (38)$$

Operate by c on (37) and by μ on (38), and transfer all operators to the denominators on the right. Then

$$\frac{c\mathbf{E}}{\mathbf{s}\mu\mathbf{E}} = \frac{\mathbf{s}}{(\mathbf{s}\mu\mathbf{s})c^{-1} - m\mu^{-1}} = \mathbf{b}_1 \quad \text{say}, \qquad (39)$$

$$\frac{\mu\mathbf{H}}{\mathbf{s}c\mathbf{H}} = \frac{\mathbf{s}}{(\mathbf{s}c\mathbf{s})\mu^{-1} - nc^{-1}} = \mathbf{b}_2 \quad \text{say}. \qquad (40)$$

(It should be noted that, in thus transferring operators, care should be taken to do it properly, otherwise it had better not be done at all. Thus, we have by (37),

$$\mathbf{b}_1 = c\frac{\mu\mathbf{s}}{(\mathbf{s}\mu\mathbf{s})\mu - mc}, \qquad \text{or} \qquad \mathbf{b}_1 = c\{(\mathbf{s}\mu\mathbf{s})\mu - mc\}^{-1}\mu\mathbf{s},$$

and the left c and the right μ are to go inside the { }. Operate by c^{-1} and then again by $\{\}^{+1}$, thus cancelling the $\{\}^{-1}$, giving

$$\mu\mathbf{s} = \{(\mathbf{s}\mu\mathbf{s})\mu - mc\}c^{-1}\mathbf{b}_1.$$

Here we can move c^{-1} inside, giving

$$\mu\mathbf{s} = \{(\mathbf{s}\mu\mathbf{s})\mu c^{-1} - m\}\mathbf{b}_1;$$

and now operating by μ^{-1}, it may be moved inside, giving

$$\mathbf{s} = \{(\mathbf{s}\mu\mathbf{s})c^{-1} - m\mu^{-1}\}\mathbf{b}_1,$$

as in (39).)

We can now, by (39) and (40), get as many forms of the index-equation as we please. We know that the displacement is perpendicular to the normal, and so is the induction. Hence

$$\mathbf{s}\mathfrak{b}_1 = 0, \qquad \mathbf{s}\mathfrak{b}_2 = 0, \quad \text{.......................(41)}$$

where $\mathfrak{b}_1$ and $\mathfrak{b}_2$ are the above vectors, in (39) and (40), are two equivalent equations of the index-surface.

Also, operate on (39) by $\mathbf{s}\mu c^{-1}$, and on (40) by $\mathbf{s}c\mu^{-1}$, and the left members become unity, by the conjugate property; hence

$$\mu\mathbf{s}c^{-1}\mathfrak{b}_1 = 1, \qquad c\mathbf{s}\mu^{-1}\mathfrak{b}_2 = 1 \quad \text{..................(42)}$$

are two other forms of the index-equation. (41) and (42) are the simplest forms. More complex forms are created with that surprising ease which is characteristic of these operators; but we do not want any more. When expanded, the different forms look very different, and no one would think they represented the same surface. This is also true of the corresponding Fresnel surface, which is comparatively simple in expression. In any equation we may exchange the operators μ and c.

Put $\mathbf{s} = \mathbf{N}v^{-1}$ in any form of index-equation, and we have the velocity-equation, a quadratic in v^2 giving the two velocities of the wave-front. And if we put $\mathbf{N}v = \mathbf{p}$, thus making $\mathbf{p}$ a vector parallel to the normal of length equal to the velocity, it will be the vector of the surface which is the locus of the foot of the perpendicular from the origin upon the tangent-plane to the wave-surface.

By (33a), remembering that $\mathbf{s}$ is parallel to the normal, we see that

$$\left.\begin{array}{llllll} & c\mathbf{E}, & \mu\mathbf{E}, & \text{and} & \mu\mathbf{N} & \text{are in one plane;} \\ \text{or} & \mathbf{E}, & \mathbf{N}, & \text{and} & \mu^{-1}c\mathbf{E} & \text{are in one plane.}\end{array}\right\}\text{...........(43)}$$

And by (34a),

$$\left.\begin{array}{llllll} & \mu\mathbf{H}, & c\mathbf{N}, & \text{and} & c\mathbf{H} & \text{are in one plane;} \\ \text{or} & \mathbf{H}, & \mathbf{N}, & \text{and} & c^{-1}\mu\mathbf{H} & \text{are in one plane.}\end{array}\right\}\text{...........(44)}$$

These conditions expanded, give us the directions of the electric force and displacement, the magnetic force and induction, for a given normal. We may write the second of (43) thus,

$$\mathbf{N}\mathrm{V}\frac{\mathbf{D}}{c}\frac{\mathbf{D}}{\mu} = 0; \quad \text{..............................(45)}$$

and the second of (44) thus,

$$\mathbf{N}\mathrm{V}\frac{\mathbf{B}}{c}\frac{\mathbf{B}}{\mu} = 0; \quad \text{..............................(46)}$$

and as these differ only in the substitution of $\mathbf{B}$ for $\mathbf{D}$, we see that the induction of either ray is parallel to the displacement of the other; that is, the two directions of induction in the wave-front are the two directions of displacement.

The Wave-Surface.—Since the velocity-surface with the vector $\mathbf{p} = v\mathbf{N}$ is the locus of the foot of the perpendicular on the tangent-plane to the wave-surface, we have, if $\mathbf{r}$ be the vector of the wave-surface,

$$\mathbf{pr} = \mathbf{p}^2. \quad \text{.................................(47)}$$

But $\mathbf{s}$ the vector of the index-surface being $= \mathbf{N}v^{-1} = \mathbf{p}v^{-2}$, we have, by (47), dividing it by v^2,

$$\mathbf{sr} = 1. \qquad \text{.........................(48)}$$

To find the wave-surface, we must therefore let $\mathbf{s}$ be variable and eliminate it between (48) and any one of the index-equations. This is not so easy as it may appear.

General considerations may lead us to the conclusion that the equation to the wave-surface and that to the index-surface may be turned one into the other by the simple process of inverting the operators, turning c into c^{-1} and μ into μ^{-1}. Although this will be verified later, any form of index-equation giving a corresponding form of wave by inversion of operators, yet it must be admitted that this requires proof. That it is true when one of the operators c or μ is a constant does not prove that it is also true when we have the inverse compound operator $\{(\mathbf{s}c\mathbf{s})\mu^{-1} - nc^{-1}\}^{-1}$ containing both c and μ, neither being constant. I have not found an easy proof. This will not be wondered at when the similar investigations of the Fresnel surface are referred to. Professor Tait, in his "Quaternions," gives two methods of finding the wave-surface; one from the velocity-equation, the other from the index-equation. The latter is rather the easier, but cannot be said to be very obvious, nor does either of them admit of much simplification. The difficulty is of course considerably multiplied when we have the two operators to reckon with. I believe the following transition from index to wave cannot be made more direct, or shorter, except of course by omission of steps, which is not a real shortening.

Given

$$\frac{c\mathbf{E}}{\mathbf{s}\mu\mathbf{E}} = \mathbf{b}_1 = \frac{\mathbf{s}}{(\mathbf{s}\mu\mathbf{s})c^{-1} - m\mu^{-1}}, \qquad \text{..........(49)} = \text{(39) } bis$$

$$\mathbf{sb}_1 = 0, \qquad \text{..........(50)} = \text{(41) } bis$$

$$\mathbf{rs} = 1. \qquad \text{..........(51)} = \text{(48) } bis$$

Eliminate $\mathbf{s}$ and get an equation in $\mathbf{r}$. We have also

$$\mu\mathbf{s}c^{-1}\mathbf{b}_1 = 1, \qquad \text{..........(52)} = \text{(42) } bis$$

which will assist later.

By (49) we have

$$\mathbf{s} = (\mathbf{s}\mu\mathbf{s})c^{-1}\mathbf{b}_1 - m\mu^{-1}\mathbf{b}_1. \qquad \text{..........(53)}$$

Multiply by $\mathbf{b}_1$ and use (50); then

$$0 = (\mathbf{s}\mu\mathbf{s})(\mathbf{b}_1 c^{-1}\mathbf{b}_1) - m(\mathbf{b}_1\mu^{-1}\mathbf{b}_1). \qquad \text{..........(54)}$$

By differentiation, $\mathbf{s}$ being variable, and therefore $\mathbf{b}_1$ also,

$$0 = 2(d\mathbf{s}\mu\mathbf{s})(\mathbf{b}_1 c^{-1}\mathbf{b}_1) + 2(\mathbf{s}\mu\mathbf{s})(d\mathbf{b}_1 c^{-1}\mathbf{b}_1) - 2m(d\mathbf{b}_1\mu^{-1}\mathbf{b}_1). \qquad \text{......(55)}$$

Also, differentiating (53),

$$d\mathbf{s} = 2(d\mathbf{s}\mu\mathbf{s})c^{-1}\mathbf{b}_1 + (\mathbf{s}\mu\mathbf{s})dc^{-1}\mathbf{b}_1 - md\mu^{-1}\mathbf{b}_1;$$

and multiplying this by $2\mathbf{b}_1$ gives

$$2\mathbf{b}_1 d\mathbf{s} = 4(d\mathbf{s}\mu\mathbf{s})(\mathbf{b}_1 c^{-1}\mathbf{b}_1) + 2(\mathbf{s}\mu\mathbf{s})(d\mathbf{b}_1 c^{-1}\mathbf{b}_1) - 2m(d\mathbf{b}_1\mu^{-1}\mathbf{b}_1). \qquad \text{...(56)}$$

Subtract (55) from (56) and halve the result; thus obtaining

$$\mathbf{b}_1 d\mathbf{s} = (d\mathbf{s}\mu\mathbf{s})(\mathbf{b}_1 c^{-1}\mathbf{b}_1),$$

or

$$\{\mathbf{b}_1 - (\mathbf{b}_1 c^{-1}\mathbf{b}_1)\mu\mathbf{s}\} d\mathbf{s} = 0. \quad \text{.......................(57)}$$

In the last five equations it will be understood that $d\mathbf{s}$ and $d\mathbf{b}_1$ are differential vectors, and that $d\mathbf{s}\mu\mathbf{s}$ is the scalar product of $d\mathbf{s}$ and $\mu\mathbf{s}$, etc.; also in getting (56) from the preceding equation we have

$$\mathbf{b}_1 dc^{-1}\mathbf{b}_1 = \mathbf{b}_1 c^{-1} d\mathbf{b}_1 = d\mathbf{b}_1 c^{-1}\mathbf{b}_1, \quad \text{etc.}$$

Equation (57) is the expression of the result of differentiating (50),

$$d(\mathbf{s}\mathbf{b}_1) = d\mathbf{s}\mathbf{b}_1 + \mathbf{s}d\mathbf{b}_1 = 0,$$

with $d\mathbf{b}_1$ eliminated.

Now (57) shows that the vector in the {} is perpendicular to $d\mathbf{s}$, the variation of $\mathbf{s}$. But by (51) we also have, on differentiation,

$$\mathbf{r}d\mathbf{s} = 0. \quad \text{.................................(58)}$$

Hence $\mathbf{r}$ and the {} vector in (57) must be parallel. This gives

$$h\mathbf{r} = \mathbf{b}_1 - (\mathbf{b}_1 c^{-1}\mathbf{b}_1)\mu\mathbf{s}, \quad \text{.........................(59)}$$

where h is a scalar. If we multiply this by $c^{-1}\mathbf{b}_1$ and use (52), we obtain

$$\mathbf{r}c^{-1}\mathbf{b}_1 = 0; \quad \text{.............................(60)}$$

or, by (49), giving $\mathbf{b}_1$ in terms of $c\mathbf{E}$,

$$\mathbf{r}\mathbf{E} = 0, \quad \text{.................................(61)}$$

a very important landmark. The ray is perpendicular to the electric force.

Similarly, if we had started from—instead of (49), (50), and (52)—the corresponding $\mathbf{H}$ equations, viz.,

$$\frac{\mu\mathbf{H}}{\mathbf{s}c\mathbf{H}} = \mathbf{b}_2 = \frac{\mathbf{s}}{(\mathbf{s}c\mathbf{s})\mu^{-1} - nc^{-1}}, \qquad \mathbf{s}\mathbf{b}_2 = 0, \qquad c\mathbf{s}\mu^{-1}\mathbf{b}_2 = 1,$$

with of course the same equation (51) connecting $\mathbf{r}$ and $\mathbf{s}$, we should have arrived at

$$h'\mathbf{r} = \mathbf{b}_2 - (\mathbf{b}_2\mu^{-1}\mathbf{b}_2)c\mathbf{s}; \quad \text{.........................(62)}$$

h' being a constant, corresponding to (59); of this no separate proof is needed, as it amounts to exchanging μ and c and turning $\mathbf{E}$ into $\mathbf{H}$, to make (39) become (40). And from (62), multiplying it by $\mu^{-1}\mathbf{b}_2$, we arrive at

$$\mathbf{r}\mu^{-1}\mathbf{b}_2 = 0, \qquad \text{or} \qquad \mathbf{r}\mathbf{H} = 0, \quad \text{....................(63)}$$

corresponding to (61). The ray is thus perpendicular both to the electric and to the magnetic force. The first half of the demonstration is now completed, but before giving the second half we may notice some other properties.

Thus, to determine the values of the scalar constants h and h'. Multiply (59) by $\mathbf{s}$, and use (50) and (51); then

$$h = -(\mathbf{b}_1 c^{-1}\mathbf{b}_1)(\mathbf{s}\mu\mathbf{s}) = -m(\mathbf{b}_1\mu^{-1}\mathbf{b}_1),$$

the second form following from (54). Insert in (59), then

$$\mathrm{r} = \frac{\mu \mathrm{s}}{\mathrm{s}\mu\mathrm{s}} - \frac{\mathrm{b}_1}{m(\mathrm{b}_1\mu^{-1}\mathrm{b}_1)} \quad \text{..........................(64)}$$

gives r explicitly in terms of μs and b_1, the latter of which is known in terms of the former by (49). Multiply this by $\mu^{-1}\mathrm{b}_1$, using (50); then

$$\mathrm{r}\mu^{-1}\mathrm{b}_1 = -m^{-1}. \quad \text{..........................(65)}$$

Similarly we shall find

$$h' = -n(\mathrm{b}_2c^{-1}\mathrm{b}_2), \quad \text{..........................(66)}$$

giving

$$\mathrm{r} = \frac{c\mathrm{s}}{\mathrm{s}c\mathrm{s}} - \frac{\mathrm{b}_2}{n(\mathrm{b}_2c^{-1}\mathrm{b}_2)}; \quad \text{..........................(67)}$$

and, corresponding to (65), we shall have

$$\mathrm{r}c^{-1}\mathrm{b}_2 = -n^{-1}. \quad \text{..........................(68)}$$

Now to resume the argument, stopped at equation (63). Up to equation (59) the work is plain and straightforward, according to rule in fact, being merely the elimination of the differentials, and the getting of an equation between r and s. What to do next is not at all obvious. From (59), or from (64), the same with h eliminated, we may obtain all sorts of scalar products containing r and b_1, and if we could put b_1 explicitly in terms of r, (60) or (65) would be forms of the wave-surface equation. From the purely mathematical point of view no direct way presents itself; but (61) and (63), considered physically as well as mathematically, guide us at once to the second half of the transformation from the index- to the wave-equation. As, at the commencement, we found the induction and the displacement to be perpendicular to the normal, so now we find that the corresponding forces are perpendicular to the ray. There was no difficulty in reaching the index-equation before, when we had a single normal with two values of v the normal velocity, and two rays differently inclined to the normal. There should then be no difficulty, by parallel reasoning, in arriving at the wave-surface equation from analogous equations which express that the ray is perpendicular to the magnetic and electric forces, considering two parallel rays travelling with different ray-velocities with two differently inclined wave-fronts.

Now, as we got the index-equation from

$$\mathrm{VNH} = -vc\mathbf{E}, \quad \text{..........................(25) } bis$$

$$\mathrm{VNE} = v\mu\mathrm{H}, \quad \text{..........................(26) } bis$$

we must have two corresponding equations for one ray-direction. Let **M** be a unit vector defining the direction of the ray, and w be the ray-velocity, so that

$$\mathrm{r} = w\mathbf{M}. \quad \text{..........................(69)}$$

Operate on (25) and (26) by V**M**, giving

$$\mathrm{VMVNH} = -v\mathrm{VM}c\mathbf{E},$$

$$\mathrm{VMVNE} = v\mathrm{VM}\mu\mathrm{H}.$$

Now use the formula of transformation (18), giving

$$\mathbf{N}(\mathbf{HM}) - \mathbf{H}(\mathbf{MN}) = -v\mathrm{V}\mathbf{M}c\mathbf{E},$$
$$\mathbf{N}(\mathbf{EM}) - \mathbf{E}(\mathbf{MN}) = v\mathrm{V}\mathbf{M}\mu\mathbf{H}.$$

But $\mathbf{HM}=0$ and $\mathbf{EM}=0$, as proved before. Also $v=w(\mathbf{MN})$, or the wave-velocity is the normal component of the ray-velocity. Hence

$$\mathbf{H} = w\mathrm{V}\mathbf{M}c\mathbf{E}, \qquad \text{(70)}$$
$$-\mathbf{E} = w\mathrm{V}\mathbf{M}\mu\mathbf{H}, \qquad \text{(71)}$$

which are the required analogues of (25) and (26). Or, by (69),

$$\mathbf{H} = \mathrm{V}\mathbf{r}c\mathbf{E}, \qquad \text{(72)}$$
$$-\mathbf{E} = \mathrm{V}\mathbf{r}\mu\mathbf{H} \qquad \text{(73)}$$

are the analogues of (28) and (29). The rest of the work is plain. Eliminating $\mathbf{E}$ and $\mathbf{H}$ successively, we obtain

$$0 = \mathbf{E} + \mathrm{V}\mathbf{r}\mu\mathrm{V}\mathbf{r}c\mathbf{E},$$
$$0 = \mathbf{H} + \mathrm{V}\mathbf{r}c\mathrm{V}\mathbf{r}\mu\mathbf{H};$$

and, using the theorem (17), these give

$$0 = \mathbf{E} + m\mathrm{V}\mathbf{r}\mathrm{V}\mu^{-1}\mathbf{r}\mu^{-1}c\mathbf{E},$$
$$0 = \mathbf{H} + n\mathrm{V}\mathbf{r}\mathrm{V}c^{-1}\mathbf{r}c^{-1}\mu\mathbf{H};$$

which, using the transformation-formula (18), become

$$0 = \mathbf{E} + m\mu^{-1}\mathbf{r}(\mu^{-1}\mathbf{r}c\mathbf{E}) - \mu^{-1}c\mathbf{E}(\mathbf{r}\mu^{-1}\mathbf{r})m,$$
$$0 = \mathbf{H} + nc^{-1}\mathbf{r}(c^{-1}\mathbf{r}\mu\mathbf{H}) - c^{-1}\mu\mathbf{H}(\mathbf{r}c^{-1}\mathbf{r})n;$$

or, rearranging, after operating by μ and c respectively,

$$\{(\mathbf{r}\mu^{-1}\mathbf{r})mc - \mu\}\mathbf{E} = m\mathbf{r}(\mu^{-1}\mathbf{r}c\mathbf{E}),$$
$$\{(\mathbf{r}c^{-1}\mathbf{r})n\mu - c\}\mathbf{H} = n\mathbf{r}(c^{-1}\mathbf{r}\mu\mathbf{H}).$$

Or

$$\frac{\mathbf{E}}{\mu^{-1}\mathbf{r}c\mathbf{E}} = \frac{\mathbf{r}}{(\mathbf{r}\mu^{-1}\mathbf{r})c - m^{-1}\mu} = \mathbf{g}_1, \text{ say}, \qquad \text{(74)}$$

$$\frac{\mathbf{H}}{c^{-1}\mathbf{r}\mu\mathbf{H}} = \frac{\mathbf{r}}{(\mathbf{r}c^{-1}\mathbf{r})\mu - n^{-1}c} = \mathbf{g}_2, \text{ say}. \qquad \text{(75)}$$

These give us the four simplest forms of equation to the wave. For, since $\mathbf{rE}=0=\mathbf{rH}$, we have

$$\mathbf{rg}_1 = 0, \qquad \mathbf{rg}_2 = 0. \qquad \text{(76)}$$

Also, operating on (74) by $\mu^{-1}\mathbf{r}c$ and on (75) by $c^{-1}\mathbf{r}\mu$ we get

$$\mu^{-1}\mathbf{r}c\mathbf{g}_1 = 1, \qquad c^{-1}\mathbf{r}\mu\mathbf{g}_2 = 1, \qquad \text{(77)}$$

two other forms.

$\mathbf{g}_1$ and $\mathbf{g}_2$ differ from $\mathbf{b}_1$ and $\mathbf{b}_2$ merely in the change from $\mathbf{s}$ to $\mathbf{r}$, and in the inversion of the operators. The two forms of wave (76) are analogous to (41), and the two forms (77) analogous to (42), inverting operators and putting $\mathbf{r}$ for $\mathbf{s}$.

Similarly, if the wave-surface equation be given and we require that of the index-surface, we must impose the same condition $\mathbf{rs}=1$ as before, and eliminate $\mathbf{r}$. This will lead us to

$$\mathbf{s}c\mathbf{g}_1=0, \qquad \mathbf{s}\mu\mathbf{g}_1=-m, \qquad\qquad (78)$$

corresponding to (60) and (65); and

$$\mathbf{s}\mu\mathbf{g}_2=0, \qquad \mathbf{s}c\mathbf{g}_2=-n, \qquad\qquad (79)$$

corresponding to (63) and (68); and the first of (78) and (79) are equivalent to

$$\mathbf{s}c\mathbf{E}=0, \qquad \mathbf{s}\mu\mathbf{H}=0;$$

or the displacement and the induction are perpendicular to the normal. This completes the first half of the process; the second part would be the repetition of the already given investigation of the index-equation.

The vector rate of transfer of energy being $\mathbf{VEH}/4\pi$ in general, when a ray is solitary, its direction is that of the transfer of energy. It seems reasonable, then, to define the direction of a ray, whether the wave is plane or not, as perpendicular to the electric and the magnetic forces. On this understanding, we do not need the preliminary investigation of the index-surface, but may proceed at once to the wave-surface by the investigation (69) to (77), following equations (25) and (26).

The following additional useful relations are easily deducible :—From (25) and (26) we get

$$\mathbf{s}=\frac{\mathbf{V}c\mathbf{E}\mu\mathbf{H}}{\mathbf{E}c\mathbf{E}}; \qquad\qquad (80)$$

and from (72) and (73),

$$\mathbf{r}=\frac{\mathbf{VEH}}{\mathbf{E}c\mathbf{E}}. \qquad\qquad (81)$$

Also, from either set,

$$\mathbf{E}c\mathbf{E}=\mathbf{H}\mu\mathbf{H}, \qquad\qquad (82)$$

expressing the equality of the electric to the magnetic energy per unit volume (strictly, at a point).

Some Cartesian Expansions.—In the important case of parallelism of the principal axes of capacity and permeability, the full expressions for the index- or the wave-surface equations may be written down at once from the scalar product abbreviated expressions. Thus, taking any equation to the wave, as the first of (76), for example, $\mathbf{rg}_1=0$, $\mathbf{g}_1$ being given in (74), take the axes of coordinates parallel to the common principal axes of c and μ; so that we can employ c_1, c_2, c_3, the principal capacities, and μ_1, μ_2, μ_3 the principal permeabilities in the three components of $\mathbf{g}_1$. We then have, x, y, z being the coordinates of $\mathbf{r}$,

$$\frac{x^2}{(\mathbf{r}\mu^{-1}\mathbf{r})c_1-m^{-1}\mu_1}+\frac{y^2}{(\mathbf{r}\mu^{-1}\mathbf{r})c_2-m^{-1}\mu_2}+\frac{z^2}{(\mathbf{r}\mu^{-1}\mathbf{r})c_3-m^{-1}\mu_3}=0, \qquad (83)$$

where

$$\mathbf{r}\mu^{-1}\mathbf{r}=\frac{x^2}{\mu_1}+\frac{y^2}{\mu_2}+\frac{z^2}{\mu_3}.$$

In (83) we may exchange the c's and μ's, getting the second of (76). Similarly the first of (77) gives

$$\frac{\mu_1^{-1}c_1x^2}{(\mathbf{r}\mu^{-1}\mathbf{r})c_1-m^{-1}\mu_1}+\frac{\mu_2^{-1}c_2y^2}{(\mathbf{r}\mu^{-1}\mathbf{r})c_2-m^{-1}\mu_2}+\frac{\mu_3^{-1}c_3z^2}{(\mathbf{r}\mu^{-1}\mathbf{r})c_3-m^{-1}\mu_3}=1 \qquad (84)$$

as another form, in which, again, the μ's and c's may be exchanged (not forgetting to change m into n) to give a fourth form.

These reduce to the Fresnel surface if either $\mu_1 = \mu_2 = \mu_3$ or $c_1 = c_2 = c_3$.

Let $x = 0$ to find the sections in the plane y, z. The first denominator in (83) gives

$$\left(\frac{y^2}{\mu_2} + \frac{z^2}{\mu_3}\right)c_1 - \frac{1}{\mu_2\mu_3} = 0, \qquad \text{or} \qquad y^2 c_1 \mu_3 + z^2 c_1 \mu_2 = 1,$$

representing an ellipse, semiaxes

$$v_{13} = (c_1\mu_3)^{-\frac{1}{2}} \qquad \text{and} \qquad v_{12} = (c_1\mu_2)^{-\frac{1}{2}}.$$

The other terms give

$$\left(\frac{y^2}{\mu_2} + \frac{z^2}{\mu_3}\right)(c_3 y^2 + c_2 z^2) = \frac{y^2}{\mu_1\mu_2} + \frac{z^2}{\mu_1\mu_3}.$$

Or $$y^2\mu_1 c_3 + z^2 \mu_1 c_2 = 1,$$

an ellipse, semiaxes $v_{31} = (c_3\mu_1)^{-\frac{1}{2}}$ and $v_{21} = (c_2\mu_1)^{-\frac{1}{2}}$. Similarly, in the plane z, x the sections are ellipses whose semiaxes are v_{21}, v_{23}, and v_{12}, v_{32}, where for brevity $v_{rs} = (c_r\mu_s)^{-\frac{1}{2}}$; and in the plane x, y, the ellipses have semiaxes v_{31}, v_{32}, and v_{13}, v_{12}.

In one of the principal planes two of the ellipses intersect, giving four places where the two members of the double surface unite.

If $c_1/\mu_1 = c_2/\mu_2 = c_3/\mu_3$, we have a single ellipsoidal wave-surface whose equation is

$$\frac{x^2}{v_{23}^2} + \frac{y^2}{v_{31}^2} + \frac{z^2}{v_{12}^2} = 1. \qquad \text{..........................(85)}$$

Now, of course, $v_{12} = v_{21}$, etc.

When the μ and c axes are not parallel, we cannot immediately write down the full expansion of the wave-surface equation. Proceed thus :— Taking $\mathbf{r}\mathbf{g}_1 = 0$ as the equation, let

$$R = m(\mathbf{r}\mu^{-1}\mathbf{r}), \qquad \text{and} \qquad \mathbf{a} = m^{-1}\mathbf{g}_1;$$

then, by (74) and (76),

$$\mathbf{r}\frac{\mathbf{r}}{Rc - \mu} = 0, \qquad \text{or} \qquad \mathbf{r}\mathbf{a} = 0,$$

where $$\mathbf{r} = (Rc - \mu)\mathbf{a}. \qquad \text{..............................(86)}$$

R is a scalar. If a_1, a_2, a_3 are the three components of $\mathbf{a}$ referred to any rectangular axes, and x, y, z the components of $\mathbf{r}$, we have, by (86) and (12),

$$x = (Rc_{11} - \mu_{11})a_1 + (Rc_{12} - \mu_{12})a_2 + (Rc_{13} - \mu_{13})a_3,$$
$$y = (Rc_{21} - \mu_{21})a_1 + (Rc_{22} - \mu_{22})a_2 + (Rc_{23} - \mu_{23})a_3,$$
$$z = (Rc_{31} - \mu_{31})a_1 + (Rc_{32} - \mu_{32})a_2 + (Rc_{33} - \mu_{33})a_3;$$

from which a_1, a_2, a_3 may be solved in terms of x, y, z; thus

$$a_1 = a_{11}x + a_{12}y + a_{13}z,$$
$$a_2 = a_{21}x + a_{22}y + a_{23}z,$$
$$a_3 = a_{31}x + a_{32}y + a_{33}z;$$

where, by using (15),

$$a_{11}=\frac{(Rc_{22}-\mu_{22})(Rc_{33}-\mu_{33})-(Rc_{23}-\mu_{23})^2}{\Delta},$$

$$a_{12}=\frac{(Rc_{13}-\mu_{13})(Rc_{23}-\mu_{23})-(Rc_{12}-\mu_{12})(Rc_{33}-\mu_{33})}{\Delta};$$

and the rest by symmetry. Then, since

$$\mathbf{ra}=xa_1+ya_2+za_3=0,$$

we get the full expansion. Δ need not be written fully, as it goes out. The equation may be written symmetrically, thus,

$$0=1+mn(\mathbf{r}\mu^{-1}\mathbf{r})(\mathbf{r}c^{-1}\mathbf{r})-\Big\{x^2(c_{22}\mu_{33}+c_{33}\mu_{22}-2c_{23}\mu_{23})+\dots$$
$$+2xy(c_{13}\mu_{23}+c_{23}\mu_{13}-c_{12}\mu_{33}-c_{33}\mu_{12})+\dots\Big\}, \quad \dots\dots(87)$$

where the coefficients of y^2, z^2, yz, and zx are omitted. Here $m=\mu_1\mu_2\mu_3$ and $n=c_1c_2c_3$; whilst

$$\mathbf{r}c^{-1}\mathbf{r}=c'_{11}x^2+c'_{22}y^2+c'_{33}z^2+2c'_{12}xy+2c'_{23}yz+2c'_{31}zx,$$

where c'_{11}, ..., are the inverse coefficients. See equation (15). The expansion of $\mathbf{r}\mu^{-1}\mathbf{r}$ is exactly similar, using the inverse μ coefficients.

If in (87) we for every c or μ write the reciprocal coefficients, we obtain the equation to the index-surface; that is, supposing x, y, z then to be the components of $\mathbf{s}$ instead of $\mathbf{r}$. And, since $\mathbf{s}v=\mathbf{N}$, the unit wave-normal, we have the velocity-equation as follows, in the general case,

$$0=\frac{\mathbf{N}\mu\mathbf{N}}{m}\cdot\frac{\mathbf{N}c\mathbf{N}}{n}+v^4-v^2\Big\{N_1^2(c'_{22}\mu'_{33}+c'_{33}\mu'_{22}-2c'_{23}\mu'_{23})+\dots$$
$$+2N_1N_2(c'_{13}\mu'_{23}+c'_{23}\mu'_{13}-c'_{12}\mu'_{33}-c'_{33}\mu'_{12})+\dots\Big\}, \quad \dots\dots(88)$$

in which N_1, N_2, N_3 are the components of $\mathbf{N}$, or the direction-cosines of the normal. To show the dependence of v^2 upon the capacity and permeability perpendicular to $\mathbf{N}$, take $N_1=1$, $N_2=0$, $N_3=0$, which does not destroy generality, because in (88) the axes of reference are arbitrary. Then (88) reduces to

$$v^4-(c'_{22}\mu'_{33}+c'_{33}\mu'_{22}-2c'_{23}\mu'_{23})v^2+(c'_{22}c'_{33}-c'^2_{23})(\mu'_{22}\mu'_{33}-\mu'^2_{23})=0.$$

When the μ and c axes are parallel, and their principal axes are those of reference, we have

$$0=\frac{\mathbf{N}\mu\mathbf{N}}{m}\cdot\frac{\mathbf{N}c\mathbf{N}}{n}+v^4-v^2\Big\{N_1^2(v_{23}^2+v_{32}^2)+N_2^2(v_{31}^2+v_{13}^2)+N_3^2(v_{12}^2+v_{21}^2)\Big\}, \quad (89)$$

where $$\mathbf{N}\mu\mathbf{N}=\mu_1N_1^2+\mu_2N_2^2+\mu_3N_3^2,$$

with a similar expression for $\mathbf{N}c\mathbf{N}$, and $v_{23}=(c_2\mu_3)^{-\frac{1}{2}}$, etc., as before. The solution is

$$v^2=\tfrac{1}{2}N_1^2(v_{23}^2+v_{32}^2)+\tfrac{1}{2}N_2^2(v_{31}^2+v_{13}^2)+\tfrac{1}{2}N_3^2(v_{12}^2+v_{21}^2)\pm\tfrac{1}{2}\sqrt{X}, \quad \dots\dots(90)$$

where $X=N_1^4u_1^2+N_2^4u_2^2+N_3^4u_3^2-2(N_1^2N_2^2u_1u_2+N_2^2N_3^2u_2u_3+N_3^2N_1^2u_1u_3)$,

in which $u_1=v_{23}^2-v_{32}^2, \qquad u_2=v_{31}^2-v_{13}^2, \qquad u_3=v_{12}^2-v_{21}^2. \quad \dots\dots(91)$

Take $u_1 = 0$, or $c_2/\mu_2 = c_3/\mu_3$; the two velocities (squared) are then

$$N_1^2 v_{23}^2 + N_2^2 v_{31}^2 + N_3^2 v_{12}^2, \qquad \text{and} \qquad N_1^2 v_{23}^2 + N_2^2 v_{13}^2 + N_3^2 v_{21}^2,$$

reducing to one velocity v_{23} when $N_1 = 1$.

If, further, $u_2 = 0$, or $u_3 = 0$, making $c_1/\mu_1 = c_2/\mu_2 = c_3/\mu_3$, $X = 0$ always, and

$$v^2 = N_1^2 v_{23}^2 + N_2^2 v_{31}^2 + N_3^2 v_{12}^2 \quad \text{.......................(92)}$$

is the single value of the square of velocity of wave-front.

Directions of **E**, **H**, **D**, *and* **B**.—We may expand (45) to obtain an equation for the two directions of the induction and displacement. Thus, since

$$\frac{\mathbf{D}}{c} = \mathbf{i}(c'_{11}D_1 + c'_{12}D_2 + c'_{13}D_3) + \mathbf{j}(c'_{21}D_1 + c'_{22}D_2 + c'_{23}D_3) + \mathbf{k}(c'_{31}D_1 + c'_{32}D_2 + c'_{33}D_3),$$

$$\frac{\mathbf{D}}{\mu} = \mathbf{i}(\mu'_{11}D_1 + \mu'_{12}D_2 + \mu'_{13}D_3) + \mathbf{j}(\mu'_{21}D_1 + \mu'_{22}D_2 + \mu'_{23}D_3)$$
$$+ \mathbf{k}(\mu'_{31}D_1 + \mu'_{32}D_2 + \mu'_{33}D_3),$$

$$\mathbf{N} = \mathbf{i}N_1 + \mathbf{j}N_2 + \mathbf{k}N_3,$$

the determinant of the coefficients of **i**, **j**, **k** equated to zero gives the required equation. When the principal axes of μ and c are parallel, the equation greatly simplifies, being then

$$0 = \frac{N_1 u_1}{D_1} + \frac{N_2 u_2}{D_2} + \frac{N_3 u_3}{D_3}, \quad \text{.........................(93)}$$

where $u_1, \ldots,$ are the same differences of squares of principal velocities as in (91). For D_1, etc., write B_1, etc.; and we have the same equation for the induction directions. For D_1, etc., write $c_1 E_1$, etc., and the resulting equation gives the directions of **E**. For D_1, etc., write $\mu_1 H_1$, etc., and the resulting equation gives the directions of **H**.

Note on Linear Operators and Hamilton's Cubic. (June 12th, 1892.)

[The reason of the ease with which the transformations concerned in the above can usually be effected is, it will be observed, the symmetrical property $\mathbf{A}c\mathbf{B} = \mathbf{B}c\mathbf{A}$ of the scalar products. But when a linear operator, say c, is not its own conjugate, some change of treatment is required. Thus, let

$$D_1 = c_{11}E_1 + c_{12}E_2 + c_{13}E_3, \qquad D'_1 = c_{11}E_1 + c_{21}E_2 + c_{31}E_3,$$
$$D_2 = c_{21}E_1 + c_{22}E_2 + c_{23}E_3, \qquad D'_2 = c_{12}E_1 + c_{22}E_2 + c_{32}E_3,$$
$$D_3 = c_{31}E_1 + c_{32}E_2 + c_{33}E_3, \qquad D'_3 = c_{13}E_1 + c_{23}E_2 + c_{33}E_3,$$

where the nine c's are arbitrary. We may then write

$$\mathbf{D} = c\mathbf{E}, \qquad \mathbf{D}' = c'\mathbf{E},$$

where the operator c' only differs from c in the exchange of c_{12} and c_{21}, etc. It is now $\mathbf{D}'$ that is conjugate to $\mathbf{D}$, whilst c' is the operator conjugate to c. It may be readily seen that

$$\mathbf{D} = f\mathbf{E} + \mathrm{V}\mathbf{eE}, \qquad \mathbf{D}' = f\mathbf{E} - \mathrm{V}\mathbf{eE},$$

where f is the self-conjugate operator obtained by replacing c_{12} and c_{21}, etc., in c by half their sums, and $\mathbf{e}$ is a certain vector whose components are half their differences. Thus,

$$f\mathbf{E} = \tfrac{1}{2}(\mathbf{D} + \mathbf{D}'), \qquad \mathrm{V}\mathbf{eE} = \tfrac{1}{2}(\mathbf{D} - \mathbf{D}'),$$

$$\mathbf{e} = \tfrac{1}{2}\mathbf{i}(c_{32} - c_{23}) + \tfrac{1}{2}\mathbf{j}(c_{13} - c_{31}) + \tfrac{1}{2}\mathbf{k}(c_{21} - c_{12}).$$

The conjugate property of scalar products is now

$$\mathbf{A}c\mathbf{B} = \mathbf{B}c'\mathbf{A}.$$

That is, in transferring the operator from $\mathbf{B}$ to $\mathbf{A}$, we must simultaneously change it to its conjugate. Another way of regarding the matter is as follows:—If we put

$$\mathbf{c}_1 = \mathbf{i}c_{11} + \mathbf{j}c_{12} + \mathbf{k}c_{13}, \qquad \mathbf{c}_2 = \mathbf{i}c_{21} + \mathbf{j}c_{22} + \mathbf{k}c_{23}, \qquad \mathbf{c}_3 = \mathbf{i}c_{31} + \mathbf{j}c_{32} + \mathbf{k}c_{33},$$

we see, by the above, that

$$\mathbf{D} = c\mathbf{E} = \mathbf{i}.\mathbf{c}_1\mathbf{E} + \mathbf{j}.\mathbf{c}_2\mathbf{E} + \mathbf{k}.\mathbf{c}_3\mathbf{E} = (\mathbf{i}.\mathbf{c}_1 + \mathbf{j}.\mathbf{c}_2 + \mathbf{k}.\mathbf{c}_3)\mathbf{E},$$

$$\mathbf{D}' = c'\mathbf{E} = \mathbf{c}_1.\mathbf{iE} + \mathbf{c}_2.\mathbf{jE} + \mathbf{c}_3.\mathbf{kE} = (\mathbf{c}_1.\mathbf{i} + \mathbf{c}_2.\mathbf{j} + \mathbf{c}_3.\mathbf{k})\mathbf{E},$$

from which we see that $c'\mathbf{E}$ is the same as $\mathbf{E}c$, and $c\mathbf{E}$ the same as $\mathbf{E}c'$. In the case of $\mathbf{A}c\mathbf{B}$, therefore, we may regard it either as the scalar product of $\mathbf{A}$ and $c\mathbf{B}$, or as the scalar product of $\mathbf{A}c$ and $\mathbf{B}$. This is equivalent to Professor Gibbs's way of regarding linear operators. That is (converted to my notation),

$$c = \mathbf{i}.\mathbf{c}_1 + \mathbf{j}.\mathbf{c}_2 + \mathbf{k}.\mathbf{c}_3$$

is the type of a linear operator. It assumes the utmost generality when $\mathbf{i}$, $\mathbf{j}$, $\mathbf{k}$ stand for any three independent vectors, instead of a unit rectangular system. Professor Gibbs has considerably developed the theory of linear operators in his Vector Analysis.

The generalised form of (17) is got thus:—Let $\mathbf{v}$ and $\mathbf{w}$ be any vectors, then, as before, we have

$$0 = \mathbf{v}\mathrm{V}\mathbf{vw} = \mathbf{v}cc^{-1}\mathrm{V}\mathbf{vw},$$

$$0 = \mathbf{w}\mathrm{V}\mathbf{vw} = \mathbf{w}cc^{-1}\mathrm{V}\mathbf{vw},$$

where the last forms assert that $c^{-1}\mathrm{V}\mathbf{vw}$ is perpendicular to $\mathbf{v}c$ and $\mathbf{w}c$, or parallel to $\mathrm{V}\mathbf{v}c\mathbf{w}c$; that is,

$$m\mathrm{V}\mathbf{vw} = c\mathrm{V}c'\mathbf{v}c'\mathbf{w}; \qquad \text{(A)}$$

from which, by multiplying by a third vector $\mathbf{u}$, we find

$$m = \frac{c'\mathbf{u}\mathrm{V}c'\mathbf{v}c'\mathbf{w}}{\mathbf{u}\mathrm{V}\mathbf{vw}}, \qquad \text{(B)}$$

which is an invariant.

Hamilton's cubic equation in c is obtained by observing that since (A) is an identity, c being any linear operator, it remains an identity when c is changed to $c-g$, which changes c' to $c'-g$, where g is a scalar constant. For $c-g$ is also a linear operator. Making this substitution in (A) and expanding, we obtain

$$(m - m_1 g + m_2 g^2 - g^3)\mathrm{V}\mathbf{vw}$$
$$= \frac{\mathrm{V}\mathbf{vw}}{\mathbf{u}\mathrm{V}\mathbf{vw}}\Big\{c'\mathbf{u}\mathrm{V}c'\mathbf{v}c'\mathbf{w} - g(\mathbf{u}\mathrm{V}c'\mathbf{v}c'\mathbf{w} + \mathbf{v}\mathrm{V}c'\mathbf{w}c'\mathbf{u} + \mathbf{w}\mathrm{V}c'\mathbf{u}c'\mathbf{v})$$
$$+ g^2(c'\mathbf{u}\mathrm{V}\mathbf{vw} + c'\mathbf{v}\mathrm{V}\mathbf{wu} + c'\mathbf{w}\mathrm{V}\mathbf{uv}) - g^3\mathbf{u}\mathrm{V}\mathbf{vw}\Big\}$$
$$= c\mathrm{V}c'\mathbf{v}c'\mathbf{w} - g(\mathrm{V}c'\mathbf{v}c'\mathbf{w} + c\mathrm{V}\mathbf{v}c'\mathbf{w} + c\mathrm{V}c'\mathbf{vw})$$
$$+ g^2(c\mathrm{V}\mathbf{vw} + \mathrm{V}\mathbf{v}c'\mathbf{w} + \mathrm{V}c'\mathbf{vw}) - g^3\mathrm{V}\mathbf{vw},$$

where m, m_1, m_2 are the coefficients of g^0, $-g$, and g^2 in the expansion of the left member of m given by (B). Comparing coefficients we see that g^0 and g^3 go out. The others give (remembering that we are dealing with an identity),

$$\mathrm{V}c'\mathbf{v}c'\mathbf{w} + c(\mathrm{V}\mathbf{v}c'\mathbf{w} + \mathrm{V}c'\mathbf{vw}) = m_1\mathrm{V}\mathbf{vw},$$
$$c\mathrm{V}\mathbf{vw} + (\mathrm{V}\mathbf{v}c'\mathbf{w} + \mathrm{V}c'\mathbf{vw}) = m_2\mathrm{V}\mathbf{vw}.$$

Operate on the first by c and second by c^2, and subtract. This eliminates the vector in the brackets, and leaves

$$c\mathrm{V}c'\mathbf{v}c'\mathbf{w} - c^3\mathrm{V}\mathbf{vw} = m_1 c\mathrm{V}\mathbf{vw} - m_2 c^2\mathrm{V}\mathbf{vw},$$

where the first term on the left is $m\mathrm{V}\mathbf{vw}$. So we have

$$m - m_1 c + m_2 c^2 - c^3 = 0, \qquad \text{(C)}$$

which is Hamilton's cubic.

If we start instead with the conjugate operator c' we shall arrive at

$$m'\mathrm{V}\mathbf{vw} = c'\mathrm{V}c\mathbf{v}c\mathbf{w}, \qquad \text{where} \qquad m' = \frac{c\mathbf{u}\mathrm{V}c\mathbf{v}c\mathbf{w}}{\mathbf{u}\mathrm{V}\mathbf{vw}};$$

and then, later, to the cubic

$$m' - m_1'c' + m_2'c'^2 - c'^3 = 0,$$

where m', etc., come from m, etc., by exchanging c and c'. But it may be easily proved that $m = m'$, and we may infer from this that $m_1 = m_1'$ and $m_2 = m_2'$, on account of the invariantic character of m being preserved when c becomes $c-g$. In fact, putting $c = f + \mathrm{V}\mathbf{e}$ and $c' = f - \mathrm{V}\mathbf{e}$, where f is self-conjugate, we may independently show that

$$m = m' = \frac{f\mathbf{u}\mathrm{V}f\mathbf{v}f\mathbf{w}}{\mathbf{u}\mathrm{V}\mathbf{vw}} + \mathbf{e}f\mathbf{e} = \frac{c\mathbf{u}\mathrm{V}c\mathbf{v}c\mathbf{w}}{\mathbf{u}\mathrm{V}\mathbf{vw}} = \frac{c'\mathbf{u}\mathrm{V}c'\mathbf{v}c'\mathbf{w}}{\mathbf{u}\mathrm{V}\mathbf{vw}},$$

$$m_1 = m_1' = \frac{\mathbf{u}\mathrm{V}f\mathbf{v}f\mathbf{w} + \mathbf{v}\mathrm{V}f\mathbf{w}f\mathbf{u} + \mathbf{w}\mathrm{V}f\mathbf{u}f\mathbf{v}}{\mathbf{u}\mathrm{V}\mathbf{vw}} + \mathbf{e}^2$$

$$= \frac{\mathbf{u}\mathrm{V}c\mathbf{v}c\mathbf{w} + \mathbf{v}\mathrm{V}c\mathbf{w}c\mathbf{u} + \mathbf{w}\mathrm{V}c\mathbf{u}c\mathbf{v}}{\mathbf{u}\mathrm{V}\mathbf{vw}} = \text{same with } c',$$

$$m_2 = m_2' = \frac{f\mathbf{u}\mathrm{V}\mathbf{vw} + f\mathbf{v}\mathrm{V}\mathbf{wu} + f\mathbf{w}\mathrm{V}\mathbf{uv}}{\mathbf{u}\mathrm{V}\mathbf{vw}} = \text{same with } c = \text{same with } c'.$$

So in Hamilton's cubic (C) we may change c to c', leaving the m's unchanged; or else in the m's only; or make the change in both the c's and the m's, without affecting its truth.

If the passage from (A) to (C) above be compared with the corresponding transition in Tait's Quaternions (3rd edition, §§ 158 to 160) it will be seen that that rather difficult proof is simplified (as done above) by omitting altogether the inverse operations ϕ^{-1} and $(\phi - g)^{-1}$ and the auxiliary operator χ; especially χ, perhaps. One is led to think from Professor Tait's proof that the object of the investigation is to solve the problem of inverting ϕ. But the mere inversion can be done by elementary methods. In Gibbs's language, if **a**, **b**, **c** is one set of vectors, the reciprocal set is **a**′, **b**′, **c**′, given by

$$\mathbf{a}' = \frac{\mathrm{V}\mathbf{bc}}{\mathbf{a}\mathrm{V}\mathbf{bc}}, \qquad \mathbf{b}' = \frac{\mathrm{V}\mathbf{ca}}{\mathbf{b}\mathrm{V}\mathbf{ca}}, \qquad \mathbf{c}' = \frac{\mathrm{V}\mathbf{ab}}{\mathbf{c}\mathrm{V}\mathbf{ab}}.$$

On this understanding, we may expand any vector **d** in terms of **a**, **b**, **c** thus:—

$$\mathbf{d} = \mathbf{a}.\mathbf{a}'\mathbf{d} + \mathbf{b}.\mathbf{b}'\mathbf{d} + \mathbf{c}.\mathbf{c}'\mathbf{d}.$$

Similarly, if **l**′, **m**′, **n**′ is the set reciprocal to **l**, **m**, **n**, we have

$$\mathbf{r} = \mathbf{l}'.\mathbf{lr} + \mathbf{m}'.\mathbf{mr} + \mathbf{n}'.\mathbf{nr}.$$

If, then, it be given that

$$\mathbf{d} = \phi(\mathbf{r}) = \mathbf{a}.\mathbf{lr} + \mathbf{b}.\mathbf{mr} + \mathbf{c}.\mathbf{nr},$$

we see that $\mathbf{lr} = \mathbf{a}'\mathbf{d}$, etc., so that

$$\mathbf{r} = \phi^{-1}(\mathbf{d}) = \mathbf{l}'.\mathbf{a}'\mathbf{d} + \mathbf{m}'.\mathbf{b}'\mathbf{d} + \mathbf{n}'.\mathbf{c}'\mathbf{d}$$

inverts ϕ. (This is equivalent to Tait, § 173.)

We see by (A) and (B) that the inverts of **u**, **v**, **w** are $c' \times$ inverts of $c\mathbf{u}$, $c\mathbf{v}$, $c\mathbf{w}$; or $c \times$ inverts of $c'\mathbf{u}$, $c'\mathbf{v}$, $c'\mathbf{w}$. The cubic (C) may be written

$$\frac{c\mathbf{u}\mathrm{V}c\mathbf{v}c\mathbf{w}}{\mathbf{u}\mathrm{V}\mathbf{vw}}\left\{c^{-1} - (\mathbf{u}c^{-1}\mathbf{u}' + \mathbf{v}c^{-1}\mathbf{v}' + \mathbf{w}c^{-1}\mathbf{w}')\right\} = c\left\{c - (\mathbf{u}c\mathbf{u}' + \mathbf{v}c\mathbf{v}' + \mathbf{w}c\mathbf{w}')\right\},$$

if **u**′, **v**′, **w**′ are the inverts of **u**, **v**, **w** (or the reciprocal set). In this identity the operators c and c^{-1} may be inverted. When that is done we see that the m of c is the reciprocal of the m of c^{-1}.]

Note on Modification of Index-equation when c *and* μ *are Rotational.*

[Let c' and μ' be the conjugates to c and μ. Then, by (A), (B), in last note,

$$m\mathrm{V}\mathbf{vw} = \mu'\mathrm{V}\mu\mathbf{v}\mu\mathbf{w} = \mu\mathrm{V}\mu'\mathbf{v}\mu'\mathbf{w},$$

where

$$m = \mu_1\mu_2\mu_3 + \mathbf{e}\mu_0\mathbf{e},$$

if μ_1, μ_2, μ_3 are the principal permeabilities of μ_0, the self-conjugate operator such that $\mu = \mu_0 + \mathrm{V}\mathbf{e}$. With this extension of meaning, we shall have (treating c and n similarly),

$$-\mathbf{E} = c^{-1}\mathrm{V}\mathbf{s}\mathbf{H}, \qquad -n\mathbf{E} = \mathrm{V}c'\mathbf{s}c'\mathbf{H}, \qquad -m\mathbf{E} = c^{-1}\mathrm{V}\mathbf{s}\mathrm{V}\mu'\mathbf{s}\mu'\mathbf{E},$$

$$\mathbf{H} = \mu^{-1}\mathrm{V}\mathbf{s}\mathbf{E}, \qquad m\mathbf{H} = \mathrm{V}\mu'\mathbf{s}\mu'\mathbf{E}, \qquad -n\mathbf{H} = \mu^{-1}\mathrm{V}\mathbf{s}\mathrm{V}c'\mathbf{s}c'\mathbf{H},$$

where the first pair replace (28), (29), the second pair (31), (32), and the third pair (33), (34). Then

$$-mc\mathbf{E}=\mu'\mathbf{s}(\mathbf{s}\mu'\mathbf{E})-\mu'\mathbf{E}(\mathbf{s}\mu'\mathbf{s}),$$
$$-n\mu\mathbf{H}=c'\mathbf{s}(\mathbf{s}c'\mathbf{H})-c'\mathbf{H}(\mathbf{s}c'\mathbf{s})$$

replace (33a) and (34a), and

$$\frac{\mathbf{E}}{\mathbf{s}\mu'\mathbf{E}}=\frac{\mu'\mathbf{s}}{(\mathbf{s}\mu'\mathbf{s})\mu'-mc},\qquad \frac{\mathbf{H}}{\mathbf{s}c'\mathbf{H}}=\frac{c'\mathbf{s}}{(\mathbf{s}c'\mathbf{s})c'-n\mu}$$

replace (37a), (38a); from which two forms of index-equation corresponding to (41) are

$$\mathbf{s}\frac{\mathbf{s}}{(\mathbf{s}\mu'\mathbf{s})c^{-1}-m\mu'^{-1}}=0,\qquad \mathbf{s}\frac{\mathbf{s}}{(\mathbf{s}c'\mathbf{s})\mu^{-1}-nc'^{-1}}=0.$$

We obtain impossible values of the velocity for certain directions of the normal. That is, there could not be a plane wave under the circumstances.]

XXXII. NOTES ON NOMENCLATURE.

[*The Electrician*, Note 1, Sep. 4, 1885, p. 311; Note 2, Jan. 26, 1886, p. 227; Note 3, Feb. 12, 1886, p. 271.]

NOTE 1. IDEAS, WORDS, AND SYMBOLS.

HOWEVER desirable it may be that writers on electrotechnics should use a common notation, at least as regards the frequently recurring magnitudes concerned—which notation should not be a difficult matter to arrange, provided it be kept within practical limits—it is perhaps more desirable that they should adopt a common language, within the same practical limits, of course. For whilst the use of certain letters for certain magnitudes requires no more explanation than, for instance, "Let us call the currents C_1, C_2, etc.," it is otherwise with the language used when speaking of the magnitudes, as more elaborate explanations are needed to identify the ideas meant to be expressed.

As regards electric conduction currents, there is a tolerably uniform usage, and a fairly good terminology. It is seldom that any doubt can arise as to a writer's meaning, unless he be an ignoramus or a paradoxist, or have unfortunately an indistinct manner of expressing himself. I would, however, like to see the word "intensity," as applied to the electric current, wholly abolished. It was formerly very commonly used, and there was an equally common vagueness of ideas prevalent. It is sufficient to speak of the current in a wire (total) as "the current," or "the strength of current," and when referred to unit area, the current-density. (In three dimensions, on the other hand, when everything is referred to the unit volume, and the current-density is meant as a matter of course, it is equally sufficient to call *it* the current.)

It is a matter of considerable practical advantage to have single words for names, instead of groups of words, and it is fortunate that the existing conduction-current terminology admits of very practical adaptation this way. Thus, "specific resistance" may be well called "resistivity," and specific conductance "conductivity," referring to the unit volume. Resistivity is the reciprocal of conductivity, and resistance of conductance. When wires are in parallel, their conductances may be more easy to manage than their resistances. We have also the convenient adjectives "conductive" and "resistive," to save circumlocution.

Passing to the subject of magnetic induction, there is considerable looseness prevailing. There is a definite magnitude called by Maxwell "the magnetic induction," which may well be called simply "the induction." It is related to the magnetic force in the same manner as current-density to the electric force. ($B=\mu H$.) The ratio μ is the "magnetic permeability." This may be simply called the permeability, since the word is not used in any other electrical sense. Induction and permeability may not be the best names, but (apart from their being understood by mathematical electricians) they are infinitely better than the long-winded "number of lines of force" (meaning magnetic) and "conductivity for lines of force," the use of which, though defensible enough in merely popular explanations, becomes almost absurd when the electrotechnical user actually goes so far as to give them quantitative expression. Conductivity should not be used at all, save in pointing out an analogy. It has its own definite meaning.

"Permeability," however, does not admit of such easy adaptation to different circumstances as conductivity. Permeability referring to the unit volume, the word permeance is suggested for a mass, analogous to conductance. We have also the adjective "permeable." By adding, moreover, the prefix "im," we get "impermeable," "impermeability," and "impermeance," for the reciprocal ideas, sometimes wanted. Thus impermeability, the reciprocal of μ, would stand for the long-winded "specific resistance to lines of magnetic force." (The permeance of a coil would be $L/4\pi$, if L is its coefficient of self-induction. In the expression $T=\frac{1}{2}LC^2$ for the magnetic energy of current C in the coil, 4π does not appear, whilst it does in the form $T=\frac{1}{2}$ magnetomotive force × total induction through the circuit ÷ 4π. It is $4\pi C$ that is the magnetomotive force, and LC the induction through the circuit. Thus we have oppositely acting 4π's. I may here remark that it would be not only a theoretical but a great practical improvement to have the electric and magnetic units recast on a rational basis. But I suppose there is no chance of such an extensive change.) It must be confessed, however, that these various words are not so good as the corresponding conduction-current words.

But now, if, thirdly, we pass to electric displacement, the analogue of magnetic induction (noting by the way that it had better not be called the electric induction, on account of our already appropriating the word induction, but be called the displacement), the existing terminology is extremely unsatisfactory; and, moreover, does not readily admit of adaptation and extension. Corresponding to conductivity and permeability we have "specific inductive capacity," or "dielectric constant," or whatever it may be called. I usually call it the electric capacity, or the capacity. It refers to the unit volume. But here it is very unfortunate that it is not this specific capacity c (say), but $c/4\pi$, that is the capacity of a unit cube condenser (such that charge = difference of potential × capacity). D, the displacement, is the charge (+ or −, according to the end), and we have $D = cE/4\pi$, E being the electric force. We may get over this trouble by putting it thus, $D = sE$, and calling s (or $c/4\pi$) the specific capacity. Then the capacity in bulk is got in the same manner as conductance from conductivity.

Supposing we have done this, there is still the trouble that capacity gives the extremely awkward inverse "incapacity," and the adjectives "capacious" and "incapacious," besides not giving us any words for use in bulk, like conductance and resistance. And, in addition, the word capacity is itself rather objectionable, as likely to give beginners entirely erroneous notions as to the physical quality involved. It is not that one dielectric absorbs electricity more readily than another. Electric displacement is an *elastic* phenomenon: one dielectric is more yielding (electrically) than another. The reciprocal of s above is the electric elasticity, measuring the electric force required to produce the unit displacement. Thus s should have a name to express the idea of elastic yielding or distortion, and its reciprocal also a name (not strings of words), and they should be readily adaptable, like conductivity, etc. (Perhaps also a better word than permeability might be introduced, although, as we see, it is tolerably accommodative.) Displacement itself might also be replaced by another word less suggestive of bodily translation; although, on the other hand, it harmonises well with "current," the displacement being the accumulated current, or the current the time-variation of the displacement.

All these things will get right in time, perhaps. Ideas are of primary importance, scientifically. Next, suitable language. As for the notation, it is an important enough matter, but still only takes the third place.

NOTE 2. ON THE RISE AND PROGRESS OF NOMENCLATURE.

In the beginning was the word. The importance of nomenclature was recognised in the earliest times. One of the first duties that devolved upon Adam on his installation as gardener and keeper of the zoological collection was the naming of the beasts.

The history of the race is repeated in that of the individual. This grand modern generalisation explains in the most scientific manner the fondness for calling names displayed by little children.

Passing over the patriarchal period, the fall of the Tower of Babel and its important effects on nomenclature, the Egyptian sojourn, the wanderings in the desert, the times of the Kings, of the Babylonian captivity, of the minor prophets, of early Christianity, of those dreadful middle ages of monkish learning and ignorance, when evolution worked backwards, and of the Elizabethan revival, and coming at once to the middle of the 19th century, we find that Mrs. Gamp was much impressed by the importance of nomenclature. "Give it a name, I beg. Sairey, give it a name!" cried that esteemed lady. She even went so far as to give a name to an entirely fictitious personage—Mrs. Harris, to wit—who has many scientific representatives.

Having thus fortified ourselves by quoting both ancient and modern instances, let us consider the names of the electrical units.

A really practical name should be short, preferably monosyllabic, pronounced in nearly the same way by all civilised peoples, and not mistakable for any other scientific unit. If, in addition, it be the name, or a part of the name, of an eminent scientist, so much the better. This is quite a sentimental matter; but if it does no harm, it is needless to object to it. But we should never put the sentiment in the first place, and give an unpractical name to a unit on account of the sentiment.

Ohm and volt are admirable; farad is nearly as good (but surely it was unpractical to make it a million times too big—the present microfarad should be the farad); erg and dyne please me; watt is not quite so good, but is tolerable. But what about those remarkable results of the Paris Congress, the ampère and the coulomb? Speaking entirely for myself, they are very unpractical. Coulomb may be turned into coul, and is then endurable; this unit is, however, little used. But ampère shortened to am or amp is not nice. Better make it père; then it will do. Now an additional bit of sentiment comes in to support us. Was not Ampère the father of electrodynamics?

It seems rather unpractical for the B.A. Committee to have selected 10^8 c.g.s. as the practical unit of E.M.F., instead of 10^9. This will hardly be appreciated except by those who make theoretical calculations; the awkward thing is that the père is one tenth of the c.g.s. unit of current. I suppose it was because the present volt was an approximation to the E.M.F. of a Daniell; that is, however, a very strong reason for making the practical unit much smaller; because the E.M.F. of a cell has now to be given in volts and tenths, or hundredths also. How awkward it would have been if the ohm had been made 10^{10} c.g.s., so as to approximate to the resistance of a mile of iron telegraph wire. The ohm and volt should be the same multiple of the c.g.s. units, both 10^9 for example. Then use the millivolt or centivolt when speaking of the E.M.F. of cells. The present 1·12 volt would be 112 millivolts. Speaking from memory, Sir W. Thomson did object to the 10^8 volt at the Paris Congress.

Mac, tom, bob, and dick are all good names for units. Tom and mac (plural, max), have sentimental reasons for adoption; bob and dick may also at some future time. I have used tom myself (no offence, I hope) for six years past to denote 10^9 c.g.s. units of self or mutual electromagnetic induction coefficient. (Some reform is wanted here. Coefficient of self-induction, or of electromagnetic capacity, is too lengthy.) The advantage is that L toms divided by R ohms gives L/R seconds of time. But it is too big a unit for little coils; then use the millitom; or even the microtom for very small coils. This applies to fine-wire coils. The c.g.s. unit itself would be most suitable for coils of a few turns of thick wire. If it is called the tom, then the kilotom or megatom will come in useful for fine-wire coils.

A name should certainly be given to a unit of this quantity, whether it be tom, or mac, or any other practical name. Also, names to a unit of magnetic force (intensity of), and of magnetic induction.

There is also the question of the names, not of the units, but of the physical magnitudes of which they are the units, but it is too large a question to discuss here except in the most superficial manner. It is engrained in the British nature to abbreviate, to make one word do for two or three, or a short for a long word. And quite right too. We have much to be thankful for; in the application of this general remark, consider what frightful names might have been given to the electrical units by the Germans. But, on account of this national, and also rational tendency to cut and clip, it is in the highest degree desirable that as many as possible of the most important physical magnitudes should be known, not by a long string of words, but by a single word, or the smallest number possible.

Thus, I find myself frequently saying force, when I mean magnetic force, and even then, I mean the intensity of magnetic force. The context will generally make the meaning plain. But it is necessary to be very careful when there are more forces than one in question. (This use of force as an abbreviation is, of course, quite distinct from the frequent positive misuse of the word force, to indicate it may be momentum, or energy, or activity, or, very often, nothing in particular, the misuser not being able to say exactly what he means; nor does it much matter.) It would be decidedly better if such a quantity as "intensity of magnetic force" had a one-word name, for people will abbreviate, and sometimes confusion may step in. This remark applies to most of the electromagnetic magnitudes.

There is an important magnitude termed the magnetic induction. I call it often simply "the induction"; but in doing so, carefully avoid calling any other quantity "the induction" (sometimes the electric displacement is called the electric induction). But there is an unfortunate thing here, which somewhat militates against "the induction," or even "the magnetic induction" being a thoroughly good name for the magnitude in question. This is, that besides being a name of a physical magnitude, the word induction has a widespread use, in a rather vague manner, in connection with transient states in general, whether of the electric or of the magnetic field, exemplified, to take an extreme example, when a man explains something complex by saying it is caused by "induction," and so settling the matter. If this vague qualitative use of induction were got rid of, then as a name for a physical magnitude it would be unobjectionable. As it is, it is a question whether the physical magnitude should not have a name for itself alone.

"Resistivity" for specific resistance, and "conductance" for what is sometimes called the conductibility of a wire, *i.e.*, not its conductivity (specific conductance), but the reciprocal of its resistance, are, I think, as I have remarked before, quite practical names.

NOTE 3. THE INDUCTANCE OF A CIRCUIT.

IN my first note, amongst other things, I remarked that whilst the conduction-current terminology admitted of the words resistivity and conductance being coined to make it more complete, the terminology in the allied cases of magnetic induction and electric displacement was unsatisfactory.

As regards the former, the following appears to me to be practical. First, abolish the word permeability, and substitute Inductivity. We then have $B=\mu H$, when B is the Induction, and μ the Inductivity, showing how the Induction is related to the magnetic force H by the specific quality of the medium at the place, its inductivity.

Now conductivity and conductance are mathematically related in the same manner (except as regards a 4π) as inductivity and what it is naturally suggested to call Inductance.

The Inductance of a circuit is what is now called its coefficient of self-induction, or of electromagnetic capacity.

Thus the quantities induction, inductivity, and inductance are happily connected in a manner which is at once concise and does justice to their real relationship. When the mutual coefficient of induction of two circuits is to be referred to, it will of course be the mutual inductance.

XXXIII. NOTES ON THE SELF-INDUCTION OF WIRES.

[*The Electrician*, 1886; Note 1, April 23, p. 471; Note 2, May 7, p. 510.]

NOTE 1. We read in the pages of history of a monarch who was "*supra grammaticam.*" All truly great men are like that monarch. They have their own grammars, syntaxes, and dictionaries. They cannot be judged by ordinary standards, but require interpretation. Fortunately the liberty of private interpretation is conserved.

No man has a more peculiar grammar than Prof. Hughes. Hence, he is liable, in a most unusual degree, to be misunderstood, as I venture to think he has been by many, including Mr. W. Smith, whose interesting letter appears in *The Electrician*, April 16, 1886, p. 455, and Prof. H. Weber, p. 451.

The very first step to the understanding of a writer is to find out what he means. Before that is done there cannot possibly be a clear comprehension of his utterances. One may, by taking his language in its ordinary significance, hastily conclude that he has either revolutionised the science of induction, or that he is talking nonsense. But to do this would not be fair. We must not judge by what a man says if we have good reason to know that what he means is quite different. To be quite fair, we must conscientiously endeavour to translate his language and ideas into those we are ourselves accustomed to use. Then, and then only, shall we see what is to be seen.

When Prof. Hughes speaks of the resistance of a wire, he does not *always* mean what common men, men of ohms, volts, and farads, mean by the resistance of a wire—only sometimes. He does not exactly define what it is to be when the accepted meaning is departed from. But by a study of the context we may arrive at some notion of its new meaning. It is not a definite quantity, and must be varied to suit circumstances. Again, there is his "inductive capacity" of a wire. We can only find roughly what that means by putting together this, that, and the other. It, too, is not a definite quantity, but must be varied to suit circumstances. It is not the coefficient of self-induction, nor is it any quantity defining a specific quality of the wire, like conductivity, or inductivity. It is a complex quantity, depending on a great many things, but which may, to a first rough approximation, be taken to be proportional to the time-constant of the wire, the quotient of its coefficient of self-induction by its resistance. Bearing these two things in mind, we shall be able to approximate to Prof. Hughes's meaning.

Owing to the mention of discoveries, apparently of the most revolutionary kind, I took great pains in translating Prof. Hughes's language into my own, trying to imagine that I had made the same experiments *in the same manner* (which could not have happened), and then asking what are their interpretations? The discoveries I looked for vanished for the most part into thin air. They became well known facts when put into common language. The satisfaction of getting verifications, however, even in so roundabout and rough a manner, is some compensation for the disappointment felt. I venture to think that Prof. Hughes does not do himself justice in thus deceiving us, however unwittingly, and that possibly there has been also some misapprehension on his part as to what the laws of self-induction are generally supposed to be.

I have failed to find any departure from the known laws of electromagnetism. In saying this, however, I should make a reservational remark. There may be lying latent in Prof. Hughes's results dozens of discoveries, but it is impossible to get at them. For consider what the mere existence of ohms, volts, and farads means? It means that, even before they were made, the laws of induction in linear circuits were known, and very precisely. To get, then, at new discoveries requires very accurate comparison of experiment with theory, by methods which enable us to see what we are doing and measuring, in terms of the known electromagnetic quantities. This is practically impossible, on the basis of Prof. Hughes's papers. We can only make very rough verifications. I have had myself, for many years past, occasional experience with induction balances of an exact nature—true balances of resistance and induction—and always found them work properly. But, in the modification made by Prof. Hughes, the balance is generally of a mixed nature, neither a true resistance nor a true induction balance, and has to be set right by a foreign impressed force, viz., induction between the battery and telephone branches. By using a strictly simple harmonic E.M.F., as of a rotating coil, we may exactly formulate the conditions of the false balance, and then, noting all the resistances, etc., concerned, derive, though in a complex manner, exact information. Or, if we use true balances, any kind of E.M.F. will answer.

To illustrate the falsity of Prof. Hughes's balances and the difficulty of getting at exact information, he finds the comparative force of the extra-currents in two similar coils in series to be 1·74 times that of a single coil. From the context it would appear that this "comparative force of the extra-currents" is the same thing as the former "inductive capacity" of wires. Now, the coefficient of self-induction of two similar coils in series, not too near one another, is double that of either, whilst the time-constant of the two is the same as of either. This can be easily verified by true balances.

The most interesting of the experiments are those relating to the effect of increased diameter on what Prof. Hughes terms the "inductive capacity" of wires. My own interpretation is roughly this. That the time-constant of a wire first increases with the diameter, and then later decreases rapidly; and that the decrease sets in the sooner the higher the conductivity and the higher the inductivity (or magnetic permeability) of the wires. If this be correct, it is exactly what I should have expected and predicted. In fact, I have already described the phenomenon substantially in *The Electrician*; or, rather, the phenomenon I described contains in itself the above interpretation. In *The Electrician* for January 10, 1885, I described how the current starts in a wire. It begins on its boundary and is propagated inward. Thus, during the rise of the current it is less strong at the centre than at the boundary. As regards the manner of inward propagation, it takes place according to the same laws as the propagation of magnetic force and current into cores from an enveloping coil, which I have described in considerable detail in *The Electrician* [Reprint, vol. 1, Art. 28. See especially § 20]. The retardation depends on the conductivity, on the inductivity, and on the section, under similar boundary conditions. If the conductivity be high enough, or the inductivity or the section be large enough, to make the central current appreciably less than the boundary current during the greater part of the time of rise of the current, there will be an apparent reduction in the time-constant. Go to an extreme case. Very rapid short currents, and large retardation to inward transmission. Here we have the current in layers, strong on the boundary, weak in the middle. Clearly, then, if we wish to regard the wire as a mere linear circuit, which it is not, and as we can only do to a first approximation, we should remove the central part of the wire—that is, increase its resistance, regarded as a line, or reduce its time-constant. This will happen the sooner the greater the inductivity and the conductivity, as the section is continuously increased. It is only thin wires that can be treated as mere lines, and even they, if the speed be only great enough, must be treated as solid conductors. I ought also to mention that the influence of external conductors, as of the return conductor, is of importance, sometimes of very great importance, in modifying the distribution of current in the transient state. I have had for years in MS. some solutions relating to round wires, and hope to publish them soon.

As a general assistance to those who go by old methods—a rising current inducing an opposite current in itself and in parallel conductors—this may be useful. Parallel currents are said to attract or repel, according as the currents are together or opposed. This is, however, mechanical force on the conductors. The distribution of current is not affected by it. But when currents are increasing or decreasing, there is an apparent attraction or repulsion between them. Oppositely going currents repel when they are decreasing, and attract when they are increasing. Thus, send a current into a loop, one wire the return to the other, both being close together. During the rise of the current it will be denser on the sides of the wires nearest one another than on the remote sides. It is an apparent force, not between currents (on the distance-action and real motion of electricity views), but between their accelerations.

NOTE 2. I did not expect to return to the subject, and do so because Prof. Hughes has apparently misunderstood my statements. On p. 495 of *The Electrician* for April 30, 1886, he says:—"Mr. Oliver Heaviside points out that upon a close examination it will be found that all the effects which I have described are well known to mathematicians, and consequently old." A regard for accuracy compels me to point out that I did not make the statement he credits me with; nor, to avoid any hypercriticism, is the above a correct summary of the many things that I pointed out.

I said, "The discoveries I looked for vanished, for the most part, into thin air. They became well-known facts when put into common language." Observe here my "for the most part" as against Prof. Hughes's "all"; and that I said not a word about mathematicians in the whole letter. An immediate consequence of my statement is another, namely, that some, although a minority, of the results were not well known. There is a material difference between what I said and what Prof. Hughes makes me say. In another place I said that I had "failed to find any departure from the known laws of electro-magnetism," and then proceeded to give my reasons for it. This statement includes the well-known facts as well as those which are not well known.

It may be as well that I should illustrate the difference between well-known facts and those that are less known, or only known theoretically. The influence of the form of a thin wire (a linear conductor), and of its length, diameter, conductivity, and inductivity on the phenomena of self-induction is well known. The various relations involved form the A B C of the subject. So are the effects of concentration of the current, and of dividing it, or spreading it out in strips, well known. There is another influence that is well known, that is scarcely touched upon by Prof. Hughes. The self-induction depends upon the distribution of inductivity, that is, in another form, of inductively magnetisablė matter, outside the current, as well as in it, in a manner which is quite definite when the magnetic properties of the matter are known.

It is not to be inferred that verifications of well-known facts are of no value—that depends upon circumstances. To be of any use, we must know what we are measuring and verifying. The theory of self and mutual induction in linear circuits is almost a branch of pure mathematics, so simply are the quantities related, and so exactly. It furnishes a most remarkable example of the dependence of complex phenomena on a very small number of independent variables, by ignoring minute dielectric phenomena. In getting verifications, then, it is first necessary to employ a correct method. I have elsewhere [*The Electrician,* April 30, 1886, p. 489; the next Art. 33] shown the approximate character of Prof. Hughes's method of balancing, and pointed out exact methods. Next, it is necessary to put results in terms of the quantities in the electromagnetic theory which is founded upon the well-known facts; how else can we know what we are doing, and see how near our verifications go?

Coming now to results that are not well known, there is the thick-wire effect, depending on size, conductivity, inductivity, place of return current, etc. This is, in my opinion, the really important part of Prof. Hughes's researches, as it, in some respects, goes beyond what was already experimentally known. Having been, so far as I know, the first to correctly describe (*The Electrician,* Jan. 10, 1885, p. 180) [Reprint, vol. I. pp. 439, 440] the way the current rises in a wire, viz., by diffusion from its boundary, and the consequent approximation, under certain circumstances, to mere surface conduction; and believing Prof. Hughes's researches to furnish experimental verifications of my views, it will be readily understood that I am specially interested in this effect; and I can (in anticipation) return thanks to Prof. Hughes for accurate measures of the same, expressed in an intelligible form, to render a comparison with theory possible if it be practicable. I send with this a first instalment of my old core investigations applied to a round wire with the current longitudinal. [Section 26 of "Electromagnetic Induction," later.]

There are also intermediate matters where one can hardly be said to be either making verifications, except roughly, or discoveries; for instance, the self-induction of an iron-wire coil. Theory indicates in the plainest manner that the self-induction coefficient will be a much smaller multiple of that of a similar copper-wire coil than if the wires were straightened. Magnetic circuits are now getting quite popularly understood, by reason of the commercial importance of the dynamo. But there is really no practical way of carrying out the theory completely, as the mathematical difficulties are so great. Hence, actual measurements of the precise amounts in various cases of magnetic circuits are of value, if they be accompanied by the data necessary for comparisons.

There is, however, this little difficulty in the way when transient currents are employed. Iron, by reason of its high inductivity, is pre-eminently suited for showing the thick-wire effect. We may not, therefore, be always measuring what we want, but something else.

XXXIV. ON THE USE OF THE BRIDGE AS AN INDUCTION BALANCE.

[*The Electrician*, April 30, 1886, p. 489.]

IN connection with a paper "On Electromagnets, etc.," that I wrote about six years ago [Reprint, Art. xvii., vol. 1, p. 95], which paper dealt mainly with the question of the influence of the electromagnetic induction of the lines and instruments on the magnitude of the signalling currents, an influence which is of the greatest importance on short lines, and which (of the instruments) is, even on long lines, where electrostatic induction is prominent, of importance as a retarding factor, I made a great many experiments on self-induction, amongst which were measurements of the inductances of various telegraph instruments, with a view to ascertaining their practical values, and also the multiplying powers of the iron cores. It was my intention to write a supplementary paper giving the results and also further investigations; but, having got involved, in the course of the experiments, in the difficult subject of magnetic inductivity, it was postponed, and then dropped out of mind.

I used, first of all, the Bridge and condenser method described by Maxwell, with reversals, and a telephone for current indicator. This was to get results at once, or by simple calculations, in electromagnetic units. Next, I discarded the condenser, and used the simple Bridge, balancing coils against standard coils. Thirdly, I have used a differential telephone with the same object, in a similar manner. The two last are very sensitive methods, and the verifications of the theory of induction in linear conductors that I have made by them are numerous.

The whole of this journal would be required to give anything like a full investigation of the various ways of using the Bridge as an induction balance. I can, therefore, only touch lightly on the subject of exact balances, especially as I have to remark upon faulty methods, approximate balances, and absolutely false balances. Prof. Hughes's balance is sometimes fairly approximate, sometimes quite false.

Put a telephone in the branch 5, battery and interrupter in 6. Then, r standing for resistance, l for inductance (coefficient of self-induction), and x for l/r, the time-constant of a branch, the conditions of a true and perfect balance, however the impressed force in 6 vary, are three in number, namely,

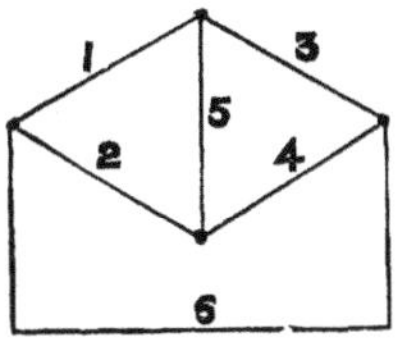

$$r_1 r_4 = r_2 r_3, \quad \text{.............................(1)}$$

$$x_1 + x_4 = x_2 + x_3, \quad \text{.............................(2)}$$

$$l_1 l_4 = l_2 l_3. \quad \text{.............................(3)}$$

Their interpretations are as follows:—If the first condition is fulfilled there will be no final current in 5 when a steady impressed force is put in 6. This is the condition for a true resistance balance.

If, in addition to this, the second condition be also satisfied, the integral extra-current in 5 on making or breaking 6 is zero, besides the steady current being zero. (1) and (2) together therefore give an approximate induction balance with a true resistance balance.

If, in addition to (1) and (2), the third condition is satisfied, the extra-current is zero at every moment during the transient state, and the balance is exact, however the impressed force in 6 vary.

Practically, take

$$r_1 = r_2, \quad \text{and} \quad l_1 = l_2, \quad \text{.......................(4)}$$

that is, let branches 1 and 2 be of equal resistance and inductance. Then the second and third conditions become identical; and, to get perfect balances, we need only make

$$r_3 = r_4, \quad \text{and} \quad l_3 = l_4. \quad \text{.......................(5)}$$

This is the method I have generally used, reducing the three conditions to two, whilst preserving exactness. It is also the simplest method. The mutual induction, if any, of 1 and 2, or of 3 and 4, does not influence the balance when this ratio of equality, $r_1 = r_2$, is employed (whether $l_1 = l_2$ or not). So branches 1 and 2 may consist of two similar wires wound together on the same bobbin to keep their temperatures equal.

The sensitiveness of the telephone has been greatly exaggerated. Altogether apart from the question of referring the sensitiveness to the human ear rather than to the telephone, it is certainly, under ordinary circumstances, often unable to appreciate the differences of the second order, which vanish when the third condition is satisfied. Thus (1) and (2) satisfied, but with (3) unsatisfied, will give silence. Take, for instance, $r_1 = r_2$ and $r_3 = r_4$, but l_1 different from l_2 and l_3 from l_4, then silence is given by

$$(l_1 - l_2)/r_1 = (l_3 - l_4)/r_3; \quad \text{.......................(6)}$$

that is, by making the differences of the inductances on the two sides of 5 proportional to the resistances. We can therefore get silence by varying the inductance of any one or more of the four branches 1, 2, 3, 4, to suit equation (6). It is certain that we do get silence this way, but it does not follow that silence is given by *exactly* satisfying (6), (and (1) of course), because it is only a balance of integral extra-currents, and other balances of this kind are certainly quite false sometimes. To avoid any doubt, it is of course best to keep to the legitimate and simpler previously-described method.

There are some other ways of using the Bridge as an induction balance in an exact manner, but they are less practically useful than theoretically interesting. Pass, therefore, to other approximate, and to false balances. Suppose we start with a true balance, and then upset it by increasing the inductance of the branch 4. It is clear that we should never alter the already truly established resistance balance. Now, besides by the exact ways, we can get approximate silence by allowing mutual induction between 5 and any of the other five branches, or between 6 and any of the other five branches, that is nine ways, not counting combinations. (Put test coils in 5 and 6 with long leading wires, so that they may be carried about from one branch to another.) These approximate balances are all of the integral extra-current only, and therefore imperfect, however nearly there may be silence. But the silences are of very different values.

I find, using fine-wire coils, that mutual induction between 6 and 4 or between 6 and 3 gives silence (to my ear) with the true resistance balance, just like the approximate balance of equation (6) in which no mutual induction is allowed.

These are only two out of the nine ways. All the rest are bad. If the difference in the inductance of 3 and 4 be small, there is very nearly silence on using any of the other seven ways; but, the larger this difference is made, the louder becomes the "silence," and sometimes it is even a very loud noise, quite comparable with the original sound that was to be destroyed, even when the combinations 6 and 4 or 6 and 3, and the formerly-mentioned method give a silence that can be felt, with the true resistance balance.

It is certainly a rather remarkable thing that the one method out of these seven faulty ways which gave the very loudest sound was the 5 and 6 combination, which is Professor Hughes's method. I do not say that it is always the worst, although it was markedly so in my experiments to test the trustworthiness of the method. And sometimes it is quite fair. In fact, when the sound to be destroyed is itself weak, all the seven faulty methods are apparently alike, nearly true. But when we exaggerate the inequality of inductance between 3 and 4, whilst the 6 to 4 and 6 to 3 combinations keep good, the others get rapidly worse, and differences appear between them.

I found that by increasing the resistance of the branch whose inductance was the smaller, the sound was diminished greatly, *i.e.*, in the seven faulty methods. The coil of greater inductance had apparently the higher resistance. That is, with a false resistance balance we may approximate to silence. Such a balance is condemned for scientific purposes.

Although mutual induction between 6 and 4 or 6 and 3 gave silence, with true resistance balances, the experiments were not sufficiently extended to prove their general trustworthiness. There is, however, some reason to be given for their superiority. For, since the disturbance in the telephone arises from the inequality of the momenta of the currents in the branches 3 and 4, and of the electric impulses arising in them when contact is broken in branch 6 (considering the break only for simplicity), we go nearest to the root of the evil by generating an additional impulse in 3 or 4 themselves from the battery branch, of the right amount.

The following is an outline of the theory of these approximate balances. Let $r_1r_4 = r_2r_3$ first; so that, C standing for current, we have, in the steady state,

$$C_1 = C_3 = C_4r_4/r_3, \qquad C_2 = C_4, \qquad C_6 = C_4(1 + r_4/r_3). \quad \text{.....(7)}$$

The momentum of the current in branch 1 is l_1C_1, that in 2 is l_2C_2, and so on. Consider the break, and the integral extra-current that then arises from l_1C_1. It is

$$l_1C_1 \div \{r_1+r_2+r_5(r_3+r_4)/(r_3+r_4+r_5)\},$$

and $(r_3+r_4)/(r_3+r_4+r_5)$ is the fraction of this that goes through 5; so that the integral current in 5 due to l_1C_1 is

$$l_1C_1(r_3+r_4) \div \{(r_1+r_2)(r_3+r_4)+r_5(r_1+r_2+r_3+r_4)\},$$

or

$$C_4r_4x_1 \div \{r_3+r_4+r_5+r_3r_5/r_1\},$$

by making use of equations (1) and (7).

Treat the others similarly. The total extra-current in 5 is

$$r_4C_4(x_1+x_4-x_2-x_3) \div \{r_3+r_4+r_5+r_3r_5/r_1\}, \quad \text{............(8)}$$

without any mutual induction. So

$$x_1+x_4=x_2+x_3$$

gives approximate balance. This was mentioned before, and becomes an exact balance with makes and breaks when a ratio of equality is taken.

Now let there be mutual induction between 6 and 4, 5 and 4, and 5 and 6, the mutual inductances being M_{64}, etc. Treating these similarly to before, we shall find the total extra-current in 5 on the break taking place to be

$$\{r_4(x_1+x_4-x_2-x_3)+M_{64}(1+r_4/r_3)+M_{54}(1+r_3/r_1) + M_{56}(1+r_3/r_1)(1+r_4/r_3)\}C_4 \div (r_3+r_4+r_5+r_3r_5/r_1). \quad \text{.........(9)}$$

The theory of the make leads to the same result—that is, as regards the *integral* extra-current. Otherwise they are different. So, using M_{56} (Hughes's method) the zero integral current is when

$$r_4(x_1+x_4-x_2-x_3)+M_{56}(1+r_3/r_1)(1+r_4/r_3)=0. \quad \text{.........(10)}$$

Using M_{45} we have

$$r_4(x_1+x_4-x_2-x_3)+M_{45}(1+r_3/r_1)=0. \quad \text{..............(11)}$$

Using M_{64} we have

$$r_4(x_1+x_4-x_2-x_3)+M_{64}(1+r_4/r_3)=0. \quad \text{..............(12)}$$

Practically employ a ratio of equality $r_1=r_2$, $l_1=l_2$; that is, make branches 1 and 2 equal fixtures. Then these three equations become

$$l_4-l_3+2M_{56}(1+r_3/r_1)=0, \quad \text{...................(10a)}$$

$$l_4-l_3+\ M_{54}(1+r_3/r_1)=0, \quad \text{...................(11a)}$$

$$l_4-l_3+2M_{46}=0. \quad \text{...................(12a)}$$

Thus the M_{46} system has the simplest formula, as well as being practically perfect. It is the same with M_{63}. Either of these must equal half the difference of the inductances of 3 and 4.

As (10*a*), or, more generally, (10) contains resistances, we cannot get any definite results from Prof. Hughes's numbers without a knowledge of the resistances concerned. Note, also, that (10) and (11) are faulty balances; to improve them, destroy the resistance balance; of course then the formula will change, and is likely to become very complex.

It will be understood that when I speak of false resistance balances in this paper I do not in any way refer to the thick-wire phenomenon, mentioned in my letter [p. 30], which, from its very nature, requires the resistance balance to be upset, or be different from what it would be if the wire were thin, but of the same real [*i.e.*, steady] resistance. The resistance balance must be upset in a perfect arrangement. Nor can there be a true balance got, but only an approximate one, unless a similar thick wire be employed to produce balance.

What I refer to here is the upsetting of the true resistance balance when there is no perceptible departure whatever from the linear theory. The two effects may be mixed.

To use the Bridge to speedily and accurately measure the inductance of a coil, we should have a set of proper standard coils, of known inductance and resistance, together with a coil of variable inductance, *i.e.*, two coils in sequence, one of which can be turned round, so as to vary the inductance from a minimum to a maximum. (The scale of this variable coil could be calibrated by (12*a*), first taking care that the resistance balance did not require to be upset.) This set of coils, in or out of circuit according to plugs, to form say branch 3, the coil to be measured to be in branch 4. Ratio of equality. Branches 1 and 2 equal. Of course inductionless, or practically inductionless resistances are also required, to get and keep the resistance balance.

The only step to this I have made (this was some years ago) in my experiments, was to have a number of little equal unit coils, and two or three multiples; and get exact balance by allowing induction between two little ones, with no exact measurement of the fraction of a unit.

So long as we keep to coils we can swamp all the irregularities due to leading wires, etc., or easily neutralise them, and therefore easily obtain considerable accuracy. With short wires, however, it is a different matter. The inductance of a circuit is a definite quantity. So is the mutual inductance of two circuits. Also, when coils are connected together, each forms so nearly a closed circuit that it can be taken as such, so that we can add and subtract inductances, and localise them definitely as belonging to this or that part of a circuit. But this simplicity is, to a great extent, lost when we deal with short wires, unless they are bent round so as to make nearly closed circuits. We cannot fix the inductance of a straight wire, taken by itself. It has no meaning, strictly speaking. The return current has to be considered. Balances can always be got, but as regards the interpretation, that will depend upon the configuration of the apparatus. [See Section xxxviii. of "Electromagnetic Induction," later.]

Speaking with diffidence, having little experience with short wires, I should recommend 1 and 2 to be two equal wires, of any convenient length, twisted together, joined at one end, of course slightly separated at the other, where they join the telephone wires, also twisted. The exact arrangement of 3 and 4 will depend on circumstances. But always use a long wire rather than a short one (experimental wire). If this is in branch 4, let branch 3 consist of the standard coils (of appropriate size), and adjust *them*, inserting if necessary, coils in series with 4 also. Of course I regard the matter from the point of view of getting easily interpretable results.

The exact balance (1), (2), (3) above is quite special. If the branches 1 and 3 consist of any combination of conductors and condensers, with induction in masses of metal allowed, and branches 2 and 4 consist of an exactly equal combination, in every respect, there will never be any current in 5 due to impressed force in 6. And, more generally, 2 + 4 may be only a copy of 1 + 3, on a reduced scale, so to speak.

P.S.—(April 27, 1886.) The great exactness with which, when a ratio of equality is used, the M_{64} and M_{63} methods conform to the true resistance balance, as above mentioned, together with the almost persistent departure of the M_{65} (Hughes's) method from the true resistance balance, led me to suspect that, as in the use of the simple Bridge method, with no mutual induction, the three conditions of a true balance are reduced to two by a ratio of equality, the same thing happens in the M_{64} and M_{63} methods, but not in the M_{65}. This I have verified.

In Hughes's system the three conditions are

$$r_1 r_4 = r_2 r_3, \quad \text{.........(13)}$$

$$r_4(x_1 + x_4 - x_2 - x_3) + M_{56}(1 + r_3/r_1)(1 + r_4/r_3) = 0, \quad \text{.........(14)}$$

$$l_1 l_4 - l_2 l_3 + M_{56}(l_1 + l_2 + l_3 + l_4) = 0. \quad \text{.........(15)}$$

Now take $l_1 = l_2$, $r_1 = r_2$, $r_3 = r_4$; then the second and third are equivalent to

$$l_4 - l_3 + 2M_{56}(1 + r_3/r_1) = 0, \qquad 2x_1/x_3 = 1 + l_4/l_3.$$

The second of these is a *special relation* that must hold before the first is true. Hence the sound with a true resistance balance, and the necessity of a false balance to get rid of it.

But in the M_{64} method the conditions are

$$r_1 r_4 = r_2 r_3, \quad \text{.........(16)}$$

$$r_4(x_1 + x_4 - x_2 - x_3) + M_{64}(1 + r_2/r_1) = 0, \quad \text{.........(17)}$$

$$l_1 l_4 - l_2 l_3 + M_{64}(l_1 + l_2) = 0. \quad \text{.........(18)}$$

Take $l_1 = l_2$, $r_1 = r_2$, $r_3 = r_4$, as before, and *now* the second and third conditions become identical, viz.,

$$l_4 - l_3 + 2M_{64} = 0,$$

agreeing with the previously obtained equation (12*a*).

Thus, whilst Hughes's method is inaccurate, sometimes greatly so, we may employ the M_{64} and M_{63} methods without any hesitation, provided a ratio of equality be kept to. They will be as accurate as the simple Bridge method, and the choice of the methods will be purely a matter of convenience.

I have verified experimentally that the Hughes system requires a false resistance balance when, instead of coils, short wires are used, the branch of greater inductance having apparently the greater resistance. I have also verified that this effect is mixed with the thick-wire effect, which last is completely isolated by using the proper M_{64} method or the simple Bridge. Its magnitude can now be exactly measured, free from the errors of a faulty method. That is, it can be estimated for any particular speed of intermittences or reversals, for it is not a constant effect. Balance a very thin against a very thick wire, so that the effect occurs only on *one* side.

XXXV. ELECTROMAGNETIC INDUCTION AND ITS PROPAGATION. (SECOND HALF.)

[*The Electrician*, 1886-7. Section XXV., April 23, 1886, p. 469; XXVI., May 14, p. 8 (vol. 17); XXVII., June 11, p. 88; XXVIII., June 25, p. 128; XXIX., July 23, p. 212; XXX., August 6, p. 252; XXXI., August 20, p. 296; XXXII., August 27, p. 316; XXXIII., November 12, p. 10 (vol. 18); XXXIV., December 24, 1886, p. 143; XXXV., January 14, 1887, p. 211; XXXVI., February 4, p. 281; XXXVII., March 11, p. 390; XXXVIII., April 1, p. 457; XXXIXa., May 13, p. 5 (vol. 19); XXXIXb., May 27, p. 50; XL., June 3, p. 79; XLI., June 17, p. 124; XLII., July 1, p. 163; XLIII., July 15, p. 206; XLIV., August 12, p. 295; XLV., August 26, p. 340; XLVI., October 7, p. 459; XLVII., December 30, 1887, p. 189 (vol. 20).]

SECTION XXV. SOME NOTES ON MAGNETISATION.

ALTHOUGH it is generally believed that magnetism is molecular, yet it is well to bear in mind that all our knowledge of magnetism is derived from experiments on masses, not on single molecules, or molecular structures. We may break up a magnet into the smallest pieces, and find that they, too, are little magnets. Still, they are not molecular magnets, but magnets of the same nature as the original; solid bodies showing magnetic properties, or intrinsically magnetised. We are nearly as far away as ever from a molecular magnet. To conclude that molecules are magnets because dividing a magnet always produces fresh magnets, would clearly be unsound reasoning. For it involves the assumption that a molecule has the same magnetic property as a mass, *i.e.*, a large collection of molecules, having, by reason of their connection, properties not possessed by the molecules separately. (Of course, I do not define a molecule to be the smallest part of a substance that has *all* the properties of the mass.) If we got down to a mass of iron so small that it contained few molecules, and therefore certainly not possessing all the properties of a larger mass, what security have we that its magnetic property would not have begun to disappear, and that their complete separation would not leave us without any magnetic field at all surrounding them of the kind we attribute to intrinsic magnetisation. That there would be magnetic disturbances round an isolated molecule in motion through a medium, and with its parts in relative motion, it is difficult not to believe in view of the partial co-ordination of radiation and electromagnetism made by Maxwell. But it might be quite different from the magnetic field of a so-called magnetic molecule—that is, the field of any small magnet. This evident magnetisation might be essentially conditioned by structure, not of single molecules, but of a collection, together with relative motions connected with the structure, this structure and relative motions conditioning that peculiar state of the medium in which they are immersed, which, when existent, implies intrinsic magnetisation of the collection of molecules, or the little mass. However this be, two things are deserving of constant remembrance. First, that the molecular theory of magnetism is a speculation which it is desirable to keep well separated from theoretical embodiments of known facts, apart from hypothesis. And next, that as the act of exposing a solid to magnetising influence is, it is scarcely to be doubted, always accompanied by a changed structure, we should take into account and endeavour to utilise in theoretical reasoning on magnetism which is meant to contain the least amount of hypothesis, the elastic properties of the body, speaking generally, and without knowing the exact connection between them and the magnetic property.

Hooke's law, *Ut tensio, sic vis,* or strain is proportional to stress, implies perfect elasticity, and is the first approximate law on which to found the theory of elasticity. But beyond that, we have imperfect elasticity, elastic fatigue, imperfect restitution, permanent set.

When we expose an unmagnetised body to the action of a magnetic field of unit inductivity, it either draws in the lines of induction, in which case it is a paramagnetic, is positively magnetised inductively, and its inductivity is greater than unity; or it wards off induction, in which case it is a diamagnetic, is negatively magnetised inductively, and its inductivity is less than unity; or, lastly, it may not alter the field at all, when it is not magnetised, and its inductivity is unity.

Regarding, as I do, the force and the induction—not the force and the induced magnetisation—as the most significant quantities, it is clear that the language in which we describe these effects is somewhat imperfect, and decidedly misleading in so prominently directing attention to the induced magnetisation, especially in the case of no induced magnetisation, when the body is still subject to the magnetic influence, and is as much the seat of magnetic stress and energy as the surrounding medium. We may, by coining a new word provisionally, put the matter thus. All bodies known, as well as the so-called vacuum, can be inductized. According to whether the inductization (which is the same as "the induction," in fact) is greater or less than in vacuum (the universal magnetic medium) for the same magnetic force (the other factor of the magnetic energy product), we have positive or negative induced magnetisation.

To the universal medium, which is the primary seat of the magnetic energy, we attribute properties implying the absence of dissipation of energy, or, on the elastic solid theory, perfect elasticity. (Dissipation in space is scarcely within a measurable distance of measurement.) But that the ether, resembling an elastic solid in some of its properties, is one, is not material here. Inductization in it is of the elastic or quasi-elastic character, and there can be no intrinsic magnetisation. Nor evidently can there be intrinsic magnetisation in gases, by reason of their mobility, nor in liquids, except of the most transient description. But when we come to solids the case is different.

If we admit that the act of inductization produces a structural change in a body (this includes the case of no induced magnetisation), and if, on removal of the inducing force, the structural change disappears, the body behaves like ether, so far, or has no inductive retentiveness. Here we see the advantage of speaking of inductive rather than of magnetic retentiveness. But if, by reason of imperfect elasticity, a portion of the changed structure remains, the body has inductive retentiveness, and has become an intrinsic magnet. As for the precise nature of the magnetic structure, that is an independent question. If we can do without assuming any particular structure, as for instance, the Weber structure, which is nothing more than an alignment of the axes of molecules, a structure which I believe to be, if true at all, only a part of the magnetic structure, so much the better. It is the danger of a too special hypothesis, that as, from its definiteness, we can follow up its consequences, if the latter are partially verified experimentally we seem to prove its truth (as if there could be no other explanation), and so rest on the solid ground of nature. The next thing is to predict unobserved or unobservable phenomena whose only reason may be the hypothesis itself, one out of many which, within limits, could explain the same phenomena, though, beyond those limits, of widely diverging natures.

The retentiveness may be of the most unstable nature, as in soft iron, a knock being sufficient to greatly upset the intrinsic magnetisation existing on first removing the magnetising force, and completely alter its distribution in the iron; or of a more or less permanent character, as in steel. But, whether the body be para- or diamagnetic, or neutral, the residual or intrinsic magnetisation, if there be any, must be always of the same character as the inducing force. That is, any solid, if it have retentiveness, is made into a magnet, magnetised parallel to the inducing force, like iron.

Until lately only the magnetic metals were known to show retentiveness. Though we should theoretically expect retentiveness in all solids, the extraordinary feebleness of diamagnetic phenomena might be expected to be sufficient to prevent its observation. But, first, Dr. Tumlirz has shown that quartz is inductively retentive, and next, Dr. Lodge (*Nature*, March 25th, 1886) has published some results of his experiments on the retentiveness of a great many other substances, following up an observation of his assistant, Mr. Davies.

The mathematical statement of the connections between intrinsic magnetisation and the state of the magnetic field is just the same whether the magnet be iron or copper, para- or dia-magnetic, or is neutral. In fact, it would equally serve for a water or a gas magnet, were they possible. That is,

$$\text{curl}\,(\mathbf{H}-\mathbf{h})=4\pi\Gamma, \qquad \text{div}\,\mathbf{B}=0\,;$$

$\mathbf{H}$ being the magnetic force according to the equation $\mathbf{B}=\mu\mathbf{H}$, where $\mathbf{B}$ is the induction and μ the inductivity, Γ the electric current, if any, and $\mathbf{h}$ the magnetic force of the intrinsic magnetisation, or the impressed magnetic force, as I have usually called it in previous sections where it has occurred, because it enters into all equations as an impressed force, distinct from the force of the field, whose rotation measures the electric current. It is **h** and μ that are the two data concerned in intrinsic magnetisation and its field; the quantity **I**, the intensity of intrinsic magnetisation, only gives the product, viz., $\mathbf{I} = \mu\mathbf{h}/4\pi$. It would not be without some advantage to make **h** and μ the objects of attention instead of **I** and μ, as it simplifies ideas as well as the formulæ. The induced magnetisation, an extremely artificial and rather unnecessary quantity, is $(\mu - 1)(\mathbf{H} - \mathbf{h})/4\pi$.

It will be understood that this system, when united with the corresponding electric equations, so as to completely determine transient states, requires **h** to be *given*, whether constant or variable with the time. The act of transition of elastic induction into intrinsic magnetisation, when a body is exposed to a strong field, cannot be traced in any way by our equations. It is not formulated, and it would naturally be a matter of considerably difficulty to do it.

In a similar manner, we may expect all solid dielectrics to be capable of being intrinsically electrized by electric force, as described in a previous section. I do not know, however, whether any dielectric has been found whose dielectric capacity is less than that of vacuum, or whether such a body is, in the nature of things, possible.

As everyone knows nowadays, the old-fashioned rigid magnet is a myth. Only one datum was required, the intensity of magnetisation **I**, assuming μ to be unity in as well as outside the magnet. It is a great pity, regarded from the point of view of mathematical theory, which is rendered far more difficult, that the inductivity of intrinsic magnets is not unity. But we must take nature as we find her, and although Prof. Bottomley has lately experimented on some very unmagnetisable steel, which may approximate to $\mu = 1$, yet it is perfectly easy to show that the inductivity of steel magnets in general is not 1, but a large number, though much less than the inductivity of soft iron, and we may use a hard steel bar, whether magnetised intrinsically or not, as the core of an electromagnet with nearly the same effects, as regards induced magnetisation, except as regards the amount, as if it were of soft iron.

Regarding the measure of inductivity, especially in soft iron, this is really not an easy matter, when we pass beyond the feeble forces of telegraphy. For all practical purposes μ is a constant when the magnetic force is small, and Poisson's assumption of a linear relation between the induced magnetisation and the magnetic force is abundantly verified. It is almost mathematically true. But go to larger forces, and suppose for simplicity we have a closed solenoid with a soft iron core, and we magnetise it. Let **F** be the magnetic force of the current. Then, if the induction were completely elastic, we should have the induction $\mathbf{B} = \mu\mathbf{F}$. But in reality we have $\mathbf{B} = \mu(\mathbf{F} + \mathbf{h}) = \mu\mathbf{H}$. If we assume the former of these equations, that is, take the magnetic force of the current as *the* magnetic force, we shall obtain too large an estimate of the inductivity, in reckoning which **H** should be taken as the magnetic force. This may be several times as large as **F**. For, the softer the iron the more imperfect is its inductive elasticity, and the more easily is intrinsic magnetisation made by large forces; although the retentiveness may be of a very infirm nature, yet whilst the force **F** is on, there is **h** on also. This over-estimate of the inductivity may be partially corrected by separately measuring **h** after the original magnetising force has been removed, by then destroying **h**. But this **h** may be considerably less than the former. For one reason, when we take off **F** by stopping the coil-current, the molecular agitation of the heat of the induced currents in the core, although they are in such a direction as to keep up the induced magnetisation whilst they last, is sufficient to partially destroy the intrinsic magnetisation, owing to the infirm retentiveness. We should take off **F** by small instalments, or slowly and continuously, if we want **h** to be left.

Another quantity of some importance is the ratio of the increment in the elastic induction to the increment in the magnetic force of the current. This ratio is the same as μ when the magnetic force is small, but is, of course, quite different when it is large.

As regards another connected matter, the possible existence of magnetic friction, I have been examining the matter experimentally. Although the results are not yet quite decisive, yet there does appear to be something of the kind in steel. That is, during the act of inductively magnetising steel by weak magnetic force, there is a reaction on the magnetising current very closely resembling that arising from eddy currents in the steel, but produced under circumstances which would render the real eddy currents of quite insensible significance. In soft iron, on the other hand, I have failed to observe the effect. It has nothing to do with the intrinsic magnetisation, if any, of the steel. But as no hard and fast line can be drawn between one kind of iron and another, it is likely, if there be such an effect in steel, where, by the way, we should naturally most expect to find it, that it would be, in a smaller degree, also existent in soft iron. Its existence, however, will not alter the fact materially that the dissipation of energy in iron when it is being weakly magnetised is to be wholly ascribed to the electric currents induced in it.

P.S. (April 13, 1886.)—As the last paragraph, owing to the hypothesis involved in magnetic friction, may be somewhat obscure, I add this in explanation. The law, long and generally accepted, that the induced magnetisation is simply proportional to the magnetic force, when small, is of such importance in the theory of electromagnetism, that I wished to see whether it was minutely accurate. That is, that the curve of magnetisation is, at the origin, a straight line inclined at a definite angle to the axis of abscissæ, along which magnetic force is reckoned. I employed a differential arrangement (differential telephone) admitting of being made, by proper means, of considerable sensitiveness. The law is easily verified roughly. When, however, we increase the sensitiveness, its accuracy becomes, at first sight, doubtful; and besides, differences appear between iron and steel, differences of kind, not of mere magnitude. But as the sensitiveness to disturbing influences is also increased, it is necessary to carefully study and eliminate them. The principal disturbances are due to eddy currents, and to the variation in the resistance of the experimental coil with temperature. For instance, as regards the latter, the approach of the hand to the coil may produce an effect larger than that under examination. The general result is that the law is very closely true in iron and steel, it being doubtful whether there is any effect that can be really traced to a departure from the law, when rapidly intermittent currents are employed, and that the supposed difference between iron and steel is unverified.

Of course it will be understood by scientific electricians that it is necessary, if we are to get results of scientific definiteness, to have true balances, both of resistance and of induction, and not to employ an arrangement giving neither one nor the other. He will also understand that, quite apart from the question of experimental ability, the theorist sometimes labours under great disadvantages from which the pure experimentalist is free. For whereas the latter may not be bound by theoretical requirements, and can employ himself in making discoveries, and can put down numbers, really standing for complex quantities, as representing the specific this or that, the former is hampered by his theoretical restrictions, and is employed, in the best part of his time, in the poor work of making mere verifications.

SECTION XXVI. THE TRANSIENT STATE IN A ROUND WIRE WITH A CLOSE-FITTING TUBE FOR THE RETURN CURRENT.

The propagation of magnetic force and of electric current (a function of the former) in conductors takes place according to the mathematical laws of diffusion, as of heat by conduction, allowing for the fact of the electric quantities being vectors. This conclusion may perhaps be considered very doubtful, as depending upon some hypothesis. Since, however, it is what we arrive at immediately by the application of the laws for linear conductors to infinitely small circuits (with a tacit assumption to be presently mentioned), it seems to me more necessary for an objector to show that the laws are not those of diffusion, rather than for me to prove that they are.

We may pass continuously, without any break, from transient states in linear circuits to those in masses of metal, by multiplying the number of, whilst diminishing the section of, the "linear" conductors indefinitely, and packing them closely. Thus we may pass from linear circuits to a hollow core; from ordinary linear differential equations to a partial differential equation; from a set of constants, one for each circuit, to a continuous function, viz., a compound of the J_0 function and its complementary function containing the logarithm. This I have worked out. Though very interesting mathematically, it would occupy some space, as it is rather lengthy. I therefore start from the partial differential equation itself.

Our fundamental equations are, in the form I give to them,

$$\text{curl}\,\mathbf{H} = 4\pi\mathbf{C}, \qquad -\text{curl}\,\mathbf{E} = \mu\dot{\mathbf{H}}, \qquad \mathbf{C} = k\mathbf{E}, \quad \ldots\ldots\ldots(1b)$$

E and **H** being the electric and magnetic forces, **C** the conduction current, k and μ the conductivity and the inductivity. The assumption I referred to is that the conductor has no dielectric capacity. Bad conductors have. We are concerned with good conductors, whose dielectric capacity is quite unknown.

We are concerned with a special application, and therefore choose the suitable coordinates. All equations referring to this matter will be marked *b*. The investigations are almost identical with those given in my paper on "The Induction of Currents in Cores," in *The Electrician* for 1884. [Reprint, vol. I., p. 353, art. XXVIII.] The magnetic force was then longitudinal, the current circular; now it is the current that is longitudinal, and the magnetic force circular.

The distribution of current in a wire in the transient state depends materially upon the position of the return conductor, when it is near. The nature of the transient state is also dependent thereon. Now, if the return conductor be a wire, the distributions in the two wires are rendered unsymmetrical, and are thereby made difficult of treatment. We, therefore, distribute the return current equally all round the wire, by employing a tube, with the wire along its axis. This makes the distribution symmetrical, and renders a comparatively easy mathematical analysis possible. At the same time we may take the tube near the wire or far away, and so investigate the effect of proximity. The present example is a comparatively elementary one, the tube being supposed to be close-fitting. As I entered into some detail on the method of obtaining the solutions in "Induction in Cores," I shall not enter into much detail now. The application to round wires with the current longitudinal was made by me in *The Electrician* for Jan. 10, 1885, p. 180, so far as a general description of the phenomenon is concerned. See also my letter of April 23, 1886. [Reprint, vol. I., p. 440; vol. II., p. 30.]

Let there be a wire of radius a, surrounded by a tube of outer radius b, and thickness $b-a$. In the steady state, if the current-density is Γ in the wire, it is $-\Gamma a^2/(b^2-a^2)$ in the tube, if both be of uniform conductivity, and the tube or sheath be the return conductor of the wire. Let H_1 be the intensity of magnetic force in the wire, and H_2 in the tube. The direction of the magnetic force is circular about the axis in both, and the current is longitudinal. We shall have

$$H_1 = 2\pi\Gamma r, \qquad H_2 = -2\pi\Gamma a^2(r^2-b^2)/r(b^2-a^2), \quad \dots\dots\dots(2b)$$

where r is the distance of the point considered from the axis. Test by the first of equations (1*b*). We have

$$\text{curl} = \frac{1}{r}\frac{d}{dr}r,$$

when applied to H.

Now let this steady current be left to itself, without impressed force to keep it up, so that the "extra-current" phenomena set in, and the magnetic field subsides, the circuit being left closed. At the time t later, if the current-density be γ at distance r from the axis, it will be represented by

$$\gamma = \Sigma AJ_0(nr)\epsilon^{pt}, \quad \dots\dots\dots\dots\dots\dots\dots\dots(3b)$$

where Σ is the sign of summation. The actual current is the sum of an infinite series of little current distributions of the type represented, in which A, n, and p are constants, and $J_0(nr)$ is the Fourier cylinder function. We have

$$\frac{1}{r}\frac{d}{dr}r\frac{d\gamma}{dr} = 4\pi\mu k\dot{\gamma}. \quad \text{......................(4}b\text{)}$$

Let $d/dt = p$, a constant, then n is given in terms of p by

$$n^2 = -4\pi\mu kp. \quad \text{......................(5}b\text{)}$$

We suppose that k and μ are the same in the wire as in the sheath. Differences will be brought in in the subsequent investigation with the sheath at any distance.

In (3b) there are two sets of constants, the A's fixing the size of the normal systems, and the n's or p's, since these are connected by (5b). To find the n's, we ignore dielectric displacement, since it is electromagnetic induction that is in question. This gives the condition

$$H_2 = 0, \qquad \text{at} \qquad r = b; \quad \text{......................(6}b\text{)}$$

i.e., no magnetic force outside the tube. This gives us

$$J_1(nb) = 0, \quad \text{......................(7}b\text{)}$$

as the determinantal equation of the n's, which are therefore known by inspection of a Table of values of the J_1 function.

Find the A's by the conjugate property. Thus,

$$A = \frac{\Gamma\int_0^a J_0(nr)rdr - \int_a^b \Gamma a^2 J_0(nr)rdr/(b^2-a^2)}{\int_0^b J_0^2(nr)rdr} = \frac{2a\Gamma J_1(na)}{n(b^2-a^2)J_0^2(nb)}. \quad \text{...(8}b\text{)}$$

The full solution is, therefore,

$$\gamma = \frac{2a\Gamma}{b^2-a^2}\sum\frac{J_1(na)J_0(nr)\epsilon^{pt}}{nJ_0^2(nb)}, \quad \text{......................(9}b\text{)}$$

giving the current at time t anywhere.

The equation of the magnetic force is obtained by applying the second of equations (1b); it is

$$H = \frac{8\pi a\Gamma}{b^2-a^2}\sum\frac{J_1(na)J_1(nr)\epsilon^{pt}}{n^2J_0^2(nb)}; \quad \text{......................(10}b\text{)}$$

and the expression for the vector-potential of the current (for its scalar magnitude A_0, that is to say, as its direction, parallel to the current, does not vary, and need not be considered), is

$$A_0 = \frac{8\pi\mu a\Gamma}{b^2-a^2}\sum\frac{J_1(na)J_0(nr)\epsilon^{pt}}{n^3J_0^2(nb)}. \quad \text{......................(11}b\text{)}$$

This may be tested by

$$\text{curl}\,\mathbf{A} = \mu\mathbf{H}; \quad \text{......................(12}b\text{)}$$

curl being now $= -d/dr$. In the steady state (initial), $t=0$,

$$\left.\begin{aligned} A_0 &= \mu\pi\Gamma\left(-r^2+\frac{2a^2b^2\log(b/a)}{b^2-a^2}\right), \\ A_0 &= \frac{\mu\pi\Gamma a^2}{b^2-a^2}\left(-b^2+r^2+2b^2\log\frac{b}{r}\right), \end{aligned}\right\} \quad \ldots\ldots\ldots\ldots\ldots(13b)$$

in the wire, and (second equation) in the sheath. Test by (12*b*) applied to (13*b*) to obtain (2*b*).

The magnetic energy being $\mu H^2/8\pi$ per unit volume, the amount in length l of wire and sheath is, by (10*b*),

$$T=\frac{\mu l}{8\pi}\left(\frac{8\pi a\Gamma}{b^2-a^2}\right)^2\sum\int\frac{J_1^2(na)J_1^2(nr)}{n^4J_0^4(nb)}2\pi r\,dr\,\epsilon^{2pt}.$$

To verify, this should equal the space-integral of $\frac{1}{2}A_0\gamma$, using (11*b*) and (9*b*). This need not be written. They are identical because

$$\int_0^b J_0^2(nr)\,r\,dr=\int_0^b J_1^2(nr)\,r\,dr=\tfrac{1}{2}b^2J_0^2(nb),$$

so that we may write the expression for T thus,

$$T=\tfrac{1}{2}\mu l\left(\frac{4\pi a\Gamma b}{b^2-a^2}\right)^2\sum\frac{J_1^2(na)\epsilon^{2pt}}{n^4J_0^2(nb)}. \quad \ldots\ldots\ldots\ldots\ldots(14b)$$

The dissipativity being γ^2/k per unit volume, the total heat in length l of wire and sheath is, if $\rho=k^{-1}$, the resistivity, and the complete variable period be included,

$$Q=\rho\{2a\Gamma/(b^2-a^2)\}^2\sum\frac{J_1^2(na)\frac{1}{2}b^2J_0^2(nb)2\pi l}{n^2J_0^4(nb)(-2p)}. \quad \ldots\ldots\ldots(15b)$$

When $t=0$, either by (14*b*) or by easy direct investigation, the initial magnetic energy in length l is

$$T_0=\mu l(\tfrac{1}{2}\pi\Gamma a^2)^2\left\{1+\frac{b^2+a^2}{b^2-a^2}-\frac{4b^2}{b^2-a^2}+\frac{4b^4}{(b^2-a^2)^2}\log\frac{b}{a}\right\}, \quad \ldots..(16b)$$

giving the inductance of length l as

$$L=\mu l\left(-\frac{b^2}{b^2-a^2}+\frac{2b^4\log(b/a)}{(b^2-a^2)^2}\right), \quad \ldots\ldots\ldots\ldots\ldots(17b)$$

which may be got in other ways. This refers to the steady state. In the transient state there cannot be said to be a definite inductance, as the distribution varies with the time. The expression in (15*b*) for the total heat may be shown to be equivalent to that in (16*b*) for the initial magnetic energy, thus verifying the conservation of energy in our system.

I should remark that it is the same formula (9*b*) that gives us the current both in the wire and tube, and the same formula (10*b*) that gives us the magnetic force. They are distributed continuously in the variable period. It is at the first moment only that they are discontinuous, requiring then separate formulæ for the wire and tube, *i.e.*, separate *finite* formulæ, although only a single infinite series.

The first term of (9*b*) is, of course, the most important, representing the normal system of slowest subsidence. In fact, there is an extremely rapid subsidence of the higher normal systems; only three or four need be considered to obtain almost a complete curve; and, at a comparatively early stage of the subsidence, the first normal system has become far greater than the rest. In fact, on leaving the current without impressed force, there is at first a rapid change in the distribution of the current (and magnetic force), besides a rapid subsidence. It tends to settle down to be represented by the first normal system; a certain nearly fixed distribution, subsiding according to the exponential law of a linear circuit.

To see the nature of the rapid change, and of the first normal system, refer to *The Electrician* of Aug. 23, 1884 [vol. I., p. 387], where is a representation of the J_0 and J_1 curves. In Fig. 1, take the distance OC_2 to be the outer radius of the tube, O being on the axis. Then the curve marked J_1 is the curve of the magnetic force, showing its comparative strength from the centre of the wire to the outside of the tube, in the first normal system. And, to correspond, the curve *vv* from O up to C_2 is the curve of the current, showing its distribution in the first normal system.

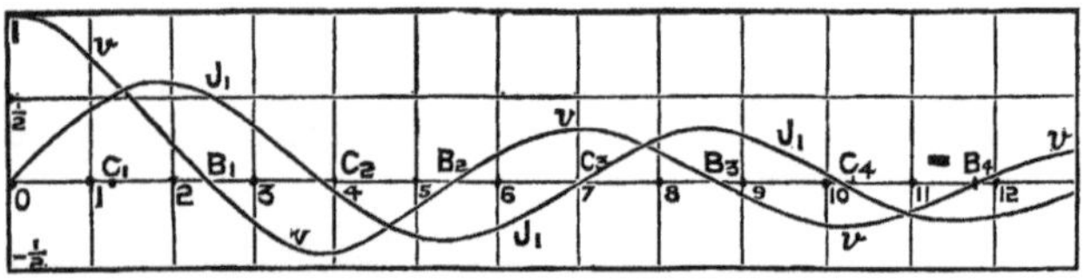

We see that the position of the point B_1 with respect to the inner radius of the sheath determines whether the current is transferred from the wire to the sheath, or *vice versâ*, in the early part of the subsidence. If the sheath is very thin, so that the radius of the wire extends nearly up to C_2, there is transfer of the sheath-current (initial) from the sheath a long way into the wire. On the other hand, if the wire be of small radius compared with the outer radius of the tube, so that the tube's depth extends from C_2 nearly up to O, there is a transfer of the original wire-current a long way into the thick sheath. In Fig. 2 [vol. I., p. 388] are shown the first four normal systems, all on the same scale as regards the vertical ordinate, but we are not concerned with them at present.

Since $$-p^{-1} = 4\pi\mu k b^2/(nb)^2,$$

by (5*b*), and $-p^{-1}$ is the time-constant of subsidence of a normal system, we have, for the value of the time-constant of the first system,

$$-p_1^{-1} = \cdot 273\, \pi\mu k b^2,$$

because the value of the first nb, say $n_1 b$, is 3·83. Compare this with the linear-theory time-constant L/R, where L is given by (17*b*), and R is the resistance of length l of the wire and sheath (sum of resistances, as the current is oppositely directed in them). Let $a = \frac{1}{2}b$. Then

$$L = 1 \cdot 128\, \mu l.$$

We have also

$$R = 16l/3\pi kb^2, \qquad \text{therefore} \qquad L/R = \cdot 211\, \pi\mu kb^2,$$

so that the time-constant of the first normal system is to that of the current in wire and tube on the linear theory as ·27 to ·21. But it is only after the first stage of the subsidence is over that this larger time-constant is valid.

We may write the expression for L thus. Let $x = b/a$, then

$$L = \frac{\mu l x^2}{x^2 - 1}\left(\frac{2x^2 \log x}{x^2 - 1} - 1\right),$$

nearly the same as $2\mu l \log x$ when x is large. The minimum is when $b = a$; then $L = \frac{1}{2}\mu l$. This is the least value of the inductance of a round wire, viz., when it has a very thin and close-fitting sheath for the return current, so that the magnetic energy is confined to the wire.

When b/a is only a little over unity,

$$L = \frac{\mu l b^2}{b^2 - a^2} \cdot \frac{3b^2 - a^2 - 2ab}{(b + a)^2}.$$

We have also $\qquad R = lb^2/\pi ka^2(b^2 - a^2),$

and therefore $\qquad L/R = \pi k\mu a^2\left(\frac{2b^2 \log(b/a)}{b^2 - a^2} - 1\right).$

Irrespective of b/a being only a little over unity, we have,

with $a/b = \frac{1}{10}$,	$L/R = \cdot 009(4\pi\mu kb^2)$,
" $\frac{1}{2}$,	" ·053 " ,
" $\frac{10}{11}$,	" ·020 " ;

whilst the time-constant of the first normal system in all three cases is 068 $(4\pi\mu kb^2)$.

The maximum of L/R with b/a variable is when

$$(x^2 - 1)(x^2 - \tfrac{1}{2})/x^4 = \log x,$$

x being b/a. This value of x is not much different from the ratio of the nodes in the first normal system, or the ratio of the value of nr making $J_1(nr) = 0$ for the first time, to that making $J_0(nr) = 0$. For the latter value makes $\log x = 465$, and makes the other side of the last equation be ·486.

In the subsidence from the steady state, the central part of the wire is the last to get rid of its current. But the steady state has to be first set up. Then it is the central part of the wire that is the last to get its full current. To obtain the equations showing the rise of the current and of the magnetic force in the wire and the tube, we have to reverse or negative the preceding solutions, and superpose the final steady states. As these are discontinuous, there are two solutions, one for the wire, the other for the sheath; but the transient part of them, which ultimately disappears, is the same in both. There is no occasion to write these out.

If the steady state is not fully set up before the impressed force is removed, we see that the central part of the wire is less useful as a conductor than the outer part, as the current is there the least. If there are short contacts, as sufficiently rapid reversals, or intermittences, the central part of the wire is practically inoperative, and might be removed, so far as conducting the current is concerned. Immediately after the impressed force is put on, there is set up a positive current on the outside of the wire, and a negative on the inside of the sheath, which are then propagated inward and outward respectively. If the sheath be thin, the initial (surface) wire-current is of greater and the initial sheath-current of less density than the values finally reached by keeping on the impressed force; whilst if it be the sheath that is thick the reverse behaviour obtains.

This case of a close-fitting tube is rather an extreme example of departure from the linear theory; the return current is as close as possible and wholly envelops the wire-current. Except as regards duration, the distributions of current and magnetic force are independent of the dimensions, *i.e.*, in the smallest possible round wire closely surrounded by the return current the phenomena are the same as in a big wire similarly surrounded, except as regards the duration of the variable period. The retardation is proportional to the conductivity, to the inductivity, and to the square of the outer radius of the tube.

When, as in our next Section, we remove the tube to a distance, we shall find great changes.

SECTION XXVII. THE VARIABLE PERIOD IN A ROUND WIRE WITH A CONCENTRIC TUBE AT ANY DISTANCE FOR THE RETURN CURRENT.

The case considered in the last Section was an extreme one of departure from the linear theory. This arose, not from mere size, but from the closeness of the return to the main conductor, and to its completely enclosing it. Practically we must separate the two conductors by a thickness of dielectric. The departure from the linear theory is then less pronounced; and when we widely separate the conductors it tends to be confined to a small portion only of the variable period. The size of the wire is then also of importance.

Let there be a straight round wire of radius a_1, conductivity k_1, and inductivity μ_1, surrounded by a non-conducting dielectric of specific capacity c and inductivity μ_2 to radius a_2, beyond which is a tube of conductivity k_3, and inductivity μ_3, inner radius a_2 and outer a_3. The object of taking c into account, temporarily, will appear later.

Let the current be longitudinal and the magnetic force circular. Then, by (1*b*), if γ is the current-density at distance r from the axis, we shall have

$$\frac{1}{r}\frac{d}{dr}r\frac{d\gamma}{dr} = 4\pi\mu k\dot{\gamma}, \qquad \text{or} \qquad = c\mu\ddot{\gamma}, \quad \ldots\ldots\ldots\ldots\ldots(18b)$$

in the conductors, and in the dielectric respectively; the latter form being got by taking $\gamma = c\dot{E}/4\pi$, the rate of increase of the elastic displacement.

A normal system of longitudinal current-density may therefore be represented by

$$\left.\begin{aligned}\gamma_1&=A_1J_0(n_1r), && \text{from } r=0 \text{ to } a_1\\ \gamma_2&=A_2J_0(n_2r)+B_2K_0(n_2r), && \quad ,, \quad r=a_1 \text{ to } a_2\\ \gamma_3&=A_3J_0(n_3r)+B_3K_0(n_3r), && \quad ,, \quad r=a_2 \text{ to } a_3\end{aligned}\right\}\text{............} (19b)$$

in the wire, in the insulator, and in the sheath, respectively, at a given moment. In subsiding, free from impressed force, each of these expressions, when multiplied by the time-factor ϵ^{pt}, gives the state at the time t later.

$J_0(nr)$ is the Fourier cylinder function, and $K_0(nr)$ the complementary function. [For their expansions see vol. I., p. 387, equations (70) and (71)]. The A's and B's are constants, fixing the size of the normal functions; the n's are constants showing the nature of the distributions, and p determines the rapidity of the subsidence.

By applying (18b) to (19b) we find

$$n_1^2=-4\pi\mu_1k_1p, \qquad n_2^2=-\mu_2cp^2, \qquad n_3^2=-4\pi\mu_3k_3p; \text{ ......}(20b)$$

expressing all the n's in terms of the p.

Corresponding to the expressions (19b) for the current, we have the following for the magnetic force:—

$$\left.\begin{aligned}H_1&=-(n_1/\mu_1k_1p)A_1J_1(n_1r),\\ H_2&=-(4\pi n_2/\mu_2cp^2)\{A_2J_1(n_2r)+B_2K_1(n_2r)\},\\ H_3&=-(n_3/\mu_3k_3p)\quad\{A_3J_1(n_3r)+B_3K_1(n_3r)\},\end{aligned}\right\}\text{..............}(21b)$$

where, as is usual, the negative of the differential coefficient of $J_0(z)$ with respect to z is denoted by $J_1(z)$; and, in addition, the negative of the differential coefficient of $K_0(z)$ with respect to z is denoted by $K_1(z)$. These equations (21b) are got by the second and third equations (1b), in the case of H_1 and H_3; and in the case of H_2, by using, instead of Ohm's law, the dielectric equation, giving

$$E=4\pi\gamma/cp,$$

in the dielectric, E being the electric force. Of course $d/dt=p$, in a normal system.

We have next to find the relations between the five A's and B's, to make the three solutions fit one another, or harmonize. This we must do by means of the boundary conditions. These are nothing more than the surface interpretations of the ordinary equations referring to space distributions. In the present case the appropriate conditions are continuity of the magnetic and of the electric force at the boundaries, because the two forces are tangential; the conditions of continuity of the normal components of the electric current and of magnetic induction are not applicable, because there are no normal components in question. If the magnetic or the electric force were discontinuous, we should have electric or magnetic current-sheets.

Thus H_1 and H_2 are equal at $r=a_1$, and H_2 and H_3 are equal at $r=a_2$. These give, by (21b),

$$(A_1n_1/\mu_1k_1p)J_1(n_1a_1)=(4\pi n_2/\mu_2cp^2)\{A_2J_1(n_2a_1)+B_2K_1(n_2a_1)\}, \quad (22b)$$

and $$(4\pi n_2/\mu_2 cp^2)\{A_2 J_1(n_2a_2)+B_2K_1(n_2a_2)\}$$
$$=(n_3/\mu_3 k_3 p)\{A_3J_1(n_3a_2)+B_3K_1(n_3a_2)\}. \quad \ldots\ldots(23b)$$

Similarly, E_1 and E_2 are equal at $r=a_1$, and E_2 and E_3 are equal at $r=a_2$. These give, by (19*b*),

$$(A_1/k_1)J_0(n_1a_1)=(4\pi/cp)\{A_2J_0(n_2a_1)+B_2K_0(n_2a_1)\}, \quad \ldots\ldots(24b)$$

and

$$(4\pi/cp)\{A_2J_0(n_2a_2)+B_2K_0(n_2a_2)\}=k_3^{-1}\{A_3J_0(n_3a_2)+B_3K_0(n_3a_2)\}. \quad (25b)$$

Thus, starting with A_1 given, (22*b*) and (24*b*) give A_2 and B_2 in terms of A_1, and then (23*b*) and (25*b*) give A_3 and B_3 in terms of A_1. Similarly we might carry the system further, by putting more concentric tubes of conductors and dielectrics, or both, outside the first tube, using similar expressions for the magnetic and electric forces; every fresh boundary giving us two boundary conditions of continuity to connect the solution in one tube with that in the next. But at present we may stop at the first tube. Ignore the dielectric displacement beyond it, *i.e.*, put $c=0$ beyond $r=a_3$, because our tube is to be the return conductor to the wire inside it. We may merely remark in passing that although when such is the case, there is, in the steady state, absolutely no magnetic force outside the tube, yet this is not exactly true in a transient state. To make it true, take $c=0$ beyond $r=a_3$; requiring $H_3=0$ at $r=a_3$. This gives, by (21*b*),

$$A_3J_1(n_3a_3)+B_3K_1(n_3a_3)=0. \quad \ldots\ldots(26b)$$

Now A_3 and B_3 are, by the previous, known in terms of A_1. Make the substitution, and we find, first, that A_1 is arbitrary, so that it, when given, fixes the size of the whole normal system of electric and magnetic force; and next, that the n's are subject to the following equation :—

$$\frac{\dfrac{n_3}{\mu_3}J_0(n_2a_2)\dfrac{J_1(n_3a_2)K_1(n_3a_3)-J_1(n_3a_3)K_1(n_3a_2)}{J_0(n_3a_2)K_1(n_3a_3)-J_1(n_3a_3)K_0(n_3a_2)}-\dfrac{n_2}{\mu_2}J_1(n_2a_2)}{\dfrac{n_3}{\mu_3}K_0(n_2a_2)\times\;.\;.\;.\;.\;.\;.\;.\;.\;.\;.\;.\;-\dfrac{n_2}{\mu_2}K_1(n_2a_2)}$$
$$=\frac{(n_1/\mu_1)J_1(n_1a_1)J_0(n_2a_1)-(n_2/\mu_2)J_1(n_2a_1)J_0(n_1a_1)}{(n_1/\mu_1)J_1(n_1a_1)K_0(n_2a_1)-(n_2/\mu_2)K_1(n_2a_1)J_0(n_1a_1)}, \quad \ldots\ldots(27b)$$

where, on the left side, to save trouble, the dots represent the same fraction that appears in the numerator immediately over them.

Now, the n's are known in terms of p, hence (27*b*) is the determinantal equation of the p's, determining the rates of subsidence of all the possible normal systems. We have, therefore, all the information required in order to solve the problem of finding how any initially given state of circular magnetic force and longitudinal electric force in the wire, insulator, and sheath subsides when left to itself. We merely require to decompose the initial states into normal systems of the above types, and then multiply each term by its proper time-factor ϵ^{pt} to let it subside at its proper rate. To effect the decomposition, make use of the universal conjugate property of the equality of the mutual potential and the mutual kinetic energy of two complete normal systems, $U_{12} = T_{12}$ [vol. I., p. 523], which results from the equation of activity. We start with a given amount of electric energy in the dielectric, and of magnetic energy in the wire, dielectric and sheath, which are finally used up in heating the wire and sheath, according to Joule's law.

It would be useless to write out the expressions, for I have no intention of discussing them in the above general form, especially as regards the influence of c. Knowing from experience in other similar cases that I have examined, that the effect of the dielectric displacement on the wire and sheath phenomena is very minute, we may put $c=0$ at once between the wire and the sheath. We might have done this at the beginning; but it happens that although the results are more complex, yet the reasoning is simpler, by taking c into account.

The question may be asked, how set up a state of purely longitudinal electric force in the tube, sheath, and intermediate dielectric? As regards the wire and sheath, it is simple enough; a steady impressed force in any part of the circuit will do it (acting equally over a complete section). But it is not so easy as regards the dielectric. It requires the impressed force to be so distributed in the conductors as to support the current on the spot without causing difference of potential. There will then be no dielectric displacement either (unless there be impressed force in the dielectric to cause it). Now, if we remove the impressed force in the conductors, the subsequent electric force will be purely longitudinal in the dielectric as well as in the conductors.

But practically we do not set up currents in this way, but by means of localised impressed forces. Then, although the steady state is one of longitudinal electric force in the wire and sheath, in the dielectric there is normal or outward electric force as well as tangential or longitudinal, and the normal component is, in general, far greater than the tangential. In fact, the electrostatic retardation depends upon the normal displacement. But electrostatic retardation, which is of such immense importance on long lines, is quite insignificant in comparison with electromagnetic on short lines, and in ordinary laboratory experiments with closed circuits (no condensers allowed) is usually quite insensible. We see, therefore, that when we put $c=0$, and have purely longitudinal electric force, we get the proper solutions suitable for such cases where the influence of electrostatic charge is negligible, irrespective of the distribution of the original impressed force. Our use of the longitudinal displacement in the dielectric, then, was merely to establish a connection in time between the wire and the sheath, and to simplify the conditions.

(In passing, I may give a little bit of another investigation. Take both electric and magnetic induction into consideration in this wire and sheath problem, treating them as solids in which the current distribution varies with the time. The magnetic force is circular, so is fully specified by its intensity, say H, at distance r from the axis. Its equation is, if z be measured along the axis,

$$\frac{d}{dr}\frac{1}{r}\frac{d}{dr}rH + \frac{d^2H}{dz^2} = 4\pi\mu k\dot{H} + \mu c\ddot{H},$$

in which discard the last term when the wire or sheath is in question; or retain it and discard the previous when the dielectric is considered. The form of the normal H solution is

$$H = J_1(sr)(A \sin + B \cos)mz\,\epsilon^{pt},$$

for the wire, where $s^2 = -(4\pi\mu kp + m^2)$. The current has a longitudinal and a radial component, say Γ and γ, given by

$$\Gamma = \quad sJ_0(sr)(A \sin + B \cos)mz\,\epsilon^{pt},$$
$$\gamma = -mJ_1(sr)(A \cos - B \sin)mz\,\epsilon^{pt}.$$

In the dielectric and sheath the K_0 and K_1 functions have, of course, to be counted with the J_0 and J_1.)

Now put $c=0$ in (27*b*). We shall have

$$J_0(n_2r)=1; \quad -n_2J_1(n_2r)=0; \quad K_0(n_2r)=\log(n_2r); \quad -n_2rK_1(n_2r)=1;$$

which will bring (27*b*) down to

$$J_1(n_1a_1) = \frac{n_3}{\mu_3}\cdot\frac{J_1(n_3a_2)K_1(n_3a_3) - J_1(n_3a_3)K_1(n_3a_2)}{J_0(n_3a_2)K_1(n_3a_3) - J_1(n_3a_3)K_0(n_3a_2)}$$
$$\times\left\{(\mu_1a_2/n_1a_1)J_0(n_1a_1) - \mu_2a_2J_1(n_1a_1)\log(a_2/a_1)\right\}, \quad (28b)$$

the determinantal equation in the case of ignored dielectric displacement.

To obtain this directly, establish a rigid connection between the magnetic and electric forces at $r=a_1$ and at $r=a_2$, thus. Since there is no current in the insulating space, the magnetic force varies inversely as the distance from the axis of the wire. Therefore, instead of the second of (21*b*), we shall have

$$H_2 = -(n_1/\mu_1k_1p)A_1J_1(n_1a_1)(a_1/r),$$

by the first of (21*b*). Thus H_2 at $r=a_2$ is known, and, equated to H_3 at $r=a_2$, gives us one equation between A_1, A_3, and B_3. Next we have

$$\int_{a_1}^{r}(a_1/r)H_1dr = H_1a_1\log(r/a_1) = -(n_1a_1/\mu_1k_1p)J_1(n_1a_1)\log(r/a_1);$$

H_1 meaning, temporarily, the value of H_1 at $r=a_1$. This, when multiplied by μ_2, is the amount of induction through a rectangular portion of a plane through the axis, bounded by straight lines of unit length parallel to the axis at distances a_1 and r from it; or the line-integral of the vector-potential round the rectangle; or the excess of the vector-potential at distance r over that at distance a_1; so, when multiplied by p, it is the excess of the electric force at a_1 over that at r. Thus the electric force is known in the insulating space in terms of that at the boundary of the wire. Its value at $r=a_2$ equated to E_3 at $r=a_2$ gives us a second equation between A_1, A_3, and B_3. The third is equation (26*b*) over again, and the union of the three gives us (28*b*) again.

We now have, if γ_1 and γ_3 are the *actual* current-densities at time t in the wire and the sheath respectively,

$$\left.\begin{aligned}\gamma_1 &= \Sigma AJ_0(n_1r)\,\epsilon^{pt}\\ \gamma_2 &= \Sigma B\{J_0(n_3r) - J_1(n_3a_3)K_0(n_3r)/K_1(n_3a_3)\}\,\epsilon^{pt}\end{aligned}\right\}, \quad \ldots\ldots..(29b)$$

where $$B = A\frac{k_3}{k_1} \cdot \frac{J_0(n_1a_1) - n_1a_1(\mu_2/\mu_1)J_1(n_1a_1)\log(a_2/a_1)}{J_0(n_3a_2) - J_1(n_3a_3)K_0(n_3a_2)/K_1(n_3a_3)},$$

in which only the A requires to be found, so that when $t=0$, the initial state may be expressed. The decomposition of the initial state into normal systems may be effected by the conjugate property of the vanishing of the mutual kinetic energy, or of the mutual dissipativity of a pair of normal systems. Thus, in the latter case, writing (29*b*) thus, $\gamma_1 = \Sigma Au$, $\gamma_3 = \Sigma Av$, we shall have

$$\int_0^{a_1} u_1u_2rdr/k_1 + \int_{a_2}^{a_3} v_1v_2rdr/k_3 = 0,$$

u_1, v_1, and u_2, v_2 being a pair of normal solutions.

We can only get rid of those disagreeable customers, the K_0 and K_1 functions, by taking the sheath so thin that it can be regarded as a linear conductor—*i.e.*, neglect variations of current-density in it, and consider instead the integral current. (Except when the sheath and wire are in contact and of the same material, as in the last section.) Let a_4 be the very small thickness of the sheath, and evaluate (28*b*) on the supposition that a_4 is infinitely small, so that a_2 and a_3 are equal ultimately. The result is

$$J_0(n_1a_1) = a_1J_1(n_1a_1)\left\{(n_1\mu_2/\mu_1)\log(a_2/a_1) - k_1/n_1k_3a_2a_4\right\}, \quad \ldots .(30b)$$

the determinantal equation in the case of a round wire of radius a_1 with a return conductor in the form of a very thin concentric sheath, radius a_2. Notice that μ_3, the inductivity of the sheath itself, has gone out altogether; that is, an iron sheath for the return, if it be thin enough, does not alter the retardation as compared with a copper sheath, provided the difference of conductivity be allowed for.

We may get (30*b*) directly, easily enough, by considering that the total sheath-current must be the negative of the total wire-current, which last is, by integrating the first of (29*b*) throughout the wire,

$$= (A/n_1)2\pi a_1 \, \Sigma \, J_1(n_1a_1) \, \epsilon^{pt}.$$

This, divided by the volume of the sheath per unit length, that is, by $2\pi a_2a_4$, gives us the sheath current-density, and this, again, divided by k_3 gives us the electric force at $r=a_2$. Another expression for the electric force at the sheath is given by the previous method (the rectangle business). Equate them, and (30*b*) results.

We have now got the heavy work over, and some results of special cases will follow, in which we shall be materially assisted by the analogy of the eddy currents in long cores inserted in long solenoidal coils.

Section XXVIII. Some Special Results relating to the Rise of the Current in a Wire.

Premising that the wire is of radius a_1, conductivity k_1, inductivity μ_1; that the dielectric displacement outside is ignored; and that the sheath for the return current is at distance a_2, and is so thin that variations of current-density in it may be ignored, so that merely the total return current need be considered; that a_4 is the small thickness of the sheath, and k_3 its conductivity, we have the determinantal equation (30*b*). Let now

$$L_0 = 2\mu_2 \log(a_2/a_1), \qquad R_1 = (k_1 \pi a_1^2)^{-1}, \qquad R_2 = (2\pi a_2 a_4 k_3)^{-1}.$$

L_0 is the external inductance per unit length, *i.e.*, the inductance per unit length of *surface*-current, ignoring the internal magnetic field. R_1 and R_2 are the resistances per unit length of the wire and sheath respectively, and $\frac{1}{2}\mu_1$ is the internal inductance per unit length, *i.e.*, the inductance per unit length of uniformly distributed wire-current when the return current is on its surface, thus cancelling the external magnetic field. We can now write (30*b*) thus:—

$$J_0(n_1 a_1) = J_1(n_1 a_1)\left\{\tfrac{1}{2}n_1 a_1 (L_0/\mu_1) - (\tfrac{1}{2}n_1 a_1)^{-1}(R_2/R_1)\right\}; \quad \ldots\ldots(31b)$$

and, in this, we have

$$\left.\begin{aligned} -n_1^2 a_1^2 &= 4\pi\mu_1 k_1 p a_1^2 = 4p\mu_1/R_1, \\ J_0(n_1 a_1) &= 1 - \frac{n_1^2 a_1^2}{2^2} + \frac{n_1^4 a_1^4}{2^2 4^2} - \ldots, \\ J_1(n_1 a_1) &= \tfrac{1}{2}n_1 a_1\left(1 - \tfrac{1}{2}\frac{n_1^2 a_1^2}{2^2} + \tfrac{1}{3}\frac{n_1^4 a_1^4}{2^2 4^2} - \ldots\right). \end{aligned}\right\} \ldots\ldots\ldots\ldots(32b)$$

From (31*b*) we see that the two important quantities are the ratio of the external to the internal inductance, and the ratio of the external to the internal resistance, *i.e.*, the ratios L_0/μ_1 and R_2/R_1.

Suppose, first, the return has no resistance. Draw the curves

$$y_1 = J_0(x)/J_1(x) \qquad \text{and} \qquad y_2 = \tfrac{1}{2}(L_0/\mu_1)x,$$

the ordinates y, abscissæ x, which stands for $n_1 a_1$. Their intersections show the required values of x. The J_0/J_1 curve is something like the curve of cotangent. If L_0/μ_1 is large, the first intersection occurs with a small value of x, so small that $J_0(x)$ is very little less than unity, so that a uniform distribution of current is nearly represented by the first normal distribution, whose time-constant is a little greater than that of the linear theory. The remaining intersections will be nearly given by $J_1(x) = 0$. On the other hand, decreasing L_0/μ_1 increases the value of the first x; in the limit it will be the first root of $J_0(x) = 0$. Thus, if the wire be of copper, and the return distant (compared with radius of wire), the linear theory is approximated to. If of iron, on the other hand, it is not practicable to have the return sufficiently distant, on account of the large value of μ_1, unless the wire be exceedingly fine. Even if of copper, bringing the return closer has the same effect of rendering the first normal system widely different from representing a uniform distribution of current. It is the external magnetic field that gives stability, and reduces differences of current-density.

Next, let the return have resistance. The curve y_2 must now be

$$y_2 = \tfrac{1}{2}(L_0/\mu_1)x - 2(R_2/R_1 x).$$

The effect of increasing R_2 from zero is the opposite of that of increasing L_0. It increases the first x, and tends to increase it up to that given by $J_1(x)=0$ (not counting the zero root of this equation). Thus there is a double effect produced. Whilst on the one hand the rapidity of subsidence is increased by the resistance of the sheath, on the other the wire-current in subsiding is made to depart more from the uniform distribution of the linear theory. The physical explanation is, that as the external field in the case of sheath of no resistance cannot dissipate its energy in the sheath it must go to the wire. But when the sheath has great resistance the external field is killed by it; then the internal field is self-contained, or the wire-current subsides as if $J_1(x)=0$, with a wide departure from uniform distribution. This must be marked when the wire-circuit is suddenly interrupted, making the return-resistance infinite.

Now, let there be no current at the time $t=0$, when, put on, and keep on, a steady impressed force, of such strength that the final current-density in the wire is Γ_0. At time t the current-density Γ at distance r from the axis is given by

$$\frac{\Gamma}{\Gamma_0}=1-\sum\frac{2}{n_1a_1}\cdot\frac{(R_1+R_2)J_0(n_1r_1)/J_1(n_1a_1)R_1}{1-R_2/\mu_1p+\{J_0(n_1a_1)/J_1(n_1a_1)\}^2}\epsilon^{pt},$$

where the n_1a_1's are the roots of equation (31*b*). And the total current in the wire, say C_1, and with it the equal and opposite sheath-current, will rise thus to the final value C_0,

$$\frac{C}{C_0}=1-\sum\frac{4}{n_1^2a_1^2}\cdot\frac{(R_1+R_2)\epsilon^{pt}/R_1}{1-R_2/\mu_1p+\{J_0(n_1a_1)/J_1(n_1a_1)\}^2}.$$

It will give remarkably different results according as we take the resistance of the wire very small and that of the sheath great, or conversely, or as we vary the ratio L_0/μ_1. Infinite conductivity shuts out the current from the wire altogether, and so does infinite inductivity; the retardation to the inward transmission of the current being proportional to the product $\mu_1k_1a_1^2$. Similarly, if the sheath has no resistance, the return current is shut out from it. In either of these shutting-out cases the current becomes a mere surface-current, what it always is in the initial stage, or when we cannot get beyond the initial stage, by reason of rapidly reversing the impressed force, when the current will be oppositely directed in concentric layers, decreasing in strength with great rapidity as we pass inward from the boundary. But if both the sheath and the wire have no resistance, there will be no current at all, except the dielectric current, which is here ignored, and the two surface-currents.

The way the current rises in the wire, at its boundary, and at its centre, is illustrated in "Induction in Cores." For the characteristic equation of the longitudinal magnetic force in a core placed within a long solenoid, and that of the longitudinal current in our present case, are identical. The boundary equations are also identical. That is, (31*b*) is the boundary equation of the magnetic force in the core, excepting that the constants L_0/μ_1 and R_2/R_1 have entirely different meanings, depending upon the number of turns of wire in the coil, and its dimensions, and resistance. If, then, we adjust the constants to be equal in both cases, it follows that when any varying impressed force acts in the circuit of the wire and sheath, the current in the wire will be made to vary in identically the same manner as the magnetic force in the core, at a corresponding distance from the axis, when a similarly varying impressed force acts in the coil-circuit (which, however, must have only resistance in circuit with it, not external self-induction as well). Thus, we can translate our core-solutions into round-straight-wire solutions, and save the trouble of independent investigation, in case a detailed solution has been already arrived at in either case.

Refer to Fig. 3 [p. 398, vol. I., here reproduced]. It represents the curves of subsidence from the steady state. The "arrival" curves are got by perversion and inversion, *i.e.*, turn the figure upside down and look at it from behind. The case we now refer to is when the sheath has negligible resistance, and when we take the constant $L_0 = 2\mu_1$, which requires a near return when the wire is of copper, but a very distant one if it is iron.

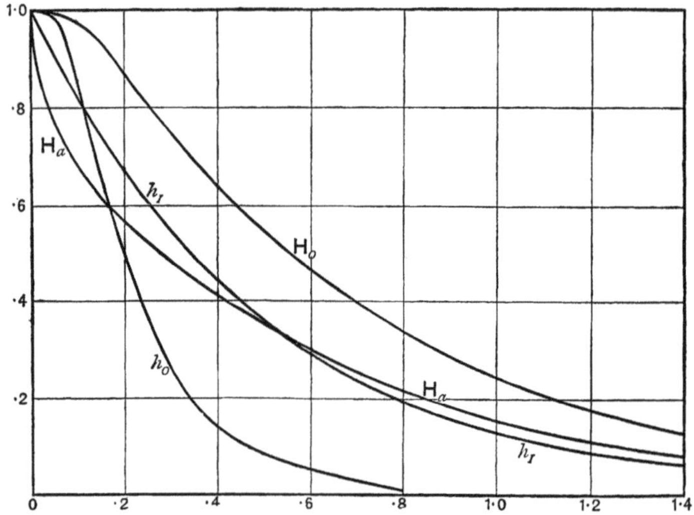

Regarding them as arrival-curves, the curve h_1h_1 is the linear-theory curve, showing how the current-density would rise in all parts of the wire if it followed the ordinarily assumed law (so nearly true in common fine-wire coils).

The curve H_aH_a shows what it really becomes, at the boundary, and near to it. The current rises much more rapidly there in the first part of the variable period, and much more slowly in the later part. From this we may conclude that, when very rapid reversals are sent, the amplitude of the boundary current-density will be far greater than according to the linear theory; whereas if they be made much slower it may become weaker. This is also verified by the separate calculation in "Induction in Cores" of the reaction on the coil-current of the core-currents when the impressed force is simple-harmonic, the amplitude of the coil-current being lowered at a low frequency, and greatly increased at high frequencies [p. 370, vol. I.].

The curve H_0H_0 shows how the current rises at the axis of the wire. It is very far more slowly than at the boundary. But the important characteristic is the preliminary retardation. For an appreciable interval of time, whilst the boundary-current has reached a considerable fraction of its final strength, the central current is infinitesimal. In fact the theory is similar to that of the submarine cable; when a battery is put on at one end, there is only infinitesimal current at the far end for a certain time, after which comes a rapid rise.

Between the axis and the boundary the curves are intermediate between H_aH_a at the boundary and H_0H_0 at the axis, there being preliminary retardation in all, which is zero at the boundary, a maximum at the axis. It is easy to understand, from the existence of this practically dead period, how infinitesimally small the axial current can be, compared with the boundary current, when very rapid reversals are sent. The formulæ will follow.

The fourth curve h_0h_0 shows the way the current rises at the axis when the return has no resistance, but when at the same time there is no external magnetic field, or $L_0/\mu_1 = 0$. The return must fit closely over the wire. We may approximate to this by using an iron wire and a close-fitting copper sheath of much lower resistance. There is preliminary retardation, after which the current rises far more rapidly than when L_0/μ_1 is finite.

That is, the effect of changing L_0/μ_1 from the value 2 to the value 0 is to change the axial arrival-curve from H_0H_0 to h_0h_0. Suppose it is a copper wire. Then $L_0 = 2$ means $\log(a_2/a_1) = 1$, or $a_2/a_1 = 2{\cdot}718$. Thus, removing the sheath from contact to a distance equal to 2·7 times the radius of the wire alters the axial arrival-curve from h_0h_0 to H_0H_0. Now this great alteration does not signify an increased departure from the linear theory (equal current-density over all the wire). It is exactly the reverse. We have increased the magnetic energy by adding the external field, and, therefore, make the current rise more slowly. But the shape of the curve H_0H_0, if the horizontal (time) scale be suitably altered, will approximate more closely to the linear-theory curve h_1h_1. By taking the sheath further and further away, continuously increasing the slowness of rise of the current, we (altering the scale) approximate as nearly as we please to the linear-theory curve, and gradually wipe out the preliminary axial retardation, and make the current rise nearly uniformly all over the section of the wire, except at the first moment. In fact, we have to distinguish between the absolute and the relative. When the sheath is most distant the current rises the most slowly, but also the most regularly. On the other hand, when the sheath is nearest, and the current rises most rapidly, it does so with the greatest possible departure from uniformity of distribution.

If the wire is of iron, say $\mu_1 = 200$, the distance to which the sheath would have to be moved would be impracticably great, so that, except in an iron wire of very low inductivity, or of exceedingly small radius, we cannot get the current to rise according to the linear theory.

The simple-harmonic solutions I must leave to another Section. We may, however, here notice the water-pipe analogy [p. 384, vol. I.]. The current starts in the wire in the same manner as water starts into motion in a pipe, when it is acted upon by a longitudinal dragging force applied to its boundary. Let the water be at rest in the first place. Then, by applying tangential force of uniform amount per unit area of the boundary we drag the outermost layer into motion instantly; it, by the internal friction, sets the next layer moving, and so on, up to the centre. The final state will be one of steady motion resisted by surface friction, and kept up by surface force.

The analogy is useful in two ways. First, because any one can form an idea of this communication of motion into the mass of water from its boundary, as it takes place so slowly, and is an everyday fact in one form or another; also, it enables us to readily perceive the manner of propagation of waves of current into wires when a rapidly varying impressed force acts in the circuit, and the rapid decrease in the amplitude of these waves from the boundary inward.

Next, it is useful in illustrating how radically wrong the analogy really is which compares the electric current in a wire to the current of water in a pipe, and impressed E.M.F. to bodily acting impressed force on the water. For we have to apply the force to the boundary of the water, not to the water itself in mass, to make it start into motion so that its velocity can be compared with the electric current-density.

The inertia, in the electromagnetic case, is that of the magnetic field, not of the electricity, which, the more it is searched for, the more unsubstantial it becomes. It may perhaps be abolished altogether when we have a really good mechanical theory to work with, of a sufficiently simple nature to be generally understood and appreciated.

In our fundamental equations of motion

$$\text{curl}\,(\mathbf{e}-\mathbf{E}) = \mu\dot{\mathbf{H}}, \qquad \text{curl}\,\mathbf{H} = 4\pi\Gamma,$$

suppose we have, in the first place, no electric or magnetic energy, so that $\mathbf{E}=0$, $\mathbf{H}=0$, everywhere, and then suddenly start an impressed force $\mathbf{e}$. The initial state is

$$\mathbf{E}=0, \qquad \mathbf{H}=0, \qquad \text{curl}\,\mathbf{e} = \mu\dot{\mathbf{H}}.$$

Thus the first effect of $\mathbf{e}$ is to set up, not electric current (for that requires there to be magnetic force), but magnetic current, or the rate of increase of the magnetic induction, and this is done, not by $\mathbf{e}$, but by its rotation, and at the places of its rotation. [A general demonstration will be given later that disturbances due to impressed $\mathbf{e}$ or $\mathbf{h}$ always have curl $\mathbf{e}$ and curl $\mathbf{h}$ for sources.]

Now, imagine $\mathbf{e}$ to be uniformly distributed throughout a wire. Its rotation is zero, except on the boundary, where it is numerically e, directed perpendicularly to the axis of the wire. Thus the first effect is magnetic current on the boundary of the wire, and this is propagated inward and outward through the conductor and the dielectric respectively. Magnetic current, of course, leads to magnetic induction and electric current.

Now, in purely electromagnetic investigations relating to wires, in which we ignore dielectric displacement, we may, for purposes of calculation, transfer our impressed forces from wherever they may be in the circuit to any other part of the circuit, or distribute them uniformly, so as to get rid of difference of potential, which is much the best plan. It is well, however, to remember that this is only a device, similar in reason and in effect to the devices employed in the statics and dynamics of supposed rigid bodies, shifting applied forces from their points of application to other points, completely ignoring how forces are really transmitted. The effect of an impressed force in one part of the circuit is assumed to be the same as if it were spread all round the circuit. It would be identically the same were there no dielectric displacement, but only the magnetic force in question. When, however, we enlarge the field of view, and allow the dielectric displacement, it is not permissible to shift the impressed forces in the above manner, for every special arrangement has its own special distribution of electric energy. The transfer of energy is, of course, always from the source, wherever it may be. The first effect of starting a current in a wire is the dielectric disturbance, directed in space by the wire, because it is a sink of energy where it can be dissipated. But the dielectric disturbance travels with such great speed that we may, unless the line is long, regard it as affecting the wire at a given moment equally in every part of its length; and this is substantially what we do when we ignore dielectric displacement in our electromagnetic investigations, distribute the impressed force as we please, and regard a long wire in which a current is being set up from outside as similar to a long core in a magnetising helix, when we ignore any difference in action at different distances along the core.

Section XXIX. Oscillatory Impressed Force at one End of a Line. Its Effect. Application to Long-Distance Telephony and Telegraphy.

Given that there is an oscillatory impressed force in a circuit, if this question be asked—what is the effect produced? the answer will vary greatly according to the conditions assumed to prevail. I therefore make the conditions very comprehensive, taking into account frictional resistance, forces of inertia, forces of elasticity, and also the approximation to surface conduction that the great frequency of telephonic currents makes of importance.

Space does not permit a detailed proof from beginning to end. The results may, however, be tested for accuracy by their satisfying all the conditions laid down, most of which I have given in the last three Sections.

The electrical system consists of a round wire of radius a_1, conductivity k_1, and inductivity μ_1; surrounded by an insulator of inductivity

μ_2 and specific dielectric capacity c, to radius a_2; surrounded by the return of conductivity k_3, inductivity μ_3, and outer radius a_3. The wire and return to be each of length l, and to be joined at the ends to make a closed conductive circuit.

Let S be the electrostatic capacity, and L_0 the inductance of the dielectric per unit length of the line. That is,

$$L_0 = 2\mu_2 \log(a_2/a_1), \qquad S = c\{2 \log(a_2/a_1)\}^{-1}. \quad \text{........(33}b\text{)}$$

We have $L_0 S = c\mu_2 = v^{-2}$; if v is the speed of undissipated waves through the dielectric.

Let V be the surface-potential of the wire, and C the wire-current, or total current in the wire, at distance x from one end, at time t. The differential equation of V is

$$\frac{d^2V}{dx^2} = (R' + L'p)SpV; \quad \text{........................(34}b\text{)}$$

where R' and L' are certain even functions of p, whose structure will be explained later, and p stands for d/dt. That of C is the same. The connection between C and V is given by

$$-\frac{dV}{dx} = (R' + L'p)C. \quad \text{..........................(35}b\text{)}$$

Both (34b) and (35b) assume that there is no impressed force at the place considered. If there be impressed force e per unit length, add e to the left side of (35b), and make the necessary change in (34b), which is connected with (35b) through the equation of continuity

$$-\frac{dC}{dx} = SpV. \quad \text{........ ....(36}b\text{)}$$

But as we shall only have e at one end of the line, we shall not require to consider e elsewhere.

Now, given (34b) and (35b), and that there is an impressed force $V_0 \sin nt$ at the $x=0$ end, find V and C everywhere. Owing to R' and L' containing only even powers of p, and to the property $p^2 = -n^2$ possessed by p in simple-harmonic arrangements, R' and L' become constants. The solution is therefore got readily enough. Let

$$\left.\begin{aligned} P &= (\tfrac{1}{2}Sn)^{\frac{1}{2}}\{(R'^2 + L'^2n^2)^{\frac{1}{2}} - L'n\}^{\frac{1}{2}}, \\ Q &= (\tfrac{1}{2}Sn)^{\frac{1}{2}}\{(R'^2 + L'^2n^2)^{\frac{1}{2}} + L'n\}^{\frac{1}{2}}. \end{aligned}\right\} \quad \text{................(37}b\text{)}$$

These are very important constants concerned. Let also

$$\left.\begin{aligned} \tan\theta_1 &= (L'nP - R'Q)/(R'P + L'nQ), \\ \tan\theta_2 &= \sin 2Ql/(\epsilon^{-2Pl} - \cos 2Ql). \end{aligned}\right\} \quad \text{...............(38}b\text{)}$$

These make θ_1 and θ_2 angles less than 90°. Then the potential V at distance x at time t is

$$V = V_0\epsilon^{-Px}\sin(nt - Qx) + V_0\frac{\epsilon^{Px}\sin(nt + Qx + \theta_2) - \epsilon^{-Px}\sin(nt - Qx + \theta_2)}{\epsilon^{Pl}(\epsilon^{2Pl} + \epsilon^{-2Pl} - 2\cos 2Ql)^{\frac{1}{2}}}. \quad (39b)$$

and the current C is

$$C = V_0 \frac{(Sn)^{\frac{1}{2}}}{(R'^2 + L'^2 n^2)^{\frac{1}{4}}} \Big[\epsilon^{-Px} \sin(nt - Qx - \theta_1)$$
$$- \frac{\epsilon^{Px} \sin(nt + Qx - \theta_1 + \theta_2) + \epsilon^{-Px} \sin(nt - Qx - \theta_1 + \theta_2)}{\epsilon^{Pl}(\epsilon^{2Pl} + \epsilon^{-2Pl} - 2\cos 2Ql)^{\frac{1}{2}}} \Big]. \quad (40b)$$

Each of these consists of the sum of three waves, two positive, or from $x = 0$ to $x = l$, and one negative, or the reverse way. If the line were infinitely long, we should have only the first wave. But this wave is reflected at $x = l$, and the result is the second term. Reflection at the $x = 0$ end produces the third and least important term.

The wave-speed is n/Q, and the wave-length $2\pi/Q$. As the waves travel their amplitudes diminish at a rate depending upon the magnitude of P. The angles θ_1 and θ_2 merely settle the phase-differences. The limiting case is wave-speed $= v$, and no dissipation.

The amplitude of the current (half its range) is important. It is

$$C_0 = \frac{V_0 (Sn)^{\frac{1}{2}}}{(R'^2 + L'^2 n^2)^{\frac{1}{4}}} \left[\frac{\epsilon^{2P(l-x)} + \epsilon^{-2P(l-x)} + 2\cos 2Q(l-x)}{\epsilon^{2Pl} + \epsilon^{-2Pl} - 2\cos 2Ql} \right]^{\frac{1}{2}},$$

at any distance x. At the extreme end $x = l$ it is

$$C_0 = \frac{2V_0 (Sn)^{\frac{1}{2}}}{(R'^2 + L'^2 n^2)^{\frac{1}{4}}} (\epsilon^{2Pl} + \epsilon^{-2Pl} - 2\cos 2Ql)^{-\frac{1}{2}}. \quad \text{..........}(41b)$$

As it is only the current at the distant end that can be utilised there, it is clear that (41*b*) is the equation from which valuable information is to be drawn.

It must now be explained how to get R' and L', and their meanings. Go back to equation (28*b*), Section XXVII. [p. 54], which is the determinantal or differential equation when dielectric displacement is ignored. We may write it

$$0 = L_0 p + \frac{\rho_1 s_1}{2\pi a_1} \frac{J_0(s_1 a_1)}{J_1(s_1 a_1)} - \frac{\rho_3 s_3}{2\pi a_2} \frac{J_0(s_3 a_2) - \frac{J_1}{K_1}(s_3 a_3) K_0(s_3 a_2)}{J_1(s_3 a_2) - \frac{J_1}{K_1}(s_3 a_3) K_1(s_3 a_2)}.$$

When p is d/dt it is the differential equation of the boundary magnetic force, or of C, since they are proportional. Separating into even and odd powers of p it will take the form, if we operate on C,

$$0 = (R' + L'p)C, \quad \text{..........................}(42b)$$

where R' and L' are functions of p^2. To suit the oscillatory state, put $-n^2$ for p^2, making R' and L' constants. They will be of the form

$$R' = R_1' + R_2', \qquad L' = L_0 + L_1' + L_2'; \quad \text{..............}(43b)$$

where R_1' depends on the wire, R_2' on the return; L_1' on the wire, L_2' on the return, and L_0 on the intermediate insulator. The forms of R_1' and L_1' have been given by Lord Rayleigh. They are, if $g^2 = \mu_1 n/R_1$, where R_1 = steady resistance of the wire per unit length,

$$\left.\begin{aligned} R_1' &= R_1\left(1+\frac{g}{12}-\frac{g^2}{12.15}+\frac{11g^3}{12.28.80}-\dots\right), \\ L_1' &= \tfrac{1}{2}\mu_1\left(1-\frac{g}{24}+\frac{13g^2}{12^2.30}-\frac{73g^3}{12^2.28.80}+\dots\right), \end{aligned}\right\} \quad \dots\dots\dots(44b)$$

to the last of which I have added an additional term. The getting of the forms of R_2' and L_2', depending upon the return, is less easy, though only a question of long division. I shall give the formulæ later. At present I give their ultimate forms at very high frequencies. Let ρ = resistivity, and q = frequency $= n/2\pi$, then

$$R_1' = \frac{(\mu_1\rho_1 q)^{\frac{1}{2}}}{a_1}, + \frac{1}{4} \qquad L_1' = R_1'/n. \quad \dots\dots\dots\dots\dots\dots(45b)$$

These are also Lord Rayleigh's. For the return we have

$$R_2' = \frac{(\mu_3\rho_3 q)^{\frac{1}{2}}}{a_2}, \qquad L_2' = R_2'/n. \quad \dots\dots\dots\dots\dots\dots(46b)$$

I express R_1' and R_2' in terms of the resistivity rather than the resistance of the wire and return because their resistances have really nothing to do with it, as we see in especial from the R_2' formula. The R_2' of the tube depends upon its inner radius only, no matter how thick it may be, that is to say upon extent of conducting surface, varying inversely as the area, which is $2\pi a_2$ per unit length. The proof of (46b) will follow.

Now, as regards the meanings. Let us call the ratio of the impressed force to the current in a line when electrostatic induction is ignorable the Impedance of the line, from the verb impede. It seems as good a term as Resistance, from resist. (Put the accent on the middle e in impedance.) When the flow is steady, the impedance is wholly conditioned by the dissipation of energy, and is then simply the resistance Rl of the line. This is also sensibly the case when the frequency is very low; but with greater frequency inertia becomes sensible. Then $(R^2+L^2n^2)^{\frac{1}{2}}l$ is the impedance. Here R and L are, in the ordinary sense, the resistance and inductance of unit length of line, including wire and return. When, further, differences of current-density are sensible, the impedance is $(R'^2+L'^2n^2)^{\frac{1}{2}}l$. This is greater or less than the former, according to the frequency, becoming ultimately less, especially if the wire is of iron, owing to the then large reduction in the value of L' as compared with L.

Now, when we further take electrostatic induction into account we shall have the above equations (34b) and (35b), in which R' and L' are the same as if there were no static charge. The proof of this I must also postpone. It is the only thing to be proved to make the above quite complete, excepting (46b), which is a mere matter of detail. The proof arises out of the short sketch I gave in Section XXVII. of the general electrostatic investigation, used there for illustration.

The impedance is made variable; it is no longer the same all along the line, simply because the current-amplitude decreases from the place of impressed force, where it is greatest, to the far end of the line, where it is least. The question arises whether we shall confine impedance according to the above definition to the place of impressed force, or extend its meaning. If we confine the use, a new word must be invented. I therefore, at least temporarily, extend the meaning to signify the ratio of V_0 to C_0 anywhere.

It is very convenient to express impedance in ohms, whatever may be its ultimate structure. Thus the greatest impedance of a line is what its resistance would have to be in order that in steady-flow the current should equal that arriving at the far end under the given circumstances. It will usually be far greater than the resistance. But there is this remarkable thing about the joint action of forces of inertia and elasticity. The impedance may be far less than in the electromagnetic theory. That is, V_0/C_0 according to (41*b*) may be far less than $(R'^2+L'^2n^2)^{\frac{1}{2}}l$. This is clearly of great importance in connection with the future of long-distance telegraphy and telephony.

(In passing I will give an illustration of reduction of impedance produced by inertia. Let an oscillatory current be kept up in a submarine cable and in the receiving coils. Insert an iron core in them. The result is to *increase* the amplitude of the current-waves. More fully, increasing the inductance of the coil continuously from zero, whilst keeping its resistance constant, increases the amplitude up to a certain point, after which it decreases. The theory will follow.)

To get the submarine cable formulæ, ignoring inertia, take $L'=0$ and $R'=R$. To get the more correct formulæ, not allowing for variations of current-density, but including inertia, take $L'=L$ the steady inductance, and $R'=R$. To get the linear magnetic theory formulæ, take $S=0$, and $L'=L$, $R'=R$. Finally, using R' and L', but with $S=0$, we have the complete magnetic formulæ suitable for short lines. Thus $S=0$ in (41*b*) brings it to

$$C_0=V_0\div(R'^2+L'^2n^2)^{\frac{1}{2}}.$$

Equations (34*b*) to (36*b*) are true generally, that is, with R' and L' the proper functions of d/dt. The solution in the case of steady impressed force will follow, including the interior state of the wire. Also the interior state in the oscillatory case.

A great deal may be dug out of (41*b*). In the remainder of this Section, however, we may merely notice the form it takes at very high frequencies, so high as to bring surface conduction into play, and show how much less the impedance is than according to the magnetic theory. Let n be so great as to make $R'/L'n$ small. Then we may also take $L'=L_0$. Then

$$P=R'/2L_0v, \qquad Q=n/v.$$

Also, if ϵ^{-Pl} is small, as it will be on increasing the frequency, we need only consider the first term under the radical sign in (41*b*), which becomes

$$C_0=\frac{2V_0}{L_0v}\epsilon^{-R'l/2L_0v}.$$

Take for R' its ultimate form

$$R'=2(\mu\rho q)^{\frac{1}{2}}/a,$$

got from (45*b*) and (46*b*) by supposing wire and sheath of the same material, and $2/a = 1/a_1 + 1/a_2$.

Then the impedance is

$$V_0/C_0 = \tfrac{1}{2}L_0 v \times \exp\frac{l(\mu\rho q)^{\frac{1}{2}}}{L_0 va},$$

where exp is defined by $\epsilon^x = \exp x$, convenient when x is complex. Here L_0 is a numeric, and $v = 30$ ohms (*i.e.*, when we reckon the impedance in ohms); $\rho = 1600$ and $\mu = 1$, if the conductors are copper; and $l = 10^5 l_1$, if l_1 is the length of the line in kilom.; therefore

$$V_0/C_0 = 15L_0 \times \exp(4l_1 q^{\frac{1}{2}}/30^4 aL_0).$$

To see how it works out, take $L_0 = 1$, $a = 1$ cm, and $q = 10^4$; then

$$V_0/C_0 = 15 \times \exp 4l_1/300 \quad \text{ohms}.$$

If the line is 100 kilom., Pl is made $1\frac{1}{3}$, which is too small for our approximate formula. If 1,000 kilom., it is made $13\frac{1}{3}$, which is rather large. $Pl = 10$ is large. If it is 500 kilom., then

$$V_0/C_0 = 15 \times \exp 6\tfrac{2}{3} = 1{,}178 \text{ ohms}.$$

So the impedance is only 1,178 ohms at 500 kilom. distance at the enormous frequency of 10,000 waves per second. It is of course much less at a lower frequency, but the more complete formula will have to be used if it be much lower.

Now compare this real impedance with the resistance of the line in the steady state, its effective resistance according to the magnetic theory, and the impedance according to the same. The resistance of the line we may take to be twice that of the wire, by choosing the return of a proper thickness, or

$$Rl = 2 \times 500 \times 10^5 \times 1600/\pi = 50 \text{ ohms, say}.$$

L will be a little more than $1\frac{1}{2}$, say 1·6, therefore

$$Lln = \cdot 8 \times 2\pi \times 10^4 = 5060,$$

so that the linear-theory impedance is nearly 5,100 ohms.

But, owing to the high frequency, we should use R' and L' instead of R and L; here take $L' = L_0 + R'/n$, then

$$R'l = 400 \text{ ohms}.$$

This large increase of resistance is more than counterbalanced by the reduction of inductance, so that the impedance is brought down from the above 5,100 to about 3,500 ohms, the magnetic theory impedance; and this is about three times the real impedance at its greatest, viz., at the distant end of the line.

It is further to be noted that the wire and return need not be solid, as we see from the value of R' compared with R. What is needed at very high frequencies is two conducting sheets of small thickness, of the highest conductivity and lowest inductivity; *i.e.*, of copper.

SECTION XXX. IMPEDANCE FORMULÆ FOR SHORT LINES. RESISTANCE OF TUBES.

In the case of a short line, a very high frequency is needed in general to make it necessary to take electrostatic induction into account in estimating the impedance. Keeping below such a frequency, the impedance per unit length is simply

$$(R'^2 + L'^2 n^2)^{\frac{1}{2}}.$$

This is greater than the common $(R^2 + L^2 n^2)^{\frac{1}{2}}$ at first, when the frequency is low, equal to it at some higher frequency, and less than it for still higher frequencies. Thus, for simplicity, let the return contribute nothing to the resistance or the inductance; then, using (44*b*), we shall have

$$(R'^2 + L'^2 n^2) - (R^2 + L^2 n^2)$$
$$= \frac{R^2 g}{6} - \frac{R^2 g^2}{24}\left(\frac{L}{\mu} + \frac{1}{10}\right) + \frac{R^2 g^3}{48.90}\left(13\frac{L}{\mu} + \frac{79}{56}\right) - \dots; \quad (47b)$$

R and L being the steady resistance and inductance of the line per unit length (the latter to include L_0 for the external medium), R' and L' the real values at frequency $n/2\pi$ per second, μ the inductivity of the wire, and $g = (\mu n/R)^2$.

Thus the first increase in the square of the impedance over that of the linear theory is $\frac{1}{6}\mu^2 n^2$, independent of resistance; large in iron, small in copper. But as the frequency is raised, the g^2 term becomes sensible; being negative, it puts a stop to the increase. We can get a rough idea of the frequency required to bring the impedance down to that of the linear theory by ignoring the g^3 term. This gives

$$n^2 = 4R^2 \div \mu^2\left(\frac{L}{\mu} + \frac{1}{10}\right). \quad \dots\dots\dots\dots(48b)$$

The real frequency required must be greater than this, and taking the g^3 term into account, we shall obtain, as a higher limit,

$$n^2 = 4\tfrac{2}{13}R^2 \div \mu^2\left(\frac{L}{\mu} + \frac{3}{28}\right), \quad \dots\dots\dots\dots(49b)$$

approximately. We see that the simpler (48*b*) is near enough.

If the wire is of copper of a resistance of 1 ohm per kilom., making $R = 10^4$, we shall have, using (48*b*),

$$n = 20^4 \div \left(L_0 + \frac{3}{5}\right)^{\frac{1}{2}}.$$

If the return is distant, we can easily have $L_0 = 9$. Then the frequency required is about 100 waves per second. This is a low telephonic frequency, so that we see that telephonic signalling is somewhat assisted by the approximation to surface conduction.

If the wire is of iron, then, on account of the large value of μ, a much lower frequency is sufficient to reduce the impedance below that of the linear theory; that is, an iron wire is not by any means so disadvantageous, compared with a copper wire of the same diameter, as its higher resistivity and far higher inductivity would lead one to expect.

But it is not to be inferred that there is any advantage in using iron, electrically speaking, from the fact that the impedance is so easily made much less than that of the linear theory. Copper is, of course, the best to use, in general, being of the highest conductivity, and lowest inductivity. Nor is any great importance to be attached to the matter in any case, for, on a short line, to which we at present refer, it will usually happen that the telephones themselves are of more importance than the line in retarding changes of current.

We also see that in electric-light mains with alternating currents there may easily be a reduction of impedance if the wires be thick and the returns not too close. On the other hand, the closer they are brought the less is the impedance, according to the ordinary formula. It should be borne in mind that we are merely dealing with a correction, not with the absolute value of the impedance, which is really the important thing.

Now take the frequency midway between 0 and the second frequency which gives the linear-theory impedance. Then $R^2+L^2n^2$ becomes

$$R^2+n^2\left(L^2+\frac{1}{24}\mu^2\right),$$

wherein use the value of n^2 given by (48*b*). The increase of impedance is not, therefore, in a copper wire, anything of a startling nature.

Impedances are not additive, in general. We cannot say that the impedance of a wire is so much, that of a coil so much more, and then that their sum is the impedance when they are put in sequence, at the same frequency.

In passing, I may as well caution the reader against the false idea somewhere prevalent. The increased resistance of a wire is not in any way caused or evidenced by the weakness of the current in the variable period compared with its final strength, a result due to the back E.M.F. of inertia. No matter how great the inertia, and how slowly it makes the current rise, there is no change of resistance, unless there be changed distribution of current. There must always be some change, but it is usually negligible. When, however, as notably in the case of iron, the central part of the wire is inoperative, of course this changed distribution of current means a large increase of resistance, though not of impedance, which is reduced. It is a hollow tube, not a solid wire, that must, to a first approximation, be regarded as the conductor. There cannot be said to be any definite resistance unless the current distribution is definite.

Thus, in the rise of the current from zero to the steady state there is, presuming that there is large departure from the regular final distribution, no definite resistance, and it is clearly not possible to balance a wire in which the above takes place against a thin wire, a conclusion that is easily verified. But the case of simple-harmonic impressed force is peculiar. The distribution of current, though not constant, goes through the same regular changes over and over again in such a manner that the total current at every moment is the same as if a true linear circuit of definite resistance and inductance were substituted. This is very considerably departed from when mere rapid makes and breaks are employed.

Consider now the resistance of a tube at a given frequency. It depends materially upon whether the return-current be within it or outside it. Let there be two tubes, a_0 and a_1 the inner and outer radii of the inner, and a_2 and a_3 of the outer. By an easy extension of equation (28*b*), the form quoted in the last Section, the differential equation of the total current is

$$0 = L_0 p + \frac{\rho_1 s_1}{2\pi a_1} \frac{J_0(s_1 a_1) - \frac{J_1}{K_1}(s_1 a_0) K_0(s_1 a_1)}{J_1 \ldots\ldots - \ldots\ldots\ldots K_1 \ldots\ldots}$$

$$- \frac{\rho_3 s_3}{2\pi a_2} \frac{J_0(s_3 a_2) - \frac{J_1}{K_1}(s_3 a_3) K_0(s_3 a_2)}{J_1 \ldots\ldots - \ldots\ldots\ldots K_1 \ldots\ldots}, \qquad \ldots\ldots\ldots\ldots\ldots (50b)$$

the dots indicating repetition of what is above them. The first term is for the insulator between tubes, the second for the inner tube, the third for the outer. Or,

$$0 = L_0 p + (R_1' + L_1' p) + (R_2' + L_2' p),$$

where R_1', R_2', L_1', L_2' are functions of p^2, and therefore constants when the current is simple-harmonic. The division of the numerators by the denominators, a simple matter in the case of a solid wire, becomes a very complex matter in the tube case. I give the results as far as p^2. It is not necessary to do the work separately for the two tubes, for, if we compare the expressions carefully, we shall see that they only differ in the exchange of the inner and outer radii, and in changed sign of the whole.

For the inner tube we have

$$L_1' = R_1 \frac{\mu_1 \pi a_1^2}{\rho_1} \left\{ \frac{1}{2} - \frac{3}{2} \frac{a_0^2}{a_1^2} - 2 \log \frac{a_1}{a_0} \cdot \frac{a_0^4}{a_1^2 (a_1^2 - a_0^2)} \right\}, \qquad \ldots\ldots\ldots (51b)$$

where R_1 is the steady resistance per unit length. This is the coefficient of p, and is therefore nothing more than the inductance per unit length of the tube in steady flow, the first correction to which depends on p^3. This may be immediately verified by the square-of-force method.

The resistance of the inner tube per unit length is

$$R_1' = R_1 + R_1 \left(\frac{n \mu_1 \pi a_1^2}{\rho_1} \right) \left[\frac{1}{12} - \frac{2a_0^2}{3a_1^2} + \frac{7a_0^4}{12a_1^4} + \frac{2a_0^4 \log \frac{a_1}{a_0}}{a_1^2 (a_1^2 - a_0^2)} + \frac{4a_0^6 \left(\log \frac{a_1}{a_0} \right)^2}{a_1^2 (a_1^2 - a_0^2)^2} \right]. (52b)$$

To obtain, from (51*b*) and (52*b*), the corresponding expressions for the outer tube, change R_1 to R_2, ρ_1 to ρ_3, μ_1 to μ_3, a_1 to a_2, and a_0 to a_3. The change of sign is not necessary, because it is involved in the substitution of R_2 for R_1. Or, simply, (51*b*) and (52*b*) holding good when the return is outside the tube, exchange a_1 and a_0, and we have the corresponding formulæ when the return is inside it.

Let $a_0 = \frac{1}{2}a_1$. This removes a fourth part of the material from the central part of a solid wire of radius a_1. The return being outside, the resistance is

$$R_1' = R_1 + R_1(n\mu_1\pi a_1^2/\rho_1)^2 \times \cdot 012.$$

If solid, the ·012 would be ·083; or the correction is reduced seven times by removing only a fourth part of the material.

But if the return is inside, all else being the same, the resistance is

$$R_1' = R_1 + R_1(n\mu_1\pi a_0^2/\rho_1) \times \cdot 503 = R_1 + R_1(n\mu_1\pi a_1^2/\rho_1) \times \cdot 031,$$

so now the correction is reduced less than three times instead of seven times, as when the return was outside.

This difference will be, of course, greatly magnified when the ratio a_1/a_0 is large; for instance, consider a solid wire surrounded by a very thick tube for return; the steady resistance of the return will be only a small fraction of that of the wire, but the percentage increase of resistance of the outer conductor will be a large multiple of that of the wire. Thus the earth's resistance, which, in spite of the low conductivity, is so small to a steady current, will be largely multiplied when the current is a periodic function of the time.

Now, as regards the resistance of the tube at high frequencies. If the return is outside it is

$$R_1' = (\mu_1\rho_1 q)^{\frac{1}{2}} \div a_1, \quad \text{..........................} (53b)$$

q being the frequency. But if the return is inside, it is

$$R_1' = (\mu_1\rho_1 q)^{\frac{1}{2}} \div a_0, \quad \text{..........................} (54b)$$

thus depending upon the inner radius when the return is inside, and on the outer when it is outside, for an obvious reason, when the position of the magnetic field where the primary transfer of energy takes place is considered.

Suppose we fix the outer radius, and then thin the tube from a solid wire down to a mere skin. In doing so we increase the steady resistance as much as we please. But the high-frequency formula (53b) remains the same. Now, as it would involve an absurdity for the resistance to be less than that in steady flow, it is clear that (53b) cannot be valid until the frequency is so high as to make R_1' much greater than R_1, which is itself very great when the tube is thin. That is to say, removing the central part of a wire, when the return is outside it, makes it become more a linear conductor, so that a much higher frequency is required to change its resistance; and when the tube is very thin the frequency must be enormous. Practically, then, a thin tube is always a linear conductor, although it is only a matter of raising the frequency to make (53b) or (54b) applicable.

To get them, use in (50b) the appropriate $J_0(x)$, etc., formulæ when x is very large. They are

$$\left.\begin{aligned} J_0(x) &= -K_1(x) = (\sin x + \cos x) \div (\pi x)^{\frac{1}{2}}, \\ J_1(x) &= K_0(x) = (\sin x - \cos x) \div (\pi x)^{\frac{1}{2}}. \end{aligned}\right\} \quad \text{................} (55b)$$

These, used in (50*b*), putting the circular functions in the exponential forms, reduce it to

$$0 = L_0 p + \frac{\rho_1 s_1 i}{2\pi a_1} + \frac{\rho_3 s_3 i}{2\pi a_2},$$

where $i = (-1)^{\frac{1}{2}}$. Here

$$-s_1^2 a_1^2 = 4\pi\mu_1 k_1 a_1^2 p, \qquad \text{therefore} \qquad \tfrac{1}{2} s_1 a_1 i = (\pi\mu_1 k_1 p)^{\frac{1}{2}} a_1,$$

and similarly for s_3; so we get

$$0 = L_0 p + (\mu_1 p/\pi k_1)^{\frac{1}{2}}/a_1 + (\mu_3 p/\pi k_3)^{\frac{1}{2}}/a_2.$$

Here, since $p^2 = -n^2$, $p^{\frac{1}{2}} = (\frac{1}{2}n)^{\frac{1}{2}}(1+i)$; which brings us to

$$0 = (R_1' + R_2') + (L_0 + L_1' + L_2')p, \qquad \text{(56}b\text{)}$$

where

$$\left.\begin{aligned} R_1' &= (\mu_1 \rho_1 q)^{\frac{1}{2}} \div a_1, & L_1' &= R_1'/n, \\ R_2' &= (\mu_3 \rho_3 q)^{\frac{1}{2}} \div a_2, & L_2' &= R_2'/n, \end{aligned}\right\} \qquad \text{(57}b\text{)}$$

as before given, except that the inner tube was a solid wire.

If, however, the frequency were really so high as to make these high-frequency formulæ applicable when the conductors are thin tubes, it is clear that we should, by reason of the high frequency, need, at least in general, to take electrostatic induction into account even on a short line, and therefore not estimate the impedance by the magnetic formulæ, but by the more general of the last Section, in which the same R' and L' occur. As for long lines, it is imperative to consider electrostatic induction. There is no fixed boundary between a "short" and a "long" line; we must take into account in a particular case the circumstances which control it, and judge whether we may treat it as a short or a long-line question. To the more general formula I shall return in the following Section, merely remarking at present that there is a curious effect arising from the to-and-fro reflection of the electromagnetic waves in the dielectric, which causes the impedance to have maxima and minima values as the speed continuously increases; and that when the period of a wave is somewhere about equal to the time taken to travel to the distant end and back, the amplitude of the received current may easily be greater than the steady current from the same impressed force. And, in correction of the definition in Section XXIX. of V as the surface potential of the wire, substitute this definition, $Q = SV$, where Q is the charge and S the electrostatic capacity, both per unit length of wire.

SECTION XXXI. THE INFLUENCE OF ELECTRIC CAPACITY. IMPEDANCE FORMULÆ.

Let us now return to the more general case of Section XXIX., the amplitude of the current due to a simple-harmonic impressed force at one end of a line. Although the formula (41*b*) for the amplitude at the distant end is very compact, yet the exponential form of the functions does not allow us to readily perceive the nature of the change made by lengthening the line, or making any other alteration that will cause the effect of the electric charge to be no longer negligible, by causing the magnetic formula to be sensibly departed from. Let us, therefore, put (41*b*) in the form $V_0/C_0 =$ etc., and then expand the right member in an infinite series of which the first term shall be the magnetic impedance itself, whilst the others depend on the electric capacity as well as on the resistance and inductance.

On expanding the exponentials and the cosine in (41*b*), we obtain a series in which the quantities $P^4 - Q^4$, $P^6 - Q^6$, etc., occur, all divided by $P^2 + Q^2$.

To put these in terms of the resistance, etc., we have, by (37*b*),

$$P^2 + Q^2 = SnI, \qquad 2PQ = SnR', \qquad Q^2 - P^2 = Sn^2L', \quad \text{...(58b)}$$

where

$$I = (R'^2 + L'^2n^2)^{\frac{1}{2}}, \quad \text{.........................(59b)}$$

I being the short-line impedance per unit length. Using these, we convert (41*b*) to the following form,

$$V_0/C_0 = Il\left[1 - \frac{l^2}{3}Sn^2L' + \frac{2l^4}{45}(Sn)^2(\tfrac{1}{4}R'^2 + L'^2n^2) - \dots\right]^{\frac{1}{2}}. \quad \text{.....(60b)}$$

Here we may repeat that V_0 and C_0 are the amplitudes of the impressed force at one end and of the current in the wire at the other end of the double wire of length l, whose "constants" are R', L', and S, the resistance, inductance, and electric capacity per unit length, R' and L' being functions of the frequency already given. I do not give more terms than are above expressed, owing to the complexity of the coefficients of the subsequent powers of S. To go further, it will be desirable to modify the notation, and also to entirely separate the terms depending upon resistance in the [ ] from the others. Let

$$SL' = v^{-2}, \qquad f = (R'/L'n)^2, \qquad h = nl/v. \quad \text{.........(61b)}$$

Here v is a velocity, f and h numerics. The least value of the velocity is $(SL)^{-\frac{1}{2}}$, at zero frequency, L being the full steady inductance per unit length, as before. As the frequency increases, so does v. Its limiting value is $(SL_0)^{-\frac{1}{2}}$ or $(\mu_2 c_2)^{-\frac{1}{2}}$, the speed of undissipated waves through the dielectric. The ratio f falls from infinity at zero frequency, to zero at infinite frequency. See equations (43*b*) to (46*b*). The ratio h is such that $h/2\pi$ is the ratio of the time a wave travelling at speed v takes to traverse the line, to the wave-period.

In terms of I, f, and h, our formula (41*b*), or rather (60*b*), when extended, becomes

$$V_0/C_0 = Il\left[\left(\frac{\sin h}{h}\right)^2 + \frac{fh^4}{90}\left\{1 - \frac{1}{7}h^2 + \frac{1}{105}h^4\left(1 + \frac{1}{12}f\right)\right.\right.$$
$$\left.\left. - \frac{4}{105.99}h^6\left(1 + \frac{3}{16}f\right) + \frac{10}{105.99.91}h^8\left(1 + \frac{3}{10}f + \frac{1}{80}f^2\right) - \dots\right\}\right]^{\frac{1}{2}}. \quad (62b)$$

From this, seeing that in the [ ], resistance appears in f only, we see that the corresponding no-resistance formula is simply

$$V_0/C_0 = Il\frac{\sin h}{h} = L_0 v \sin\frac{nl}{v}, \quad \text{.....................(63b)}$$

where, of course, v is the speed corresponding to L_0, or the speed of undissipated waves. The sine must be reckoned positive always. To check (63b), derive it immediately from (41b) by taking $R=0$. We shall find the following form of (41b) in terms of f and h useful later:—

$$V_0/C_0=\tfrac{1}{2}L'v(1+f)^{\frac{1}{4}}\{\epsilon^{2Pl}+\epsilon^{-2Pl}-2\cos 2Ql\}^{\frac{1}{2}}, \quad \text{..........(64}b)$$

where
$$\left.\begin{aligned} Pl&=h(\tfrac{1}{2})^{\frac{1}{2}}\{(1+f)^{\frac{1}{2}}-1\}^{\frac{1}{2}}, \\ Ql&=h(\tfrac{1}{2})^{\frac{1}{2}}\{(1+f)^{\frac{1}{2}}+1\}^{\frac{1}{2}}. \end{aligned}\right\} \quad \text{.....................(65}b)$$

Let us now dig something out of the above formulæ. This arithmetical digging is dreadful work, only suited for very robust intellects. I shall therefore be glad to receive any corrections the following may require, if they are of any importance.

It will be as well to commence with the unreal, but easily imaginable case of no resistance. Let the wire and return be of infinite conductivity. We have then merely wave propagation through the dielectric, without any dissipation of energy, at the wave-speed $v=(\mu_2c_2)^{-\frac{1}{2}}$, which is, in air, that of light-waves. Any disturbances originating at one end travel unchanged in form; but owing to reflection at the other end, and then again at the first end, and the consequent coexistence of oppositely travelling waves, the result is rather complex in general. Now, if we introduce a simple-harmonic impressed force at one end, and adjust its frequency until the wave-period is nearly equal to the time taken by a wave to travel to the other end and back again at the speed v, it is clear that the amplitude of the disturbance will be enormously augmented by the to-and-fro reflections nearly timing with the impressed force. This will explain (63b), according to which the distant-end impedance falls to zero when

$$nl/v=\pi, \quad \text{or } 2\pi, \quad \text{or } 3\pi, \qquad \text{etc.}$$

Here $2\pi/n$ is the wave-period, and $2l/v$ the time of a to-and-fro journey. The current-amplitude goes up to infinity.

If, next, we introduce only a very small amount of resistance, we may easily conclude that, although the impedance can never fall to zero, yet, at particular frequencies, it will fall to a minimum, and, at others, go up to a maximum; and that the range between the consecutive maximum and minimum impedance will be very large, if only the resistance be low enough.

Increasing the resistance will tend to reduce the range between the maximum and minimum, but cannot altogether obliterate the fluctuations in the value of the impedance as the frequency continuously increases. In practical cases, starting from frequency zero, and raising it continuously, the impedance, which is simply Rl, the resistance of the line, in the first place, rises to a maximum, then falls to a minimum, then rises to a second maximum greater than the first, and falls to a second minimum greater than the first, and so on, there being a regular increase in the impedance on the whole, if we disregard the fluctuations, whilst the fluctuations themselves get smaller and smaller, so that the real maxima and minima ultimately become false, or only tendencies towards maxima and minima at certain frequencies.

By this to-and-fro reflection, or electrical reverberation or resonance, the amplitude of the received current may be made far greater than the strength of the steady current from the same impressed force, even when the electrical data are not remote from, but coincide with, or resemble, what may occur in practice. To show this, let us work out some results numerically.

As this matter has no particular concern with variations of current-density in the conductors, ignore them altogether; or, what comes to the same thing, let the conductors be sheets, so that $R' = R$, the steady resistance, and $L' = L_0$ very nearly, the dielectric inductance, both per unit length. Then, in (64*b*), let

$$f = 1, \qquad Ql = \pi, \qquad v = 30 \text{ ohms.} \quad \text{............(66}b\text{)}$$

Then, by the second of (65*b*), we find that

$$h = 2{\cdot}85\,;$$

and, by (64*b*), that

$$V_0/C_0 = \tfrac{1}{2}L_0 v . 2^{\frac{1}{4}}[\epsilon^{\cdot 8284\pi} + \epsilon^{-\cdot 8284\pi} - 2]^{\frac{1}{2}} = 60{\cdot}6\, L_0 \text{ ohms.} \quad \text{......(67}b\text{)}$$

The ratio of the distant-end impedance to the resistance is therefore

$$\frac{60{\cdot}6 \times 10^9 L_0}{Rl} = \frac{60{\cdot}6 \times 10^9}{nl} = \frac{20{\cdot}2}{10h} = \frac{202}{285}, \quad \text{............(68}b\text{)}$$

by making use of the data (66*b*). That is, the amplitude of the received current is 42 per cent. greater than the steady current, when (66*b*) is enforced.

But let $Ql = \frac{1}{4}\pi$, then

$$V_0/C_0 = \tfrac{1}{2}L_0 v . 2^{\frac{1}{4}}[\epsilon^{\cdot 2071\pi} + \epsilon^{-\cdot 2071\pi}]^{\frac{1}{2}} = 28\, L_0 \text{ ohms;}$$

and the ratio of impedance to resistance is

$$\frac{28}{60{\cdot}6} \times \frac{202}{285} \times 4 = \frac{4}{3} \text{ nearly,}$$

or the amplitude of current is only 3/4 of the steady current.

And if $Ql = \frac{1}{2}\pi$, we shall find

$$V_0/C_0 = 43{\cdot}5 \text{ ohms,}$$

and that the impedance is slightly greater than the resistance. Whilst, if $Ql = \frac{3}{4}\pi$, we shall have

$$V_0/C_0 = 47{\cdot}8 \text{ ohms,}$$

and find the ratio of impedance to resistance to be 63/85, making the received current 35 per cent. stronger than the steady current.

The above data of $f = 1$, and $Ql = \frac{1}{4}\pi$, $\frac{1}{2}\pi$, $\frac{3}{4}\pi$, and π, have been chosen in order to get near the first maximum and minimum of impedance. The range, it will be seen, is very great. Let us next see how these data resemble practical data in respect to resistance, etc. Remember that 1 ohm per kilom. makes $R = 10^4$, (resistance per cm. of double conductor). Also, that $f = 1$ means $R = nl = 10^5 nl_1$, if l_1 is in kilometres. Then, in the case to which (66*b*) to (68*b*) refer, we shall have, first assuming a given value of R, then varying L_0, and deducing the values of n and l_1, the following results:—

	$L_0 = 1$,	$L_0 = 10$,
$R = 10^3$,	$n = 10^3$,	$n = 10^2$,
	$l_1 = 856$,	$l_1 = 8568$.

This is an excessively low resistance, $\frac{1}{10}$ ohm per kilom.; the frequencies are rather low, and the lengths great. Next, 1 ohm per kilom. :—

$$
\begin{array}{llll}
 & L_0=1, & L_0=10, & L_0=100, \\
R=10^4, & n=10^4, & n=10^3, & n=10^2, \\
 & l_1=85. & l_1=856. & l_1=8568.
\end{array}
$$

The $L_0=100$ case is extravagant, requiring such a very distant return current (therefore very low electric capacity). Next, 10 ohms per kilom. :—

$$
\begin{array}{llll}
 & L_0=1, & L_0=10, & L_0=100, \\
R=10^5, & n=10^5, & n=10^4, & n=10^3, \\
 & l_1=8\cdot5. & l_1=85. & l_1=856.
\end{array}
$$

Lastly, very high resistance of 100 ohms per kilom. :—

$$R=10^6, \qquad L_0=10, \qquad n=10^5, \qquad l_1=8\cdot5.$$

In all these cases the amplitude of received current is 42 per cent. greater than the steady current.

In the next case, $Ql=\frac{1}{4}\pi$, the quantity nl/v has a value one-fourth of that assumed in the above; hence, with the same R and L_0, and same frequency, the above values of l_1 require to be quartered. Then, in all cases, the current-amplitude will be three-fourths of the steady current. Similarly, to meet the $Ql=\frac{1}{2}\pi$ case, use the above figures, with the l_1's halved; and in the $Ql=\frac{3}{4}\pi$ case, with the l_1's multiplied by $\frac{3}{4}$.

A consideration of the above figures will show that there must be, in telephony, a good deal of this reinforcement of current strength sometimes; not merely that the electrostatic influence tends to increase the amplitude all round, from what it would be were only magnetic induction concerned, but that there must be special reinforcement of certain tones, and weakening of others. It will be remembered that *good* reproduction of human speech is not a mere question of getting the lower tones transmitted well, but also the upper tones, through a long range; the preservation of the latter is required for good articulation. The ultimate effect of electrostatic retardation, when the line is long enough, is to kill the upper tones, and convert human speech into mere murmuring.

The formula (62b) is the most useful if we wish to see readily to what extent the magnetic formula is departed from. In this, two quantities only are concerned, f and h, or $(R'/L'n)^2$ and nl/v; and if both f and h are small, it is readily seen that the *first* form of (63b) applies, the factor by which the magnetic impedance is multiplied being $(\sin h)/h$. Even when h is not small the f terms in (62b) may be negligible, and the first form of (63b) apply. For example, suppose $h=\frac{1}{3}$, and f small, then $(\sin h)/h = 3\times\cdot3272=\cdot9816$, showing a reduction of 2 per cent. from the magnetic impedance.

Now, this $h=\frac{1}{3}$ means $nl_1=10^5$, or the high frequency of $10^5/2\pi$ on a line of one kilom., $10^4/2\pi$ on 10 kilom., and so on, down to $10/2\pi$ on 10,000 kilom., always provided the f terms are still negligible. This may easily be the case when the line is short, but will cease to be true as the line is lengthened, owing to the n in f getting smaller and smaller. Thus, in the just-used example, if the resistance is 10 ohms per kilom., and $L=10$, we shall have $f=\frac{1}{100}$ on the line of 1 kilom., and $f=1$ on 10 kiloms. So far, the f terms are negligible, and the first form of (63b) applies. But f becomes 100 on 100 kiloms., which will make an appreciable, though not large, difference; and $f=10{,}000$ on 1,000 kilom. will make a large difference and cause the first (63b) formula to fail. It is remarkable, however, that this formula should have so wide a range of validity.

In the above we have always referred to the distant-end impedance. But at the seat of impressed force there is a large increase of current on account of the "charge." Thus, at $x=0$, by the formula preceding (41b), we have

$$\frac{V_0}{C_0}=L'v(1+f)^{\frac{1}{4}}\left[\frac{\epsilon^{2Pl}+\epsilon^{-2Pl}-2\cos 2Ql}{\ldots\ldots\ldots+\ldots\ldots\ldots}\right]^{\frac{1}{2}}. \qquad \text{(69}b\text{)}$$

The term impedance is of course strictly applicable at the seat of impressed force. As the frequency is raised, this impedance tends to be represented by

$$V_0/C_0=L'v(1+f)^{\frac{1}{4}},$$

and, ultimately, by $\quad V_0/C_0=L_0v=30\,L_0$ ohms, $\qquad$ (70b)

if the dielectric be air. L_0 is usually a small number.

Section XXXII. The Equations of Propagation along Wires. Elementary.

In another place (*Phil. Mag.*, Aug., 1886, and later) the method adopted by me in establishing the equations of V and C, Section XXIX., was to work down from a system exactly fulfilling the conditions involved in Maxwell's scheme, to simpler systems nearly equivalent, but more easily worked. Remembering that Maxwell's is the only complete scheme in existence that will work, there is some advantage in this; also, we can see the degree of approximation when a change is made. In the following I adopt the reverse plan of rising from the first rough representation of fact up to the more complete. This plan has, of course, the advantage of greater intelligibility to those who have not studied Maxwell's scheme in its complete form; besides being, from an educational point of view, the more natural plan.

Whenever the solution of a so-called physical problem has been obtained, according to which, under such or such conditions, such or such effects *must* happen, what has really been done has been to solve another problem, which resembles the real one more or less in those features we wish to study, which we regard as essential, whilst it is of such a greatly simplified nature that its solution is, in comparison with that of the real problem, quite elementary. This remark, which is of rather an obvious nature, conveys a lesson that is not always remembered; that the difference between theory and empiricism is only one of degree, even when the word theory is used in its highest sense, and is applied to legitimate deductions from laws which are known to be very true indeed, within wide limits.

It is quite possible to imagine the solution of the general problem of the universe. There does not seem to be anything against it except its possible infinite extent. Stop the extension of the universe somewhere; then, if its laws be fully known, and be either invariable or known to vary in some definite manner, and if its state be known at a given moment, it is difficult to see how it can be indefinite at any later time, even in the minutest particulars in the history of nations or of animalculæ, or in the development of a human soul (which is certainly immortal, for the good and evil worked by a soul in this life live for ever, in the permanent impress they make on the future course of events).

But if this be imagined to be all done, and the universe made a machine, no one would be a bit the wiser as to the reason why of it. (Even if we ask what we mean by the reason why, we shall in all probability get into a vicious circle of reasoning, from which there is no escape.) All that would be done would be the formulation of facts in a complete manner. This naturally brings us to the subject of the equations of propagation, for they are merely the instruments used in attempts to formulate facts in a more or less complete manner.

The first to solve a problem in the propagation of signals was Ohm, whose investigation is a very curious chapter in the history of electricity, as he arrived at results which are, under certain conditions, nearly correct, by entirely erroneous reasoning. Ohm followed the theory of the conduction of heat in wires, as developed by Fourier. Up to a certain point there is a resemblance between the flow of heat and the electric conduction current, but after that a wide dissimilarity.

Let a wire be surrounded by a non-conductor of heat, in imagination; let the heat it contains be indestructible when in the wire, and be in a state of steady flow along it. If C is the heat-current across a given section, and V the temperature there, C will be proportional to the rate of decrease of V along the wire. Or

$$-dV/dx = RC,$$

if x be length measured along the wire. The ratio R of the fall of temperature per unit length, to the current, is the "resistance" per unit length, and is, more or less, a constant. Or, the current is proportional to the difference of temperature between any two sections, and is the same all the way between.

The law which Ohm discovered and correctly applied to steady conduction currents in wires is similar to this. Make C the electric current in the wire, and V the potential at a certain place. The current, which is the same all the way between any two sections, is proportional to their difference of potential. The ratio of the fall of potential to the current is the electrical resistance, and is constant (at the same temperature). But V is, in Ohm's memoir, an indistinctly defined quantity, called electroscopic force, I believe. Even using the modern equivalent potential, there is not a perfect parallel between the temperature V and the potential V. For a given temperature appears to involve a definite physical state of the conductor at the place considered, whereas potential has no such meaning. The real parallel is between the temperature gradient, or slope, and the potential slope.

Now, returning to the conduction of heat, suppose that the heat-current is not uniform, or that the temperature-gradient changes as we pass along the wire. If the current entering a given portion of the wire at one end be greater than that leaving it at the other, then, since the heat cannot escape laterally, it must accumulate. Applying this to the unit length of wire, we have the equation of continuity,

$$-dC/dx = \dot{q},$$

t being the time, and q the quantity of heat in the unit length. But the temperature is a function of q, say

$$q = SV,$$

where S is the capacity for heat per unit length of wire, here regarded, for simplicity of reasoning, as a constant, independent of the temperature. This makes the equation of continuity become

$$-dC/dx = S\dot{V}.$$

Between this and the former equation between C and the variation of V, we may eliminate C and obtain the characteristic equation of the temperature,

$$d^2V/dx^2 = RS\dot{V},$$

which, when the initial state of temperature along the wire is known, enables us to find how it changes as time goes on, under the influence of given conditions of temperature and supply of heat at its ends.

Ohm applied this theory to electricity in a manner which is substantially equivalent to supposing that electricity (when prevented from leaving the wire) flows like heat, and so must accumulate in a given portion of the wire if the current entering at one end exceeds that leaving at the other. The quantity q is the amount of electricity in the unit length, and is proportional to V, their ratio S being the capacity per unit length. With the same formal relations we arrive, of course, at the same characteristic equation, now of the potential, so that electricity diffuses itself along a wire, by difference of potential, in the same way as heat by difference of temperature.

A generation later, Sir W. Thomson arrived at a system which is formally the same, but having a quite different physical significance. Between the times of Ohm and Thomson great advances had been made in electrical science, both in electrostatics and electromagnetism, and the quantities in the system of the latter are quite distinct. We have

$$\left.\begin{aligned} -\frac{dV}{dx} &= RC, \qquad q = SV, \\ -\frac{dC}{dx} &= \frac{dq}{dt} = S\frac{dV}{dt}, \end{aligned}\right\} \quad \frac{d^2V}{dx^2} = RS\frac{dV}{dt}, \quad \ldots\ldots\ldots\ldots\ldots(71b)$$

where on the left appear the elementary relations, and on the right the resultant characteristic equation of V.

Here C is the current in the wire, R its resistance per unit length, and V the electrostatic potential. So far there is little change. But S is the electrostatic capacity per unit length of the condenser formed by the dielectric outside the wire, whose two coatings are the surface of the wire and that of some external conductor, as water, for instance, which serves as the return conductor. Thus S, from being in Ohm's theory a hypothetical quantity depending upon the nature of the conducting wire, its size and shape, has become a definitely known quantity depending on the nature of the dielectric, and its size and shape. Here is the first step towards getting out of the wire into the dielectric, to be followed up later. The equation $q = SV$ is the electrostatic law expressing the relation between the charge of a condenser and its potential-difference, q being the charge on the wire per unit length, and V its potential. It is assumed that $V = 0$ at the outer conductor, which requires that its resistance must be very small, theoretically nothing. This makes V definitely the potential at the surface of the wire, and it must be the potential all over its section at a given distance x, if the current is uniformly distributed across the section.

The meaning of the equation of continuity is now, that when the current entering a given length of wire on one side is greater than that leaving it on the other, the excess is employed in increasing the charge of the condenser formed by the given length of wire, the dielectric, and the outer conductor. In the wire, therefore, comparing the electric current to the motion of a fluid, such fluid must be incompressible. It can, however, accumulate on the boundary of the wire, where it makes the surface-charge. This is exceedingly difficult to understand. But in any case, whether electricity accumulates in the wire or only on its boundary, is quite immaterial as regards the form of the equation of continuity, and of the characteristic equation. (Of course it is the equations which give rise to it, and their interpretation, that are of the greatest importance.)

There is very little hypothesis in this system. We unite the condenser-law with Ohm's law of the conduction current, on the hypothesis, which is supported by experiments with condensers and conductors, that the equation of continuity is of the kind supposed. But it is assumed that the electric force is entirely due to difference of potential. As, when the current is changing in strength, this is not true, there being then also the electric force of inertia, or of magnetic induction, this should also be taken into account in the Ohm's law equation, making a corresponding change in the characteristic equation. What difference this will make in the manner of the propagation will depend upon the relative magnitude of the electric force of inertia and of the charge, and materially upon the length of the line. The necessary change will be made in the next Section. At present we may only remark that electrostatic induction is most important on long submarine cables, and that the (71*b*) equations are those to be used for them for general purposes, as the first approximate representation of the facts of the case.

Now, as regards the accumulation difficulty. This is entirely removed in a beautifully simple manner in Maxwell's theory. The line-integral of the magnetic force round a wire measures the current in it, a fact that cannot be too often repeated, until it is impressed upon people that the electric current is a function of the magnetic field, which is in fact what we generally make observations upon, the electricity in motion through the wire being a pure hypothesis. Maxwell made this the universal definition of electric current anywhere. There is no difference between a current in a conductor or in a dielectric as a function of the magnetic field, though there is great difference in the effect produced, according to the nature of the matter. All currents are closed, either in conductors alone or in dielectrics alone, or partly in one and partly in the other. In a conductor heat is the universal result of electric current, and energy is wasted; in a dielectric, on the other hand, the energy which would be wasted were it conducting is stored temporarily, becoming the electric energy, which is recoverable. In a conductor, the time-integral of the current is not a quantity of any physical significance; but in a dielectric it is a very important quantity, the electric displacement, which can only be removed by an equal reverse current. The electric displacement involves a back electric force, which will cause the displacement to subside when it is permitted by the removal of the cause that produced it. Put a condenser in circuit with a conductor and battery. The current goes right through the condenser. But it cannot continue, on account of the back force of the displacement; when this equals the impressed force of the battery, there is equilibrium. Remove the battery, and leave the circuit closed. The back force of the displacement can now act, and discharges the condenser. As for the positive and negative charges, they are numerically equal to the total displacement through the condenser. They are located at the places of, and measure the amount of discontinuity of the elastic displacement, and that is all.

If we must have a fluid to assist (keep it well in the back-ground), then this fluid must be everywhere, and be incompressible, and accumulate nowhere. I am no believer in this fluid. Its only utility is to hang facts together. But when one has obtained an accurate idea of the facts it has to hang together, it has served its purpose. A fluid has mass, and when in motion, momentum and kinetic energy. But the facts of electromagnetism decidedly negative the idea that the electric current *per se* has momentum or energy, or anything of that kind; these really belong to the magnetic field. It is therefore well to dispense with the fluid behind the scenes.

But when one thinks of the old fluids (of surprising vitality), and of their absurd and wholly incomprehensible behaviour, their miraculous powers of attracting and repelling one another, of combining together and of separating, and all the rest of that nonsense, one is struck with the extremely rational behaviour of the Maxwell fluid. When, further, one thinks of the greatly superior simplicity of the manner in which it hangs the facts together (it is remarkably good in advanced electrostatics, impressed forces in dielectric, etc.), one wonders why it does not take the place of the commonly used two-fluid hypothesis, merely *as* a working hypothesis, and nothing more.

Returning to the wire. It is important to remember that there are two conductors, not one only, with a dielectric between. When we put an impressed force in the wire we send current across the dielectric as well as round the conducting circuit. The dielectric current ceases as soon as the back force of the elastic displacement supplies that difference of potential which is appropriate to the distribution of impressed force (which difference of potential depends entirely on the conductivity conditions). The equation of continuity means that when the current entering a unit length of wire on one side is greater than that leaving it on the other, the excess goes across the dielectric to the outer conductor, in which there is a precisely equal variation in the current. The time-integral of this dielectric current $\dot{q}$ is q, which is the total displacement outward per unit length of wire. The quantity V is the back E.M.F. of the displacement. On removing the impressed force, there is left the electric energy of the displacement, which is $\frac{1}{2}Vq$ per unit length of wire; the back forces act, discharge the dielectric, and this energy is used up as heat in the conductors.

We can now make some easy extensions of the system (71*b*). R must be the sum of the resistances of the wire and return, per unit length, thus removing the restriction that the return has no resistance. S, of course, remains the same. But V cannot be the potential of the wire, because V cannot $=0$ all along the return. We may, however, call V the difference of potential (although that is not exactly true, on account of inertia, unless we agree to include a part of the E.M.F. of inertia in V). It is, however, definitely the E.M.F. of the condenser, given by $q=SV$. We need not restrict ourselves, in these first approximations, to round wires, or to symmetrically-arranged returns. The return may be a parallel wire. Of course the proper change must then be made in the value of S.

SECTION XXXIII. THE EQUATIONS OF PROPAGATION. INTRODUCTION OF SELF-INDUCTION.

The next step to a correct formulation of the laws of propagation along wires is, obviously, to take account of the electric force of inertia in the expression of Ohm's law. This appears to have been first attempted by Kirchhoff in 1857. According to J. J. Thomson ("Electrical Theories," *The Electrician*, June 25, 1886, p. 138) this was his system. Let

$$e=X\sin ns,$$

where e is the charge per unit length, and s is length measured along the wire. The equation of X is

$$\frac{c^2}{2}\frac{d^2X}{ds^2}=\frac{rc^2}{16l\gamma}\frac{dX}{dt}+\frac{d^2X}{dt^2},$$

where r is the resistance of the wire in electrostatic units, l its length, $\gamma=\log(l/a)$, where a is its radius, and c is a quantity occurring in Weber's hypothesis, the velocity with which two particles of electricity must move in order that the electrostatic repulsion and the electromagnetic attraction may balance.

As it stands, I can make neither head nor tail of it. But, by extensive alterations, it may be converted to something intelligible. Turn X into e, in the second equation; or, what will come to the same thing, take V as the variable, since e and V are proportional. Then ignore the first equation altogether. Turn s into our variable x.

Then
$$\frac{d^2V}{dx^2}=\frac{r}{8l\gamma}\frac{dV}{dt}+\frac{2}{c^2}\frac{d^2V}{dt^2}.$$

Clearly this should reduce to (71b) by ignoring the last term. Therefore
$$r/8l\gamma=RS.$$

Here r/l is the resistance per unit length. Therefore $(8\gamma)^{-1}$ should be the capacity per unit length, or $\{8\log(l/a)\}^{-1}$. This is clearly wrong. The l should be a_2, the resistance of the return, a far smaller quantity than l; and the 8 should be 2, if the dielectric is air. This last correction may, however, be merely required by a change of units. Making it, we get this result
$$\frac{d^2V}{dx^2}=RS\frac{dV}{dt}+L_0S\frac{d^2V}{dt^2},$$
in our previous notation, with the addition that L_0 is the inductance per unit length of the dielectric only. That is,
$$L_0=2\mu\log(a_2/a_1),$$
with unit inductivity; a_2 distance of return, a_1 radius of wire. This estimate of the inductance is, of course, too low. The change of units makes it doubtful whether L_0 or some multiple of it was meant, but it is clearly a wrong estimate. Notice that L_0S is the reciprocal of the square of a velocity, which is numerically equal to the ratio of the electromagnetic and electrostatic units, and is the velocity of light, or close to it.

It is clear that there is room for considerable improvement here in several ways, such as the establishment of the equations independently of such a very special hypothesis as Weber's; also in the estimation of L; and, in interpretation, to modernise it in accordance with Maxwell's ideas. Having observed that Maxwell, in his treatise, described the system (71b) of the last section, with no allowance for self-induction, and knowing this system to be quite inapplicable to short lines, I (in ignorance of Kirchhoff's investigation) made the necessary change of bringing in the electric force of inertia (*Phil. Mag.*, August, 1876), [vol. I., p. 53], converting the system (71b) to the following:—

$$\left.\begin{array}{ll}-\dfrac{dV}{dx}=RC+L\dfrac{dC}{dt}, & q=SV,\\ -\dfrac{dC}{dx}=\dfrac{dq}{dt}=S\dfrac{dV}{dt}, & \end{array}\right\}\qquad \frac{d^2V}{dx^2}=RS\frac{dV}{dt}+LS\frac{d^2V}{dt^2}. \quad \ldots(72b)$$

The equations on the left side show the elementary relations, and that on the right the resultant equation of V.

The difference from (71*b*) is only in the first equation of electric force, and in the characteristic equation of V. To the electric force due to V is added the electric force of inertia $-L\dot{C}$, where L is the inductance of the circuit per unit length, according to Maxwell's system of coefficients of electromagnetic induction. That is, L consists of three parts, say L_0 for the dielectric, L_1 for the wire, and L_2 for the return. Their expressions will vary according to the size and shape of the conductors and their distance apart. In case of symmetry about an axis, their determination is very easy by the square-of-force method. The magnetic energy per unit length is $\frac{1}{2}LC^2$. It is also $\Sigma \mu \mathrm{H}^2/8\pi$, if H is the magnetic force, and the summation extends over the region of space belonging to the unit length. As H is a simple function of C and of the distance from the axis, the integration is very easily effected.

L is calculated on the hypothesis that the current-density has always the steady distribution, just as R is the steady resistance. As it is, strictly speaking, impossible to have the Faraday-law of induction true in all parts of the conductors without some departure from the steady distributions, it is satisfactory to know that more exhaustive investigation shows that L, not L_0, should be used in a first approximation.

In connection with this matter I may mention that, rather singularly, just as I was investigating it, my brother, Mr. A. W. Heaviside, called my attention to certain effects observed on telegraph lines, which could be explained by the combined action of the electrostatic and electromagnetic induction, causing electrical oscillations which made the pointers of the old alphabetical indicators jump several steps instead of one. When freed from practical complications, and worked down to the simplest form, the matter reduced to this, that the discharge of a condenser through a coil is of an oscillatory character, under certain circumstances, and I described the theory in the paper I have mentioned. It had been given by Sir W. Thomson in 1853, but it is a singular circumstance that this very remarkable and instructive phenomenon should not be so much as mentioned in the whole of Maxwell's treatise (first edition), though it is scarcely possible that he was unacquainted with it; if for no other reason, because it is so *simple* a deduction from his equations. I lay stress on the word simple, because it is not to be supposed that Maxwell was fully acquainted with the whole of the consequences of his important scheme.

Mr. Webb, the author of a suggestive little book on "Electrical Accumulation and Conduction," had very early practical experience of electrical oscillations in submarine cables, when they were coiled up on board ship, ceasing, more or less, as they were submerged.

It is far more difficult to obtain a satisfying mental representation of the electric force of inertia $-LC$ than of that due to the potential, or $-dV/dx$, as described in the last section. The water-pipe analogy is, however, simple enough. Let L be the mass of the fluid per unit length, C its velocity, then $\frac{1}{2}LC^2$ is its kinetic energy, LC its momentum, LC the force that must be applied to increase it, $-L\dot{C}$ the force of reaction. A mental representation of many of the phenomena connected with electrical oscillations is also very simply got by the use of the fluid analogy. It is, however, certainly wrong, as we find by carrying it out more fully into detail. Remark, however, that, as $\frac{1}{2}LC^2$ is the magnetic energy per unit length, LC is the *generalised* momentum corresponding to C as a generalised velocity, $L\dot{C}$ the generalised externally applied force, an electric force, of course, and $-L\dot{C}$ the force of reaction—that is, the electric force of inertia. This is by the simple principles of dynamics, disconnected from any hypothesis as to the mechanism concerned.

The magnetic energy must be definitely localised in space, to the amount $\frac{1}{2}\mu\mathbf{H}^2/4\pi$ per unit volume, and be regarded as the kinetic energy of some kind of motion in the magnetic field. When steady, there is no force of inertia. But when $\mathbf{H}$ changes, and with it C, since these are rigidly connected (in our first approximation) there is necessarily a force of inertia, which, reckoned as an electric force appropriate to C as a generalised velocity, is $-L\dot{C}$ per unit length.

In the discharge of a condenser through a coil, if we start with a charge, but no current, there is in the first place only the potential energy of the displacement in the condenser. The discharge cannot take place without setting up a magnetic field, proportional in intensity to the current at any moment, so that the original electrical energy is employed in heating the wire, and also in setting up the magnetic energy. When the condenser is wholly discharged, the inertia of the magnetic field keeps the current going, and it will continue until the whole energy of the magnetic field is restored to the condenser (less the part wasted in the wire) in the form of the energy of the negative displacement there produced. Except that the charge is smaller, and of the opposite sign, everything is now as when we started, so that we may begin again and have a reverse current, continuing until the condenser is again charged in the same sense as at first, with no magnetic field. This is the course of a complete oscillation. But if the resistance be of or above a certain amount, depending on the capacity of the condenser and the inductance of the coil, the oscillations cease, and the discharge is completed in a single current which does not reverse itself.

Similar effects take place, in general, in any circuit, when a change is made which involves a redistribution of electric displacement, or its total discharge, but the full theory is usually very difficult to follow in detail. The so-called "false discharge" of a submarine cable is, however, easily comprehensible by the last paragraph.

If, in the characteristic equation of V in (72*b*), we take $L=0$, reducing it to that of (71*b*), we have simple diffusion of the static charge. If, for instance, the ends of the lines be insulated, any initial state of charge will settle down to be a uniform distribution, in a non-oscillatory manner, the smaller inequalities (smaller as regards length of line over which they extend) being wiped out rapidly, the larger more slowly; the law being that similar distributions subside similarly, but in times which are proportional to the squares of the lengths concerned.

If, on the other hand, we take $R=0$ in the characteristic equation of V in (72b), we have an entirely different order of events. As there is no waste in the wire, it is clear that the total energy of any initial state, electric and magnetic, remains undiminished. We can definitely divide the initial state into two distinct states travelling in the manner of waves in opposite directions, and being continuously reflected at the ends. Or, more simply, set up a charge at a single point of the line. It will divide into two, which will go on travelling backwards and forwards for ever. But into details of this kind we must not be tempted to enter at present, the immediate object being to lay the foundations for a more general theory.

When both terms on the right side of the characteristic equation are counted, propagation takes place by a mixture of diffusion and wave-transfer. A wave sent from one end of the line which would, were there no resistance, travel unchanged in form, and be reflected over and over again at the ends, in reality spreads out or diffuses itself, as well as, to a certain extent, being carried forward as a wave. The length of the line is an important factor. Wave characteristics get rapidly wiped out in the transmission of signals on a very long submarine cable, so that the manner of variation of the current at the distant end approximates to what it would be in the case of mere diffusion.

On the other hand, coming to a very short line, there are, every time a signal is made, immensely rapid dielectric oscillations, before the steady state is reached, due to to-and-fro reflection. As a general rule, this oscillatory phenomenon is unobservable, but it is none the less existent. It is customary to ignore it altogether in formulation, regarding the matter as one in which magnetic induction alone is concerned. Of course the magnetic energy is then far more important than the electric, and the current in the wire rises nearly in accordance with the magnetic theory.

The immense rapidity of the dielectric vibrations is one reason why they are unobservable, except indirectly, and under peculiar circumstances. Sometimes, however, they become prominent, especially when a circuit is suddenly interrupted, when we shall have large differences of potential. Mr. Edison discovered a new force. The enthusiasm displayed by his followers in investigating its properties was most edifying, and thoroughly characteristic of a vigorous and youthful nation. But it was only the dielectric oscillations, it is to be presumed; unless indeed it be really true, as has been reported, that the renowned inventor has kept the new force concealed on his person ever since.

How is it, it may be asked, that in the rise of the current in a short wire, according to the simple magnetic theory, the potential at any point in the wire is regarded as a constant, viz., its final value when the current has reached the steady state? Thus, as we have

$$\frac{e}{l}-\frac{dV}{dx}=RC+L\frac{dC}{dt}, \qquad \text{and} \qquad \frac{d^2V}{dx^2}=0,$$

if e is the total impressed force in the circuit, and l the length, the potential variation dV/dx must be constant. Supposing then e to exist only at $x=0$, the current will rise thus:—

$$C=\frac{e}{Rl}(1-\epsilon^{-Rt/L}),$$

and the value of $-dV/dx$ must be e/l, from the very moment e is started, and so long as it is kept on.

When we seek the interpretation, in the more general theory, we find that although the current oscillations become so insignificant on shortening the line that the well-known last formula becomes valid, practically, yet the potential oscillations remain in full force during the variable period. A wave of potential travels to and fro at the velocity $(LS)^{-\frac{1}{2}}$, making the potential at any one spot rapidly vibrate between a higher and a lower limit, though not according to the S. H. law, but in such a manner that its mean value is the final value, whilst the limits between which the vibration occurs continuously approach one another; the vibration, on the whole, subsiding according to the exponential law, with $2L/R$ as time-constant. The quantity e/l, which in the above rudimentary theory is taken to be the actual potential variation, is really the mean value of the real rapidly vibrating potential variation, at every point of the circuit and during the whole variable period, at whose termination, on subsidence of the vibrations, it becomes the real potential variation. [See vol. I., pp. 57 and 132 for details.]

To get rid of this vibration, we have merely to distribute the impressed force so as to do away with the potential variation.

Having now got the elementary relations established, we can proceed to the simplest manner of extending them to include the phenomena attending the propagation of current into the conductors from the dielectric.

Section XXXIV. Extension of the Preceding to Include the Propagation of Current into a Wire from its Boundary.

The first step to getting out of the wire into the dielectric occurs in Sir W. Thomson's theory, Section XXXII. We certainly get as far as the boundary of the wire. To some extent we make progress in adopting (same Section) Maxwell's idea of the continuity of the conduction and the dielectric current, when the conduction current is discontinuous itself. Further progress is made (Section XXXIII.) in introducing the electric force of inertia and the magnetic energy, so far as dependent on the first differential coefficient of the current with respect to the time, assuming the magnetic field to be fixed by the single quantity C, the wire-current, just as the electric field is fixed by the single quantity V, the potential-difference of the two wires at a given distance.

But the magnetic machinery does not move in rigid connection with the wire-current, as is implied in the specifications of the magnetic energy by $\frac{1}{2}LC^2$, like that of the electric energy by $\frac{1}{2}SV^2$, L and S being the inductance and the electric capacity, per unit length of line. In going further, I believe the following to be the most elementary method possible, as well as being pretty comprehensive. To fix ideas, and simplify the nature of the magnetic field, let the line consist of two concentric tubes, separated by a dielectric nonconducting tube. The dielectric is to occupy our attention mainly, in the first place. Let a_1 and a_2 be its inner and outer radii, a_0 the inner radius of the inner tube, and a_3 the outer radius of the outer. Find the connection between the longitudinal electric force at the inner and outer boundaries of the dielectric tube, and the E.M.F. of the condenser, and the E.M.F. of inertia, so far as it depends upon the magnetic field in the dielectric.

Let ABCD in the figure be a rectangle in a plane through the common axis of the tubes, AB being on the inner and CD on the outer boundary of the dielectric, both of unit length. Let the current be from A to B in the inner tube, in which direction x is measured, and therefore from C to D in the outer tube. These currents are not precisely equal under all circumstances, but are so nearly equal that we can ignore the longitudinal current in the dielectric in comparison with them; then the current C in the inner necessitates the same current C in the outer tube. The lines of magnetic force are directed upward through the paper, and the intensity of force is $2C/r$ at distance r from the common axis of the tubes.

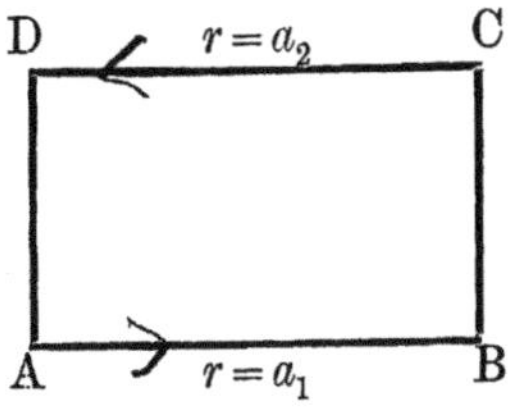

The total induction through the rectangle is therefore

$$\mu_2\int_{a_1}^{a_2}\frac{2C}{r}dr=\left(2\mu_2\log\frac{a_2}{a_1}\right)C=L_0C,$$

if μ_2 be the inductivity of the dielectric, and L_0 the inductance of the dielectric per unit length of line.

Now, the rate of decrease of the induction with the time, or $-L_0\dot{C}$, is the E.M.F. of inertia in the circuit ABCD in the order of the letters. But if E and F are the longitudinal electric forces in AB and DC, and V and W the radial forces in BC and AD, another expression for the E.M.F. in the circuit is $E-F+V-W$. But as AB and CD are of unit length, $V-W=dV/dx$. Hence

$$E-F+dV/dx=-L_0\dot{C}, \qquad \text{or} \qquad -dV/dx=L_0\dot{C}+E-F. \qquad (73b)$$

Next, let Γ_1 and Γ_2 be the longitudinal current-densities at the boundaries of the conductors, ρ_1 and ρ_2 their resistivities, and e_1, e_2 the impressed forces, if any, in them. Then, by Ohm's law,

$$e_1+E=\rho_1\Gamma_1, \qquad e_2+F=\rho_2\Gamma_2;$$

and therefore $\qquad E-F=\rho_1\Gamma_1-\rho_2\Gamma_2-e, \qquad \text{.......................}(74b)$

if $e=e_1-e_2$. Thus e is the impressed force in the circuit per unit length, irrespective of how it is divided between the inner and the outer conductor. Also, e is supposed to be longitudinal.

Now use (74*b*) in (73*b*), making it become

$$e - dV/dx = L_0\dot{C} + \rho_1\Gamma_1 - \rho_2\Gamma_2. \quad \text{.................. (75b)}$$

We now require to connect Γ_1 and Γ_2, the current-densities at the boundaries of the conductors, with the total currents in them. Representing these connections thus,

$$\rho_1\Gamma_1 = R_1''C, \qquad -\rho_2\Gamma_2 = R_2''C, \quad \text{...... .... (76b)}$$

we require to find the forms of R_1'' and R_2'', one for the inner, the other for the outer conductor. If this be imagined to be done, and we put

$$R'' = L_0(d/dt) + R_1'' + R_2,$$

the equation (75*b*) becomes

$$e - dV/dx = R''C = L_0\dot{C} + R_1''C + R_2''C, \quad \text{.............. (77b)}$$

wherein R'' is known. The complete scheme will therefore be,

$$e - dV/dx = R''C, \qquad q = SV, \qquad -dC/dx = dq/dt = S\dot{V}, \quad (78b)$$

which should be compared with (71*b*) and (72*b*). As for the equations of V and of C, they may be obtained by elimination, but it is unnecessary to write them at present.

We have supposed R_1'' and R_2'' to be known. The question is, then, how to find them. We know that in steady-flow they must be R_1 and R_2, the steady resistances of the conductors. We know, further, that they are $R_1 + L_1(d/dt)$ and $R_2 + L_2(d/dt)$, when only the first derivative $\dot{C}$ of the current is allowed for. Now, we know that, under all ordinary circumstances, the length of a wire must be a very large multiple of its diameter before the influence of the electric charge becomes sensible. When it does become sensible, the current is of a different strength in different parts of the line during the setting up of a steady current. But in a section of the line which, though long compared with the diameter of the wire, is short compared with its length, the current changes insensibly, even when the change is very great between the current-strength in that section and in another, which, by contrast, may be called distant from the first.

It is, clear, therefore, that we shall come exceedingly near the truth if, in the investigation of the function R_1'' we altogether disregard the change in strength of the current in passing along the line. This amounts to ignoring the small radial component of the current in the conductors, and making the current quite longitudinal. This is only done for purposes of simplification, and does not involve any physical assumption in contradiction of the continuity of the current; for we join on the dielectric current to that in the conductors, by means of the equation of continuity, the third of (78*b*).

The determination of R_1'' and R_2'' is thus made a magnetic problem, of which I have already given the solution. See equation (50*b*), Section XXX., where the first big fraction represents R_1'' for the inner conductor, and the second R_2'' for the outer. The separation of these into even and odd differential coefficients, thus,

$$R_1'' = R_1' + L_1'p,$$

is of principal utility in the periodic applications. It may, perhaps, be as well pointed out that the first equation (78*b*) should, in strictness, be cleared of fractions to obtain the rational differential equation. But the advantages of the form (78*b*) are too great to be lightly sacrificed to formal accuracy.

We have now the means of fully investigating the transmission of disturbances along the line, including the retardation to inward transmission from the dielectric into the conductors as well as the effects of the electrostatic charge. The system is a practical working one; for, the electrical variables being V and C, we are enabled to submit the line to any terminal conditions arising from the attachment of apparatus, the effect of which is fully determinable, because the differential equation of the apparatus itself is one between V and C. Both the ratio of V to C and their product are important quantities. The first is, in steady-flow, a mere resistance. In variable states it becomes a complex operator of great importance in the theoretical treatment. The second, VC, is the energy-current, concerning which more in the next Section.

In the meantime I will briefly indicate the nature of the changes made when we go further towards a complete representation of Maxwell's electric and magnetic connections. First, as regards the small radial component of current in the conductors. The quantity s that appears in the expression for R_1'' is given by

$$-s_1^2 = 4\pi\mu_1 k_1 p,$$

μ_1 being the inductivity and k_1 the conductivity of the inner conductor, whilst p is, when we are dealing with a normal system of subsidence, a constant; thus, ϵ^{pt} is the time-factor showing how it subsides, p being always negative in an electromagnetic problem, and also always negative in an electrostatic problem, whilst in a combined electrostatic and magnetic case it is either negative and real, or negative with an imaginary part, when its term must be paired with a companion to make a real oscillatorily subsiding system. Now the simplest form of terminal condition possible is $V=0$ at both ends of the line, *i.e.*, short-circuits. Then

$$V = A \sin(j\pi x/l),$$

where j is any integer, represents a V system, satisfying the condition of vanishing at both ends. Let the factor of x, which is $j\pi/l$, be denoted by m. Only the first few j's are of much importance, 1, 2, 3, etc. Now, if we change the connection between s_1 and p above-given to

$$s_1^2 = -4\pi\mu_1 k_1 p + m_1^2,$$

we shall be able to take the radial component of current in the conductors into account; but the change made is usually very insignificant. There are four other cases in which we can work similarly—viz., when the line is insulated at both ends, or $C=0$; when it is insulated at either end and short-circuited at the other—two cases; and when the line is closed upon itself, each conductor making a closed circuit without interposed resistances, etc. In all except the last case, when the line has no ends, the quantity VC vanishes at both ends of the line, either V or C being zero at these places, so that no energy can enter or leave the line (dielectric and two conductors). Nor can this happen in the last case. But if we join on terminal apparatus, thus making VC finite at one or both ends, the system breaks down, and we require to fall back upon the preceding.

But if we keep to the five cases mentioned, we may make a further refinement, by taking the longitudinal current in the dielectric into account, which we have previously considered negligible in comparison with the current C. We cannot do this in terms of V, which is inadequate to express the electric energy. But we may do it in terms of the electric and magnetic forces, and then obtain a full representation of Maxwell's connections, instead of an approximate. But even in this it is assumed that there is no magnetic disturbance outside the outer conducting tube or inside the inner, which there must really be, for we must have continuity of the tangential electric force, which necessitates electric force, and therefore also electric displacement and current and magnetic force, outside the outer tube and inside the inner, having some minute disturbing effect on the current in the conductors.

We may, however, leave these refinements to take care of themselves, and return to the V and C system of representation. The advantage of dealing with concentric tubes is due to the circularity of the lines of magnetic force, which produces considerable mathematical simplifications, as well as physical. Suppose, however, the tubes are not concentric, although the dielectric is still shut in by them. Here, clearly, to a first approximation, we have merely to give changed values to the constants S and L, whilst R is unchanged. But to go further, the determination of R_1'' and R_2'' will present great difficulties. This, however, is clear: that the full L' will have for its minimum value, approached with very rapid oscillations, L_0, such that $SL_0 = v^{-2}$, where v is the speed of propagation of undissipated disturbances through the dielectric. This follows by regarding the conductors as infinitely conducting, so that there is no waste in them, when the equation of V becomes

$$\frac{d^2V}{dx^2} = \mu_2 c \frac{d^2V}{dt^2}, \qquad \text{or} \qquad = SL_0 \frac{d^2V}{dt^2}, \quad \text{..............}(79b)$$

showing wave propagation with velocity v.

But if the two conductors be parallel solid wires or tubes (not concentric), and be placed at a sufficient distance from one another, the lines of magnetic force in and close round the conductors will be very nearly circles, so that we may regard R_1'' and R_2'' as known by the preceding; and we can therefore go beyond the approximate method of representation founded upon R, S, and L only. Even if we bring the conductors so close that there is considerable disturbance from the assumed state, we should still, in reckoning R_1'' and R_2'' in the same way, go a long distance in the direction required, especially in the case of iron wires, in which, by reason of the high inductivity, the magnetic retardation is so great.

The effect of leakage has not been allowed for in the preceding. The making of the necessary changes is, however, quite an elementary matter in comparison with those connected with magnetic retardation. We require to change the form of the equation of continuity. If there be a leakage-fault on an otherwise perfectly insulated line, we have the line divided into two sections, in each of which the former equations hold good; whilst at the place of the leak there is continuity of V and discontinuity of C, the current arriving at the leak on the one side exceeding that leaving it on the other by the current in the leak itself, which is the quotient of V by the resistance of the leak, if it be representable as a resistance merely. But when the leakage is widely distributed it must be allowed for in the line-equations. Even in the case of leakage over the surface of the insulators of a suspended wire, the proper and rational course is to substitute continuously distributed leakage for the large number of separate leaks; which amounts to the same thing as substituting a continuous curve for a large number of short straight lines joined together so as to closely resemble the curve. The equation of continuity becomes

$$-dC/dx = KV + S\dot{V}, \qquad \text{........................(80}b\text{)}$$

where the fresh quantity K is the conductance, or reciprocal of the insulation resistance, per unit length of line. That is, the true current leaving the line is the sum of the former $S\dot{V}$, the condenser-current, and of KV, the leakage-current, both of which co-operate to make the current in the line vary along its length, although in the steady state it is the leakage alone that thus operates. But as regards retardation, their effects are opposed. The setting up of the permanent state is greatly facilitated by leakage, as is most easily seen by considering the converse, viz., the subsidence of the previously set-up steady state to zero when the impressed force is removed. If, then, we wish to increase the clearness of definition of current-changes at the distant end of a line on which electrostatic retardation is important, we can do it by lowering the insulation-resistance as far as is practicable.

SECTION XXXV. THE TRANSFER OF ENERGY AND ITS APPLICATION TO WIRES. ENERGY-CURRENT.

When the sage sits down to write an elementary work he naturally devotes Chapter I. to his views concerning the very foundation of things, as they present themselves to his matured intellect. It may be questioned whether this is to the advantage of the learner, who may be well advised to "skip the Latin," as the old dame used to say to her pupils when they came to a polysyllable, and begin at Chapter II. If this be done, Prof. Tait's "Properties of Matter" is such an excellent scientific work as might be expected from its author. But Chapter I. is metaphysics. There are only two Things going, Matter and Energy.

Nothing else is a thing at all; all the rest are Moonshine, considered as Things.

However this be, the transfer of energy is a fact well known to all, even when we put the statement in such a form that the energy seems to lose its thinginess, by calling it the transfer of the power of doing work. Thus, after transfer of energy from the sun ages ago, followed by long storage underground and convection to the stove or furnace, we set free the imprisoned energy, to be generally diffused by the most varied paths. The transfer from place to place can be, in great measure, traced so far as quantity and time are concerned; but it does not seem possible to definitely follow the motion of an atom of energy, so to speak, or to give a fixed individuality to any definite quantity of energy.

Whenever the dynamical connections are known, the transfer of energy can be found, subject to a certain reservation. In the elementary case of a force, F, acting on a particle of mass m and velocity v, F is measured by the rate of acceleration of momentum, or $F=m\dot{v}$; and, to obtain the equation of activity, we merely multiply this equation by the velocity, getting $Fv=mv\dot{v}=\dot{T}$, the rate of increase of the kinetic energy T, or $\frac{1}{2}mv^2$, which is the amount of work the particle can do against resistance in coming to rest. Where the energy came from is here left unspecified. In a case of impact, we may clearly understand that the transfer of kinetic energy is from one of the colliding bodies to the other through the forces of elasticity brought into play, thus making potential energy an intermediary, though what the potential energy may be, and whether it is not itself kinetic, or partly kinetic, we are not able to decide.

It is much more difficult in the case of gravity. As the stone falls to the ground, it acquires kinetic energy truly; and if energy moves continuously, as its indestructibility seems to imply, it must receive its energy from the surrounding medium; or the energy of gravitation must be in space generally, wholly or in part, and be transferred through space by definite paths through stresses in the medium, by which means Maxwell endeavoured to account for gravitation. In general, we have only to frame the equations of motion of a continuous system of forces, and it stands to reason that the transfer of energy is to be got by forming the equation of activity, not of the system as a whole, but of a unit volume.

Now, in the admirable electromagnetic scheme framed by Maxwell, continuous action through space is involved, and the kinetic and potential energies (or magnetic and electric) are definitely located, as well as the seat and amount of dissipation of energy. We therefore need only form the equation of activity to find the transfer-of-energy vector. Of course impressed forces are subject to the energy definition. No other is possible in a dynamical system.

But if we take Maxwell's equations and endeavour to immediately form the equation of activity (like $Fv=\dot{T}$ from $F=m\dot{v}$), it will be found to be impossible. They will not work in the manner proposed. But we may consider the energy, electric and magnetic, entering and leaving a given space, and that dissipated within it, and by laborious transformations evolve the expression for the vector transfer. This was first done by Prof. Poynting for a homogeneous isotropic medium (*Phil. Trans.*, 1884). In my independent investigation of this matter, I also followed this method in the first place (*The Electrician*, June 21, 1884) [p. 377, vol. I.] in the case of conductors. But the roundabout nature of the process to obtain what ought to follow immediately from the equations of motion, led me to remodel Maxwell's equations in some important particulars, as in the commencing Sections of this Article (Jan., 1885) with the result of producing important simplifications, and bringing to immediate view useful analogies which are in Maxwell's equations hidden from sight by the intervention of his vector-potential. This done, the equation of activity is at once derivable from the two cross-connections of electric force and magnetic current, magnetic force and electric current, in a manner analogous to $Fv = \dot{T}$, without roundabout work, and applicable without change to heterogeneous and heterotropic media, with distinct exhibition of what are to be regarded as impressed forces, electric and magnetic. (*Electrician*, Feb. 21, 1885) [p. 449, vol. I.]

Knowing the electric field and the magnetic field everywhere, the transfer of energy becomes known. The vector transfer at any place is perpendicular to both the electric and the magnetic forces there, not counting impressed forces. Its amount per unit area equals the product of the intensities of the two forces and the sine of their included angle.

But I mentioned that there is a reservation to be made. It is like this. If a person is in a room at one moment, and the door is open, and we find that he is gone the next moment, the irresistible conclusion is that he has left the room by the door. But he might have got under the table. If you look there you can make sure. But if you are prevented from looking there, then there is clearly a doubt whether the person left the room by the door or got under the table hurriedly. There is a similar doubt in the electromagnetic case in question, and in other cases. Thus, we can unhesitatingly conclude from the properties of the magnetic field of magnets that the mechanical force on a complete closed circuit supporting a current is the sum of the electromagnetic forces per unit volume (vector-product of current and induction), but it does not follow strictly that the so-called electromagnetic force is the force really acting per unit volume, for any system of forces might be superadded which cancel when summed up round a closed circuit.

So, in the transfer-of-energy case, there may be any amount of circulation of energy in closed paths going on (as pointed out in another manner by Prof. J. J. Thomson), besides the obviously suggested transfer, provided this superposed closed circulation is without dissipation of energy. Or, if **W** be the vector energy-current density, according to the above-mentioned rule, we may add to it another vector, say **w**, provided **w** have no convergence anywhere. The existence of **w** is possible, but there does not appear to be any present means of finding whether it is real, and how it is to be expressed.

Its consideration may seem quite useless, in fact. But it is forced upon us in quite another way, by the fact that, when $\mathbf{w} = 0$, we are sometimes led to the circuital flux of energy. Let, for instance, a magnet be placed in the field of an electrified body; or, more simply, let a magnet be itself electrified. There is no waste of energy; hence the flux of energy caused by the coexistence of the two fields, electric and magnetic, is entirely circuital. *E.g.*, in the case of a spherical uniformly magnetised body, uniformly superficially electrified, it takes place in circles in parallel planes perpendicular to the axis of magnetisation, the circles being centred on this axis. This circuital flux is entirely through the air or other dielectric. What is the use of it? On the other hand, what harm does it do? And if the medium is really strained by coexistent electric and magnetic stresses, why should there not be this circuital flux? But, if we like, we may cancel it by introducing the auxiliary $\mathbf{w}$.

There is yet another kind of closed circulation, according to $\mathbf{W}$ alone, not existing by itself, but set going by impressed forces causing a useful transfer of energy, and ceasing when the useful transfer ceases. If, for instance, we close a conductive circuit containing a battery, we set up a useful transfer from the battery to all parts of the wire, through the dielectric usually. Suppose there is also impressed electric force in the dielectric, or electrification, or any stationary electric field. If the battery does not work there is no transfer of energy. But when it does, there is, besides the regular first-mentioned transfer from the battery to the wire, a closed circulation due to the coexistence of the stationary electric field and the magnetic field of the wire-current, the resultant transfer being got by superposing the regular flux and the closed circulation. Here again, by introducing $\mathbf{w}$, we may reduce it to the regular undisturbed transfer. It is clear, then, in considering the nature of the transfer in a useful problem, that it is of advantage to entirely ignore the useless transfer, and confine our attention to the undisturbed.

A general description of the transfer along a straight wire was given in Section II. [vol. I., p. 434]. It takes place, in the vicinity of the wire, very nearly parallel to it, with a slight slope towards the wire, as there described. Prof. Poynting, on the other hand (Royal Society, *Transactions*, February 12, 1885), holds a different view, representing the transfer as nearly perpendicular to a wire, *i.e.*, with a slight departure from the vertical. This difference of a quadrant can, I think, only arise from what seems to be a misconception on his part as to the nature of the electric field in the vicinity of a wire supporting electric current.

The lines of electric force are nearly perpendicular to the wire. The departure from perpendicularity is usually so small that I have sometimes spoken of them as being perpendicular to it, as they practically are, before I recognised the great physical importance of the slight departure. It causes the convergence of energy into the wire. To estimate the amount of departure, we may compare the normal and tangential components of electric force. Let there be a steady current in a straight wire, and the fall of potential from beginning to end be $V_0 - V_1$; the tangential component is then $(V_0 - V_1) \div l$, if l be the length of wire. On the other hand, the fall of potential from the wire to its return—of no resistance, for simplicity—at any distance from the beginning of the line, is V, which is V_0 at one end and V_1 at the other. It is clear at once that the tangential is an exceedingly small fraction of the normal component of electric force, if the wire be long, and that it is only under quite exceptional circumstances anything but a small fraction. Prof. Poynting should therefore, I think, make his tubes of displacement stick nearly straight up as they travel along the wire, instead of having them nearly horizontal, unless I have greatly misunderstood him.

But if we distribute the impressed force uniformly throughout the circuit, so that there shall be, in the steady state, no difference of potential and no transfer of energy, owing to the impressed force at any place being just sufficient to support the current there then, on starting the impressed force, the transfer of energy will be perpendicular to the wire outward, ceasing when the steady state is reached; and, on the other hand, on stopping the impressed force the transfer will be perpendicular to the wire inward, the magnetic energy travelling back again (assisted by temporary longitudinal electric force, which has no existence in the steady state) to be dissipated in the wire. But this case, though imaginable, is not practically realisable.

In the vicinity of the wire the radial electric force varies inversely as the distance, and so does the intensity of magnetic force. The density of the energy-current therefore varies inversely as the square of the distance approximately. This does not continue indefinitely. Thus, if the return be a parallel wire the middle distance is the place of minimum density of the energy-current, in the plane of the two wires. As regards the total energy-current, this is VC, the product of the fall of potential from one wire to the other into the current in each. One factor, V, is the line-integral of the electric force across the dielectric. The other, C, is the line-integral $(\div 4\pi)$ of the magnetic force round either wire.

In the figure, AB and CD are the two wires, enormously shortened in length compared with their distance apart, joined through terminal

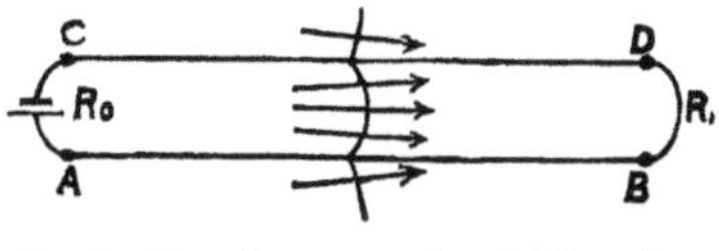

resistances R_0 and R_1, in the former of which alone is the impressed force e. The fall of potential from A to C is V_0, from B to D is V_1, and at any intermediate distance is V. The total activity of the source is eC, of which $(e - V_0)C$ is wasted in R_0. What is left, or V_0C, is the energy-current at AC, entering the line. By regular waste into the wires, its strength falls to V_1C at BD, where the line is left, and the terminal arrangement entered, to be wasted in frictional heat-generation R_1C^2 therein, or otherwise disposed of. The curved lines and arrows perpendicular to them show lines of electric force and the direction of the energy-flux at a certain place, the inclination of the lines of force to the perpendicular being greatly exaggerated, as well as that of the lines of flux of energy to the horizontal, in order to show the convergence of energy upon the wires, there to be wasted. Its further transfer belongs to another science.

The rate of decrease of VC as we travel along the line is the waste per unit length. Thus,

$$-\frac{d}{dx}(VC) = -C\frac{dV}{dx} = (R_1+R_2)C^2, \quad \text{.................(81}b\text{)}$$

R_1 and R_2 being the resistances of the wires per unit length. This is in steady-flow, with no leakage. But if there be leakage, we have the equation of continuity

$$-(dC/dx) = KV,$$

making

$$-\frac{d}{dx}(VC) = KV^2 + (R_1+R_2)C^2, \quad \text{..................(82}b\text{)}$$

where KV^2 is the waste-heat per second due to the leakage-resistance.

But when the state is not steady, we have the equation of continuity

$$-\frac{dC}{dx} = KV + S\dot{V}, \quad \text{.....................(80}b\text{) }\textit{bis.}$$

and the equation of electric force

$$-\frac{dV}{dx} = L_0C + E - F, \quad \text{................(73}b\text{) }\textit{bis.}$$

so that

$$-\frac{d}{dx}(VC) = KV^2 + \frac{d}{dt}(\tfrac{1}{2}SV^2) + \frac{d}{dt}(\tfrac{1}{2}L_0C^2) + EC - FC. \quad \text{......(83}b\text{)}$$

Here we account for the leakage-heat, for the increase of electric energy, and for the increase of magnetic energy in the dielectric by the first, second, and third terms on the right side. EC, the fourth term, represents the energy entering the first wire per second, E being the tangential electric force; and $-FC$, the last term, represents the energy entering the second wire per second, F being the tangential electric force at its boundary reckoned the same way as E. The energy-flux is now perpendicular to the current, *i.e.*, after entering the wires, ceasing when the axes are reached. And,

$$EC = Q_1 + \dot{T}_1, \qquad -FC = Q_2 + \dot{T}_2, \quad \text{...............(84}b\text{)}$$

if Q_1, Q_2, are the dissipativities, T_1 and T_2 the magnetic energies in the two wires, per unit length of line.

If the impressed force is a S.H. function of the time, so is the current, etc., everywhere, and

$$E = R_1'C + L_1'\dot{C}, \qquad -F = R_2'C + L_2'\dot{C}, \quad \text{............(85}b\text{)}$$

where R_1', R_2', L_1', and L_2' are constants depending upon the frequency, reducing to the steady resistances and inductances when the frequency is infinitely low. In this S.H. case

$$Q_1 = \tfrac{1}{2}R_1'C_0^2, \qquad Q_2 = \tfrac{1}{2}R_2'C_0^2, \qquad T_1 = \tfrac{1}{4}L_1'C_0^2, \qquad T_2 = \tfrac{1}{4}L_2'C_0^2,$$

are the *mean* dissipativities and magnetic energies in the wires, C_0 being the amplitude of the current; the halving arising from the mean value of the square of a sinusoidal function being half the square of its amplitude. But in no other case is there anything of the nature of a definite resistance, although, if the magnetic retardation to inward transmission is small, we may ignore it altogether, and drop the accents in (85*b*).

SECTION XXXVI. RESISTANCE AND SELF-INDUCTION OF A ROUND WIRE WITH CURRENT LONGITUDINAL. DITTO, WITH INDUCTION LONGITUDINAL. THEIR OBSERVATION AND MEASUREMENT.

When the effective resistance to sinusoidal currents is not much greater than the steady resistance, we may employ the formulae (44*b*) [p. 64], to estimate the effective resistance and inductance. On the other hand, when it is a considerable multiple of the steady resistance, we may employ the simple formulae (45*b*). But in intermediate cases, neither pair of formulae is suitable, and it therefore happens that in some practically realisable cases we require the fully developed formulae which are equivalent to (44*b*), but are always convergent.

Let R be the steady resistance per unit length of round wire of radius a, conductivity k, inductivity μ; and R' its effective resistance to sinusoidal currents of frequency $q = n/2\pi$. Let also

$$z = \mu n/R = \pi\mu k n a^2 = 2\pi^2\mu k a^2 q. \qquad \text{(86}b\text{)}$$

Then the formula required for R' is

$$\frac{R'}{R} = \frac{1 + \frac{z^2}{6}\left(1 + \frac{z^2}{2^3.10}\left(1 + \frac{z^2}{3^3.14}\left(1 + \frac{z^2}{4^3.18}\left(1 + \frac{z^2}{5^3.22}\left(1 + \dots\right.\right.\right.\right.\right.}{1 + \frac{z^2}{12}\left(1 + \frac{z^2}{3.2^2.10}\left(1 + \frac{z^2}{4.3^2.14}\left(1 + \frac{z^2}{5.4^2.18}\left(1 + \dots\right.\right.\right.\right.}. \qquad \text{(87}b\text{)}$$

The law of formation of the terms is plainly shown, so that the series may be continued as far as is necessary to ensure accuracy. But so far as is written is quite sufficient up to $z = 10$.

The corresponding formula for L', what the L of the wire becomes at the frequency q, is

$$\frac{L'}{L} = \frac{1 + \frac{z^2}{2^2.6}\left(1 + \frac{z^2}{2.3^2.10}\left(1 + \frac{z^2}{3.4^2.14}\left(1 + \frac{z^2}{4.5^2.18}\left(1 + \dots\right.\right.\right.\right.}{\text{Same denominator as in (87}b\text{)}}. \qquad \text{(88}b\text{)}$$

Here $L = \tfrac{1}{2}\mu$, simply. R'/R increases continuously, and L'/L decreases continuously, as the frequency increases.

The following are the values of R'/R for values of z from $\frac{1}{2}$ to 10:—

z.	R'/R.	z.	R'/R.
$\frac{1}{2}$	1·02	6	2·01
1	1·08	7	2·14
2	1·26	8	2·27
3	1·48	9	2·39
4	1·68	10	2·51
5	1·85		

The curve, whose ordinate is $R'/R-1$ and abscissa z, is convex to the axis of abscissae up to about $z=2\frac{1}{2}$, and then concave later.

Let us take the case of an iron wire of one-eighth of an inch in radius (about No. 4 B.W.G.), of resistivity 10,000, and inductivity 100. These data give us $z=q/51$, by (86b). Take, then, $z=q/50$. Each unit of z means 50 vibrations per second. Then $q=50$ makes $R'/R=1{\cdot}08$; $q=500$ makes $z=10$ and $R'/R=2{\cdot}51$, or the effective resistance $2\frac{1}{2}$ times the steady.

To obtain similar results in copper, with $\mu=1$, $k^{-1}=1600$, making μk to be $\frac{1}{16}$ part of its former value, we require the radius to be four times as great, or the wire to be 1 in. in diameter. But if it be of the same diameter, $q=500$ will only make $z=\frac{10}{16}$, and there will be only a slight increase in the effective resistance.

In the present notation the very-high-frequency formulae are

$$R'=L'n=R(\tfrac{1}{2}z)^{\frac{1}{2}}\,; \qquad \text{(89}b\text{)}$$

and, by comparison with the table, we shall be able to see how large z must be before these are sensibly true. Using (89b), $z=4$ gives $R'/R=1{\cdot}41$, much less than the real value; $z=8$ gives 2 instead of 2·274; $z=10$ gives 2·234 instead of 2·507. On the other hand, (89b) makes L' too big, but not so much as it makes R' too small. Thus $q=10$ makes $L'n/R=2{\cdot}234$ instead of 2·21, which is what the correct formula (88b) gives.

Probably $z=20$ would make (89b) fairly well represent the resistance, as it nearly does the inductance when $z=10$. In the case of the iron wire above mentioned, $z=50$, or $q=2500$, will make the effective resistance five times the steady.

If the wire be exposed to sinusoidal variations of longitudinal magnetic force by insertion within a long solenoidal coil, the effect, when small, on the coil-current, is the same as if the resistance of the coil-circuit were increased by the amount lR_1', given by

$$R_1'=L_1\times\tfrac{1}{2}\pi\mu kn^2a^2=L_1n\times\tfrac{1}{2}z. \qquad \text{(90}b\text{)}$$

[Reprint, vol. I., p. 369, the last equation. Also p. 364, equation (36).] Here l is the length of the core and coil, having N turns of wire per unit length, and

$$L_1=(2\pi aN)^2\mu$$

is the steady inductance, due to the core only, per unit of its length.

If C_0 be the amplitude of the coil-current, the mean rate of generation of heat in the core is $\frac{1}{2}R_1'C_0^2$, per unit of its length.

When the effect is large, use the formula

$$\frac{R_1'}{L_1 n} = \frac{1}{2}z . \frac{1 + \frac{z^2}{2^2.6}\left(1 + \frac{z^2}{2.3^2.10}\left(1 + \frac{z^2}{3.4^2.14}\left(1 + \dots\right.\right.\right.}{1 + \frac{z^2}{2}\left(1 + \frac{z^2}{2^3.6}\left(1 + \frac{z^2}{3^3.10}\left(1 + \dots\right.\right.\right.}. \quad \dots (91b)$$

[vol. I., p. 364, equation (36), and the next one.] (I have slightly changed the notation to suit present convenience, and show the law of formation of the terms. The old y equals the new $16z^2$.)

I did not give any separately developed expression for the L_1' corresponding to L_1; being only a portion of the L of the circuit it was merged in the expression for the tangent of the phase-difference. [Vol. I., pp. 369 to 374, §§ 16, 17.] Exhibiting now L_1' by itself, we have this formula :—

$$\frac{L_1'}{L_1} = \frac{1 + \frac{z^2}{6}\left(1 + \frac{z^2}{2^3.10}\left(1 + \frac{z^2}{3^3.14}\left(1 + \frac{z^2}{4^3.18}\left(1 + \dots\right.\right.\right.\right.}{\text{Same denominator as in (91}b\text{)}}. \quad \dots\dots (92b)$$

Notice that the numerators in (91b) and (88b) are the same, and that those of (92b) and (87b) are the same.

At the frequency 500, using the same iron wire above described, we have, taking $z = 10$ in (91b) and (92b),

$$R_1' = 188\, L_1 n, \qquad L_1' = \cdot 225\, L_1. \quad \dots\dots\dots\dots\dots (93b)$$

Or, with a little development,

$$R_1' = 622\, L_1 = 243{,}000\, N^2, \quad \dots\dots\dots\dots\dots\dots\dots (94b)$$

i.e., the extra resistance is 243 microhms multiplied by the length of the core, and by the square of the number of windings per unit length. At this particular frequency the amplitude of the magnetic force oscillations at the axis of the core is only one-fourteenth of the amplitude at the boundary. When it is the current that is longitudinal, it is the current-density at the axis that is only $\frac{1}{14}$ its boundary-value.

Now, as cores may be so easily taken thicker, it is also desirable to have the high-frequency formulæ corresponding to (91b) and (92b), which I now give. They are

$$R_1' = L_1' n = \frac{L_1 n}{(2z)^{\frac{1}{2}}}. \quad \dots\dots\dots\dots\dots\dots\dots\dots (95b)$$

The value $z = 10$ is scarcely large enough for their applicability. Thus (95b) give (same iron wire),

$$R_1' = L_1' n = \cdot 223\, L_1 n, \quad \dots\dots\dots\dots\dots\dots\dots\dots (96b)$$

instead of (93b), making R_1' too big, and L_1' too small, although the latter is nearly correct.

In one respect the reaction of metal in the magnetic field on a coil-current is far simpler than the reaction on itself when it contains the impressed force in its own circuit. If we have a sinusoidal current in a coil, subject to

$$e = RC + L\dot{C},$$

e being the sinusoidal impressed force, C the current, R and L the steady resistance and inductance of the circuit; and we, by putting metal in its magnetic field, induce currents in it, and waste energy there, we know that the new state is also sinusoidal, subject to

$$e = R'C + L'\dot{C},$$

where R' and L' have some other values. So far is elementary. This, however, is also elementary, that R' must be greater than R. For the heat in the coil per second is $\frac{1}{2}RC_0^2$, and the total heat per second is $\frac{1}{2}R'C_0^2$. As the latter includes the heat externally generated, R' is necessarily greater than R. But this simple reasoning, without any appeal to abstrusities, breaks down when it is the wire itself in which the change from R to R' takes place, and we then require to use reasoning based upon the changed distribution of current.

To observe these changes qualitatively is easy enough. But to do so quantitatively and accurately is another matter. It cannot be done with intermittences. A convenient little machine giving a strictly sinusoidal impressed force of good working strength, adjustable from zero up to very high frequencies, is a thing to be desired. But we may employ very rapid intermittences with an approximation to the theoretical results. I have obtained the best results with a microphonic contact, without interruptions, but it was difficult to keep it going uniformly. Slow intermittences give widely erroneous results, *i.e.*, according to the sinusoidal theory, which does not apply, making the changes in resistance and induction much too large. Here, of course, the silence—the best minimum to be got—is a loud sound.

I should observe, by the way, that a correct method of balancing is presumed. In Prof. Hughes's researches, which led him to such remarkable conclusions, the method of balancing was not such as to ensure, save exceptionally, either a true resistance or a true induction balance. Hence, the complete mixing up of resistance and induction effects, due to false balances. And hidden away in the mixture was what I termed the "thick-wire effect," causing a true change in resistance and inductance [vol. II., p. 30]. In fact, if I had not, in my experiments on cores and similar things, been already familiar with real changes in resistance and inductance, and had not already worked out the theory of the phenomenon of approximation to surface conduction [first general description in vol. I., Art. 30, p. 440; vol. II., p. 30], on which these effects in a wire with the current longitudinal depend, it is quite likely that I should have put down all anomalous results to the false balances.

Of course, we should separate inductance from resistance. Perhaps the simplest way is that I described [vol. II., p. 33, Art. XXXIV.] of using a ratio of equality, reducing the three conditions to two, ensuring independence of the mutual induction of sides 1 and 2, and also of sides 3 and 4 (allowing us to wind wires 1 and 2 together, and so remove the source of error due to temperature inequality which is so annoying in fine work), and requiring us merely to equalise the resistances and the inductances of sides 3 and 4, varying the inductance to the required amount by means of a coil of variable inductance, consisting of two coils joined in sequence, one of which is movable with respect to the other, thus varying the inductance from a minimum to a maximum—an arrangement which I now call an Inductometer, since it is for the measurement of induction. The oddly-named Sonometer will do just as well, if of suitable size, and its coils be joined in sequence. The only essential peculiarity of the inductometer is the way it is joined and used. This method of equal ratio was adopted by Prof. Hughes in his later researches (Royal Society, May 27, 1886); he, however, varies his induction by a flexible coil, which I hardly like. Lord Rayleigh has also adopted this method of separating induction from resistance, and of varying the inductance. (*Phil. Mag.*, Dec., 1886.) I found that the calibration could be expeditiously effected with a condenser, dividing the scale into intervals representing equal amounts of inductance. Lord Rayleigh does, indeed, seem to approve somewhat of Prof. Hughes's method, with its extraordinary complications in theoretical interpretation (very dubious at the best, owing to intermittences not being sinusoidal). But if it be wished to employ mutual induction between two branches to obtain a balance, there is the M_{63} or M_{64} method I described [vol. II., Art. XXXIV.], which is, like the method of equal ratio, exact in its separation of resistance and inductance, with simple interpretation. I have since found that there are no other ways than these, except the duplications which arise from the exchange of the source of electricity and the current indicator. Using any of these methods, we completely eliminate the false balances; now we shall have perfect silences, independent of the manner of variation of the currents, whenever the side 4 [in figure, p. 33, vol. II.], containing the experimental arrangement, is equivalent to a *coil*, with the two *constants* R and L, and can therefore equalise a coil in side 3 (presuming that the equal-ratio method is employed). But if in the equation $V=ZC$ of the experimental wire, Z is not reducible to the form of $R+L(d/dt)$, it is not possible to make the currents vary in the same manner in the sides 3 and 4, and so secure a balance. That is, we cannot balance merely by resistance and self-induction, the departure of the nearest approach to a balance from a true balance being little or great, as the manner of variation of the current in side 4 differs little or much from that of the current in its ought-to-be equivalent side 3. The difference is great when a coil with a big core is compared with a coil without a core; and, as in all similar cases, as before remarked, at a moderate rate of intermittence, we must not apply the sinusoidal theory to the interpretation. If we want to have true balances when there is departure from coil-equivalence, we must specialise the currents, making them sinusoidal. Then we can have silences, and correctly interpret results. We appear to have false balances. But they are quite different from the before-mentioned false balances, as they indicate true changes in resistance and inductance, owing to the reduction of Z to the required form, in which, however, the two "constants" are functions of the frequency.

SECTION XXXVII. GENERAL THEORY OF THE CHRISTIE BALANCE. DIFFERENTIAL EQUATION OF A BRANCH. BALANCING BY MEANS OF REDUCED COPIES.

It is not easy to find a good name for Mr. S. H. Christie's differential arrangement. There are objections to all the names bridge, balance, lozenge, parallelogram, quadrangle, quadrilateral, and *pons asinorum*, which have been used. It seems to be a nearly universal rule for words, used correctly in the first place, to gradually change their meaning, and finally cause us to talk nonsense, according to their original signification. Thus the Bridge is the conductor which bridges across two others. But it has become usual to speak of the differential arrangement as a whole as the Bridge; and then we have the four sides of the bridge, which is absurd. Quadrilateral is the latest fashion. It has four sides, truly. But there are six conductors concerned; so we should not call the differential arrangement itself the quadrilateral. I propose to simply call it the Christie, without any addition, just as telegraphers speak of the Morse, or the Wheatstone, meaning the apparatus taken as a whole. Thus we can refer to the Christie, the quadrilateral, and the bridge, the latter two being parts of the former. This will suppress the farrago.

In the usual form of the Christie we have four points, A, B_1, B_2, C, united by six conductors, numbered from 1 to 6 in the figure. The quadrilateral has the four sides, 1, 2, 3, 4. The bridge-wire is 5, joining B_1 to B_2, and 6 is the battery-wire. The battery-current goes from A to C by the two distinct routes AB_1C and AB_2C. Some of it crosses the bridge, up or down; except under special circumstances, when the bridge-wire is free from current, which is the useful property.

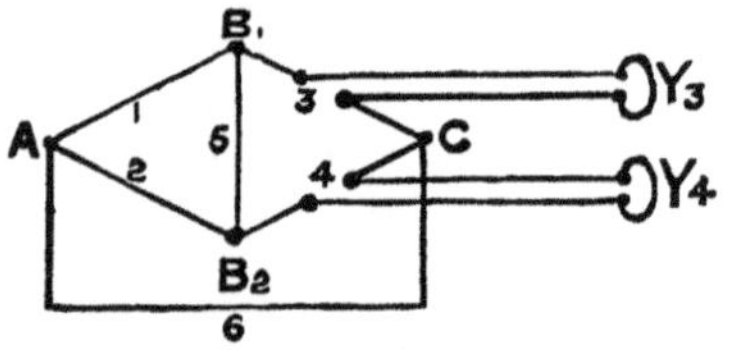

Let us generalise the Christie thus:—Let the sole characteristics of a branch be that the current entering it at one end equals that leaving it at the other, with the additional property that the electromagnetic conditions prevailing in it are stationary, so that the branch becomes quite definite, independent of the time.

Thus, all six branches may be any complex combinations of conductors and condensers satisfying these conditions. The communication between the two ends of a branch need not be conductive at all; for example, a condenser may be inserted. As an example of a complex combination, let branch 3 consist of a long telegraphic circuit, symbolised by the two parallel lines starting from 3 and ending at Y_3, where they are connected through terminal apparatus. This branch then consists of a long series of small condensers, whose + poles are all connected together by one wire, and the – poles by the other wire. There is also conductive connection (by leakage) between the two wires. There is also electromagnetic induction all along the line. But, as the current entering the line from B_1 to 3, and that leaving it, from 3 to C, are equal, the telegraphic line comes under our definition, provided it be stationary in its properties. Observe that this does not exclude the presence of other conductors, between which and the line in branch 3 there is mutual induction, providing this does not disturb our fundamental property of a branch. We may, indeed, remove the original restriction, but then it will no longer be the Christie, for more than *four* points will be in question. Suppose, for example, there is mutual induction of the electrostatic kind between branches 1 and 2, which is most simply got by connecting the middles of 1 and 2, taken as resistances, through a condenser. Then there are *six* points, or junctions, concerned, and a slight enlargement of the theory is required.

Let us now inquire into the general condition of a balance, or of no current in the bridge-wire due to current in 6, which, therefore, enters the quadrilateral at A and leaves at C, and which may arise from impressed force in 6 itself, or be induced in it by external causes. First, as regards the self-induction balance in the extended sense. This does not mean that each side of the quadrilateral must be equivalent to a coil, but merely that the four sides are independent of one another in every respect, except in being connected at A, B_1, B_2, C. Thus we can have electrostatic and electromagnetic induction in all six branches, but independently of one another. Under these circumstances it is always possible to write the differential equation of a branch in the form $V=ZC$, where C is the current (at the ends), V the fall of potential from end to end, and Z a differential operator in which time is the independent variable. When the branch is a mere resistance R, then $Z=R$, simply. When it is a coil, independent of all other conductors, then

$$Z=R+Lp,$$

where L is the inductance of the coil, and p stands for d/dt. When it is a condenser, then $Z=(Sp)^{-1}$, where S is the capacity. If the condenser have also conductance K, or be shunted by a mere resistance, then

$$Z=(K+Sp)^{-1}.$$

These are merely the simplest cases. In general, Z is a function of p, p^2, etc., and electrical constants.

Now let the positive direction of current be from left to right in sides 1, 2, 3, 4, and suppose we know their differential equations

$$V_1=Z_1C_1, \qquad V_2=Z_2C_2, \quad \text{etc.}$$

To have a balance, so far as the current from 6 is concerned, the potentials at B_1 and B_2 must be always equal, except as regards inequalities arising from impressed forces in other branches than 6, with which we are not concerned. Therefore

$$\left.\begin{array}{lll} V_1=V_2, & \text{and} & V_3=V_4, \\ \text{or,} \quad Z_1C_1=Z_2C_2, & \text{and} & Z_3C_3=Z_4C_4. \end{array}\right\} \qquad (1c)$$

But, $C_1 = C_3$, $C_2 = C_4$.

So, using these in (1c), we get

$$\left.\begin{aligned} Z_1C_1 = Z_2C_2, \\ Z_3C_1 = Z_4C_2. \end{aligned}\right\} \qquad (2c)$$

Eliminate the currents by cross-multiplication, and we get

$$Z_1Z_4 = Z_2Z_3, \qquad (3c)$$

which is the condition required. It has to be identically satisfied, so that, on expansion, the coefficient of every power of p must vanish.

If we take $Z = R + Lp$ (as when each side is a coil, or equivalent to one), we obtain the three conditions given in my paper "On the Use of the Bridge as an Induction Balance, equations (1), (2), (3) [vol. II., Art. XXXIV., p. 33].

As another example, take $Z = (K + Sp)^{-1}$ (shunted condensers), and we obtain three similar conditions. But it is needless to multiply examples here. We have only to find the forms of the four Z's, expand equation (3c), and equate to zero separately the coefficient of every power of p. It does not follow that a balance is possible in a particular case, but our results will always tell us how to make it possible, as by giving zero values to some of the constants concerned, when one branch is too complex to be balanced by simpler arrangements in other branches.

The theory of a balance of self and mutual electromagnetic induction I propose to give by a different and very simple method in the next Section. At present, in connection with the above generalised self-induction balance, let us inquire how to balance telegraph lines of different types, or when they can be simply balanced. It is clear, in the first place, that if we choose sides 1 and 3 quite arbitrarily, we have merely to make side 2 an exact copy of side 1, and 4 an exact copy of side 3, in order to ensure a perfect balance. Imagine the bridge-wire to be removed; then we have points A and C joined by two identical arrangements. The disturbances produced in these by the current from 6 must be equal in similar parts; hence, if B_1 and B_2 be corresponding points, their potentials will be always equal, so that no current will pass in the bridge-wire when they are connected. But we can also get a true balance when the "line" AB_2C is not a full-sized, but a reduced copy of the line AB_1C. It is not the most general balance, of course, but is still a great extension upon the balance by means of full-sized copies. The general principle is this:—

Starting with sides 1 and 3 arbitrary, make 2 and 4 copies of them, first simply qualitatively, as it were; thus, a resistance for a resistance, a condenser for a condenser, and so on. This is like constructing an artificial man with all organs complete, but in no particular proportion. Then, make every resistance in sides 2 and 4 any multiple, say s times the corresponding resistance in sides 1 and 3. Make every condenser in sides 2 and 4 have, not s times, but s^{-1} times the capacity of the corresponding condenser in sides 1 and 3. And, lastly, make every inductance in sides 2 and 4 be s times the corresponding inductance in sides 1 and 3. This done, s being any numeric, AB_2C is made a reduced (or enlarged) copy of AB_1C, and there will be a true balance. That is, the *potentials* at corresponding points will be equal, so that the bridge-wire may connect any pair of them, without causing any disturbance.

Now let a telegraph line be defined by its length l, and by four electrical constants R the resistance, S the electrostatic capacity, L the inductance, and K the leakage-conductance, all per unit length. It is not by any means the most general way of representation of a telegraph line, but is sufficient for our purpose. Let C be the current, and V the potential-difference at distance x from its beginning. We require the form of Z in $V=ZC$ at its beginning. This will depend somewhat upon the terminal conditions at the distant end, so, in the first place let $V=0$ there. Take

$$C=\cos mx.A+\sin mx.B, \qquad \text{(4c)}$$

$$V=-(R+Lp)m^{-1}(\sin mx.A-\cos mx.B), \qquad \text{(5c)}$$

$$-m^2=(K+Sp)(R+Lp), \qquad \text{(6c)}$$

p standing for d/dt as before. These are general, subject to no impressed forces in the line. A and B are arbitrary so far. But at the end $x=l$, we have $V=0$ imposed, which gives, by (5c),

$$B/A=\tan ml, \qquad \text{(7c)}$$

so that at the $x=0$ end, we have, by (4c), (5c), and (7c),

$$\frac{V}{C}=\frac{R+Lp}{m}\cdot\frac{B}{A}; \qquad \text{or} \qquad Z=(R+Lp)l.\frac{\tan ml}{ml}. \qquad \text{(8c)}$$

This is the Z required. From the form of m^2, we see that if the total resistance Rl and total inductance Ll in one line be, say, s times those in a second, whilst the total capacity Sl and total leakage-conductance Kl in the second line are s times those in the first, then the values of ml are identical for the two lines. If these lines be in branches 3 and 4, we therefore have

$$\frac{Z_3}{Z_4}=\frac{(R_3+L_3p)l_3}{(R_4+L_4p)l_4}=s, \qquad \text{(9c)}$$

so that we may balance by making sides 1 and 2 resistances whose ratio R_1/R_2 is s; or, if coils be used, by having, additionally, $L_1/L_2=s$; or, if condensers are used, $(K_2+S_2p)/(K_1+S_1p)=s$; and so on.

But if there be apparatus at the distant end of the line, it must also be allowed for. Let $V=YC$ be the equation of the terminal apparatus; that is, this equation connects (4c) and (5c) when $x=l$. Using it, instead of the former $V=0$, we shall arrive at

$$Z=\frac{(R+Lp)}{m}\cdot\frac{\tan ml+mY/(R+Lp)}{1-\tan ml.mY/(R+Lp)}, \qquad \text{(10c)}$$

instead of (8c). Now, just as before, adjust the constants of lines 3 and 4, so that $m_3l_3=m_4l_4$, and, in addition, make $Y_3/Y_4=s$. Then, supposing each side of the quadrilateral to be a telegraph line, the full conditions of balance by this kind of reduced copies are

$$\left.\begin{aligned} \frac{R_1 l_1}{R_2 l_2} = \frac{L_1 l_1}{L_2 l_2} = \frac{S_2 l_2}{S_1 l_1} = \frac{K_2 l_2}{K_1 l_1} = \frac{Y_1}{Y_2}, \\ = \frac{R_3 l_3}{R_4 l_4} = \frac{L_3 l_3}{L_4 l_4} = \frac{S_4 l_4}{S_3 l_3} = \frac{K_4 l_4}{K_3 l_3} = \frac{Y_3}{Y_4}. \end{aligned}\right\} \qquad (11c)$$

The difference from the former case is that we now have in sides 2 and 4 reduced copies of the terminal apparatus of lines 1 and 3. It will be observed that the equalities in the first line of (11c) make side 2 a reduced copy of side 1, and that those in the second line make side 4 a reduced copy of 3, whilst the equalisation of the two lines of (11c) makes the scale of reduction the same, so that AB_2C is made a reduced copy of AB_1C.

If one of the four sides, say side 3, of the quadrilateral be a telegraph line, we must have at least one other telegraph line, or imitation line, namely, in side 4. But, of course, sides 1 and 2 may be electrical arrangements of a quite different type. Further, notice that only two of the sides, either 1 and 2, or 3 and 4, can be single wires with return through earth, so that if the other two are also to be telegraph lines they must be looped, or double wires. In certain cases precisely the same form of Z as that above used will be valid, but this is quite immaterial as regards balancing by means of a reduced copy.

The balance expressed by equation (3c) is exact—that is, it is independent of the manner of variation of the current. The balance by means of reduced copies is also exact, but is only a special case of the former. But there is always, in addition, the periodic or S.H. balance, when the currents are undulatory. Then merely two conditions are required, to be got by putting $p^2 = -n^2$, where $n/2\pi$ is the frequency, in $Z_1 Z_4 - Z_2 Z_3$, which will reduce it to the form $a + bp$, in which a and b contain the frequency. Now, $a = 0$ and $b = 0$ specify this peculiar kind of balance, which is, generally speaking, useless. Whilst, however, the balance of AB_1C and AB_2C by making the latter a reduced copy of the former is, when applied to the Christie, only a special case of (3c), it is, in another respect, far more general; for it will be observed that any pair of corresponding points may be joined by the bridge-wire, although the result may be an arrangement which is not the Christie.

SECTION XXXVIII. THEORY OF THE CHRISTIE AS A BALANCE OF SELF AND MUTUAL ELECTROMAGNETIC INDUCTION. FELICI'S INDUCTION BALANCE.

As promised in the last Section, I now give a simple, and, I believe, the very simplest, investigation of the conditions of balance when all six branches of the Christie have self and mutual induction. Referring to the same figure (in which we may ignore the extensions of branches 3 and 4 to Y_3 and Y_4), we see that as there are six branches there are twenty-one inductances, viz., six self and fifteen mutual.

This looks formidable. But since there are only three independent currents possible there can really be only six independent inductances concerned, viz., three self and three mutual, each of which is a combination of those of the branches separately.

Thus, let C_1, C_3, and C_6 be the currents that are taken as independent, and let them exist in the three circuits AB_1B_2A, CB_2B_1C, and AB_2CA (*via* branch 6), with right-handed circulation when positive. Then the other three real currents C_2, C_4, and C_5 are given by

$$C_2 = C_6 - C_1, \qquad C_4 = C_6 - C_3, \qquad C_5 = C_1 - C_3;$$

if the positive direction be from left to right in sides 1, 2, 3, and 4, from right to left in 6, and down in 5, which harmonises with the positive directions of the cyclical currents C_1, C_3, and C_6.

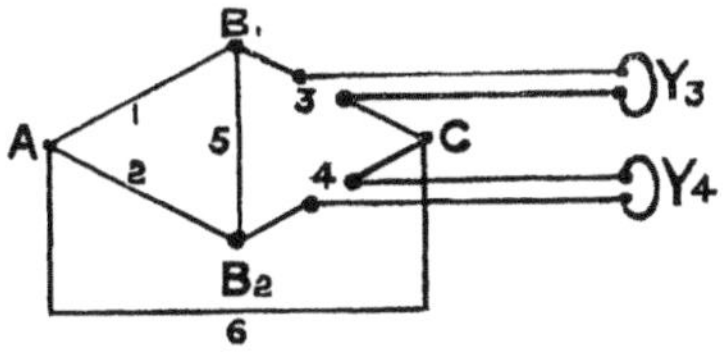

Next, let m_1, m_3, m_6, and m_{13}, m_{36}, m_{61} be the inductances, self and mutual, of the three circuits. Thus, m_1 = induction through AB_1B_2A due to unit current in this circuit; and m_{13} = the induction through CB_2B_1C due to the same, etc. We have to find what relations must exist amongst the resistances and the inductances in order that there may never be any current in the bridge-wire, provided there be no impressed forces in 1, 2, 3, 4 or 5.

We obtain them by writing down the equations of E.M.F. in the two circuits AB_1B_2A and CB_2B_1C on the assumption that there is no current in the bridge-wire, which requires $C_1 = C_3$; and this we do by equating the E.M.F. of induction in a circuit, or the rate of decrease of the induction through the circuit, to the E.M.F supporting current, which is the sum of the products of the real currents into the resistances, taken round the circuit.

Thus,

$$\left.\begin{aligned} -p(m_1C_1 + m_{13}C_3 + m_{16}C_6) &= R_1C_1 - R_2(C_6 - C_1), \\ -p(m_3C_3 + m_{31}C_1 + m_{63}C_6) &= R_3C_3 - R_4(C_6 - C_3), \end{aligned}\right\} \qquad (12c)$$

where p stands for d/dt. But $C_1 = C_3$, which, substituted, makes

$$\left.\begin{aligned} \{(R_1 + R_2) + (m_1 + m_{13})p\}C_1 &= (R_2 - m_{16}p)C_6, \\ \{(R_3 + R_4) + (m_3 + m_{13})p\}C_1 &= (R_4 - m_{36}p)C_6, \end{aligned}\right\} \qquad (13c)$$

which have to be identically satisfied. Eliminate the currents by cross-multiplication, and then equate to zero separately the coefficients of the powers of p. This gives us

$$\left.\begin{aligned} R_1R_4 &= R_2R_3, \\ (m_1 + m_{13} + m_{16})R_4 - m_{36}R_1 &= (m_3 + m_{31} + m_{36})R_2 - m_{16}R_3, \\ (m_1 + m_{13})m_{36} &= (m_3 + m_{13})m_{16}, \end{aligned}\right\} \qquad (14c)$$

which are the conditions required. First the resistance balance; next the vanishing of integral extra-current due to putting on a steady impressed force in branch 6; and the third condition to wipe out all trace of current, and make branches 5 and 6 perfectly conjugate under all circumstances.

If the Christie consists of short wires, which are not nearly closed in themselves, then, as I pointed out before [vol. II., Art. 34, p. 37], the theory of the balance expressed in terms of the self and mutual inductances of the different branches becomes meaningless, because the inductances themselves are meaningless. Under these circumstances, equations (14c) are *the* conditions of a balance, from which alone can accurate deductions be made. Even if we have the full equations in terms of the twenty-one inductances of the branches, they will express no more than (14c) do. We could not, for instance, generally assume any one of the inductances to vanish, as it would produce an absurdity, viz., the consideration of the amount of induction passing through an open circuit. Hence it is quite possible that (14c) may be useful in certain experiments, in which such short wires are used that terminal connections become not insignificant.

At the same time it is to be remarked that such cases are quite exceptional. I would not think, for example, of measuring the inductance of a wire a few inches long, in which case (14c) would, at least in part, be applicable, if I could get a long wire and swamp the terminal connections. Still, however, equations (14c) and the way they are established are useful in another respect. In general, I have not found any particular advantage in Maxwell's method of cycles.* It has seemed to me to often lead to very roundabout ways of doing simple work, from what I have seen of it. This applies both when the steady distribution of current in a network of conductors is considered, due to steady impressed forces, as in the original application; and also when the branches are not treated as mere resistances, but transient states are considered, provided the branches be independent, so that, as I remarked before, the equation of a branch may be represented by $V=ZC$, where Z takes the place of R, the resistance in the elementary case. But in our present problem there is such a large number of inductances that there is a real advantage in using the above method, an advantage which is non-existent in a problem relating to steady states. We greatly simplify the preliminary work by reducing the number of inductances from 21 to 6. But, of course, on ultimate expansion of results we shall come to the same end.

If we use the first and third of (14c) in the second, it becomes

$$\left\{m_1 + m_{31} + m_{61}\left(1 + \frac{R_1}{R_2}\right)\right\} \cdot \left(\frac{R_3}{R_1} - \frac{m_{63}}{m_{61}}\right) = 0 ; \quad \dots\dots\dots(15c)$$

and, as either of these factors may vanish, we have in general two entirely distinct solutions. If the second factor vanish, the whole

* [Not given in his treatise, but described by Dr. Fleming in the *Phil. Mag.*]

set of conditions may be written

$$\frac{R_1}{R_3}=\frac{R_2}{R_4}=\frac{m_{16}}{m_{36}}=\frac{m_1+m_{13}}{m_3+m_{13}}; \quad \text{......................(16c)}$$

whilst, if it be the first factor that vanishes, we shall have

$$-\frac{R_1}{R_2}=-\frac{R_3}{R_4}=1+\frac{m_1+m_{13}}{m_{61}}=1+\frac{m_3+m_{13}}{m_{63}}, \quad \text{...........(17c)}$$

expressing the full conditions. Both (16*c*) and (17*c*) are included in (14*c*).

Suppose now that we make the branches long wires, or coils of wire, or many coils in sequence, etc., and can therefore localise inductances in and between the branches. We require to expand the six m's. Their full expressions will vary according to circumstances. When all the twenty-one inductances are counted, they are given by

$$\left.\begin{aligned}
m_1 &= L_1+L_5+L_2+2(M_{15}-M_{25}-M_{12}),\\
m_3 &= L_3+L_4+L_5+2(M_{45}-M_{34}-M_{35}),\\
m_6 &= L_6+L_2+L_4+2(M_{62}+M_{64}+M_{34}),\\
m_{13} &= -L_5+(M_{13}-M_{14}-M_{15}-M_{23}+M_{24}+M_{25}+M_{53}-M_{54}),\\
m_{16} &= -L_2+(M_{12}+M_{14}+M_{16}-M_{24}-M_{26}+M_{52}+M_{54}+M_{56}),\\
m_{36} &= -L_4+(M_{32}+M_{34}+M_{36}-M_{42}-M_{46}-M_{52}-M_{54}-M_{56}).
\end{aligned}\right\} \text{(18c)}$$

Here L stands for the inductance of a branch, and M for the mutual inductance of two branches. These are got by inspection of the figure, with careful attention to the assumed positive directions of both the cyclical and the real currents.

In the use of these, for insertion in (14*c*), we shall of course equate to zero all negligible inductances. As an example of a very simple case, let coils be put in branches 4 and 6, between which there is mutual induction, and let the other four branches be double-wound or of negligible inductance. Then all except L_4, L_6, and M_{46} are zero, giving

$$m_1=0, \qquad m_3=L_4, \qquad m_6=L_4+2M_{64},$$
$$m_{13}=0, \qquad m_{16}=0, \qquad m_{36}=-L_4-M_{46}.$$

Insert these in the second of (14*c*), and we get

$$R_1(L_4+M_{46})=-M_{46}R_2, \qquad \text{or} \qquad -L_4=(1+R_2/R_1)M_{46}. \quad \text{...(19c)}$$

The third condition is nugatory. Hence (19*c*), with a resistance balance, but without the need of measuring R_3 (or, equivalently, R_4), gives us the ratio of the M of two coils to the L of one of them in terms of the ratio of two resistances.

As another example, let all the M's be zero except M_{12} and M_{34}, whilst all the L's are finite. We shall then have, besides the resistance balances, the two conditions

$$\left.\begin{aligned}
0&=(L_1L_4-L_2L_3)+(L_2-L_1)M_{34}+(L_3-L_4)M_{12},\\
0&=(L_1R_4+L_4R_1-L_2R_3-L_3R_2)+(R_3-R_4)M_{12}+(R_2-R_1)M_{34}.
\end{aligned}\right\} \text{(20c)}$$

If we now take $R_1=R_2$, $L_1=L_2$; that is, let sides 1 and 2 be equal, we reduce the three conditions (14*c*) to $R_3 = R_4$, $L_3 = L_4$. This is obvious enough in the absence of mutual induction; but we also see that induction between sides 1 and 2, and between 3 and 4, does not in the least interfere with the self-induction balance Whilst remarkable, this property is of great utility. For it allows us to have the equal wires 1 and 2 close together, preferably twisted, and then this double wire may be doubled on itself, and the result wound on a bobbin. We ensure the equality of the wires at all times, doing away with the troublesome source of error arising from the disturbance of the resistance balance from temperature changes, which occur when 1 and 2 are separated, and also doing away with interferences from induction between 1 and 2 and the rest. We also do away with the necessity of keeping coils 3 and 4 widely separated from one another.

Passing to a connected matter, Maxwell, Vol. II., Art. 536, describes the well-known mutual induction balance with which Felici made such instructive experiments, that may be made the basis of the science of electromagnetic induction. It is very simple and obvious. The figure explains itself. If the M of the two circuits is *nil*, there is no current in the secondary on making or breaking the primary. This is secured when the M of coils 1 and 2 is cancelled by the M of coils 3 and 4, presuming that the pair 1, 2 is well-removed from the pair 3, 4.

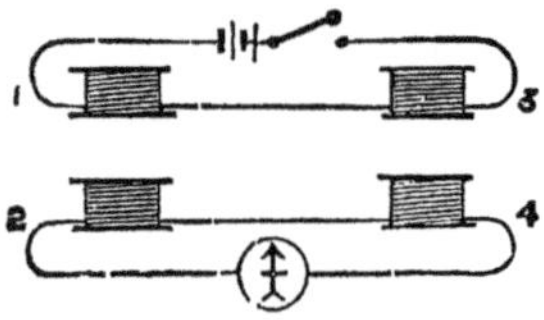

The balance is independent of the self-inductions of the four coils, and also of the resistance of the two circuits, and may be made very sensitive. In fact, Felici's balance is unique, and should be used whenever possible. To exhibit its merits fully, we should use a telephone and automatic intermitter, giving a steady tone. It is then doubly unique, and it is difficult to imagine anything better. Compared with the galvanometer, the use of the telephone is a real pleasure. It is science made amusing.

But if we want not merely to balance M_{12} against M_{34}, but to know the value of the M_{12} of a *given* pair of coils, M_{34} should be both variable and known. Coils 3 and 4 may be the coils of an inductometer [vol. II., p. 101] calibrated once for all. There are many ways of doing it, in terms of the capacity of a condenser, or in terms of the inductance of a coil, etc., none of which methods has the merits of Felici's balance. Suppose it done by Maxwell's condenser method (using a telephone, of course). It is, perhaps, as good as any (certainly better than many) for the particular purpose, as we have only to give particular values to the time-constant of the condenser—a series of values with a common difference—and get silence at once by moving the pointer to a series of particular places, which is very different from dodging about to find the value of the time-constant when the M of the coils is fixed. We should also measure the L of each coil by itself, and it is well to previously adjust the coils to have equal L and R. But the present use of the inductometer is not to measure self-induction, but mutual induction. Therefore make 1 and 2 the coils whose M is wanted, 3 and 4 the coils of the inductometer. If within range (it is well to have inductometers of different sizes, for various purposes), we immediately measure M_{12}, and have the full advantages of Felici's balance.

But if there is metal about the coils 1 and 2 (of course there should be none, or very little, about the inductometer, or it should be carefully divided), we cannot get telephone balances. If the departure from balance is serious, and it is not practicable to remove the metal, we may give up the telephone and use a suitable galvanometer, one whose needle will not move till all the current due to a make has passed, and then move if it can. But if the metal be iron, and we want to measure the steady M in presence of the iron (not finely divided), of course we must not remove the iron and measure something else than what we want to know. Then the galvanometer is indispensable. We lose the advantage of the telephone, but Felici's balance has still its peculiar merits left, in a very great measure.

Apart from the question of measurement, Felici's balance is highly instructive, as to which see Maxwell's treatise, to which we should add that the telephone should always be used if possible. Besides the experiments referred to, the balance is useful for studying the influence of iron in the field on the M of two coils, increasing or decreasing it, according to position. Use non-conducting iron [vol. II., Art. 36, later]. Here we have another proof to that there mentioned, that there is no appreciable waste of energy in finely divided iron when the range of the magnetic force is moderate, although very *perfect* silences, like those when there are no F. currents, and no iron, are not always obtainable.

As regards Felici's balance when employed for observing differential effects, *e.g.*, Prof. Hughes's magical experiments with coins, and so forth, I cannot recommend it, for several reasons. The theory is complex, in the first place, so that scientific interpretation of results is difficult. Next, considerable accuracy in adjustment of the coils, in two equal pairs, similarly placed, is required. Lastly, the independence of resistance, etc., ceases when there are F. currents to disturb; and as we are not able to trace the variations of resistance, we may, in sensitive arrangements, when balancing one set of F. currents and reactions against another set, be interfered with by unknown temperature variations.

Perhaps the easiest way is to take a long wire, double it on itself and then double again, giving four equal wires. Wind two side by side to make one pair of coils (1 and 2), and the others in the same manner, to make the other pair. Of course we have increased sensitiveness by the closeness of the wires.

But it is far better not to use four coils, but only two, viz., coils 3 and 4 in the equal-sided *self*-induction balance, with 1 and 2 made permanently equal, as before described. The temperature error is then under constant observation, and we know at once when the resistance balance of coils 3 and 4 (apart from F. currents) is upset. Interpretation is also an easier matter, both in general reasoning and in calculations.

SECTION XXXIXa. FELICI'S BALANCE DISTURBED, AND THE DISTURBANCE EQUILIBRATED.

Referring to the last figure, in which imagine the galvanometer to be replaced by a telephone, and the key by an automatic intermitter, let us start with a perfect balance due to the M of one pair of coils being cancelled by the M of the other pair, and consider the nature of the effects produced by the presence of metal in or near either pair of coils.

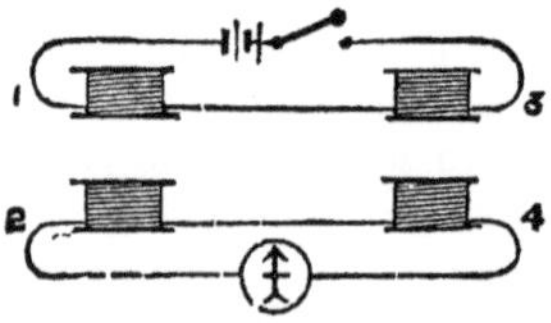

First, let 3 and 4 be the coils of an inductometer, and 1, 2 other coils of any kind, separate from one another. The simplest action is that caused by non-conducting iron. It acts to increase or decrease the M of either or both pairs of coils according to its position with respect to them, and its effect can be perfectly balanced by a suitable increase or decrease of the M (mutual inductance) of the inductometer coils. Suppose, for example, the disturber is a non-conducting iron bullet, and is brought into the field of the coils 1, 2. If it be inserted in either coil, it increases their M. This is mainly because it increases the L of the coil in which it is inserted. If the two coils have their axes coincident, as in the figure, the bullet will cause their M to be increased by placing it anywhere on the axis, or near it. But if the bullet be brought between the coils laterally, so as to be, for instance, between the numerals 1 and 2 in the figure, the result is a decreased M. Here the L of each coil is little altered, and the decrease of M results from the lateral diversion of the magnetic induction by the bullet from its normal distribution. By pushing it in towards the axis a position of minimum M is reached, after which further approach to the axis causes M to increase, ending finally on the axis with being greater than the normal amount.

If the disturber be a non-conducting core (round cylinder), the greatest increase of M is, of course, when it is pushed through both coils, which are themselves brought as close together as possible, and when the core itself is several times as long as the depth of the coils. M is then multiplied about four times when the coils are about of the shape shown, with internal aperture about $\frac{1}{3}$ the diameter of the coils. If the coils be wound parallel on the same bobbin, the increase is much greater. If the whole space surrounding the coils be embedded in iron to a considerable distance, we shall approach the maximum M possible. The effective inductivity of the non-conducting iron is considerably less than that of solid iron, which counterbalances the freedom from F. currents.

Using solid iron, no silence is possible, owing to the F. currents, although there is a more or less distinctly marked minimum sound for a particular value of M. The substitution of a bundle of iron wires reduces this minimum sound, and when the wires are very fine, it is brought to comparative insignificance; but only by very fine division of iron are the F. currents rendered of insensible effect. It will be, of course, remembered that the range of the magnetic force variations must be moderate, so as to render the variations in the magnetic induction strictly proportional to them, otherwise no perfect balance is possible with non-conducting iron.

On the other hand, non-conducting (*i.e.*, very finely divided) brass (or presumably any other non-magnetic metal) does nothing. Diamagnetic effects are insensible. The above remarks apply, for the most part, equally well to the self-induction balance, except that iron always increases the L of a coil.

So far is very simple. It is the effect of the conductivity (in mass) of the disturbing matter that makes the interpretation of results troublesome. If the disturber be non-magnetic, we have a secondary current due to the action on the secondary circuit of the current induced in the disturber by the primary current; at least I suppose that this is the way it might be popularly explained. If the disturber be not too big, the M of the inductometer which gives the least sound (instead of silence) is sensibly the old value which gave silence before its introduction. If it be magnetic, there is usually increased M also. Changing the M of the inductometer to suit this, the minimum sound is still far louder than with an equally large non-magnetic disturbing mass (metallic) because the F. currents are so much stronger in iron. To this an exception is Prof. Bottomley's manganese-steel of nearly unit inductivity, in which the F. currents should be, and no doubt are, far weaker than in copper, on account of the comparatively low conductivity. If this be not so, then it must be found out why not. Again, if the iron be independently magnetised so intensely as to reduce the effective inductivity sufficiently, then, as I pointed out in 1884, the F. currents should be made less than in copper.

To obtain an idea of the disturbance in the secondary circuit due to a conducting mass, let it be a simple linear circuit, and call it the tertiary. Let the suffixes $_1$ and $_2$ refer to the primary and secondary circuits, and $_3$ to the tertiary. Then the equations of E.M.F. are

$$\left.\begin{aligned} e &= Z_1C_1 + M_{13}pC_3, \\ 0 &= Z_2C_2 + M_{23}pC_3, \\ 0 &= M_{31}pC_1 + M_{32}pC_2 + Z_3C_3; \end{aligned}\right\} \qquad (21c)$$

where $Z = R + Lp$, and e is the impressed force in the primary. Here M_{12} is missing, it being supposed to be properly adjusted to be zero. From these,

$$C_2 = \frac{M_{13}M_{23}p^2e}{Z_1(Z_2Z_3 - M_{23}^2p^2) - Z_2M_{13}^2p^2} \qquad (22c)$$

is the secondary current's equation. The secondary current therefore varies as the product of the M of the tertiary and primary into the M of the tertiary and secondary. It is therefore made greatest by making coils 1 and 2 in the figure coincident (practically) by double-winding, and putting the disturber in their centre. In this case, let R and L be the resistance and inductance of the primary and also of the secondary circuit, r and l those of the tertiary, and m the former M_{13} or M_{23}, now equal. Then (22c) becomes, if $z = r + lp$,

$$C_2 = \frac{m^2p^2e}{zZ^2 - 2Zm^2p^2}. \qquad (23c)$$

But m is very small compared with L, so

$$C_2 = \frac{m^2p^2e}{zZ^2}. \qquad (24c)$$

Let the impressed force be sinusoidal; then $p^2 = -n^2$, making

$$C_2 = -\frac{m^2n^2e}{(r+lp)\{(R^2 - L^2n^2) + 2RLp\}}. \qquad (25c)$$

Let $R = Ln$, which condition is readily reached approximately. Then

$$C_2 = \frac{m^2}{2RL} \cdot \frac{ln^2 + rp}{r^2 + l^2n^2} \cdot e \qquad (26c)$$

gives the secondary current in amplitude, $(m^2n/2RL)(r^2 + l^2n^2)^{-\frac{1}{2}}$ per unit impressed force, and phase. If the tertiary could have no resistance, the secondary current would be of amplitude $m^2/2RLl$ per unit impressed force, and in the same phase with it.

Now seek the conditions of balance by means of a fourth linear circuit placed between coils 3 and 4 in the figure, supposed to be exactly like coils 1 and 2. Let the suffix $_4$ relate to this fourth circuit. Then (21c) become

$$\left.\begin{aligned} e &= Z_1C_1 \qquad\qquad\qquad\qquad + M_{13}pC_3 + M_{14}pC_4, \\ 0 &= \qquad\qquad\qquad Z_2C_2 + M_{23}pC_3 + M_{24}pC_4, \\ 0 &= M_{31}pC_1 + M_{32}pC_2 + \quad Z_3C_3 \qquad\qquad\quad , \\ 0 &= M_{41}pC_1 + M_{42}pC_2 \qquad\qquad\qquad + \quad Z_4C_4. \end{aligned}\right\} \qquad (27c)$$

Here, besides M_{12}, M_{34} is also missing, because of the distance between the two disturbers. From these,

$$\Delta C_2 = (M_{23}M_{31}Z_4 + M_{24}M_{41}Z_3)e \qquad (28c)$$

is the equation of C_2, where Δ is the determinant of the coefficients in (27c). For a balance, the coefficient of e must vanish. This gives

$$\frac{r_3}{r_4} = \frac{L_3}{L_4} = -\frac{M_{31}M_{32}}{M_{41}M_{42}}. \qquad (29c)$$

If the coils of each "transformer" are coincident and equal, $M_{31} = M_{32}$, and $M_{41} = -M_{42}$; and, the M's being small, (28c) becomes

$$C_2 = \frac{M_{31}^2Z_4 - M_{41}^2Z_3}{Z^2Z_3Z_4}e, \qquad (30c)$$

where Z is that of either the primary or secondary circuit.

We do not need to balance the disturber in one pair of coils by means of a precise copy of it in the other pair, similarly placed. It may be a reduced copy, according to (29*c*).

SECTION XXXIX*b*. THEORY OF THE BALANCE OF THICK WIRES, BOTH IN THE CHRISTIE AND FELICI ARRANGEMENTS. TRANSFORMER WITH CONDUCTING CORE.

This brings me to the subject of balancing rods against one another, either in the Christie or in the Felici differential arrangements, when placed in long solenoids; and to the similar question of balancing thick wires in the Christie, when the current in them is longitudinal. As I pointed out before [vol. II., p. 37], if a wire be so thick that the effect of diffusion is sensible, it cannot be balanced in the Christie against a fine wire, but requires another thick wire in which the diffusion effect also occurs. I refer to true balances, independent of the manner of variation of the current, in which, therefore, the resistance of the one wire, though different at every moment, is yet precisely that of the other wire (or any constant multiple of it). Perhaps the best way to define the resistance is by Joule's RC^2. In the sinusoidal case a mean value is taken. According to this heat-generation formula, there always is a definite resistance *at a particular moment*, but what it may be will require elaborate calculation to find. This definition of the resistance to suit the instantaneous value of the dissipativity does not agree precisely with the sinusoidal R', which represents a mean value; but the sinusoidal R' has important recommendations which outweigh this disadvantage.

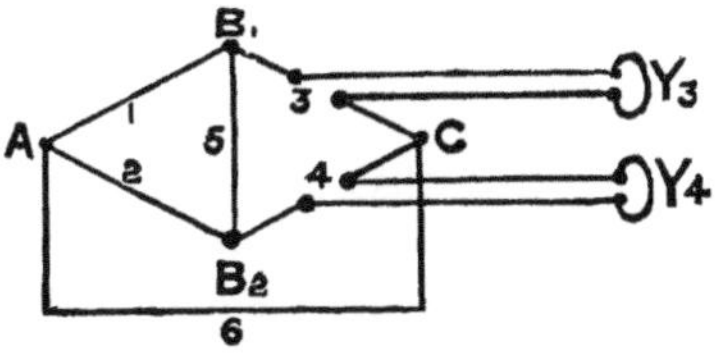

Suppose, now, we want to balance an iron wire against a copper wire, the wires being straight and long, though not so long as to require the consideration of electrostatic capacity. For simplicity, first let the ratio be one of equality, so that sides 1 and 2 in the Christie are any precisely equal admissible arrangements, which may be mere resistances. Let the iron wire be in side 3, the copper wire in side 4. We have to make side 3 an electrical full-sized copy of side 4. For definiteness, imagine Y_3 and Y_4 to be short-circuits, that one of the two parallel lines leading to either is the wire under test, whilst the other is a return tube, thin and concentric.

First, in accordance with the description of how to make copies [vol. II., p. 104], make the resistances of the two returns equal. Next, make the inductances due to the magnetic field in the space between wires and returns equal, by proper distance of returns, or by inductometers in sequence with sides 3 and 4. There is now left only the wires themselves to be equalised. First, their steady resistances require to be equal. Next, their steady inductances ($\frac{1}{2}\mu \times$ length). These two conditions will give balance to infinitely slow variations of current, and can be satisfied with wires of all sorts of sizes and lengths. But we require to make them balance during rapid variations of any kind. For instance, a very short impulse will cause a mere surface current in the wires, that is, in appreciable strength, if they be thick; and still the wires must balance. The full balance is secured by a third condition, viz., that the time-constants of diffusion shall be equal. This time-constant is $\mu k\pi c^2$, where μ is the inductivity, k the conductivity, and c the radius of a wire. Or, $\mu l/R$, the quotient of the inductivity by the resistance per unit length (or any multiple that we may find convenient of this quotient).

Thus, if the iron has inductivity 100, that of copper being 1, whilst k for copper is about six times the value for iron, the copper wire must have a radius of about four times that of the iron. This is indispensable. Fixing thus the relative diameters, the rest is easy, by properly choosing the lengths. In a similar manner, we may have the resistances in any proportion; as, for instance, to obviate the necessity of having wires of very different lengths, keeping, however, the proper ratio of diameters.

The following will be more satisfactory as a demonstration. If Z is the V/C operator, then $Z_1/Z_2 = Z_3/Z_4$ is the condition of balance [vol. II., p. 104]. So we have merely to examine the form of the Z of a straight wire. This is [vol. II., p. 63].

$$Z = L_0 p + Rf, \qquad \text{.........................(31c)}$$

where f is the operator given by

$$f = \tfrac{1}{2}sc\frac{J_0(sc)}{J_1(sc)}, \qquad s^2 = -4\pi\mu kp. \qquad \text{................(32c)}$$

L_0 is the inductance other than that due to the wire itself, and R is its steady resistance. Using this form of Z in our general equation of balance, we see that if we take $s_3c_3 = s_4c_4$, that is, make the diffusion time-constants equal, we make $f_3 = f_4$, so that the balance is given by

$$\frac{Z_1}{Z_2} = \frac{R_3}{R_4} = \frac{L_{03}}{L_{04}} = \frac{r_3}{r_4}, \qquad \text{........................(33c)}$$

where the additional r_3 and r_4 are for the two return-sheaths, or other resistances that may be in sides 3 and 4. Of course Z_1 and Z_2 may be R_1 and R_2, the resistances of sides 1 and 2, when they are mere resistances. In virtue of the equality of the diffusion time-constants, we may express the full conditions by adding to (33c) this:—

$$= \frac{\mu_3 l_3}{\mu_4 l_4}, \qquad \text{..............................(34c)}$$

where l_3 and l_4 are the lengths of the two wires.

Although this balance is true, yet there will be one practical difficulty in the way. As is very easily shown by sliding a coil along an iron wire or rod, the inductivity often varies from place to place. But if the wire be made homogeneous, the evil is cured.

Next, let it be required to balance a long iron against a long copper rod in long magnetising solenoids forming sides 3 and 4. Here the form of Z for the circuit of the solenoid is

$$Z = R + L_0 p + Lpf^{-1}, \qquad (35c)$$

where R is the total resistance (as ordinarily understood) of the circuit of the solenoid, L_0 the total inductance ditto, due to the magnetic field everywhere except in the core, L that due to the core itself when the field is steady, and f as before, in (32c).

To balance the iron against the copper we therefore require, first, the equality of time-constants of diffusion, or the iron rod should be one-fourth the radius of the copper; this being done,

$$\frac{Z_1}{Z_2} = \frac{R_3}{R_4} = \frac{L_{03}}{L_{04}} = \frac{L_3}{L_4} \qquad (36c)$$

will complete the balance. The value of L (*i.e.*, L_3 or L_4) is

$$L = (2\pi c N)^2 \mu l, \qquad (37c)$$

if N is the number of turns per unit length, and l the length of the solenoid. As for L_0, that is adjustible *ad lib.* nearly. The only failure will be due to want of homogeneity.

Lastly, balance two rods, one of iron, the other of copper, against one another in Felici's arrangement, when each pair of coils consists of long coaxial solenoids, making two primaries and two secondaries, properly connected together. Let R_1, R_2 be the total resistances of the primary and the secondary circuits; L_{01}, L_{02}, the total inductances, not counting the parts due to cores; M_0 the total mutual inductance, not counting the parts due to cores; L_1, L_2, and M those parts of the inductances, self and mutual, of the first pair of coils, due to the cores; and l_1, l_2, m the same for the second pair. The equations of E.M.F. in the primary and secondary are then, if F and f are the two core-operators, as per (32c), and C_1, C_2 the primary and secondary currents,

$$\left.\begin{aligned} e &= R_1C_1 + L_{01}pC_1 + M_0pC_2 + F^{-1}p(L_1C_1 + MC_2) + f^{-1}p(l_1C_1 + mC_2), \\ 0 &= R_2C_2 + L_{02}pC_2 + M_0pC_1 + F^{-1}p(L_2C_2 + MC_1) + f^{-1}p(l_2C_2 + mC_1). \end{aligned}\right\} \qquad (38c)$$

The first terms on the right are the E.M.F.'s used in the solenoid circuits against their resistance; the two following terms taken negatively the E.M.F.'s of induction not counting cores; and the last two taken negatively those due to the cores. To have a balance, C_2 must vanish. The second equation then gives

$$M_0 + MF^{-1} + mf^{-1} = 0. \qquad (39c)$$

So $M_0 = 0$, or the mutual inductance of the circuits due to other causes than the cores, must vanish. Then, further,

$$F = f, \qquad \text{and} \qquad M = -m. \qquad (40c)$$

So the diffusion time-constants of the cores must be equal, and the steady mutual inductance of one pair be cancelled by that of the other pair of coils, so far as depends on the cores, as well, as before said, as depends on the rest of the system. (When not counting cores is spoken of, it is not meant that *air* must be substituted. *Nothing* must be substituted.) The latter part is capable of external balancing. The balancing of the former part requires the value of

$$M=(2\pi c)^2 N_1 N_2 \mu l, \qquad \text{(41c)}$$

where N_1, N_2 are the turns per unit length in the two coils of a transformer of length l, to be the same for the two transformers.

The condition (39c) of course makes the primary equation independent of the secondary. It is then the same as if the secondary coils were removed.

This leads us to show the modification made in the equation of a transformer by the conductivity of its core. In (38c) we have merely to ignore the f terms, thus confining ourselves to one transformer, when the equations are given by the first lines. Now if the solenoids be of small depth, and there be no L externally, L_{01} and L_{02} become insignificant, and also M_0, provided the cores fill the coils. We have then

$$\left.\begin{aligned} e &= R_1C_1 + F^{-1}p(L_1C_1+MC_2), \\ 0 &= R_2C_2 + F^{-1}p(L_2C_2+MC_1), \end{aligned}\right\} \qquad \text{(42c)}$$

which only differ from the equations when cores are non-conducting by the introduction of F. The first approximation to F is unity (when very slow variations take place). It may be written thus:—

$$F^{-1}=A-Bp, \qquad \text{(43c)}$$

when A and B are positive functions of p^2, whose initial values are $A=1$, $B=0$. When the impressed force is sinusoidal, $p^2=-n^2$, and A and B are constants. Then (42c) become

$$\left.\begin{aligned} e &= R_1C_1 + (L_1C_1+MC_2)Bn^2 + Ap(L_1C_1+MC_2), \\ 0 &= R_2C_2 + (L_2C_2+MC_1)Bn^2 + Ap(L_2C_2+MC_1). \end{aligned}\right\} \qquad \text{(44c)}$$

From these, by elimination, we have

$$\frac{e}{C_1}=R_1+L_1R_2\frac{Bn^2(R_2+L_2Bn^2)+A^2L_2n^2+AR_2p}{(R_2+L_2Bn^2)^2+(L_2An)^2}, \qquad \text{(45c)}$$

showing the effective resistance and inductance of the primary as modified by the secondary *and* conducting core.

But it is very easy with iron cores, without excessive frequency, to make simpler formulæ suit. Let $z=\pi c^2 k\mu n$; then, if this is 10 or over [see vol. II., p. 99], we have

$$A=Bn=(2z)^{-\frac{1}{2}} \qquad \text{(46c)}$$

approximately, which may be used in (44c), (45c) at once.

In an iron rod of only 1 cm. radius, and $\mu=100$, $k=1/10{,}000$, the value of z is one-fifth of the frequency. If of 10 cm. radius, it equals twenty times the frequency. With large values of z we have

$$\frac{e}{C_1}=R_1+\frac{l_1 n}{1+2l_2^2n^2\{R_2(R_2+2l_2n)\}^{-1}}+\frac{R_2^2 l_1 p}{(R_2+l_2n)^2+l_2^2n^2}, \qquad \text{(47c)}$$

if L_1, L_2, and M, when divided by $(2z)^{\frac{1}{2}}$, become l_1, l_2, and m. This gives the primary current. And

$$\frac{e}{C_2} = -\left(\frac{R_1R_2}{2mn} + \frac{R_1L_2 + R_2L_1}{M}\right) + \frac{R_1R_2}{2mn^2}p \quad \text{.............}(48c)$$

gives the secondary current.

We can predict beforehand what these should lead to ultimately, from the general property that a secondary circuit, at sufficiently high frequencies, shuts out induction, or tends to bring $L_2C_2 + MC_1$ to zero, giving the ratio of the currents at every moment. The coefficients of p in (47c) and (48c) tend to zero, and the current in the primary to be the same as if its resistance were increased by the amount R_2L_1/L_2. The core need not be solid. A cylinder will do as well, since the magnetisation does not penetrate deep. It should, however, be remembered that although at low frequencies it is the core that contributes the greater part of the inductance, so that the rest is then negligible, yet when that due to the core actually becomes negligible, the rest becomes relatively important, and should therefore be allowed for.

SECTION XL. PRELIMINARY TO INVESTIGATIONS CONCERNING LONG-DISTANCE TELEPHONY AND CONNECTED MATTERS.

Although there is more to be said on the subject of induction-balances, I put the matter on the shelf now, on account of the pressure of a load of matter that has come back to me under rather curious circumstances. In the present Section I shall take a brief survey of the question of long-distance telephony and its prospects, and of signalling in general. In a sense, it is an account of some of the investigations to follow.

Sir W. Thomson's theory of the submarine cable is a splendid thing. His paper on the subject marks a distinct step in the development of electrical theory. Mr. Preece is much to be congratulated upon having assisted at the experiments upon which (so he tells us) Sir W Thomson based his theory; he should therefore have an unusually complete knowledge of it. But the theory of the eminent scientist does not resemble very closely that of the eminent practician.

But all telegraph circuits are not submarine cables, for one thing; and, even if they were, they would behave very differently according to the way they were worked, and especially as regards the rapidity with which electrical waves were sent into them. It is, I believe, a generally admitted fact that the laws of Nature are immutable, and everywhere the same. A consequence of this fact, if it be granted, is that all circuits whatsoever always behave in exactly the same manner. This conclusion, which is perfectly correct when suitably interpreted, appears to contradict a former statement; but further examination will show that they may be reconciled. The mistake made by Mr. Preece was in arguing from the particular to the general. If we wish to be accurate, we must go the other way to work, and branch out from the general to the particular. It is true, to answer a possible objection, that the want of omniscience prevents the literal carrying out of this process; we shall never know the most general theory of anything in Nature; but we may at least take the general theory so far as it is known, and work with that, finding out in special cases whether a more limited theory will not be sufficient, and keeping within bounds accordingly. In any case, the boundaries of the general theory are not unlimited themselves, as our knowledge of Nature only extends through a limited part of a much greater possible range.

Now a telegraph circuit, when reduced to its simplest elements, ignoring all interferences, and some corrections due to the diffusion of current in the wires in time, still has no less than four electrical constants, which may be most conveniently reckoned per unit length of circuit—viz., its resistance, inductance, permittance, or electrostatic capacity, and leakage-conductance. These connect together the two electric variables, the potential-difference and the current, in a certain way, so as to constitute a complete dynamical system, which is, be it remembered, not the real but a simpler one, copying the essential features of the real. The potential-difference and the permittance settle the electric field, the current and the inductance settle the magnetic field, the current and resistance settle the dissipation of energy in, and the leakage-conductance and potential-difference that without the wires. Now, according to the relative values of these four constants it is conceivable, I should think, by the eminent engineer, that the results of the theory, taking all these things into account, will, under different circumstances, take different forms. The greater includes the lesser, but the lesser does not include the greater.

In the case of an Atlantic cable it is only possible (at present) to get a small number of waves through per second, because, first, the attenuation is so great, and next it increases so fast with the frequency, thus leading to a most prodigious distortion in the shape of irregular waves as they travel along. Of course we may *send* as many waves as we please per second, but they will not be utilisable at the distant end. This distortion is a rather important matter. Mere attenuation, if not carried too far, would not do any harm. Now the distortion and the attenuation, though different things, are intimately connected. The more rapidly the attenuation varies with the frequency, the greater is the distortion of arbitrary waves; and if the attenuation could be the same for all frequencies, there would be no distortion. This can be realised, very nearly, as will appear later.

Now when there are only a very few waves per second, the influence of inertia in altering the shape of received signals becomes small, and this is why the cable-theory of Sir W. Thomson, which wholly ignores inertia, works as a substituted approximate theory. But suppose we shorten the cable continuously, and at the same time raise the frequency. Inertia becomes more and more important; the theory which ignores it will not suffice; and carrying this further, we at length arrive at a state of things in which the old cable-theory gives results which have no resemblance whatever to the real. This is *usually* the case in telephony, as I have before proved. It is always partly the case, viz., for the very high frequencies, and it may be true, and practically is sometimes, down to the low frequencies also. I have shown that the attenuation tends to *constancy* as the frequency is raised, except in so far as the resistance of the wire increases, and that at the same time the speed of the waves tends to approximate to the speed of light, or to a speed of the same order of magnitude, which is the only speed which can, I think, be said, even in a restricted sense, to be the "speed of electricity." But if the dielectric be solid, there must be some uncertainty about what this speed is, for obvious reasons, with very high frequencies. The speed of the current is *never* proportional to the square of the length of the line.

Within the limits of approximately constant attenuation the distortion is small. This is what is wanted in telephony, to be good. Lowering the resistance is perhaps the most important thing of all. Other means I will mention later. What the limiting distance of long-distance telephony may be, who can tell? We must find out by trial. We know that human speech admits of an extraordinary amount of distortion (never mind the attenuation) before it becomes quite unrecognisable. The "perfect articulation," "even different voices could be distinguished," etc., etc., mean really a large amount of distortion, of which little may be due to the circuit. There is the transmitter, the receiver, and several transformations between the speaker and the listener, besides the telephone line. What additional amount of distortion is permissible clearly must depend upon what is already existent due to other causes. Even if that be fixed, I see no legitimate way of fixing its amount by theoretical principles; the matter is too involved, and includes too many unknown data, including "personal equation." But this is certain, in my opinion—that good telephony is possible through a circuit whose electrostatic time-constant, the product of the total resistance into the total permittance, is several times as big as the recent estimate of Mr. W. H. Preece, and I shall give my reasons for this conclusion.

Increasing the inductance is another way of improving things. Hang your wires wider apart. The longer the circuit, the wider apart they should be; besides this, they may be advantageously raised higher. You can then telephone further, with similar attenuation and distortion. There is a critical value of the inductance for minimum attenuation-ratio. It is from $L = Rl/2v$ to $L = Rl/v$, according to circumstances to be later explained; L being the inductance and R the resistance per unit length, l the length, and v the speed of waves which are not, or are only slightly dissipated, which is $(LS)^{-\frac{1}{2}}$, if S be the permittance per unit length. The resulting attenuation may be an enlargement, as I have before explained, due to to-and-fro reflections. This is to be avoided. I shall explain its laws, and how to prevent it. By this method, carrying it out to an impracticable extent, however, we could make the amplitude of sinusoidal currents received at the distant end of an Atlantic cable greater than the greatest possible steady current from the same impressed force—an unbelievable result. And, without altering the permittance or the resistance, we could make the distortion quite small.

There is some experimental evidence in favour of increasing the inductance (apart from lessening the permittance); though, owing to want of sufficient information, I do not wish to magnify its importance. I refer to the statement that excellent results have been obtained in long-distance telephony with copper-covered steel wires. Here the copper covering practically decides the greatest resistance of the wire; what current penetrates into the steel lowers the resistance and increases the inductance. Clearly, we should magnify this effect, and, electrically speaking, it would seem that a bundle of soft-iron wires with a covering of copper is the thing, as this will allow the current to penetrate more readily, lower the resistance the most, and increase the inductance the most. But it is too complex a matter for hasty decision. We also see that the iron sheathing of a cable may be beneficial.

When we have little distortion, we get into the regions of radiation. The dielectric should be the central object of attention, the wires subsidiary, determining the rate of attenuation. The waves are waves of light, in all save wave-length, which is great, and gradual attenuation as they travel, by dissipation of energy in the wires. There is the electric disturbance and the magnetic disturbance keeping time with it, and perpendicular to it, and both perpendicular to the transfer of energy, which is parallel to the wire, very nearly. A tube of energy-current may be regarded as a ray of light (dark, of course).

It is to such *long* waves that I attribute the magnetic disturbances that come from the sun occasionally, and simultaneously show themselves all over the world; arising from violent motions of large quantities of matter, giving shocks to the ether, and causing the passage from the sun of waves of enormous length. On such a wave passing the earth, there are immediately induced currents in the sea, earth's crust, telegraph lines, etc.

But to return to the circuit. The attenuation-ratio per unit length is represented by $\epsilon^{-R/2Lv}$, this being the ratio of the transmitted to the original intensity of the wave. This is when the insulation is perfect. These waves are subject to reflection, refraction, absorption, etc., according to laws I shall give. Of these the simplest cases are reflection by short-circuiting, when the potential-difference is reversed by reflection, but not the current, and in the act of reflection the former is annulled, the latter doubled. Also reflection by insulation, when it is the current that is reversed, and potential-difference unchanged; or, in the act of reflection, the first cancelled, the second doubled. But there are many other cases I have investigated.

I have also examined leakage. This is an old subject with me. An Atlantic cable is worked under the worst conditions (electrical) possible with high insulation; there is the greatest possible distortion. One megohm per mile or less instead of hundreds or thousands would vastly accelerate signalling. The attenuation-factor is now $\epsilon^{-R/2Lv}.\ \epsilon^{-K/2Sv}$, if K be the leakage-conductance, and S the permittance per unit length. The attenuation is increased, but the distortion is reduced. This has led me to a theoretically perfect arrangement. Make $R/L=K/S$, and the distortion is annihilated (save corrections for increased resistances, etc.). The solution is so simple I may as well give it now. Let V and C be the potential-difference and current at distance x, subject to

$$-\frac{dV}{dx}=RC+L\dot{C}, \qquad -\frac{dC}{dx}=KV+S\dot{V};$$

then, with equality of time-constants as described, the complete solution consists of two oppositely travelling trains of waves, of which we need only write one; thus,

$$V=f(x-vt)\epsilon^{-Rt/L},$$

where $f(x)$ is the state when $t=0$. The current is $C=V/Lv$. The energy is half electric, half magnetic; the dissipation is half in the wire, half outside. Change the sign of v in a negative wave. There is a perfect correspondence of properties, when this unique state of things is not satisfied, between V solutions with $K=0$, and C solutions with $R=0$. This perfect system would require very great leakage in an Atlantic cable, and cause too much attenuation; but this perfect state may be aimed at, and partly reached. Are there really any hopes for Atlantic telegraphy? Without any desire to be over sanguine, I think we may expect great advances in the future. Thus, without reducing the resistance or reducing the permittance (obvious ways of increasing speed), increase the leakage as far as is consistent with other things, and increase the inductance greatly. One way is with my non-conducting iron, which I have referred to more than once, an insulator impregnated with plenty of iron-dust. Use this to cover the conductor. It will raise the inductance greatly, and so greatly diminish the attenuation; whilst the insulation-resistance will be lowered, somewhat increasing the attenuation, but assisting to diminish the distortion, which the increased inductance does. The change in the permittance must also be allowed for. But I shall show that we can have practical approximations to almost negligible distortion in telephony, and that it is the reduction of R/L that is most important.

I have also examined the question of apparatus. We must stop the reflection, if possible, to prevent interference. In the perfect system this is also quite easy. The receiver must have resistance Lv and zero inductance. All waves arriving are then wholly absorbed. Similarly, to make the transmitted waves agree with the impressed force, Lv should be the resistance there, (or else zero). Another remarkable property is that if the receiving coil be fixed in size and shape, whilst its resistance varies, then this same Lv is the resistance that makes the magnetic force of the coil a maximum. We cannot imagine anything more perfect. No distortion, and maximum effect. I shall show that these things may be fairly approximated to in telephony. It should be understood that in the perfect system we have nothing to do with what the frequency may be, whilst in telephony it is the high frequency that allows us to approximate to the ideal state.

Then there is the matter of bridges, and the nature of the reflected, transmitted, and absorbed waves. The phenomena formally resemble those due to the insertion of resistance in the main circuit, except that the potential-difference and the current change places. Thus if R_1 be an inserted resistance, when there is no leakage and no resistance in the line $(1+R_1/2Lv)^{-1}$ is the ratio of transmitted to incident wave. Now let there be no resistance inserted, but a bridge of conductance K_1; then the substitution of K_1 for R_1, and S for L gives us the corresponding formula. In the first case the reflected current is reversed, in the second case it is the potential-difference of the reflected wave that is reversed. Now let there be both a resistance inserted and a conducting bridge, and choose $R_1/L=K_1/S$; then the reflected wave is abolished. Part of the original wave is absorbed in the bridge, and the rest is transmitted unchanged. This explains the perfect system above described.

I have also examined the changes made when the state is not perfect. The result is that a wave throws out a long slender tail behind it; and whilst the nucleus goes forward at speed v, the tail goes backwards at this speed. In time, if the line be long enough, the nucleus, which changes shape as it progresses, diminishes so as to come to be a part of the tail itself. It is then all tail. I will give the equation of the nucleus and tail. It is the mixing up of these tails that causes arbitrary waves to be distorted as they travel from beginning to end of the line. (But I have, in the above, usually referred to distortion as the change in the shape of the curve of current at a single spot.) There is residual reflection due to the self-induction of the receiver, even when the resistance is of the proper amount. The effect of diffusion in the wires is to make a wave with an abrupt front, which would continue abrupt, have a curved front, and thus mitigate that perfection which only exists on paper. I shall also describe graphical methods of following the progress of waves, and of calculating arrival-curves of various kinds, the submarine cable and oscillatory; approximate only, but very easy to follow. Other matters, perhaps more practical, but certainly duller, will find their place, if space allow.

SECTION XLI. NOMENCLATURE SCHEME. SIMPLE PROPERTIES OF THE IDEALLY PERFECT TELEGRAPH CIRCUIT.

To explain the word "permittance" that I used in the last Section, I may remark that in stating my views in 1885 in several communications to this journal on the subject of a systematic and convenient electrical nomenclature based upon the explicit recognition of the three fluxes, conduction-current, magnetic induction, and electric displacement, proposing several new words, some of which have found partial acceptance, I remarked upon the unadaptable character of the word "capacity." It must be the capacity of something or other, as of permitting displacement. I did not then go further in connection with the flux displacement than to use "elastance," for the reciprocal of electrostatic capacity. The following shows the scheme so far as it is at present developed :—

Flux.		Force/Flux.	Flux/Force.	Force.
Conduction-Current	{	Resistance. Resistivity.	Conductance. Conductivity.	} Electric.
Induction*	{	? ?	Inductance. Inductivity.	} Magnetic.
Displacement	{	Elastance. Elastivity.	Permittance. Permittivity.	} Electric.

Why elastivity? Maxwell called the reciprocal of the permittance of a unit cube "the electric elasticity." By making it simply elastivity, we first get rid of the qualifying adjective; next, we avoid confusion with any other sort of elasticity; and, thirdly, we produce harmony with the rest of the scheme. There are now only two gaps. "Resistance to lines of force," or "magnetic resistance," now used, will not do for permanent employment. Besides the above, there is Impedance, to express the ratio of force to flux in the very important case of sinusoidal current. Impedance is at present known by various names that seem to be founded upon entirely false ideas. The impedance (which, derived from impede, need not be mispronounced) of a coil is the ratio of the amplitude of the impressed force to that of the current. A coil used for impeding may be called an impeder. The same definition obviously applies in any case that admits of reduction to *one circuit* (even though parts of it may be multiple), *e.g.*, any number of coils in sequence, in sequence with any number in parallel (to be regarded as one), in sequence with a condenser, or arrangement reducible to a condenser. The impedance is always reducible to $(R^2+L^2n^2)^{\frac{1}{2}}$, where R is the effective *resistance*, which is real, and L the effective inductance, or sometimes *quasi*-inductance. It is not necessary to exclude inductive action on *other* circuits, although the heat corresponding to R may be partly in them. As for resistance, it is very desirable to confine its use to the established meaning in connection with Joule's law.

Now let R, L, S and K be the resistance, inductance, permittance and leakage-conductance per unit length of a circuit; and let V and C be the potential-difference (an awkward term) and current at distance x. We have the following fundamental equations of connection:—

$$-\frac{dV}{dx}=(R+Lp)C, \qquad -\frac{dC}{dx}=(K+Sp)V, \quad \text{...........} \ (1d)$$

p standing for d/dt. Observe that the space-variation of C is related to V in the same manner (formally) as the space-variation of V is related to C, so that we can translate solutions in an obvious manner by exchanging V and C, R and K, L and S, which are reciprocally related, in a manner.

To fix ideas, the circuit may be the common pair of parallel wires. There is one case in which the four constants are all finite that is characterised by such extreme simplicity that it is desirable to begin

* [The two blanks were filled up later by the words Reluctance and Reluctivity or Reluctancy.]

with it, especially as it casts a flood of light upon all the other cases, which may be simpler in appearance, and yet are immensely more complex in results. Let

$$R/L = K/S = s. \qquad \text{and} \qquad LSv^2 = 1. \qquad \text{(2d)}$$

The number of circuit-constants is now virtually three, owing to the fixing of the fourth constant. The equation of V is now

$$v^2\frac{d^2V}{dx^2} = (s+p)^2V, \qquad \text{(3d)}$$

or, which is equivalent,

$$v^2\frac{d^2u}{dx^2} = p^2u, \qquad \text{(4d)}$$

if $V = u\epsilon^{-st}$. Since (4*d*) is the equation of undissipated waves, with constant speed v, whose solution consists of two oppositely travelling arbitrary waves, the complete solution of (3*d*) consists of such waves attenuated as they progress at the rate s (logarithmic). Thus,

$$V = f(x - vt)\epsilon^{-st} \qquad \text{(5d)}$$

is the complete expression of the positive wave, if $f(x)$ be the state when $t = 0$. Shift the wave bodily a distance vt to the right, and attenuate it from 1 to ϵ^{-st}, and we obtain the state at time t. The corresponding current is

$$C = V/Lv = SvV, \qquad \text{(6d)}$$

in every part of the wave. To express a negative wave, change the sign of v in (5*d*) and (6*d*). The second form of (6*d*) says that a charge Q moving at speed v is equivalent to a current Qv.

Since V is an E.M.F., it is convenient to reckon Lv in ohms, as was done before; v is 30 ohms, in air, when it has its greatest value (speed of light, 30 earth-quadrants per second) and L is a convenient numeric. $L = 20$ is a common value (copper suspended wires); in this case our "resistance" is 600 ohms. But it is not "ohmic" or "joulic" resistance; the current and E.M.F. are perpendicular. V is the line-integral of the electric force across the dielectric from wire to wire, and C is the line-integral ($\div 4\pi$) of the magnetic force round either wire. The electric and magnetic forces are perpendicular, and so are V and C regarded as vectors, [*i.e.*, their elements **E** and **H** are perpendicular]. The product VC is the energy-current; their ratio is the important quantity Lv, the impedance.

In a positive wave V and C are similarly signed, and in a negative wave are oppositely signed. Thus, if the electrification be positive, the direction of the current is the direction of motion of the wave; whilst if it be negative, the current is against the motion of the wave.

When oppositely travelling waves meet, the resultant V is the sum of the two V's, and the resultant C the sum of the two C's.

Thus, if the waves be so shaped as to fit, then, on coincidence, V is doubled and C is annulled. The energy is then all electric. But if the electrifications be opposite, V is annulled and C is doubled, on coincidence.

The energy is then all magnetic. On emergence, however, the two waves are unaltered, save in the attenuation that is always going on.

The electric energy is $\frac{1}{2}SV^2$ per unit length of circuit, and the magnetic energy is $\frac{1}{2}LC^2$. From this, by (6*d*) and the second of (2*d*), we see that the electric and magnetic energies are equal in a solitary wave, either positive or negative. The dissipativity in the wires is RC^2, and outside them KV^2, per unit length of circuit. These are also equal, for the same reason.

Should the disturbance be given arbitrarily, *i.e.*, V and C any functions of x, the division into the positive wave V_1 and the negative wave V_2 is effected thus:—

$$V_1 = \tfrac{1}{2}(V + LvC), \qquad V_2 = \tfrac{1}{2}(V - LvC). \quad \text{..........} \ (7d)$$

Notice that $SV_1V_2 = -LC_1C_2$, so that the total energy per unit length is always

$$S(V_1^2 + V_2^2) = L(C_1^2 + C_2^2). \quad \text{..................} \ (8d)$$

Similarly, the total dissipativity is always

$$R(C_1^2 + C_2^2) = K(V_1^2 + V_2^2). \quad \text{..................} \ (9d)$$

Similarly the total energy-current is always

$$V_1C_1 + V_2C_2, \quad \text{........................} \ (10d)$$

since

$$V_1C_2 = -V_2C_1.$$

If, at a given moment, $V = V_0$ through unit distance anywhere, with no C, this immediately breaks into two equally big waves, one positive, the other negative, which at once separate. If initially there be no V, but only C, the same is true for the current-waves; *i.e.*, the result is two equal but oppositely signed V waves, which at once separate.

What happens when disturbances reach the end of the circuit depends upon the nature of the terminal connections there. At present only one case—the simplest—will be noticed. Let there be a resistance of amount Lv at the distant end B of the circuit. The terminal condition is then $V = LvC$. But this is the property of a positive wave. Hence all waves travelling towards B are immediately *absorbed* on reaching B. The electricity is all gobbled up at once, so to speak. Similarly, if there be a resistance Lv at the end A (where $x = 0$) it imposes the condition $V = -LvC$, which is the property of a negative wave, so that all disturbances on arrival at A are absorbed immediately. Thus, given the circuit in any state of electrification and current, without impressed force, it is wholly cleared in the time l/v at the most, l being the length of the circuit.

Now, let the circuit be short-circuited at A, and have a resistance Lv at B. Insert an impressed force e at A momentarily, producing $V = e$ through unit distance, say. This will travel towards B at speed v, attenuating as it goes, and on arrival at B, what is left will be at once absorbed. This being true for every momentary impressed force, we see that if it be put on at time $t = 0$, and kept steadily on thereafter, the full solution is

$$V = e.\epsilon^{-Rx/Lv}, \quad \text{..........................} \ (11d)$$

from $x=0$ to $x=vt$, and zero beyond. Thus the steady state at a given point is instantly assumed the moment the wave-front reaches it. After that, there is still transfer of energy going on there, viz., to supply the waste in the part of the wave that has passed the spot under consideration, and to increase the energy at the front of the wave. The current is V/Lv, as before. On reaching B, the current is

$$C=\frac{e}{Lv}\epsilon^{-Rl/Lv}=\frac{e}{Rl}\times\frac{Rl}{Lv}\epsilon^{-Rl/Lv}. \quad \text{.................. (12d)}$$

If we let Rl, the resistance of the circuit, be 3,000 ohms, which is 5 times the before-assumed value of Lv, then the received current is

$$C=\frac{e}{150\,Lv}=\frac{e}{90{,}000}=\frac{e}{30\,Rl}. \quad \text{.................. (13d)}$$

The attenuation is such that the current is one-thirtieth part of the full steady current with perfect insulation.

The electrostatic time-constant of the circuit is

$$RSl^2=\frac{l}{v}\times\frac{Rl}{Lv}; \quad \text{......................... (14d)}$$

or, in our example, five times the time of a journey from A to B. It may have any value we please. If we want it to be ·1 second, l/v must be ·02 second, and therefore $l=6{,}000$ kilometres, which requires $R=$ ·5 ohm per kilom. This is lower than that of any telephone line yet erected. But to make the electrostatic time-constant ·05 second, with the same attenuation, it must be 3,000 kilom. at 1 ohm per kilom.

If e vary in any manner at A, the current at B is given by (12*d*), in which e varies in the same way at a time l/v later. As there is no distortion, it becomes a question of suitable instruments. With proper instruments, no doubt the permissible attenuation could be much greater, and the circuit much longer. Again, if we raise the insulation we lessen the attenuation. We bring on distortion, but a good deal is allowable, so that again we can work further. The insulation-resistance should be 36 megohm per kilom. in the 3,000 kilom. example; the product of the resistance of any portion of the circuit (wires) into the insulation-resistance of the corresponding part is $(Lv)^2$. In the 6,000 kilom. example it should be ·72 megohm per kilom. But if it be not arbitrary waves, but only waves of high frequency that are in question, then we may approximate to the distortionless transmission without attending to the exactly-required leakage.

SECTION XLII. SPEED OF THE CURRENT. EFFECT OF RESISTANCE AT THE SENDING END OF THE LINE. OSCILLATORY ESTABLISHMENT OF THE STEADY STATE WHEN BOTH ENDS ARE SHORT-CIRCUITED.

Although the speed of the current is not quite so fast as the square of the length of the line, yet, on the other hand, it is not quite so slow as the inverse-square of the length, as a writer in a contemporary (*Electrical Review*, June 17, 1887, p. 569) assures us has been proved by recent researches. However, if we strike a sort of mean, not an arithmetic mean, nor yet a harmonic mean, but what we may call a scienticulistic mean (whatever *that* may mean), and make the speed of the current altogether independent of the length of the line, we shall probably come as near to the truth as the present state of electromagnetic science will allow us to go. But, apart from this, there is some *à priori* evidence to be submitted. Is it possible to conceive that the current, when it first sets out to go, say, to Edinburgh, *knows* where it is going, how long a journey it has to make, and where it has to stop, so that it can adjust its speed (scienticulistic speed) accordingly? Of course not; it is infinitely more probable that the current has no choice at all in the matter, that it goes just as fast as the laws of Nature, preordained from time immemorial, will let it; and if the circuit be so constructed that the conditions prevailing are constant, there is every reason to expect that the speed will be constant, whether the line be long or short. Q.E.D.

Now, a great and striking thing about the distortionless system, whose elementary properties were discussed in the last Section, is the distinct manner in which it brings the speed of the current into full view. Another and very important thing is this. When the leakage is not so adjusted as to remove the distortion altogether, solutions become difficult of interpretation, owing to the almost necessary employment of Fourier or other transcendental series to express results. But by a proper adjustment of the leakage so as to abolish the tailing, which is the cause of the mathematical difficulties, we are enabled to follow with ease the whole course of events, say, in the setting up of the final state, due to a steady impressed force, without laborious calculations. And, although the state of things supposed to exist in the distortionless system is rather an ideal one, yet it allows us to obtain a very fair idea of what happens when there is distortion, *e.g.*, in the oscillatory establishment of the steady state in a well-insulated circuit.

When we speak of a charge travelling along a wire at speed v, it should be always remembered what this implies. There are two conductors, parallel to one another, and the positive charge on the one is accompanied by its complementary negative charge on the other (corrections due to parallel wires, etc., are ignored here). The two charges move together. More comprehensively, the whole electromagnetic field, of which the charges are a feature only, is moving along at speed v, in the space between the wires, into which it also penetrates to a greater or less extent. In the distortionless system this penetration is assumed to be perfect and instantaneous, so that the resistance and the inductance are strictly constants; and, by the ratio R/L being made equal to K/S, we make any isolated disturbance travel on without spreading out behind. In travelling it attenuates by loss of energy in the conductors and by leakage in such a way that if it attenuate from 1,000 to 900 in the first 50 kilometres, it will attenuate to 810 in the second, to 729 in the third, and so on; multiplying by 9/10 in every 50 kilometres.

In the last Section was considered the uniquely simple case of a short-circuit at A, the beginning of the circuit, where any impressed force is placed, sending any-shaped waves into the circuit, travelling undistorted, with uniform attenuation, and completely absorbed on arrival at the distant end B by a terminal resistance of amount Lv. Of course this complete absorption at B of all waves arriving there is independent of the nature of the terminal arrangements at A. But these will materially influence the magnitude of the waves leaving A. Keeping at present entirely to simple cases, if we insert a resistance Lv at A we can make a safe guess that the current will be just halved, because when there is a short-circuit there, the line itself behaves just as if it were a resistance Lv. That is, the current at A is then e/Lv, however e may vary, provided there be a resistance Lv at B; or, which is equivalent, the circuit be continued indefinitely beyond B unchanged in its properties. This guess may be easily justified. That the current is zero when we insulate, or insert an infinite resistance at A, is also evident. In general, the insertion of a resistance R_0 at A causes the potential-difference V_0 there, due to an impressed force e, to be

$$V_0 = eLv(R_0 + Lv)^{-1}, \quad \text{......................} (15d)$$

and the current to match to be V_0/Lv. The transmission to the distant end, and the attenuation are as before.

But if the place of e be shifted along the circuit from A, interferences will result whenever the resistance at A has not the value Lv. Imagine e to be at distance x_1 from A. When put on, the result is to send a positive wave $\frac{1}{2}e$ to the right, and a negative wave $-\frac{1}{2}e$ to the left, both travelling at speed v, and attenuating similarly. Thus the circuit behaves towards e as a resistance $2Lv$, half to the right, half to the left. Now, when the negative wave arrives at A, if there be a resistance Lv there to absorb it, there will be no interference with the positive wave, which will go on to B and be absorbed there. The current at B will therefore be

$$C_B = \tfrac{1}{2}(e/Lv)\epsilon^{-R(l-x_1)/Lv}, \quad \text{......................} (16d)$$

the value of e to be taken at a given moment being that at x_1, at the time $(l-x_1)/v$ earlier. But if there be a resistance at A of any other amount than Lv, there will be a reflected wave from A, which will run after the original positive wave, and so make every signal at B have a double or familiar following it after an interval of time $2x_1/v$, which is that required to go from x_1 to A, and back again. Now the closer the seat of e is shifted towards A, the more closely will the familiar follow the original positive wave; and when e is at A itself, they will be coincident in front. Now, the current at A corresponding to (16d) is

$$C_A = \tfrac{1}{2}(e/Lv)\epsilon^{-Rx_1/Lv}; \quad \text{......................} (17d)$$

and (as will be explained in the Section on Reflections) the reflected wave is got by multiplying by ρ_0, where

$$\rho_0 = (R_0 - Lv)(R_0 + Lv)^{-1}. \quad \text{....................} (18d)$$

Now make $x_1 = 0$, and we shall verify (15d), and, by the union of the positive and the reflected (also positive) wave, show that V at x at time t due to $e = f(t)$, any function of t, at A, is

$$V = f(t - x/v) \times Lv(R_0 + Lv)^{-1} \times \epsilon^{-Rx/Lv}, \quad \text{.............} (19d)$$

and the current there is V/Lv.

The most simple case after these of complete absorption at B, with complete absorption, or short-circuit, or any resistance at A, is perhaps that in which we short-circuit at both A and B. If a charge be then moving towards B, it is wholly reflected with reversal of electrification. We must have $V=0$ at B, and this requires every disturbance arriving at B to be at once reversed and sent back again. The same thing happens at the short-circuit at A. Perhaps, however, the easiest way to follow events is to imagine the two charges, positive and negative, which always travel together, to pass through one another when they come to the short-circuit, so as to exchange wires. Thus one charge goes round and round the circuit one way, whilst the other, just opposite, goes round and round the other way. There is the usual attenuation. On this view of the matter, we may imagine the effect of a terminal resistance Lv to be simply to bring the charges to rest against friction. It need scarcely be said, however, that the day has gone by for any such fanciful explanation to be taken seriously.

Since the current in a negative wave (from B to A) is of the opposite sign to the electrification, there is no reversal of current by reflection at a short-circuit. As, therefore, the reflected wave is to be superimposed upon the incident wave, we see that the current is doubled at B from what it would be were the circuit to be continued beyond B, or the critical resistance Lv were inserted in place of the continuation.

The process of setting up the permanent state due to a steady e at A is now this:—First the positive wave

$$V_1 = e.\epsilon^{-Rx/Lv}, \qquad \text{..........................(20}d\text{)}$$

if $x<vt$, which would be the complete solution were there no reflection at B. Now B is reached by V_1 in the time l/v, and the value of V_1 at B just on arrival is $e\rho$, if $\rho=\epsilon^{-Rl/Lv}$, which is the attenuation in the circuit. The reflected wave V_2 now begins. This is

$$V_2 = -e\rho^2.\epsilon^{Rx/Lv}, \qquad \text{..........................(21}d\text{)}$$

which travels towards A at speed v. In the meantime the first wave V_1 is still going on, for the battery at A does not know what is going on at B. Thus, from $t=l/v$ to $t=2l/v$, the state of the circuit is given by the sum of V_1 and V_2 so far as V_2 has reached, and by V_1 alone in the rest. On arrival of V_2 at A it is attenuated to $-e\rho^2$, and reflection then produces a positive wave

$$V_3 = e\rho^2.\epsilon^{-Rx/Lv}, \qquad \text{..........................(22}d\text{)}$$

which is a copy of V_1, only smaller to the extent produced by the multiplication by ρ^2. This wave reaches B when $t=3l/v$, and then there commences the reflected wave, V_4, given by

$$V_4 = -e\rho^4.\epsilon^{Rx/Lv}, \qquad \text{..........................(23}d\text{)}$$

going from B to A. This is a copy of V_2. And so on. Thus we have an infinite series of reflected waves, coming into existence one after the other; the state at any moment is expressed by the sum of the waves already existent; the final state is the sum of them all. Since the sizes of the positive waves form a geometrical series, and also those of the negative waves, they are easily summed. The positive waves V_1, V_3, etc., come to

$$e(1-\rho^2)^{-1}\epsilon^{-Rx/Lv}, \quad\text{......................}(24d)$$

and the negative come to

$$-e\rho^2(1-\rho^2)^{-1}\epsilon^{Rx/Lv}; \quad\text{......................}(25d)$$

so that the sum of (24*d*) and (25*d*) expresses the final V of the circuit. And, since the current is got by dividing by Lv in a positive wave and by $-Lv$ in a negative, the final current is the excess of (24*d*) over (25*d*), divided by Lv. Notice that whilst it is a process of settling down to the final state of electrification, it is a process of rising up to the final state of current. More strictly, whilst the potential-difference at any spot oscillates about its final value, being alternately above and below it, the excursions getting smaller and smaller as time goes on, the current-increments are all positive, though they get smaller and smaller. Now if the time l/v of a journey be exceedingly small, so that there may be thousands of journeys performed in getting up to say 99 per cent. of the final current, the current will appear to rise continuously, and the potential-difference to have its final value from the first moment, which is in reality its mean value during the oscillatory period. This is the explanation I have before given of how it comes about that there is no sign of oscillation in any purely electromagnetic formulæ, such as are universally employed when such short circuits are in question that the current seems to have the same strength (when no leakage) everywhere. It is really rising by little jumps, and differently timed at different places, but the jumps are too small to be perceived, and too rapidly executed. And the electrification at any spot is really (unless the vibrations are specially checked) vibrating about its mean value, which is its final value, though this mean value is assumed (in electromagnetic formulæ) to be the actual value. But if the resistance in circuit be great, so that the final current is small, we have an oscillatory settling down of the current, instead of a rise.

The solution (24*d*), (25*d*) is what we may at once get by considering the differential equation of the steady state and its solution to satisfy the terminal conditions. But our solution gives us the whole history of the establishment of this final state, and allows us to readily follow the oscillatory phenomenon into minute detail. When there is distortion there is difference in detail, which is then difficult to follow; but there is no substantial difference in the general results. We cannot make or break a circuit without a similar action in general. But we cannot expect to be able to formularise the results simply when the circuit is of an irregular type, *e.g.*, a laboratory circuit.

Section XLIII. Reflection due to any Terminal Resistance, and Establishment of the Steady State. Insulation. Reservational Remarks. Effect of varying the Inductance. Maximum Current.

If there be a resistance R_1 at the end B of a distortionless circuit, its presence imposes the condition $V=R_1C$ at B permanently. If, then, there be a wave travelling towards B, we find the nature of the reflected wave from B by applying the above terminal condition to the actual V and C, which are the sum of V_1, V_2, the potential-differences in corresponding portions of the incident and reflected waves, and of C_1, C_2 the currents in these portions. Thus we have

$$V_1 = LvC_1, \qquad V_1 + V_2 = V, \qquad V = R_1 C,$$
$$-V_2 = LvC_2, \qquad C_1 + C_2 = C,$$

to represent the full connections. From these we find

$$V_2/V_1 = (R_1 - Lv)(R_1 + Lv)^{-1} = \rho_1 \quad \text{say}, \qquad \text{......(26}d\text{)}$$

giving the reflected in terms of the incident wave. This ratio is positive if R_1 be greater, and negative if it be less than the critical Lv. In the former case there is reversal of current, in the latter of electrification, produced by the reflection. The three most striking cases are when $R_1 = 0$, ∞, or Lv, *i.e.* short-circuit, insulation, and the critical resistance of complete absorption, making $\rho_1 = -1$, $+1$, or zero. There is partial absorption and loss of energy whenever R_1 is finite, but none whatever in the two extreme cases. The loss of energy is accounted for by the Joule-heat in the terminal resistance.

In a similar manner, if there be a resistance R_0 at the near end A, the transforming factor is

$$\rho_0 = (R_0 - Lv)(R_0 + Lv)^{-1}. \qquad \text{..................(27}d\text{)}$$

If there be given an isolated charge moving towards B at a certain time, it will, after reflection at B, be replaced by another charge moving towards A, which may be of the same or of the opposite kind, according as the reflecting resistance is greater or less than the critical. On arrival at A it is transformed into a third charge moving towards B, and so on. There is the usual attenuation ρ in each journey, where $\rho = \epsilon^{-Rl/Lv}$. If there be complete insulation at both ends, there is no other attenuation than this due to the circuit; and, similarly, if the ends be short-circuited; but in all other cases it has to be remembered that the act of reflection attenuates, besides causing a reversal of either the electrification or the current.

The complete history of the establishment of the steady state due to a steady impressed force at A is now expressible in terms of the three constants ρ_0, ρ, and ρ_1; with, of course, x the distance, t the time, and e the impressed force. There is first the positive wave

$$V_1 = \tfrac{1}{2}e(1 - \rho_0)\epsilon^{-Rx/Lv}, \qquad \text{......................(28}d\text{)}$$

due, as mentioned in the last Section, to the union of the initial positive wave of half strength and of the positive wave which is the reflection of the initial negative wave of half strength, which latter is rendered visible by shifting the seat of e towards B. The solution (28d) applies to all values of x less than vt, which is the extreme distance reached by the wave at time t after starting. On arrival at B we have to introduce the transforming factor ρ_1, above defined. The reflected wave is therefore

$$V_2 = \tfrac{1}{2}e(1 - \rho_0)\rho^2\rho_1 \,.\, \epsilon^{Rx/Lv}, \qquad \text{....................(29}d\text{)}$$

which is to be superimposed on the former wave to obtain the real state during the second journey, from B to A. The region over which V_2 extends grows at a uniform rate with the time, from B to A. On arrival of V_2 at A we must introduce the transforming factor ρ_0 to obtain the third wave, which is

$$V_3 = \tfrac{1}{2}e(1-\rho_0)\rho^2\rho_1\rho_0 \,.\, \epsilon^{-Rx/Lv}. \qquad (30d)$$

This reaches B at time $t = 3l/v$, when the fourth wave commences, which is to be found by introducing the transforming factor ρ_1; thus

$$V_4 = \tfrac{1}{2}e(1-\rho_0)\rho^4\rho_1^2\rho_0 \,.\, \epsilon^{Rx/Lv}. \qquad (31d)$$

It is unnecessary to proceed further, as it would only produce repetitions. The positive waves V_1, V_3, etc., have the common ratio $\rho^2\rho_1\rho_0$, and are otherwise similar. Their sum is therefore

$$\tfrac{1}{2}e(1-\rho_0)(1-\rho^2\rho_1\rho_0)^{-1} \,.\, \epsilon^{-Rx/Lv}. \qquad (32d)$$

Similarly the sum of the negative waves is

$$\tfrac{1}{2}e(1-\rho_0)\rho^2\rho_1(1-\rho^2\rho_1\rho_0)^{-1} \,.\, \epsilon^{Rx/Lv}. \qquad (33d)$$

The final state of V is therefore expressed by the sum of (32*d*) and (33*d*). In all the positive waves the current is from A to B, and in the negative from B to A; hence the excess of (32*d*) over (33*d*), divided by Lv, expresses the final state of current.

The solution of the above problem by means of Fourier-series is extremely difficult. It expresses the whole history of the variable period by a single formula. But this exceedingly remarkable property of comprehensiveness, which is also possessed by an infinite number of other kinds of series, has its disadvantages. The analysis of the formula into its finite representatives, so that during one period of time it shall represent (28*d*), then in another period represent the sum of (28*d*) and (29*d*), and so on, *ad inf.*, is trying work. And the getting of the formula itself is not child's play. Considering this, and also the fact that a large number of other cases besides the above can be fully solved by common algebra (with a little common-sense added), the importance of a full study of the distortionless system will, I think, be readily admitted by all who are dissatisfied with official views on the subject of the speed of the current. The important thing is to let in the daylight on a subject which it was difficult to believe could ever be freed from mathematical complications.

There is a rather important remark to be made concerning the two extremes, $R_1 = 0$ and $R_1 = \infty$, at the end of the line, in the above solution. Although described as short-circuiting and insulation, they do not really represent the state of things existent when we actually terminate a long circuit of two parallel wires by a thick cross-wire (the short-circuit) or leave the ends disconnected in the air. Every theory that ever was made is more or less a paper theory; we must simplify the real conditions to make a theory workable. Now a theory may very closely represent reality (when pursued into numerical detail) through a wide range, and yet go quite wrong at extremes. The justification for making the constants of the circuit independent of its length is that the length is an enormous multiple of the distance between the wires. But if we terminate the circuit somewhere, it is no longer true that the permittance and the inductance per unit length are constants, near and at the termination. The theory, to be correct, must wholly change its nature, as may be seen at once on thinking of the changed nature of the electromagnetic field as the termination is reached. Now our theory says that when the circuit is insulated at B, every charge arriving there is at once sent back again unchanged; and that during the period of reflection, the potential-difference is doubled and the current annulled. The doubling of the potential-difference is obviously due to there being a double charge with the same assumed permittance. But the permittance is not the same, nor anything like the same, at the termination as it is far away from it. The theory therefore wholly fails to represent the case of insulation, so far as the potential-difference at the termination is concerned, though there does not seem to be any reason to suppose that this will affect matters elsewhere; for when the reflected wave gets away from the termination, the old state of things is restored. There is a similar want of correspondence between the theory and reality when we make a real short-circuit, which we have supposed to be represented by $R_1 = 0$.

Now the question may suggest itself: Since this failure is due to the assumption that the permittance and inductance continue constants right up to the termination, and this assumption being made in all cases, may there not also be a failure when R_1 is finite? The following reasoning will show that this is not to be expected. For if the terminal resistance (although it may be small) be equal to that of a considerable length of the circuit, the influence of this resistance on the course of events must be much greater than that due to the changed nature of the circuit near its end. We therefore swamp the terminal corrections, which become so important themselves when the terminal resistance is quite negligible.

The general principle that may be recognised is this. If the transfer of energy between the circuit and the terminal apparatus (of any kind) be of sensible amount, we may wholly disregard the fact that the circuit changes its nature as the termination is approached. But should it be insensible, then we fail to represent matters correctly at and near the termination.

Again, if the ends of the circuit, supposed insulated, be brought sufficiently close together, there may be a spark or disruptive discharge there when a charge arrives, involving a loss of energy and attenuation. It is scarcely necessary to remark that effects of this kind have no place in the theory.

In the same connection it may be remarked that when we are following the history of an isolated charge, which may, in the theory, be confined to the shortest piece of the circuit imaginable, we should really spread it over a length which is several times as big as the distance between the two wires. This is to make the element of length have the same properties as a great length. Similar assumptions are made (though seldom, if ever, mentioned) in most theories in mathematical physics. An element of volume, for instance, must be large enough to contain such an immense number of molecules as to impart to it the properties of the mass.

Returning now to the study of the properties of the circuit, let us examine the effect of varying the constants. For simplicity, insert the critical resistance at B, and let there be none at A, where the impressed force is. The current at B is then

$$C_B = (e/Lv)\epsilon^{-Rl/Lv} = (e/Rl)y\epsilon^{-y}, \quad \text{.....................(34}d)$$

if $y = Rl/Lv$. The value of e to be taken in the formula at a given moment should be that at A at the time l/v earlier. Now, with the resistance of the circuit kept constant, vary y to make the current a maximum. We require $y = 1$, or the critical resistance should equal the resistance of the circuit (without leakage). It then also equals the insulation-resistance $(Kl)^{-1}$. If the resistance at A be any constant multiple of Lv, we shall have the same property $y = 1$ to get maximum current. (But should the resistance at A be kept constant, we shall have $y^2(R_0/Rl) + y = 1$, which it is unnecessary to discuss.) The received current is therefore

$$C_B = e(2{\cdot}718\,Rl)^{-1}, \quad \text{..........................(35}d)$$

when no resistance at A; and if there be resistance of amount zLv, we must divide the right side of (35d) by $(1+z)$ to obtain the current at B. Thus the result is the same as if the circuit were a mere resistance whose value is a small multiple of the true resistance, with abolition of the leakage, permittance, and inductance, but with a retardation of amount l/v. This is not the electrostatic retardation, of course; it merely means the interval of time that elapses between sending and receiving, whereas electrostatic retardation, as formerly understood, is quite another thing. Neither is it the speed of the current; that is v. But singularly enough, the value of the electrostatic time-constant RSl^2 is now l/v itself, proportional to the first power of the length, and inversely proportional to the speed of the current.

Example. 1,200 kilometres at 2 ohms per kilom. Lv should be 2,400. If it be an air-circuit, of copper, with v practically = 30 ohms (the formulæ for permittance, inductance, etc., will be given later), we require $L = 80$. This is much too great. The inductance must be artificially increased, if we are to have so little attenuation as above on a circuit of that length. Or the resistance may be reduced. If 1 ohm per kilom., $L = 40$ is wanted. If $\frac{1}{2}$ ohm per kilom, $L = 20$.

The shorter the circuit, the smaller is the value of L needed to get the maximum current; and the longer the circuit, the greater L should be. If L could be made large enough, without altering the resistance, the circuit could be of any length we pleased. The lower the resistance of the circuit, the less leakage is needed to prevent distortion, and the less attenuation there is. The higher the resistance, the more leakage is needed, and the greater is the attenuation. We see, by inspection of (34d), that without varying either the resistance or the permittance, but solely by increasing L (remembering that $Lv = (L/S)^{\frac{1}{2}}$), we could make Atlantic fast-speed telegraphy possible, with little attenuation and distortion. But the speed of the current would be very low. This I shall return to in connection with the sinusoidal solution.

SECTION XLIV. ANY NUMBER OF DISTORTIONLESS CIRCUITS RADIATING FROM A CENTRE, OPERATED UPON SIMULTANEOUSLY. EFFECT OF INTERMEDIATE RESISTANCE: TRANSMITTED AND REFLECTED WAVES. EFFECT OF A CONTINUOUS DISTRIBUTION OF RESISTANCE. PERFECTLY INSULATED CIRCUIT OF NO RESISTANCE. GENESIS AND DEVELOPMENT OF A TAIL DUE TO RESISTANCE. EQUATION OF A TAIL IN A PERFECTLY INSULATED CIRCUIT.

If the ends of the two conductors of a distortionless circuit at its termination at A be caused to have a difference of potential V_0, varying in any manner with the time, and if there be an absorbing resistance inserted at the other termination B, we know that the impedance of the circuit to V_0 is Lv, a constant, at every moment, so that the current there is V_0/Lv. We also know how the potential-difference and current are transmitted, attenuating to $V_0\rho$ and $V_0\rho/Lv$ on arrival at B.

If there be a second distortionless circuit starting from A, and we simultaneously maintain the same difference of potential V_0 on it, we know what happens on it, viz., as above described, merely changing, if necessary, the values of ρ and Lv. That is, if the circuit be not of the same type as the first one, and of the same length, we require to use different values of ρ and Lv.

This obviously leads to the working of any number of distortionless circuits in parallel by a common impressed force at A. Call the wires of a circuit the right and the left wires, merely for distinction. Join all the right wires to one terminal A_1, and all the left wires to another, A_2, and then maintain a difference of potential V_0 between A_1 and A_2. Then, provided every circuit has its proper absorbing resistance at the distant end, we know what happens. The reciprocal of the sum of the reciprocals of the impedances of the various circuits is the effective impedance to V_0. Next, V_0 divided by the effective impedance (say I) is the total current. Finally the total current divides amongst the circuits in the inverse ratio of their impedances. The current at the distant end B of any circuit is the current entering it at A at the time l/v earlier, multiplied by the attenuation-factor ρ of the circuit. I do not write out the equations, as the description is fully equivalent.

In order that V_0 should be strictly proportional to an impressed force e in the branch joining the two common terminals A_1, A_2 of the circuits, it is necessary that it should be a mere resistance, which may have any value. Let it be R_0; then, R_0 added to the previous effective impedance to V_0, is the impedance to e; so that the total current is $e/(R_0+I)$, and the value of V_0 is $eI/(R_0+I)$. In practice, it is not possible to fully realise this simplicity. Suppose, for instance, the secondary of the transformer, in the circuit of whose primary a microphone is placed, is joined across the common terminals of the circuits.

Even if the circuits be distortionless, we see that there must be terminal distortion, or V_0 will not vary as it should for the accurate transmission of speech. There are several causes of distortion here. At the distant end, one cause of further distortion will be the inductance of the receiving telephone, and an additional and very important one will be the mechanical troubles that will prevent the disc from copying accurately, in its motion, the magnetic-force variations.

After this example of a complex arrangement of circuits admitting of simple treatment, let us return to a single circuit. Examine the effect of inserting any resistance r intermediately. This should be put half in each wire, if the circuit consist of a pair of equal wires, to prevent interferences. Let there be a wave travelling from left to right towards r. Let V_1, V_2, V_3 be the potential-differences in corresponding portions of the incident, reflected, and transmitted waves, so that, at a certain moment, they are coincident, viz. at r itself, where let V be the actual potential-difference on the left side of r. Then we have

$$\left.\begin{array}{lll} V_1 = LvC_1, & V_2 = -LvC_2, & V_3 = LvC_3, \\ V_1 + V_2 = V = rC_3 + V_3, & & C_1 + C_2 = C_3. \end{array}\right\} \quad \text{.........(36}d\text{)}$$

These are the full connections. From them,

$$\frac{V_2}{V_1} = \frac{r}{r+2Lv}, \qquad \frac{V_3}{V_1} = \frac{2Lv}{r+2Lv} = \sigma, \text{ say.} \quad \text{...........(37}d\text{)}$$

Particularly notice that

$$V_1 = V_2 + V_3, \quad \text{..........................(38}d\text{)}$$

as this is an important property. Every element of electrification in the incident wave arriving at the resistance is split into two (without any loss), one part σV_1 (in terms of potential-difference) is transmitted, the remainder is reflected.

As we have, by (37d),

$$r = 2Lv\,(\sigma^{-1} - 1), \quad \text{..........................(39}d\text{)}$$

we see that if 1 per cent. of the incident wave be reflected, and 99 per cent. transmitted, we require $r = \frac{2}{99}Lv$. If 10 per cent. be reflected and 90 per cent. transmitted, then $r = \frac{2}{9}Lv$. There is no transmitted wave if r be infinite. Half is transmitted and half reflected when $r = 2Lv$.

There is always a loss of energy by this division of the charge, which is accounted for by the Joule-heat in the resistance. This is rC_3^2 per second; and since a wave of unit length takes v^{-1} second to pass, rC_3^2/v is the loss of energy per unit length of the incident wave, which loss, if added to the sum of the energies in the reflected and transmitted waves, makes up the energy per unit length in the incident. Another expression for the loss of energy is given by

$$\frac{rC_3^2}{v} = \frac{4V_1^2 r}{v\,(r+2Lv)^2}. \quad \text{..........................(40}d\text{)}$$

There is the greatest possible loss of energy when $r = 2Lv$, making $\sigma = \frac{1}{2}$, and the loss $= \frac{1}{2}SV_1^2$. That is, when the intermediate resistance is twice the critical, and the incident wave is consequently half transmitted, half reflected, then half the energy is wasted in the resistance.

As the resistance is further increased, the transmitted wave gets smaller, and when it is infinite, we fall back upon the case already considered of total reflection without reversal of electrification or loss of energy.

If we have the absorbing resistance at A and at B, and any resistance r at an intermediate point C, we have a very simple result when any waves are sent from A to B, or from B to A. Suppose e acts at A, and that ρ_1, ρ_2 are the attenuations in the two sections AC and CB. Then $V_1 = e$ at A becomes $V_1 = e\rho_1$ on arriving at C. The reflected wave is $V_2 = e\rho_1(1-\sigma)$, where σ is given by (37d). On arrival (multiplied by ρ_1) at A it is absorbed, so there is an end of it. The transmitted wave at C is $V_3 = e\rho_1\sigma$, which attenuates to $V_3 = e\rho_1\sigma\rho_2 = e\sigma\rho$ on arrival at B, where it is absorbed. The last equation therefore gives the potential-difference at B in terms of that at A at the time l/v earlier. In the first section of the circuit V is the sum of two oppositely travelling waves, and the current is their difference divided by Lv; but in the second section there is but one wave.

We are also able to solve by algebra alone the following problem. Given a distortionless circuit with any terminal resistances and any intermediate resistances at different places, find the effect due to a steady impressed force inserted anywhere in the circuit (half in each wire, pointing oppositely in space, to avoid interferences). For we have the circuit divided into sections, for each of which the attenuation is known (*i.e.*, $\rho_1 = \epsilon^{-Rx_1/Lv}$ in a section of length x_1); we also know the transforming factors of the terminal resistances (ρ_0 and ρ_1 of the last Section); and we also now know the factors σ and $1-\sigma$ for any intermediate resistance, by which we express how a wave divides there. So, starting when e is first put on, with the initial waves $\frac{1}{2}e$ to the right, and $-\frac{1}{2}e$ to the left, we can follow the whole course of events until we arrive (asymptotically) at the steady state. But it is no part of my intention to enter into the details, as nothing new would be contained therein.

But the effect of a great number of equal intermediate resistances equidistantly situated is of importance. Let ρ_1 be the attenuation due to the circuit between two consecutive resistances, and σ the attenuation due to each resistance, that is, the attenuation of the transmitted wave. Let an isolated disturbance go from A to B. If it be initially V_0, it becomes $V_0\rho_1\sigma$ one section further on, $V_0(\rho_1\sigma)^2$ after another section is passed, and so on, becoming $V_0\rho_1^n\sigma^n$ after passing n sections. If these n sections make up the whole circuit, then $\rho_1^n = \rho$, the attenuation in the circuit due to itself only, as before, so that in passing through the circuit, V_0 is attenuated to $V_0\rho\sigma^n$.

Now let the sum of the inserted resistances be $nr = R_1$. Increase n indefinitely, whilst reducing r in the same ratio, thus keeping R_1 constant. In the limit the resistance R_1 becomes uniformly distributed in the circuit, and the attenuation due to it becomes, by (37d),

$$\sigma^n = (1 + R_1/2Lvn)^{-n}, \quad \text{with} \quad n = \infty,$$
$$= \epsilon^{-R_1/2Lv}. \qquad \text{...}(41d)$$

Observe the presence of the 2. From this we may conclude certainly (as will be shown later), that if this uniformly distributed resistance R_1, in addition to the original Rl, be accompanied by uniformly distributed leakage-conductance of total amount K_1, such that $R_1/L = K_1/S$, the attenuation due to both R_1 and K_1 together is expressed by the square of (41d). For what we do is to make the circuit distortionless again, by the additional leakage to compensate the additional resistance of the wires.

But the simplest way of viewing the matter is to start with a perfectly insulated circuit of no resistance. This is a distortionless circuit, of course, since it obeys the law $R/L = K/S$. The only difference from a real distortionless circuit is that there is no attenuation at all. All the preceding results therefore apply, remembering that $\rho = 1$, or any waves are transmitted, not merely undistorted, but also unattenuated. They are, in fact, purely plane waves of light (very long waves practically) travelling through a perfectly non-conducting dielectric. They are merely guided through space in a definite manner by the conductors, imagined to have no resistance, so that, to use a very gross simile, the electricity slips along like greased lightning. There is no penetration of the electromagnetic field into the conductors, but purely surface-conduction, where we may use the word in a popular sense (conduct = to lead). Some curious consequences of the absence of resistance I will notice later; at present I may observe that owing to the relative simplicity produced by the absence of attenuation, the imaginary circuit of no resistance is useful for investigating the effect of inserting resistances, bridges, etc., and the action of a real distortionless circuit itself.

Thus, imagine an isolated charge moving from left to right in the circuit of no resistance. Introduce anywhere a resistance r; this will cause an attenuation from 1 to σ in passing the resistance (equation (37d)), and the remainder $1 - \sigma$ will be reflected back. Next let there be a great number of equidistant small equal resistances; every one of these will attenuate in the ratio $1 : \sigma$, and throw back the fraction $1 - \sigma$. The result is that the original isolated charge, as it travels along, becomes a nucleus with a long slender tail behind it; the nucleus travelling forward at speed v and attenuating in the manner described; the tail stretching out the other way at speed v. If these isolated resistances be packed together very closely, and be each very small, we approximate to the effect of continuously distributed resistance, that is, the resistance of the wires in a real circuit. In the limit, the result is, by (41d), that the nucleus, if originally represented by $V_0 a$, that is, the potential-difference V_0 through the very small distance a, with current to match, viz., V_0/Lv through the same distance a, and therefore moving entirely to the right at that particular moment, becomes attenuated to

$$V_0 a \epsilon^{-Rx/2Lv} = \epsilon^{-Rt/2L} V_0 a \quad \text{.......................(42}d\text{)}$$

in the time $t = x/v$, during which it has moved through the distance x to the right, if the resistance per unit length be R.

Since there is here no leakage, the rest of the original charge must be in the tail. The amount of electricity in the tail is therefore

$$S \times V_0 a(1 - \epsilon^{-Rt/2L}), \quad \text{.........................(43}d\text{)}$$

when the circuit is perfectly insulated. The length of the tail is $2x$, half being to the right and half to the left of the position of the original isolated charge, it being of course supposed that neither the head nor the tail has suffered any extraneous operations, as terminal reflections, etc.

In a similar manner, if initially the isolated charge SV_0a be without current, so that it would, were there no resistance, at once divide into equal halves, travelling in opposite directions without attenuation, what will really happen will be an immediate splitting into halves and separation of two nuclei, travelling in opposite directions at speed v, attenuating as they progress according to (42d), and joined by a band, consisting of the two tails superimposed. The equation of this double-tail is

$$V = \tfrac{1}{2} V_0 a \,.\, \epsilon^{-Rt/2L} \,.\, \frac{1}{v}\left(\frac{R}{2L} + \frac{d}{dt}\right) J_0\left\{\frac{R}{2Lv}(x^2 - v^2t^2)^{\frac{1}{2}}\right\}, \quad \text{........(44}d\text{)}$$

in a finite form (as usually understood, by a convention that a solution in terms of a sine or J_0 function, etc., is in a finite form, though it is really an infinite series), true from $x = -vt$ to $x = +vt$, it being supposed that the origin of x was the original position of the charge. At the ends of this tail the two nuclei, each represented by

$$V = \tfrac{1}{2} V_0 \epsilon^{-Rt/2L} \quad \text{...........................(45}d\text{)}$$

through the very small distance a, must be placed, to make up the complete solution. I shall later illustrate this graphically, and also explain the other kind of tail.

Section XLV Effect of a Single Conducting Bridge on an Isolated Wave. Conservation of Current at the Bridge. Maximum Loss of Energy in Bridge-Coil, with Maximum Magnetic Force. Effect of any Number of Bridges, and of Uniformly Distributed Leakage. The Negative Tail. The Property of the Persistence of Momentum.

Let a distortionless circuit be bridged across anywhere by a wire whose conductance is k, and let us examine its effect on a wave passing along the circuit. In the first place, we may remark that we have already solved one bridge-problem, viz., the result due to an impressed force in the bridge itself, this being made a special case of the first part of the last Section, by limiting the number of radial circuits to two of the same type.

Now let V_1, V_2, and V_3 be the potential-differences in corresponding parts of an incident, reflected, and transmitted wave; V_1 going from left to right on the left side of the bridge, V_2 from right to left on the same side, and V_3 from left to right on the further side of the bridge.

At a certain moment these are coincident, viz., at the bridge itself. Then, by the properties of positive and negative waves and elementary principles, we have the following full connections:—

$$\left.\begin{array}{lll} V_1 = LvC_1, & V_2 = -LvC_2, & V_3 = LvC_3, \\ V_1 + V_2 = V_3, & C_1 + C_2 = C_3 + kV_3. & \end{array}\right\} \quad \text{(46d)}$$

From these we find

$$-\frac{V_2}{V_1} = \frac{C_2}{C_1} = \frac{k}{k+2Sv}, \qquad \frac{V_3}{V_1} = \frac{C_3}{C_1} = \frac{2Sv}{k+2Sv}. \quad \text{(47d)}$$

Particularly notice that

$$C_1 = C_2 + C_3, \quad \text{(48d)}$$

which, though extremely simple, is not by any means obvious at first sight, whilst it is an extremely important property. It is an example of the persistence of momentum; though this may not be immediately recognised, it will be made plain enough later on.

These equations should be compared with (36*d*), (37*d*), the corresponding ones relating to the effect of a resistance r inserted in the circuit. We see that this resistance is replaced by the conductance of the bridge, that L becomes S, and that V and C change places in the expressions for the ratios of the transmitted and reflected waves to the incident.

If we fix our attention upon the current, we see that every element of current, when it arrives at the bridge, is split into two, in the ratio of k to $2Sv$, or of $\frac{1}{2}Lv$ to k^{-1}, half the critical resistance to the resistance of the bridge. The first part is reflected, increasing the current on the left side, and lowering the potential-difference; whilst the other part is transmitted. The electrification in the reflected wave is negative, if that in the incident wave be positive; and conversely.

It may be as well here to remind the reader that from left to right is the arbitrarily assumed positive direction along the circuit, which is the direction of motion of a positive wave (therefore so-called); whilst a negative wave goes from right to left. Also, that the sign of the current, whether positive or negative, is a quite different thing. That is, the current in a positive wave may be negative, and the current in a negative wave may be positive, or the reverse. What is a possible source of some preliminary confusion is the fact that the vector we term the current, and the vector direction of motion of a wave, are in the same straight line, one way or the other. These connections are all summed up in $V_1 = LvC_1$, the property of a positive, and $V_2 = -LvC_2$, the property of a negative wave. If the first of these relations be true, the wave must move from left to right, whether V and C be both positive or both negative; whilst if the second be true, the wave must move from right to left. I can also recommend the reader to take the advice before given to fix his attention upon the electromagnetic field which is implied by a stated V and a stated C, viz., a field of electric displacement across the dielectric from one conductor to the other, and a field of magnetic induction round the conductors. A very useful purpose may perhaps be served by a careful study of the properties of the distortionless circuit, viz., to assist in abolishing the time-honoured but (in my opinion) essentially vicious practice of associating the electric current in a wire with the motion through the wire of a hypothetical *quasi*-substance, which is a pure invention that may well be dispensed with.

Returning to the effect of a bridge, notice that by the union of (48*d*) with the last of (46*d*), we produce

$$2C_2 = kV_3. \qquad \text{.............................(49}d\text{)}$$

That is, the current in the bridge equals twice the current in the reflected wave. The corresponding property when it is a resistance r inserted in the circuit that is in question is, by (36*d*), (37*d*),

$$2V_2 = rC_3; \qquad \text{...........................(50}d\text{)}$$

that is, the fall of potential through the resistance equals twice the difference of potential of the reflected wave.

If the bridge have no resistance, making a short-circuit (subject to reservations that need not be repeated), there is no transmitted wave. In fact, the case becomes identical with that of a terminal short-circuit, producing total reflection with reversal of electrification. If, on the other hand, the bridge have no conductance, it does nothing. If the conductance of the bridge be $2Sv$, or its resistance be $\frac{1}{2}Lv$, the transmitted wave is half the incident, or the attenuation due to the bridge is $\frac{1}{2}$. Then, by superimposition, the current on the left side is increased in the ratio 2 to 3, and is therefore made three times the transmitted current.

The current in the bridge being kV_3, and the corresponding heat per second divided by v being the heat due to the bridge per unit length of the incident wave, this amounts to

$$kV_3^2/v = 4S^2V_1^2kv/(k+2Sv)^2, \qquad \text{...................(51}d\text{)}$$

by (47*d*). If k be variable, we make the quantity in question a maximum when $k = 2Sv$, which is the above case of attenuation $\frac{1}{2}$. The heat in the bridge per unit length of the incident wave is then $\frac{1}{2}SV_1^2$, which is half its energy; the other half is equally divided between the transmitted and reflected waves.

If this bridge-wire be a coil of a given size and shape, the variation of k implies a variation of the thickness of the wire and of the number of turns. Whence, in a well-known manner, the magnetic force of the coil varies as the current in it and as the square root of its resistance; in another form, the square of the magnetic force varies as the product of the resistance of the coil into the square of the current, that is, as the heat per second. Hence, by what has just been said, the magnetic force is also a maximum when the resistance of the coil is $\frac{1}{2}Lv$. Notice that this is the impedance of the circuit as viewed from the coil itself. A correction is required for the inductance of the coil. It ought not, however, to be a very large correction, if it be a telephone that is in question, and of a really good type, having the smallest possible time-constant consistent with other necessary conditions. We require the magnetic force to be a maximum (*i.e.*, due to the current coming from the circuit) to make the stress-variations the greatest possible, and act most strongly on the disc. [*See* "Theory of Telephone," Art. XXXVI., vol. II.] Allowing for the inductance of the coil, if the currents be sinusoidal, we require equality of its impedance to that external to it, which is the general law.

Now let there be any number of bridges at different parts of the circuit, and let the ratio V_3/V_1 of a transmitted to an incident wave be denoted by s, its value being given by (47*d*), separately for each bridge. Let also ρ_1, ρ_2, etc., be the attenuations due to the circuit in the different sections into which it is divided by the bridges, and start with an isolated positive wave V_1 at A, the beginning of the first section. On arrival at the first bridge, it has attenuated to $V_1\rho_1$. What passes the bridge (not what crosses it) is $V_1\rho_1s_1$, which attenuates to $V_1\rho_1\rho_2s_1$ on arrival at the second bridge. Then there is another sudden attenuation, to $V_1\rho_1\rho_2s_1s_2$, followed by a gradual attenuation in the third section, to $V_1\rho_1\rho_2\rho_3s_1s_2$; and so on, to the end of the circuit, at B. The disturbance is then attenuated to $V_1\rho s_1s_2 \ldots s_n$; where ρ is the product of all the former ρ's, or the attenuation due to the circuit from A to B, and s_n is the last s, belonging to the bridge next to B. If the absorbing resistance Lv be put at B, it will at once absorb the wave just described; but after that there will come dribbling in and be absorbed the dregs of the original disturbance at A, arising from the complex system of small reflected waves due to the bridges across the circuit, much attenuated by the many to-and-fro journeys. But if there be but one bridge, and the absorbing resistance be put at A, to get rid of the wave reflected from the bridge, then there is no dribbling in at B.

However many bridges there be, there is, by (48*d*), no attenuation of current due to them, when its integral amount is considered, but only a redistribution of current. This exactly corresponds to the absence of any alteration of the total charge by inserting resistances in the circuit. They merely redistribute the charge.

If there be n bridges in the distance x, each of conductance k, the total attenuation produced by them is, by (47*d*),

$$s^n = \{1 + k/2Sv\}^{-n}. \qquad \text{(52}d\text{)}$$

Now place the bridges at equal distances apart, and increase the number n in the distance x indefinitely, keeping the total conductance constant, $=K_1x$, say. In the limit we shall arrive at a uniform distribution of leakage, K_1 being its conductance per unit length, and the attenuation due to it will be the limit of

$$\{1 + K_1x/2Svn\}^{-n}, \qquad \text{with} \qquad n = \infty,$$

$$= \epsilon^{-K_1x/2Sv}. \qquad \text{(53}d\text{)}$$

This is therefore the attenuation of the nucleus, when an initially isolated disturbance travels through the distance x, due to the extra leakage K_1 per unit length. There is, in addition, the regular attenuation due to the circuit. Disregard this for the present, by letting the circuit have no resistance and no leakage, that is, no leakage before the leakage represented by K_1 was introduced. Then we see that if there be initially an isolated disturbance represented by $V_0 = LvC_0$, extending through the very small distance a, it becomes, at the time x/v later, removed a distance x to the right, attenuated to (writing K for the leakage-conductance per unit length)

$$V = LvC = V_0\epsilon^{-Kx/2Sv}, \quad \text{..........} (54d)$$

extending through the distance a, with a tail of length $2x$ behind it. This tail is of the negative kind, the electrification being opposite in kind to that in the head, and is such that the line-integral of the current in it amounts to

$$C_0a(1 - \epsilon^{-Kx/2Sv}), \quad \text{..........} (55d)$$

because this, when added to the corresponding line-integral for the head, according to (54*d*), makes up C_0a, the initial value of the line-integral.

This tail is, as regards current, of the same shape as the corresponding tail due to resistance, as regards electrification, so its equation may be derived from (44*d*). But I shall consider the tails all together in a later Section.

The property involved in (48*d*), which leads to the deduction of (55*d*) from (54*d*), is worthy of notice. It is the persistence (or conservation) of momentum. If a circuit have no resistance, then, as Maxwell showed, we cannot change its momentum, the amount of induction passing through it. This was a linear circuit, with the current of the same strength all round it. Now our example is a remarkable extension of this property. Our circuit is linear and of no resistance, but it has any number of leaks, or conducting bridges, as well as what is equivalent to a series of condensers. The current in the circuit may be varied indefinitely in its distribution, but we cannot change its momentum. The line-integral of LC expresses the momentum, but since L is here a constant, of course the line-integral of C cannot change either. This property only continues true so long as there is no resistance bounding the magnetic field; therefore, if the circuit be of finite length, we must not insert resistances at the terminals. For instance, short-circuit at A and B, and we can at once say what will ultimately happen due to any initial distribution of current. It will settle down to uniformity of distribution, *i.e.*, making a uniform magnetic field, so that the strength of current will equal the original total momentum divided by the total inductance. There is, of course, a loss of energy in the settling down, due to the leakage. If the circuit be infinitely long, so that the disturbance can spread out infinitely, the total energy will decrease asymptotically to zero, in spite of the persistence of the momentum, which indeed tends to zero in any finite length, but keeps its total amount unchanged.

If the circuit have resistance, the *total* momentum decreases according to the time-factor $\epsilon^{-Rt/L}$, whatever be the initial distribution, if it be short-circuited at A and B, or be infinitely long. On the other hand, the total charge subsides according to the time-factor $\epsilon^{-Kt/S}$, if the circuit be *insulated* at A and B, or else be infinitely long. The meaning of terminal short-circuit or of insulation may clearly be extended to various other cases not involving loss of charge in the latter case (*e.g.* a terminal condenser) or of momentum in the former, with appropriate corresponding changes in the measure of S or L respectively.

SECTION XLVI. CANCELLING OF REFLECTION BY COMBINED RESISTANCE AND BRIDGE. GENERAL REMARKS. TRUE NATURE OF THE PROBLEM OF LONG-DISTANCE TELEPHONY. HOW NOT TO DO IT. NON-NECESSITY OF LEAKAGE TO REMOVE DISTORTION UNDER GOOD CIRCUMSTANCES, AND THE REASON. TAILS IN A DISTORTIONAL CIRCUIT. COMPLETE SOLUTIONS.

Having in Sections XLIV and XLV discussed in some detail the effects due to resistances inserted in, and also those due to conducting bridges across, a distortionless circuit, which are of fundamental importance, and which lead to the development of a positive tail by a continuous distribution of resistance in excess of the distortionless amount, and of a negative tail by an excess of leakage, the full investigation of the case of resistance and leakage combined in any proportions presents no difficulty.

Start with a circuit having no resistance and no leakage, which is therefore both distortionless and conservative (or characterised by the absence of attenuation), and let there be an isolated disturbance going from left to right, defined by $V_1 = LvC_1$. Also, let there be, at a certain place X, a bridge across the circuit, of conductance k; and, at the same place, a resistance r inserted in the circuit. When our incident wave V_1 arrives at X, there result a reflected wave represented by $V_2 = -LvC_2$, and a transmitted wave $V_3 = LvC_3$.

Now, considering the moment when these are all at X together (corresponding elements, of course), we have the following two equations connecting the three V's :—

$$V_1 + V_2 = V_3(1 + r/Lv), \qquad (56d)$$

$$C_1 + C_2 = C_3(1 + k/Sv + rk). \qquad (57d)$$

The first is simply the expression of Ohm's law applied to the resistance r, and the second expresses the continuity of the current at X. (Remember that Lv and Sv are reciprocal, so that the sum of the second and third terms on the right of (57d) expresses the bridge-current.) The equation (57d) may also be written

$$V_1 - V_2 = V_3(1 + k/Sv + rk), \qquad (58d)$$

so that, by adding this to (56d) first, and then subtracting it, we obtain the desired ratios. Thus,

$$V_1/V_3 = 1 + r/2Lv + k/2Sv + \tfrac{1}{2}rk, \qquad (59d)$$

$$V_2/V_3 = r/2Lv - k/2Sv - \tfrac{1}{2}rk, \qquad (60d)$$

when written in the simplest manner. Of course the ratio V_2/V_1, if wanted, is the quotient of (60d) by (59d).

We see that the reflected wave may be either of the same or of the opposite electrification to the incident; and that, in order to completely abolish the reflected wave, we require, by (60d),

$$r/Lv = k/Sv + rk, \qquad (61d)$$

and that we then have, by (59*d*),

$$V_1/V_3 = 1 + r/Lv \quad \text{.........................} \quad (62d)$$

simply. The reciprocal V_3/V_1 expresses the attenuation suffered by the incident wave in passing X.

The above equations are not in any way altered when we start with a real distortionless circuit instead of an imaginary one of no resistance. But by adopting the latter course we are directed to the nearest approach to a physical explanation of the properties of the real distortionless circuit itself. For, in the case of the circuit of no resistance we are dealing merely with progressive waves in a conservative medium, and we cannot expect to come to anything simpler than this. They simply carry their energy and all their properties forward at speed v unchanged, this speed being $(\mu c)^{-\frac{1}{2}}$, if μ be the inductivity and c the permittivity of the medium; which expression is equivalent to the other, $(LS)^{-\frac{1}{2}}$, where L is the inductance and S the permittance, which is more convenient in the practical application concerned. Except in the matter of wave-length, these waves are identical with light-waves, with the peculiarity that the two (supposed) perfect conductors of our circuit prevent the waves from spreading in space generally, by guiding them definitely along the circuit. (The simplest case is that of a tubular dielectric bounded by perfect conductors, say an internal wire and an external sheath.) Now we prove by elementary principles, (Ohm's law, etc.) that an inserted resistance, causing tangential dissipation of energy, produces a reflected wave of the positive kind, involving a redistribution, without loss, of the electrification on the bounding conductors; and a redistribution, with loss, of the corresponding magnetic quantity, the momentum. On the other hand, we show that a bridge causes a reflected wave of the negative kind, involving a redistribution, without loss, of the momentum; and a redistribution, with loss, of the electrification. (In speaking of redistribution, the mere translatory motion of waves is disregarded.) And by having both the bridge and the inserted resistance so proportioned as to make the loss of energy in each be of the same amount (when small enough), we abolish the reflected wave, so that there is no redistribution, but merely attenuation produced by the resistance and bridge. This applies to any number of resistances inserted in the main circuit, each with its corresponding bridge; so that when we pack them infinitely closely together to represent continuously distributed resistance and leakage, we arrive at a real circuit, along which waves are propagated unchanged except in size. Thus any circuit (apart from interferences) may be made distortionless by adding a suitable amount of leakage. This amount is usually too great for practical purposes. Nor is it required. In the very important problem of long-distance telephony, employing circuits of *low* resistance (which are the only proper things to use), making the well-known ratio R/Ln of the two components of the electromagnetic impedance small, say $\frac{1}{2}$ or $\frac{1}{4}$, which may be easily done without using an extravagant amount of copper, we tend naturally, by bringing the inductance into relative importance, or equivalently, reducing the importance of the factor resistance, to a state of things resembling that which obtains in the truly distortionless circuit (independent of frequency of variations), and approximate to distortionless transmission. These statements may be proved by an inspection of the sinusoidal solutions I have given, but it would enlarge the subject too greatly to discuss them at present. I may, however, repeat that the problem of long-distance telephony is very remote from that of a long submarine cable which can only be worked slowly, unless we should unknowingly create a parallelism by employing quite unsuitable conductors; as, for instance, was done by the Post Office a few years since when they put down conductors having a resistance of 45 ohms per mile of circuit, combined with large permittance and small inductance; and then, to make the violation of electromagnetic principles more complete, put the intermediate apparatus in sequence, so as to introduce as much additional impedance as possible. The proper place for intermediate apparatus is in bridge, removing all their impedance completely. This method was invented and introduced into the Post Office by Mr. A. W. Heaviside. It makes a wonderful difference in the capabilities of a circuit, as is now pretty well known.

The theory of tails allows us to give an intelligible physical explanation of how it comes to pass that a perfectly insulated circuit violating the distortionless condition completely, will yet tend to behave in a distortionless manner to waves of great frequency, provided the circuit be of a suitable nature, as above described. For let the circuit be so long that we can get several waves into it at once, when telephoning. They divide the circuit into regions of opposite electrification, each of which may (very roughly) represent what I have termed an isolated disturbance. Every one of them has its tail, but as they are alternately of opposite kinds, their residual effect in producing distortion becomes quite small. We can see clearly that the greater the frequency the less is the distortion, unless the increased frequency should bring with it increased resistance, which is very much to be avoided, and is what renders iron wire so unsuitable for *long*-distance telephony. By this mutual cancelling of the effects of the tails, we simulate the effect of the leakage which would wholly remove distortion, even of the biggest waves, without the disadvantage of the extra attenuation thereby introduced. I am induced to make these remarks rather out of their proper place, as they illustrate the importance of the distortionless circuit from the scientific point of view, in casting light upon the obscurities of distortional circuits.

From (59*d*) we can get some results relating to the tails of waves in a distortional circuit. Thus, let there be n bridges in the distance x, equidistantly placed, and each of conductance Kx/n, with a corresponding resistance Rx/n in the main circuit. Let a disturbance pass from beginning to end of the length x. If σ be the attenuation at each bridge, the total attenuation of the head of the disturbance produced by all the bridges and resistances is σ^n. Now make n infinite, keeping R and K finite. The total attenuation becomes, by (59*d*),

$$\sigma^n = \{1 + Rx/2Lvn + Kx/2Svn + RKx^2/2n^2\}^{-n} = \epsilon^{-Rx/2Lv}\epsilon^{-Kx/2Sv} \qquad (63d)$$

This is therefore the attenuation of the head suffered by every element in traversing the distance x, when R and K are the resistance and the leakage-conductance per unit length in any uniform circuit.

It will now be convenient to introduce a simpler mode of expressing the exponentials. Let

$$f=R/2L, \qquad g=K/2S, \qquad h=f-g, \qquad q=f+g, \quad \text{......(64}d\text{)}$$

all four being reciprocals of time-constants. Now (63d) becomes ϵ^{-qt} simply, if $t=x/v$ be the time of the journey over the length x. If, therefore, we have initially a disturbance $V_0=LvC_0$ extending through the small distance a, possessing the charge SV_0a and the momentum LC_0a, then, at the time t later, when the disturbance extends over the distance $2x$, half on each side of its initial position, being a nucleus of length a and a tail of length $2x$, the charge and momentum in the nucleus become

$$SV_0a\,\epsilon^{-qt} \quad \text{and} \quad LC_0a\,\epsilon^{-qt}. \quad \text{...............(65}d\text{)}$$

We have next to examine to what extent the total charge has attenuated by the leakage, and the total momentum by the resistance. This we can ascertain by (59d) and (60d), applied to find the loss of electrification caused by a single bridge, and of momentum by a single resistance. Those equations give

$$\frac{V_2+V_3}{V_1}=\frac{1+r/2Lv-k/2Sv-rk/2}{1+r/2Lv+k/2Sv+rk/2}, \quad \text{...........(66}d\text{)}$$

$$\frac{C_2+C_3}{C_1}=\frac{V_3-V_2}{V_1}=\frac{1+k/2Sv+\frac{1}{2}rk-r/2Lv}{1+k/2Sv+\frac{1}{2}rk+r/2Lv}. \quad \text{...........(67}d\text{)}$$

These fractions, multiplied into the values of the charge and momentum respectively before the splitting, give their *total* values after the splitting. We can, therefore, apply the previous method of equidistant resistances and bridges, to ascertain the method of subsidence of the total charge and momentum, in the infinitely numerous splittings that occur in a finite time, when we pass to the limit and have uniform R and K. Putting $r=Rx/n$, etc., as before, and finding the limit of the nth powers of (66d) and (67d), we arrive at $\epsilon^{-Rt/L}$ and $\epsilon^{-Kt/S}$ respectively.

We thus see that a moving charge, no matter how it redistributes itself, subsides at the same rate as if it were at rest; for, obviously, S/K is the time-constant of the circuit regarded as a condenser, when uniformly charged and insulated at its terminations. It is *as if* electricity were atomic, so that we could follow the course of every particle. Then, no matter how it moves about, it *shrinks* at the same rate as if it were at rest. Similarly as regards the momentum of the moving disturbance. Could we identify *its* elements, each would shrink in a manner independent of its translatory motions along the circuit. Notice, also, that the attenuation of the total charge equals the square of the attenuation of the nucleus due to leakage alone; whilst the attenuation of the total momentum equals the square of the attenuation of the nucleus due to resistance alone.

Thus, corresponding to (65*d*), we have

$$SV_0a.\epsilon^{-gt}(\epsilon^{-gt}-\epsilon^{-ft}) \qquad \text{and} \qquad LC_0a.\epsilon^{-ft}(\epsilon^{-ft}-\epsilon^{-gt}) \quad \ldots(68d)$$

to express the charge and momentum in the tail; since these, when added to (65*d*), make up the actual values otherwise found, viz.,

$$SV_0a.\epsilon^{-2gt} \qquad \text{and} \qquad LC_0a.\epsilon^{-2ft}.$$

If $f>g$, or the resistance be in excess, the current in the tail is from head to tip, if that in the head be positive. But as time goes on, if the circuit be long enough, the head attenuates practically to nothing, leaving the big tail to work with. The region of positive current now extends from the vanishing nucleus a long way towards the middle of the tail; and, in the limit, the disturbance tends to become symmetrically arranged with respect to the origin from which it started as a positive wave, tailing off on both sides, with the current positive on one side and negative on the other.

But if $f<g$, or the leakage be in excess, a quite anomalous state of affairs occurs, which may be inferred from the preceding by changing V to C, etc.

The full solutions of all tail-problems (shape, growth, etc.) are contained in the following four equations. Let a charge SV_0a be at the origin at time $t=0$, without any current. At time t we shall have, if $y=(h/v)(x^2-v^2t^2)^{\frac{1}{2}}$,

$$V=\tfrac{1}{2}\frac{V_0a}{v}\epsilon^{-qt}\left(h+\frac{d}{dt}\right)J_0(y), \quad \ldots\ldots(69d)$$

$$C=\tfrac{1}{2}\frac{V_0a}{Lv}\epsilon^{-qt}\left(-\frac{d}{dx}\right)J_0(y), \quad \ldots\ldots(70d)$$

to express the double-tail or band connecting the two nuclei at its ends, which are already known. Similarly, if there be initially a current at the origin, of momentum LC_0a, without charge, then at time t we shall have

$$C=\tfrac{1}{2}\frac{C_0a}{v}\epsilon^{-qt}\left(-h+\frac{d}{dt}\right)J_0(y), \quad \ldots\ldots(71d)$$

$$V=\tfrac{1}{2}\frac{C_0a}{Sv}\epsilon^{-qt}\left(-\frac{d}{dx}\right)J_0(y). \quad \ldots\ldots(72d)$$

As before, put on the two nuclei at the ends. Since the J_0 function is a simple one, viz.,

$$J_0(y)=1-\frac{y^2}{2^2}+\frac{y^4}{2^24^2}-\frac{y^6}{2^24^26^2}+\ldots,$$

it is quite easy to follow the changes of shape by these formulæ, except when t has become large and the nuclei small, when other formulæ may be derived from the above which will approximately suit. [For further information, see Part VIII. of Art. XL., Part I. of "Electromagnetic Waves," and "The General Solution of Maxwell's Equations."]

SECTION XLVII. TWO DISTORTIONLESS CIRCUITS OF DIFFERENT TYPES IN SEQUENCE. PERSISTENCE OF ELECTRIFICATION, MOMENTUM, AND ENERGY. ABOLITION OF REFLECTION BY EQUALITY OF IMPEDANCES. DIVISION OF A DISTURBANCE BETWEEN SEVERAL CIRCUITS. CIRCUIT IN WHICH THE SPEED OF THE CURRENT AND THE RATE OF ATTENUATION ARE VARIABLE, WITHOUT ANY TAILING OR DISTORTION IN RECEPTION.

If two distortionless circuits of different types be joined in sequence, a wave passing along one of them will, on arrival at the junction, be usually split into two, a transmitted and a reflected wave. Let, in the former notation, V_1, V_2, V_3 denote the potential-differences in corresponding elements of the incident wave in the first circuit, the reflected wave in the same, and the transmitted wave in the second circuit. The sole conditions at the junction are that V and C shall not change in passing through it. Thus,

$$V_1 + V_2 = V_3, \qquad C_1 + C_2 = C_3. \quad \text{................(73}d\text{)}$$

Now let L_1v_1 and L_2v_2 be the impedances of the two circuits, L_1 and L_2 being the inductances per unit length, and v_1, v_2 the speeds of the current. Put the first of (73d) in terms of the currents. Thus,

$$L_1v_1(C_1 - C_2) = L_2v_2C_3 \, ; \quad \text{......................(74}d\text{)}$$

showing that the momentum of the incident disturbance equals the sum of the momenta of the reflected and transmitted disturbances. Corresponding lengths are compared, of course, proportional to the speed of the current. The condition of continuity of V is therefore identical with that of persistence of momentum.

Next, put the second of (73d) in terms of potential-differences. Thus,

$$S_1v_1(V_1 - V_2) = S_2v_2V_3 \, ; \quad \text{......................(75}d\text{)}$$

which expresses that the electrification suffers no loss by the splitting. The condition of continuity of C is therefore equivalent to that of the persistence of electrification.

Multiply the first of (73d) into (75d); the second of (73d) into (74d); the two members of (73d) together; and (74d) into (75d). The results are

$$\left.\begin{array}{c} S_1v_1(V_1^2 - V_2^2) = S_2v_2V_3^2, \qquad L_1v_1(C_1^2 - C_2^2) = L_2v_2C_3^2, \\ V_1C_1 + V_2C_2 = V_3C_3 \, ; \end{array}\right\} \text{.....(76}d\text{)}$$

which are equivalent expressions of the fact of persistence of energy, while the last of (76d) is the equation of transfer of energy. That it should be equivalent to the others will be understood on remembering that the energy is transferred at speed v_1 or v_2, according to position.

We have, therefore, three things that persist, electrification, momentum, and energy, and these are expressed most simply by the two equations (73d) and by their product. If the continuity of V could be violated at the surface across the dielectric common to the two circuits at their junction, there would be a surface magnetic-current; and if the continuity of C could be violated, there would be a surface electric-current. These statements are implied in the general equations

$$-\operatorname{curl}\mathbf{E}=4\pi\mathbf{G}, \qquad \operatorname{curl}\mathbf{H}=4\pi\Gamma, \qquad \text{(77d)}$$

where **E** and **H** are the electric and magnetic forces, Γ and **G** the electric and magnetic currents. That is, tangential continuity of **E** implies normal continuity of **G** (or of the induction, since it, like **G**, can have no divergence); and tangential continuity of **H** implies normal continuity of Γ, and therefore, in our special case, of electrification. In fact (73*d*) express the same facts as (77*d*) do generally.

Now the continuity of V and C is violated at the boundaries of an isolated disturbance (*e.g.*, V = constant in a certain part of the circuit, and zero before and behind). Then we do have the surface electric and magnetic currents on the front and back of the disturbance. It should, however, be stated that the conception of an isolated disturbance is merely employed for convenience of description and argument. Practically, there cannot be abrupt discontinuities; we must make them gradual. Then the surface-currents become real, with finite volume-densities.

The ratio of the reflected to the incident wave is given by

$$V_2/V_1=(L_2v_2-L_1v_1)/(L_2v_2+L_1v_1), \qquad \text{(78d)}$$

and is positive or negative according as the impedance of the second circuit is greater or less than that of the first. The abolition of reflection is therefore secured by equality of impedances, irrespective of any change of type that does not conflict with this equality. Every element of the transmitted wave therefore carries forward, in passing the junction, its potential-difference, current, electrification, momentum and energy unchanged, but is changed in length in the same ratio (inversely) as the speed of the current is changed.

In a similar manner, we can determine fully what happens when a disturbance travelling along one distortionless circuit is caused to divide between any number of others, of any types. We have merely to ascertain the magnitude of the reflected wave in the first circuit. Let V_1 and U_1 be the incident and reflected waves. Then, corresponding to (78*d*), we shall have

$$U_1/V_1=(I-L_1v_1)/(I+L_1v_1), \qquad \text{(79d)}$$

where I is the resultant impedance of all the other circuits (instead of L_2v_2, that of one only), viz. the reciprocal of the sum of the reciprocals of their separate impedances. Knowing thus U_1 in terms of V_1, we know their sum. But this is the common potential-difference in all the transmitted waves, which are therefore known, since by dividing by the impedance of any circuit we find the current. As regards the attenuation as the disturbances travel away from the junction, that must be separately reckoned for each circuit, according to the value of R/L, in the way before described. There will be found to be the previously-mentioned persistences, provided all the waves are counted, including the reflected in the first circuit.

Now put any number of distortionless circuits in sequence. If their impedances be equal, we know, by the above, that a disturbance will travel from end to end without any reflection at the junctions. It will vary in its length and in its speed, and also in the rate at which it attenuates, but there will be no tailing, however many changes there may be in the values of R and L. By pushing this to the limit, we arrive at a circuit in which R and L vary in an arbitrary manner (functions of x), whilst K varies in the same way as R, and S in the same way as L. The impedance is a constant, but the rate of attenuation and the speed vary in different parts of the circuit.

If we start an isolated disturbance at one end, it will travel to the other without tailing. But it will be distorted on the journey, owing to the variable speed of its different parts and the variable attenuation. But as regards the reception of the wave, there is no distortion whatever. For, on arrival at the distant end, where we may place the absorbing resistance, every element of the wave has gone through the same ordeal precisely, passing over the same resistances in the same sequence and at the same speed at corresponding places, so as to arrive at the distant end in the same time, attenuated to the same extent. Similarly there is no intermediate distortion as regards the succession of values of V and C at any one spot. There is only distortion when it is the wave as a whole that is looked at, comparing its state at one instant with that at another. And if we should cause this wave to start in a uniform circuit, then pass into an irregular one as just described, and finally emerge in a uniform circuit again, it will then have recovered its original shape, every part being attenuated to the same extent.

As regards the time taken to pass over a distance x in the variable circuit, we have to solve the kinematical problem: given the path of a particle, and its speed at every point, find the time t taken. Thus,

$$t = \int \frac{dx}{v},$$

taken between the proper limits, wherein v is to be a function of x. The attenuation suffered in this journey is more easily expressed. Go back to the former case of any number of uniform distortionless circuits of equal impedance joined in sequence. The attenuation produced in passing through any number of them is the product of their separate attenuations, *i.e.*,

$$\epsilon^{-R_1/Lv} \times \epsilon^{-R_2/Lv} \times \ldots = \epsilon^{-\Sigma R/Lv}, \qquad \text{.....................(80}d\text{)}$$

where R_1, R_2, ..., are the resistances of the separate sections, and Lv the common value of the impedances. As this is independent of the number of sections or their closeness, we see that in our variable circuit the attenuation in any distance is expressed by the right member of (80d), wherein ΣR represents the total resistance of the circuit in that distance, or $\int R dx$ between the proper limits, R being a function of x.

The above-given demonstration of the properties of the variable distortionless circuit, which is rather a curiosity, depends entirely upon our previous proof that the abolition of reflection at the junction of a pair of simple distortionless circuits is obtained by equality of impedances, irrespective of any change that may take place in the resistances. The following is also of some use. Go back to the fundamental equations

$$-\nabla V=(R+Lp)C, \qquad -\nabla C=(K+Sp)V; \quad \text{.........(81}d\text{)}$$

wherein ∇ means d/dx, and p means d/dt. Now *assume* $V=LvC$, making them become

$$\left.\begin{aligned}-\nabla(LvC)&=L\,(R/L+p)C,\\ -\nabla C\quad&=S\,(K/S+p)(LvC).\end{aligned}\right\}\quad\text{..................(82}d\text{)}$$

If our assumption can be justified, these equations must become identical. They do become identical if $R/L=K/S$, and $Lv=$ constant; becoming

$$-v\nabla V=(R/L+p)V. \quad \text{.......................(83}d\text{)}$$

This is for the positive wave. The assumption $V=-LvC$ again makes (81d) identical under the same conditions, the resulting equation being (83d) with the sign of v changed. The necessary conditions may be written

$$R/K=L/S=(Lv)^2=\text{constant}; \quad \text{.................(84}d\text{)}$$

and since we have made no assumption as to the constancy of R, L, K, and S, we see that R and L are left arbitrary, any functions of x. Or, what comes to the same thing, R/L and v are arbitrary, making the attenuation and the speed variable, but without any tailing.

A third way is to examine what happens when we place a bridge of conductance k across the junction of two distortionless circuits of different types, but of the same impedance, along with a resistance r in the circuit at the same place. The two conditions, using the former notation, are

$$\left.\begin{aligned}V_1+V_2&=(1+r/Lv)V_3,\\ V_1-V_2&=\{1+(k/Sv)(1+r/Lv)\}V_3;\end{aligned}\right\}\quad\text{..............(85}d\text{)}$$

from which,

$$\left.\begin{aligned}V_1/V_3&=1+r/2Lv+(k/2Sv)(1+r/Lv),\\ V_2/V_3&=r/2Lv-(k/2Sv)(1+r/Lv),\end{aligned}\right\}\quad\text{............(86}d\text{)}$$

which give the ratios of incident and reflected to transmitted wave. We destroy the reflection by

$$r/Lv=k/Sv+rk,$$

and then the attenuation is

$$V_3/V_1=(1+r/Lv)^{-1},$$

due to r and k. An infinite number of these r's and k's in succession, placed infinitely close together, leads to the expression (80d).

We can also go a little way towards finding what occurs when the only condition is $Lv=$ constant, so that there is tailing. For we then have, at a single junction,

$$V_3/V_1=(1+r/2Lv)^{-1}(1+k/2Sv)^{-1};$$

and therefore, when the distribution of r and k is made continuous, the attenuation of the head of a disturbance in passing through any distance is

$$\epsilon^{-R_1/2Lv} \times \epsilon^{-K_1/2Sv},$$

if R_1 be the total resistance and K_1 the total conductance of the leakage in that part of the circuit. But we cannot similarly estimate to what extent the total charge and momentum have attenuated, as we could when the circuit was uniform, because the attenuation now occurs at a different rate in different parts of the tail, and we are not able to trace the paths followed by the different parts of a charge as it splits up repeatedly. The determination of the exact shape of the tail is of course an infinitely more difficult matter. But an approximation may be obtained by easy numerical calculations, if we concentrate the resistance and leakage in a succession of points.

NOTE (Nov. 30, 1887).—The author much regrets to be unable to continue these articles in fulfilment of Section XL., having been requested to discontinue them.

XXXVI. SOME NOTES ON THE THEORY OF THE TELEPHONE, AND ON HYSTERESIS.

[*The Electrician*, Feb. 11, 1887, p. 302.]

As was found in the early days of the telephone, its cores need to be permanently magnetised before it becomes efficient. I refer, of course, to the ordinary magnetic telephone, in which an iron disc is attracted by an electromagnet, which does not differ essentially from a common Morse instrument with a flexible armature, with the important addition that the electromagnet is permanently polarised. The permanent magnetisation may be communicated by a permanent current in the circuit, or, in the usual way, by employing a permanent magnet on whose pole or poles the coils are placed. But the permanent magnetisation, except of the iron disc, is not essential. Thus we may abolish the magnet and core from the telephone, leaving only the coil and disc, and produce the necessary permanent field of force by means of an external magnet suitably placed. The efficiency is then greatly increased by inserting a soft-iron core in the coil. Similarly, we may destroy the efficiency of a complete telephone by an external magnet, or we may increase it, by suitably placing the external magnet so as to, in the first place decrease, and in the second increase the strength of the permanent magnetic field. And if we carry the destruction of the magnetic field by the external magnet so far as to reverse it, and bring it on again strongly enough, we restore the efficiency of the telephone. That is, the permanent polarity may be of either kind. The disc is strongly magnetically attracted in either case, and that is the really essential thing. Most of these facts, if not all, are pretty well known, but it appears to be different as regards their explanation.

A good many years ago I read in Mr. Prescott's work on "The Telephone," an article by Mr. Elisha Gray on the subject, containing some of the above facts, and, in particular, describing the effect of a permanent current in the circuit. He looked upon the necessity of a permanent field of force as a great mystery, and suggested some reasons for its necessity that appeared to me to be unwarranted and inadequate. I now observe that Professor S. P. Thompson, in his recent paper, "Telephonic Investigations," remarks upon this question (*The Electrician*, Feb. 4, 1887, pp. 290, 291). Whilst not explaining the necessity of a permanent field, he brings in to complicate the thing such matters as hysteresis and the curve of induction referred to magnetic force, which do not appear to be materially concerned. I have very little acquaintance with telephonic literature, and, therefore, it may happen that the following explanation has been already well threshed out, and accepted or proved to be erroneous, as the case may be; but the perusal of the remarks of the above authority has suggested to me that the following explanation may be not only generally useful, but even absolutely novel to many of my readers.

The stress between the iron disc and the poles of the electromagnet varies, under similar circumstances, as the square of the intensity of magnetic force in the space between them. There is no occasion to consider the relative intensity in different places, or to perform integrations, as we have merely to deal with the fundamental fact of the stress on the diaphragm varying as the square of the magnetic force. Now, as we cause this diaphragm to execute forced vibrations by varying the stress upon it, we should make the variations of stress as great as possible in order to obtain the greatest amplitude of vibration, and the greatest intensity of sound from it.

Suppose, then, that there is a permanent field of intensity H, producing a steady stress proportional to H^2, and that we vary the stress by means of the magnetic force of undulatory currents in the coils. Let h be the amplitude of the undulations of magnetic force, small in comparison with H, so that we vary the real magnetic force from $H-h$ to $H+h$, through the range $2h$. This is quite independent of H, so that if it were a mere question of the intensity of magnetic force, we could just as well do without the permanent field, except for a reason to be mentioned later. But the stress varies from being proportional to $(H-h)^2$ to $(H+h)^2$; or the range is $4Hh$, not troubling about any constant multiplier. That is, the stress-variation is proportional to the *product* of the intensity of the permanent magnetic force into that of the undulatory magnetic force. This contains the explanation.

We see at once that it is in at least approximate agreement with facts. For, with the same weak undulatory current passing, which keeps h constant, we know that the intensity of sound continuously increases as we increase the intensity of the permanent field. And, keeping the permanent field the same, we know that the intensity of sound continuously increases as we increase the amplitude of the current-undulations, and therefore h. The question of exact proportionality is an independent one. We have got already what appears to be the main explanation.

Now to consider some other points. It has been assumed for simplicity that H was several times h. In a telephone H is a very large multiple of h under ordinary circumstances. But as H is reduced, or h increased sufficiently, the effects change. Thus, if $H=h$, the magnetic force varies from 0 to $2h$, and the stress from 0 to $(2h)^2$. And if H is less than h, the magnetic force varies from a negative to a positive value, whilst the stress varies from a positive value through zero to another positive value. In the extreme, when the permanent field is altogether abolished, whilst the magnetic force varies from $-h$ to $+h$, the stress varies from h^2 through zero to h^2 again. The disc is therefore urged to execute vibrations of double the frequency of the current-undulations. It is similar to sending reversals through a Morse instrument, when the armature will make a rap for every current, positive or negative, or two raps for every complete wave. This alone would be, I think, a serious hindrance to getting good speech from a magnetic telephone without a permanent field. But, with ordinary speaking currents, the double vibrations, in the absence of the permanent field, are insensible. On the other hand, when we put on the strong permanent field they are non-existent, *i.e.*, in the stress-variations, as there is no reversal of the magnetic field, but only a change in its intensity.

But we may easily examine the effect of h alone, or in combination with H of a similar strength, by means of a vibrating microphone sensitively set, producing a very large variation of current in the circuit of battery, microphone, and telephone. Here the current is equivalent to the co-existence of a permanent current and of an undulatory current, and the latter may be made not insignificant compared with the former, but even $\frac{1}{4}$ or $\frac{1}{3}$ its strength. It is not a matter of indifference now which way the current goes. In one case the permanent current increases, and in the other it decreases the permanent magnetic field of the magnet, producing corresponding changes in the intensity of the sound. We may cancel the permanent field by an external strong magnet, approximately, or make H small compared with h. Then the disc is attracted both when the current is above and when it is below its mean strength.

We cannot increase the efficiency of a telephone indefinitely by multiplying the intensity of the permanent field. In the first place, the disc becomes stiffened under strong attraction, so that ultimately a large increase in the stress makes little difference in its displacement. Again, when the core is very strongly magnetised, we may expect that the effective inductivity of the core, so far as variations in the magnetic force are concerned, will be reduced, so that undulations of current of given amplitude will not continue to produce stress-variations proportional to the amplitude of the current.

There are many other things concerned, of course, between the stress-variation and the intensity of sound, especially mechanical; as, for instance, the multiplication in the intensity of certain tones, especially the fundamental of the disc, which has also the disagreeable result of keeping up a sound after it should have ceased.

The application of the preceding is not merely to the telephone, but to various electromagnetic instruments. I frequently make use of the multiplying power of a permanent magnetic field. For example, to make a trembler-bell go with a weak current; or to make an electromagnetic intermitter go firmly with a current that, unassisted, would do nothing. Then a strong permanent magnet takes the place of a strong permanent current. It should be so placed as to increase the strength of field due to the electromagnet.

In the other way of getting power, by having a movable coil in a strong permanent field, first done, I believe, by Mr. Gott in 1877 (*Journal* S.T.E., Vol. V., p. 500), the action is different, as it is the electromagnetic force on the moving coil that is operative. There is no stress on the coil when no current passes in it. But when a current passes, the torque may be taken to be proportional to the intensity of the permanent field and to the current passing, as in the other case.

In conclusion, a few words, from my own point of view, of course, on the subject of the hysteresis which has lately become prominent, and which has been, perhaps, rather overdone by some writers. It is, substantially, an old thing in a new dress. Iron exposed to magnetising force usually, perhaps always, more or less, becomes magnetised intrinsically as well as elastically, just as ductility is probably always in action to some extent in a strained elastic spring. Thus, in changing the elastic magnetisation, which does not involve any recognised or as yet recognisable dissipation of energy, we change the intrinsic magnetisation, which does. But that there is no sensible dissipation of energy in an iron core placed in a rapidly intermittent or undulatory magnetic field of moderate strength I assured myself of experimentally some years ago, as I mentioned in *The Electrician* for June 14, 1884 [vol. I., p. 370]. I repeated the experiments in a far more effective form last year (*The Electrician*, April 23, 1886), [vol. II., p. 43]. The method is very simple and obvious, being merely to show that iron, when sufficiently divided, is exactly equivalent to self-induction. Use the differential telephone, or the Bridge. The former is a handy little thing, but the latter is much more adaptable and generally useful. Take two coils of the same resistance but of widely different inductances, and complete the balance by making up the deficit with iron. If sufficiently divided, the changed resistance due to dissipation in the iron vanishes or becomes exceedingly small. I formerly used a bundle of the finest iron wires I could get, and the residual effect was small.

In the repetition I used iron dust, worked up with wax into solid cores (1 wax to 5 or 6 iron by bulk), and the residual effect is far smaller, scarcely recognisable. But if the magnetising force be made stronger there is a small increased resistance, which can hardly be due to the Foucault or Farrago currents in the insulated dust. It is possibly due to hysteresis. But at the same time the variation in the inductivity is recognisable, so that the effect is complex. It is clear that in the case of telephone-speaking currents, dissipation (except F.) is nowhere, whether the core be permanently magnetised or not.

We require strong forces to make hysteresis important. Even then it is probable that when the variations of force are very rapid (undulatory, not with jerks) dissipation due to hysteresis may be considerably reduced, and the results of Ewing and Hopkinson not be applicable.

XXXVII. ELECTROSTATIC CAPACITY OF OVERGROUND WIRES.

[*The Electrician*, Sept. 25, 1885, p. 375.]

In the late Prof. F. Jenkin's "Electricity and Magnetism" (p. 332, first edition) is a formula for the capacity of an overhead wire. Owing to the remark there made, that experiment gave results nearly double as great as the formula, which was attributed by him to induction between the wires and the posts and insulating supports, and thinking that the presence of neighbouring wires should have a marked influence in increasing the capacity, owing to the neighbouring wires being earthed, I verified this by working out the theoretical formulæ for the capacities (self and mutual) of overground parallel wires, and applying them numerically in a special case. [Vol. I., Art. XII., p. 42.] With one additional parallel wire the increase of the capacity of the first was 11 per cent.; with three additionals it was 24 per cent. As to further increase by more wires, it would not be *very* great, as they would be practically much further away. As a guess, it might run up to 50 or 60 per cent., with a large number of wires, but of course it would depend materially upon their mutual distances and height above the ground.

The recent measurements of capacities of wires in the North of England supply some definite information. Taking the case of a wire 20 feet above the ground, of diameter ·08 inch, the calculated capacity, supposing there to be no other wires (nor trees, etc.), is ·0095 mcf. per mile. The average result observed is given in Mr. Preece's paper (*The Electrician*, Sept. 18, 1885, p. 348) as ·0120 with the other wires insulated, and ·0142 when earthed. And for the iron wire, ·171 inch diameter, supposed 20 feet above the ground, the similar three results are ·0103, ·0131, and ·0169. I take $v = 30^{10}$ instead of the 2880^8 centim. used in the paper referred to [vol. I., p. 44].

In both cases we may observe that the experimental result with wires insulated is about midway between the calculated result and the experimental result with wires earthed; so that it would appear that the influence of surrounding objects (other than neighbouring wires earthed) in increasing the capacity was about equal to that of the neighbouring wires themselves. This might, of course, be true in some particular case, but we cannot safely conclude it from the above, on account of leakage, as may be seen thus. If the wire experimented on were perfectly insulated from earth through the poles, whilst the other wires (though insulated at the ends) were so very badly insulated at the poles that they could be considered as connected to earth, it is clear that a measurement of capacity of the first wire would give the highest result. And this would be true with fair insulation, if the total charge could be observed. But when the observation is made by throw of needle, only a part of the charge is observed, the remainder (due to the leakage of the neighbouring wires) going in slowly, or coming out slowly when discharge is taken. In any case, however, the effect of the imperfect insulation of the neighbouring wires is to make the apparent capacity greater, and so reduce the difference between the capacity with wires insulated and to earth. Thus, bettering the insulation would shift the middle results above given towards the lower.

How far this operates might perhaps be experimentally determined by charging the first wire with the others insulated, then waiting a little, and observing the extra charge produced by suddenly earthing the other wires. If the insulation be bad, the extra charge will be *nil*; if first-rate, it might amount to nearly the full difference.

XXXVIII. MR. W. H. PREECE ON THE SELF-INDUCTION OF WIRES.

[Sept. 24, 1887; but now first published.]

A VERY remarkable paper "On the Coefficient of Self-Induction of Iron and Copper Telegraph Wires" was read at the recent meeting of the B. A. by William Henry Preece, F.R.S., the eminent electrician. This paper will be found in *The Electrician*, Sept. 16, 1887, p. 400. It contains an account of the latest researches of this scientist on this important subject, and of his conclusions therefrom. The fact that it emanates from one who is—as the *Daily News* happily expressed it in its preliminary announcement of Mr. Preece's papers—one of the acknowledged masters of his subject, would alone be sufficient to recommend this paper to the attention of all electricians. But there is an additional reason of even greater weight. The results and the reasoning are of so surprising a character that one of two things must follow. Either, firstly, the accepted theory of electromagnetism must be most profoundly modified; or, secondly, the views expressed by Mr. Preece in his paper are profoundly erroneous. Which of these alternatives to adopt has been to me a matter of the most serious and even anxious consideration. I have been forced finally to the conclusion that electromagnetic theory is right, and consequently, that Mr. Preece is wrong, not merely in some points of detail, but radically wrong, generally speaking, in methods, reasoning, results, and conclusions. To show that this is the case, I propose to make a few remarks on the paper.

It will be remembered that Mr. Preece, in spite of the well-known influence of resistance in lowering the speed of signalling, was formerly an advocate of thin wires of high resistance for telephony; but that, perhaps taught by costly failures in his own department, and by the experience of more advanced Americans and Continentals who had signally succeeded with wires of low resistance, he recently signified his conversion. Along with this, however, it will be remembered that, although it had been previously shown how very different the theory of the rapid undulatory currents of telephony is from the electrostatic theory of the submarine cable, he adopted rather pronouncedly what should, it appears, be understood to be the electrostatic theory, with full application to telephony. It is not to be presumed that Mr. Preece meant to deny the existence of magnetic induction, but that he meant to assert that it was of so little moment as to be negligible. It will also be remembered that his views were rather severely criticised by Prof. S. P. Thompson, and that Prof. Ayrton and others pointed out that he had not treated the telephonic problem at all. More recently still, it may be remembered by the readers of this journal that it has been endeavoured to explain how and why the electrostatic theory has so limited an application to telephony. (E. M. I. and its P., Section XL. *et seq.*) [vol. II., pp. 119 to 155.] Nothing daunted, however, Mr. Preece now, although to some extent modifying his views as regards iron wires, maintains that self-induction is negligible in copper-wire circuits; and in fact, on the basis of his latest researches, asks us to believe that the inductance of a copper circuit is *several hundred times smaller* than what it is maintained to be by experimental theorists, and is really quite negligible in consequence. His paper is devoted to proving this. It is necessary to examine it in detail.

(1). Mr. Preece finds the inductance of a certain iron wire to be ·00504 macs per mile. The unit employed is inconveniently large. It is so large that, even for use with coils, I have proposed that $\frac{1}{1000}$ part, or 10^6 centim. would be a convenient size. As regards straight wires, however, I find that it saves much useless figuring to reckon the inductance per centim. simply, with the result that we have a conveniently-sized numeric to deal with. Thus, in the present case, we have $L = 31$, if L be the inductance per centim.

Now Mr. Preece tells us that the inductance of a copper circuit will be approximately got by dividing by μ, the inductivity of the iron, which he reckons at from 300 to 1000. This gives

$$L = \cdot 1 \quad \text{to} \quad \cdot 031 \quad \text{in copper circuits.}$$

Let us compare with theory. The least value of the L of a copper wire of radius r at height h above the ground is

$$L = 2 \log (2h/r),$$

on the assumption that the return-current is on the surface of the ground, and that the wire-current is on *its* surface, so that the real value of L is *greater* than this formula states. The value ranges from 10 to 30, roughly speaking, according to radius and height. Thus, as a copper wire of 6·3 ohms per kilom. must be of radius ·091 centim.. if it be only 318 centim. above the ground, the inductance is 17·7 per centim. This is 177 times as big as Mr. Preece's *biggest* estimate. Even if we assume $\mu = 100$, which is more in accordance with my own measurements, Mr. Preece's estimate would be 60 times too small. In the presence of such stupendous errors it is of course useless to take account of the small corrections to which the above formula is subject. A proof will be found in my paper "On Electromagnets," Journal S. T. E. and E., vol. VII., p. 303 [vol. I., p. 101]. It is derived from Maxwell's formula for the inductance of a pair of parallel wires by the method of images.

(2). Mr. Preece does not seem to have observed that in measuring the permittance of his copper-circuits he was virtually measuring their inductance, though very roughly. Thus, if L and S be the inductance and the permittance of a solitary suspended copper wire, per unit length, and v be the speed of light in air, or 30 ohms, then on the assumption of return-current on the surface of the ground, we have $LSv^2 = 1$. This gives $L = (9s)^{-1}$, if s be the permittance per kilom. in microfarads. Since Mr. Preece's copper wire was 7·44 microf. per 261 miles, or 420 kilom., we have $s = {\cdot}018$, and therefore $L = 6{\cdot}2$. Although it is a considerable underestimate, yet we see that Mr. Preece's enormous error has disappeared. Why it is underestimated is mainly because the permittance is so greatly increased by the presence of neighbouring wires, as is explained in my paper "On the Electrostatic Capacity of Suspended Wires," Journal S. T. E. and E., vol. IX., p. 115 [vol. I., p. 46]. Allowing for this influence, we shall certainly come near to the true magnitude of L. It is possible that very carefully executed measurements by correct methods might reveal some quite new correction, but we cannot expect anything amounting to several hundred cent. per cent.

(3). Let us now briefly examine Mr. Preece's methods. First, he tried to measure the L of a copper circuit by a differential arrangement, and could not find that there was any to measure. But it will be clear to those who are acquainted with the properties of electrical balances that he did not go the right way to work. He supposed that the balancing resistance balances the quantity he calls the throttling or spurious resistance $(R^2 + L^2n^2)^{\frac{1}{2}}$, if R be the resistance, L the inductance, and $n/2\pi$ the frequency. This would be the impedance of the circuit if the effect of its permittance were ignorable. But it was not, as the permittance was, say 7 microfarads, so that the impedance formula is quite different. But, in any case, it is not the impedance that is balanced by resistance, but the resistance of the circuit. It is well known that the resistance of a copper wire is not sensibly increased, unless the undulations be excessively rapid, or the wire be very thick. And it was not increased. Whilst corroborating theory to some extent therefore, Mr. Preece's argument fails completely, as his experiment proved nothing about the impedance or the inductance, except in the indirect way I mentioned in (2) above, which is wholly against his conclusion that $L = 0$ nearly.

I should remark, however, that the proper way to observe and measure the inductance of a copper-wire circuit is to shorten it until the effect of its permittance is insensible. The L will be found to be about what I have stated. Why it should be shortened in this way will be obvious when it is remembered what a very rough business the P. O. *duplex* balancing with condensers is. In fact, no attempt seems to have been made to balance the L, nor would it be practicable under the circumstances.

(4). "It is, however, quite another matter with iron," as Mr. Preece remarks. It is known that the resistance of, say, a No. 4 iron wire can easily be 2 or 3 times its steady value, when currents of telephonic frequency are passed. But, as before, Mr. Preece supposed that he was measuring the impedance, or rather, what it would have been had there been no permittance, which makes a material difference. Consequently Mr. Preece's results are wrong. The value of L deduced is not related to the quantity observed in the manner he supposes. It is not a question of small corrections, but of an entire change of method.

(5). Coming next to the "direct measurement of the time-constant L/R," we are involved in further mysteries. How the chronograph was made to indicate the values of L/R is not stated. But let us assume that it did do this, and that ·0044 sec. and ·00667 sec. were really the values of L/R for the copper and the iron circuits. Now one is half as great again as the other. The resistances, too, are not widely different. It follows that the L's are of the same order of magnitude. But Mr. Preece argues in quite another manner. He assumes that self-induction is negligible first, and then reasons that the time-constant of the iron circuit would have been less than the measured ·00667 sec. in the ratio of the electrostatic time-constant of the iron to that of the copper circuit, and should therefore have been ·00624 sec.; and that the difference ·00043 was due to self-induction in the iron wire; from which he finds L. It is scarcely necessary to say that there is no warrant for this singular reasoning from the point of view of electromagnetic theory. These questions have been pretty fully worked out, but there is no resemblance to be found between Mr. Preece's methods and those which are, I believe, generally admitted to be correct.

The values of L come out 277 copper, 540 iron, per centim., taking the given ·0044 and ·0066 sec. as the values of L/R given by the chronograph. These values of L and L/R are much too great. It is suggested that the chronograph figures represent something quite different from L/R. If they represent the time of transit, the reasoning is equally erroneous.

(6). Mr. Preece next gives a table of the values of the impedance on the assumptions of no permittance, and that L had the value he had erroneously deduced, and that it was a constant. The table is quite inapplicable, because there *is* permittance, a great deal. If there were not, the figures would not represent the resistance. Nor do they represent the impedance, which does not run up in the way Mr. Preece makes it do as the frequency is raised. In fact, I may remark that Mr. Preece employs such entirely novel and unintelligible methods, that it would surely be right that he should give some reason for the faith that is in him.

(7). In conclusion, I would point out what is perhaps the most striking thing of all, in its ultimate consequences. Mr. Preece wants to prove that L is negligible in copper circuits, being under the idea that self-induction is prejudicial to long-distance telephony (and also very rapid telegraphy, of course, *if* rapid enough). Mr. Preece has spoken through a copper circuit of 270×2 miles with a clearness of articulation that is "entirely opposed to the idea of any measurable magnitude of L." As regards the speaking, it has been done over a thousand (1000×2) miles in America. But the important thing is the vital error involved in the reasoning. So far from being prejudicial, precisely the contrary is the case, as I have proved in considerable detail in this journal. [*The Electrician* is referred to.] Increasing L increases the amplitude and diminishes the distortion, and therefore renders long-distance telephony possible under circumstances that would preclude possibility were there no inductance.

The following examples will serve to show the importance of this matter. Take a circuit 100 kilom. long, 4 ohms and $\frac{1}{4}$ microf. per kilom., and *no* inductance in the first place. Short-circuit at both ends. Introduce at end A a sinusoidal impressed force, and calculate the current-amplitude at the other end B by the formula of the electrostatic theory which Mr. Preece believes in. Let the ratio of the full steady current to the amplitude of the actual current be ρ, and let the frequency range through 4 octaves, from $n = 1250$ to $n = 20{,}000$, where $n = 2\pi \times$ frequency. The values of ρ are

1·723, 3·431, 10·49, 58·87, 778.

It is barely credible that any kind of speaking would be possible, owing to the extraordinarily rapid increase of attenuation with the frequency. Nothing but murmuring would result.

Now introduce the additional datum that L has the very low value of $2\frac{1}{2}$ per centim., without other change, and calculate the corresponding results. They are

1·567, 2·649, 5·587, 10·496, 16·607.

The change is marvellous. It is by the preservation of the currents of great frequency that good articulation is possible, and we see that a very little inductance immensely improves matters. There is no "dominant" frequency in telephony. What is wanted is to have currents of *all* frequencies reproduced at the distant end in proper proportion, attenuated as nearly as may be to the same degree.

Change L to 5, which is a more probable value. Results :—

1·437, 2·251, 3·176, 4·169, 4·670.

We see that *good* telephony is now possible, though much distortion remains.

Finally, increase L to 10. Results :—

1·235, 1·510, 1·729, 1·825, 1·854,

showing splendid articulation. In fact we have approximated very considerably towards a distortionless circuit.

Now, this is all done by the inductance which Mr. Preece dreads so much, and would make out to be 0. It is the very essence of good long-distance telephony that inductance should *not* be negligible. R/Ln must be made small, a fraction. The bigger L is the better (*cæteris paribus*). It is proved, not merely by theory but by the experimental facts, especially with copper wires of low resistance. It is not the inductance of iron that is prejudicial, nor yet its impedance, but its high resistance. R is increased whilst L is reduced, which is exactly the opposite to what is required for good articulation over long circuits.

But it is impossible to treat these questions by the electrostatic theory. Nor yet, as Mr. Preece attempts, by a mixed process, a little bit of the electromagnetic theory put into the electrostatic The true theory takes both the static and the magnetic effects into consideration simultaneously. No particular exactness need be attributed to the above figures. What is important is the nature of the effect of self-induction, and that it is, without entering into refined calculations, of great magnitude. The permittance has been purposely chosen large.

XXXIX. NOTES ON NOMENCLATURE.

[*The Electrician*; Note 4, June 24, 1887, p. 143; Note 5, May 11, 1888, p. 27.]

NOTE 4. MAGNETIC RESISTANCE, ETC.

As there is at the present time at least a possibility of the various words I have proposed coming into general use, I take the opportunity of making a few casual remarks upon the subject supplementary to those of 1885 and since. First, I observe (*The Electrician*, June 17, 1887, p. 114) it mentioned that I disapprove of "magnetic resistance." This is only a part of the fact. To illustrate this, I may say that were I investigating the theory of the dynamo, I think I should make use of the term myself, provisionally. What is really my objection is to its permanent use. There must always be a certain latitude allowed to investigators who do not find words ready to meet their wants. Were it an isolated question, there would be little difficulty in finding a perfectly unobjectionable word; but it is not an isolated question. My aim has been to make a scheme which shall be at once theoretically defensible and yet thoroughly practical. Bearing this in mind, I prefer to leave a blank in the place of "magnetic resistance" at present [vol. II. p. 125].

To illustrate the difficulties connected with nomenclature I may mention that, last summer, I was extremely in want of a term which should be an extension of impedance. The impedance of a circuit at a given frequency (under stated external conditions) is quite definite (with occasional departures due to want of proportionality between forces and fluxes), if it be a simple circuit, or reducible to a simple circuit, so that the strength of the current does not vary in different parts. But when it does, we can certainly only apply the term impedance legitimately at the seat of the impressed force, if at a single spot; or else, if it be wholly localised in a part of the main circuit in which the current does not vary, then the term impedance is again applicable. Now, I used impedance in an extended sense, but expressly stated that it was only done provisionally [vol. II., p. 65]. I have since found a far better way of expressing results, viz., in terms of "attenuation" and "distortion," both very important things. The idea of attenuation, expressed in a more roundabout manner in terms such as "diminution of amplitude," and so forth, is nothing new; the *word* "attenuation" I found Lord Rayleigh use, and at once adopted it myself as the very thing I wanted. "Distortion," on the other hand, I chose myself as preferable to "mutilation" and similar words. Its meaning is obvious. Make current-variations in a certain way at one place. If the current-variations at another place are similar, no matter how much attenuated they may be, there is no distortion. The extremest kind of distortion is to be found on Atlantic cables. Drawn on the same scale, there is little resemblance between the curves at one end and at the other. Telephony would obviously be impossible even were the frequency allowable to be sufficiently great, which is of course out of the question under present conditions. But, only make the distortion reasonably small at a sufficiently great frequency, and telephony is at once possible, provided the attenuation be not of unreasonable amount. (Frequency is Lord Rayleigh's word for "pitch," number of waves per second.)

Referring to magnetic resistance again. A certain person once declared that $E=RC$, to express Ohm's law, was nonsense; it must be $C=E/R$. This eminent scienticulist could not see the force of Maxwell's argument, that electricity could not be a form of energy because it was only one of the factors of energy. Now, however, by the development of the electric light rendering energy a marketable commodity through electric agency, there is little fear of converts being made to these views. So we may return to $E=RC$, or $C=KE$, if K be the conductance. One is just as good as the other, theoretically, and is just as meaningful. Which to use (including the ideas) is purely a matter of convenience in the particular application that is in question. As a general rule, resistances are more useful, because we usually deal with wires *in sequence*. But if they be in parallel, conductances are the proper things to use. With condensers, on the other hand, permittances are more useful; should, however, we join in sequence, then elastances are the proper things. In theoretical investigations disconnected from special applications, the unit-volume properties conductivity, inductivity, and permittivity, are generally much more useful than their reciprocals, resistivity the [reluctivity], and elastivity. Now, in late years, there has been some development of practical applications in connection with the flux magnetic induction; in theory, inductivity would be the more convenient basis; but several practicians find that the reciprocal ideas, say, provisionally, "magnetic resistivity" and "magnetic resistance," are more useful. I think their choice has been a wise one, whilst at the same time I recognise the difficulties with which they have to contend, through "magnetic leakage," and so forth. It is for the practicians to find practical ways of getting a round peg to fit a square hole. They know best what they want, and whether empirical formulæ will not suit them better than more elaborate empiricism, which could, perhaps, be scientifically better defended. For it is clear that, beyond the region of proportionality of force to flux, the science of magnetic induction must continue very empirical for some time to come. I do not think the time has yet arrived for laying down the law by conventions or committees in this matter (as it may have come in more definite parts of electrical science); but that practical and theoretical investigators should be allowed to develop their ideas freely. In short, Conventions or Committees should not meddle with matters (save very lightly) which are in a provisional stage. And I may add that, just as treaties are made to be broken, so the laws of Conventions will be broken as soon as ever it is found inconvenient to obey them. The introduction of anything of the nature of officialism into scientific matters should be strenuously opposed—in *this* country. It would be as bad as the passport system. The utility of a Convention seems to consist in the formation of a temporary consensus of opinion from which to make fresh departures. There cannot be any finality.

Mac.—Here we are on firmer ground. There cannot, I think, be any question that this is the right name for the practical unit of inductance, in honour of the man who knew something about self-induction, and whose ideas on the subject are not yet fully appreciated. This was very much his own fault. He had the most splendid and thoroughly philosophical ideas on electromagnetism all round, but kept them too much in the background. Maxwell's treatise requires to be *studied*, not read, before the inner meaning of his scheme can be appreciated. Had he lived, he would probably, in some future edition, have brought his views prominently forth *ab initio*, and developed the whole treatise on their basis exclusively. Should the mac be 10^9 or 10^6 centimetres? If 10^9, which has great recommendations, then millimac will be practically wanted, to avoid decimals. It is quite a euphonious and unobjectionable word.

Inductometer.—Naturally, in accordance with induction, inductivity, and inductance, this is a measurer of inductance (self or mutual) in terms of units of inductance—macs, or millimacs. I would apply the term to any instrument that measured inductance at once in terms of known inductances, as resistances are compared with known resistances. Some practical acquaintance with self and mutual induction, desultory, but of long continuance, has gradually forced upon me the idea (not to be easily displaced) that really practical ways of measuring inductances should be in terms of standard inductances—or, which is the same thing, by a properly calibrated inductometer—and not absolute measurements. What particular method of making the comparisons is best I do not know, nor yet how best to calibrate the inductometer. If it were a mere question of coils of fine wire, nothing is simpler, or more expeditious, or more accurate, or more sensitive, than the immediate balancing of the self-induction against that of an inductometer of *variable* inductance, using the telephone [vol. II., p. 37 and p. 100]. The advantages and the simplicity are so great that I think practical men might well turn their attention to practical ways of extending the method to cases other than those in which mere coils are alone concerned.

"Absolutism."—The most absolute of all ways of finding the inductance of a coil is with a tape. Herein lies a moral of very wide application.

NOTE 5. MAGNETIC RELUCTANCE.

There is a tendency at the present time among some writers to greatly extend the application of the word resistance in electromagnetism, so as to signify cause/effect. This seems a pity, because the term resistance has already become thoroughly specialised in electromagnetism in strict relationship to frictional dissipation of energy. What the popular meaning of resistance may be is beside the point; ditto dimensions, etc.

I would suggest that what is now called magnetic resistance be called the magnetic reluctance; and when referred to unit volume, the reluctancy [or reluctivity].

XL. ON THE SELF-INDUCTION OF WIRES.

[*Phil. Mag.*, 1886-7. Part 1, August, 1886, p. 118; Part 2, Sept., 1886, p. 273; t 3, Oct., 1886, p. 332; Part 4, Nov., 1886, p. 419; Part 5, Jan., 1887, p. 10; t 6, Feb., 1887, p. 173; Part 7, July, 1887, p. 63; Part 8, now first published.]

PART I.

Remarks on the Propagation of Electromagnetic Waves along Wires outside them, and the Penetration of Current into Wires. Tendency to Surface Concentration. Professor Hughes's experiments.

A SERIES of experiments made some years ago, in which I used the Wheatstone-bridge and the differential telephone as balances of induction as well as of resistance, led me to undertake a theoretical investigation of the phenomena occurring when conducting-cores are placed in long solenoidal coils, in which impressed electromotive force is made to act, in order to explain the disturbances of balance which are produced by the dissipation of energy in the cores. The simpler portions of this investigation, leaving out those of greater mathematical difficulty and less practical interest, relating to hollow cores and the effect of allowing dielectric displacement, were published in *The Electrician*, May 3, 1884, and after [vol. I., Art. xxviii., p. 353].

This investigation led me to the mathematically similar investigation of the transmission of current into wires. I say *into* wires, instead of through wires, because the current is really transmitted by diffusion from the boundary into a wire from the external dielectric, under all ordinarily occurring circumstances. In the case of a core placed in a coil, the magnetic force is longitudinal and the current circular; in the case of a straight round wire, the current is longitudinal and the magnetic force circular. The transmission of the longitudinal current into the wire takes place, however, exactly in the same manner as the transmission of the longitudinal magnetic force into the core within the coil, when the boundary conditions are made similar, which is easily realizable. Similarly, we may compare the circular electric current in the core with the circular magnetic flux in the wire.

I also found the transfer of energy to be similar in both cases, viz., radially inward or outward, to or from the axis of the core or the wire. It was therefore necessary to consider the dielectric, in order to complete the course of the transfer of energy from its source, say a voltaic cell, to its sink, the wire or the core where it is finally dissipated in the form of heat, with temporary storage as electric and magnetic energy in the field generally, including the conductors.

Terminating the paper above referred to, having so much other matter, I started a fresh one under the title of "Electromagnetic Induction and its Propagation," [vol. I., Art. XXX., p. 429; and vol. II., Art. XXXV., p. 39]. Having, according to my sketched plan, to get rid of general matter first, before proceeding to special solutions, I took occasion near the commencement of the paper to give a general account of some of my results regarding the propagation of current, in which the following occurs, describing the way the current rises in a wire, and the consequent approximation, under certain circumstances, to mere surface-conduction. It was meant to illustrate the previously-mentioned stoppage of current-conduction by high conductivity. After an account of the transfer of energy through the dielectric (concerning which I shall say a few words later) I continue [vol. I., p. 440]:—

"Since, on starting a current, the energy reaches the wire from the medium without, it may be expected that the electric current is first set up in the outer part, and takes time to penetrate to the middle. This I have verified by investigating some special cases.

"Increase the conductivity enormously, still keeping it finite, however. Let it, for instance, take minutes to set up a current at the axis. Then ordinary rapid signalling 'through the wire' would be accompanied by a surface-current only, penetrating to but a small depth. The disturbance is then propagated parallel to the wire in the manner of waves, with reflection at the end, and hardly any tailing off. With infinite conductivity, there can be no current set up in the wire at all. There is no dissipation; wave-propagation is perfect. The wire-current is wholly superficial, an abstraction, yet it is nearly the same with very high conductivity. This illustrates the impenetrability of a perfect conductor to magnetic induction (and similarly to electric current) applied by Maxwell to the molecular theory of magnetism. ..."

Attention has recently been forcibly directed towards the phenomenon above described of the inward transmission of current into wires by Professor Hughes's Inaugural Address to the Society of Telegraph

Engineers and Electricians, January, 1886. This paper was, for many reasons, very remarkable. It was remarkable for the ignoration of well-known facts, thoroughly worked out already; also for the mixing up of the effects due to induction and to resistance, and the author's apparent inability to separate them, or to see the real meaning of his results; one might indeed imagine that an entirely new science of induction was in its earliest stages. It was remarkable that the great experimental skill of the author should have led him to employ a method which was in itself objectionable, being capable of giving, in general, neither a true resistance nor a true induction-balance (as may be easily seen by simple experiments with coils, without mathematical examination of the theory) —a method which does not therefore admit of exact interpretation of results without full particulars being given and subjected to laborious calculations. Finally, it was remarkable as containing, so far as could be safely guessed at, many verifications of the approximation towards mere surface-conduction in wires. This is, after all, the really important matter, against which all the rest is insignificant.

As regards the method employed, I have shown its inaccuracy in my paper "On the Use of the Bridge as an Induction-balance" [vol. II., p. 33], wherein I also described correct methods, including the simple Bridge without mutual induction, and also methods in which mutual induction is employed to get balance, giving the requisite formulæ, which are of the simplest character.

As regards the interpretation of Professor Hughes's thick-wire results, showing departure from the linear theory, by which I mean the theory that ignores differences in the current-density in wires, I have before made the following remarks [vol. II., p. 30]. After commenting upon the difficulty of exact interpretation, I proceed:—

"The most interesting of the experiments are those relating to the effect of increased diameter on what Prof. Hughes terms the inductive capacity of wires. My own interpretation is roughly this. That the time-constant of a wire first increases with the diameter" (this is of course what the linear theory shows), "and, then, later, decreases rapidly; and that the decrease sets in the sooner the higher the conductivity and the higher the inductivity (or magnetic permeability) of the wires. If this be correct, it is exactly what I should have expected and predicted. In fact, I have already described the phenomenon in this Journal; or, rather, the phenomenon I described contains in itself the above interpretation. In *The Electrician* for January 10, 1885, I described how the current starts in a wire. It begins on its boundary, and is propagated inward. Thus, during the rise of the current it is less strong at the centre than at the boundary. As regards the manner of inward propagation, it takes place according to the same laws as the propagation of magnetic force and current into cores from an enveloping coil, which I have described in considerable detail in *The Electrician* [vol. I., Art. xxviii.; see especially § 20]. The retardation depends upon the conductivity, upon the inductivity, and upon the section, under similar boundary-conditions. If the conductivity be high enough, or the inductivity, or the section, be large enough to make the central current appreciably less than the boundary-current during the greater part of the time of rise of the current, there will be an apparent reduction in the time-constant. Go to an extreme case—very rapid short currents, and large retardation to inward transmission. Here we have the current in layers, strong on the boundary, weak in the middle. Clearly then, if we wish to regard the wire as a mere linear circuit, which it is not, and as we can only do to a first approximation, we should remove the central part of the wire—that is, increase its resistance, regarded as a line, or reduce its time-constant. This will happen the sooner, the greater the inductivity and the conductivity, as the section is continuously increased. It is only thin wires that can be treated as mere lines, and even they, if the speed be only great enough, must be treated as solid conductors. I ought also to mention that the influence of external conductors, as of the return conductor, is of importance, sometimes of very great importance, in modifying the distribution of current in the transient state. I have had for years in manuscript some solutions relating to round wires, and hope to publish them soon.

"As a general assistance to those who go by old methods, a rising current inducing an opposite current in itself and in parallel conductors, this may be useful. Parallel currents are said to attract or repel, according as the currents are together or opposed. This is, however, mechanical force on the conductors. The distribution of current is not affected by it. But when currents are increasing or decreasing, there is an apparent attraction or repulsion between them. Oppositely-going currents repel when they are decreasing and attract when they are increasing. Thus, send a current into a loop, one wire the return to the other, both being close together. During the rise of the current it will be denser on the sides of the wires nearest one another than on the remote sides. ..."

An iron wire, through which rapid reversals are sent, should afterwards be found, by reason of its magnetic retentiveness, magnetized in concentric cylindrical shells, of alternately positive and negative magnetization. This would only occur superficially. The thickness of the layers would give information regarding the amount of retardation, from which the inductivity could be deduced. The case is similar to that of the superficial layers of magnetization produced in a core placed in a coil through which reversals are sent, the magnetization being then, however, longitudinal instead of circular.

The linear theory is departed from in the most extreme manner, when the return-current closely envelops the wire. The theory of the rise of the current in this case I have given before [vol. II., p. 44], and also the case of the return-current at any distance [vol. II., p. 50]. The investigation following in this paper is more comprehensive, taking into account both electrostatic and magnetic induction, working down to the magnetic theory on the one hand, and approximating towards the electrostatic theory (long submarine cable) on the other; with this difference, that inertia is not so wholly ignorable in the long-line case as is elastic yielding in the case of a short wire. Nor is the variation of current-density wholly ignorable.

New (Duplex) Method of Treating the Electromagnetic Equations. The Flux of Energy.

But first as regards the transfer of energy in the electromagnetic field. This is a very important matter theoretically. It is a necessity of a rationally intelligible scheme (even if it be only on paper) that the transfer of energy should be explicitly definable. It is the absence of this definiteness that makes the German methods so repulsive to a plain man who likes to see where he is going and what he is doing, and hates metaphysics in science.

I found that I had been anticipated by Prof. Poynting [*Phil. Trans.*, 1884] in the deduction of the transfer-of-energy formula appropriate to Maxwell's electromagnetic scheme, in the main. It is, therefore, only as having given the equation of activity in a more general form, the most general that Maxwell's scheme admits of, and having deduced it in a simple manner, that I can attach myself to the matter. In connection with it, however, there is another matter of some importance, viz., the use of a certain fundamental equation. That I should have been able to arrive at the most general form, taking into account intrinsic magnetization, as well as not confining myself to media homogeneous and isotropic as regards the three quantities conductivity, inductivity, and dielectric capacity, in a simple and direct manner, without any volume-integrations or complications, arose from my method of treating the general equations. I here sketch out the scheme, in the form I give it.

Let $\mathbf{H}_1$ be the magnetic force and Γ the current. (Thick letters here for vectors. The later investigation is wholly scalar.) Then, "curl" denoting the well-known rotatory operator, Maxwell's fundamental current-equation is

$$\operatorname{curl} \mathbf{H}_1 = 4\pi\Gamma, \qquad \text{.............................(1)}$$

and is his definition of electric current in terms of magnetic force. It necessitates closure of the electric current, and, at a surface, tangential continuity of $\mathbf{H}_1$ and normal continuity of Γ. The electric current may be conductive, or the variation of the elastic "displacement," say

$$\Gamma = \mathbf{C} + \dot{\mathbf{D}},$$

where $\mathbf{C}$ is the conduction-current, and $\mathbf{D}$ the displacement, linear functions of the electric force $\mathbf{E}$, thus,

$$\mathbf{C} = k\mathbf{E}_1, \qquad \mathbf{D} = c\mathbf{E}_1/4\pi;$$

k being the conductivity, and c the dielectric capacity (or $c/4\pi$ the condenser-capacity per unit-volume). Equation (1) thus connects the electric and the magnetic forces one way. But this is not enough to make a complete system. A second relation between $\mathbf{E}_1$ and $\mathbf{H}_1$ is wanted.

Maxwell's second relation is his equation of electric force in terms of two highly artificial quantities, a vector and a scalar potential, say $\mathbf{A}$ and P, thus

$$\mathbf{E}_1 = -\dot{\mathbf{A}} - \nabla P, \qquad \text{...........................(2)}$$

ignoring impressed force for the present. From **A** we get down to $\mathbf{H}_1$ again, thus,

$$\text{curl}\,\mathbf{A}=\mathbf{B}, \qquad \mathbf{B}=\mu\mathbf{H}_1;$$

B being the magnetic induction, and μ the inductivity. (Here we ignore intrinsic magnetization.)

The equation (2) is arrived at through a rather complex investigation. From these equations are deduced the general equations of electromagnetic disturbances in vol. ii., art. 783. They contain both **A** and P. One or other must go before we can practically work the equations, which are, independently of this, rather unmanageable, although they are not really general, for impressed forces are omitted, and the intrinsic magnetization must be zero, and the medium isotropic. Again—and this is an objection of some magnitude—the two potentials **A** and P, if given everywhere, are *not sufficient* to specify the state of the electromagnetic field. Try it; and fail.

Even without using these complex general equations referred to, but those on which they are based, (1) and (2), the very artificial nature of **A** and P greatly obscures and complicates many investigations. Not being able to work practically in terms of **A** and P in a general manner, and yet knowing there was nothing absolutely wrong, I went to the root of the evil, and cured it, thus:—

As a companion to equation (1) use this,

$$-\text{curl}\,\mathbf{E}_1=4\pi\mathbf{G}; \qquad \text{.........................(3)}$$

where **G** is the magnetic current, or $\dot{\mathbf{B}}/4\pi$. That this may be derived at once from (2) is obvious. But what is of greater importance in view of the difficult establishment of (2), is that (3) can be got immediately independently, and that (2) is its consequence. Equation (3) is, in fact, the mathematical expression of the Faraday law of induction, that the electromotive force of induction in any closed circuit is to be measured by the rate of decrease of the induction through it.

Now make (1) and (3) the fundamental equations, and ignore (2) altogether, except for special purposes. There are several great advantages in the use of (3). First, the abolition of the two potentials. Next, we are brought into immediate contact with $\mathbf{E}_1$ and $\mathbf{H}_1$, which have physical significance in really defining the state of the medium anywhere (k, μ, and c of course to be known), which **A** and P do not, and cannot, even if given over all space. Thirdly, by reason of the close parallelism between (1) and (3), electric force being related to magnetic current, as magnetic force to electric current, we are enabled to perceive easily many important relations which are not at all obvious when the potentials **A** and P are used, and (3) ignored. Fourthly, we are enabled with considerable ease, if we have obtained solutions relating to variable states in which the lines of $\mathbf{E}_1$ and $\mathbf{H}_1$ are related in one way, to at once get the solutions of problems of quite different physical meaning, in which $\mathbf{E}_1$ and $\mathbf{H}_1$, or quantities directly related to them, change places. For example, the variation of magnetic force in a core placed in a coil, and of electric current in a round wire; and many others.

That the advantages attending the use of (3) as a fundamental equation are not imaginary, I have repeatedly verified. The establishment of the general equation of activity, however, which I now reproduce [vol. I., p. 449], shows that (3) is really the proper and natural fundamental equation to use. But we must first introduce impressed forces, allowing energy to be taken in by the electric and magnetic currents. In (1) and (3), $\mathbf{E}_1$ and $\mathbf{H}_1$ are not the effective electric and magnetic forces concerned in producing the fluxes conduction-current, displacement, and induction, but require impressed forces, say $\mathbf{e}$ and $\mathbf{h}$, to be added. Let $\mathbf{E}=\mathbf{E}_1+\mathbf{e}$, and $\mathbf{H}=\mathbf{H}_1+\mathbf{h}$; then we shall have

$$\mathbf{B}=\mu\mathbf{H}, \qquad \mathbf{C}=k\mathbf{E}, \qquad \mathbf{D}=c\mathbf{E}/4\pi, \quad \text{................}(4)$$

as the three linear relations between forces and fluxes; two equations,

$$\Gamma=\mathbf{C}+\dot{\mathbf{D}}, \qquad \mathbf{G}=\dot{\mathbf{B}}/4\pi, \quad \text{......................}(5)$$

showing the structure of the currents; and two equations of cross-connection,

$$\text{curl}\,(\mathbf{H}-\mathbf{h})=4\pi\Gamma, \quad \text{...........................}(6)$$

$$-\text{curl}\,(\mathbf{E}-\mathbf{e})=4\pi\mathbf{G}. \quad \text{...........................}(7)$$

Next, let Q be the dissipativity, U the electric energy, and T the magnetic energy per unit volume, defined thus:

$$\mathrm{Q}=\mathbf{EC}, \qquad \mathrm{U}=\tfrac{1}{2}\mathbf{ED}, \qquad \mathrm{T}=\tfrac{1}{2}\mathbf{HB}/4\pi, \quad \text{.............}(8)$$

(according to the notation of scalar products used in my paper in the *Philosophical Magazine*, June, 1885 [vol. ii., p. 4]; c, k, and μ are in general the operators appropriate to linear connection between forces and fluxes). Then we get the full equation of activity at once, by multiplying (6) by $\mathbf{E}$, and (7) by $\mathbf{H}$, and adding the results. It is

$$\left.\begin{aligned}\mathbf{e}\Gamma+\mathbf{hG}&=\mathbf{E}\Gamma+\mathbf{HG}+\text{div}\,\mathrm{V}(\mathbf{E}-\mathbf{e})(\mathbf{H}-\mathbf{h})/4\pi,\\ &=Q+\dot{U}+\dot{T}+\text{div}\,\mathrm{V}(\mathbf{E}-\mathbf{e})(\mathbf{H}-\mathbf{h})/4\pi,\end{aligned}\right\}\text{............}(9)$$

where div stands for divergence, the negative of Maxwell's convergence. The left side showing the energy taken in per second per unit volume by reason of impressed forces, and $Q+\dot{U}+\dot{T}$ being expended on the spot in heating, and in increasing the electric and magnetic energies, we see that $\mathrm{V}(\mathbf{E}-\mathbf{e})(\mathbf{H}-\mathbf{h})/4\pi$ is the vector flux of energy per unit area per second, or the energy-current density. The appropriateness of (7) as a companion to (6) is very clearly shown.

The scheme expressed by (4), (5), (6), (7) is, however, in one respect too general. The magnetic current is closed, by (7); but that does not necessitate the closure of the magnetic induction, which is necessary to avoid having unipolar magnets. Hence

$$\text{div}\,\mathbf{B}=0 \quad \text{..................................}(10)$$

is required to meet facts, in addition to (4), (5), (6), (7). There is no magnetic conduction-current with dissipation of energy, analogous to the electric conduction-current.

As regards the meanings of $\mathbf{e}$ and $\mathbf{h}$, in the light of dynamics they define themselves in the equation of activity; that is, so far as the mere measure of impressed forces is concerned, apart from physical causation. Thus, **e** is the amount of energy taken in by the electromagnetic field per second per unit volume per unit electric current, and **h** is similarly related to magnetic current. Under **e** have to be included the recognised voltaic and thermoelectric forces. But besides them, **e** has to include the impressed electric force due to motion in a magnetic field, or $\mathrm{V}\mathbf{vB}$, if **v** is the vector velocity, necessitating a mechanical force $\mathrm{V}\Gamma\mathbf{B}$. It has also to include intrinsic electrization, the state which is set up in solid dielectrics under the continued application of electric force. Thus,

$$\mathbf{J} = c\mathbf{e}/4\pi$$

connects the intensity of intrinsic electrization **J** with the corresponding **e**.

I can find only two kinds of **h**. First, due to motion in an electric field, viz., $4\pi\mathrm{V}\mathbf{Dv}$, necessitating a mechanical force $4\pi\mathrm{V}\mathbf{DG}$; and, secondly, much more importantly, intrinsic magnetization **I**, connected with the corresponding **h** thus,

$$\mathbf{I} = \mu\mathbf{h}/4\pi.$$

As regards potentials, there are, to match the two electric potentials **A** and P, two magnetic potentials, say **Z** and Ω; Ω being the single-valued scalar magnetic potential, and **Z** the vector-potential of the magnetic current, some of whose properties in relation to dielectric and conductive displacement I have worked out in the paper referred to before.

As regards the general equations of disturbances, like Maxwell's (7), chapter xx. vol. ii., they are far more a hindrance than an assistance in general investigations. But when we come to a special investigation, and need to know the forms of the functions involved, then we may eliminate either **E** or **H** between (6) and (7), and use the suitable coordinates.

Application of the General Equations to a Round Wire with Coaxial Return-Tube. The Differential Equations and Normal Solutions. Arbitrary Initial State.

We may make use of the above equations at the start, in passing to the question of the propagation of disturbances along a wire, after which the investigation will be wholly scalar. Put $\mathbf{e} = 0$ in (7); then we see that we cannot alter the magnetic force at a point without giving rotation to the electric force. Now, as in a steady state the electric force has no rotation (away from the seat of impressed force), it follows that under no circumstances (except by artificial arrangements of impressed force) can we set up the steady state in a conductor strictly according to the linear theory. We may approximate to it very closely throughout the greater part of the variable period, but it will be widely departed from in the very early stages.

Let there be a straight round wire of radius a_1, conductivity k_1, inductivity μ_1, and dielectric capacity c_1; surrounded up to radius a_2 by a dielectric of conductivity k_2, inductivity μ_2, and dielectric capacity

c_2; in its turn surrounded to radius a_3 by a conductor of k_3, μ_3, and c_3. This might be carried on to any extent; but we stop at $r=a_3$, r being distance from the axis of the wire, as the outer conductor is to be the return to the inner wire.

Let the magnetic lines be such as would be produced by longitudinal impressed electric force, viz. circles in planes perpendicular to the axis of the wire, and centred thereon. Let H be the intensity of magnetic force at distance r from the axis, and distance z along it from a fixed point. Use (6), with $\mathbf{h}=0$, to find the electric current. It has two components, say Γ longitudinal, or parallel to $\mathbf{z}$, and γ radial, or parallel to $\mathbf{r}$, given by

$$4\pi\Gamma = \frac{1}{r}\frac{d}{dr}rH, \qquad 4\pi\gamma = -\frac{dH}{dz}. \quad \text{.............(11)}$$

We have also $\mathbf{E}=\rho\Gamma$, if ρ is a generalised resistivity, or

$$\rho^{-1} = k + \frac{c}{4\pi}\frac{d}{dt}. \quad \text{.............(12)}$$

Now use equation (7), with $\mathbf{e}=0$. The curl of the longitudinal and of the radial electric force are both circular, like $\mathbf{H}$, giving

$$\mu\dot{H} = \rho\left(\frac{d\Gamma}{dr} - \frac{d\gamma}{dz}\right). \quad \text{.............(13)}$$

In this use (11), and we get the H equation, which is

$$\frac{d}{dr}\frac{1}{r}\frac{d}{dr}rH + \frac{d^2H}{dz^2} = 4\pi\mu k\dot{H} + \mu c\ddot{H}. \quad \text{.............(14)}$$

The suffixes $_1$, $_2$, and $_3$ are to be used, according as the wire, dielectric, or sheath, is in question.

In a normal state of free subsidence, $d/dt=p$, a constant. Let also $d^2/dz^2 = -m^2$, where m^2 is a constant, depending upon the terminal conditions. Also, let

$$-s^2 = 4\pi\mu kp + \mu cp^2 + m^2. \quad \text{.............(15)}$$

Then (14) becomes

$$\frac{d}{dr}\frac{1}{r}\frac{d}{dr}rH + s^2H = 0; \quad \text{.............(16)}$$

which is the equation of the $J_1(sr)$ and its complementary function, which call $K_1(sr)$. Thus, for reference,

$$\left.\begin{aligned}
J_0(sr) &= 1 - \frac{s^2r^2}{2^2} + \frac{s^4r^4}{2^24^2} - \dots, \\
J_1(sr) &= -\frac{d}{d(sr)}J_0(sr) = \tfrac{1}{2}sr\left(1 - \tfrac{1}{2}\frac{s^2r^2}{2^2} + \tfrac{1}{3}\frac{s^4r^4}{2^24^2} - \dots\right), \\
K_0(sr) &= J_0(sr).\log sr + \frac{s^2r^2}{2^2} - (1+\tfrac{1}{2})\frac{s^4r^4}{2^24^2} + \dots, \\
K_1(sr) &= -\frac{dK_0(sr)}{d(sr)} = -\frac{J_0(sr)}{sr} + J_1(sr)\log sr - \tfrac{1}{2}sr\left\{1 - \tfrac{3}{4}\frac{s^2r^2}{2^2} + \dots\right\}.
\end{aligned}\right\} \quad (17)$$

We have therefore the following sets of solutions, in the wire, dielectric, and sheath respectively, the A's and B's being constants :—

$$\left.\begin{aligned}
H_1 &= A_1J_1(s_1r)\cos(mz+\theta)\epsilon^{pt},\\
4\pi\gamma_1 &= A_1J_1(s_1r)m\sin(mz+\theta)\epsilon^{pt},\\
4\pi\Gamma_1 &= A_1J_0(s_1r)s_1\cos(mz+\theta)\epsilon^{pt},\\
H_2 &= \{A_2J_1(s_2r)+B_2K_1(s_2r)\}\cos(mz+\theta)\epsilon^{pt},\\
4\pi\gamma_2 &= \{A_2J_1(s_2r)+B_2K_1(s_2r)\}m\sin(mz+\theta)\epsilon^{pt},\\
4\pi\Gamma_2 &= \{A_2J_0(s_2r)+B_2K_0(s_2r)\}s_2\cos(mz+\theta)\epsilon^{pt},\\
H_3 &= \{A_3J_1(s_3r)+B_3K_1(s_3r)\}\cos(mz+\theta)\epsilon^{pt},\\
4\pi\gamma_3 &= \{A_3J_1(s_3r)+B_3K_1(s_3r)\}m\sin(mz+\theta)\epsilon^{pt},\\
4\pi\Gamma_3 &= \{A_3J_0(s_3r)+B_3K_0(s_3r)\}s_3\cos(mz+\theta)\epsilon^{pt}.
\end{aligned}\right\}\quad\ldots\ldots\ldots\ldots(18)$$

To harmonise these, we have the boundary conditions of continuity of tangential electric and magnetic forces, and of normal electric and magnetic currents (or of magnetic induction). Thus, $\gamma_1=\gamma_2$ and $\rho_1\Gamma_1=\rho_2\Gamma_2$, at $r=a_1$, give us

$$\left.\begin{aligned}
(A_2/A_1)(J_1K_0-J_0K_1)(s_2a_1) &= J_1(s_1a_1)K_0(s_2a_1)\\
&\quad-(\rho_1s_1/\rho_2s_2)J_0(s_1a_1)K_1(s_2a_1),\\
(B_2/A_1)(J_1K_0-J_0K_1)(s_2a_1) &= (\rho_1s_1/\rho_2s_2)J_0(s_1a_1)J_1(s_2a_1)\\
&\quad-J_1(s_1a_1)J_0(s_2a_1).
\end{aligned}\right\}\quad\ldots\ldots\ldots\ldots(19)$$

As there is to be no current beyond the sheath, $\gamma_3=0$, or $H_3=0$, at $r=a_3$. This gives

$$B_3=-A_3\frac{J_1}{K_1}(s_3a_3).\quad\ldots\ldots\ldots\ldots\ldots\ldots\ldots\ldots(20)$$

This, and the conditions $\gamma_3=\gamma_2$, and $\rho_3\Gamma_3=\rho_2\Gamma_2$, at $r=a_2$, give us

$$\left.\begin{aligned}
(A_2J_1+B_2K_1)(s_2a_2) &= A_3\left\{J_1(s_3a_2)-\frac{J_1}{K_1}(s_3a_3)K_1(s_3a_2)\right\},\\
(\rho_2s_2/\rho_3s_3)(A_2J_0+B_2K_0)(s_2a_2) &= A_3\left\{J_0(s_3a_2)-\frac{J_1}{K_1}(s_3a_3)K_0(s_3a_2)\right\};
\end{aligned}\right\}\quad\ldots(21)$$

whence, eliminating A_3 by division, and putting for A_2 and B_2 their values in terms of A_1, through (19), we obtain the determinantal equation of the p's for a particular value of m^2. It is

$$\frac{\rho_3s_3}{\rho_2s_2}\cdot\frac{J_0(s_3a_2)K_1(s_3a_3)-J_1(s_3a_3)K_0(s_3a_2)}{J_1(s_3a_2)K_1(s_3a_3)-J_1(s_3a_3)K_1(s_3a_2)}$$

$$=\frac{\dfrac{J_1(s_1a_1)K_0(s_2a_1)-(\rho_1s_1/\rho_2s_2)J_0(s_1a_1)K_1(s_2a_1)}{(\rho_1s_1/\rho_2s_2)J_0(s_1a_1)J_1(s_2a_1)-J_1(s_1a_1)J_0(s_2a_1)}J_0(s_2a_2)+K_0(s_2a_2)}{\ldots\ldots\ldots\ldots\ldots\ldots\ldots\ldots\ldots\ldots J_1(s_2a_2)+K_1(s_2a_2)},\quad\ldots(22)$$

where the dots indicate repetition of the fraction immediately over them.

Before proceeding to practical simplifications, we may in outline continue the process of finding the complete solution to correspond to any given initial state. The m's must be found from the terminal conditions. Suppose, for example, that the wire, of length l, forms a closed circuit, and that the sheath and the dielectric are similarly closed on themselves. Then, clearly, we shall have Fourier periodic series, with

$$m=0, \qquad 2\pi/l, \qquad 4\pi/l, \qquad 6\pi/l, \quad \text{etc.}$$

If, again, we desire to make the sheath the return to the wire, without external resistance, join them at the end $z=0$ by a conducting-plate of no resistance, placed perpendicular to the axis; and do the same at the other end, where $z=l$. This will make

$$\gamma=0 \quad \text{at} \quad z=0, \quad \text{and at} \quad z=l;$$

will make the θ's vanish, and make

$$m=0, \qquad \pi/l, \qquad 2\pi/l, \qquad 3\pi/l, \quad \text{etc.}$$

Each of these m's has its infinite series of p's, by the equation (22).

Now, as regards the initial state, the electric field and the magnetic field must be both given. For, although the quantity H, fully expressed, alone settles the complete state of the system after the first moment, yet at the first moment (when the previously acting impressed forces finally cease) the electric field and the magnetic field are independent. The energy which is dissipated according to Joule's law has two sources, the electric and the magnetic energies. Now we may, by longitudinal impressed force, set up a certain distribution of magnetic energy, without electric energy. Or, having set up a certain magnetic and a certain electric field by a particular distribution of impressed force, we may alter it in various ways, so as to keep the magnetic field the same whilst we vary the electric field. So both fields require to be known, or equivalent information given.

We may then decompose them into the proper normal systems by means of the universal conjugate property derived from the equation of activity, that of the equality of the mutual electric energy of two complete normal systems to their mutual magnetic energy [vol. I., p. 523.] Thus, if U_{11} and T_{11} are the doubles of the complete electric and magnetic energies of any normal system, and U_{01} is the mutual electric energy of the initial electric field and the normal electric field in question, and T_{01} is the mutual magnetic energy of the initial magnetic field and the normal magnetic field, we shall have

$$A_1=\frac{U_{01}-T_{01}}{U_{11}-T_{11}} \qquad \text{.............................(23)}$$

as the expression for the value of the coefficient A_1, which settles the actual size of the normal system in question. Equal roots require further investigation. This would complete the theoretical treatment. It is best to use the electric and magnetic forces as initial data in the general case. As regards potentials, we cannot express the electric energy in terms of merely the electric potential and the electrification, but require to use also the vector-potential **Z** and the magnetic current.

Simplifications. Thin Return Tube of Constant Resistance. Also Return of no Resistance.

Now there are several important practical simplifications. Suppose, first, that the thickness of the sheath is only a small fraction of its distance from the axis. Then it may be treated as if it were infinitely thin, making the sheath a linear conductor; of course its resistance may remain the same as if of finite thickness. Let a_4 be the very small thickness of the sheath, then the big fraction on the left side of (22) will become

$$\frac{(J_0+s_3a_4J_1)K_1-(K_0+s_3a_4K_1)J_1}{(J_1+s_3a_4J_2)K_1-(K_1+s_3a_4K_2)J_1}(s_3a_3)=-\frac{1}{s_3a_4}\frac{J_1K_0-J_0K_1}{J_2K_1-J_1K_2}(s_3a_3)=-\frac{1}{s_3a_4};$$

wherein J_2 and K_2 are derived from J_1 and K_1 as the latter are derived from J_0 and K_0. So the left side of (22) will become

$$-\frac{\rho_3s_3}{\rho_2s_2}\frac{1}{s_3a_4}=-\frac{\rho_3}{\rho_2a_4s_2}. \qquad \text{.....................(24)}$$

The inductivity of the sheath is now of no importance. Being on the outer edge of the magnetic field, the thinness of the sheath makes its contribution to the magnetic energy be diminished indefinitely.

Again, in important practical cases, the resistance of the return is next to nothing in comparison with that of the wire. Then put $\rho_3=0$ in (22). This makes the left side vanish, and then we sweep away the denominator on the right side, and get the determinantal or differential equation

$$0=\frac{J_1(s_1a_1)K_0(s_2a_1)-(\rho_1s_1/\rho_2s_2)J_0(s_1a_1)K_1(s_2a_1)}{(\rho_1s_1/\rho_2s_2)J_0(s_1a_1)J_1(s_2a_1)-J_1(s_1a_1)J_0(s_2a_1)}J_0(s_2a_2)+K_0(s_2a_2). \quad \text{...(25)}$$

Although we may have the return of nearly no resistance and yet of low conductivity (as in the case of the earth), yet it cannot be quite zero without infinite conductivity, which is what is here assumed. The result is that we shut out the return-conductor from participation, except superficially, in the phenomena. (25) will result from the condition $\rho_2\Gamma_2=0$, or $\Gamma_2=0$, at $r=a_2$; that is, no tangential current, or electric force, in the dielectric close to the sheath. If there could be any, it would involve infinite current-density in the sheath. As it is, there is none, and the return-current has become a mere abstraction, to be measured by the tangential magnetic force divided by 4π, and turned round through a right angle on the inner boundary of the sheath. In a similar manner, if we make the wire infinitely conducting (or of infinitely great inductivity * either) the wire will be shut out. Then the magnetic and electric fields are confined to the dielectric only, and we shall have purely wave-propagation, unless it be a conductor as well.

Now, with the return of no resistance, let the dielectric be non-conducting and the wire non-dielectric, or $c_1=0$, $k_2=0$. The most important simplification arises from the smallness of s_2a_2. For we have

$$-s_2^2=\mu_2cp^2+m^2$$

* [The case, parenthetically mentioned, of infinite inductivity, though resembling that of infinite conductivity in excluding magnetic disturbances from the body of the conductors, differs widely from it in other respects. Considering here only the effect on a train of waves sent along the conductors, the effect of increasing conductivity with constant inductivity is a tendency to surface-concentration and also to a state of perfect slip, without attenuation. But the effect of increasing inductivity is a tendency to surface-concentration together with large attenuation in transit. The S.H. solutions will give more details on this point.]

If the length l of the line is a large multiple of the greatest transverse length a_2 we are concerned with, m^2 is made a small quantity—very small when the line is miles in length, except in case of the insignificant terms involving large multiples of π in $m = n\pi/l$. Again, $(\mu_2 c)^{-\frac{1}{2}}$ is the speed of light through the dielectric, so that unless p be extravagantly large $\mu_2 cp^2$ is exceedingly small also. Thus, with moderate distance of return-current, $s_2 a_2$ is in general exceedingly small.

Therefore, in the expressions (17), take first terms only, making

$$\left.\begin{array}{ll} J_0(s_2 r) = 1, & J_1(s_2 r) = \tfrac{1}{2}s_2 r, \\ K_0(s_2 r) = \log s_2 r, & K_1(s_2 r) = -(s_2 r)^{-1}. \end{array}\right\} \quad \ldots\ldots\ldots\ldots(26)$$

These, used in (25), bring it down to

$$\log\frac{a_2}{a_1} \cdot J_1(s_1 a_1) = \frac{cps_1}{4\pi k_1 a_1 s_2^2} J_0(s_1 a_1)\,; \quad \ldots\ldots\ldots\ldots\ldots\ldots(27)$$

concerning which, so far as substantial accuracy is concerned, the only assumption made is that the return has no resistance.

We have now the following complete normal system :—

$$\left.\begin{aligned} H_1 &= AJ_1(s_1 r)\cos(mz+\theta)\epsilon^{pt}, \\ 4\pi\gamma_1 &= AJ_1(s_1 r)m\sin(mz+\theta)\epsilon^{pt}, \\ 4\pi\Gamma_1 &= AJ_0(s_1 r)s_1\cos(mz+\theta)\epsilon^{pt}, \\ H_2 &= B(s_2^2 r)^{-1}\cos(mz+\theta)\epsilon^{pt}, \\ 4\pi\gamma_2 &= B(s_2^2 r)^{-1}m\sin(mz+\theta)\epsilon^{pt}, \\ 4\pi\Gamma_2 &= B\log(a_2/r)\cos(mz+\theta)\epsilon^{pt}, \end{aligned}\right\} \quad \ldots\ldots\ldots\ldots\ldots\ldots(28)$$

where $$B = A(\rho_1 s_1/\rho_2)J_0(s_1 a_1) \div \log(a_2/a_1).$$

The longitudinal current and electric force in the dielectric vary as the logarithm of the ratio a_2/r, vanishing at $r = a_2$. The radial components vary inversely as the distance. Numerically considered, the longitudinal electric force is negligible against the radial, which is important as causing the electrostatic retardation on long lines. But, theoretically, the longitudinal component of the electric force is very important when we look to the physical actions that take place, as it determines the passage of energy from the dielectric, its seat of transmission along the wire, into the conductor, where it is dissipated.

Regarding (28), however, it is to be remarked that, on account of the approximations, the dielectric solutions do not satisfy the fundamental equation (6). Applying it, we get $\Gamma = 0$. But the other fundamental (7) is satisfied. To satisfy (6), take

$$K_1(s_2 r) = -(s_2 r)^{-1} + \tfrac{1}{2}s_2 r(\log s_2 r - 1)\,;$$

leading to the determinantal equation

$$\log\frac{a_2}{a_1} \cdot J_1(s_1 a_1) = \frac{\rho_1 s_1}{\rho_2} J_0(s_1 a_1)\left\{\frac{1}{s_2^2 a_1} + \tfrac{1}{2}a_1\left(\log\frac{a_2}{a_1} + 1\right)\right\},$$

and requiring us to substitute

$$(s_2^2 r)^{-1} + \tfrac{1}{2}r\log(a_2/r) + \tfrac{1}{2}r$$

for $(s_2^2 r)^{-1}$ in the H_2 and γ_2 formulæ in (28). Then (6) is nearly satisfied, and is quite satisfied if we change the last term in the last expression to $\frac{1}{4}r$. But the other fundamental is violated.

Ignored Dielectric Displacement. Magnetic Theory of Establishment of Current in a Wire. Viscous Fluid Analogy.

Now take $m=0$ in (27), making $-s_2^2=\mu_2 cp^2$, and bringing (27) down to

$$\tfrac{1}{2}s_1a_1J_0(s_1a_1) = -\frac{L_0}{R_0}pJ_1(s_1a_1); \qquad\qquad (29)$$

where $$L_0 = 2\mu_2\log(a_2/a_1),$$

the coefficient of self-induction of the surface-current, and

$$R_0 = (\pi k_1 a_1^2)^{-1},$$

the resistance of the wire, both per unit length of wire; so that L_0/R_0 is the time-constant of the linear theory, on the supposition that the resistance of the wire fully operates, although the current is confined to the surface. This case of $m=0$ is appropriate when the line is so short that the electrostatic induction is really negligible in its effects on the wire-current. In fact we shall arrive at (29) from purely electromagnetic considerations, with $c=0$ everywhere. But it is also the proper equation in the $m=0$ case when the electrostatic retardation is not negligible. It must be taken into account, for instance, in the subsidence of an initially steady current, independently of the electrostatic charge.

Expanding (29) in powers of p, by means of $\frac{1}{4}s_1^2a_1^2 = -\mu_1 p/R_0$, we get

$$1+\frac{\mu_1 p}{R_0}+\frac{1}{4}\left(\frac{\mu_1 p}{R_0}\right)^2+\ldots = -\frac{L_0 p}{R_0}\left(1+\frac{\mu_1 p}{2R_0}+\ldots\right) \qquad (30)$$

Taking first powers only, we get

$$-p^{-1} = (\mu_1 + L_0)/R_0;$$

which is greater than the linear-theory time-constant of the wire by the amount $\frac{1}{2}\mu_1/R_0$, since $\frac{1}{2}\mu_1$ is the inductance per unit length of wire when the return-current is upon its surface.

But taking second powers as well, we get, if $L=\frac{1}{2}\mu_1+L_0$,

$$-p^{-1} = L/R_0 \quad\text{and}\quad \tfrac{1}{2}\mu_1/R_0,$$

of which the first is exactly the linear-theory value. The real time-constant of the first normal system of current, therefore, exceeds the linear-theory value by an amount which is less than $\frac{1}{2}\mu_1/R_0$, when the return is so distant, or the retardation $(\mu_1 k_1 a_1^2)$ of the wire is so small that a steady current subsides with very nearly uniform current-density, being very slightly less at the boundary than at the axis. It is not, however, to be inferred that the subsidence of the "current in the wire" is delayed. It is accelerated, at least at first.

Equation (29) may be written

$$(\mu_1/L_0)J_0(s_1a_1) = \tfrac{1}{2}s_1a_1J_1(s_1a_1), \qquad\qquad (31)$$

the appropriate form when a full investigation is desired. Draw the curves y_1 = right member, and y_2 = left member, the abscissa being s_1a_1. Their intersections will give the values of s_1a_1 satisfying (31). The first root has been already considered, when μ_1/L_0 is very small. The rest, under the same circumstances, will be nearly those of $J_1(s_1a_1)=0$. But if the wire is of iron, μ_1/L_0 may be very large, and there will be no approach to the linear theory. Many normal systems must be taken into account to get numerical solutions. Similarly if the sheath be close to the wire, whether it be magnetic or not.

Electrostatic charge being ignored, join the wire and sheath to make a closed circuit, in which insert a steady impressed force e at time $t=0$. Let Γ be the current at distance r from the axis at time t. (There is no γ now.) The rise of Γ to the final steady value, say Γ_0, is given by

$$\Gamma=\Gamma_0\left\{1-\sum\frac{2}{s_1a_1}\frac{J_0(s_1r)\epsilon^{pt}}{J_1(s_1a_1)(1+s_1^2q^2)}\right\}, \qquad\qquad (32)$$

where $q=L_0a_1/2\mu_1$. The values of s_1a_1 are to be got by (31).

The total current C, or the current in the wire, in ordinary language, rises thus to its final value C_0:—

$$C=C_0\left\{1-\sum\left(\frac{2}{s_1a_1}\right)^2\frac{\epsilon^{pt}}{1+s_1^2q^2}\right\}. \qquad\qquad (33)$$

The boundary-condition of Γ is that, at $r=a_1$,

$$\Gamma+q\frac{d\Gamma}{dr}=0, \qquad \text{therefore} \qquad \frac{J_0}{J_1}(s_1a_1)=s_1q. \qquad\qquad (34)$$

Considering the first term only in the summation in (33), as may be done when the linear theory is nearly followed, that is, after the first stage of the rise, put $-p^{-1}=(L+L_1)/R_0$, where L_1 must be very small compared with L; then

$$C=C_0\left\{1-\frac{(L+L_1)^2\epsilon^{pt}}{L^2+\mu_1L_1+\frac{1}{4}\mu_1^2}\right\}.$$

When the current is started, by a steady impressed force in the coil-circuit, in a long solenoidal coil of small thickness, containing a solid conducting core, the magnetic force in the core rises in the same manner as the current in the wire, according to (32); because the boundary-condition of the magnetic force is of the same form as (34), q being then a function of the number of windings, etc.

There is also the water-pipe analogy, which is always turning up. This I have before made use of [vol. I., p. 384]. Water in a round pipe is started from rest and set into a state of steady motion by the sudden and continued application of a steady longitudinal dragging or shearing-force applied *to its boundary*, according to the equation (32). This analogy is useful because every one is familiar with the setting of water in motion by friction on its boundary, transmitted inward by viscosity.

Graphically representing (32), abscissæ the time, and ordinates Γ, at the centre, intermediate points, and the boundary, by what we may call the arrival-curves of the current, and comparing them with

$$\Gamma=\Gamma_0(1-\epsilon^{-R_0t/L}),$$

the linear theory arrival curve at all parts of the wire, we may notice these characteristics. The current rises much more rapidly at the boundary than according to the linear theory, at first, but much more slowly in the later stages. Going inward from the boundary we find that an inflection is produced in the arrival-curve near its commencement; the rapid rise being delayed for an appreciable interval of time. This dead period is, of course, very marked at the axis of the wire, there being practically no current at all there until a certain time has elapsed. That the central part of the wire is nearly inoperative when rapid reversals are sent is easily understood from this, or perhaps more easily by the use of the water-pipe analogue. Some curves of (32), for two special values of q, I have already given [vol. I., p. 398; vol. II., p. 58].

Magnetic Theory of S.H. Variations of Impressed Voltage and resulting Current.

Let there be a simple-harmonic impressed force $e \sin nt$ in the circuit of wire and sheath, with no external resistance, making a total circuit-resistance R. (I translate the core-solution into the wire-solution.) The boundary condition is

$$\frac{e \sin nt}{R\pi a_1^2} = \Gamma + q\frac{d\Gamma}{dr}; \qquad \text{.........(35)}$$

and the solution is

$$\Gamma = \frac{e}{R\pi a_1^2}(P_0^2 + Q_0^2)^{-\frac{1}{2}}\Big\{(P_0 M + Q_0 N)\sin nt + (P_0 N - Q_0 M)\cos nt\Big\}; \quad \text{...(36)}$$

where M and N are the following functions,

$$\left.\begin{aligned} M &= \tfrac{1}{2}J_0(xr\sqrt{i}) + \tfrac{1}{2}J_0(xr\sqrt{-i}), \\ N &= \tfrac{1}{2}iJ_0(xr\sqrt{i}) - \tfrac{1}{2}iJ_0(xr\sqrt{-i}), \end{aligned}\right\} \qquad \text{.........(37)}$$

i standing for $\sqrt{-1}$, and x for $\sqrt{4\pi\mu_1 k_1 n}$. Also

$$P = M + qM', \qquad Q = N + qN', \qquad \text{.........(38)}$$

the $'$ denoting differentiation to r. In (36), M and N have the values at distance r, and P_0, Q_0 the values at $r = a_1$, the boundary.

We have

$$P^2 + Q^2 = M^2 + N^2 + 2q(MM' + NN') + q^2(M'^2 + N'^2). \qquad \text{.......(39)}$$

If $y = (xr)^4 = (4\pi\mu_1 k_1 r^2 n)^2$, we have the following series :—

$$\left.\begin{aligned} M^2 + N^2 &= 1 + \frac{y}{2.4^2}\Big(1 + \frac{1}{2}\frac{y}{6.8^2}\Big(1 + \frac{1}{3}\frac{y}{10.12^2}\Big(1 + \frac{1}{4}\frac{y}{14.16^2}\Big(1 + \dots, \\ M'^2 + N'^2 &= \frac{y}{4r^2}\Big(1 + \frac{3y}{4^2 6^2}\Big(1 + \frac{3\frac{1}{3}y}{8^2 10^2}\Big(1 + \frac{3\frac{1}{2}y}{12^2 14^2}\Big(1 + \frac{3\frac{3}{5}y}{16^2 18^2}\Big(1 + \dots, \\ MM' + NN' &= \frac{y}{16r}\Big(1 + \frac{6y}{6^2 8^2}\Big(1 + \frac{5y}{10^2 12^2}\Big(1 + \frac{4\frac{2}{3}y}{14^2 16^2}\Big(1 + \frac{4\frac{1}{2}y}{18^2 20^2}\Big(1 + \dots \end{aligned}\right\} (40),$$

These are suitable for calculating the amplitude of Γ or of C when y is not a very large quantity. The wire-current C is given by

$$C = \frac{ea_1}{2l\mu_1 n}\left(\frac{M'^2+N'^2}{P^2+Q^2}\right)^{\frac{1}{2}} \sin\left(nt - \tan^{-1}\frac{QN'+PM'}{PN'-QM'}\right), \quad \text{.....(41)}$$

where P, Q, M, N, M', N' have the boundary values. As for M and N themselves, their expansions are

$$\left.\begin{aligned} M &= 1 - \frac{y}{2^2 4^2} + \frac{y^2}{2^2 4^2 6^2 8^2} - \dots, \\ M &= \frac{y^{\frac{1}{2}}}{2^2} - \frac{y^{\frac{3}{2}}}{2^2 4^2 6^2} + \frac{y^{\frac{5}{2}}}{2^2 4^2 6^2 8^2 10^2} - \dots . \end{aligned}\right\} \quad \text{..................(42)}$$

But these series are quite unsuitable when y is very large. Then use the approximate formulæ

$$\left.\begin{aligned} J_0(sr) &= \left(\frac{2}{\pi sr}\right)^{\frac{1}{2}} \cos\left(sr - \frac{\pi}{4}\right), \\ J_1(sr) &= \left(\frac{2}{\pi sr}\right)^{\frac{1}{2}} \cos\left(sr - \frac{3\pi}{4}\right), \end{aligned}\right\} \quad \text{..........................(43)}$$

which make, if $f = y^{\frac{1}{4}}$,

$$\left.\begin{aligned} M^2 + N^2 &= J_0(f\sqrt{i})J_0(f\sqrt{-i}) = \epsilon^{f\sqrt{2}}/2\pi f, \\ M'^2 + N'^2 &= f\epsilon^{f\sqrt{2}}/2\pi r^2, \\ MM' + NN' &= \epsilon^{f\sqrt{2}}/2\pi r\sqrt{2}. \end{aligned}\right\} \quad \text{...............(44)}$$

In the extreme, very high frequency, or large retardation, or both combined, making y very great, the amplitude of the wire-current C tends to be represented by

$$e/L_0 ln; \quad \text{.....................................(45)}$$

showing that the current is stronger than according to the linear theory, and far stronger in the case of an iron wire, or very close return.

The amplitude of the current-density at the axis, under the same circumstances, with $r = a_1$ in f, is

$$\frac{e}{R\pi a_1^2 L_0} \cdot \frac{2\mu}{a_1} \cdot \left(\frac{2\pi a_1^2}{f\epsilon^{f\sqrt{2}}}\right)^{\frac{1}{2}}, \quad \text{......................(46)}$$

which is of course excessively small. On the other hand, the boundary current-density amplitude is

$$\frac{(2\mu_1/a_1)e}{R\pi a_1^2 L_0 (4\pi\mu_1 k_1 n)^{\frac{1}{2}}} = \frac{e}{L_0 l a_1}\left(\frac{\mu_1 k_1}{\pi n}\right)^{\frac{1}{2}}, \quad \text{..................(47)}$$

which may be greater than the linear-theory amplitude.

Analogous to this, the amplitude of the current in a coil due to a S.H. impressed force in the coil-circuit is greatly increased by allowing dissipation of energy by conduction in a core placed in the coil, when the corresponding y is great, a large core, high inductivity, etc.; that is, the inertia or retarding-power of the electromagnet is greatly reduced, so far as the coil-current is concerned. This is, in a great measure, done away with by dividing the core to stop the electric currents, when the linear theory is approximated to.

If $y=1600$, the axial is about one-fourteenth of the boundary-current amplitude. To get this in a thick copper wire of 1 centim. radius, a frequency of about 850 waves per second would be required. But in an iron rod of the same size, if we take $\mu_1=500$, only about $8\frac{1}{2}$ waves per second would suffice.

Returning to the former expressions, if we go only as far as n^6, the amplitude C_0 of the wire-current is given by $C_0=e/R''l$; where the square of R'', which is the "apparent resistance," or the impedance, per unit length of wire, is given by

$$R''^2=R_0^2+L^2n^2+\frac{R_0^2g}{6}-\frac{R_0^2g^2}{24}\left(\frac{L}{\mu_1}+\frac{1}{10}\right)+\frac{R_0^2g^3}{48.90}\left(13\frac{L}{\mu_1}+\frac{79}{56}\right)-\ldots, \tag{48}$$

where $g=(\mu_1 n/R_0)^2$, and R_0 and L have the former meanings.

When only the total current is under investigation, the method followed by Lord Rayleigh (*Phil. Mag.*, May, 1886) possesses advantages. I find it difficult, however, to understand how the increased resistance can become of serious moment. For, above a certain frequency, the current-amplitude is increased; whilst, below that frequency, its reduction, from that given by the linear theory, appears to be, in copper wires, quite insignificant in general [vol. II., p. 67].

PART II.

Extension of General Theory to two Coaxial Conducting Tubes.

In Part I. the inner conductor was solid. Let now the central portion be removed, making it a hollow tube of outer radius a_1 and inner a_0. The reason for this modification is that the theory of a tube is not the same when the return-conductor is outside as when it is inside it; that is to say, it depends upon the position of the dielectric, the primary seat of the transfer of energy. The expression for H_1, the magnetic force at distance r from the axis, will now be

$$H_1=\{J_1(s_1r)-(J_1/K_1)(s_1a_0)K_1(s_1r)\}A_1; \tag{49}$$

instead of the former $A_1J_1(s_1r)$, of the first of equations (18); if we impose the condition $H_1=0$ at the inner boundary of the wire (as we may still call the inner tube). This means that there is to be no current from $r=0$ to $r=a_0$; we therefore ignore the minute longitudinal dielectric-current in this space, just as we ignored that beyond $r=a_3$ previously. If we wish to necessitate that this shall be rigidly true, we may suppose that within $r=a_0$, and beyond $r=a_3$, we have not merely $k=0$, but also $c=0$, thus preventing current, either conducting or dielectric. In any case, with only $k=0$, the dielectric disturbance must be exceedingly small. On this point I may mention that my brother, Mr. A. W. Heaviside, experimenting with a wire and outer tube for the return, using a (for telegraphic purposes) very strong current, rapidly interrupted, and a sensitive telephone in circuit with a parallel outer wire, could not detect the least sign of any inductive action outside the tube, at least when the source of energy (the battery) was kept at a distance from the telephone. In explanation of the last remark, we need only consider that, although the transfer of energy is from the battery along the tubular space between the wire and return, yet, before getting to this confined space, there is a spreading out of the disturbances, so that in the neighbourhood of the battery the disk of a telephone may be strongly influenced by the variations of the magnetic field. On the other hand, the induction between parallel wires whose circuits are completed through the earth, is perceptible with the telephone at hundreds of miles distance, or practically at any distance, if the proper means be taken which theory points out. His direct experiments have, so far, only gone as far as forty miles, quite recently; but this distance may easily be extended.

Corresponding to (49) we shall have

$$4\pi\Gamma_1 = s_1\{J_0(s_1r) - (J_1/K_1)(s_1a_0)K_0(s_1r)\}A_1; \quad \text{...........(50)}$$

omitting, in both, the z and t factors. Now, to obtain the corresponding development of the general equation (22), we have only to change the $J_0(s_1a_1)$ in it to the quantity in the {} in (50), and the $J_1(s_1a_1)$ to that in the {} in (49), with $r=a_1$ in both cases.

Electrical Interpretation of the Differential Equations. Practical Simplification in Terms of Voltage V *and Current* C.

The method by which (22) was got was the simplest possible, reducing to mere algebra the work that would otherwise involve much thinking out; and, in particular, avoiding some extremely difficult reasoning relating to potentials, scalar and vector, that would occur were they considered *ab initio*. But, having got (22), the interpretation is comparatively easy. Starting with the inner tube, (49) is the general solution of (14), with the limitation $H_1=0$ at $r=a_0$; if, in s, given by

$$-s_1^2 = 4\pi\mu_1k_1p + m^2,$$

we let p mean d/dt and m^2 mean $-d^2/dz^2$, instead of the constants in a normal system of subsidence, and let A_1 be an arbitrary function of z and t. Similarly, (50) gives us the connection between Γ_1 and A_1. From it we may see what A_1 means. For, put $r=a_0$ in (50); then, since

$$(J_0K_1 - J_1K_0)(x) = -x^{-1},$$

we see that
$$A_1 = -4\pi a_0K_1(s_1a_0)\Gamma_0,$$

if Γ_0 is the current-density at $r=a_0$. When the tube is solid, $A_1=4\pi\Gamma_0/s_1$. But, without knowing A_1, (49) and (50) connect H_1 and Γ_1 directly, when A_1 is eliminated by division. Also, $H_1=C_1\times(2/r)$, if C_1 be the total longitudinal current from $r=a_0$ to r; hence

$$\Gamma_1 = \frac{s_1}{2\pi r}\,\frac{J_0(s_1r) - (J_1/K_1)(s_1a_0)K_0(s_1r)}{J_1\ldots\ldots - \ldots\ldots\ldots\ldots\ldots K_1\ldots\ldots}C_1 \quad \text{..............(51)}$$

connects the current-density and the integral current.

Now pass to the outer tube. Quite similarly, remembering that $H_3=0$ at $r=a_3$, we shall arrive at

$$\Gamma_3=\frac{s_3}{2\pi r}\frac{J_0(s_3r)-(J_1/K_1)(s_3a_3)K_0(s_3r)}{J_1\ldots\ldots-\ldots\ldots\ldots\ldots\ldots K_1\ldots\ldots}C_3, \qquad (52)$$

connecting Γ_3, the longitudinal current-density at distance r in the outer tube, with C_3, the current through the circle of radius r in the plane perpendicular to the axis.

Next, let there be longitudinal impressed electric forces in the wire and return, of uniform intensities e_1 and e_2 over the sections of the two conductors. We shall have

$$\rho_1\Gamma_1=e_1+E_1, \qquad \rho_3\Gamma_3=e_3+E_3\,; \qquad (53)$$

if E_1 and E_3 are the longitudinal electric forces "of the field." Therefore

$$e_1-e_3=e=\rho_1\Gamma_1-\rho_3\Gamma_3-(E_1-E_3), \qquad (54)$$

where e is the impressed force per unit length *in the circuit* at the place considered: the positive direction in the circuit being along the wire in the direction of increasing z, and oppositely in the return.

If we take $r=a_1$ in (51), and $r=a_2$ in (52), and use them in (54), then, since C_1 becomes C, the wire-current, and C_3 becomes the same *plus* the longitudinal dielectric-current, we see that if we agree to ignore the latter, and can put E_1-E_3 in terms of C, (54) will become an equation between e and C.

To obtain the required E_1-E_3, consider a rectangular circuit in a plane through the axis, two of whose sides are of unit length parallel to z at distances a_1 and a_2 from the axis, and the other two sides parallel to r, and calculate the E.M.F. of the field in this circuit in the direction of the circular arrow. If z be positive from left to right, the positive direction of the magnetic force through the circuit is upward through the paper. Therefore, if V be the line-integral of the radial electric force from $r=a_1$ to $r=a_2$, so that dV/dz is the part of the E.M.F. in the rectangular circuit due to the radial force, we shall have

$$E_1-E_3+\frac{dV}{dz}=-\int_{a_1}^{a_2}\mu_2\dot{H}_2\,dr,$$

by the Faraday law, or equation (7); H_2 being the magnetic force in the dielectric. This being $2C/r$, on account of our neglect of Γ_2, we get, on performing the integration, $-L_0\dot{C}$, on the right side, where L_0 is the previously-used inductance of the dielectric per unit length. This brings (54) to

$$e-\frac{dV}{dz}=L_0pC+\frac{\rho_1s_1}{2\pi a_1}\frac{J_0(s_1a_1)-(J_1/K_1)(s_1a_0)K_0(s_1a_1)}{J_1\ldots\ldots-\ldots\ldots\ldots\ldots\ldots K_1\ldots\ldots}C$$

$$-\frac{\rho_3s_3}{2\pi a_2}\frac{J_0(s_3a_2)-(J_1/K_1)(s_3a_3)K_0(s_3a_2)}{J_1\ldots\ldots-\ldots\ldots\ldots\ldots\ldots K_1\ldots\ldots}C, \qquad (55)$$

which, for brevity, write thus,

$$e-\frac{dV}{dz}=L_0pC+R_1''C+R_2''C, \qquad \text{(56)}$$

where R_1'' and R_2'' define themselves in (55). They are generalised resistances of wire and return respectively, per unit length. But of their structure, later. Equation (56) is what we get from (22) by treating s_2r as a small quantity and using (26); remembering also the extension from a solid to a hollow wire.

By more complex reasoning we may similarly put the right member of (54) in terms of C without the neglect of Γ_2, and arrive at (22) itself, in a form similar to (55) or (56). But we may get it from (22) at once by a proper arrangement of the terms. It becomes

$$0=\left(R_1''+R_2''\frac{R_{02}''}{R_{01}''}+R_{03}''+\frac{R_1''R_2''}{R_{01}''}\right)C. \qquad \text{(57)}$$

Here R_1'' and R_2'' are as before, whilst R_{01}'' and R_{02}'' are similar expressions for the dielectric, on the assumption that $H=0$ at $r=a_1$ or at $r=a_2$ respectively; thus,

$$R_{01}''=+\frac{\rho_2 s_2}{2\pi a_2}\frac{J_0(s_2a_2)-(J_1/K_1)(s_2a_1)K_0(s_2a_2)}{J_1\ldots\ldots-\ldots\ldots\ldots\ldots K_1\ldots\ldots},$$

$$R_{02}''=-\frac{\rho_2 s_2}{2\pi a_1}\frac{J_0(s_2a_1)-(J_1/K_1)(s_2a_2)K_0(s_2a_1)}{J_1\ldots\ldots-\ldots\ldots\ldots\ldots K_1\ldots\ldots}.$$

R_{03}'' has a different structure, being given by

$$R_{03}''=-\frac{\rho_2 s_2}{2\pi a_1}\frac{J_0(s_2a_1)-(J_0/K_0)(s_2a_2)K_0(s_2a_1)}{J_1\ldots\ldots-\ldots\ldots\ldots\ldots K_1\ldots\ldots}.$$

In these take s_2r small; they will become

$$R_{01}''=R_{02}''=\frac{\rho_2}{\pi(a_2^2-a_1^2)};$$

that is, if ρ_2 be imagined to be resistivity, the steady resistance per unit length of the dielectric tube (fully, ρ_2 is the reciprocal of $k_2+c_2p/4\pi$); and, with $k_2=0$,

$$R_{03}''=-\frac{s_2^2}{cp}2\log\frac{a_2}{a_1}=L_0p+\frac{m^2}{Sp},$$

if S is the electric capacity per unit length, such that $L_0S=\mu_2c_2$. Then, introducing e, (57) reduces to

$$e=(L_0p+m^2/Sp+R_1''+R_2'')C, \qquad \text{(58)}$$

which is really the same as (56). For, by continuity, or by the second of (11),

$$-\frac{dC}{dz}=2\pi a_1\gamma_1=2\pi a_1p\sigma=SpV, \qquad \text{(59)}$$

if σ is the time-integral of the radial current at $r=a_1$, or, in other words, the electrification surface-density there, when the conductors are non-dielectric. (There is equal $-\sigma$ at the $r=a_2$ surface.) Therefore

$$-\frac{1}{Sp}\frac{d^2C}{dz^2}=\frac{m^2}{Sp}C=\frac{dV}{dz}, \qquad (60)$$

which establishes the equivalence.

Particular attention to the meaning of the quantity V is needed. It is the line-integral of the radial force in the dielectric from $r=a_1$ to $r=a_2$. Or it may be defined by

$$SV=2\pi a_1\sigma=Q,$$

if Q be the charge per unit length of wire. But it is not the electric potential at the surface of the wire. It is not even the excess of the potential at the wire-boundary over that at the inner boundary of the return. For, as it is the line-integral of the electric force from end to end of the tubes of displacement, it includes the line-integral of the electric force of inertia. It has, however, the obvious property of allowing us to express the electric energy in the dielectric in the form of a surface-integral, thus, $\frac{1}{2}V\sigma$ per unit area of wire-surface, or $\frac{1}{2}VQ$ per unit length of wire, instead of by a volume-integration throughout the dielectric. Hence the utility of V. The possibility of this property depends upon the comparative insignificance of the longitudinal current in the dielectric, which we ignore. It may happen, however, that the longitudinal displacement is far greater than the radial; but then it will be of so little moment that the problem could be taken to be a purely electromagnetic one. We need not use V at all, (58) being the equation between e and C without it. It is, however, useful in electrostatic problems, for the above-mentioned reason. Again, instead of V, we may use σ or Q, which are definitely localized.

The physical interpretation of the force $-dV/dz$, in terms of Maxwell's inimitable dielectric theory, is sufficiently clear, especially when we assist ourselves by imagining the dielectric displacement to be a real displacement, elastically resisted, or any similar elastically resisted generalized displacement of a vector character. When there is current from the wire into the dielectric there is necessarily a back electric force in it due to the elastic displacement; and if it vary in amount along the wire, its variation constitutes a longitudinal electric force.

(58) being a differential equation previously, let m^2 be a constant in it. Then R_1'' and R_2'' may be thus expressed:—

$$R_1''=R_1'+L_1'p, \qquad R_2''=R_2'+L_2'p, \qquad (61)$$

where R_1' and R_2', L_1' and L_2' are functions of p^2. The utility of this notation arises from R_1' etc. becoming mere constants in simple-harmonically vibrating systems. Let e_m, V_m, and C_m be the corresponding quantities for the particular m; then, by (56),

$$e_m-\frac{dV_m}{dz}=L_0pC_m+(R_{1m}'+L_{1m}'p)C_m+(R_{2m}'+L_{2m}'p)C_m. \qquad (62)$$

Or

$$e_m-\frac{dV_m}{dz}=(R_m'+L_m'p)C_m, \qquad (63)$$

where

$$R_m'=R_{1m}'+R_{2m}', \qquad L_m'=L_0+L_{1m}'+L_{2m}'. \qquad (64)$$

R'_m and L'_m are functions of p^2. Therefore, by (62), summing up,

$$e - \frac{dV}{dz} = \Sigma (R'_m + L'_m p) C_m. \quad \text{......................(65)}$$

Now, although R'_m and L'_m are really different functions of p^2 for every different value of m, since they contain m^2, yet if, in changing from one m to another, through a great many m's, from $m=0$ upward, they should not materially change, we may regard R'_m and L'_m as having the $m=0$ expressions, as in the purely electromagnetic case, and denote them by R' and L' simply. Then (65) becomes

$$e - \frac{dV}{dz} = (R' + L'p)C \quad \text{......................(66)}$$

simply. The equation of V is now

$$-\frac{de}{dz} + \frac{d^2V}{dz^2} = (R' + L'p)SpV; \quad \text{......................(67)}$$

and that of C_m being

$$e_m = (R'_m + L'_m p + m^2/Sp)C_m \quad \text{......................(68)}$$

in the m case, that of C becomes now simply

$$Spe + \frac{d^2C}{dz^2} = (R' + L'p)SpC. \quad \text{......................(69)}$$

The assumption above made is, in general, justifiable.

Previous Ways of treating the subject of Propagation along Wires.

Let us now compare these equations with the principal ways that have been previously employed to express the conditions of propagation of signals along wires. For simplicity, leave out the impressed force e. First, we have Ohm's system, which may be thus written:—

$$-\frac{dV}{dz} = RC, \qquad -\frac{dC}{dz} = SpV, \qquad \frac{d^2V}{dz^2} = RSpV. \quad \text{........(70)}$$

Here the first equation expresses Ohm's law. C is the wire-current, R the resistance per unit length, and V is a quantity whose meaning is rather indistinct in Ohm's memoir, but which would be now called the potential. The second equation is of continuity. Misled by an entirely erroneous analogy, Ohm supposed electricity could accumulate in the wire in a manner expressed by the second of (70), wherein S therefore depends upon a specific quality *of the conductor*. The third equation results from the two previous, and shows that V, or C, or $Q = SV$ diffuse themselves through the wire as heat does by difference of temperature when there is no surface-loss. This system has at present only historical interest. The most remarkable thing about it is the getting of equations correct in form, at least approximately, by entirely erroneous reasoning.

The matter was not set straight till a generation later, when Sir W. Thomson arrived at a system which is formally the same as (70), but in which V is precisely defined, whilst S changes its meaning entirely. V is now to be the electrostatic potential, and S is the electrostatic capacity *of the condenser* formed by the opposed surfaces of the wire and return with dielectric between. The continuity of the current in the wire is asserted; but it can be discontinuous at its surface, where electricity accumulates and charges the condenser. In short, we simply unite Ohm's law (with continuity of current in the conductor) and the similar condenser law. The return is supposed to have no resistance, and $V=0$ at its boundary.

The next obvious step is to bring the electric force of inertia into the Ohm's law equation, and make the corresponding change in that of V; that is, if we decide to accept the law of quasi-incompressibility of electricity in the conductor, which is implied by the second of (70), when Sir W Thomson's meanings of S and V are accepted. Kirchhoff seems to have been the first to take inertia into account, arriving at an equation which is reducible to the form

$$d^2V/dz^2=(R+Lp)SpV.$$

I am, unfortunately, not acquainted with his views regarding the continuity of the current, so that, translated into physical ideas, his equation may not be conformable to Maxwell's ideas, even as regards the conductor. Also, as his estimation of the quantity L was founded upon Weber's hypothesis, it may possibly turn out to be different in value from that in the next following system. In ignorance of Kirchhoff's investigation, I made the necessary change of bringing in the electric force of inertia in a paper "On the Extra Current" (*Phil. Mag.*, August, 1876), [Art. XIV., vol. I., p. 53] getting this system,

$$-\frac{dV}{dz}=(R+Lp)C, \qquad -\frac{dC}{dz}=SpV, \qquad \frac{d^2V}{dz^2}=(R+Lp)SpV, \qquad (71)$$

wherein everything is the same as in Sir W. Thomson's system, with the addition of the electric force of inertia $-LpC$, where L is the coefficient of self-induction, or, as I now prefer to call it [vol. II., p. 28], the inductance, per unit length of the wire, according to Maxwell's system, being numerically equal to twice the energy, per unit length of wire, of the unit current in the wire, uniformly distributed.

The system (71) is amply sufficient for all ordinary purposes, with exceptions to be later mentioned. It applies to short lines as well as to long ones; whereas the omission of L, reducing (71) to (70), renders the system quite inapplicable to lines of moderate length, as the influence of S tends to diminish as the line is shortened, relatively to that of L. An easily-made extension of (71) is to regard R as the sum of the steady resistances of wire and return, and V as the quantity Q/S, Q being the charge per unit length of wire. Nor are we, in this approximate system (71), obliged to have the return equidistant from the wire. It may, for instance, be the earth, or a parallel wire, with the corresponding changes in the formulæ for the electric capacity and inductance.

But there are extreme cases when (71) is not sufficient. For example, an iron wire, unless very fine, by reason of its high inductivity; a very thick copper wire, by reason of thickness and high conductivity; or, a very close return-current, in which case, no matter how fine a wire may be, there is extreme departure from uniformity of current-distribution in the variable period; or, extremely rapid reversals of current, for, no matter what the conductors may be, by sufficiently increasing the frequency we approximate to surface-conduction.

We must then, in the system (71), with the extension of meaning of R and V just mentioned, change R and L to R' and L', as in (67), and other equations. In a S.H. problem, this simply changes R and L from certain constants to others, depending on the frequency. But, in general, it would, I imagine, be of no use developing R_1'' etc. in powers of p, so that we must regard $(R_1'+L_1'p)$ etc. merely as a convenient abbreviation for the R_1'' etc. defined by (56) and (55).

A further refinement is to recognise the differences between R' and L' in one m system and another, instead of assuming $m=0$ in R_m''. And lastly, to obtain a complete development, and exact solutions of Maxwell's equations, so as to be able to fully trace the transfer of energy from source to sink, fall back upon (57), or (22), and the normal systems (18) of Part I.

The Effective Resistance and Inductance of Tubes.

Now, as regards our obtaining the expansions of R_1' etc. in powers of p^2, we have to expand the numerators and the denominators of R_1'' and R_2'' in powers of p, perform the divisions, and then separate into odd and even powers. When the wire is solid, the division is merely of $\frac{1}{2}xJ_0(x)$ by $J_1(x)$, a comparatively easy matter. The solid wire R' and L' expansions were given by Lord Rayleigh (*Phil. Mag.*, May, 1886). I should mention that my abbreviated notation was suggested by his. But in the tubular case, the work is very heavy, so, on account of possible mistakes, I go only as far as p^2, or three terms in the quotient. The work does not need to be done separately for the inner and the outer tube, as a simple change converts one R' or L' into the other. Thus, in the case of the inner tube, we shall have

$$R_1'=R_1\left[1+n^2(\mu_1k_1\pi a_1^2)^2\left\{\frac{1}{12}-\frac{2}{3}\frac{a_0^2}{a_1^2}+\frac{7}{12}\frac{a_0^4}{a_1^4}+\frac{2a_0^4\log(a_1/a_0)}{a_1^2(a_1^2-a_0^2)}-\frac{4a_0^6\{\log(a_1/a_0)\}^2}{a_1^2(a_1^2-a_0^2)^2}\right\}\right], \quad (72)$$

$$L_1'=R_1(\mu_1k_1\pi a_1^2)\left\{\frac{1}{2}-\frac{3}{2}\frac{a_0^2}{a_1^2}+2\log\frac{a_1}{a_0}\cdot\frac{a_0^4}{a_1^2(a_1^2-a_0^2)}\right\}, \quad \text{..............}(73)$$

where n^2 is written for $-p^2$, for the S.H. application.

As for L_1', it is simply the inductance of the tube per unit length (of the tube only), as may be at once verified by the square-of-force method. The first correction depends upon p^3. But R_1' gives us the first correction to R_1, which is the steady resistance, so it is of some use. To obtain R_2' and L_2' from these, change R_1 to R_2, μ_1 and k_1 to μ_3 and k_3, a_0 to a_3, and a_1 to a_2. Or, more simply, (72) and (73) being the tube-formulæ when the return is outside it, if we simply exchange a_0 and a_1 we shall get the formulæ for the *same* tube when the return is inside it.

If the tube is thin, there is little change made by thus shifting the locality of the return. But if a_1/a_0 be large, there is a large change. This will be readily understood by considering the case of a wire whose return is outside it, and of great bulk. Although the steady resistance of the return may be very low, yet the percentage correction will be very large, compared with that for the wire.

Taking $a_1/a_0 = 2$ only, we shall find

$$R_1' = R_1[1 + (\pi k_1 a_1^2 \mu_1 n)^2 \times \cdot 012]$$

when the return is outside, and

$$R_1' = R_1[1 + (\pi k_1 a_0^2 \mu_1 n)^2 \times \cdot 503]$$
$$= R_1[1 + (\pi k_1 a_1^2 \mu_1 n)^2 \times \cdot 031],$$

when the return is inside. In the case of a solid wire, the decimals are ·083, so that whilst the correction is reduced, in this $a_1/a_0 = 2$ example, the reduction is far greater when the return is outside than when it is inside.

The high-frequency tube-formulæ are readily obtained. Those for the inner tube are the same as for a solid wire, and those for the outer tube depend not on its bulk, but on its inner radius. That is, in both cases it is the extent of surface that is in question, next the dielectric, from which the current is transmitted into the conductors. Let $G_0(x) = (2/\pi)K_0(x)$, and $G_1(x) = (2/\pi)K_1(x)$; then, when x is very large,

$$\left.\begin{aligned} J_0(x) &= -G_1(x) = (\sin x + \cos x) \div (\pi x)^{\frac{1}{2}}, \\ J_1(x) &= G_0(x) = (\sin x - \cos x) \div (\pi x)^{\frac{1}{2}}. \end{aligned}\right\} \qquad (74)$$

Use these in the R_1'' fraction, and put in the exponential form. We shall obtain

$$R_1'' = (\rho_1 s_1 i)/(2\pi a_1).$$

But $\quad \frac{1}{2} s_1 a_1 i = (\pi k_1 \mu_1 p)^{\frac{1}{2}} a_1, \quad$ therefore $\quad R_1'' = (\mu_1 \rho_1 p/\pi a_1^2)^{\frac{1}{2}}.$

Also, $\quad p^2 = -n^2, \quad$ therefore $\quad p^{\frac{1}{2}} = (\frac{1}{2}n)^{\frac{1}{2}}(1+i) = (\frac{1}{2}n)^{\frac{1}{2}} + p(\frac{1}{2}n^{-1})^{\frac{1}{2}},$

so that, finally, $\quad R_1' = \dfrac{(\mu_1 \rho_1 q)^{\frac{1}{2}}}{a_1}, \qquad L_1' = \dfrac{R_1'}{n}, \qquad (75)$

where $q = n/2\pi$ is the frequency. To get R_2' and L_2', change the μ and ρ of course, and also a_1 to a_2.

It is clear that the thinner the tube, the greater must be the frequency before these formulæ can be applicable. For the steady resistance is increased indefinitely by reducing the thickness of the tube, whilst the high-frequency resistance is independent of the steady resistance, and must be much greater than it. In (75) then, q must be great enough to make R' several times R, itself very large when the tube is very thin. Consequently thin tubes, as is otherwise clear, may be treated as linear conductors, subject to the equations (71), with no corrections, except under extreme circumstances. The L may be taken as L_0, except in the case of iron.

Train of Waves due to S.H. Impressed Voltage. Practical Solution.

I will now give the S.H. solution in the general case, subject to (58). Let there be any distribution of e (longitudinal, and of uniform intensity over cross-sections). Expand it in the Fourier-series appropriate to the terminal conditions at $z=0$ and l. For definiteness, let wire and return be joined direct, without any terminal resistances. Then, $e_0 \sin nt$ being e at distance z, the proper expansion is

$$e_0 = e_{00} + e_{01} \cos m_1 z + e_{02} \cos m_2 z + \dots,$$

where $m_1 = \pi/l$, $m_2 = 2\pi/l$, etc. (It should be remembered that e is the $e_1 - e_2$ of (54) and (53). Shifting impressed force from the wire to the return, with a simultaneous reversal of its direction, makes no difference in e. Thus two e's directed the same way in space, of equal amounts, and in the same plane $z=$ constant, one in the inner, the other in the outer conductor, cancel. This will clearly become departed from as the distance of the return from the wire is increased.) Then, in the equation

$$\begin{aligned} e_m &= (R'_m + L'_m p) C_m + (m^2/Sp) C_m \\ &= R'_m C_m + (L'_m - m^2/Sn^2) p C_m, \end{aligned}$$

we know e_m; whilst R'_m and L'_m are constants. The complete solution is obtained by adding together the separate solutions for e_{00}, e_{01}, etc., and is

$$C = \frac{1}{l}\left\{\frac{e_{00}\sin(nt-\theta_0)}{(R'^2 + L'^2 n^2)^{\frac{1}{2}}} + 2\sum \frac{e_{0m}\sin(nt-\theta_m)\cos mz}{[R'^2_m + (L'_m - m^2/Sn^2)^2 n^2]^{\frac{1}{2}}}\right\}, \quad \dots\dots(76)$$

where the summation includes all the m's, and

$$\tan\theta_m = (L'_m - m^2/Sn^2)n \div R'_m.$$

A practical case is, no impressed force anywhere except at $z=0$, one end of the line, where it is $V_0 \sin nt$. Then, imagining it to be V_0/z_1 from $z=0$ to $z=z_1$, and zero elsewhere, and diminishing z_1 indefinitely, the expansion required is

$$V_0/z_1 = (V_0/l)(1 + 2\Sigma \cos j\pi z/l),$$

j going from 1, 2, ... to ∞. This makes the current-solution become

$$C = \frac{V_0}{l}\left\{\frac{\sin(nt-\theta_0)}{(R'^2+L'^2n^2)^{\frac{1}{2}}} + 2\sum\frac{\sin(nt-\theta_m)\cos mz}{\{R'^2_m + (L'_m - m^2/Sn^2)^2n^2\}^{\frac{1}{2}}}\right\}. \quad \dots.(77)$$

If the line is short, neglect the summation altogether, unless the frequency is excessive. Now (77) may perhaps be put in a finite form when R'_m is allowed to be different from R', though I do not see how to do it. But when $R'_m = R'$ and $L'_m = L'$ it can of course be done, for we may then use the finite solutions of (66) and (67). Thus, given $V = V_0 \sin nt$ at $z=0$, and no impressed force elsewhere, find V and C everywhere subject to (66) and (67) with $e=0$, and $V=0$ at $z=l$.

Let

$$\left.\begin{aligned} P &= (\tfrac{1}{2}Sn)^{\frac{1}{2}}\{(R'^2 + L'^2n^2)^{\frac{1}{2}} - L'n\}^{\frac{1}{2}}, \\ Q &= \dots\dots\{(\dots\dots\dots\dots)^{\frac{1}{2}} + \dots\}^{\frac{1}{2}}, \end{aligned}\right\} \dots\dots\dots\dots\dots\dots(78)$$

$$\left.\begin{aligned} \tan\theta_2 &= \sin 2Ql \div (\epsilon^{-2Pl} - \cos 2Ql), \\ \tan\theta_1 &= (L'nP - R'Q) \div (R'P + L'nQ); \end{aligned}\right\} \dots\dots\dots\dots\dots\dots(79)$$

then the finite V and C solutions are

$$V = V_0 \epsilon^{-Pz}\sin(nt - Qz) + V_0 \frac{\epsilon^{Pz}\sin(nt + Qz + \theta_2) - \epsilon^{-Pz}\sin(nt - Qz + \theta_2)}{\epsilon^{Pl}(\epsilon^{2Pl} + \epsilon^{-2Pl} - 2\cos 2Ql)^{\frac{1}{2}}}, \quad (80)$$

$$C = V_0 \frac{(Sn)^{\frac{1}{2}}}{(R'^2 + L'^2n^2)^{\frac{1}{4}}}\Big[\epsilon^{-Pz}\sin(nt - Qz - \theta_1) - \frac{\epsilon^{Pz}\sin(nt + Qz - \theta_1 + \theta_2) + \epsilon^{-Pz}\sin(nt - Qz - \theta_1 + \theta_2)}{\epsilon^{Pl}(\epsilon^{2Pl} + \epsilon^{-2Pl} - 2\cos 2Ql)^{\frac{1}{2}}}\Big]. \quad (81)$$

If we expand the last in cosines of mz we shall obtain (77), with $R'_m = R'$. There are three waves; the first is what would represent the solution if the line were of infinite length; but, being of finite length, there is a reflected wave (the ϵ^{Pz} term), and another reflected at $z = 0$, the third and least important.

The amplitude of C anywhere is

$$V_0 \frac{(Sn)^{\frac{1}{2}}}{(R'^2 + L'^2n^2)^{\frac{1}{4}}}\Big[\frac{\epsilon^{2P(l-z)} + \epsilon^{-2P(l-z)} + 2\cos 2Q(l-z)}{\epsilon^{2Pl} + \epsilon^{-2Pl} - 2\cos 2Ql}\Big]^{\frac{1}{2}}.$$

At the distant ($z = l$) end it is

$$C_0 = 2V_0 \frac{(Sn)^{\frac{1}{2}}}{(R'^2 + L'^2n^2)^{\frac{1}{4}}}(\epsilon^{2Pl} + \epsilon^{-2Pl} - 2\cos 2Ql)^{-\frac{1}{2}}. \quad \text{.........(82)}$$

Effects of Quasi-Resonance. Fluctuations in the Impedance.

I have already spoken of the apparent resistance of a line as its impedance (from impede). The steady impedance is the resistance. The short-line impedance is $(R^2 + L^2n^2)^{\frac{1}{2}}l$ or $(R'^2 + L'^2n^2)^{\frac{1}{2}}l$, at the frequency $n/2\pi$, according as current-density differences are, or are not, ignorable. The impedance according to the latter formula increases with the frequency, but is greater or less than that of the former formula (linear theory) according as the frequency is below or above a certain value.

But if the frequency is sufficiently increased, even on a short line, the formula ceases to represent the impedance, whilst, if the line be long, it will not do so at any frequency except zero. According to (82), we have

$$\frac{V_0}{C_0} = \frac{(R'^2 + L'^2n^2)^{\frac{1}{4}}}{2(Sn)^{\frac{1}{2}}}(\epsilon^{2Pl} + \epsilon^{-2Pl} - 2\cos 2Ql)^{\frac{1}{2}}, \quad \text{............(83)}$$

as the distant-end impedance of the line. That is, we have extended the meaning of impedance, as we must (or else have a new word), since the current-amplitude varies as we pass from beginning to end of the line. (83) will, roughly speaking, on the average, give the greatest value of the impedance. It is what the resistance of the line would have to be in order that when an S.H. impressed force acts at one end, the current-amplitude at the distant-end should be, without any magnetic and electrostatic induction, what it really is. The distant-end impedance may easily be less than the impedance according to the magnetic reckoning. What is more remarkable, however, is that it may be much less than the steady resistance of the line. This is due to the to-and-fro reflection of the dielectric waves, which is a phenomenon similar to resonance.

To show this, take $R'=0$ in the first place, which requires the conductors to be of infinite conductivity. Then $L'=L_0$, the dielectric inductance. We shall have, by (83) and (78),

$$V_0/C_0 = L_0 v \sin(nl/v), \qquad \text{.........................(84)}$$

where $v=(L_0S)^{-\frac{1}{2}}=(\mu_2c_2)^{-\frac{1}{2}}$, the speed of waves through the dielectric when undissipated. The sine is to be taken positive always. If $nl/v=\pi$, 2π, etc., the impedance is zero, and the current-amplitude infinite. Here $nl/v=\pi$ means that the period of a wave equals the time taken to travel to the distant end and back again. This accounts for the infinite accumulation, which is, of course, quite unrealizable.

Now, giving resistance to the line, it is clear that although the impedance can never vanish, it will be subject to maxima and minima values as the speed increases continuously, itself increasing, on the whole. We may transform (83) to

$$V_0/C_0$$

$$= (R'^2+L'^2n^2)^{\frac{1}{2}}l\Big[\Big(\frac{v'}{nl}\Big)^2\sin^2\Big(\frac{nl}{v'}\Big)+\Big(\frac{nl}{v'}\Big)^4\frac{h}{90}\Big\{1-\tfrac{1}{7}\Big(\frac{nl}{v'}\Big)^2+\frac{1}{105}\Big(\frac{nl}{v'}\Big)^4(1+\tfrac{1}{12}h)$$

$$-\frac{4}{105.99}\Big(\frac{nl}{v'}\Big)^6(1+\tfrac{3}{16}h)+\frac{10}{105.99.91}\Big(\frac{nl}{v'}\Big)^8(1+\tfrac{3}{10}h+\tfrac{1}{80}h^2)-\dots\Big\}\Big]^{\frac{1}{2}}, \quad (85)$$

where $\qquad v'=(L'S)^{-\frac{1}{2}}, \qquad$ and $\qquad h=(R'/L'n)^2.$

The factor outside the [ ] is the electromagnetic impedance; and, if we take only the first term within the [ ], we shall obtain the former infinite-conductivity formula (84). The effect of resistance is shown by the terms containing h.

With this v' and h notation (83) becomes

$$V_0/C_0=\tfrac{1}{2}L'v'(1+h)^{\frac{1}{4}}\{\epsilon^{2Pl}+\epsilon^{-2Pl}-2\cos 2Ql\}^{\frac{1}{2}}; \qquad \text{...........(86)}$$

where

$$Ql=(nl/v')(\sqrt{1+h}+1)^{\frac{1}{2}}\div\sqrt{2},$$

$$Pl=(nl/v')(\sqrt{1+h}-1)^{\frac{1}{2}}\div\sqrt{2}.$$

Choose Q so that $2Ql=2\pi$, and let $h=1$. This requires $nl/v'=2{\cdot}85$. Then

$$V_0/C_0=\tfrac{1}{2}L'v'\,.\,2^{\frac{1}{4}}[\epsilon^{\cdot 8284\pi}+\epsilon^{-\cdots}-2]^{\frac{1}{2}},$$

$$=60{\cdot}6\,L' \quad \text{ohms},$$

if we take $v=30^{10}$ cm. $=30$ ohms. This implies $L'=L_0$, and the dielectric air. Without making use of current-density differences, we may suppose that the conductors are thin tubes. Therefore

$$\frac{\text{Impedance}}{\text{Resistance}}=\frac{60{\cdot}6\,L'.\,10^9}{R'l}=\text{ about }\frac{202}{285},$$

by making use of the above values of h and nl/v'.

But taking $2Ql = \frac{1}{2}\pi$, or one fourth of the above value. Then

$$V_0/C_0 = 28\,L' \quad \text{ohms},$$

and
$$\frac{\text{Impedance}}{\text{Resistance}} = \text{ about } \frac{4}{3}.$$

Thus the amplitude of the current, from being less than the steady strength in the last case, becomes 42 per cent. greater than the steady current by quadrupling nl/v', and keeping $h=1$. We have evidently ranged from somewhere near the first maximum to the first minimum value of the impedance. These figures suit lines of any length, if we choose the resistances, etc., properly. The following will show how the above apply practically. Remember that 1 ohm per kilom. $= 10^4$ per cm. Then, if l_1 = length of line in kilom.,

If $R' = 10^3$, and $L' = 1$, $\therefore n = 10^3$, and $l_1 = 856$,
,, $R' = 10^3$, ,, $L' = 10$, ,, $n = 10^2$, ,, $l_1 = 8568$,
,, $R' = 10^4$, ,, $L' = 1$, ,, $n = 10^4$, ,, $l_1 = 85$,
,, $R' = 10^4$, ,, $L' = 10$, ,, $n = 10^3$, ,, $l_1 = 856$,
,, $R' = 10^4$, ,, $L' = 100$, ,, $n = 10^2$, ,, $l_1 = 8568$,
,, $R' = 10^5$, ,, $L' = 1$, ,, $n = 10^5$, ,, $l_1 = 8{\cdot}5$,
,, $R' = 10^5$, ,, $L' = 10$, ,, $n = 10^4$, ,, $l_1 = 85$,
,, $R' = 10^5$, ,, $L' = 100$, ,, $n = 10^3$, ,, $l_1 = 856$,
,, $R' = 10^6$, ,, $L' = 10$, ,, $n = 10^5$, ,, $l_1 = 8{\cdot}5$.

The resistances vary from $\frac{1}{10}$ to 100 ohms per kilom., the inductances from 1 to 100 per cm., the frequencies from $10^2/2\pi$ to $10^5/2\pi$, and the lengths from 8·5 to 8568 kilom. In all cases $\frac{2}{3}$ is the ratio of the distance-end impedance to the resistance. The common value of nl_1 is 856800.

In the other case, nl/v' has one fourth of the value just used, so that, with the same R' and L', l_1 has values one fourth of those in the above series.

Telephonic currents are so rapidly undulatory (it is the upper tones that go to make articulation, and convert mumblings and murmurs into something like human speech) that it is evident there must be a considerable amount of this dielectric resonance, if a tone last through the time of several wave-periods.

Derivation of Details from the Solution for the Total Current.

Having got the solution for C, the wire-current, we may obtain those for H, Γ, and γ from it. Thus, H_r being the same as $(2/r)C_r$, where C_r is the longitudinal current through the circle of radius r, we may first derive C_r or H_r from C, and then derive Γ and γ from either by (11). Thus, make use of (49) and (50), and the value of A_1 there given. Then we shall obtain

$$C_r = \frac{r}{a_1}\,\frac{J_1(s_1 r) - (J_1/K_1)(s_1 a_0) K_1(s_1 r)}{J_1(s_1 a_1) - (J_1/K_1)(s_1 a_0) K_1(s_1 a_1)}\,C, \qquad \text{..............(87)}$$

where, in the s_1, p and m^2 are to be d/dt and $-d^2/dz^2$. Similarly for the return-tube.

In a comprehensive investigation, the C-solution would be only a special result. As this special result is more easily got by itself, it might appear that there would be some saving of labour by first getting the C-solution and then deriving the general from it. But this does not stand examination; the work has to be done, whether we derive the special results from the general, or conversely.

In the solid-wire case

$$C_r = \frac{r J_1(s_1 r)}{a_1 J_1(s_1 a_1)} C,$$

or

$$C_r = \frac{r^2}{a_1^2}\Big\{1 + \tfrac{1}{2}(\pi\mu_1 k_1 p + \tfrac{1}{4}m^2)(r^2 - a_1^2) + \tfrac{1}{3}\frac{1}{1^2 2^2}(\pi\mu_1 k_1 p + \tfrac{1}{4}m^2)^2(r^2 - a_1^2)(r^2 - 2a_1^2)$$

$$+ \tfrac{1}{4}\frac{1}{1^2 2^2 3^2}(\pi\mu_1 k_1 p + \tfrac{1}{4}m^2)^3(r^2 - a_1^2)(r^4 - 5r^2a_1^2 + 7a_1^4)\} + \dots\Big\}C.$$

Or, use the M and N functions of Part I., equations (42). For we have

$$J_0(s_1 r) = (M + iN)(s_1 r i^{\frac{1}{2}}),$$

where $s_1 r i^{\frac{1}{2}}$ takes the place of the y in those equations. M contains the even, and N the odd powers of $(p + m^2/4\pi\mu_1 k_1)$.

We have also

$$\Gamma_r = J_0(s_1 r)\Gamma_0 = \frac{s_1}{2\pi a_1}\frac{J_0(s_1 r)}{J_1(s_1 a_1)}C,$$

Γ_0 being Γ at $r = 0$; and, since by the first of these,

$$\Gamma_{a_1} = J_0(s_1 a_1)\Gamma_0$$

connects the boundary and axial current-densities, we see that the ratio of their amplitudes in the S.H. case is

$$(M^2 + N^2)^{\frac{1}{2}},$$

using the $r = a_1$ expressions, with $m = 0$.

Note on the Investigation of Simple-Harmonic States. (July, 1892.)

[I have been asked by more than one correspondent how the above solutions (80) and (81) are obtained, and therefore add some details, giving the working rather fully, as it will serve to show the procedure in other cases.

We have an impressed force acting at one spot, and desire to know the effect produced there and elsewhere. The first step is to form the differential equation connecting the impressed force with the effect produced. Now we have

$$d^2V/dz^2 = (K + Sp)(R' + L'p)V = F^2 V \qquad \text{.................(1A)}$$

in the line generally, if we introduce K the leakage-conductance per unit length (as in Parts IV. and V.), and therefore

$$V=\epsilon^{Fz}.A+\epsilon^{-Fz}.B, \qquad\text{.........................(2A)}$$

where A and B are undetermined. To suit the present case, we find them by the terminal conditions

$$V=0 \quad \text{at} \quad z=l, \qquad \text{therefore} \qquad 0=\epsilon^{Fl}.A+\epsilon^{-Fl}.B,$$
$$V=e \quad \text{at} \quad z=0, \qquad \text{therefore} \qquad e=\quad A+B;$$

which give A and B and develop (2A) to

$$V=\frac{\epsilon^{F(l-z)}-\epsilon^{-F(l-z)}}{\epsilon^{Fl}-\epsilon^{-Fl}}.e. \qquad\text{.........................(3A)}$$

This is the differential equation connecting V at z with e at $z=0$, the latter being any function of the time. It may also be regarded as the solution of the problem of finding V due to e. For the march of V is strictly connected with that of e through the operator in (3A) and by nothing else, all indefiniteness having been removed by the previous work. But, whilst (3A) is the solution, it is (usually) in a very condensed form, needing development to more immediately interpretable forms. If, however, F be constant, as happens when $p=0$, (3A) needs no development. It then represents the ultimate steady state of V due to steady e. But the primitive solution in general requires a good deal of development. Thus, if we wish to find the ultimate simple-harmonic state of V due to simple-harmonic e of frequency $n/2\pi$, we know that $p^2=-n^2$, or $p=ni$, making $F=P+Qi$, where P and Q are given in the text (when $K=0$). This substitution made in (3A) will make it be convertible to the simple form

$$V=(a+bp)e, \qquad\text{..............................(4A)}$$

expressing V fully when e is given fully in any amplitude and phase. The work is now to turn (3A) to (4A). First put $F=P+Qi$, then (3A) becomes, when the real and imaginary parts in the numerator and denominator are separated,

$$\frac{V}{e}=\frac{(\epsilon^{P(l-z)}-\epsilon^{-P(l-z)})\cos Q(l-z)+i(\epsilon^{P(l-z)}+\epsilon^{-P(l-z)})\sin Q(l-z)}{(\epsilon^{Pl}-\epsilon^{-Pl})\cos Ql+i(\epsilon^{Pl}+\epsilon^{-Pl})\sin Ql}. \qquad\text{...(5A)}$$

To rationalise the denominator, multiply it and the numerator by the denominator with the sign of i changed, producing

$$\frac{V}{e}=\left\{\begin{bmatrix}+(\epsilon^{P(l-z)}-\epsilon^{-P(l-z)})\cos Q(l-z).(\epsilon^{Pl}-\epsilon^{-Pl})\cos Ql\\ +(\ldots\ldots+\ldots\ldots.)\sin\ldots\ldots\ldots.(\ldots+\ldots.)\sin\ldots\end{bmatrix}\right.$$
$$\left.+i\begin{bmatrix}-(\ldots\ldots-\ldots\ldots.)\cos\ldots\ldots\ldots.(\ldots+\ldots.)\sin\ldots\\ +(\ldots\ldots+\ldots\ldots.)\sin\ldots\ldots\ldots.(\ldots-\ldots.)\cos\ldots\end{bmatrix}\right\}$$
$$\div\left[(\epsilon^{Pl}-\epsilon^{-Pl})^2\cos^2 Ql+(\epsilon^{Pl}+\epsilon^{-Pl})^2\sin^2 Ql\right]. \qquad\text{...............(6A)}$$

This is in rational form, since $i=p/n$. But it can be simplified. The denominator, say D, is evidently

$$D=\epsilon^{2Pl}+\epsilon^{-2Pl}-2\cos 2Ql, \qquad\text{.....................(7A)}$$

and we may easily reduce (6A) to

$$V=\Big[\cos Qz\Big(\epsilon^{Pz}\epsilon^{-2Pl}+\epsilon^{-Pz}\epsilon^{2Pl}-(\epsilon^{Pz}+\epsilon^{-Pz})\cos 2Ql\Big)$$

$$+\sin Qz\Big(-(\epsilon^{Pz}+\epsilon^{-Pz})\sin 2Ql\Big)$$

$$+i\cos Qz\Big(+(\epsilon^{Pz}-\epsilon^{-Pz})\sin 2Ql\Big)$$

$$+i\sin Qz\Big(\epsilon^{Pz}\epsilon^{-2Pl}-\epsilon^{-Pz}\epsilon^{2Pl}-(\epsilon^{Pz}-\epsilon^{-Pz})\cos 2Ql\Big)\Big]\frac{e}{D}, \quad \text{...(8A)}$$

the full solution with e simple-harmonic, but left arbitrary in amplitude and phase. If it is $V_0 \sin nt$, then the terms in the first two lines of (8A) receive $\sin nt$ as a factor, whilst the next two lines receive $\cos nt$ (by the operation of the differentiator i on e), giving the result, after rearrangement,

$$V=\frac{V_0}{D}\Big\{\sin 2Ql\Big[\epsilon^{Pz}\cos(nt+Qz)-\epsilon^{-Pz}\cos(nt-Qz)\Big]$$

$$-\cos 2Ql\Big[\ldots\sin(\ldots\ldots)+\ldots\sin(\ldots\ldots)\Big]$$

$$+\epsilon^{2Pl}\epsilon^{-Pz}\sin(nt-Qz)+\epsilon^{-2Pl}\epsilon^{Pz}\sin(nt+Qz)\Big\}. \quad \text{......(9A)}$$

This differs in form from (80), which was arranged to show the solution for an infinitely long line (obtainable by the same process, only greatly simplified) explicitly, with the additions caused by the reflection at $z=l$ and the subsequent complex minor reflections at beginning and end of the line. To get (80) from (9A) observe the form of D in (7A), and add and substract from (9A) terms so as to isolate the solution for an infinitely long line. Thus

$$V=V_0\epsilon^{-Pz}\sin(nt-Qz)+V_0\frac{\sin 2Ql}{D}\Big[\epsilon^{Pz}\cos(nt+Qz)-\epsilon^{-Pz}\cos(nt-Qz)\Big]$$

$$+V_0\frac{\epsilon^{-2Pl}-\cos 2Ql}{D}\Big[\ldots\sin(\ldots\ldots)-\ldots\sin(\ldots\ldots)\Big]. \quad \text{(10A)}$$

The transition to the shorter form (80) is now obvious, by taking

$$\cos\theta_2=\frac{\epsilon^{-2Pl}-\cos 2Ql}{D^{\frac{1}{2}}\epsilon^{-Pl}}, \qquad \sin\theta_2=\frac{\sin 2Ql}{D^{\frac{1}{2}}\epsilon^{-Pl}}. \quad \text{..........(11A)}$$

Some of the above work may be saved, perhaps, by taking $e=V_0\epsilon^{int}$ at the beginning, that is, a special complex form of impressed force. The result is a complex solution, divisible into one due to $V_0\cos nt$ and another due to $V_0\sin nt$, either of which may be selected, or any combination made. But I find the above method more generally useful.

We may derive C from V thus,

$$C=\frac{-dV/dz}{R'+L'p}=-\frac{R'-L'p}{R'^2+L'^2n^2}\frac{dV}{dz}. \quad \text{..............(12A)}$$

This process may be applied to the final form of solution for V or to any previous form, as the primitive (3A). The easiest way will depend on circumstances. Similarly we may derive the V-solution from the C-solution, by

$$V=\frac{-dC/dz}{K+Sp}=-\frac{K-Sp}{K^2+S^2n^2}\frac{dC}{dz}. \qquad \text{(13A)}$$

The above details will also serve to illustrate the working of the problem in Part V., for it is the same problem as above, but with arbitrary terminal connections (instead of short-circuits), and is done in the same way. Its complexity arises from the reactions between the terminal apparatus and the main circuit.]

Part III.

Remarks on the Expansion of Arbitrary Functions in Series.

The subject of the decomposition of an arbitrary function into the sum of functions of special types has many fascinations. No student of mathematical physics, if he possess any soul at all, can fail to recognise the poetry that pervades this branch of mathematics. The great work of Fourier is full of it, although there only the mere fringe of the subject is reached. For that very reason, and because the solutions can be fully realised, the poetry is more plainly evident than in cases of greater complexity. Another remarkable thing to be observed is the way the principle of conservation of energy and its transfer, or the equation of activity, governs the whole subject, in dynamical applications, as regards the possibility of effecting certain expansions, the forms of the functions involved, the manner of effecting the expansions, and the possible nature of the "terminal conditions" which may be imposed.

Special proofs of the possibility of certain expansions are sometimes very vexatious. They are frequently long, complex, difficult to follow, unconvincing, and, after all, quite special; whilst there is an infinite number of functions equally deserving. Something is clearly wanted of a quite general nature, and simple in its generality, to cover the whole field. This will, I believe, be ultimately found in the principle of energy, at least as regards the functions of mathematical physics. But in the present place only a small part of the question will be touched upon, with special reference to the physical problem of the propagation of electromagnetic disturbances through a dielectric tube, bounded by conductors.

It will be, perhaps, in the recollection of some readers that Professor Sylvester, a few years since, in the course of his learned paper on the Bipotential, poked fun at Professor Maxwell for having, in his investigation of the conjugate properties possessed by complete spherical-surface harmonics, made use of Green's Theorem concerning the mutual energy of two electrified systems. He said (in effect, for the quotation is from memory) that one might as well prove the rule of three by the laws of hydrostatics—or something similar to that. In the second edition of his treatise, Prof. Maxwell made some remarks that appear to be meant for a reply to this; to the effect that although names, involving physical ideas, are given to certain quantities, yet, as the reasoning is purely mathematical, the physicist has a right to assist himself by the physical ideas.

Certainly; but there is much more in it than that. For not only the conjugate properties of spherical harmonics, but those of all other functions of the fluctuating character, which present themselves in physical problems, including the infinitely undiscoverable, are involved in the principle of energy, and are most simply and immediately proved by it, and predicted beforehand. We may indeed get rid of the principle of energy, and treat the matter as a question of the properties of quadratic functions; a method which may commend itself to the pure mathematician. But by the use of the principle of energy, and assisted by the physical ideas involved, we are enabled to go straight to the mark at once, and avoid the unnecessary complexities connected with the use of the special functions in question, which may be so great as to wholly prevent the recognition of the properties which, through the principle of energy, are necessitated.

The Conjugate Property $\mathrm{U}_{12}=\mathrm{T}_{12}$ *in a Dynamical System with Linear Connections.*

Considering only a dynamical system in which the forces of reaction are proportional to displacements, and the forces of resistance to velocities, there are three important quantities—the potential energy, the kinetic energy, and the dissipativity, say U, T, and Q, which are quadratic functions of the variables or their velocities. When there is no kinetic energy, the conjugate properties of normal systems are $U_{12}=0$ and $Q_{12}=0$; these standing for the mutual potential energy and the mutual dissipativity of a pair of normal systems. When there is no potential energy, we have $T_{12}=0$ and $Q_{12}=0$. When there is no dissipation of energy, $U_{12}=0$ and $T_{12}=0$. And in general, $U_{12}=T_{12}$, which covers all cases, and has two equivalents, $\frac{1}{2}Q_{12}+\dot{U}_{12}=0$, and $\frac{1}{2}Q_{12}+\dot{T}_{12}=0$; for, as the mutual potential and kinetic energies are equal, the mutual dissipativity is derived half from each.

Let the variables be x_1, x_2, ..., their velocities $v_1=\dot{x}_1$, ..., and the equations of motion

$$\left.\begin{aligned} F_1&=(A_{11}+B_{11}p+C_{11}p^2)x_1+(A_{12}+B_{12}p+C_{12}p^2)x_2+\ldots,\\ F_2&=(A_{21}+B_{21}p+C_{21}p^2)x_1+(A_{22}+B_{22}p+C_{22}p^2)x_2+\ldots,\\ &\ldots\ldots\ldots\ldots\ldots\ldots\ldots\ldots\ldots\ldots\ldots\ldots\ldots\ldots \end{aligned}\right\}\quad\ldots\ldots(88)$$

where F_1, F_2, ..., are impressed forces, and p stands for d/dt. Forming the equation of total activity, we obtain

$$\Sigma\, Fv=Q+\dot{U}+\dot{T}\,;\quad\ldots\ldots\ldots\ldots\ldots\ldots\ldots\ldots(89)$$

where

$$\left.\begin{aligned} 2U&=A_{11}x_1^2+2A_{12}x_1x_2+A_{22}x_2^2+\ldots,\\ Q&=B_{11}v_1^2+2B_{12}v_1v_2+B_{22}v_2^2+\ldots,\\ 2T&=C_{11}v_1^2+2C_{12}v_1v_2+C_{22}v_2^2+\ldots. \end{aligned}\right\}\quad\ldots\ldots\ldots\ldots(90)$$

So far will define, in the briefest manner, U, T, Q, and activity.

Now let the F's vanish, so that no energy can be communicated to the system, whilst it can only leave it irreversibly, through Q. Then let p_1, p_2 be any two values of p satisfying (88) regarded as algebraic. Let Q_1, U_1, T_1 belong to the system p_1 existing alone; then, by (89) and (90),

$$0 = Q_1 + \dot{U}_1 + \dot{T}_1, \qquad \text{or} \qquad 0 = Q_1 + 2p_1(U_1 + T_1);$$
$$0 = Q_2 + \dot{U}_2 + \dot{T}_2, \qquad \text{or} \qquad 0 = Q_2 + 2p_2(U_2 + T_2).$$

But when existing simultaneously, so that

$$Q = Q_1 + Q_2 + Q_{12}, \qquad U = U_1 + U_2 + U_{12}, \qquad T = T_1 + T_2 + T_{12},$$

where U_{12}, T_{12}, Q_{12} depend upon products from both systems, thus :—

$$Q_{12} = 2\{B_{11}v_1v_1' + B_{22}v_2v_2' + B_{12}(v_1v_2' + v_2v_1') + \dots\},$$
$$U_{12} = A_{11}x_1x_1' + A_{22}x_2x_2' + A_{12}(x_1x_2' + x_2x_1') + \dots,$$
$$T_{12} = C_{11}v_1v_1' + C_{22}v_2v_2' + C_{12}(v_1v_2' + v_2v_1') + \dots,$$

the accents distinguishing one system from the other, we shall find, by forming the equations of mutual activity $\Sigma Fv' = \dots$, and $\Sigma F'v = \dots$, that is, with the F's of one system, and the v's of the other, in turn,

$$0 = \tfrac{1}{2}Q_{12} + p_2U_{12} + p_1T_{12},$$
$$0 = \tfrac{1}{2}Q_{12} + p_1U_{12} + p_2T_{12};$$

adding which, there results the equation of mutual activity,

$$0 = Q_{12} + (p_1 + p_2)(U_{12} + T_{12}), \qquad \text{or} \qquad 0 = Q_{12} + \dot{U}_{12} + \dot{T}_{12};$$

and, on subtraction, there results

$$0 = (p_1 - p_2)(U_{12} - T_{12}), \qquad \text{.......................(91)}$$

giving $U_{12} = T_{12}$, if the p's are unequal. But this property is true whether the p's be equal or not; that is, $U_{11} = T_{11}$ when p_1 is a repeated root. I have before discussed various cases of the above, with special reference to the dynamical system expressed by Maxwell's electromagnetic equations. [Vol. i., pp. 520 to 531.]

Application to the General Electromagnetic Equations.

The following applies to Maxwell's system, using the equations (4) to (10) of Part. I. [vol. ii., p. 174]. A comparison with the above is instructive. Let $\mathbf{E}_1$, $\mathbf{H}_1$ and $\mathbf{E}_2$, $\mathbf{H}_2$ be any two systems satisfying these equations, with no impressed forces, or $\mathbf{e} = 0$, $\mathbf{h} = 0$. Then the energy entering the unit volume per second by the action of the first system on the second is

$$\text{conv}\ \mathrm{V}\mathbf{E}_1\mathbf{H}_2/4\pi = (\mathbf{E}_1 \,\text{curl}\, \mathbf{H}_2 - \mathbf{H}_2 \,\text{curl}\, \mathbf{E}_1)/4\pi,$$
$$= \mathbf{E}_1\Gamma_2 + \mathbf{H}_2\mathbf{G}_1,$$
$$= \mathbf{E}_1\mathbf{C}_2 + \mathbf{E}_1\dot{\mathbf{D}}_2 + \mathbf{H}_2\dot{\mathbf{B}}_1/4\pi \qquad \text{................(92)}$$

Similarly, by the action of the second system on the first,

$$\text{conv}\ \mathrm{V}\mathbf{E}_2\mathbf{H}_1/4\pi = \mathbf{E}_2\mathbf{C}_1 + \mathbf{E}_2\dot{\mathbf{D}}_1 + \mathbf{H}_1\dot{\mathbf{B}}_2/4\pi. \qquad \text{.............(93)}$$

Addition gives the equation of mutual activity. And, subtracting (93) from (92), we find

$$\text{conv}\,(\mathbf{VE}_1\mathbf{H}_2 - \mathbf{VE}_2\mathbf{H}_1)/4\pi = (\mathbf{E}_1\dot{\mathbf{D}}_2 - \mathbf{E}_2\dot{\mathbf{D}}_1) - (\mathbf{H}_1\dot{\mathbf{B}}_2 - \mathbf{H}_2\dot{\mathbf{B}}_1)/4\pi\,; \quad ..(94)$$

since $\mathbf{E}_1\mathbf{C}_2 = \mathbf{E}_1 k\mathbf{E}_2 = \mathbf{E}_2 k\mathbf{E}_1 = \mathbf{E}_2\mathbf{C}_1$, if there be no rotatory power, or $\mathbf{C}$ be a symmetrical linear function of $\mathbf{E}$. Similarly for $\mathbf{D}$ and $\mathbf{E}$, and $\mathbf{B}$ and $\mathbf{H}$. Hence, if the systems are normal, making $d/dt = p_1$ in one, and p_2 in the other, (94) becomes

$$\text{conv}\,(\mathbf{VE}_1\mathbf{H}_2 - \mathbf{VE}_2\mathbf{H}_1)/4\pi = (p_2 - p_1)(\mathbf{E}_1\mathbf{D}_2 - \mathbf{H}_1\mathbf{B}_2/4\pi). \quad ....(95)$$

Therefore, by the well-known theorem of Convergence, if we integrate through any region, and U_{12}, T_{12} be the mutual electric energy and the mutual magnetic energy of the two systems in that region, we obtain

$$U_{12} - T_{12} = \sum \frac{\mathbf{N}(\mathbf{VE}_2\mathbf{H}_1 - \mathbf{VE}_1\mathbf{H}_2)/4\pi}{p_1 - p_2}, \quad ....(96)$$

where $\mathbf{N}$ is the unit normal drawn inward from the boundary of the region, over which the summation extends. And if the region include the whole space through which the systems extend, the right member will vanish, giving $U_{12} = T_{12}$, when these are complete.

From (96) we obtain, by differentiation, the value of twice the excess of the electric over the magnetic energy of a single normal system in any region; thus

$$2(U - T) = \sum \mathbf{N}\left(\mathbf{VE}\frac{d\mathbf{H}}{dp} - \mathbf{V}\frac{d\mathbf{E}}{dp}\mathbf{H}\right)/4\pi. \quad ....(97)$$

This formula, or some special representative of the same, is very useful in saving labour in investigations relating to normal systems of subsidence.

Application to any Electromagnetic Arrangements subject to V = ZC.

The quantity that appears in the numerator in (96) is the excess of the energy entering the region through its boundary per second by the action of the second system on the first, over that similarly entering due to the action of the first on the second system. Bearing this in mind, we can easily form the corresponding formula in a less general case. Suppose, for example, we have two fine-wire terminals, a and b, that are joined through any electromagnetic and electrostatic combination which does not contain impressed forces, nor receives energy from without except by means of the current, say C, entering it at a and leaving it at b. Let also V be the excess of the potential of a over that of b. Then VC is the energy-current, or the amount of energy added per second to the combination through the terminal connections with, necessarily, some other combination. (In the previous thick-letter vector investigation V was the symbol of vector product. There will, however, be no confusion with the following use of V, as in Part II., to express the line-integral of an electric force. One of the awkward things about the notation in Prof. Tait's "Quaternions" is the employment of a number of most useful letters, as S, T, U, V, K, wanted for other purposes, as mere symbols of operations, putting another barrier in the way of practically combining vector methods with ordinary scalar methods, besides the perpetual negative sign before scalar products.) The combination need not be of mere linear circuits, in which differences of current-density are insensible; there may, for example, be induction of currents in a mass of metal either connected conductively or not with a and b; but in any case it is necessary that the arrangement should terminate in fine wires at a and b, in order that the two quantities V and C may suffice to specify, by their product, the energy-current at the terminals. Even in this we completely ignore the dielectric currents and also the displacement, in the neighbourhood of the terminals, *i.e.*, we assume $c=0$, to stop displacement. This is, of course, what is always done, unless specially allowed for.

Now, supposing the structure of the combination to be given, we can always, by writing out the equations of its different parts, arrive at the characteristic equation connecting the terminal V and C. For instance,

$$V=ZC, \quad \text{.................................(98)}$$

where Z is a function of d/dt. In the simplest case Z is a mere resistance. A common form of this equation is

$$f_0V+f_1\dot{V}+f_2\ddot{V}+\ldots=g_0C+g_1\dot{C}+g_2\ddot{C}+\ldots,$$

where the f's and g's are constants. But there is no restriction to such simple forms. All that is necessary is that the equation should be linear, so that Z may be a function of p. If, for example, $(dC/dt)^2$ occurred, we could not do it.

Now this combination must necessarily be joined on to another, however elementary, to make a complete system, unless V is to be zero always. The complete system, without impressed forces in it, has its proper normal modes of subsidence, corresponding to definite values of p. Consequently, by (96),

$$U_{12}-T_{12}=(V_2C_1-V_1C_2)\div(p_1-p_2), \quad \text{.................(99)}$$

if V_1, C_1 belong to p_1, and V_2, C_2 to p_2, whilst the left member refers to the combination given by $V=ZC$. Or,

$$U_{12}-T_{12}=C_1C_2\left(\frac{V_1}{C_1}-\frac{V_2}{C_2}\right)\div(p_2-p_1)=C_1C_2\frac{Z_1-Z_2}{p_2-p_1}, \quad \text{......(100)}$$

and the value of $2(U-T)$ in a single normal system is

$$2(U-T)=V\frac{dC}{dp}-C\frac{dV}{dp}=-C^2\frac{d}{dp}\frac{V}{C}=-C^2\frac{dZ}{dp}. \quad \text{.. ......(101)}$$

In a similar manner we can write down the energy-differences for the complementary combination, whose equation is, say, $V=YC$; remembering that $-VC$ is the energy entering it per second, we get

$$C_1C_2\frac{Y_1-Y_2}{p_1-p_2}, \quad \text{and} \quad C^2\frac{dY}{dp}, \quad \text{respectively.}$$

By addition, the complete $U_{12} - T_{12}$ is

$$C_1 C_2 \frac{Y_1 - Y_2 - Z_1 + Z_2}{p_1 - p_2} = 0 = C_1 C_2 \frac{\phi_1 - \phi_2}{p_1 - p_2}; \quad \text{.............(102)}$$

and the complete $2(U-T)$ is

$$C^2 \frac{d}{dp}(Y-Z), \qquad \text{or} \qquad C^2 \frac{d\phi}{dp}, \quad \text{.................(103)}$$

where $\phi = 0$, or $Y - Z = 0$, is the determinantal equation of the complete system (both combinations which join on at a and b, where V and C are reckoned), expressed in such a form that every term in ϕ is of the dimensions of a resistance.

Determination of Size of Normal Systems of V *and* C *to express Initial State. Complete Solutions obtainable with any Terminal Arrangements provided* R, S, L *are Constants.*

If the complete system depends only upon a finite number of variables, it is clear that the number of independent normal systems is also finite, and there is no difficulty whatever in understanding how any possible initial state is decomposable into the finite number of normal states; nor is any proof needed that it is possible to do it. The constant A_1, fixing the size of a particular normal system p_1, will be given by

$$A_1 = \frac{U_{01} - T_{01}}{U_{11} - T_{11}} = \frac{U_{01} - T_{01}}{2(U_1 - T_1)} = \frac{U_{01} - T_{01}}{C_1^2 \dfrac{d\phi}{dp_1}}, \quad \text{........... (104)}$$

by the previous, if U_{01} be the mutual electric energy of the given initial state and the normal system, and T_{01}, similarly, the mutual magnetic energy.

And, when we increase the number of variables infinitely, and pass to partial differential equations and continuously varying normal functions, it is, by continuity, equally clear that the decomposition of the initial state into the now infinite series of normal functions is not only possible, but necessary. Provided always, that we have the whole series of normal functions at command. Therein lies the difficulty, when there is any.

In such a case as the system (71) of Part II., involving the partial differential equation

$$\frac{d^2V}{dz^2} = RS\frac{dV}{dt} + LS\frac{d^2V}{dt^2}, \quad \text{......................(105)}$$

wherein R, S, and L are constants, to hold good between the limits $z=0$ and $z=l$, subject to

$$V = Z_0 C \quad \text{at} \quad z=0, \qquad \text{and} \qquad V = Z_1 C \quad \text{at} \quad z=l,$$

there is no possible missing of the true normal functions which arise by treating d/dt as a constant; so that we can be sure of the possibility of the expansions. Thus, denoting $RSp + LSp^2$ by $-m^2$, we may take the normal V-function as

$$u = \sin(mz+\theta), \qquad (106)$$

and the corresponding normal C-function as

$$w = +\frac{Sp}{m^2}\frac{du}{dz} = +\frac{Sp}{m}\cos(mz+\theta). \qquad (107)$$

Here θ will be determined by the terminal conditions

$$\frac{u}{w} = Z_0 \quad \text{at} \quad z=0, \qquad \frac{u}{w} = Z_1 \quad \text{at} \quad z = l, \qquad (108)$$

and the complete V and C solutions are

$$V = \Sigma A u \epsilon^{pt}, \qquad C = \Sigma A w \epsilon^{pt} \qquad (109)$$

at time t; where any A is to be found from the initial state, say V_0, C_0, functions of z, by

$$A = \frac{\int_0^l (SV_0 u - LC_0 w)dz}{\left[w^2 \frac{d}{dp}\left(\frac{u}{w} - Z\right)\right]_0^l}, \qquad (110)$$

provided there be no energy initially in the terminal arrangements. If there be, we must make corresponding additions to the numerator, without changing the denominator of A. The expression to be used for u/w is, by (106) and (107),

$$\frac{u}{w} = \frac{m}{Sp}\tan(mz+\theta), \qquad (111)$$

remembering that m is a function of p. There are four components in the denominator of (110), as there are three electrical systems; viz., the terminal arrangements, which can only receive energy from the "line," and the line itself, which can receive or part with energy at both ends.

Complete Solutions obtainable when R, S, L *are Functions of* z, *though not of* p. *Effect of Energy in Terminal Arrangements.*

In a similar manner, if we make R, S, and L any single-valued functions of z, subject to the elementary relations of (71), Part II., or

$$-\frac{dV}{dz} = RC + L\dot{C}, \qquad -\frac{dC}{dz} = S\dot{V}, \qquad (112)$$

getting this characteristic equation of C,

$$\frac{d}{dz}\left(S^{-1}\frac{dC}{dz}\right) = \left(R + L\frac{d}{dt}\right)\frac{dC}{dt}, \qquad (113)$$

and, after putting w for C and p for $\frac{d}{dt}$, this equation for the current-function,

$$\frac{d}{dz}\left(S^{-1}\frac{dw}{dz}\right) = (R + Lp)pw, \qquad (114)$$

and finding the u-function by the second of (112), giving

$$-Spu=\frac{dw}{dz}, \qquad \text{.............................(115)}$$

we see that the expansions of the initial states V_0 and C_0 can be effected, subject to the terminal conditions (108). For the normal potential- and current-functions will be perfectly definite (singularities, of course, to receive special attention), given by (115) and (114), each as the sum of two independent functions, and the terminal conditions will settle in what ratio they must be taken. (109) and (110) will constitute the solution, except as regards the initial energy beyond the terminals.

It is, however, remarkable, that we can often, perhaps universally, find the expression for the part of the numerator of (110) to be added for the terminal arrangements, except as regards arbitrary multipliers, from the mere form of the Z-functions, without knowing in detail what electrical combinations they represent. This is to be done by first decomposing the expression for $C^2(dZ/dp)$ into the sum of squares, for instance,

$$C^2\frac{dZ}{dp}=r_1\{f_1(p)\}^2+r_2\{f_2(p)\}^2+\dots, \qquad \text{...............(116)}$$

where r_1, r_2, ... are constants. The terminal arbitraries are then $\Sigma Af_1(p)$, $\Sigma Af_2(p)$, etc.: calling these E_1, E_2, ..., the additions to the numerator of (110) are

$$-\{E_1r_1f_1(p)+E_2r_2f_2(p)+\dots\}, \qquad \text{................(117)}$$

wherein the E's may have any values. This must be done separately for each terminal arrangement. The matter is best studied in the concrete application, which I may consider under a separate heading.

It is also remarkable that, as regards the obtaining of correct expansions of functions, there is no occasion to impose upon R, S, and L the physical necessity of being positive quantities, or real. This will be understandable by going back to a finite number of variables, and then passing to continuous functions. [See Art. XX., vol. I., p. 141, for examples.]

Case of Coaxial Tubes when the Current is Longitudinal. Also when the Electric Displacement is Negligible.

Let us now proceed to the far more difficult problems connected with propagation along a dielectric tube bounded by concentric conducting tubes, and examine how the preceding results apply, and in what cases we can be sure of getting correct solutions. Start with the general system, equations (11) to (14), Part I., with the extension mentioned at the commencement of Part II. from a solid to a tubular inner conductor. Suppose that the initial state is of purely longitudinal electric force, independent of z, so that the longitudinal E and circular H are functions of r only. How can we secure that they shall, in subsiding, remain functions of r only, so that any short length is representative of the whole? Since E is to be longitudinal, there must be no longitudinal energy current, or it must be entirely radial. Therefore no energy must be communicated to the system at $z=0$ or $z=l$, or leave it at those places. This seems to be securable in only five cases. Put infinitely conducting plates across the section at either or both ends of the line. This will make $V=0$ there, if V is the line-integral of the radial electric force across the dielectric. Or put nonconducting and non-dielectric plates there similarly. This will make $C=0$. Or, which is the fifth case, let the inner and the outer conductors be closed upon themselves. In any of these cases, the electric force will remain longi tudinal during the subsidence, which will take place similarly all along the line. By (14), the equation of H will be

$$\frac{d}{dr}\frac{1}{r}\frac{d}{dr}rH = 4\pi k\mu\dot{H} + \mu c\ddot{H};$$

and it is clear that the normal functions are quite definite, so that the expansion of the initial state of E and H can be truly effected. In the already-given normal functions, take $m=0$.

But if we were to join the conductors at one end of the line through a resistance, we should, to some extent, upset this regular subsidence everywhere alike. For energy would leave the line; this would cause radial displacement, first at the end where the resistance was attached, and later all along the line. (By "the line" is meant, for brevity, the system of tubes extending from $z=0$ to $z=l$.)

Now in short-wire problems the electric energy is of insignificant importance, as compared with the magnetic. It is usual to ignore it altogether. This we can do by assuming $c=0$. This necessitates equality of wire- and return-current, for one thing; but, more importantly, it prevents current leaving the conductors, so that C and H, and Γ the current-density, are independent of z. There will be no radial electric force in the conductors, in which, therefore, the energy-current will be radial. But there will be radial force in the dielectric, and therefore longitudinal energy-current. Since the radial electric force and also the magnetic force in the dielectric vary inversely as the distance from the axis, the longitudinal energy-current density will vary inversely as the square of the distance. But, on account of symmetry, we are only concerned with its total amount over the complete section of the dielectric. This is

$$\frac{1}{4\pi}\int_{a_1}^{a_2}\frac{2C}{r}\,.\,E_r\,.\,2\pi r\,dr = VC, \quad\text{.........................(118)}$$

if V is the line-integral of E_r the radial force, and C the wire-current. It is clear, then, that we can now allow terminal connections of the form $V/C=Z$ before used, and still have correct expansions of the initial magnetic field, giving correct subsidence-solutions.

But it is simpler to ignore V altogether. For the equation of E.M.F. will be

$$e_0 = (Z_0 + Z_1 + lL_0p + lR_1'' + lR_2'')C, \quad\text{....................(119)}$$

if e_0 is the total impressed force in the circuit, R_1'' and R_2'' the wire- and sheath-functions of equations (55) and (56), Part II., on the assumption

$m=0$, and Z_0, Z_1 the terminal functions, such that $V/C=Z_1$ at $z=l$, and $=-Z_0$ at $z=0$. It does not matter how e_0 is distributed so far as the magnetic field and the current are concerned. Let it then be distributed in such a way as to do away with the radial electric field, for simplicity of reasoning. The simple-harmonic solution of (119) is obviously to be got by expanding Z_0 and Z_1 in the form $R+Lp$, where R and L are functions of p^2, and adding them on to the $l(R'+L'p)$ equivalent of $l(L_0p+R_1''+R_2'')$, as in equation (66), Part II.

Regarding the free subsidence, putting $e_0=0$ in (119) gives us the determinantal equation of the p's; and as the normal H-functions are definitely known, the expansion of the magnetic field can be effected. The influence of the terminal arrangements must not be forgotten in reckoning A.

Coaxial Tubes with Displacement allowed for. Failure to obtain Solutions in Terms of V *and* C, *except when Terminal Conditions are* VC = 0, *or when there are no Terminals, on account of the Longitudinal Energy-Flux in the Conductors.*

In coming, next, to the more general case of equation (56), but without restriction to exactly longitudinal current in the conductors, it is necessary to consider the transfer of energy more fully. In the dielectric the longitudinal energy-current is still VC. The rate of decrease of this quantity with z is to be accounted for by increase of electric and magnetic energy in the dielectric, and by the transfer of energy into the conductors which bound it. Thus,

$$-\frac{d}{dz}VC=-\frac{dV}{dz}C-\frac{dC}{dz}V.$$

But here,

$$-\frac{dC}{dz}=S\dot{V}, \qquad \text{and} \qquad -\frac{dV}{dz}=L_0\dot{C}+E-F, \qquad \text{.........(120)}$$

by (59) and (56), Part II., E and F being the longitudinal electric forces at the inner and outer boundaries of the dielectric (when there is no impressed force). So

$$-\frac{d}{dz}VC=SV\dot{V}+L_0C\dot{C}+EC-FC. \qquad \text{...............(121)}$$

The first term on the right side is the rate of increase of the electric energy, the second term the rate of increase of the magnetic energy in the dielectric, the third is the energy entering the inner conductor per second, the fourth that entering the outer conductor; all per unit length.

If the electric current in the conductors were exactly longitudinal, the energy-transfer in them would be exactly radial, and EC and $-FC$ would be precisely equal to the Joule-heat per second *plus* the rate of increase of the magnetic energy, in the inner and outer conductor, respectively. But as there is a small radial current, there is also a small longitudinal transfer of energy in the conductors. Thus, E_r and E_z being the radial and longitudinal components of the electric force, in the inner conductor, for example, the longitudinal and the radial components of the energy-current per unit area are

$$E_r H/4\pi \quad \text{and} \quad E_z H/4\pi,$$

the latter being inward. Their convergences are

$$-\frac{d}{dz}\frac{E_r H}{4\pi}, \quad \text{and} \quad \frac{1}{r}\frac{d}{dr}r\frac{E_z H}{4\pi},$$

or
$$\frac{E_r}{4\pi}\left(-\frac{dH}{dz}\right)-\frac{H}{4\pi}\frac{dE_r}{dz}, \quad \text{and} \quad \frac{E_z H}{4\pi}+\frac{E_z}{4\pi}\frac{dH}{dr}+\frac{H}{4\pi}\frac{dE_z}{dr},$$

or
$$E_r\Gamma_r-\frac{H}{4\pi}\frac{dE_r}{dz}, \quad \text{and} \quad E_z\Gamma_z+\frac{H}{4\pi}\frac{dE_z}{dr},$$

if Γ_r and Γ_z are the components of the electric current-density. The sum of the first terms is clearly the dissipativity per unit volume; and that of the second terms is, by equation (13), Part. I., $H\mu\dot{H}/4\pi$, the rate of increase of the magnetic energy.

The longitudinal transfer of energy in either conductor per unit area is also expressed by $-(4\pi k)^{-1}H(dH/dz)$; or, by $-(4\pi k\mu)^{-1}(dT_1/dz)$ across the complete section, if T_1 temporarily denotes the magnetic energy in the conductor per unit length.

Now let E_1, F_1, C_1, V_1, and E_2, F_2, C_2, V_2, refer to two distinct normal systems. Then, if we could neglect the longitudinal transfer in the conductors, we should have

$$U_{12}-T_{12}=\frac{d}{dz}(V_1C_2-V_2C_1)\div(p_1-p_2), \quad \text{..............(122)}$$

the left side referring to unit length of line; and, in the whole line,

$$U_{12}-T_{12}=[V_1C_2-V_2C_1]_0^l\div(p_1-p_2). \quad \text{................(123)}$$

Similarly, for a single normal system,

$$2(U-T)=\frac{d}{dz}C^2\frac{d}{dp}\frac{V}{C}, \quad \text{........................(124)}$$

per unit length; and, in the whole line,

$$2(U-T)=\left[C^2\frac{d}{dp}\frac{V}{C}\right]_0^l. \quad \text{....................(125)}$$

We have to see how far these are affected by the longitudinal transfer. We have

$$-\frac{d}{dz}V_1C_2=SV_1\dot{V}_2+L_0C_2\dot{C}_1+(E_1-F_1)C_2,$$

$$-\frac{d}{dz}V_2C_1=SV_2\dot{V}_1+L_0C_1\dot{C}_2+(E_2-F_2)C_1;$$

therefore, if the systems are normal,

$$\frac{d}{dz}(V_1C_2-V_2C_1)=(p_1-p_2)(SV_1V_2-L_0C_1C_2)-(E_1-F_1)C_2+(E_2-F_2)C_1.$$

It will be found that we cannot make the parts depending upon E and F exactly represent the $U_{12}-T_{12}$ in the conductors except when

m^2 is the same in both systems p_1 and p_2. In that case, the parts $(E_r)_1(H)_2$ and $(E_r)_2(H)_1$ of the longitudinal transfer of energy in the conductors, depending upon the mutual action of the two systems, are equal; $(E_r)_1$ and $(E_r)_2$ being proportional to $\sin mz$, and H_1 and H_2 proportional to $\cos mz$. So, in case p_1 and p_2 are values of p belonging to the same m^2, the influence of the longitudinal energy-transfer in the conductors goes out from (122) and (123), which are therefore true in spite of it. Similarly, provided the m's can be settled independently of the p's, equations (124) and (125) are true.

Now the normal V and C functions, say u and w, as before, may be taken to be

$$\left.\begin{aligned} u &= \frac{m}{Sp}\Big(\tfrac{1}{2}a_1J_1(s_1a_1) - \tfrac{1}{2}a_1(J_1/K_1)(s_1a_0)K_1(s_1a_1)\Big)\sin(mz+\theta), \\ w &= \Big(\dots\dots\dots\dots\dots\dots\dots\dots\dots\dots\dots\Big)\cos(mz+\theta), \end{aligned}\right\} \quad (126)$$

so that $V = Au\epsilon^{pt}$, $C = Aw\epsilon^{pt}$, and

$$\frac{V}{C} = \frac{u}{w} = \frac{m}{Sp}\tan(mz+\theta); \quad \dots\dots\dots\dots(127)$$

and the complete equations for the determination of m, θ, and p are

$$\frac{m}{Sp}\tan\theta = Z_0, \qquad \frac{m}{Sp}\tan(ml+\theta) = Z_1, \qquad 0 = \frac{m^2}{Sp} + R'_m + L'_m p; \quad (128)$$

the first two of these being the terminal conditions, and $R'_m + L'_m p$ being merely a convenient way of writing the real complex expressions; (equation (68), with $e_m = 0$). It is clear that the only cases in which the m's become clear of the p's are the before-mentioned five cases, equivalent to Z_0 and Z_1 being zero or infinite, and the line closed upon itself, which is a sort of combination of both. Considering only the four, they are summed up in this, $VC = 0$ at the terminals, or the line cut off from receiving or losing energy at the ends. We have then the series of m's, 0, π/l, $2\pi/l$, etc.; or $\frac{1}{2}\pi/l$, $\frac{3}{2}\pi/l$, $\frac{5}{2}\pi/l$, etc.; and every m^2 has its own infinite series of p's through the third equation (128). These, though very special, are certainly important cases, as well as being the most simple. We can definitely effect the expansions of the initial states in the normal functions, and obtain the complete solutions in every particular.

Verification by Direct Integrations. A Special Initial State.

Although rather laborious, it is well to verify the above results by direct integration of the proper expressions for the electric and magnetic energies of normal systems throughout the whole line. Thus, let

$$\frac{d}{dr}\frac{1}{r}\frac{d}{dr}rH_1 + s_1^2H_1 = 0, \qquad \text{where} \qquad -s_1^2 = 4\pi\mu_1k_1p_1 + m_1^2,$$

$$\frac{d}{dr}\frac{1}{r}\frac{d}{dr}rH_2 + s_2^2H_2 = 0, \qquad \text{where} \qquad -s_2^2 = 4\pi\mu_1k_1p_2 + m_2^2,$$

in the inner conductor. We shall find

$$(s_1^2 - s_2^2)\int_{a_0}^{a_1} H_1 H_2 r dr = 8\pi(C_1\Gamma_2 - C_2\Gamma_1),$$

as $H_1 = 0 = H_2$ at $r = a_0$; Γ_1 and Γ_2 being the longitudinal current-densities at $r = a_1$. Similarly, for the outer conductor,

$$(s_1'^2 - s_2'^2)\int_{a_2}^{a_3} H_1' H_2' r dr = -8\pi(C_1\Gamma_2' - C_2\Gamma_1'),$$

if C_1, C_2 still be the currents in the inner conductor; the accents merely meaning changes produced by the altered μ and k in the outer conductor. We have $H_1' = 0 = H_2'$, at $r = a_3$, in this case. Then, thirdly, for the intermediate space,

$$\int_{a_1}^{a_2} H_1'' H_2'' r dr = C_1 C_2 \times 4 \log \frac{a_2}{a_1}.$$

Therefore the total mutual magnetic energy of the two distributions per unit length is

$$\frac{\mu_1}{4\pi}\int_{a_0}^{a_1} H_1 H_2 \,.\, 2\pi r dr + \frac{\mu_2}{4\pi}\int_{a_1}^{a_2} H_1'' H_2'' \,.\, 2\pi r dr + \frac{\mu_3}{4\pi}\int_{a_2}^{a_3} H_1' H_2' \,.\, 2\pi r dr,$$

which, by using the above expressions, becomes, provided $m_1^2 = m_2^2$,

$$L_0 C_1 C_2 - \frac{C_1(E_2 - F_2)}{p_1 - p_2} + \frac{C_2(E_1 - F_1)}{p_1 - p_2}, \quad \text{.................(126}a\text{)}$$

E and F being Γ/k, or the longitudinal electric forces at $r = a_1$ or $r = a_2$. But

$$E - F = R''C,$$

where $R'' =$ the $R_1'' + R_2''$ of equation (56), Part II.; and

$$0 = \frac{m^2}{Sp} + L_0 p + R'' = \frac{m^2}{Sp} + R' + L'p,$$

so (126) becomes

$$\left(C_1\frac{dV_2}{dz} - C_2\frac{dV_1}{dz}\right) \div (p_1 - p_2), \quad \text{or} \quad \frac{m^2 C_1 C_2}{Sp_1 p_2}. \quad \text{.......(127}a\text{)}$$

The mutual electric energy is obviously SV_1V_2 per unit length. By summation with respect to z from 0 to l, subject to $VC = 0$ at both ends, we verify that the total mutual magnetic energy equals the total mutual electric energy. The value of $2T$ in a single normal system is, by (126a) and the next equation,

$$L_0 C^2 + C^2 \frac{dR''}{dp} = C^2 \frac{d}{dp}(R' + L'p) \quad \text{................(128}a\text{)}$$

per unit length; and that of $2U$ is SV^2. Hence, per unit length,

$$2(U - T) = SV^2 - C^2\frac{d}{dp}(R' + L'p). \quad \text{..................(129)}$$

In this use $V=u$ and $C=w$, equations (126), and we shall obtain, for the complete energy-difference in the whole line,

$$-\left\{\frac{a_1}{2}J(s_1a_1)-\dots\right\}^2\frac{l}{2}\frac{d}{dp}\left(\frac{m^2}{Sp}+R'+L'p\right)=M \quad \text{say}, \qquad \dots\dots\dots(130)$$

which is the expanded form of

$$\left[w\frac{du}{dp}-u\frac{dw}{dp}\right]_0^l \qquad \text{or} \qquad \left[w^2\frac{d}{dp}\left(\frac{u}{w}-Z\right)\right]_0^l,$$

as may be verified by performing the differentiations, using the expression for u/w in (127), remembering that m^2 in it is a function of p; or, more explicitly, put $\sqrt{-Sp(R'+L'p)}$ for m, and then differentiate to p.

Given, then, the initial state to be $V=V_0$, a function of z, and $H=H_{01}$ in the inner conductor, H_{02} in the dielectric, and H_{03} in the outer conductor, functions of r and z, and that the system is left without impressed force, subject to $VC=0$ at both ends, the state at time t later will be given by

$$V=\Sigma\, Au\epsilon^{pt}, \qquad C=\Sigma\, Aw\epsilon^{pt};$$

the summations to include every p, with similar expressions for H, Γ, γ, etc., the magnetic force and two components of current, by substituting for u or w the proper corresponding normal functions; the coefficient A being given by the fraction whose denominator is the expression M in (130), and whose numerator is the excess of the mutual electric energy of the initial and the normal system over their mutual magnetic energy, expressed by

$$\frac{m}{p}C'\int_0^l V_0 \sin(mz+\theta)dz$$
$$-\int_0^l \cos(mz+\theta)\,dz\left\{\int_{a_0}^{a_1}\mu_1H_{01}C_1'dr+\int_{a_1}^{a_2}\mu_2H_{02}C'dr+\int_{a_2}^{a_3}\mu_3H_{03}C_3'dr\right\}, \quad \dots(131)$$

where
$$C'=\frac{a_1}{2}\Big(J_1(s_1a_1)-(J_1/K_1)(s_1a_0)K_1(s_1a_1)\Big);$$

and C_1' is the same with r put for a_1, and C_3' is the same with r put for a_1, a_3 for a_0, and s_3 for s_1. It should not be forgotten that in the case $m=0$, the denominator (130) requires to be doubled, $\frac{1}{2}l$ becoming l. Also that R'', or $R'+L'p$, contains m^2, and must not be the $m=0$ expression for the same.

To check, take the initial state to be $e_0(1-z/l)$, with no magnetic force, and let $V=0$ at both ends. We find immediately, by (130) and (131), that at time t,

$$V=\frac{2e_0}{l}\sum\frac{1}{m}\sin mz\sum\frac{(m^2/Sp^2)\epsilon^{pt}}{-\frac{d}{dp}\left(\frac{m^2}{Sp}+R'+L'p\right)}, \qquad \dots\dots\dots(132)$$

where the m's are to be π/l, $2\pi/l$, $3\pi/l$, etc.; the first summation being with respect to m, and the second for the p's of a particular m.

But, initially,

$$V=e_0\left(1-\frac{z}{l}\right)=\frac{2e_0}{l}\sum\frac{1}{m}\sin mz.$$

Therefore we must have

$$1 = \sum \frac{m^2/Sp^2}{-\frac{d}{dp}\left(\frac{m^2}{Sp} + R' + L'p\right)}$$

Simplified, it makes this theorem :—

$$-\frac{1}{\phi(0)} = \sum\left(p\frac{d\phi}{dp}\right)^{-1},$$

if the p's are the roots of $\phi(p) = 0$. This is correct.

The Effect of Longitudinal Impressed Electric Force in the Circuit. The Condenser Method.

To determine the effect of longitudinal impressed force, keeping to the case of uniform intensity over the cross-section of either conductor. Let a steady impressed force of integral amount e_0 be introduced in the line at distance z_1; it may be partly in one and partly in the other conductor, as in Part II. By elementary methods, we can find the steady state of V, C it will set up. If, then, we remove e_0, we can, by the preceding, find the transient state that will result. Let V_0 be the steady state of V set up, and V_1 what it becomes at time t after removal of e_0; then $V_0 - V_1$ represents the state at time t after e_0 is put on. So, if ΣAu represent the V set up by the unit impressed force at z_1,

$$V = V_0 - e_0 \Sigma Au\epsilon^{pt}$$

will give the distribution of V at time t after e_0 is put on, being zero when $t = 0$, and V_0 when $t = \infty$. No zero value of p is admissible here.

From this we deduce that the effect of e_0, lasting from $t = t_1$ to $t = t_1 + dt_1$, at the later time t, is

$$-\Sigma Aupe_0 dt_1 \epsilon^{p(t-t_1)};$$

therefore, by time-integration, the effect due to an impressed force e_0 at one spot, variable with the time, starting at time t_0, is

$$V = -\sum Aup\epsilon^{pt}\int_{t_0}^{t} e_0\epsilon^{-pt_1}dt_1,$$

in which e_0 is a function of t_1.

By integrating along the line, we find the effect of a continuously distributed impressed force, e per unit length, to be

$$V = -\sum up\epsilon^{pt}\int_0^l\int_{t_0}^{t} Ae\epsilon^{-pt_1}dz_1dt_1, \quad \text{................(133)}$$

wherein e is a function of both z_1 and t_1, and starts at time t_0; whilst A is a function of z_1, the position of the elementary impressed force edz_1.

To find A as a function of z_1, we might, since ΣAu is the V set up by unit e at z_1, expand this state by the former process of integration. But the following method, though unnecessary for the present purpose, has the advantage of being applicable to cases in which VC is not zero at the terminals, but $V = ZC$ instead. It is clear that the integration process, including the energy in the terminal apparatus, would be very lengthy, and would require a detailed knowledge of the terminal combinations. This is avoided by replacing the impressed force at z_1 by a charged condenser; when, clearly, the integration is confined to one spot. Let S_1 be the capacity, and V_0 the difference of potential, of a condenser inserted at z_1. If we increase S_1 infinitely it becomes mathematically equivalent to an impressed force V_0, without the condenser.

Suppose $\Sigma Aw'\epsilon^{pt}$ is the current at z at time t after the introduction of the condenser, of finite capacity; then, since $-S_1\dot{V}$ is the current leaving the condenser, or the current at z_1, we have

$$-S_1\dot{V} = \Sigma Aw_1'\epsilon^{pt},$$

w_1 being the value of w' at z_1. The expansion of V_0 is therefore

$$V_0 = -\Sigma Aw_1'/S_1 p,$$

initially; and the mutual potential energy of the initial charge of the condenser and of the normal u' corresponding to w' must be

$$S_1 V_0(-w_1'/S_1 p) = -V_0 w_1'/p.$$

But since there is, initially, electric energy only at z_1, and magnetic energy nowhere at all, the only term in the numerator of A will be that due to the condenser, or this $-V_0 w_1'/p$; hence

$$A = -V_0 w_1/pM,$$

where M is the $2(U-T)$ of the complete normal system, as modified by the presence of the condenser, is the value of A in $V = \Sigma Au'\epsilon^{pt}$, making

$$V = -V_0 \Sigma (w_1'/pM)u'\epsilon^{pt},$$

expressing the effect at time t after the introduction of the condenser, and due to its initial charge.

So far S_1 has been finite, and consequently u', w', M, and p depend on its capacity as well as on the line and terminal conditions. But on infinitely increasing its capacity, u' and w' become u and w, the same as if the condenser were non-existent. Therefore

$$V = -\Sigma V_0(w_1/pM)u\epsilon^{pt} \quad \text{.....................(134)}$$

expresses the effect due to the steady impressed force V_0 at z_1, at time t after it was started. This will have a term corresponding to a zero p (due to the infinite increase of S_1 in the previous problem), expressing the final state. Hence, leaving out this term, the summation (134), with sign changed, and $t=0$, expresses the final state itself. Thus, taking $V_0 = 1$,

$$\Sigma Au = \Sigma w_1 u/pM$$

is the expansion required to be applied to (133). Put $A = w_1/pM$ in it, and it becomes

$$V = -\Sigma\left(\frac{u}{M}\right)\epsilon^{pt}\int_0^l\int_{t_0}^t w_1 e\,\epsilon^{-pt_1}\,dz_1 dt_1, \quad \text{................(135)}$$

fully expressing the effect at z, t, due to the impressed force e, a function of z_1 and t_1, starting at time t_0. To obtain the current, change u to w

outside the double integral. The M, when the condition $VC=0$ at the ends is imposed, is that of (130); the u and w expressions those of (126). But if we regard S, R', and L' as constants (or functions of z), then (135) holds good when terminal conditions $V=ZC$ are imposed, provided the impressed force be in the line only, as supposed in (135).

Special Cases of Impressed Force.

When the impressed force is steady, and is confined to the place $z=0$, and is of integral amount e_0, (135) gives

$$V=e_0\Sigma uw_0/pM-e_0\Sigma uw_0\epsilon^{pt}/pM, \quad \text{.............. (136)}$$

w_0 being the value of w at $z=0$, as the effect at time t after starting e_0. The first summation expresses the state finally arrived at.

Again, in (135) let the impressed force be a simple-harmonic function of the time. I have already given the solution in this case, so far as the formula for C is concerned, in the case $V=0$ at both ends, in equation (76), Part II., which may be derived from (135) by using in it w instead of u at its commencement, putting $e=e_0\sin nt$, and effecting some reductions. The V-formula may be got in a similar manner to that used in getting (76), but it is instructive to derive it from (135), as showing the inner meaning of that formula. Let $e=e_0\sin(nt+\alpha)$ in it, where e_0 is a function of z. Effect the t_1 integration, with $t_0=0$ for simplicity. The result is

$$V=-\Sigma\frac{u\epsilon^{pt}}{M}\left(\frac{p\sin\alpha+n\cos\alpha}{p^2+n^2}\right)\int_0^l w_1e_0dz_1$$
$$+\Sigma\frac{u}{M}\left(\frac{p\sin(nt+\alpha)+n\cos(nt+\alpha)}{p^2+n^2}\right)\int_0^l w_1e_0dz_1. \quad \text{......(137)}$$

The first summation cancels the second at the first moment, and ultimately vanishes, leaving the second part to represent the final periodic solution. Take $\alpha=0$; and use the u, w, M expressions of (126) and (130), and let ϕ_m stand for $m^2+Sp(R_m'+L_m'p)$, so that $\phi_m=0$ gives the p's for a particular m^2. Then we obtain, (with $V=0$ at both ends),

$$V=\frac{d}{dz}\Sigma\frac{\cos mz\int_0^l\cos mz_1\,.\,e_0dz_1\,.\,(p\sin nt+n\cos nt)}{\frac{1}{2}l\frac{d\phi_m}{dp}(p^2+n^2)}$$

$$=\frac{d}{dz}\frac{2}{l}\Sigma\frac{\cos mz\int_0^l\cos mz_1\,.\,e_0\sin nt\,.\,dz_1}{\left(\frac{d}{dt}-p\right)\frac{d\phi_m}{dp}}, \quad \text{..................(138)}$$

because $d^2/dt^2=-n^2$. But, if $e_0=\Sigma e_m$, the equation of V_m is

$$-\phi_mV_m=\frac{de_m}{dz}\sin nt,$$

(by (60) and (63), Part II.), so that

$$V_m = -\phi_m^{-1}\frac{de_m}{dz} = -\frac{d}{dz}\sum\frac{e_m \sin nt}{\left(\frac{d}{dt}-p\right)\frac{d\phi}{dp}}, \qquad \dots\dots\dots\dots\dots(139)$$

by a well-known algebraical theorem, the summation being with respect to the p's which are the roots of $\phi_m = 0$, considered as algebraic. We have also

$$e_0 = \frac{2}{l}\sum \cos mz \int_0^l \cos mz_1\, e_0 dz_1, \qquad \dots\dots\dots\dots\dots(140)$$

the summation being with respect to m.

Uniting (139) and (140), there results the previous equation (138), in which the summation is with respect to all the p's belonging to all the m's. In the case $m=0$, the $2/l$ must be halved. In the form of a summation with respect to m, similar to (77) for C, the corresponding V-solution is

$$V = -\frac{2V_0}{Snl}\sum\frac{m \sin mz\{(L_m' - m^2/Sn^2)n \sin nt + R_m' \cos nt\}}{R_m^2 + (L_m - m^2/Sn^2)^2 n^2},$$

the impressed force being $V_0 \sin nt$, at $z=0$. This, on the assumption $R_m' = R'$, $L_m' = L'$, will be found to be the expansion of the form (80), Part II.

How to make a Practical Working System of V *and* C *Connections.*

Now to make some remarks on the impossibility of joining on terminal apparatus without altering the normal functions, the terminal arrangements being made to impose conditions of the form $V = ZC$. It is clear, in the first place, that if the quantity VC at $z=0$ and $z=l$ really represents the energy-transfer in or out of the line at those places, then the equation

$$A_1 = \frac{U_{01} - T_{01}}{\left[w^2 \frac{d}{dp}\left(\frac{u}{w} - Z\right)\right]_0^l}$$

will be valid, provided u and w be the correct normal functions. But to make VC be the energy-transfer at the ends, requires us to stop the longitudinal transfer in the conductors there, or make the current in the conductors longitudinal. This condition is violated when the current-function w is proportional to $\cos(mz+\theta)$, as in the previous, except in the special cases, because the radial current γ in the conductors is proportional to $\sin(mz+\theta)$, and γ has to vanish. Not in the dielectric, but merely in the conductors.

We can ensure that VC is the energy-transfer at the ends, by coating the conductors over their exposed sections with infinitely conducting material, and joining the terminal apparatus on to the latter. The current in the conductors will be made strictly longitudinal, close up to the infinitely conducting material, and γ will vanish in the conductors. But γ in the dielectric at the same place will be continuous with the radial surface-current on the infinitely conducting ends, due to the sudden discontinuity in the magnetic force. Thus the energy-transfer, at the ends, is confined to the dielectric.

It is clear, however, that the normal current-functions in the two conductors must be such as to have no radial components at the terminals, so that they cannot be what have been used, such that $d^2/dz^2 =$ constant. They require alteration, of sensible amount, it may be, only near the terminals, but, theoretically, all along the line. It would therefore appear that only the five cases of $V=0$ at either or both ends, or $C=0$ ditto, or the line closed upon itself, admit of full solution in the above manner. The only practical way out of the difficulty is to abolish the radial electric current in the conductors, making (66) the equation of V, and VC the longitudinal energy-transfer, with full applicability of the $V=ZC$ terminal conditions.

PART IV.

Practical Working System in terms of V *and* C *admitting of Terminal Conditions of the Form* V = ZC.

As mentioned at the close of Part III., it would appear that the only practicable way of making a workable system, which will allow us to introduce the terminal conditions that always occur in practice, in the form of linear differential equations connecting C and V, the current and potential-difference at the terminals, is to abolish the very small radial component of current in the conductors. This does not involve the abolition of the radial dielectric current which produces the electric displacement, or alter the equation of continuity to which the total current in the wires is subject. The dielectric current, which is $S\dot{V}$ per unit length of line, and which must be physically continuous with the radial current in the conductors at their boundaries, may, when the latter is abolished, be imagined to be joined on to that part of the longitudinal current in the conductors that goes out of existence by some secret method with which we are not concerned.

We assume, therefore, that the propagation of magnetic induction and electric current into the conductors takes place, at any part of the line, as if it were taking place in the same manner at the same moment at all parts (as when the dielectric displacement is ignored, making it only a question of inertia and resistance), instead of its being in different stages of progress at the same moment in different parts of the line. This requires that a small fraction of its length, along which the change in C is insensible, shall be a large multiple of the radius of the wire. The current may be widely different in strength at places distant, say, a mile, and yet the variation in a few yards be so small that this section, so far as the propagation of magnetic induction into it is concerned, may be regarded as independent of the rest of the line; the variation of the boundary magnetic-force, or of C, fully determining the internal state of the conductors, exactly as it would do were there no electrostatic induction.

In a copper wire, in which $\mu=1$, and $k=1/1700$, the value of the quantity $4\pi\mu kp$ is $p/135$. On the other hand, the quantity m in $-s^2=4\pi\mu kp+m^2$ has values 0, π/l, $2\pi/l$, etc., or a similar series, in which l is the length of the line in centimetres, so that $j\pi/l$ is a minute fraction, unless j be excessively large. But then it would correspond to an utterly insignificant normal system. We may therefore take

$$-s^2=4\pi\mu kp.$$

It will be as well to repeat the system that results, from Part II. The line-integral of the radial electric force across the dielectric being V, from the inner to the outer conductor (concentric tubes), and the line-integral of the magnetic force round the inner conductor being $4\pi C$, so that C is the total current in it, accompanied by an oppositely directed current of equal strength in the outer conductor, V and C are connected by two equations, one of continuity of C, the other the equation of electric force, thus :—

$$-\frac{dC}{dz}=S\dot{V}, \qquad e-\frac{dV}{dz}=L_0\dot{C}+R_1''C+R_2''C. \quad \text{..........(141)}$$

Here e is impressed force, S the electric capacity, and L_0 the inductance of the dielectric, all per unit length of line ; and R_1'' and R_2'' are certain functions of d/dt and constants such that $R_1''C$ and $-R_2''C$ are the longitudinal electric forces of the field at the inner and outer boundaries of the dielectric, which, when only the first differential coefficient dC/dt is counted, become

$$R_1''=R_1+L_1\frac{d}{dt}, \qquad R_2''=R_2+L_2\frac{d}{dt},$$

respectively, where R_1, L_1, and R_2, L_2 are the steady resistances and inductances of the two conductors.

Extension to a Pair of Parallel Wires, or to a Single Wire.

The forms of R_1'' and R_2'' are known when the conductors are concentric circular tubes, of which the inner may be solid, making it an ordinary round wire. Now if the return-conductor be a parallel wire or tube externally placed, it is clear that we may regard R_1'' and R_2'' as known in the same manner, provided their distance apart be sufficiently great to make the departure of the distribution of current in them from symmetry insensible. We have merely to remember that it is now the inner boundary of the return-tube that corresponds to the former outer boundary, *i.e.* when it surrounded the inner wire concentrically.

The quantity V will still be the line-integral of the electric force across the dielectric by any path that keeps in one plane perpendicular to the axes of the conductors, in which plane lie the lines of magnetic force. Also, the product VC will still represent the total longitudinal transfer of energy per second in the dielectric at that plane, or, in short, the energy-current. As regards the modified forms of S and L_0, there is, in strictness, some little difficulty, on account of the dielectric being necessarily bounded by other conductors than the pair under consideration, in which others energy is wasted, to a certain extent. This can only be allowed for by the equations of mutual induction of the various conductors, which are not now in question. But if our pair, for instance, be suspended alone at a uniform height above the ground, so that only the very small dissipation of energy in the earth interferes, it would seem, so far as the wire-current is concerned, to be an unnecessary refinement to take the earth into consideration. There are, then, two or three practical courses open to us; as to suppose the earth to be a perfect nonconductor and behave as if it were replaced by air, or to treat it as a perfect conductor. In neither case will there be dissipation of energy except in our looped wires, which have no connection with the earth, but there will be a different estimation of the quantities L_0 and S required. For, when we suppose the earth is perfectly conducting, we shut it out from the magnetic field as well as from the electric field. The electric capacity S is that of the condenser formed by the two wires and intermediate dielectric, as modified by the presence of the earth (the method of images gives the formula at once), and the value of L_0 is such that $L_0S = \mu c = v^{-2}$, where v is the velocity of undissipated waves through the dielectric; that is, as before, L_0 is simply the inductance of the dielectric, per unit length of line. On the ground, there will be both electrification and electric current, due to the discontinuity in the electric displacement and the magnetic force respectively; but with these we have no concern. In the other case, with extension of the magnetic and electric fields, the product L_0S still equals v^{-2}. Neither course is quite satisfactory; perhaps it would be best to sacrifice consistency and let the magnetic field extend unimpeded into the earth, considered as nonconducting, with consequently no electric current and waste of energy, whilst, as regards the external electric field, we treat it as a conductor. We must compromise in some way, unless we take the earth into account fully as an ordinary conductor. Similarly, if the line consist of a single wire whose circuit is completed through the earth, by regarding it as infinitely conducting we replace the true variably distributed return-current by a surface-current, and, terminating the magnetic field there, have $L_0S = v^{-2}$; but if we allow the magnetic field to extend into it, though with insignificant loss of energy by electric current, we shall no longer have this property.

Effect of Perfect Conductivity of Parallel Straight Conductors. Lines of Electric and Magnetic Force strictly Orthogonal, irrespective of Form of Section of Conductors. Constant Speed of Propagation.

The property is intimately connected with the influence of perfect conductivity on the state of the dielectric. For perfect conductivity will make the lines of electric force normal to the conducting boundaries, will make them cut perpendicularly the magnetic-force lines, which lie in the planes $z =$ const. and are tangential at the boundaries, and will make $L_0S = v^{-2}$, irrespective of the shape of section of the conductors. Now, at the first moment of putting on an impressed force, wires always behave as if they were infinitely conducting, so that, by the above, the initial effect is simply a dielectric disturbance, travelling along the dielectric, guided by the conductors, with velocity v, irrespective of the form of section. Of course dissipation of energy in the conductors immediately begins, and finally completely alters the state of things, which would be, in the absence of dissipation, the to-and-fro passage of a wave through the dielectric for ever. Except the extension to other than round conductors, this does not add to the knowledge already derived from their study. The effect of alternating currents in tending to become mere surface-currents as the frequency is raised (Part I.) may be derived from, or furnish itself a proof of, the property above mentioned—that at the first moment there is merely a dielectric disturbance. For in rapid alternations of impressed force, we are continually stopping the establishment of the steady state at its very commencement, and substituting the establishment of a steady state of the opposite kind, to be itself immediately stopped, and so on.

When the dielectric is unbounded—not enclosed within conductors—there is also the outward propagation of disturbances to be considered; but it would appear, by general reasoning, that this is, relatively to the main effect, or propagation parallel to the wires, a secondary phenomenon.

Extension of the Practical System to Heterogeneous Circuits, with "Constants" varying from place to place. Examination of Energy Properties.

It is clear that the same principles apply to conductors having other forms of section than circular, when V and C are made the variables, provided the functions R_1'' and R_2'' can be properly determined. The quantity VC being in all cases the energy-current, its rate of decrease as we pass along the line is accounted for (as in Part III.), thus, by making use of (141), with $e=0$,

$$-\frac{d}{dz}(VC)=\frac{d}{dt}(\tfrac{1}{2}SV^2+\tfrac{1}{2}L_0C^2)+CR_1''C+CR_2''C; \quad \text{........(142)}$$

that is, in increasing the electric and magnetic energies in the dielectric, and in transfer of energy into the conductors, to the amounts $CR_1''C$ and $CR_2''C$ per second respectively; which are, in their turn, accounted for by the rate of increase of the magnetic energy, and the dissipativity, or Joule-heat per second in the two conductors; or

$$CR_1''C=Q_1+\dot{T}_1, \qquad CR_2''C=Q_2+\dot{T}_2, \quad \text{..........(143)}$$

Q being the dissipativity and T the magnetic energy per unit length of conductor.

These equations (143) must therefore contain the enlarged definition of the meaning of the functions R_1'' and R_2''. For it is no longer true that $R_1''C$ is, as it was in the tubular case, the longitudinal electric force at the boundary of the conductor to which R_1'' belongs. It is a sort of mean value of the longitudinal electric force. Thus, we must have

$$\int\frac{EH}{4\pi}\,.\,ds=CR_1''C, \quad \text{........................(144)}$$

if E be the longitudinal electric force and H the component of the magnetic force along the line of integration, which is the circuital boundary of the section of the conductor perpendicular to its length. But no extension of the meaning of V is required from that last stated.

Let us, then, assume that R_1'' and R_2'' can be found, their actual discovery being the subject of independent investigation. We can always fall back upon round wires or tubes if required. They are functions of d/dt and constants, if the line is homogeneous. But, as we have got rid of the radial component of current in the conductors, and its difficulties, the constancy of the constants in R_1'' and R_2'' (as the conductivity and the inductivity, or the steady resistance, or the diameter) need no longer be preserved. Provided the conductors may be regarded as homogeneous along any few yards of length, they may be of widely different resistances, etc., at places miles apart. Then R_1'', R_2'' become functions of z as well as of d/dt, and S a function of z. Let our system be

$$-\frac{dC}{dz}=S''V, \qquad e-\frac{dV}{dz}=R''C, \qquad \text{..............(145)}$$

where both R'' and S'' are functions of d/dt and z. As regards S'', it is simply $S(d/dt)$ when the dielectric is quite nonconducting. But when leakage is allowed for, it becomes $K+S(d/dt)$, where K is the conductance, or reciprocal of the resistance, of the dielectric across from one conductor to the other. Then both K and S are functions of z. The conduction-current is KV, and the displacement-current $S\dot{V}$, whilst their sum, or $S''V$, is the true current across the dielectric per unit length of line. We have now, by (145), with $e=0$,

$$-\frac{d}{dz}(VC)=VS''V+CR''C=KV^2+\frac{d}{dt}\tfrac{1}{2}SV^2+CR''C. \qquad \text{....(146)}$$

The additional quantity KV^2 is the dissipativity in the dielectric per unit length, whilst now $CR''C$ includes the whole magnetic-energy increase, and the dissipativity (rate of dissipation of energy) in the conductors.

Let V_1, C_1, and V_2, C_2 be two systems satisfying (145) with $e=0$. Then

$$-\frac{d}{dz}(V_1C_2)=V_1S''V_2+C_2R''C_1, \qquad -\frac{d}{dz}(V_2C_1)=V_2S''V_1+C_1R''C_2;$$

from which we see that if the systems be normal, d/dt becoming p_1 and p_2 respectively, we shall have

$$\frac{d}{dz}(V_1C_2-V_2C_1)=(p_1-p_2)\left\{SV_1V_2-\frac{R_1''-R_2''}{p_1-p_2}C_1C_2\right\}, \qquad \text{......(147)}$$

R_1'' and R_2'' being what R'' becomes with p_1 and p_2 for d/dt. As the quantity in the $\{\}$ is the $U_{12}-T_{12}$ of Part III., and the first term is U_{12}, we see that the mutual magnetic energy is

$$T_{12}=C_1C_2(R_1''-R_2'')\div(p_1-p_2). \qquad \text{................(148)}$$

The division by p_1-p_2 can be effected, and the right member of (148) put in the form

$$C_1f(p_1)\times C_2f(p_2).$$

When this is done, we can find the mutual magnetic energy of any magnetic field (proper to our system) and a normal field, in terms of the total current in the wire and its differential coefficients with respect to t; so that, in the expansion of an arbitrary initial state, C, $\dot{C}$, $\ddot{C}$, etc., may be the data of the magnetic energy, instead of the magnetic field itself.

We see also, from (148), that if T be the magnetic energy of any normal system per unit length of line, then

$$2T = C^2 \frac{dR''}{dp}; \quad \text{.........................(149)}$$

and therefore, if Q be the dissipativity in the conductors,

$$Q = R''C^2 - \dot{T} = C^2\left(R'' - p\frac{dR''}{dp}\right). \quad \text{................(150)}$$

Now consider the connection of the two solutions for the normal functions. Since the equation of C in general is, by (145),

$$\frac{d}{dz}\left(\frac{1}{S''}\frac{dC}{dz}\right) = R''C - e, \quad \text{......................(151)}$$

the normal C-function, say w, is to be got from

$$\frac{d}{dz}\left(\frac{1}{S''}\frac{dw}{dz}\right) = R''w, \quad \text{......................(152)}$$

with $d/dt = p$ in R'' and S'', making them functions of z and p. Let X and Y be the two solutions, making

$$w = X + qY, \quad \text{...........................(153)}$$

where q is a constant. The normal V-function, say u, is got from w by the first of (145), giving

$$u = -\frac{1}{S''}\frac{dw}{dz} = -\frac{1}{S''}(X' + qY'), \quad \text{.................(154)}$$

if $\qquad X' = dX/dz, \qquad Y' = dY/dz.$

In X and Y, which together make up the w in (153), p has the same value. Therefore, in (147), supposing C_1 to be X and C_2 to be Y, we have disappearance of the right member, making

$$\frac{d}{dz}(V_1C_2 - V_2C_1) = 0, \qquad \text{or} \qquad V_1C_2 - V_2C_1 = \text{constant},$$

or $\qquad XY' - YX' = S'' \times \text{constant} = hS''$, say, ...(155)

leading to the well-known equation

$$Y = X\int\frac{hS''}{X^2}dz,$$

connecting the two solutions of the class of equations (152); which we see expresses the reciprocity of the mutual *activities* of the two parts into which we may divide the electromagnetic state represented by a single normal solution.

Also, by (147), integrating with respect to z from 0 to l,

$$\int_0^l Su_1u_2dz - \int_0^l \frac{R_1'' - R_2''}{p_1 - p_2}w_1w_2dz = \frac{[u_1w_2 - u_2w_1]_0^l}{p_1 - p_2}, \quad \text{.........(156)}$$

either member of which represents the complete $U_{12} - T_{12}$ of the line. The negative of this quantity, as in Part III., is the corresponding $U_{12} - T_{12}$ in the terminal arrangements; so that the value of $2(U-T)$ in a complete normal system, including the apparatus, is

$$2(U-T) = \int_0^l Su^2dz - \int_0^l \frac{dR''}{dp}w^2dz - w_1^2\frac{dZ_1}{dp} + w_0^2\frac{dZ_0}{dp}, \quad \text{......(157)}$$

if $V/C = Z_1$ at $z=l$ and Z_0 at $z=0$, (these being functions of p and constants), and w_1, w_0 are the values of w at $z=l$ and 0. Or, which is the same,

$$2(U-T) = \left[w^2\frac{d}{dp}\left(\frac{u}{w} - Z\right)\right]_0^l, \quad \text{..................(158)}$$

as before used.

The Solution for V *and* C *due to an Arbitrary Distribution of* e, *subject to any Terminal Conditions.*

There is naturally some difficulty in expressing the state at time t in this form :—

$$V = \Sigma Au\epsilon^{pt}, \qquad C = \Sigma Aw\epsilon^{pt},$$

due to an arbitrary initial state, on account of the difficulty connected with

$$(R_1'' - R_2'') \div (p_1 - p_2),$$

and the unstated form of R''. But when the initial state is such as can be set up by any steadily-acting distribution of longitudinal impressed force (e an arbitrary function of z), so that whilst V is arbitrary, C is only in a very limited sense arbitrary, and $\dot{C}$, $\ddot{C}$, etc., are initially zero, and certain definite distributions of electric and magnetic energy in the terminal apparatus are also necessarily involved; in this case we may readily find the full solutions, and therefore also determine the effect of any distribution of e varying anyhow with the time. In fact, by the condenser-method of Part III., we shall arrive at the solution (135); we have merely to employ the present u and w, and let M be the value of the right member of (158). The following establishment, however, is quite direct, and less mixed up with physical considerations.

To determine how V and C rise from zero everywhere to the final state due to a steadily-acting arbitrary distribution of e put on at the time $t=0$. Start with e_2 at $z=z_2$ and none elsewhere, and let $(X+q_0Y)A_0$ and $(X+q_1Y)A_1$ be the currents on the left (nearest $z=0$) and right sides of the seat of impressed force. We have to find q_0, q_1, A_0, and A_1. The condition $V=Z_0C$ at $z=0$ gives us, by (153), 154),

$$-(X_0' + q_0Y_0') \div S_0'' = Z_0(X_0 + q_0Y_0);$$

therefore $$q_0 = -(X_0' + S_0''Z_0X_0) \div (Y_0' + S_0''Z_0Y_0). \quad \text{...........(159)}$$

Similarly, $V=Z_1C$ at $z=l$, gives us

$$q_1 = -(X_1'+S_1''Z_1X_1) \div (Y_1'+S_1''Z_1Y_1). \qquad \text{.........(160)}$$

Here the numbers $_0$ and $_1$ mean that the values of X, etc., and S'' at $z=0$ and at $z=l$ are to be taken.

Now, at the place $z=z_2$, the current is continuous, whilst the V rises by the amount e_2 suddenly in passing through it. These two conditions give us

$$(X_2+q_0Y_2)A_0=(X_2+q_1Y_2)A_1,$$

$$S_2''e_2+(X_2'+q_0Y_2')A_0=(X_2'+q_1Y_2')A_1,$$

where the $_2$ means that the values at $z=z_2$ are to be taken. These determine A_0 and A_1 to be

$$A_0 \text{ or } A_1 = \frac{(X_2+q_1Y_2)e_2 \text{ or } (X_2+q_0Y_2)e_2}{(S_2'')^{-1}(X_2Y_2'-Y_2X_2')(q_0-q_1)}. \qquad \text{.........(161)}$$

Now use (155), making the denominator in (161) be $h(q_0-q_1)$. We have then, if C_0 and C_1 are the currents on the left and right sides of the seat of impressed force,

$$C_0 = \frac{(X+q_0Y)(X_2+q_1Y_2)}{h(q_0-q_1)}e_2, \qquad C_1 = \frac{(X+q_1Y)(X_2+q_0Y_2)}{h(q_0-q_1)}e_2. \qquad (162)$$

These are, when the p is throughout treated as d/dt, the ordinary differential equations of C_0 and C_1 arising out of the partial differential equation of C by subjecting it to the terminal conditions and to the impressed-force discontinuity.

Now make use of the algebraical expansion *

$$\frac{f(p_0)}{\phi(p_0)} = \sum \frac{f(p)}{(p_0-p)\dfrac{d\phi}{dp}}, \qquad \text{.....................(163)}$$

* [The limitations to which this expansion is subject render its use in the above manner undesirable even when it gives correct results, and, of course, when it gives incorrect results, as when the initial C is not zero, the manner of application should necessarily be changed. We should rather proceed thus:—Let

$$C=\frac{f(p_0)}{\phi(p_0)}e \qquad \text{..(1)}$$

be the differential equation connecting C with e, where p_0 stands for d/dt, and $\phi(p)=0$ is the determinantal equation of the system, that is, $\phi(p)$ may be either the characteristic function in fully developed form, or the same multiplied by any function that does not conflict with its use in the determinantal equation. Then we shall have, by the algebraical theorem,

$$\frac{1}{\phi(p_0)} = \sum \frac{1}{(p_0-p)\phi'}, \qquad \text{..................................(2)}$$

where ϕ' means $d\phi/dp$, and the summation includes all the roots of $\phi(p)=0$. Therefore, by (1), using (2) and integrating,

$$C=f(p_0)\sum \frac{e}{(p_0-p)\phi'} = ef(p_0)\sum \frac{\epsilon^{pt}-1}{p\phi'}, \qquad \text{.........................(3)}$$

e being zero before, and constant after $t=0$. But also, by (2),

$$\frac{1}{\phi_0} = \sum \frac{1}{-p\phi'}, \qquad \text{..(4)}$$

the summation being with respect to the p's which are the roots of $\phi(p)=0$, without inquiring too curiously into its strict applicability, or troubling about equal roots. Here p_0 has to be d/dt, and the p's the roots of

$$\phi = h(q_0 - q_1) = 0 ;$$

so that (162) expands to

$$C = \sum \frac{(X+qY)(X_2+qY_2)}{h\frac{d}{dp}(q_0-q_1)} \frac{e_2}{\frac{d}{dt}-p}, \quad \text{.........(164)}$$

where the single q takes the place of the previous q_0 or q_1, which have now equal values, and C has the same expression on both sides of the seat of impressed force. But e_2 is constant with respect to t, whilst C is initially zero; hence

$$\frac{e_2}{d/dt - p} = \frac{e_2(1-\epsilon^{pt})}{-p},$$

where ϕ_0 means ϕ with $p=0$, so that (3) becomes

$$C = ef(p_0)\frac{1}{\phi_0} + ef(p_0)\sum \frac{\epsilon^{pt}}{p\phi'}. \quad \text{.........(5)}$$

Now perform the operations indicated by $f(p_0)$ and we get

$$C = e\frac{f_0}{\phi_0} + e\sum \frac{f(p)}{p\phi'} \cdot \epsilon^{pt}, \quad \text{.........(6)}$$

where f_0 means f with $p=0$. (See also the investigation at the end of the (later) paper on "Resistance and Conductance Operators.") Here ef_0/ϕ_0 is the final steady current, when there is such a thing.

Thus, if we take $\phi = h(q_0 - q_1)$, and use (6), (162) lead to

$$C_0 = \frac{w_0 w_2}{\phi_0}(p=0) + \sum \frac{ww_2}{p\phi'} e_2 \cdot \epsilon^{pt}, \quad \text{.........(7)}$$

$$C_1 = \frac{w_1 w_2}{\phi_0}(p=0) + \sum \frac{ww_2}{p\phi'} e_2 \cdot \epsilon^{pt}, \quad \text{.........(8)}$$

instead of (165). In the first terms the $p=0$ values must be taken, with

$$w_2 = X_2 + q_1 Y_2 \text{ in (7), and } w_2 = X_2 + q_0 Y_2 \text{ in (8).}$$

Here q_0 and q_1 are not the same, but they are the same in the summation; because then $\phi=0$. We may write (165) thus:—

$$C = C_0 + \sum \frac{ww_2}{p\phi'} e_2 \cdot \epsilon^{pt}, \quad \text{.........(9)}$$

where C_0 is the final steady current, to be got direct from the first or second of (162) as the case may be. Therefore (166) should be

$$C = C_0 + \sum w \int_0^l \frac{ewdz}{p\phi'} \cdot \epsilon^{pt}, \quad \text{.........(10)}$$

where C_0 is the final steady current at z due to the whole impressed force.

In accordance with the above (167) is not always applicable, and in accordance with the text (168) is incorrect. But the substituted method of finding C_0, viz. (169), will do when (164) is applicable, and fail otherwise. The result (170), however, is independent of this restriction, as it is immediately obtainable from the differential equations (162).

So up to (162) inclusive the text is correct. Then pass on to (170), (172), as the next clear results. Between these places modify the method as in the present note.]

which brings (164) to

$$C=\sum\frac{(X+qY)(X_2+qY_2)}{-p\frac{d\phi}{dp}}e_2(1-\epsilon^{pt}),\quad\text{.............}(165)$$

which is the complete solution. By integration with respect to z we find the effect due to a steady arbitrary distribution of e put on at $t=0$; thus

$$C=\sum\frac{w\int_0^l ewdz}{-p\phi'}(1-\epsilon^{pt}),\quad\text{.....................}(166)$$

where $\phi'=d\phi/dp$, and w is the normal current-function $X+qY$. To express the V-solution, turn the first w into u. The extension to e variable with t, as in Part III., is obvious. But as the only practical case of e variable with t is the case of periodic e, whose solution can be got immediately from the equations (162) by putting $p^2=-n^2$, constant, the extension is useless. Note that q_0 and q_1 are not equal in (162), and therefore in the periodic solution obtained from (162) direct they must be both used.

The quantity $-\phi'$ which occurs here is identical with the former complete $2(U-T)$ of the line and terminal apparatus of (157) or (158).

Let C_0 be the finally-reached steady current. By (166) it is

$$C_0=\sum\left(-\frac{w}{p\phi'}\right)\int_0^l ewdz.\quad\text{.....................}(167)$$

To this apply (163), with $p_0=0$. Then a finite expression for C_0 is

$$C_0=\frac{w_0}{\phi_0}\int_0^l ew_0dz,\quad\text{.........................}(168)$$

where w_0 and ϕ_0 are what w and ϕ become when $p=0$ in them. Or, rather, it would be so if q_0 and q_1 taken as identical could be consistent with $p=0$. But this is not generally true, so that (168) is wrong. To suit our present purpose, we must write, by (162),

$$C_0=\sum\frac{1}{-p\phi'}\left\{(X+q_1Y)\int_0^z e(X+q_0Y)dz+(X+q_0Y)\int_z^l e(X+q_1Y)dz\right\}$$

$$=\sum(-p\phi')^{-1}\left\{w_1\int_0^z ew_0dz+w_0\int_z^l ew_1dz\right\};\quad\text{.........................}(169)$$

the q_0 being used in w_0, and the q_1 in w_1. Now we can take $p=0$, and get the correct formula to replace (168), viz.

$$C_0=\frac{1}{\phi_0}\left\{w_{10}\int_0^z ew_{00}dz+w_{00}\int_z^l ew_{10}dz\right\};\quad\text{............}(170)$$

the second $_0$ meaning that $p=0$ in w_0 and w_1.

If there is no leakage ($K=0$ in S''), C_0 becomes a constant, given by

$$C_0=\int_0^l edz\div\left\{\int_0^l Rdz+R_0+R_1\right\},\quad\text{.................}(171)$$

where the numerator is the total impressed force, and the denominator the total steady resistance; R, R_0, and R_1 being what R'', $-Z_0$, and Z_1 become when $p=0$ in them.

But when there is leakage (170) must be used; it would require a very special distribution of impressed force to make C_0 the same everywhere. To find the corresponding distribution of V, say V_0, in the steady state, we have then

$$-dC_0/dz = KV_0,$$

so that a single differentiation applied to (170) finds V_0.

Knowing thus C_0 finitely, we may write (166) thus,

$$C = C_0 - \sum\left(-\frac{w}{p\phi'}\right)\int_0^l ew\,dz.\epsilon^{pt}, \quad \text{..................(172)}$$

where C_0 is given in (170). The summation here, with $t=0$, is therefore the expansion of C_0.

The internal state of the wire is to be got by multiplying the first w by such a function of r, distance from the axis, and of whatever other variables may be necessary, as satisfies the conditions relating to inward propagation of magnetic force, and whose value at the boundary is unity. In the simple case of a round solid wire, (172) becomes, by (87), Part II.,

$$C_r = C_{0r} - \sum\frac{r}{a_1}\frac{J_1(s_1r)}{J_1(s_1a_1)}\frac{w\int ewdz}{(-p\phi')}\epsilon^{pt}. \quad \text{...............(173)}$$

This gives C_r, the current through the circle of radius r, less than a_1 the radius of the wire, C_{0r} being the final value. The value of s_1 is $(-4\pi\mu_1k_1p)^{\frac{1}{2}}$. Here of course we give to μ_1, k_1, and a_1 their proper values for the particular value of z. As before remarked, they must only vary slowly along z.

In the case of a wire of elliptical section it is naturally suggested that the closed curves taking the place of the concentric circles defined by $r=$ constant in (173) are also ellipses; and that in a wire of square section they vary between the square at the boundary and the circle at the axis. The propagation of current into a wire of rectangular section, to be considered later, may easily be investigated by means of Fourier-series, at least when the return-current closely envelops it.

Explicit Example of a Circuit of Varying Resistance, etc. Bessel Functions.

As an explicit example of the previous, let us, to avoid introducing new functions, choose the electrical data so that the current-functions X and Y are the J_0 and K_0 functions. This can be done by letting R'' be proportional and S'' inversely proportional to the distance from one end of the line. Let there be no leakage, and

$$R'' = R_0''z, \qquad S = S_0z^{-1};$$

where S_0 is a constant, and R_0'' a function of d/dt, but not of z. The electromagnetic and electrostatic time-constants do not vary from one part of the line to another. The equation of the current-function is

$$\frac{1}{z}\frac{d}{dz}\left(z\frac{dw}{dz}\right)=R_0''S_0pw; \quad \text{.......................(152a)}$$

from which we see that

$$X=J_0(fz), \qquad Y=K_0(fz), \qquad \text{where} \qquad f=(-R_0''S_0p)^{\frac{1}{2}}.$$

But, owing to the infinite conductivity at the $z=0$ end of the line, making $K_0(fz)=\infty$ there, we shall only be concerned with the J_0 function, that is, on the left side of the impressed force, in the first place. Since V is made permanently zero at $z=0$, the terminal condition there is nugatory. So

$$w=J_0(fz), \qquad \text{and} \qquad w=J_0(fz)+q_1K_0(fz);$$

$$u=(f/Sp)J_1(fz), \qquad \text{and} \qquad u=(f/Sp)\{J_1(fz)+q_1K_1(fz)\};$$

on the left and right sides of an impressed force, say at $z=z_2$. The value of q_1, got from the $V=Z_1C$ condition at $z=l$, is

$$q_1=\frac{(fl/S_0p)J_1(fl)-Z_1J_0(fl)}{Z_1K_0(fl)-(fl/S_0p)K_1(fl)}. \quad \text{..............(160a)}$$

We have also

$$\frac{XY'-YX'}{S''p}=\frac{f}{S''p}(J_1K_0-J_0K_1)(fz)=\frac{1}{S_0p}; \quad \text{........(155a)}$$

and the C-solution (166) becomes *

$$C=\Sigma(-p\phi')^{-1}J_0(fz)\int_0^l eJ_0(fz)dz.(1-\epsilon^{pt}), \quad \text{........ (166a)}$$

where $\phi=-q_1/S_0p$, and q_1 is given by (160a).

If we short-circuit at $z=l$, making $Z_1=0$, we introduce peculiarities connected with the presence of the series of p's belonging to $f=0$. The expression of q_1 is then, by (160a), $q_1=-J_1(fl)/K_1(fl)$. It seems rather singular that we should have anything to do with the K_1 function, seeing that C and V are expanded in series of the J_0 and J_1 functions. But on performing the differentiation of ϕ with respect to p it turns out to be all right, the denominator in (166a) becoming

$$-p\phi'=-\tfrac{1}{2}l^2J_0^2(fl)\frac{d}{dp}(R_0''p)$$

in general; whilst in the $f=0$ case, which makes $\phi=\frac{1}{2}R_0''l^2$, we have

$$-p\phi'=-\tfrac{1}{2}pl^2\frac{dR_0''}{dp}.$$

The value of ϕ when $p=0$ in it is, by inspection of the expansions of J_1 and K_1, simply $\frac{1}{2}R_0l^2$, the steady resistance of the line; R_0 being the

* [In accordance with the remarks in the footnote on page 226, we should write the equation (166a) thus:—

$$C=C_0+\sum J_0(fz)\int_0^l\frac{eJ_0(fz)dz}{p\phi'}.\epsilon^{pt},$$

where C_0 is the expression for the steady current at z due to e.]

constant that R_0'' becomes with $p=0$. We may therefore write (166a) thus :—

$$C=\int_0^l\frac{edz}{\frac{1}{2}R_0l^2}-\sum\int_0^l\frac{edz.\epsilon^{pt}}{-\frac{1}{2}pl^2\frac{dR_0''}{dp}}-\sum_f J_0(fz)\int_0^l\frac{J_0(fz)edz}{\frac{1}{2}l^2J_0^2(fl)}\sum_p\frac{\epsilon^{pt}}{-\frac{d}{dp}(pR_0'')},\quad(172a)$$

where the first term is C_0, the finally-reached current; the following summation, extending over the p's belonging to $f=0$, is its expansion, and therefore cancels the first term at the first moment; and the third part is a double summation, extending over all the f's except $f=0$, each f-term having its following infinite series of p-terms. This quantity (the third part) is zero initially as well as finally. If there were no elastic displacement permitted ($S_0=0$), the solution would be represented by the remainder of (172a), for we should then have C independent of z, and

$$\int_0^l edz=\int_0^l R''dz.C=\tfrac{1}{2}R_0''l^2.C$$

for the differential equation of C, whose solution is plainly given by the first two terms. The third part of (172a) is therefore entirely due to the combined action of the electrostatic and magnetic induction.

When the impressed force is entirely at $z=l$, and of such strength as to produce the steady current C_0, and if we take $R_0''=R+Lp$, where R and L are constants, there will be only two p's to each f, given by $f^2=-S_0p(R+Lp)$. The subsidence from the steady state, on removal of the impressed force, is represented by

$$C=C_0\epsilon^{-Rt/L}-\sum\frac{RC_0}{R+2Lp}\frac{J_0(fz)}{J_0(fl)}\epsilon^{pt},\qquad V=-\sum\frac{RC_0}{R+2Lp}\frac{J_1(fz)}{J_0(fl)}\frac{fz}{S_0p}\epsilon^{pt};$$

where the summations range over the p's, not counting the $p=-R/L$ whose C-term is exhibited separately; there is no corresponding V-term. A comparatively simple solution of this nature may be of course independently obtained in a more elementary manner. On the other hand, great power is gained by the use of more advanced symbolical methods, which, besides, seem to give us some view of the inner meaning of the expansions and of the operations producing them, that is wanting in the treatment of a special problem on its own merits, by the easiest way that presents itself.

Homogeneous Circuit. Fourier Functions. Expansion of Initial State to suit the Terminal Conditions.

Leaving, now, the question of variable electrical constants, let the line be homogeneous from beginning to end, so that R'' and S'' are functions of p, but not of z. The normal current-functions are then simply

$$X=\cos mz,\qquad Y=\sin mz,$$

where m is the function of p given by $-m^2=R''S''$, so that

$$w=\cos mz+q\sin mz,\qquad u=(m/S'')(\sin mz-q\cos mz).\quad(174)$$

Let there be a single impressed force e_2 at $z=z_2$; then the differential equations of the currents on the left and right sides of the same, corresponding to (162), will be

$$\left.\begin{aligned} C_0&=(\cos mz+q_0\sin mz)\frac{\cos mz_2+q_1\sin mz_2}{(m/S'')(q_0-q_1)}e_2,\\ C_1&=(\cos mz+q_1\sin mz)\frac{\cos mz_2+q_0\sin mz_2}{(m/S'')(q_0-q_1)}e_2,\end{aligned}\right\}\quad\text{.........(162b)}$$

where q_0 and q_1 are given by

$$q_0=-\frac{S''}{m}Z_0,\qquad q_1=\frac{(m/S'')\sin ml-Z_1\cos ml}{(m/S'')\cos ml+Z_1\sin ml}.\quad\text{......(160b)}$$

As before, in the case of an arbitrary distribution of e we are led to the solution (165), wherein for w (and for u in the corresponding V-formula) use the expressions (174), in which q is to be the common value of the q_0 and q_1 of (160b), and

$$\phi=(m/S'')(q_0-q_1)=0\quad\text{........................(175)}$$

is the determinantal equation of the p's.

Use (170) to find the final steady current-distribution. Thus, now,

$$C_0=\Big[(\cos mz+q_1\sin mz)\int_0^z(\cos mz+q_0\sin mz)e\,dz$$
$$+(\cos mz+q_0\sin mz)\int_z^l(\cos mz+q_1\sin mz)e\,dz\Big]\div\frac{m}{S''}(q_0-q_1),\quad(176)$$

in which m, q_0, q_1, and S'' have the $p=0$ values. They are, if $i=(-1)^{\frac{1}{2}}$,

$$S''=K,\qquad m=(-RK)^{\frac{1}{2}}=gi\quad\text{say,}$$

if R is the steady resistance of line (both conductors), and K is the conductance of the insulator, both per unit length of line;

$$q_0=(K/m)R_0=-KR_0i/g,$$

if R_0 = effective steady resistance at the $z=0$ terminals, and

$$q_1=\frac{gi\sin gli-KR_1\cos gli}{gi\cos gli+KR_1\sin gli},$$

if R_1 = effective steady resistance at the $z=l$ terminals.

The expression on the right side of (176) is, of course, real in the exponential form, and the steady distribution of V is got by

$$KV_0=-dC_0/dz.$$

Using the thus-obtained expressions, we reach the (172) form of C-solution, and the corresponding

$$V=V_0-\sum\frac{u}{(-p\phi')}\int_0^l ew\,dz.\epsilon^{pt}.\quad\text{................(176a)}$$

The value of ϕ' here, got by differentiation with respect to p, may be written in many ways, of which one of the most useful, for expansions in Fourier series, is the following. Let

$$w=(1+q^2)^{\frac{1}{2}}\cos(mz+\theta);$$

then $$\frac{d\phi}{dp}=\frac{m}{S''\cos^2\theta}\frac{d}{dp}\left\{\tan^{-1}\left(\frac{m}{S''}\frac{Z_1-Z_0}{(m/S'')^2+Z_1Z_0}\right)-ml\right\}$$

$$=\frac{l}{2S''\cos^2\theta}\frac{dm^2}{dp}\left\{\cos^2 ml\frac{d}{d(ml)}\left(\frac{m}{S''}\frac{Z_1-Z_0}{(m/S'')^2+Z_1Z_0}\right)-1\right\}. \quad (177)$$

Corresponding to this,

$$\tan ml=\frac{m}{S''}\frac{Z_1-Z_0}{(m/S'')^2+Z_1Z_0} \quad \text{......................}(178)$$

finds the angles ml; it is got by the union of

$$\tan\theta=S''Z_0/m, \qquad \tan(ml+\theta)=S''Z_1/m, \quad \text{..........}(179)$$

which are equivalent to (160*b*).

For example, if we take $R''=R$, constant, thus abolishing inertia, and $S''=Sp$, no leakage, and S constant (R and S not containing p, that is to say), the expansion of V_0 (an arbitrary function of z) is [see also vol. I., p. 123, and p. 152]

$$V_0=\sum\frac{\sin(mz+\theta)\int_0^l V_0\sin(mz+\theta)dz}{\frac{l}{2}\left\{1-\cos^2 ml\frac{d}{d(ml)}\frac{m}{Sp}\frac{Z_1-Z_0}{(m/Sp)^2+Z_1Z_0}\right\}}, \quad \text{.......}(180)$$

subject to (178). Here $p=-m^2/RS$, so that the state of the line at time t after it was V_0, when left to itself, is got by multiplying each term in the expansion by $\epsilon^{-m^2t/RS}$. The corresponding current is given by $RC=-dV/dz$. But the solution thus got will usually only be correct, although (180) is correct, when there is, initially, no energy in the terminal apparatus. If there be, additional terms in the numerator of (180) are required, to be found by the energy-difference method of Part III. They will not alter the value of the right member of (180) at all; they only come into effect after the subsidence has commenced. Similar remarks apply whatever be the nature of the line. It is, however, easy to arrange matters so that the energy in the terminal apparatus shall produce no effect in the line. For example, join the two conductors at one end of the line through two equal coils in parallel; if the currents in these coils be equal and similarly directed in the circuit they form by themselves, they will not, in subsiding, affect the line at all.

Returning to (177), or other equivalent expression, it is to be observed that particular attention must be paid to the roots $ml=0$, which may occur, or to the series of roots p belonging to the $m=0$ case, when we are working down from the general to the special, and happen to bring in $m=0$. Take $Z_1=0$ for instance, making, by (175) and (160*b*),

$$\phi=-Z_0-\frac{m}{Sp}\tan ml,$$

where $m^2=-SpR''$. Then

$$\frac{d\phi}{dp}=-\frac{dZ_0}{dp}-\frac{\tan ml}{2m}\left(\frac{dR''}{dp}-\frac{R''}{p}\right)+\frac{l}{2}\sec^2 ml\left(\frac{dR''}{dp}+\frac{R''}{p}\right). \quad \text{......}(181)$$

Now, as long as Z_0 is finite, m cannot vanish; but when Z_0 is zero, giving ml = any integral multiple of π, $m=0$ is one case. Then we have, when m is finite,

$$\frac{d\phi}{dp}=\frac{l}{2}\left(\frac{dR''}{dp}+\frac{R''}{p}\right),\qquad\text{and}\qquad p\frac{d\phi}{dp}=\frac{l}{2}\frac{d}{dp}(pR''); \quad \text{......(182)}$$

but when m is zero the middle term on the right of the preceding equation becomes finite, making

$$d\phi/dp=l(dR''/dp).$$

The result is that the current-solution contains a term, or infinite series, apparently following a different law to the rest, with no corresponding terms in the V-solution. This merely means that the mean current subsides without causing any electric displacement across the dielectric, when the ends are short-circuited ($Z=0$); so that if, in the first place, the current is steady, and there is no displacement, there will be none during the subsidence.

Transition from the Case of Resistance, Inertia, and Elastic Yielding to the same without Inertia.

The transition from the combined inertia-and-elasticity solutions to elasticity alone is very curious. Thus, let $Z=0$ at both ends, and $R''=R+Lp$, where R and L are constants not containing p. The rise of current due to e is shown by

$$C=\int_0^l\frac{edz.(1-\epsilon^{-Rt/L})}{Rl}+\frac{2}{l}\sum\cos mz\int_0^l\frac{e\cos mz\,dz}{R+2Lp}\epsilon^{pt}, \quad \text{.......(183)}$$

the m's in the summation being π/l, $2\pi/l$, etc.; and each having two p's, given by

$$0=m^2+RSp+LSp^2.$$

The $m=0$ part is exhibited separately, and is what the solution would be if e were a constant (owing to the constancy of R). But, whatever e be, as a function of z, the summation comes to nothing initially, on account of the doubleness of the p's, just as in (172a) the double summation vanishes by reason of every p-summation vanishing when $t=0$.

Now, in (183), let L be exceedingly small. The two p's approximate to $-m^2/RS$, the electrostatic one, and to $-R/L$, the magnetic one, which goes up to ∞, the storehouse for roots. The current then rises thus:—

$$C=\int_0^l\frac{edz.(1-\epsilon^{-Rt/L})}{Rl}+\frac{2}{Rl}\sum\cos mz\int_0^l e\cos mz\,dz.(1-\epsilon^{-Rt/L})$$
$$-\frac{2}{Rl}\sum\cos mz\int_0^l e\cos mz\,dz.(1-\epsilon^{-m^2t/RS}). \quad \text{........(184)}$$

But the first line on the right side is equivalent to

$$(e/R)(1-\epsilon^{-Rt/L}),$$

and here the exponential term vanishes instantly, on L being made exactly zero, so that (184) becomes

$$C=\frac{e}{R}-\frac{2}{Rl}\sum\cos mz\int_0^l e\cos mz\,dz.(1-\epsilon^{-m^2t/RS}),\quad\text{.........(185)}$$

except at the very first moment, when it gives $C=e/R$, which is quite wrong, although the preceding formula, giving $C=0$ at the first moment, is correct. Or, (185) is equivalent to

$$C=\frac{1}{R}\left(e-\frac{dV}{dz}\right),$$

from which inertia has disappeared. Here V is given by (188) below. The process amounts to taking one half the terms of the summation in (183), and joining them on to the preceding term to make up e/R, which is quite arbitrary. An alternative form of (185) is

$$C=\frac{1}{Rl}\int_0^l edz+\frac{2}{Rl}\sum\cos mz\int_0^l e\cos mz\,dz\,.\,\epsilon^{-m^2t/RS}.\quad\text{........(186)}$$

On the other hand, there is no such peculiarity connected with the V-solution in the act of abolishing inertia. The $m=0$ term is

$$-\frac{1}{Rl}\left(\frac{m}{Sp}\sin mz\int_0^l edz\right),\quad\text{which }=0,$$

because m is zero and p finite. Therefore V rises thus,

$$V=\frac{2}{l}\sum\frac{m\sin mz\int_0^l e\cos mz\,dz}{-Sp(R+2Lp)}(1-\epsilon^{pt}),\quad\text{..............(187)}$$

before abolition of inertia. But as L is made zero, the denominator becomes m^2 for the electrostatic p, and ∞ for the other; thus one half the terms vanish, leaving

$$V=\frac{2}{l}\sum\frac{\sin mz}{m}\int_0^l e\cos mz\,dz(1-\epsilon^{-m^2t/RS}),\quad\text{...........(188)}$$

when $L=0$, without any of the curious manipulation to which the current-formula was subjected.

Transition from the Case of Resistance, Inertia, and Elastic Yielding to the same without Elastic Yielding.

Next, let us consider the transition from the combined elasticity-and-inertia solution to inertia alone (of course with resistance in both cases, as in the preceding transition). It is usual to wholly ignore electrostatic induction in investigations relating to linear circuits. This is equivalent to taking $S=0$, stopping elastic displacement, and compelling the current to keep *in* the wires always, *i.e.* when the insulation is perfect, as will be here assumed. We then have, by (145),

$$-\frac{dC}{dz}=0,\qquad e-\frac{dV}{dz}=R''C.\quad\text{.................(189)}$$

By integrating the second of these with respect to z we get rid of V, and obtain the differential equation of C,

$$\int_0^l e\,dz = \left\{\int_0^l R''dz + Z_1 - Z_0\right\}C = \phi_1 C, \quad \text{say}, \quad \text{...........(190)}$$

whence follows this manner of rise of the current, when e is steady and put on everywhere at the time $t=0$, reaching the final value C_0,

$$C = C_0 - \sum\left(-p\frac{d\phi_1}{dp}\right)^{-1}\int_0^l e\,dz \,.\, \epsilon^{pt}, \quad \text{..............(191)}$$

$\phi_1 = 0$ finding the p's. We can find V at distance z by integrating the second of (189) with respect to z from 0 to z; thus,

$$V = \int_0^z e\,dz + \left(Z_0 - \int_0^z R''dz\right)C, \quad \text{..................(192)}$$

wherein C is to be the right member of (191). This finds V by differentiations with respect to t performed on C. In the final state put R_0'' for R'', and $-R_0$ for Z_0, steady resistances. V will usually vary with the time until the steady state is reached; but if the line is homogeneous, with only the two constants R and L, and if also Z_0 and Z_1 are zero, V will be independent of t, and instantly assume its final distribution.

Then, on these assumptions, we shall have

$$C = \left[\int_0^l \frac{e\,dz}{Rl}\right](1 - \epsilon^{-Rt/L}), \qquad V = \int_0^z e\,dz - \left(\frac{z}{l}\right)\int_0^l e\,dz, \quad \text{.....(193)}$$

showing the current to rise independently of the distribution of e, and V to have its final distribution from the first moment, which, when the impressed force is wholly at $z=0$, of amount e_0, is $e_0(1 - z/l)$. This infinitely rapid propagation of V is common-sense according to the prescribed conditions, but absolute nonsense physically considered, especially in view of the transfer of energy. The question then arises, How does V really set itself up, when the line is so short that the current rises sensibly according to the magnetic theory?

To examine this, let the line-constants be R, S, L (independent of d/dt), and $Z_1 = Z_0 = 0$. Put on e_0 at $z=0$ at time $t=0$. V and C will rise thus (a special case of (183) and (187)),

$$\left.\begin{aligned} C &= \frac{e_0}{Rl}(1 - \epsilon^{-Rt/L}) + \frac{2e_0}{Rl}\epsilon^{-Rt/2L}\sum\frac{2}{m'}\cos mz \sin\frac{Rm't}{2L}, \\ V &= e_0\left(1 - \frac{z}{l}\right) - \frac{2e_0}{l}\epsilon^{-Rt/2L}\sum\frac{\sin mz}{m}\left(\cos + \frac{\sin}{m'}\right)\frac{Rm't}{2L}, \end{aligned}\right\} \text{......(194)}$$

where m has the values π/l, $2\pi/l$, etc., and

$$m' = (4m^2L/R^2S - 1)^{\frac{1}{2}}.$$

It is clear that when S is made to vanish, making $m' = \infty$, the current-oscillations wholly vanish, reducing the C-solution to the first of (193). But the V-oscillations remain in full force, though of infinitely short period, and subside at a definite rate. This means that the mean value of V at any place has to be taken to represent its actual value, and this mean value is its final value. That is, if $\overline{V}$ denote the mean value about which V oscillates, we have

$$\overline{V}=e_0(1-z/l)=V_0.$$

Introduce $LS=v^{-2}$, where v is constant, making

$$m'=2mLv/R$$

very nearly, when the line is short; then the second of (194) becomes

$$V=e_0\left(1-\frac{z}{l}\right)-\frac{2e_0}{l}\epsilon^{-Rt/2L}\sum\frac{\sin mz}{m}\cos mvt, \quad \text{..........(195)}$$

which must very nearly show the subsidence of the oscillations. First ignore the subsidence-factor, replacing it by unity, then (195) represents a wave of V travelling to and fro at velocity v, as thus expressed,

$$\left.\begin{array}{lll} V=e_0 & \text{from} \quad z=0 & \text{to} \quad z=vt, \\ V=0 & \text{beyond} \quad z=vt, & \end{array}\right\} \text{when} \quad vt<l.$$

When $vt=l$, the whole line is charged to $V=e_0$. The wave then moves back in the same manner as it advanced, so that the state of things at time $t=l/v\pm\tau$ is the same, until t reaches $2l/v$, when we have $V=0$ as at first. This would be repeated over and over again if there were no resistance, which, through the exponential factor, causes the range of the oscillations of V at any place about the final value to diminish according to the time-constant $2L/R$. Also, the resistance has the effect of rounding off the abrupt discontinuity in the wave of V.

I have given a fuller description of this case elsewhere [vol. I., p. 132], and only bring it in here in connection with the interpretation according to my present views regarding the transfer of energy. As it is clear that this oscillatory phenomenon is, primarily, a dielectric phenomenon, and only affects the conductor secondarily, it is necessary that the L in the above should not at the beginning be the full L of dielectric and wires, but only L_0, that of the dielectric, making v the velocity of undissipated waves, although as the oscillations subside the velocity must diminish, tending towards $v=(LS)^{-\frac{1}{2}}$, which may, however, be far from being reached, especially in the case of an iron wire. The nature of the dielectric wave is far more simply studied graphically than by means of Fourier series, on the assumption of infinite conductivity, which allows us to represent things by means of two oppositely travelling waves. To this I may return in the next Part.

On Telephony by Magnetic Influence between Distant Circuits.

I will conclude the present Part with a brief outline of the reasoning which guided me six months ago, when my brother's experiments on induction between distant circuits (mentioned in Part II.) in the north of England commenced, to the conclusion that long-distance signalling (*i.e.* hundreds of miles) was possible by induction, a conclusion which has been somewhat supported by results, so far as the experiments have yet gone. Recognising the great complexity of the problem, and the difficulty of hitting the exact conditions, I made no special calculations, but preferred to be guided by general considerations; for, in the endeavour to be precise when the data are uncertain and very variable, one is in great danger of swallowing the camel.

One may be fairly well acquainted with electromagnetism, and also with the capabilities of the telephone, and yet receive the idea of signalling by induction long distances with utter incredulity, or at least in the same way as one might accept the truth of the statement, that when one stamps one's foot the universe is shaken to its foundations. Quite true, but insensible a few yards away. The incredulity will probably be based upon the notion of rapid decrease with distance of inductive effects. This, however, leaves out of consideration an important element, namely the size of the circuits.

The coefficients of electromagnetic induction of linear circuits are proportional to their linear dimensions. If, then, we increase the size of two circuits n times, and also their distance apart n times, the mutual inductance M is increased n times. Let R_1 and R_2 be the resistances of primary and secondary. The induced current (integral) in the secondary due to starting or stopping a current C_1 in the primary is MC_1/R_2, or Me_1/R_1R_2, if e_1 be the impressed force in the primary. Now increasing the linear dimensions, and the distance, in the ratio n (with the same kind of wire) increases M, R_1, and R_2 all n times. So only e_1 remains to be increased n times to get the same secondary-current impulse. We can therefore ensure success in long-distance experiments on the basis of the success of short-distance experiments, with elements of uncertainty arising from new conditions coming into operation at the long distances.

But practically the result must be far more favourable to the long than to the short distances than the above asserts. For no one, when multiplying the distance and size of circuits, say ten times, would think of putting ten telephones in circuit to keep rigidly to the rule. Thus it may be that only a slight increase of e_1 is required, on account of M being multiplied in a far greater ratio than the resistances, or the self-inductances. Thus, it is not uncommon for the R and L of a telephone to be 100 ohms and 12 million centim. These form the principal parts of the R and L of a circuit of moderate size, and of course do not increase when we enlarge the circuit. It is therefore certain that we can signal long distances on the above basis, with a margin in favour of the long distances, which will be large or small according as the circuits are small or large.

Again, if e_1 in the primary be periodic, of frequency $n/2\pi$, the ratio of the amplitude of the current in the secondary to that in the primary will be

$$Mn \div (R_2^2 + L_2^2n^2)^{\frac{1}{2}}.$$

Now, without any statement of the magnitude of the current in the primary, if it be largely in excess of requirements for signalling in the primary, so that $\frac{1}{100}$ part, say, would be sufficient for the purpose, then we shall have enough current in the secondary if the above ratio is only

$\frac{1}{100}$. But, without going to precise formulæ, it may be easily seen that the above ratio may be made quite a considerable fraction, in comparison with $\frac{1}{100}$, with closed metallic circuits whose linear dimensions and distance are increased in the same ratio. But we should expect a rapid decrease of effect when the mean distance between the circuits exceeds their diameter, keeping the circuits unchanged. (It should be understood that squares, circles, etc., are referred to.)

The theory seems so very clear (though it is only the first approximation to the theory), that it would be matter for wonder and special inquiry if we found that we could not signal long distances by induction between closed metallic circuits, starting on the basis of a short-distance experiment, and following up the theory.

As a matter of fact, my brother found it was possible to speak by telephone between two metallic circuits of $\frac{1}{4}$ mile square, $\frac{1}{2}$ mile between centres, using two bichros with the microphone.

Now, coming to metallic lines whose circuits are closed through the earth, the theory is rendered far more difficult on account of there being a conduction-current from the primary to the secondary due to the earth's imperfect conductivity. We therefore have, to say nothing of electrostatic induction, a superposition of effects due to induction and conduction, the latter being far more difficult to theoretically estimate than the former. But the reasoning regarding the magnetic induction is not very greatly changed, although not so favourable to long-distance signalling. If the return currents diffused themselves uniformly in all directions from the ends of the line, the same property of n-fold increase of M with n-fold lengthening of the lines and their distance would still be true. But the diffusion is one-sided only, and is even then only partial, especially when exceedingly rapid alternations of current take place. But we have the power of counterbalancing this by the multiplication of the variations of current in the primary that we can get by making and breaking the circuit, with a considerable battery-power if necessary, getting something enormous compared with the feeble variations of current in the microphonic circuit, or that can work a telephone. Electrostatic induction also comes in to assist, as it increases the activity of the battery, and therefore the current in the secondary also.

But, as regards wires connected to earth, this does not profess to be more than the very roughest reasoning, though in my opinion quite plain enough to show that we may ascribe the signalling across 40 miles of country between lines about 50 miles long mainly to induction, as we should be necessitated to do if we carried the experiment further and closed the circuits metallically by roundabout courses, for then the plain arguments relating to induction will become valid. Experiments of this kind are of the greatest value from the theoretical point of view, and it is to be hoped that they will be greatly extended.

PART V.

St. Venant's Solutions relating to the Torsion of Prisms applied to the Problem of Magnetic Induction in Metal Rods, with the Electric Current longitudinal, and with close-fitting Return-Current.

The mathematical difficulties in the way of the discovery of exact solutions of problems concerning the propagation of electromagnetic disturbances into wires of other than circular section—or, even if of circular section, when the return-current is not equidistantly distributed as regards the wire, or is not so distant that its influence on the distribution of the wire-current throughout its section may be disregarded—are very considerable. As soon as we depart from the simple type of magnetic field which occurs in the case of a straight wire of circular section, we require at least two geometrical variables in place of the one, distance from the axis of the wire, which served before; and we may have to supplement the magnetic force "of the current," as usually understood, by a polar force, or a force which is the space-variation of a single-valued scalar, the magnetic potential, in order to make up the real magnetic force.

There are, however, some simplified cases which can be fully solved, viz., when the external magnetic field, that in the dielectric, is abolished, by enclosing the wire in a sheath of infinite conductivity. It is true that we must practically separate the wire from the sheath by some thickness of dielectric, in order to be able to set up current in the circuit by means of impressed force, so that we cannot entirely abolish the external magnetic field; but we may approximate in a great measure to the state of things we want for purposes of investigation. The wire, of course, need not be a wire in the ordinary sense, but a large bar or prism. The electrostatic induction will be ignored, requiring the wire to be not of great length; thus making the problem a magnetic one.

Consider, then, a straight wire or rod or prism of any symmetrical form of section, so that, when a uniformly distributed current passes through it, its axis is the axis of the magnetic field, where the intensity of force is zero. Let a steady current exist in the wire, longitudinal of course, and let the return-conductor be a close-fitting infinitely-conducting sheath. This stops the magnetic field at the boundary of the wire. The sudden discontinuity of the boundary magnetic-force is then the measure and representative of the return-current.

The magnetic energy per unit length is $\frac{1}{2}LC^2$, where C is the current in the wire and L the inductance per unit length. As regards the diminution of the L of a circuit in general, by spreading out the current, as in a strip, instead of concentrating it in a wire, that is a matter of elementary reasoning founded on the general structure of L. If we draw apart currents, keeping the currents constant, thus doing work against their mutual attraction, we diminish their energy at the same time by the amount of work done against the attraction. Thus the quantity $\frac{1}{2}LC^2$ of a circuit is the amount of work that must be done to take a current to pieces, so to speak; that is, supposing it divided into infinitely fine filamentary closed currents, to separate them against their attractions to an infinite distance from one another. We do not need, therefore, any examination of special formulæ to see that the inductance of a flat strip is far less than that of a round wire of the same sectional area; their difference being proportional to the difference of the amounts of the magnetic energy per unit current in the two cases. The inductance of a circuit can, similarly, be indefinitely increased by fining the wire; that of a mere line being infinitely great. But we can no more have a finite current in an infinitely thin wire than we can have a finite charge of electricity at a point, in which case the electrostatic energy would also be infinitely great, for a similar reason; although by a useful and almost necessary convention we may regard fine-wire circuits as linear, whilst their inductances are finite.

Now, as regards our enclosed rod with no external magnetic field, we can in several cases estimate L exactly, as the work is already done, in a different field of Physics. The nature of the problem is most simply stated in terms of vectors. Thus, let **h** be the vector magnetic force when the boundary of the section perpendicular to the length is circular, and **H** what it becomes with another form of boundary; then

$$\mathbf{H} = \mathbf{h} + \mathbf{F}, \qquad \text{and} \qquad \mathbf{F} = -\nabla\Omega. \quad \text{.......................(1}a\text{)}$$

That is, the field of magnetic force differs from the simple circular type by a polar force **F**, whose potential is Ω. This must be so because the curl of **H** and of **h** are identical, requiring the curl of **F** to be zero. To find **F** we have the datum that the magnetic force must be tangential to the boundary, and therefore have no normal component; or, if **N** be the unit vector-normal drawn outward,

$$-\mathbf{FN} = \mathbf{hN} \quad \text{..............................(2}a\text{)}$$

is the boundary-condition. This gives **F**, when it is remembered that **F** must have no convergence within the wire.

In another form, since we have **h** circular about the axis, and of intensity $2\pi r\Gamma_0$ at distance r from it, the current-density being Γ_0; or

$$\mathbf{h} = 2\pi\Gamma_0 \mathrm{V}\mathbf{kr}, \quad \text{..............................(3}a\text{)}$$

if **r** is the vector distance from the axis in a plane perpendicular to it, and **k** a unit vector parallel to the current; we have

$$\mathbf{hN} = (2\pi\Gamma_0)(\mathbf{N}\mathrm{V}\mathbf{kr}) = (2\pi\Gamma_0)(\mathbf{r}\mathrm{V}\mathbf{Nk}) = -\pi\Gamma_0\frac{d(r^2)}{ds}, \quad \text{........(4}a\text{)}$$

if s be length measured along the bounding curve, in the direction of the magnetic force. The boundary-condition (2a) therefore becomes, in terms of the magnetic potential,

$$-\frac{d\Omega}{dp_1} = \pi\Gamma_0\frac{d(r^2)}{ds}, \quad \text{..............................(5}a\text{)}$$

which, with $\nabla^2\Omega = 0$, finds the magnetic potential. Here p_1 is length measured outward along the normal to the boundary.

Or, we may use the vector-potential **A**. It is parallel to the current, and consists of two parts; thus,

$$\mathbf{A} = \mathbf{A}' - (\mu\pi\Gamma_0 r^2)\mathbf{k}, \qquad \text{(6a)}$$

where the second part on the right side is, except as regards a constant, what it would be if the boundary were circular, its curl being $\mu\mathbf{h}$. To find $\mathbf{A}'$, let its tensor be A'; then

$$\nabla^2 A' = 0, \qquad \text{and} \qquad A' = \mu\pi\Gamma_0 r^2, \qquad \text{(7a)}$$

the latter being the boundary-condition, expressing that **A** is zero at the boundary. Comparing with (5*a*), we see that (7*a*) is the simpler.

The magnetic energy per unit length of rod, say T, is

$$T = \Sigma\,\mu\mathbf{H}^2/8\pi = \Sigma\,\mu(\mathbf{h}+\mathbf{F})^2/8\pi, \qquad \text{(8a)}$$

the summation extending over the section. But $\Sigma\,\mathbf{FH} = 0$, because **F** is polar and **H** is closed; so that

$$T = \Sigma\,\mu\mathbf{h}^2/8\pi - \Sigma\,\mu\mathbf{F}^2/8\pi = \Sigma\,\mu\mathbf{h}^2/8\pi + \Sigma\,\mu\mathbf{hF}/8\pi. \qquad \text{(9a)}$$

Or, in Cartesian coordinates, let H_1 and H_2 be the x and y components of the magnetic force **H**, z being parallel to the current; then

$$H_1 = -2\pi y\Gamma_0 - \frac{d\Omega}{dx}, \qquad H_2 = 2\pi x\Gamma_0 - \frac{d\Omega}{dy} \qquad \text{(10a)}$$

express (1*a*), and (8*a*) is represented by

$$T = \frac{\mu}{8\pi}\Sigma\,(H_1^2 + H_2^2) = \frac{\mu\pi}{2}\Gamma_0^2\Sigma\,(x^2+y^2) - \frac{\mu\Gamma_0}{4}\sum\left(x\frac{d\Omega}{dy} - y\frac{d\Omega}{dx}\right), \qquad \text{(11a)}$$

the latter form expressing (9*a*).

It will be observed that the mathematical conditions are identical with those existing in St. Venant's torsion problems. Thus, if $\mathfrak{a}$ and $\mathfrak{b}$ are the y and x tangential strain-components in the plane x, y in a twisted prism, and γ the longitudinal displacement along z, parallel to the length of the prism, we have

$$\mathfrak{b} = -\tau y + \frac{d\gamma}{dx}, \qquad \mathfrak{a} = \tau x + \frac{d\gamma}{dy}, \qquad \text{(12a)}$$

where τ is the twist (Thomson and Tait, Part II., § 706, equation (9)). The corresponding forces are n times as great, if n is the rigidity (*loc. cit.* equation (10)); so that the energy per unit length is

$$\tfrac{1}{2}n\Sigma\,(\mathfrak{a}^2 + \mathfrak{b}^2) \quad \text{over section.} \qquad \text{(13a)}$$

Also, to find γ, we have

$$\nabla^2\gamma = 0, \qquad \frac{d\gamma}{dp_1} = \tfrac{1}{2}\tau\frac{dr^2}{ds}, \qquad \text{(14a)}$$

(*loc. cit.* equations (12) and (18)). Comparing (14*a*) with (5*a*), (12*a*) with (10*a*), and (13*a*) with the first of (11*a*), we see that there is a perfect correspondence, except, of course, as regards the constants concerned. The lines of tangential stress in the torsion-problem and the lines of magnetic force in our problem are identical, and the energy is similarly reckoned. We may therefore make use of all St. Venant's results.

It will be sufficient here to point out that the ratio of the inductance of wires of different sections is the same as the ratio of their torsional rigidities. Thus, as $L=\frac{1}{2}\mu$ in the case of a round wire, that of a wire of elliptical section, semiaxes a and b, is $L=\mu ab/(a^2+b^2)$; when the section is a square, it is $\cdot 4417\mu$; when it is an equilateral triangle, $\cdot 3627\mu$, etc. [Remember the limitation of close-fitting return, above mentioned.] That of a rectangle will be given later in the course of the following subsidence-solution.

Subsidence of Initially Uniform Current in a Rod of Rectangular Section, with close-fitting Return-Current.

Consider the subsidence from the initial state of steady flow to zero, when the impressed force that supported the current is removed, in a prism of rectangular section. Let $2a$ and $2b$ be its sides, parallel to x and y respectively, the origin being taken at the centre. Let H_1 and H_2 be the x and y components of the magnetic force at the time t. Let E be the intensity of the magnetic-force vector $\mathbf{E}$, which is parallel to z; then the two equations of induction ((6), (7), Part I.), or

$$\operatorname{curl}\mathbf{H}=4\pi\Gamma, \qquad -\operatorname{curl}\mathbf{E}=\mu\dot{\mathbf{H}},$$

are reduced to

$$-\frac{dE}{dy}=\mu\dot{H}_1, \qquad \frac{dE}{dx}=\mu\dot{H}_2, \quad \dots\dots\dots(15a)$$

$$\frac{dH_2}{dx}-\frac{dH_1}{dy}=4\pi kE=4\pi\Gamma; \quad \dots\dots\dots(16a)$$

if Γ is the current-density, k the conductivity, μ the inductivity. (I speak of the intensity of a "force" and of the "density" of a flux, believing a distinction desirable.) The equation of Γ is therefore

$$\left(\frac{d^2}{dx^2}+\frac{d^2}{dy^2}\right)\Gamma=4\pi\mu k\dot{\Gamma}, \quad \dots\dots\dots(17a)$$

of which an elementary solution is

$$\Gamma=\cos mx\cos ny\,\epsilon^{pt}, \quad \dots\dots\dots(18a)$$

if

$$4\pi\mu kp=-(m^2+n^2). \quad \dots\dots\dots(19a)$$

At the boundary we have, during the subsidence, $E=0$, or $\Gamma=0$; therefore

$$\cos mx\cos ny=0 \quad \text{at the boundary},$$

or

$$\cos ma=0, \qquad \cos nb=0, \quad \dots\dots\dots(20a)$$

or $ma=\frac{1}{2}\pi$, $\frac{3}{2}\pi$, $\frac{5}{2}\pi$, etc.; $nb=$ ditto. The general solution is therefore the double summation over m and n,

$$\Gamma=\Sigma\Sigma A\cos mx\cos ny\,\epsilon^{pt},$$

if we find A to make the right member represent the initial state. This has to be $\Gamma=\Gamma_0$, a constant. Now

$$1=\Sigma(2/ma)\sin ma\cos mx, \quad \text{from} \quad x=-a \quad \text{to} \quad +a,$$
$$1=\Sigma(2/nb)\sin nb\cos ny, \quad \text{from} \quad y=-b \quad \text{to} \quad +b.$$

Hence the required solution is

$$\Gamma = \frac{4}{ab}\Gamma_0 \sum \frac{\sin ma}{m} \cos mx\, \epsilon^{-\frac{m^2t}{4\pi\mu k}} . \sum \frac{\sin nb}{n} \cos ny\, \epsilon^{-\frac{n^2t}{4\pi\mu k}},$$

or
$$\Gamma = \frac{4}{ab}\Gamma_0 \sum\sum \frac{\sin ma \sin nb}{mn} \cos mx \cos ny\, \epsilon^{pt}. \quad \text{.................(21}a\text{)}$$

From this derive the magnetic force by (15a). Thus

$$\left.\begin{aligned} H_1 &= -\frac{16\pi}{ab}\Gamma_0 \sum\sum \frac{\sin ma}{m} \sin nb \cos mx \sin ny \frac{\epsilon^{pt}}{m^2+n^2}, \\ H_2 &= \frac{16\pi}{ab}\Gamma_0 \sum\sum \frac{\sin nb}{n} \sin ma \sin mx \cos ny \frac{\epsilon^{pt}}{m^2+n^2}. \end{aligned}\right\} \text{.....(22}a\text{)}$$

The total current in the prism, say C, is given by

$$4\pi C = 2\int_{-b}^{b} H_2 dy_{(x=a)} - 2\int_{-a}^{a} H_1 dx_{(y=b)} = \frac{64\pi}{ab}\Gamma_0 \sum\sum \frac{\epsilon^{pt}}{m^2n^2},$$

by line-integration round the boundary. Or

$$C = \frac{4}{a^2b^2} C_0 \sum\sum \frac{\epsilon^{pt}}{m^2n^2}, \quad \text{.......................(23}a\text{)}$$

if $C_0 = 4ab\Gamma_0$, the initial current in the prism.

Since the current is longitudinal, and there is no potential-difference, the vector-potential is given by $\mathbf{E} = -\dot{\mathbf{A}}$; or, A being the tensor of $\mathbf{A}$, A is got by dividing the general term in the Γ-solution (21a) by $-pk$; giving

$$A = \frac{16\pi\mu}{ab} \sum\sum \frac{\sin ma \sin nb}{mn(m^2+n^2)} \cos mx \cos ny\, \epsilon^{pt}. \quad \text{.........(24}a\text{)}$$

Since the magnetic energy is to be got by summing up the product $\frac{1}{2}A\Gamma$ over the section, we find, by integrating the square of Γ, that the amount per unit length is

$$T = \frac{2\pi\mu C_0^2}{a^3b^3} \sum\sum \frac{\epsilon^{2pt}}{m^2n^2(m^2+n^2)}. \quad \text{..................(25}a\text{)}$$

By the square-of-the-force method the same result is reached, of course. We may also verify that $Q + \dot{T} = 0$ during the subsidence, Q being the dissipativity per unit length of prism.

The steady inductance per unit length is the L in $T = \frac{1}{2}LC_0^2$, which (25a) becomes when $t = 0$; this gives

$$L = 4\pi\mu \sum\sum \frac{1}{(ma)^2(nb)^2\left\{\frac{a}{b}(nb)^2 + \frac{b}{a}(ma)^2\right\}}. \quad \text{..........(26}a\text{)}$$

The lines of magnetic current are also the lines of equal electric current-density. That is, a line drawn in the plane x, y through the points where Γ has the same value is a line of magnetic current. For, if $\mathbf{s}$ be any line in the plane x, y,

$$\frac{dE}{ds} = \text{component of } \mu\dot{\mathbf{H}} \text{ perpendicular to } \mathbf{s},$$

so that $\dot{\mathbf{H}}$ is parallel to $\mathbf{s}$, when $dE/ds=0$. The transfer of energy is, as usual, perpendicular to the lines of magnetic force and electric force.

The above expression (26a) for L may be summed up either with respect to ma or to nb, but not to both, by any way I know. Thus, writing it

$$L=4\pi\mu\sum\frac{1}{(ma)^2}\sum\frac{1}{(nb)^2}\frac{1}{\frac{a}{b}(nb)^2+\frac{b}{a}(ma)^2}, \qquad (27a)$$

we may effect the second summation, with respect to nb, regarding ma as constant in every term. Use the identity

$$\frac{l-x}{h^2}-\frac{\epsilon^{h(l-x)}-\epsilon^{-h(l-x)}}{h^3(\epsilon^{hl}+\epsilon^{-hl})}=\frac{2}{l}\sum\frac{\cos(i\pi x/2l)}{(i\pi/2l)^2\{(i\pi/2l)^2+h^2\}},$$

where i has the values 1, 3, 5, etc. Take $x=0$, $i\pi/2l=nb$, $h=(b/a)(ma)$, $l=1$, and apply to (27a), giving

$$L=4\pi\mu\sum\frac{1}{(ma)^2}\left\{\frac{a}{2b}\frac{1}{(ma)^2}\left(1-\frac{a}{b}\cdot\frac{1}{ma}\cdot\frac{\epsilon^{\frac{b}{a}(ma)}-\epsilon^{-\frac{b}{a}(ma)}}{\epsilon^{\frac{b}{a}(ma)}+\epsilon^{-\frac{b}{a}(ma)}}\right)\right\}, \qquad (28a)$$

where the quantity in the {} is the value of the second Σ in (27a). The first part of (28a) is again easily summed up, and the result is

$$L=4\pi\mu\frac{a}{b}\left\{\frac{1}{12}-\frac{1}{2}\frac{a}{b}\sum\frac{1}{(ma)^5}\frac{\epsilon^{\frac{2b}{a}(ma)}-1}{\epsilon^{\frac{2b}{a}(ma)}+1}\right\}, \qquad (29a)$$

in which summation, we may repeat, ma has the values $\frac{1}{2}\pi$, $\frac{3}{2}\pi$, $\frac{5}{2}\pi$, etc. The quantities a and b may be exchanged; that is, a/b changed to b/a, without altering the value of L. This follows by effecting the ma summation in (26a) instead of the nb, as was done.

When the rod is made a flat sheet, or a/b is very small, we have $L=\frac{1}{3}\pi\mu(a/b)$.

Compare (29a) with Thomson and Tait's equation (46) § 707, Part II. Turn the nab^2 outside the [ ] to nab^3, and multiply the Σ by 2. These corrections have been pointed out by Ayrton and Perry. When made, the result is in agreement with the above (29a), allowing, of course, for changed multiplier. (I also observe that the $-\tau$ in their equation (44) should be $+\tau$, and the $+\tau$ in (45), (the second τ) should be $-\tau$.) Such little errors will find their way into mathematical treatises; there is nothing astonishing in that; but a certain collateral circumstance renders the errors in their equation (46) worthy of being long remembered. For the distinguished authors pointedly called attention to the astonishing theorems in pure mathematics to be got by the exchange of a and b, such as rarely fall to the lot of pure mathematicians. They were miraculous.

Effect of a Periodic Impressed Force acting at one end of a Telegraph Circuit with any Terminal Conditions. The General Solution.

I now pass to a different problem, viz., the solution in the case of a periodic impressed force situated at one end of a homogeneous line, when subjected to any terminal conditions of the kind arising from the attachment of apparatus. The conditions that obtain in practice are very various, but valuable information may be arrived at from the study of the comparatively simple problem of a periodic impressed force, of which the full solution may always be found. In Part II. I gave the fully developed solution when the line has the three electrical constants R, L, and S (resistance, inductance, and electric capacity), of which the first two may be functions of the frequency, but without any allowance for the effect of terminal apparatus. If we take $L=0$ we get the submarine-cable formula of Sir W. Thomson's theory; but, although the effect of L on the amplitude of the current at the distant end becomes insignificant when the line is an Atlantic cable, its omission would in general give quite misleading results.

There are some *à priori* reasons against formulating the effect of the terminal apparatus. They complicate the formulæ considerably in the first place; next, they are various in arrangement, so that it might seem impracticable to formulate generally; and, again, in the case of a very long submarine cable, we may divide the expression of the current-amplitude into factors, one for the line and two more for the terminal apparatus, of which the first, for the line, is always the same, whilst the apparatus-factors vary, and are less important than the line-factor. But in other cases the terminal apparatus may be of far greater importance than the line, in their influence on the current-amplitude, whilst the resolution into independent factors is no longer possible.

The only serious attempt to formulate the effect of the terminal apparatus with which I am acquainted is that of the late Mr. C. Hockin (Journal S. T. E. and E., vol. v. p. 432). His apparatus arrangement resembled that usually occurring then in connection with long submarine cables, including, of course, many derived simpler arrangements; and from his results much interesting information is obtainable. But the results are only applicable to long submarine cables, on account of the omission of the influence of the self-induction of the line. The work must, therefore, be done again in a more general manner. It is, besides, independently of this, not easy to adapt his formulæ, in so far as they show the influence of terminal apparatus, to cases that cannot be derived from his. For instance, the effect of magnetic induction in the terminal arrangements was omitted. I have therefore thought it worth while to take a far more general case as regards the line, and at the same time have endeavoured to put it in such a form that it can be readily reduced to simpler cases, whilst at the same time the results apply to any terminal arrangements we choose to use.

The general statement of the problem is this. A homogeneous line, of length l, whose steady resistance is R, inductance L, electric capacity S, and conductance of insulator K, all per unit length of line, is acted upon by an impressed force $V_0 \sin nt$ at one end, or in the wire attached to it; whilst any terminal arrangements exist. Find the effect produced; in particular, the amplitude of the current at the end remote from the impressed force. If the line consists of two parallel wires, R must be the sum of their resistances per unit length.

Let C be the current in the line and V the potential-difference at distance z from the end where the impressed force is situated. Then

$$-\frac{dC}{dz}=\left(K+S\frac{d}{dt}\right)V, \qquad -\frac{dV}{dz}=R''C, \quad \text{.............. (1}b\text{)}$$

are our fundamental line-equations. Here $R''=R+L(d/dt)$ to a first approximation, and $=R'+L'(d/dt)$ in the periodic case, where R' and L' are what R and L become at the given frequency. Let the terminal conditions be

$$\left.\begin{aligned} V&=Z_1C \quad \text{at} \quad z=l \quad \text{end,}\\ -V_0\sin nt+V&=Z_0C \quad \text{at} \quad z=0 \quad \text{end,}\end{aligned}\right\} \quad \text{................ (2}b\text{)}$$

so that $V=Z_0C$ would be the $z=0$ terminal condition if there were no impressed force.

The solution is a special case of the second of (162b), Part IV., which we may quote. In it take

$$S''=K+Sp, \qquad R''=R'+L'p, \quad \text{................. (3}b\text{)}$$

p meaning d/dt so far. Also put $z_2=0$, $e_2=V_0\sin nt$, and

$$-m^2=F^2=(K+Sp)(R'+L'p), \quad \text{.................. (4}b\text{)}$$

and put the equation referred to in the exponential form. Thus,

$$C=\frac{(F/S''+Z_1)\epsilon^{F(l-z)}+(F/S''-Z_1)\epsilon^{-F(l-z)}}{\epsilon^{Fl}(F/S''+Z_1)(F/S''-Z_0)-\epsilon^{-Fl}(F/S''-Z_1)(F/S''+Z_0)}V_0\sin nt. \quad (5b)$$

This is the differential equation of C in the line. Now in F, S'', Z_0, and Z_1, let $d^2/dt^2=-n^2$. It is then reducible to

$$C=\frac{P'+Q'\dfrac{d}{dt}}{A'+B'\dfrac{d}{dt}}V_0\sin nt=\frac{(A'P'+B'Q'n^2)+(A'Q'-B'P')\dfrac{d}{dt}}{A'^2+B'^2n^2}V_0\sin nt, \quad (6b)$$

giving the amplitude and phase-difference anywhere; and the amplitude is

$$C_0=V_0(A'^2+B'^2n^2)^{-\frac{1}{2}}(P'^2+Q'^2n^2)^{\frac{1}{2}}. \quad \text{............... (7}b\text{)}$$

Here P' and Q' are functions of z, whilst A' and B' are constants.

Put

$$\left.\begin{aligned} F&=P+Qi,\\ Z_1&=R_1'+L_1'ni,\\ -Z_0&=R_0'+L_0'ni,\end{aligned}\right\} \quad \begin{aligned}&\text{where} \quad i=(-1)^{\frac{1}{2}},\\ &\text{or} \quad p=ni.\end{aligned} \quad \text{............... (8}b\text{)}$$

The values of P and Q are

$$\left.\begin{aligned} P&=(\tfrac{1}{2})^{\frac{1}{2}}\{(R'^2+L'^2n^2)^{\frac{1}{2}}(K^2+S^2n^2)^{\frac{1}{2}}+(KR'-L'Sn^2)\}^{\frac{1}{2}},\\ Q&=\text{.....}\{\text{...........................}-\text{.................}\}^{\frac{1}{2}},\end{aligned}\right\} \text{...... (9}b\text{)}$$

possessing the following properties, to be used later,

$$\left.\begin{aligned} P^2+Q^2&=(K^2+S^2n^2)^{\frac{1}{2}}(R'^2+L'^2n^2)^{\frac{1}{2}},\\ P^2-Q^2&=KR'-L'Sn^2,\\ 2PQ&=(R'S+KL')n.\end{aligned}\right\} \text{............... (10}b\text{)}$$

The expressions of R_0', R_1', L_0', L_1' can only be stated when the terminal conditions are fully given. Their structure will be considered later. P and Q depend only upon the line.

Let

$$\left.\begin{aligned} A &= R' - Sn^2(R_0'L_1' + R_1'L_0') + K(R_0'R_1' - L_0'L_1'n^2),\\ B &= L'n + Sn(R_0'R_1' - L_0'L_1'n^2) + Kn(R_0'L_1' + R_1'L_0'),\\ a &= P(R_0' + R_1') - Qn(L_0' + L_1'),\\ b &= Q(R_0' + R_1') + Pn(L_0' + L_1'). \end{aligned}\right\}\quad\ldots\ldots\ldots(11b)$$

The effect of making the substitutions (8*b*) in (5*b*) is to express C in terms of the P, Q of (9*b*) and the A, B, a, b of (11*b*); thus:—

$$\begin{aligned} C = \Big[&\big\{ (P - L_1'Sn^2 + KR_1')\cos Q(l-z) - (Q + R_1'Sn + KL_1'n)\sin Q(l-z)\big\}\epsilon^{P(l-z)}\\ &+\big\{ (\ldots + \ldots\ldots - \ldots\ldots)\ldots\ldots\ldots\ldots + (\ldots - \ldots\ldots - \ldots\ldots)\ldots\ldots\ldots\ldots\big\}\epsilon^{-P(l-z)}\\ &+i\big\{ (\ldots - \ldots\ldots + \ldots\ldots)\sin Q(l-z) + (\ldots + \ldots\ldots + \ldots\ldots)\cos Q(l-z)\big\}\epsilon^{P(l-z)}\\ &+i\big\{ -(\ldots + \ldots\ldots - \ldots\ldots)\ldots\ldots\ldots\ldots + (\ldots - \ldots\ldots - \ldots\ldots)\ldots\ldots\ldots\ldots\big\}\epsilon^{-P(l-z)}\Big]\\ &\times V_0 \sin nt\\ \div \Big[&\big\{(A+a)\epsilon^{Pl}\cos Ql - (A-a)\epsilon^{-Pl}\cos Ql - (B+b)\epsilon^{Pl}\sin Ql - (B-b)\epsilon^{-Pl}\sin Ql\big\}\\ &+i\big\{(B+b)\ldots\ldots\ldots - (B-b)\ldots\ldots\ldots\ldots + (A+a)\ldots\ldots\ldots + (A-a)\ldots\ldots\ldots\ldots\big\}\Big] \end{aligned}\quad(12b)$$

The dots indicate repetition of what is immediately above them. Here we see the expressions for the four quantities A', B', P', Q' of (6*b*), which we require. (12*b*) therefore fully serves to find the phase-difference, if required. I shall only develope the amplitude-expression (7*b*). It becomes, by (12*b*),

$$\begin{aligned} \left(\frac{C_0}{V_0}\right) = &\\ \Big[&\epsilon^{2P(l-z)}\big\{(P^2+Q^2) + (K^2+S^2n^2)(R'^2 + L_1'^2n^2) + 2Qn(R_1'S + KL_1') + 2P(KR_1' - L_1'Sn^2)\big\}\\ &+\epsilon^{-2P(l-z)}\big\{\ldots\ldots\ldots + \ldots\ldots\ldots\ldots\ldots\ldots\ldots\ldots - \ldots\ldots\ldots\ldots\ldots\ldots - \ldots\ldots\ldots\ldots\ldots\ldots\big\}\\ &+2\cos 2Q(l-z)\big\{(P^2+Q^2) - (K^2+S^2n^2)(R_1'^2 + L_1'^2n^2)\big\}\\ &-4\sin 2Q(l-z)\big\{Pn(R_1'S + KL_1') + Q(L_1'Sn^2 - KR_1')\big\}\Big]^{\frac{1}{2}}\\ &\div\Big[\epsilon^{2Pl}\big\{(A+a)^2 + (B+b)^2\big\} + \epsilon^{-2Pl}\big\{(A-a)^2 + (B-b)^2\big\}\\ &-2\cos 2Ql\,.\,(A^2+B^2-a^2-b^2) + 4\sin 2Ql\,.\,(Ab - aB)\Big]^{\frac{1}{2}}, \end{aligned}\quad\ldots\ldots\ldots\ldots(13b)$$

in terms of A, B, a, b of (11*b*).

Derivation of the General Formula for the Amplitude of Current at the End remote from the Impressed Force.

This referring to any point between $z=0$ and l, a very important simplification occurs when we take $z=l$ It reduces the numerator to $2(P^2+Q^2)^{\frac{1}{2}}$. It only remains to simplify the denominator as far as possible, to show as explicitly as we can the effect of the terminal apparatus, which is at present buried away in the functions of A, B, a, b occurring in (13b).

First of all, we may show that the product of the coefficients of ϵ^{2Pl} and ϵ^{-2Pl} equals one-fourth the square of the amplitude of the circular part in the denominator. This is an identity, independent of what A, B, a, b are. (13b) therefore takes the form

$$C_0 = 2V_0(P^2+Q^2)^{\frac{1}{2}} \div \left[G\epsilon^{2Pl} + H\epsilon^{-2Pl} - 2(GH)^{\frac{1}{2}}\cos 2(Ql+\theta)\right]^{\frac{1}{2}}. \quad (14b)$$

The following are the expansions of the quantities occurring in the denominator of (13b):—

Let

$$I^2 = R'^2 + L'^2n^2, \qquad I_0^2 = R_0'^2 + L_0'^2n^2, \qquad I_1^2 = R_1'^2 + L_1'^2n^2. \quad \text{.....}(15b)$$

Then

$$\left.\begin{aligned} A^2+B^2 &= I^2 + (K^2+S^2n^2)I_0^2I_1^2 + 2(R_0'R_1' - L_0'L_1'n^2)(KR'+L'Sn^2) \\ &\quad + 2(R_1'L_0' + R_0'L_1')n^2(KL' - R'S), \\ a^2+b^2 &= (P^2+Q^2)\left\{(R_0'+R_1')^2 + (L_0'+L_1')^2n^2\right\}, \\ Aa+Bb &= (R_0'+R_1')(R'P+L'nQ) + (L_0'+L_1')n(L'nP - R'Q) \\ &\quad + (R_0'I_1^2 + R_1'I_0^2)(KP+SnQ) + (L_0'I_1^2 + L_1'I_0^2)n(KQ - SnP), \\ Ab - aB &= (R_0'+R_1')(R'Q - L'nP) + (L_0'+L_1')n(R'P + L'nQ) \\ &\quad + (R_0'I_1^2 + R_1'I_0^2)(KQ - SnP) - (L_0'I_1^2 + L_1'I_0^2)n(KP + SnQ). \end{aligned}\right\} \text{...}(16b)$$

These may be used direct in the denominator of (14b), which is the same as that of (13b). But G and H may be each resolved into the product of two factors, each containing the apparatus-constants of one end only. Noting therefore that the θ in (14b) is given by

$$\tan 2\theta = \frac{2(Ab - aB)}{A^2+B^2-a^2-b^2}, \quad \text{....................}(17b)$$

whose numerator and denominator are given in (16b) [the numerator being $(GH)^{\frac{1}{2}}\sin 2\theta$, and the denominator $(GH)^{\frac{1}{2}}\cos 2\theta$], it will clearly be of advantage to develop these factors. First observe that the expansion of H is to be got from that of G, using (16b), by merely turning P to $-P$ and Q to $-Q$. We have therefore merely to split up one of them, say G. If we put $R_1'=0$, $L_1'=0$ in G it becomes

$$I^2 + (P^2+Q^2)I_0^2 + 2P(R_0'R' + L_0'L'n^2) + 2Q(L'nR_0' - R'nL_0'). \quad (18b)$$

If, on the other hand, we put $R_0'=0$, $L_0'=0$ in G, it becomes the same function of R_1', L_1' as (18b) is of R_0', L_0'. It is then suggested that G is really the product of (18b) into the similar function of R_1', L_1'; when the result is divided by I^2. This may be verified by carrying out the operation described. But I should mention that it is not immediately evident, and requires some laborious transformations to establish it, making use of the three equations (10b). When done, the final result is that (14b) becomes

$$C_0 = 2V_0\left[\frac{K^2+S^2n^2}{R'^2+L'^2n^2}\right]^{\frac{1}{2}}$$

$$\div\left[G_0G_1\epsilon^{2Pl} + H_0H_1\epsilon^{-2Pl} - 2(G_0G_1H_0H_1)^{\frac{1}{2}}\cos 2(Ql+\theta)\right]^{\frac{1}{2}}, \quad \text{........}(19b)$$

wherein G_0 and H_0 contain only constants belonging to the apparatus at $z=0$, and G_1 and H_1 those belonging to $z=l$, besides the line-constants. Only one of the four need be written; thus

$$G_0 = 1 + \frac{1}{I^2}\left[(P^2+Q^2)I_0^2 + 2P(R'R_0' + L'L_0'n^2) + 2Qn(R_0'L' - R'L_0')\right]. \quad (20b)$$

From this get H_0 by changing the signs of P and Q. Then, to obtain G_1 and H_1, the corresponding functions for the $z=l$ end, change R_0' to R_1' and L_0' to L_1'. These functions have the value unity when the line is short-circuited at the ends, ($Z_0=0$, $Z_1=0$). They may therefore be referred to as the terminal functions. Their form is invariable. We only require to find the R' and L', or the effective resistance and inductance of the terminal arrangements, and insert in (20b) and its companions.

The Effective Resistance and Inductance of the Terminal Arrangements.

Thus, let the two conductors at the $z=l$ end be joined through a coil. Then R_1' is its resistance, L_1' its inductance, the steady values, and the accents may be dropped, except under very unusual circumstances, and I_1 is its impedance at the given frequency, when on short-circuit. But if the coil contain a core, especially if it be of iron, neither R_1 nor L_1 can have the steady values, on account of the induction of currents in the core. Their approximate values at a given frequency may be experimentally determined by means of the Wheatstone Bridge. Of course R_1 and L_1 are really somewhat changed in a similar manner by allowing any induction between the coil and external conductors, the brass parts of a galvanometer, for instance; L going down and R going up, though this does not materially affect I.

If, instead of a coil, it be a condenser of capacity S_1 that is inserted at $z=l$; then, since

$$C = S_1\dot{V} = S_1pV,$$

we have
$$Z_1 = (S_1p)^{-1} = -p/(S_1n^2).$$

Therefore take $R_1' = 0$, and $L_1' = -(S_1n^2)^{-1}$.

The condenser behaves, so far as the current is concerned, as a coil of no resistance and negative inductance, the latter decreasing as the frequency is raised, and as the capacity is increased; tending to become equivalent to a short-circuit, though this would require a great frequency in general, as the *quasi*-negative inductance is large. (Thus, $n=100$, $S=10^{-15}$ = one microfarad, make $L_1' = -10^{11}$. To make the inductance of a coil be 10^{11} it must contain a very large number of turns of fine wire.) Thus, whilst the condenser stops slowly periodic or steady currents, it tends to readily pass rapidly periodic currents, a property which is very useful in telephony, as in Van Rysselberghe's system.

On the other hand, the coil passes the slowly periodic, and tends to stop the rapidly periodic, a property which is also very useful in telephony. A very extensive application of this principle occurs in the system of telephonic intercommunication invented and carried out by

Mr. A. W. Heaviside, known as the Bridge System, from the telephones at the various offices being connected up as bridges across from one to the other of the two conductors which form the line. Whilst all stations are in direct communication with one another, one important desideratum, there is no overhearing, which is another. For all stations except the two which are in correspondence at a certain time have electromagnets of high inductance inserted in their bridges, which electromagnets will not pass the rapid telephonic currents in appreciable strength, so that it is nearly as if the non-working bridges were non-existent; and, in consequence, a far greater length of buried wire can be worked through than on the Sequence system, wherein the various stations have their apparatus in sequence with the line; whilst at the same time (in the Bridge system) a balance is preserved against inductive interferences. When the two stations have finished correspondence, they insert their own electromagnets in their bridges. As these electromagnets are used as call-instruments, responding to slowly periodic currents, we have the direct intercommunication. Of course there are various other details, but the above sufficiently describes the principle.

As regards the property of the self-induction of a coil in stopping or greatly decreasing the amplitude of rapidly periodic currents, or acting as an insulation at the first moment of starting a current, its influence was entirely overlooked by most writers on telegraphic technics before 1878, when I wrote on the subject [vol. I., p. 95]. A knowledge of the important quantity $(R^2+L^2n^2)^{\frac{1}{2}}$, which is now the common property of all electrical schoolboys (especially by reason of the great impetus given to the spread of a scientific knowledge of electromagnetism by the commercial importance of the dynamo), was, before then, confined to a few theorists.

If the coil R, L, and the condenser S_1 be in parallel, we have

$$C=\left(S_1p+\frac{1}{R+Lp}\right)V,$$

or

$$\frac{V}{C}=\frac{R+\{L-S_1(R^2+L^2n^2)\}p}{(1-LS_1n^2)^2+(RS_1n)^2},$$

which show the expressions of R_1' and L_1', the second being the coefficient of p, the first the rest.

Similarly in other simple cases. And, in general, from the detailed nature of the combination inserted at the end of the line, write out the connections between the current and potential-difference in each branch, and eliminate the intermediates so as to arrive at $V=Z_1C$, the differential equation of the combination, wherein Z_1 is a function of p or d/dt. Put $p^2=-n^2$, and it takes the form $Z_1=R_1'+L_1'p$, wherein R_1' and L_1' are functions of the electrical constants and of n^2, and are the required effective R_1' and L_1' of the combination, to be used in (20*b*), or rather, in its $z=l$ equivalent G_1.

As regards the $z=0$ end, it is to be remarked that, owing to the current being reckoned positive the same way at both ends, when we write $V=Z_0C$ as the terminal equation, it is $-Z_0$ that corresponds to Z_1. Thus $-Z_0=R_0'+L_0'p$, where, in the simplest case, R_0' and L_0' are the resistance and inductance of a coil.

Special Details concerning the above. Quickening Effect of Leakage. The Long-Cable Solution, with Magnetic Induction ignored.

So far sufficiently describing how to develope the effective resistance and inductance expressions to be used in the terminal functions G and H, we may now notice some other peculiarities in connection with the solution (19*b*). First short-circuit the line at both ends, making the terminal functions unity, and $\theta=0$. The solution then differs from that given in Part II., equation (82), in the presence of the quantity K, the former Sn now becoming $(K^2+S^2n^2)^{\frac{1}{2}}$, whilst P and Q differ from the former P and Q of (78), Part II., by reason of K, whose evanescence makes them identical. If we compare the old with the new P and Q, we find that

$$\left.\begin{array}{lll} L' & \text{becomes} & L'-KR'/Sn^2, \\ R' & \text{becomes} & R'+KL'/S, \end{array}\right\} \quad \text{.....................(21}b\text{)}$$

in passing from the old to the new. Then the function

$$\frac{R'^2+L'^2n^2}{S^2n^2} \text{ becomes } \frac{(R'+KL'/S)^2+(L'-KR'/Sn^2)^2n^2}{K^2+S^2n^2}=\frac{R'^2+L'^2n^2}{S^2n^2},$$

or is unaltered by the leakage. It follows that the equation (85), Part II., is still true, with leakage, if we make the changes (21*b*) just mentioned in it, or put

$$\frac{1}{v'^2}=S\left(L'-\frac{KR'}{Sn^2}\right), \qquad h=\frac{(R'S+KL')^2}{n^2}v'^4, \quad \text{...........(22}b\text{)}$$

instead of using the v' and h expressions of Part II.

At the particular frequency given by $n^2=KR'/L'S$, we shall have

$$P=Q=(\tfrac{1}{2})^{\frac{1}{2}}(R'^2+L'^2n^2)^{\frac{1}{4}}(K^2+S^2n^2)^{\frac{1}{4}}=(\tfrac{1}{2})^{\frac{1}{2}}(R'S+KL')n, \quad \text{...(23}b\text{)}$$

making

$$\frac{V_0}{C_0}=(R'^2+L'^2n^2)^{\frac{1}{2}}l\left\{1+2\left(\frac{(2Pl)^4}{\underline{|6}}+\frac{(2Pl)^8}{\underline{|10}}+\frac{(2Pl)^{12}}{\underline{|14}}+\dots\right)\right\}^{\frac{1}{2}}. \quad \text{...(24}b\text{)}$$

If we should regard the leakage as merely affecting the amplitude of the current at the distant end of a line, we should be overlooking an important thing, viz., its remarkable effect in accelerating changes in the current, and thereby lessening the distortion that a group of signals suffers in its transmission along the line. If there is only a sufficient strength of current received for signalling purposes, the signals can be far more distinct and rapid than with perfect insulation, as I have pointed out and illustrated in previous papers. Thus the theoretical desideratum for an Atlantic cable is not high, but low insulation—the lowest possible consistent with having enough current to work with. Any practical difficulties in the way form a separate question.

Regarding this quickening effect, or partial abolition of electrostatic retardation, I have [vol. I., pp. 531 and 536] pushed it to its extreme in the electromagnetic scheme of Maxwell. In a medium whose conductivity varies in any manner from point to point, possessed of dielectric capacity which varies in the same manner (so that their ratio, or the electrostatic time-constant, is everywhere the same), but destitute of magnetic inertia ($\mu = 0$, no magnetic energy), I have shown that electrostatic retardation is entirely done away with, except as regards imaginable preexisting electrification, which subsides everywhere according to the common time-constant, without true electric current, by the discharge of every elementary condenser through its own resistance. This being over, if any impressed force act, varying in any manner in distribution and with the time, the corresponding current will everywhere have the steady distribution appropriate to the impressed force at any moment, in spite of the electric displacement and energy; and, on removal of the impressed force, there will be instantaneous disappearance of the current and the displacement. This seems impossible; but the same theory applies to combinations of shunted condensers, arranged in a suitable manner, as described in the paper referred to.

Of course this extreme state of things is quite imaginary, as we cannot really overlook the magnetic induction in such a case. If we regard it as the limiting form of a real problem, in which inertia occurs, to be afterwards made zero, we find that the instantaneous subsidence of the electrostatic problem becomes [with reflecting barriers] an oscillatory subsidence of infinite frequency but finite time-constant, about the mean value zero; which is mathematically equivalent to instantaneous non-oscillatory subsidence.

The following will serve to show the relative importance of R, S, K, and L in determining the amplitude of periodic currents at the distant end of a long submarine cable, of fairly high insulation-resistance:—

4 ohms per kilom. makes $R = 40^4$,

$\frac{1}{4}$ microf. " " $S = \dfrac{1}{40^{20}}$,

100 megohms " " $K = 10^{-22}$.

Here, it should be remembered, K is the conductance of the insulator per centim. The least possible value of L would be such that $LS = v^{-2}$, where $v = 30^{10}$; this would make $L = \frac{4}{9}$ only. But it is really much greater, requiring to be multiplied by the dielectric constant of the insulator in the first place, making $L = 2$ say. It is still further increased by the wire, and considerably by the sheath and by the extension of the magnetic field beyond the sheath, to an extent which is very difficult to estimate, especially as it is a variable quantity; but it would seem never to become a very large number, as of course an iron wire for the conductor is out of the question. But leaving it unstated, we have, by (9b), taking $R' = R$, $L' = L$,

$$P = (\tfrac{1}{2})^{\frac{1}{2}}\left\{(160^8 + L^2n^2)^{\frac{1}{2}}\left(\frac{1}{10^{44}} + \frac{n^2}{160^{40}}\right)^{\frac{1}{2}} + \left(\frac{4}{10^{18}} - \frac{Ln^2}{40^{20}}\right)\right\}^{\frac{1}{2}}$$

$$= \frac{(\frac{1}{2})^{\frac{1}{2}}}{10^{10}}\left\{(160^8 + L^2n^2)^{\frac{1}{2}}\left(\frac{1}{10^4} + \frac{n^2}{16}\right)^{\frac{1}{2}} + \left(400 - \frac{Ln^2}{4}\right)\right\}^{\frac{1}{2}}$$

Now $n/2\pi$ is the frequency, necessarily very low on an Atlantic cable. We see then that the first L^2n^2 is quite negligible in its effect upon P, even when we allow L to increase greatly from the above $L=2$. The high insulation also makes the $(RK-LSn^2)$ part negligible, making approximately

$$P=Q=(\tfrac{1}{2}n)^{\frac{1}{2}}\,.\,10^{-8},$$

P being a little greater than Q, at least when L is small. Now this is equivalent to taking $L=0$, $K=0$, when

$$P=Q=(\tfrac{1}{2}RSn)^{\frac{1}{2}}, \quad \text{.........................} (25b)$$

reducing (19*b*) to

$$C_0=2V_0(Sn/R)^{\frac{1}{2}} \div \{G_0G_1\epsilon^{2Pl}+H_0H_1\epsilon^{-2Pl}-2(G_0G_1H_0H_1)^{\frac{1}{2}}\cos 2Pl\}^{\frac{1}{2}}, \quad (26b)$$

which is, except as regards the terminal functions I introduce, quite an old formula. It is what we get by regarding the line as having only resistance and electrostatic capacity. But, still regarding the line as an Atlantic or similar cable, worked nearly up to its limit of speed, Pl is large, say 10 at most, so that we may take this approximation to (26*b*),

$$C_0=2V_0(Sn/R)^{\frac{1}{2}}\epsilon^{-Pl}\times G_0^{-\frac{1}{2}}\times G_1^{-\frac{1}{2}}, \quad \text{....................} (27b)$$

where the first of the three factors is the line-factor, the second that due to the apparatus at the $z=0$ end, and the third to that at the $z=l$ end of the line; thus, by (20*b*) and (25*b*), with $L'=0$ and $R'=R$ in the former,

$$\left.\begin{aligned} G_0&=1+\frac{1}{R^2}\Big(2PR(R_0'-L_0'n)+2P^2(R_0'^2+L_0'^2n^2)\Big), \\ G_1&=1+\frac{1}{R^2}\Big(2PR(R_1'-L_1'n)+2P^2(R_1'^2+L_1'^2n^2)\Big). \end{aligned}\right\} \text{..........} (28b)$$

This reduction to (27*b*) is of course not possible when the line is very far from being worked up to its possible limit; in fact, all three terms in the { } of (26*b*), or, more generally, of (19*b*), require to be used in general. For this reason a full examination of the effect of terminal apparatus is very laborious. Most interesting results may be got out of (19*b*), especially as regards the relative importance of the line and terminal apparatus at different speeds, complete reversals taking place as the speed is varied whilst the line and apparatus are kept the same. The general effect is that, as the speed is raised, the influence of the apparatus increases much faster than that of the line. For instance, to work a land-line of, say, 400 miles up to its limit, we must reduce the inertia of the instruments greatly to make it even possible. In fact, electromagnets seem unsuitable for the purpose, unless quite small, and chemical recording has probably a great future before it. But it would be too lengthy a digression to go into the necessarily troublesome details.

Some Properties of the Terminal Functions.

The following relates to some properties of the terminal function G, which have application when (27*b*) is valid. Consider the G_1 of (28*b*). Let it be simply a coil that is in question. Then R_1 is its resistance and L_1 its inductance, dropping the accent. Keep the resistance constant, whilst varying the inductance so as to make G_1 a minimum, and therefore the current-amplitude a maximum. The required value of L_1 is

$$L_1 = R/2Pn, \qquad \text{.............................(29b)}$$

depending only upon the line-constants and the frequency, independently of the resistance of the coil. Taking $Pl = 10$, this makes $L_1 = Rl/20n$, where Rl is the resistance of the line. The relation (29b) makes

$$G_1 = \tfrac{1}{2} + \frac{2PR_1}{R} + \frac{2P^2R_1^2}{R^2}. \qquad \text{....................(30b)}$$

If the coil had no inductance, but the same resistance, G_1 would have the same expression, but with 1 instead of $\frac{1}{2}$ in (30b). The effect of the inductance has therefore increased the amplitude of the current, and it is conceivable that G_1 could be made less than unity, though it may not be practicable.

Now the G_1/R_1 of (30b) is a minimum, with R_1 variable, when $R = 2PR_1$, and this will make $G_1 = 2$, or make the terminal factor be $G_1^{-\frac{1}{2}} = \cdot 7$. Now if we vary the number of turns of wire in the coil, keeping it of the same size and shape, the magnetic force will vary as $(R_1/G)^{\frac{1}{2}}$, so it at first sight appears that $R_1 = R/2P$ and $L_1 = R/2Pn$ make the magnetic force a maximum for a fixed size and shape of coil. There is, however, a fallacy here, because varying the size of the wire as stated varies L_1 nearly in the same ratio as R_1, whilst (30b) assumes L_1 to be a constant, given by (29b). It is perhaps conceivable to keep L_1 constant during the variation of R_1, by means of iron, and so get $(R_1/G)^{\frac{1}{2}}$ to be a maximum; but then, on account of the iron, this quantity will not represent the magnetic force.

If, on the other hand, we vary R_1 in the original G_1 of (28b), keeping L_1/R_1 constant (size and shape of coil fixed, size of wire variable), G_1/R_1 is made a minimum by

$$R_1^2 + L_1^2n^2 = R^2/2P^2, \qquad \text{..........................(31b)}$$

giving a definite resistance to the coil, of stated size and shape, to make the magnetic force a maximum. Now G_1 becomes

$$G_1 = 2 + \frac{2P}{R}(R_1 - L_1n), \qquad \text{.......................(32b)}$$

where L_1/R_1 has been constant. If this constant have the value n^{-1}, we have $G_1 = 2$ again, and R_1, L_1 have the same values as before. There is thus some magic about $G_1 = 2$.

Again, if the terminal arrangement consist of a coil R_1, L_1, and a condenser of capacity S_1 and conductance K_1, joined in sequence, we shall have

$$\begin{aligned} V/C &= (R_1 + L_1p) + (K_1 + S_1p)^{-1}, \\ &= \left(R_1 + \frac{K_1}{K_1^2 + S_1^2n^2}\right) + \left(L_1 - \frac{S_1}{K_1^2 + S_1^2n^2}\right)p, \qquad \text{.............(33b)} \\ &= R_1' + L_1'p, \quad \text{say} \end{aligned}$$

if R_1', L_1' are the effective resistance and inductance, to be used in G_1, making

$$G_1 = 1 + \frac{2P}{R}\left\{R_1 - L_1 n + \frac{K_1 + S_1 n}{K_1^2 + S_1^2 n^2}\right\}$$
$$+ \frac{2P^2}{R^2}\left\{R_1^2 + L_1^2 n^2 + \frac{1}{K_1^2 + S_1^2 n^2} + 2\frac{R_1 K_1 - S_1 L_1 n^2}{K_1^2 + S_1^2 n^2}\right\}. \quad \text{.........(34}b\text{)}$$

Variation of L_1 alone makes G_1 a minimum when

$$L_1 n = \frac{S_1 n}{K_1^2 + S_1^2 n^2} + \frac{R}{2P}; \quad \text{........................(35}b\text{)}$$

and if we take $K_1 = 0$ (condenser non-leaky, and not shunted), we have the value of G_1 given by (30*b*) again, independent of the condenser. Similarly we can come round to the same $G_1 = 2$ again. These relations are singular enough, but it is difficult to give them more than a very limited practical application to the question of making the magnetic force of the coil a maximum, although the (30*b*) relation is not subject to any indefiniteness.

Part VI.

General Remarks on the Christie considered as an Induction Balance. Full-Sized and Reduced Copies.

The most important as well as most frequent application of Mr. S. H. Christie's differential arrangement, known at various times under the names of Wheatstone's parallelogram, lozenge, balance, bridge, quadrangle, and quadrilateral, is to balance the resistances of four conductors, when supporting steady currents due to an impressed force in a fifth, and this is done by observing the absence of steady current in a sixth. But its use in other ways and for other purposes has not been neglected. Thus, Maxwell described three ways of using the Christie to obtain exact balances with transient currents (these will be mentioned later in connection with other methods); Sir W. Thomson has used it for balancing the capacities of condensers*; and it has been used for other purposes. But the most extensive additional use has been probably in connection with duplex telegraphy; and here, along with the Christie, we may include the analogous differential-coil system of balancing, which is in many respects a simplified form of the Christie.

On the revival of duplex telegraphy some fifteen years ago, it was soon recognised that "the line" required to be balanced by a similar line, or artificial line, not merely as regards its resistance, but also as regards its electrostatic capacity—approximately by a single condenser; better by a series of smaller condensers separated by resistances; and, best of all, by a more continuous distribution of electrostatic capacity along the artificial line. The effect of the unbalanced self-induction was also observed. This general principle also became clearly recognised, at least by some,—that no matter how complex a line may be,

* Journal S. T. E. and E., vol. I., p. 394.

considered as an electrostatic and magnetic arrangement, it could be perfectly balanced by means of a precisely similar independent arrangement; that, in fact, the complex condition of a perfect balance is identity of the two lines throughout. The great comprehensiveness of this principle, together with its extreme simplicity, furnish a strong reason why it does not require formal demonstration. It is sufficient to merely state the nature of the case to see, from the absence of all reason to the contrary, that the principle is correct.

Thus, if AB_1C and AB_2C [see figure on p. 263] be two identically similar independent lines (which of course includes similarity of environment in the electrical sense in similar parts), joined in parallel, having the A ends connected, and also the C ends, and we join A to C by an external independent conductor in which is an impressed force e, the two lines must, from their similarity, be equally influenced by it, so that similar parts, as B_1 in one line and B_2 in the other, must be in the same state at the same moment. In particular, their potentials must always be equal, so that, if the points B_1 and B_2 be joined by another conductor, there will be no current in it at any moment, so far as the above-mentioned impressed force is concerned, however it vary. The same applies when it is not mere variation of the impressed force e, but of the resistance of the branch in which it is placed. And, more generally, B_1 and B_2 will be always at the same potential as regards disturbances originating in the independent electrical arrangement joining A to C externally, however complex it may be.

There is, however, this point to be attended to, that might be overlooked at first. Connecting the bridge-conductor from B_1 to B_2 must not produce current in it from other causes than difference of potential; for instance, there should be, at least in general, no induction between the bridge-wire and the lines, or some special relation will be required to keep a balance. This case might perhaps be virtually included under similarity of environment.

If we had sufficiently sensitive methods of observation, the statement that one line must be an exact copy of the other would sometimes have to be taken literally. But the word copy may practically be often used to mean copy only as regards certain properties, either owing to the balance being independent of other properties, or owing to our inability to recognise the effects of differences in other properties. Thus, in the steady resistance-balance we only require AB_1 and AB_2 to have equal total resistances, and likewise B_1C and B_2C; resistances in sequence being additive. But evidently, if the balance is to be kept whilst B_1 and B_2 are shifted together from end to end of the two lines, the resistance must be similarly distributed along them.

If, now, condensers be attached to the lines, imitating a submarine cable, though of discontinuous capacity, we require that the resistance of corresponding sections shall be equal, as well as the capacities of corresponding condensers, in order that we shall have balance in the variable period as well as in the steady state; and the two properties, resistance and capacity, are the elements involved in making one line a copy of the other.

In case of magnetic induction again, if AB_1C and AB_2C each consist of a number of coils in sequence, they will balance if the coils are alike, each for each, in the two lines, and are similarly placed with respect to one another. But the lines will easily balance under simpler conditions, inductances being additive, like resistances; and it is only necessary that the total self-inductions of AB_1 and AB_2 (including mutual induction of their parts) be equal, and likewise of B_1C and B_2C. Again, if a coil a_1 in the branch AB_1 have another coil b_1 in its neighbourhood (not in either line, but independent), and a_2, in the branch AB_2, be a copy of a_1, we can complete the balance by placing a coil b_2 (which is a copy of b_1) in the neighbourhood of the coil a_2, so that the action between a_1 and b_1 is the same as that between a_2 and b_2. But it is not necessary for b_1 and b_2 to be copies of one another except in the two particulars of resistance and inductance; whilst as regards their positions with respect to a_1 and a_2, we only require the mutual inductance of a_1 and b_1 to equal that of a_2 and b_2.

On the other hand, if b_1 be not a coil of fine wire, but a piece of metal that is placed near the coil a_1, many more specifications are required to make a copy of it. The piece of metal is not a linear conductor; and, although no doubt only a small number (instead of an infinite number) of degrees of freedom allowed for, would be sufficient to make a practical balance, yet, as we have not the means of simply analyzing pieces of metal (like coils) into a few distinct elements, we must generally make a copy of b_1 by means of a similar piece, b_2, of the same metal, and place it with respect to a_2 as b_1 is to a_1, to secure a good balance. But very near balances may be sometimes obtained by using quite dissimilar pieces of metal, dissimilarly placed.

So far, copy signifies equality in certain properties. But one line need be merely a reduced copy of the other. It is only when we inquire into what makes one line a reduced copy of another, that we require to examine fully the mathematical conditions of the case in question. In the state of steady flow the matter is simple enough. If AB_1 has n times the resistance of AB_2, then must B_1C have n times the resistance of B_2C to keep the potentials of B_1 and B_2 equal. If condensers be connected to the lines, as before mentioned, we require, first, the resistance-balance of the last sentence applied to every section between a pair of condensers; and next, that the capacity of a condenser in the line AB_1C shall be, not n times (as patented by Mr. Muirhead, I believe), but $1/n$ of the capacity of the corresponding condenser in the line AB_2C [vol. I., p. 25]. If the lines are representable by resistance, inductance, electrostatic capacity, and leakage-conductance (R, L, S, K of Parts IV. and V., per unit length), one line will be a reduced copy of the other if, when R and L in the first line are n times those in the second, S and K in the second are n times those in the first, in similar parts.

Conjugacy of Two Conductors in a Connected System. The Characteristic Function and its Properties.

After these general remarks, and preliminary to a closer consideration of the Christie, let us briefly consider the general theory of the conjugacy of a pair of conductors in a connected system, when an impressed force in either can cause no current in the other, either transient or permanent. The direct way is to seek the full differential equation of the current in either, when under the influence of impressed force in the other alone. Let $V=ZC$ be the differential equation of any one branch, C being the current in it, V the fall of potential in the direction of C, and Z the differential operator concerned, according to the notation of Parts III., IV., and V. If there be impressed force e in the branch, it becomes $e+V=ZC$. We have $\Sigma V=0$ in any circuit, by the potential-property; therefore $\Sigma e=\Sigma ZC$ in any circuit. Also the currents are connected by conditions of continuity at the junctions. These, together with the former circuit-equations, lead us to a set of equations :—

$$\left.\begin{array}{l} FC_1=f_{11}e_1+f_{12}e_2+\dots, \\ FC_2=f_{21}e_1+f_{22}e_2+\dots, \\ \dots\dots\dots\dots\dots\dots, \end{array}\right\} \qquad (1c)$$

C_1, C_2, ..., being the currents, and e_1, e_2, ... the impressed forces in branches 1, 2, etc.; F being common to all, and it and the f's being differential operators. We arrive at similar equations when the differential equation of a branch is not merely between the V and C of that branch, but between those of many branches; for instance, when

$$V_1=Z_{11}C_1+Z_{12}C_2+\dots \qquad (2c)$$

is the form of the differential equation of branch 1.

Now let there be impressed force e in one branch only, and C be the current in a second, dropping the numbers as no longer necessary. We then have

$$FC=fe. \qquad (3c)$$

Conjugacy is therefore secured by $fe=0$, making C independent of e. Therefore $fe=0$ is the complex condition of conjugacy. If, for example,

$$fe=a_0e+a_1\dot{e}+a_2\ddot{e}+\dots, \qquad (4c)$$

where the a's are constants, functions of the electrical constants concerned, then, to ensure conjugacy, we require

$$a_0=0, \qquad a_1=0, \qquad a_2=0, \quad \text{etc.}, \qquad (5c)$$

separately; and if these a's cannot all vanish together we cannot have conjugacy.

What C may be then depends only upon the initial state of the system in subsiding, or upon other impressed forces that we have nothing to do with. As depending upon the initial state, the solution is

$$C=\Sigma A\epsilon^{pt}; \qquad (6c)$$

the summation being with respect to the p's which are the roots of $F(p)=0$, p being put for d/dt in F; and the A belonging to a certain p is to be obtained by the conjugate property of the equality of the mutual electric to the mutual magnetic energy of the normal systems of any pair of p's.

As depending upon e, the impressed force in the conductor which is to be conjugate to the one in which the current is C, let e be zero before time $t=0$, and constant after. Then, by (3c),

$$C=\frac{f(d/dt)e}{F(d/dt)}=\sum\frac{f(p)e}{-pF'}(1-\epsilon^{pt})=C_0-\sum\frac{f(p)e}{-pF'}\epsilon^{pt}, \quad \text{........(7c)}$$

if C_0 is the final steady current, and $F'=dF/dp$, the summation being with respect to the p's.*

If there is a resistance-balance, $a_0=0$, $C_0=0$, and

$$C=\sum\frac{f(p)e}{pF'}\epsilon^{pt} \quad \text{..............................(8c)}$$

Now, subject to (4c), calculate the integral transient current:—

$$\int_0^\infty Cdt=\sum\frac{f(p)e}{-p^2F'},$$

$$=\text{value of } f(p)e/pF(p) \quad \text{when} \quad p=0,$$

$$=ea_1/F_0, \quad \text{..(9c)}$$

if F_0 is the $p=0$ value of F. If then $a_1=0$ also, we prove that the integral transient current is zero.

Supposing both $a_0=0$, $a_1=0$, then

$$C=\sum\frac{a_2p^2+\dots}{pF'}e\,.\,\epsilon^{pt}\,.$$

therefore $$\int_0^t Cdt=\sum\frac{a_2p+\dots}{-pF'}e(1-\epsilon^{pt})=\sum\frac{a_2p+\dots}{pF'}e\,.\,\epsilon^{pt},$$

and therefore $$\int_0^\infty dt\int_0^t Cdt=\sum\frac{a_2+\dots}{-pF'}e=\frac{a_2}{F_0}e. \quad \text{.................(10c)}$$

Thus, if $a_2=0$ also, we have

$$\int_0^\infty dt\int_0^t Cdt=0. \quad \text{..............................(11c)}$$

Similarly, if $a_3=0$ also, then

$$\int_0^\infty dt\int_0^t dt\int_0^t Cdt=0, \quad \text{..........................(12c)}$$

and so on. The physical interpretation of $a_0=0$ and $a_1=0$ is obvious, but after that it is less easy.

If F contain inverse powers of p, the steady current may be zero. But in spite of that, it will be found that to secure perfect conjugacy for transient currents we must have a true resistance-balance, or that relation amongst the resistances which would make the steady current zero, if we were to allow the possibility of a steady current by changing the value of other electrical quantities concerned. I will give an example of this later.

I have elsewhere [vol. I., p. 412] pointed out these properties of the

* [In these equations (7c) to (10c) modify as in the footnote on p. 226, vol. II., if necessary.]

function F, in the case where there is no mutual induction, or $V=ZC$ is the form of the differential equation of a branch. Let n points be united by $\frac{1}{2}n(n-1)$ conductors, whose conductances are K_{12}, K_{13}, etc., it being the points that are numbered 1, 2, etc. Then the determinant

$$\begin{vmatrix} K_{11}, & K_{12}, & \dots, & K_{1n} \\ K_{21}, & K_{22}, & \dots, & K_{2n} \\ \dots & \dots & \dots & \dots \\ K_{n1}, & K_{n2}, & \dots, & K_{nn} \end{vmatrix} \quad \text{.........................(13}c\text{)}$$

is zero, and its first minors are numerically equal, if any K with equal double suffixes be the negative of the sum of the real K's in the same row or column.* Remove the last row and column, and call the determinant that is left F. It is the F required, and is the characteristic function of the combination, expressed in terms of the conductances. If every branch have self-induction, so that $R+L(d/dt)$ takes the place of K^{-1}, then $F=0$ is the differential equation of the combination, without impressed forces; and $F=0$ is always the differential equation subject to the condition of no mutual induction. In the paper referred to cores are placed in the coils, giving a special form to K.

When K is conductance merely, the characteristic function contains within itself expressions for the resistance between every two points in the combination, which can therefore be written down quite mechanically. For it is the sum of products each containing first powers of the K's, and therefore may be written

$$F=K_{12}X_{12}+Y_{12}=K_{23}X_{23}+Y_{23}=\dots, \quad \text{..............(14}c\text{)}$$

where X_{23}, Y_{23} do not contain K_{23}, and X_{12}, Y_{12} do not contain K_{12}. (It is to be understood that the diagonal K_{11}, K_{22}, ..., are got rid of.)

Then $\left.\begin{aligned} R'_{12}&=X_{12}/Y_{12}=\text{resistance between points 1 and 2,} \\ R'_{23}&=X_{23}/Y_{23}= \quad \text{"} \quad \text{"} \quad \text{" 2 and 3,} \end{aligned}\right\}$ ......(15c)

etc., it being understood that these resistances are not R_{12}, R_{23}, etc., but the resistances complementary to them, the combined resistance of the rest of the combination; thus, if e_{12} be the impressed force in the conductor 1, 2, the current (steady) in it is

$$C_{12}=\frac{e_{12}}{R_{12}+X_{12}/Y_{12}}=\frac{e_{12}}{R_{12}+R'_{12}}. \quad \text{...................(16}c\text{)}$$

The proof by determinants is rather troublesome, using the K's, but, in terms of their reciprocals, and extending the problem, it becomes simple enough. Thus, if we turn K to R^{-1} in F, and then clear of fractions, we may write $F=0$ as

$$R_{12}X'_{12}+Y'_{12}=0, \qquad R_{23}X'_{23}+Y'_{23}=0, \quad \text{etc.,} \quad \text{........(17}c\text{)}$$

where X'_{12}, Y'_{12}, do not contain R_{12}; etc. From this we see that the differential equation of the current C_{12} in 1, 2, subject to e_{12} only, is

$$e_{12}=(R_{12}+R'_{21})C_{12}, \quad \text{.........................(18}c\text{)}$$

* As in Maxwell, vol. I., art. 280.

if $R'_{21} = Y'_{12}/X'_{12}$. For this make the dimensions correct, and that is the only additional thing required, when we observe that it makes the steady current be

$$C_{12} = e_{12}/(R_{12} + R'_{21}), \quad \ldots\ldots\ldots\ldots\ldots\ldots\ldots(19c)$$

so that R'_{21} is the resistance complementary to R_{12}.

Although it is generally best to work in terms of resistances, yet there are times when conductances are preferable, and, to say nothing of conductors in parallel arc, the above is a case in point, as will be seen by the way the characteristic function is made up out of the K's. There is also less work in another way. Thus, $\frac{1}{2}n(n-1)$ conductors uniting n points give $\frac{1}{2}(n-1)(n-2)$ degrees of freedom to the currents. It is the least number of branches in which, when the currents in them are given, those in all the rest follow. Thus, if 10 conductors unite 5 points, the currents in at least 6 conductors must be given, and no four of them should meet at one point. The remaining conductors are $n-1$ in number, or one less than the number of points, and $n-1$ is the degree of the characteristic function in terms of the conductances. Now put $F=0$ in terms of the resistances, by multiplying by the product of all the resistances. It is then made of degree $\frac{1}{2}(n-1)(n-2)$ in terms of the resistances, which is the number of current-freedoms. If $n=4$, the degree is the same, viz., three, whether in terms of conductances or resistances; but if $n=5$, it is of the sixth degree in terms of resistances and only of the fourth in terms of the conductances; and if $n=6$, it is of the tenth degree in terms of the resistances, but only of the fifth in terms of the conductances; and so on, so that F becomes greatly more complex in terms of resistances than conductances.

When every branch has self-induction, $Z=R+Lp$, and the degree of p in $F=0$ is the number of freedoms, so that there are $n-1$ fewer roots than the number of branches. It is the same when there is mutual induction. The missing roots belong to terms, in the solutions for subsidence from an arbitrary initial state, which instantaneously vanish, producing a jump from the initial state to another, which subsides in time.

On the other hand, if every branch (without self-induction) is shunted by a condenser of capacity S_1, S_2, etc., K becomes $K+Sp$, so that the degree of p in $F=0$ is the same as that of K, or $\frac{1}{2}(n-1)(n-2)$ fewer than the number of condensers. [Vol. I., p. 540.]

Theory of the Christie Balance of Self-Induction.

Coming next to the Christie as a self-induction balance, let there be six conductors, 1, 2, etc., uniting the four points A, B_1, B_2, C in the figure. AB_1C and AB_2C are "the lines" referred to in the beginning. Let R be the resistance and L the inductance of a branch in which the current is C, reckoned positive in the direction of the arrow, and the fall of potential V in the same direction; thus R_1, L_1, V_1, C_1 for the first branch The six branches may be conjugate in pairs, thus: 1 and 4, or 2 and 3, or 5 and 6. In the following 5 and 6 are selected always, the battery or other source being in 6, and the telephone or other indicator in 5. Mutual inductances will be denoted by M; thus, $M_{12}C_1$ is the electromotive impulse in 2 due to the stoppage of the current C_1 in 1; similarly $M_{12}C_2$ is the impulse in 1 due to stopping C_2.

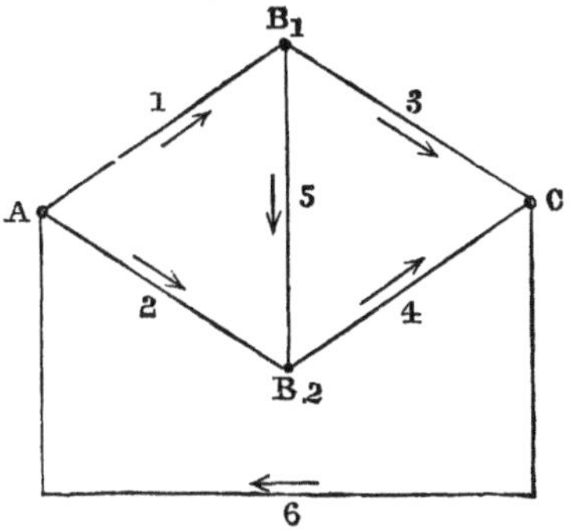

Deferring mutual induction for the present, though not confining self-induction to be of the magnetic kind only, but to include electrostatic if required, the condition of conjugacy is that the potentials at B_1 and B_2 be always equal. Therefore

$$V_1 = V_2, \quad \text{and} \quad V_3 = V_4; \quad (20c)$$

so, if $V = ZC$,

$$Z_1C_1 = Z_2C_2, \quad \text{and} \quad Z_3C_3 = Z_4C_4. \quad \text{...............(21}c\text{)}$$

But, by continuity, $C_1 = C_3$, and $C_2 = C_4$ at every moment (including equality of all their differential coefficients); so that (21c) becomes

$$Z_1C_1 = Z_2C_2, \qquad Z_3C_1 = Z_4C_2; \quad \text{...................(22}c\text{)}$$

consequently

$$Z_1Z_4 - Z_2Z_3 = 0 = f \quad \text{.........................(23}c\text{)}$$

is the complex condition of conjugacy. This function is the f of the previous investigation.

When the self-induction is of the magnetic kind, $Z = R + Lp$; so that, arranging f in powers of p,

$$0 = (R_1R_4 - R_2R_3) + (R_1L_4 + R_4L_1 - R_2L_3 - R_3L_2)p + (L_1L_4 - L_2L_3)p^2. \quad (24c)$$

Therefore, if $x = L/R$, the time-constant of a branch, we have three conditions to satisfy, namely,

$$R_1R_4 = R_2R_3, \quad \text{.................................(25}c\text{)}$$

$$x_1 + x_4 = x_2 + x_3, \quad \text{.................................(26}c\text{)}$$

$$L_1L_4 = L_2L_3. \quad \text{.................................(27}c\text{)}$$

"If the first condition is fulfilled, there will be no final current in 5 when a steady impressed force is put in 6. This is the condition for a true resistance-balance.

"If, in addition to this, the second condition is also satisfied, the integral extra-current in 5 on making or breaking 6 is zero, besides the steady current being zero; (25c) and (26c) together therefore give an approximate induction-balance with a true resistance-balance.

"If, in addition to (25c) and (26c), the third condition is satisfied, the extra-current is zero at every moment during the transient state, and the balance is exact however the impressed force in 6 vary.

"Practically, take

$$R_1 = R_2, \quad \text{and} \quad L_1 = L_2; \quad \text{.......................(28}c\text{)}$$

that is, let branches 1 and 2 be of equal resistance and inductance.

Then the second and third conditions become identical; and, to get perfect balance, we need only make

$$R_3 = R_4, \quad \text{and} \quad L_3 = L_4. \quad \text{......................}(29c)$$

"This is the method I have generally used, reducing the three conditions to two, whilst preserving exactness. It is also the simplest method. The mutual induction, if any, of 1 and 2, or of 3 and 4, does not influence the balance when this ratio of equality $R_1 = R_2$ is employed (whether $L_1 = L_2$ or not).* So branches 1 and 2 may consist of two similar wires wound together on the same bobbin, to keep their temperatures equal." [Vol. II., p. 33].

Of the eight quantities, four R's and four L's, only five can be stated arbitrarily, of which not more than three may be R's, and not more than three may be L's. We may state the matter thus:—There must first be a resistance-balance. Then, if we give definite values to two of the L's, the corresponding time-constants usually become fixed, and it is required that the other two time-constants shall be equal to them; thus

$$\text{either} \quad x_1 = x_3 \quad \text{and} \quad x_2 = x_4,$$

$$\text{or else} \quad x_1 = x_2 \quad \text{and} \quad x_3 = x_4.$$

Thus the remaining two L's become usually fixed. In fact, eliminating R_4 and L_4 from (26c) by (25c) and (27c), the second condition may be written

$$(x_1 - x_2)(x_1 - x_3) = 0.$$

Suppose R_1, R_2, R_3 given, then R_4 is fixed by (25c). Two of the inductances may then be given, fixing the corresponding time-constants. If these inductances be L_1 and L_2, then we must have (unless $x_1 = x_2$)

$$x_1 = x_3, \qquad x_2 = x_4$$

But if L_1 and L_3 be given, then we require (unless $x_1 = x_3$)

$$x_1 = x_2, \qquad x_3 = x_4.$$

These two cases present a remarkable difference in one respect. The absence of current in 5 allowing us to remove 5 altogether, we see by (18c) that the differential equation of C_6 is

$$e = \left\{ Z_6 + \frac{(Z_1 + Z_3)(Z_2 + Z_4)}{Z_1 + Z_2 + Z_3 + Z_4} \right\} C_6,$$

manipulating the Z's like resistances. The absence of branch 5 thus reduces the number of free-subsidence systems to two. [In the last equation we may eliminate one of the Z's by (23c), and then again eliminate one of the remaining three L's.] Now, if we choose $x_1 = x_2$, we shall make

$$(L_1 + L_3)/(R_1 + R_3) = (L_2 + L_4)/(R_2 + R_4),$$

* The words in the ( ) should be cancelled. The independence of M_{12} and M_{34}, which is exact when $L_1 = L_2$, $L_3 = L_4$, and sensibly true when the inequalities are small, becomes sensibly untrue when the inequalities $L_1 - L_2$ and $L_3 - L_4$ are great.

or the time-constants of the two branches $1+3$ and $2+4$ equal. Then one of the p's is

$$p_1 = -\frac{R_1+R_3}{L_1+L_3};$$

and this is only concerned in the free subsidence of current in the circuit AB_1CB_2A. Consequently the second p, which is

$$p_2 = -\frac{(R_1+R_3)R_2+R_6(R_1+R_2)}{(L_1+L_3)R_2+L_6(R_1+R_2)},$$

is alone concerned in the setting-up of current by the impressed force in 6; and the current divides between AB_1C and AB_2C in the ratio of their conductances, in the variable period as well as finally. In fact, the fraction in the above equation of C_6 will be found to contain Z_1+Z_3 as a factor in its numerator and denominator, thus excluding the p_1 root, so far as e is concerned. On the other hand, if we choose $x_1=x_3$, we do not have equality of time-constants of AB_1C and AB_2C, so that there are two p's concerned, which are not those given; and the current C_6 does not, in the variable period, divide between AB_1C and AB_2C in the ratio of their conductances, but only finally.

In the above statement it was assumed that when L_1 and L_2 were chosen, it was not so as to make $x_1=x_2$. When this happens, however, it is only the ratio of L_3 to L_4 that becomes fixed, for we have $x_3=x_4$ $=$ anything.

Similarly, when L_1 and L_3 are so chosen that $x_1=x_3$, we shall have $x_2=x_4=$ anything, so that only the ratio of L_2 to L_4 is fixed.

And if L_3, L_4 be so chosen that $x_3=x_4$, then $x_1=x_2=$ anything, only fixing the ratio of L_1 to L_2. But should x_3 not $=x_4$, then we require $x_1=x_3$ and $x_2=x_4$, thus fixing L_1 and L_2.

And if L_2, L_4 be so chosen that $x_2=x_4$, then $x_1=x_3=$ anything, only fixing the ratio of L_1 to L_3. But if so that x_2 not $=x_4$, then $x_1=x_2$ and $x_3=x_4$ fix L_1 and L_3.

There are yet two other pairs that may be initially chosen, and with somewhat different results. Let it be L_1 and L_4 that are chosen; if not so as to make $x_1=x_4$, there are two ways of fixing L_2 and L_3, viz., either by $x_1=x_3$ and $x_2=x_4$, or by $x_1=x_2$ and $x_3=x_4$; but if so that $x_1=x_4$ in the first place, then they must also $=x_2=x_3$.

Similarly the choice of L_2 and L_3 so as not to make $x_2=x_3$, gives two ways of fixing L_1 and L_4, by vertical or by horizontal equality of time-constants, as before; whilst $x_2=x_3$ produces equality all round.

The special case of all four sides equal in resistance may be also noticed. Balance is given in two ways, either by horizontal or by vertical equality in the L's.

Remarks on the Practical Use of Induction Balances, and the Calibration of an Inductometer.

Leaving the mathematical treatment for a little while, I proceed to give a short general account of my experience of induction-balances. I did not originally arrive at the method of equal-ratio just described through the general theory (20c) to (27c), but simply by means of the general principle of balancing by making one line a copy of the other, of which I obtained knowledge through duplex telegraphy, and investigated the conditions (25c) to (27c) more from curiosity than anything else, though the investigation came in useful at last. In 1881 I wished to know what practical values to give to the inductances of various electromagnets used for telegraphic purposes, and to get this knowledge went to the Christie. Not having coils of known inductance to start with, I employed Maxwell's condenser-method,* with an automatic intermitter and telephone. Let 1, 2, and 3 be inductionless resistances, and 4 a coil having self-induction. Put the telephone in 5, the battery and intermitter in 6. We require first the ordinary resistance-balance, $R_1R_4 = R_2R_3$. But the self-induction of the coil will cause current in 5 when 6 is made or broken. This will be completely annulled by shunting 1 by a condenser of capacity S_1, such that

$$R_1S_1 = L_4/R_4,$$

signifying that the time-constant of the coil on short-circuit and that of the condenser on short-circuit with the resistance R_1 are equal.

The method is, in itself, a good one. But the double adjustment is sometimes very troublesome, especially if the capacity of the condenser be not adjustable. For when we vary R_1, to approximate to the correct value of R_1S_1, we upset the resistance-balance, and have, therefore, to make simultaneous variations in some of the other resistances to restore it. But the method has the remarkable recommendation of giving us the value of the inductance of a coil at once in electromagnetic units.

In the course of these experiments I observed the upsetting of the resistance and induction-balance by the presence of metal in the neighbourhood of the coils, which is manifested in an exaggerated form in electromagnets with solid cores. So, having got the information I wanted in the first place, I discarded the condenser-method with its troublesome adjustments, and, to study these effects with greater ease, went to the equal-ratio method, with the assistance that I had obtained (by the condenser-method), the values of the inductances of various coils, to be used as standards.

"To use the Bridge to speedily and accurately measure the inductance of a coil, we should have a set of proper standard coils, of known inductance and resistance, together with a coil of variable inductance, *i.e.* two coils in sequence, one of which can be turned round, so as to vary the inductance from a minimum to a maximum.† The scale of this coil could be calibrated by (12a), first taking care that the resistance-balance did not require to be upset. This set of coils, in or out of circuit according to plugs, to form say branch 3, the coil to be measured to be in branch 4. Ratio of equality. Branches 1 and 2 equal. Of course inductionless, or practically inductionless, resistances are also

* Maxwell, vol. II., art. 778.

† Prof. Hughes's oddly named Sonometer will do just as well, if of suitable size and properly connected up. It is the manner of connection and use that give individuality to my inductometer.

required to get and keep the resistance-balance. The only step to this I have made (this was some years ago) . . . was to have a number of little equal coils, and two or three multiples; and get exact balance by allowing induction between two little ones, with no exact measurement of the fraction of a unit" [vol. II., p. 37].

Although rather out of order, it will be convenient to mention here that although I have not had a regular inductance-box made (the coils, if close together, would have to be closed solenoids), yet shortly after making these remarks, I returned to my earlier experiments by calibrating the scale of the coil of variable inductance. As it then becomes an instrument of precision, it deserves a name; and as it is for the measurement of induction it may, I think, be appropriately termed an Inductometer. Of course, for many purposes no calibration is needed.

I found that the calibration could be effected with ease and rapidity by the condenser-method more conveniently than by comparisons with coils. Thus, first ascertain the minimum and the maximum inductance, and that of the coils separately. Suppose the range is from 20 to 50 units (hundreds, thousands, millions, etc., of centimetres, according to the quite arbitrary size of the instrument). It will then be sufficient to find the places on the scale corresponding to 20, 21, 22, etc., 49, 50. Starting at 21, set the resistance-balance so that L_4 should be 21 units; turn the moveable coil till silence is reached, and mark the place 21. Then set the balance to suit 22, turn again till silence comes, and mark again; repeat throughout the whole range. Why this can be done rapidly is because the resistance-balance is at every step altered in the same manner. We have thus an instrument of constant resistance and variable known inductance, ranging from

$$l_1+l_2-2m_0 \qquad \text{to} \qquad l_1+l_2+2m_0,$$

if l_1 and l_2 are the separate inductances and m_0 the maximum mutual inductance. The calibration is thoroughly practical, as no table has to be referred to to find the value of a certain deflection.

I formerly chose 10^9 centim. as a practical unit of inductance, and called it a tom; the attraction this had for me arose from L toms $\div$ R ohms equalling L/R seconds of time. But it was too big a unit, and millitoms and microtoms were wanted. Another good name is mac. 10^6 centim. might be called a mac. Since Maxwell made the subject of self-induction his own, and described methods of correctly measuring it, there is some appropriateness in the name, which, as a mere name, is short and distinctive.

The two coils of the inductometer need not be equal; but it is very convenient to make them so, before calibration, by the equal-ratio method, which, of course, merely requires us to get a balance, not to measure the values. Let 1 and 2 be any equal coils; put one coil of the inductometer in 3, the other in 4, and balance. It happened by mere accident that my inductometer had nearly equal coils; so I made them quite equal, to secure two advantages. First, there is facility in calculations; next, the inductometer may be used with its coils in parallel or in sequence, as desired. When in parallel, the effective resistance and inductance are each one fourth of the sequence-values. Thus, let $V=ZC$ be the differential equation of the coils in parallel, C being the total current, and V the common potential-fall; it is easily shown that

$$Z=\frac{(r_1+l_1p)(r_2+l_2p_2)-m^2p^2}{r_1+r_2+(l_1+l_2-2m)p}, \quad \text{.................}(30c)$$

when the coils are unequal; r_1 and r_2 being their resistances, l_1 and l_2 their inductances, and m their mutual inductance in any position. Now make $r_1=r_2$, and $l_1=l_2$; this reduces Z to

$$Z=\tfrac{1}{2}r+\tfrac{1}{2}(l+m)p; \quad \text{.......................}(31c)$$

whilst, when in sequence, we have

$$Z=2r+2(l+m)p, \quad \text{.......................}(32c)$$

thus proving the property stated. We may therefore make one inductometer serve as two distinct ones, of low or high resistance.

There does not seem to be any other way of making the two coils in parallel behave as a single coil as regards external electromotive force. Any number of coils whose time-constants are equal will, when joined up in parallel, behave as a single coil of the same time-constant; but there must be no mutual induction. (This is an example of the property* that any linear combination whose parts have the same time-constant has only that one time-constant.) This seriously impairs the utility of the property, but the reservation does not apply in the case of the equal-coil inductometer.

Having got the inductometer calibrated, we may find the inductance of a given coil, or of a combination of coils in sequence, with or without mutual induction, nearly as rapidly as the resistance. Thus, 1 and 2 being equal, put the coil to be measured in 3, and the inductometer in 4. We have to make $R_3=R_4$ and $L_3=L_4$, or to get a resistance-balance, and then turn the inductometer till silence is reached, when the scale reading tells us the inductance. This assumes that L_3 lies within the range of the inductometer. If not, we may vary the limits as we please by putting a coil of known inductance in sequence with branch 3 or 4 as required, putting at the same time equal resistance in the other branch.

Or, the inductometer being in 4, and 1, 2 being inductionless resistances, put the coil to be measured in 3. If it has a larger time-constant than the inductometer's greatest, insert resistance along with it to bring the time-constants to equality. The conditions of silence are $R_1R_4=R_2R_3$ and $L_3/R_3=L_4/R_4$. Here a ratio of equality is not required. The method is essentially the same as one of Maxwell's †, and is a good one for certain purposes.

* This property supplies us with induction-balances of a peculiar kind. Let there be any network of conductors, every branch having the same time-constant. Set up current in the combination, and then remove the impressed force. During the subsidence all the *junctions* will be at the same potential, and any pair of them may consequently be joined by an external conductor without producing current in it.

† Maxwell, vol. ii. art. 757.

Or, 1 and 2 being any equal coils, put one coil of the inductometer in 6 and the other in 4, the coil to be measured being in 3. Then

$$L_3 = L_4 - 2M_{46} \qquad (33c)$$

gives the induction-balance, L_4 being here the inductance of the coil of the inductometer in 4, and M_{46} the mutual inductance of the two coils, in the position giving silence. This is known in all positions, because the scale-reading gives the value of $l_1 + l_2 + 2m$ (or else $2(l+m)$ if the coils are equal), and $l_1 + l_2$ is known. If the range is not suitable, we may, as before, insert other coils of known inductance.

There are other ways; but these are the simplest, and the equal-ratio method is preferable for general purposes. I have spoken of coils always, where inductances are large and small errors unimportant. When, however, it is a question of small inductances, or of experiments of a philosophical nature, needing very careful balancing, then the equal-ratio method acquires so many advantages as to become *the* method.

"So long as we keep to coils we can swamp all the irregularities due to leading wires, etc., or easily neutralize them, and can therefore easily obtain considerable accuracy. With short wires, however, it is a different matter. The inductance of a circuit is a definite quantity: so is the mutual inductance of two circuits. Also, when coils are connected together, each forms so nearly a closed circuit that it can be taken as such; so that we can add and subtract inductances, and localise them definitely as belonging to this or that part of a circuit. But this simplicity is, to a great extent, lost when we deal with short wires, unless they are bent round so as to make nearly closed circuits. We cannot fix the inductance of a straight wire, taken by itself. It has no meaning, strictly speaking. The return-current has to be considered. Balances can always be got, but as regards the interpretation, that will depend upon the configuration of the apparatus.

"Speaking with diffidence, having little experience with short wires, I should recommend 1 and 2 to be two equal wires, of any convenient length, twisted together, joined at one end, of course slightly separated at the other, where they join the telephone-wires, also twisted. The exact arrangement of 3 and 4 will depend on circumstances. But always use a long wire rather than a short one (experimental wire). If this is in branch 4, let branch 3 consist of the standard coils (of appropriate size), and adjust *them*, inserting, if necessary, coils in series with 4 also. Of course I regard the matter from the point of view of getting easily interpretable results" [vol. II. p. 37].

Some Peculiarities of Self-Induction Balances. Inadequacy of S.H. Variations to represent Intermittences.

Consider the equations (24*c*) to (27*c*). *Three* conditions have to be satisfied, in general, the resistance-balance (25*c*) and the balance of integral extra-current (26*c*) not being sufficient. To illustrate this in a simple manner, let 2 and 3 be equal coils, by previous adjustment, and 1 and 4 coils having the same resistance as the others, but of lower inductance, or else two coils whose total resistance in sequence is that of each of the others, but of lower inductance when separated. The resistance-balance is satisfied, of course. Now, if the next condition were sufficient to make an induction-balance, all we should have to do would be to make $L_1 + L_4 = 2L_3$. For instance, if L_1 is first adjusted to equal L_2 and L_3, then, by increasing either L_1 or L_4 to the right amount, silence would result. It does result when it is L_4 that is increased, but not when it is L_1. If the sound to be quenched is slight, the residual sound in the L_1 case is feeble and might be overlooked; but if it be loud, then the residual sound in the L_1 case is loud and is comparable with that to be destroyed, whilst in the L_4 case there is perfect silence.

The reason of this is that in the L_1 case we satisfy only the second condition, whilst in the L_4 case we satisfy the third as well.

Another way to make the experiment is to make 1, 2, and 3 equal, and 4 of the same resistance but of lower inductance—much lower. Then the insertion of a non-conducting iron core in 1 will lead to a loud minimum sound, but if put in 4 will bring us to silence, except as regards something to be mentioned later.

Supposing, however, we should endeavour to get silence by operating upon L_1, although we cannot do it exactly, yet by destroying the resistance-balance we may approximate to it. Thus we have a false resistance- and a false induction-balance, and the question would present itself, If we were to wilfully go to work in this way in the presence of exact methods, how should we interpret the results? As neither (25*c*) nor (26*c*) is true, it is suggested that we make use of the formula based upon the assumption that the currents are sinusoidal or pendulous, or S.H. functions of the time. Take $p^2 = -n^2$ in (24*c*), the frequency being $n/2\pi$, and we find

$$R_1R_4(x_1 + x_4) = R_2R_3(x_2 + x_3), \quad \ldots\ldots\ldots\ldots\ldots\ldots\ldots\ldots \quad (34c)$$

$$(R_1R_4 - R_2R_3) = n^2(L_1L_4 - L_2L_3) \quad \ldots\ldots\ldots\ldots\ldots\ldots\ldots (35c)$$

are the two conditions to be satisfied; and we can undoubtedly, if we take enough trouble, correctly interpret the results, if the assumption that has been made is justifiable.

I should have been fully inclined to admit (and have no doubt it is sometimes true) that, with an intermitter making regular vibrations, we might regard the residual sound as due to the upper partials, and that $n/2\pi$ could be taken as the frequency of the intermitter, and (34*c*), (35*c*) employed safely, though not with any pretensions to minute accuracy, if circumstances compelled us to ignore the exact methods of true balances, were it not for the fact that this hypothesis sometimes leads to utterly absurd results when experimentally tested. Of this I will give an illustration, and, as we have only to test that intermittences may be regarded as S.H. reversals, simplify by taking $R_1 = R_2$, $L_1 = L_2$, which makes an exact equal-ratio balance, $R_3 = R_4$, $L_3 = L_4$.

Since a steady or slowly varying current does not produce sound in the telephone, if a battery could be treated as an ordinary conductor, we could put it in one of the sides of the quadrilateral and balance it, just like a coil, in spite of its electromotive force. So, let 1 and 2 be equal coils, 3 the battery to be tested, and 4 the balancing coils. I find that a good battery can be very well balanced, though not perfectly, with intermittences, as regards resistance, which is, however, far less with rapid intermittences than with a steady current.* Thus: steady, 2½ ohms; intermittent (about 500), 1½ ohm. Another battery: steady, 166 ohms; intermittent, 126 ohms. The steady resistances are got by cutting out the intermitter, using a make-and break instead; the deflection of a galvanometer in 5 must be the same whether 6 is in or out. If we leave out the battery in 6, it becomes Mance's method. The sensitiveness is, however, far greater when the battery is not left out, although other effects are then produced

So far regarding the resistance. As regards the inductance, or apparent inductance, of batteries, that is, I find, usually negative. That is to say, after bringing the sound to a minimum by means of resistance-adjustment, the residual sound (sometimes considerable) may be quenched by inserting equal coils in branches 3 and 4, and then increasing the inductance of the one containing the battery under test. I selected the battery which showed the greatest negative inductance, about ½ mac, or 500,000 centim., got the best possible silence by adjustment of resistance and inductance, and then found the residual sound could be nearly quenched by allowing induction between the coil in 3 and a silver coin, provided, at the same time, R_4 were a little increased.

It was naturally suggested by the negative inductance and lower resistance that the battery behaved as a shunted condenser, or as a shunted condenser with resistance in sequence, or something similar; and I examined the influence of the frequency on the values of the effective resistance and inductance. The change in the latter was uncertain, owing to the complex balancing, but the apparent resistance was notably increased by increasing the frequency, viz., from 125 to 130 ohms, when the frequency was raised from about 500 to about 800, whilst there was a small reduction in the amount of the negative inductance. The effect was distinct, under various changes of frequency, but was the opposite (as regards resistance) of what I expected on the S.H. assumption. To see whereabouts the minimum apparent resistance was (being 165 steady), I lowered the frequency by steps. The resistance went down to 113 with a slow rattle, and so there was no minimum at all. The S.H. assumption had not the least application to the apparent resistance, as regards the values 165 steady, 113 slow intermittences, although it no doubt is concerned in the rise from 113 to 130 at frequency 800. The balance (approximate) was some complex compromise, but was principally due to a vanishing of the integral extra-current. Of course in such a case as this we should employ a strictly S.H. impressed force; a remark that applies more or less in all cases where the combination tested does not behave as a mere coil of constant R and L.

* I am aware that Kohlrausch employs the telephone with intermittences to find the resistance of electrolytes, but have no knowledge of how he gets at the true resistance.

The other effects, due to using a battery in branch 6 as well, are complex. It made little difference when the current in the cell was in its natural direction; but on reversal (by reversing the battery in 6) there was a rapid fall in the resistance—for instance, from 46 ohms to 18 ohms in half a minute in the case of a rather used-up battery, but a comparatively small fall when the battery was good.

Besides the advantage of independence of the manner of variation of the impressed force (in all cases where the resistance and inductance do not vary with the frequency), and the great ease of interpretation, the equal-ratio method gives us independence of the mutual induction of 1 and 2 and of 3 and 4; and this, again, leads to another advantage of an important kind. If the arrangement is at all sensitive, the balance will continually vary, on account of temperature inequalities occurring in experimenting, caused by the breath, heat of hands, lamps, etc. Now, if the four sides of the quadrilateral consist of four coils, equal in pairs, it is a difficult matter to follow the temperature-changes. To restore a resistance-balance is easy enough; but more than that is needed, viz. the preservation of the ratio of equality. But, by reason of the independence of the self-induction balance of M_{12}, we may, as before mentioned, wind them together, and thus ensure their equality at every moment. There is then only left the inequality between branches 3 and 4, which must, of course, be separated for experimental purposes, and that is very easily followed and set right. When a sound comes on, holding a coin over the coil of lower resistance will quench it, if it be slight and due to resistance-inequality, and tell us which way the inequality lies. If it be louder, the cancelling will be still further assisted by an iron wire over or in the same coil, or by a thicker iron wire alone, for reasons to be presently mentioned.

On the other hand, a small inequality in the inductance may be at once detected by a fine iron wire, quenching the sound when over or in the coil of lower inductance; and when the resistance and inductance-balances are both slightly wrong, a combination of these two ways will show us the directions of departure. These facts are usefully borne in mind when adjusting a pair of coils to equality, during which process it is also desirable to handle them as little as possible, otherwise the heating will upset our conclusions and cause waste of time. But a pair of coils once adjusted to equality, and not distorted in shape afterwards, will practically keep equal in inductance; for the effect of temperature-variation on the inductance is small, compared with the resistance-change.

Regarding the intermitter, I find that it is extremely desirable to have one that will give a pure tone, free from harsh irregularities, for two reasons: first, it is extremely irritating to the ear, especially when experiments are prolonged, to have to listen to irregular noises, or grating and fribbling sounds; next, there is a considerable gain in sensitiveness when the tone is pure.*

* *I.e.*, pure in the common acceptation, not in the scientific sense of having a definite single frequency, which is only needed in a special class of cases, when no true balance could be got without it.

Disturbances produced by Metal, Magnetic and Non-magnetic. The Diffusion-Effect. Equivalence of Nonconducting Iron to Self-Induction.

Coming now to the effects of metal in the magnetic field of a coil, the matter is more easily understood from the theoretical point of view, in the first instance, than by the more laborious course of noting facts and evolving a theory out of them—a quite unnecessary procedure, seeing that we have a good theory already, and, guided by it, have merely to see whether it is obeyed and what the departures are, if any, that may require us to modify it.

First, there is the effect of inductive magnetisation in increasing the inductance of a coil. Diamagnetic decrease is quite insensible, or masked by another effect, so that we are confined to iron and the other strongly magnetic bodies. The foundation of the theory is Poisson's assumption (no matter what his hypothesis underlying it was) that the induced magnetisation varies as the magnetic force; and when this is put into a more modern form, we see that impressed magnetic force is related to a flux, the magnetic induction, through a specific quality, the inductivity, in the same manner as impressed electric force is related to electric conduction-current through that other specific quality, the conductivity of a body. Increasing the inductivity in any part of the magnetic field of a coil, therefore, always increases the inductance L, or the amount of induction through the coil per unit current in it, and the magnetic energy, $\frac{1}{2}LC^2$. The effect of iron therefore is, in the steady state, merely to increase the inductance of a coil, without influence on its resistance. I have, indeed, speculated [vol. I. p. 441] upon the existence of a magnetic conduction-current, which is required to complete the analogy between the electric and magnetic sides of electromagnetism; but whilst there does not appear to be any more reason for its existence than its suggestion by analogy, its existence would lead to phenomena which are not observed.

But this increase of L by a determinable amount—determinable, that is, when the distribution of inductivity is known, on the assumption that the only electric current is that in the coil—breaks down when there are other currents, connected with that in the coil, such as occur when the latter is varying, the induced currents in whatever conducting matter there may be in the field. L then ceases to have any definite value. But in one case, that of S.H. variation, the mean value of the magnetic energy becomes definite, viz., $\frac{1}{4}L'C_0^2$, where L' is the effective L, and C_0 the amplitude of the coil-current, the change from $\frac{1}{2}$ to $\frac{1}{4}$ being by reason of the mean of the square of a sine or cosine being $\frac{1}{2}$. There must be this definiteness, because the variation of the coil-current is S.H., as well as that of the whole field. That L' is less than L, the steady value, may be concluded in a general though vague manner from the opposite direction of an induced current to that of an increasing primary, and its magnetic field in the region of the primary; or, more distinctly, from the power of conducting-matter to temporarily exclude magnetic induction.

In a similar manner, the resistance of a coil, if regarded as the R in

RC^2, the Joulean generation of heat per second, ceases to have a definite value when the current is varying, if C be taken to be the coil-current, on account of the external generation of heat. But in the S.H. case, as before, the mean value is necessarily a definite quantity (at a given frequency), making $\frac{1}{2}R'C_0^2$ the heat per second, where R' is the effective resistance. That R' is always greater than R is certain and obvious without mathematics; for the coil-heat is $\frac{1}{2}RC_0^2$, and there is the external heat as well. It is suggested that, in a similar manner, a non-mathematical and equally clear demonstration of the reduction of L is possible. The magnetic energy of the coil-current alone is $\frac{1}{4}LC_0^2$, and we have to show non-mathematically, but quite as clear as in the argument relating to the heat, that the existence of induced external current reduces the energy, without any reference to a particular kind of coil or kind of distribution of the external conductivity. Perhaps Lord Rayleigh's dynamical generalisation * might be made to furnish what is required.

When the matter is treated in an inverse manner, not regarding electric current as causing magnetic force, but as caused by or being an affection of the magnetic force, there is some advantage gained, inasmuch as we come closer to the facts as a whole, apart from the details relating to the reaction on the coil-current. Magnetic force, and with it electric current, a certain function of the former, are propagated with such immense rapidity through air that we may, for present purposes, regard it as an instantaneous action. On the other hand, they are diffused through conductors in quite another manner, quite slowly in comparison, according to the same laws as the diffusion of heat, allowing for their being vector magnitudes, and for the closure of the current, thus producing lateral propagation. The greater the conductivity and the inductivity, the slower the diffusion. Hence a conductor brought with sufficient rapidity into a magnetic field is, at the first moment, only superficially penetrated by the magnetic disturbance to an appreciable extent; and a certain time—which is considerable in the case of a large mass of metal, especially copper, by reason of high conductivity, and more especially iron, by reason of high inductivity more than counteracting the effect of its lower conductivity—is required before the steady state is reached, in which the magnetic field is calculable from the coil-current and the distribution of inductivity. And hence, a sufficiently rapidly oscillatory impressed force in the coil-circuit induces only superficial currents in a piece of metal in the field of the coil, the interior being comparatively free from the magnetic induction.

The same applies to the conductor forming the coil-circuit itself; it, also, may be regarded as having the magnetic disturbance diffused into its interior from the boundary, and we have only to make the coil-wire thick enough to make the effect of the approximation to surface-conduction experimentally sensible. But in common fine-wire coils it may be wholly ignored, and the wires regarded as linear circuits. There is no distinction between the theory for magnetic and for non-magnetic conductors; we pass from one to the other by changing the values of

* *Phil. Mag.*, May, 1886.

the two constants, conductivity and inductivity. Nor is there any difference in the phenomena produced, if the steady state be taken in each case as the basis of comparison. But, owing to copper having practically the same inductivity as air, there seems to be a difference in the theory which does not really exist.

A fine copper wire placed in one (say in branch 3) of a pair of balanced coils in the quadrilateral, under the influence of intermittent currents, produces no effect on the balance. Its inductivity is that of the air it replaces, so that the steady magnetic-field is the same; and it is too small for the diffusion-effect to sensibly influence the balance. On the other hand, a fine iron wire, by reason of high inductivity, requires the inductance of the balancing-coil (say in 4) to be increased. The other effect is small in comparison, but quite sensible, and requires a small increase of the resistance of branch 4 to balance it. A thicker copper wire shows the diffusion-effect; and if we raise the frequency and increase the sensitiveness of the balance, its thickness may be decreased as much as we please, if other things do not interfere, and still show the diffusion-effect. If thick, so that the disturbance is considerable, the approximate balancing of it by change of resistance is insufficient, and the inductance of coil 4 requires a slight decrease, or that of 3 a slight increase. A thick iron wire shows both effects strongly: the inductance and the resistance of branch 4 must be increased. These effects are greatly multiplied when big cores are used; then the balancing, with intermittences, at the best leaves a considerable residual sound. The influence of pole-pieces and of armatures outside coils in increasing the inductance, which is so great in the steady state, becomes relatively feeble with rapid intermittences. This will be understood when the diffusion-effect is borne in mind.

If the metal is divided so that the main induced conduction-currents cannot flow, but only residual minor currents, we destroy the diffusion-effect more or less, according to the fineness of the division, and leave only the inductivity effect. In my early experiments I was sufficiently satisfied by finding that the substitution of a bundle of iron wires for a solid iron core, with a continuous reduction in the diameter of the wires, reduced the diffusion-effect to something quite insignificant in comparison with the effect when the core was solid, to conclude that we had only to stop the flow of currents to make iron, under weak magnetising forces, behave merely as an inductor. More recently, on account of some remarks of Prof. Ewing on the nature of the curve of induction under weak forces, I immensely improved the test by making and using nonconducting cores, containing as much iron as a bundle of round wires of the same diameter as the cores. I take the finest iron filings (siftings) and mix them with a black wax in the proportion of 1 of wax to 5 or 6 of iron filings by bulk. After careful mixture I roll the resulting compound, when in a slightly yielding state, under considerable pressure, into the form of solid round cylinders, somewhat resembling pieces of black poker in appearance. ($\frac{1}{2}$ inch diameter, 4 to 6 inches long.) That the diffusion-effect was quite gone was my first conclusion. Next, that there was a slight effect, though of doubtful amount and character. The resistance-balance had to be very carefully attended to. But, more recently, by using coils containing a much greater number of windings, and thereby increasing the sensitiveness considerably, as well as the magnetising force, I find there is a distinct effect of the kind required. Though small, it is much greater than the least effect that might be detected; but whether it should be ascribed to the cause mentioned or to other causes, as dissipation of energy due to variations in the intrinsic magnetisation, or to slight curvature in the line of induction, so far as the quasi-elastic induction is concerned, is quite debateable. To show it, let 1 and 2 be equal coils wound together ($L=3$ macs, $R=47$ ohms), 3 and 4 equal in resistance ($R_3=R_4=93$ ohms), but of very unequal inductances, that of coil 3 ($L_3=24$ macs) being so much greater than that of coil 4 that the iron core must be fully inserted in the latter to make $L_4=L_3$. (Coils 3 and 4; $1\frac{1}{4}$ inch external, $\frac{1}{2}$ inch internal diameter, and $\frac{3}{8}$ inch in depth. Frequency 500.) The balancing of induction is completed by means of an external core. Resistance of branch 6 a few ohms, E.M.F. 6 volts. There is, of course, an immense sound in the telephone when the core is out of coil 3, but when it is in, there is merely a faint residual sound, which is nearly destroyed by increasing R_3 by about $\frac{1}{600}$ part, a relatively considerable change. On the other hand, pure self-induction of copper wires gives perfect silence, and so does M_{64}, a method I have shown to be exact [vol. II., p. 38]. (I may, however, here mention that in experiments with mere fine copper-wire coils there are sometimes to be found traces of variations of resistance-balance with the frequency of intermittence, of very small amount, and difficult to elucidate owing to temperature-variations.) Balancing partly by M_{64}, and partly by the iron cores, the residual sound increases from zero with M_{64} only, to the maximum with the cores only. Halving the strength of current upsets the induction-balance in this way:—the auxiliary core must be set a little closer when the current is reduced. This would indicate a slightly lower inductivity with the smaller magnetising force, and proves slight curvature in the line of induction. But, graphically represented, it would be invisible except in a large diagram.

It is confidently to be expected, from our knowledge of the variation of μ, that when the range of the magnetising force is made much greater, the ability of nonconducting iron to act merely as an increaser of inductance will become considerably modified, and that the dissipation of energy by variations in the intrinsic magnetisation will cease to be insensible. But, so far as weak magnetising oscillatory forces are concerned, we need not trouble ourselves in the least about minute effects due to these causes. Under the influence of regular intermittences, the iron gets into a stationary condition, in which the variations in the intrinsic magnetisation are insensible. It seems probable that μ must have a distinctly lower value under rapid oscillations than when they are slow. The values of μ calculated from my experiments on cores have been usually from 50 to 200, seldom higher. I should state that I define μ to be the ratio $\mathbf{B}/\mathbf{H}$, if $\mathbf{B}$ is the induction and $\mathbf{H}$ the magnetic force, which is to include $\mathbf{h}$, the impressed force of intrinsic magnetisation. (See the general equations in Part I.) It is with this μ, not with the ratio of the induction to the magnetising force as ordinarily understood, that we are concerned with in experiments of the present kind.

Inductance of a Solenoid. The Effective Resistance and Inductance of Round Wires at a given Frequency, with the Current Longitudinal; and the Corresponding Formulæ when the Induction is Longitudinal.

Knowing, then, that iron when made a nonconductor acts merely as an inductor, when we remove the insulation and make the iron a solid mass, it requires to be treated as both a conductor and inductor, just like a copper mass, in fact, of changed conductivity and inductivity. When the coil is a solenoid whose length is a large multiple of its diameter, and the core is placed axially, the phenomena in the core become amenable to rigorous mathematical treatment in a comparatively simple manner.

In passing, I may mention that on comparing the measured with the calculated value of the inductance of a long solenoid according to Maxwell's formula (vol. II., art. 678, equations (21) and (23)) in the first edition of his treatise, I found a far greater difference than could be accounted for by any reasonable error in the ohm (reputed) or in the capacity of the condenser, and therefore recalculated the formula. The result was to correct it, and reduce the difference to a reasonable one. On reference to the second edition (not published at the time referred to) I find that the formula has been corrected. I will therefore only give my extension of it. Let M be the mutual inductance of two long coaxial solenoids of length l, outer diameter c_2, inner c_1, having n_1 and n_2 turns per unit length. Then

$$M = 4\pi^2 n_1 n_2 c_2^2 (l - 2ac_1), \quad \ldots\ldots\ldots\ldots\ldots\ldots(46c)$$

where, if $\rho = c_1/c_2$,

$$2a = 1 - \frac{\rho}{8}\Big(1 + \frac{\rho}{8}\Big(1 + \frac{5\rho}{16}\Big(1 + \frac{7\rho}{16}\Big(1 + \frac{21\rho}{40}\Big(1 + \frac{33\rho}{56} + \ldots \quad \ldots\ldots(47c)$$

When

$$c_1 = c_2, \qquad 2a = 1 - \cdot 149 = \cdot 851.$$

As regards Maxwell's previous formula (22), art. 678, however, there is disagreement still.

References to authors who have written on the subject of induction of currents in cores other than, and unknown to, and less comprehensively than, myself, are contained in Lord Rayleigh's recent paper.* So far as the effect on an induction-balance is concerned, when oscillatory currents are employed, it is to be found, as he remarks, by calculating the reaction of the core on the coil-current. This I have fully done in my article on the subject. Another method is to calculate the heat in the core, to obtain the increased resistance. This I have also done. When the diffusion-effect is small, its influence on the amplitude and

* *Phil. Mag.*, December, 1886.

phase of the coil-current is the same as if the resistance of the coil-circuit were increased from the steady value R to [vol. I., p. 369]

$$\left.\begin{aligned} R' &= R + \tfrac{1}{2}L_1\pi\mu k n^2 c^2 \\ &= R + 2l\pi k(\pi N c^2 \mu n)^2 = R + R_1 \quad \text{say.} \end{aligned}\right\} \qquad (48c)$$

"Many phenomena which may be experimentally observed when rods are inserted in coils may be usefully explained in this manner." Here μ and k are the inductivity and conductivity of the core, of length l, the same as that of the coil, $n/2\pi$ the frequency, c the core's radius, and N the number of turns of wire in the coil per unit length; whilst

$$L_1 = (2\pi N c)^2 \mu l$$

is that part of the steady inductance of the coil-circuit which is contributed by the core.

The full expression for the increased resistance due to the dissipation of energy in the core is to be got by multiplying the above R_1 by Y, which is given by [vol. I., p. 364]

$$Y = \frac{1 + \dfrac{y}{6.8^2}\left(1 + \dfrac{y}{2.10.12^2}\left(1 + \dfrac{y}{3.14.16^2}\left(1 + \dfrac{y}{4.18.20^2}\left(1 + \dots\right.\right.\right.\right.}{1 + \dfrac{y}{2.4^2}\left(1 + \dfrac{y}{2.6.8^2}\left(1 + \dfrac{y}{3.10.12^2}\left(1 + \dfrac{y}{4.14.16^2}\left(1 + \dots\right.\right.\right.\right.}, \qquad (49c)$$

where $y = (4\pi\mu k n c^2)^2$. The value of R' is therefore $R + R_1 Y$. The series being convergent, the formula is generally applicable. The law of the coefficients is obvious. I have slightly changed the arrangement of the figures in the original to show it. We may easily make the core-heat a large multiple of the coil-heat, especially in the case of iron, in which the induced currents are so strong. When y is small enough, we may use the series obtained by division of the numerator by the denominator in (49c), which is

$$Y = 1 - \frac{11y}{16.24} + \frac{11.43y^2}{15.16^3.9} - \dots \qquad (50c)$$

Corresponding to this, I find from my investigation [vol. I., p. 370] of the phase-difference, that the decrease of the effective inductance from the steady value is expressed by

$$L_1 \times \frac{y}{48}\left(1 - \frac{19y}{16.40} + \frac{229y^2}{16^3.63} + \dots\right). \qquad (51c)$$

When the same core is used as a wire with current longitudinal, and again as core in a solenoid with induction longitudinal, the effects are thus connected. Let L_1 be the above steady inductance of the coil so far as is due to the core, and L_1' its value at frequency $n/2\pi$, when it also adds resistance R_1' to the coil. Also let R_2 be the steady resistance of the same when used as a wire, and R_2' and L_2' its resistance and inductance at frequency $n/2\pi$, the latter being what $\frac{1}{2}\mu$ then becomes. Then

$$\left.\begin{aligned} 4\pi\mu N^2 l^2/k = R_2 L_1 &= R_1' L_2' + R_2' L_1', \\ R_1' R_2' &= L_1' L_2' n^2. \end{aligned}\right\} \qquad (52c)$$

I did not give any separate development of the L_1' of the core, corresponding to (48*c*) and (49*c*) above for R', but merged it in the expression for the tangent of the difference in phase between the impressed force and the current in the coil-circuit. The full development of L_1' is

$$\frac{L_1'}{L_1}=\frac{1+\frac{y}{6.16}\left(1+\frac{y}{2^3.10.16}\left(1+\frac{y}{3^3.14.16}\left(1+\frac{y}{4^3.18.16}\left(1+\ldots\right.\right.\right.\right.}{\text{same denominator as in (49}c)}.$$

The high-frequency formulæ for R_1' and L_1' are

$$R_1'=L_1'n=\frac{L_1 n}{(2z)^{\frac{1}{2}}},$$

if $y=16z^2$ When z is as large as 10, this gives

$$R_1'=L_1'n=\cdot2234\ L_1 n,$$

whereas the correct values by the complete formulæ are

$$R_1'=198\ L_1 n,\qquad L_1'=\cdot225\ L_1.$$

It is therefore clear that we may advantageously use the high-frequency formulæ when z is over 10, which is easily reached with iron cores at moderate frequencies.

The corresponding fully developed formulæ for R_2' and L_2', when the current is longitudinal, are

$$\frac{R_2'}{R_2}=\frac{1+\frac{y}{6.16}\left(1+\frac{y}{2^3.10.16}\left(1+\frac{y}{3^3.14.16}\left(1+\ldots\right.\right.\right.}{1+\frac{y}{2.6.16}\left(1+\frac{y}{3.2^2.10.16}\left(1+\frac{y}{4.3^2.14.16}\left(1+\ldots\right.\right.\right.},$$

showing the laws of formation of the terms, and

$$\frac{L_2'}{\frac{1}{2}\mu}=\frac{1+\frac{y}{2^2.6.16}\left(1+\frac{y}{2.3^2.10.16}\left(1+\frac{y}{3.4^2.14.16}\left(1+\ldots\right.\right.\right.}{\cdots\cdots\cdots\cdots\cdots\cdots\cdots\cdots},$$

the denominator being as in the preceding formula. At $z=10$, or $y=1600$, these give

$$R_2'=2{\cdot}507\ R_2,\qquad L_2'=\tfrac{1}{2}\mu\times\cdot442\ ;$$

whereas Lord Rayleigh's high-frequency formulæ, which are

$$R_2'=L_2'n=R_2(\tfrac{1}{2}z)^{\frac{1}{2}},$$

make $$R_2'=2{\cdot}234\ R_2,\qquad L_2'=\tfrac{1}{2}\mu\times 447$$

This particular frequency makes the amplitude of the magnetic force in the case of the core, and of the electric current in the other case, fourteen times as great at the boundary as at the axis of the wire or core (see Part I.). As, however, we do not ordinarily have very thick wires for use with the current longitudinal, the high-frequency formulæ are not so generally applicable as in the case of cores, which may be as thick as we please, whilst by also increasing the number of windings the core-heating per unit amplitude of coil-current may be greatly increased.

If the core is hollow, of inner radius c_0, else the same, the equation of the coil-current is, if e be the impressed force and C the current in the coil-circuit whose complete steady resistance and inductance are R and L, whilst L_1 is the part of L due to the core and contained hollow (dielectric current in it ignored),

$$e = RC + (L - L_1)\dot{C} + \frac{2}{sc} \cdot \frac{J_1(sc) - qK_1(sc)}{J_0(sc) - qK_0(sc)} L_1 \dot{C}, \quad \text{..........(53c)}$$

when q depends upon the inner radius, being given by

$$q = \frac{\frac{1}{2}sc_0 J_0(sc_0) - J_1(sc_0)}{\frac{1}{2}sc_0 K_0(sc_0) - K_1(sc_0)}, \quad \text{........................(54c)}$$

(whose value is zero when the core is solid), and

$$s^2 = -4\pi\mu k(d/dt).$$

There may be a tubular space between the core and coil, and R, L may include the whole circuit. In reference to this equation (53c), however, it is to be remarked that there is considerable labour involved in working it out to obtain what may be termed practical formulæ, admitting of immediate numerical calculation. The same applies to a considerable number of unpublished investigations concerning coils and cores that I made, including the effects of dielectric displacement; the analysis is all very well, and is interesting enough for educational purposes, but the interpretations are so difficult in general that it is questionable whether it is worth while publishing the investigations, or even making them.

The Christie Balance of Resistance, Permittance, and Inductance.

Leaving now the question of cores and the balance of purely magnetic self-induction, and returning to the general condition of a self-induction balance, $Z_1 Z_4 = Z_2 Z_3$, equation (23c), let the four sides of the quadrilateral consist of coils shunted by condensers. Then R, L, and S denoting the resistance, inductance, and capacity of a branch, we have

$$Z = \{Sp + (R + Lp)^{-1}\}^{-1}; \quad \text{....................(55c)}$$

so that the conjugacy of branches 5 and 6 requires that

$$\{S_1 p + (R_1 + L_1 p)^{-1}\}\{S_4 p + (R_4 + L_4 p)^{-1}\}$$
$$= \{S_2 p + (R_2 + L_2 p)^{-1}\}\{S_3 p + (R_3 + L_3 p)^{-1}\}, \quad \text{...........(56c)}$$

wherein the coefficient of every power of p must vanish, giving seven conditions, of which two are identical by having a common factor. It is unnecessary to write them out, as such a complex balance would be useless; but some simpler cases may be derived. Thus, if all the L's vanish, leaving condensers shunted by mere resistances, we have the three conditions

$$\left.\begin{aligned} R_1R_4=R_2R_3,\\ S_1/R_4+S_4/R_1=S_2/R_3+S_3/R_2,\\ S_1S_4=S_2S_3,\end{aligned}\right\}\quad\text{(57c)}$$

which may be compared with the three self-induction conditions (25c) to (27c).

If we put $RS=y$, the time-constant, the second of (57c) may be written

$$y_1+y_4=y_2+y_3,\quad\text{(58c)}$$

which corresponds to (26c) If $S_2=0=S_4$, the single condition in addition to the resistance-balance is $y_1=y_3$. If $S_1=0=S_2$, it is $y_3=y_4$.

Next, let each side consist of a condenser and coil in sequence. Then the expression for Z is

$$Z=R+Lp+(Sp)^{-1},\quad\text{(59c)}$$

which gives rise to five conditions,

$$\left.\begin{aligned} S_1S_4=S_2S_3,\qquad y_1+y_4=y_2+y_3,\\ S_1S_4(R_1R_4-R_2R_3)=L_2S_2+L_3S_3-L_1S_1-L_4S_4,\\ \frac{1}{x_1}+\frac{1}{x_4}=\frac{1}{x_3}+\frac{1}{x_2},\qquad L_1L_4=L_2L_3.\end{aligned}\right\}\quad\text{(60c)}$$

Here it looks as if the resistance-balance were unnecessary; and, as there can be no steady current, this seems a sufficient reason for its not being required. But, in fact, the third condition, by union with the others, eliminating S_3, L_3, S_4, and L_4 by means of the other four conditions, becomes

$$0=(R_1R_4-R_2R_3)\frac{S_1S_2(R_1S_1-R_2S_2)(L_1R_2-R_1L_2)-(L_2S_2-L_1S_1)^2}{(R_3S_1-R_4S_2)(L_1R_2-R_1L_2)}.\quad\text{(61c)}$$

So the obvious way of satisfying it is by the true resistance-balance. [But see, on this point, the beginning of the next Part VII.]

If there are condensers only, without resistance-shunts, we have

$$Z=(Sp)^{-1},\quad\text{(62c)}$$

so that

$$S_1S_4=S_2S_3\quad\text{(63c)}$$

is the sole condition of balance.

If two sides are resistances, R_1 and R_2, and two are condensers, S_3 and S_4, we obtain

$$R_1/R_2=S_4/S_3\quad\text{(64c)}$$

as the sole condition. The multiplication of special kinds of balance is a quite mechanical operation, presenting no difficulties.

General Theory of the Christie Balance with Self and Mutual Induction all over.

Passing now to balances in which induction between different branches is employed, suppose we have, in the first place, a true resistance-balance, $R_1R_4=R_2R_3$, but not an induction-balance, so that there is sound produced in the telephone. Then, by means of small test-coils placed in the different branches, we find that we may reduce the sound to a minimum in a great many ways by allowing induction between different branches. If the sound to be destroyed is feeble, we may think that we have got a true induction-balance; but if it is loud, then the minimum sound is also loud, and may be comparable to the original in intensity. We may also, by upsetting the resistance-balance by trial, still further approximate to silence, and it may be a very good silence, with a false resistance-balance. The question arises, Can these balances, or any of them, be made of service and be as exact as the previously described exact balances? and are the balances easily interpretable, so that we may know what we are doing when we employ them?

There are fifteen M's concerned, and therefore fifteen ways of balancing by mutual induction when only two branches at a time are allowed to influence one another, and in every case three conditions are involved, because there are three degrees of current-freedom in the six conductors involved. Owing to this, and the fact that in allowing induction between a pair of branches we use only one condition (*i.e.* giving a certain value to the M concerned), whilst the resistance-balance makes a second condition, I was of opinion, in writing on this subject before [vol. II., p. 35], that all the balances by mutual induction, using a true resistance-balance, were imperfect, although some of them were far better than others. Thus, I observed experimentally that when a ratio of equality ($R_1 = R_2$, $L_1 = L_2$) was taken, the balances by means of M_{63} or M_{64} were very good, whilst that by M_{65} was usually very bad, the minimum sound being sometimes comparable in intensity to that which was to be destroyed.

I investigated the matter by direct calculation of the integral extra-current in branch 5 arising on breaking or making branch 6, due to the momenta of the currents in the various branches, making use of a principle I had previously deduced from Maxwell's equations [vol. I., p. 105], that when a coil is discharged through various paths, the integral current divides as in steady flow, in spite of the electromotive forces of induction set up during the discharge. This method gives us the second condition of a true balance.

But more careful observation, under various conditions, showing a persistent departure from the true resistance-balance in the M_{65} method (due to Professor Hughes), and that the M_{63} and M_{64} methods were persistently good and were not to be distinguished from true balances, led me to suspect that the second and third conditions united to form one condition when a ratio of equality was used (just as in (28*c*), (29*c*) above) in the M_{63} and M_{64} methods, but not in the M_{65} method. So I did what I should have done at the beginning; investigated the differential equations concerned, verified my suspicions, and gave the results in a Postscript [vol. II., p. 38]. I have since further found that, when using the only practical method of equal-ratio, there are no other ways than those described in the paper referred to of getting a true balance of induction by variation of a single L or M, after the resistance-balance has been secured. This will appear in the following investigation, which, though it may look complex, is quite mechanical in its simplicity.

Write down the equations of electromotive force in the three circuits $6+1+3$, $1+5-2$, and $3-4-5$, when there is impressed force in branch 6 only. They are (p standing for d/dt),

$$\left.\begin{aligned}
e_6 &= (R_6+L_6p)C_6+(R_1+L_1p)C_1+(R_3+L_3p)C_3\\
&\quad +p(M_{61}C_1+M_{62}C_2+M_{63}C_3+M_{64}C_4+M_{65}C_5)\\
&\quad +p(M_{12}C_2+M_{13}C_3+M_{14}C_4+M_{15}C_5+M_{16}C_6)\\
&\quad +p(M_{31}C_1+M_{32}C_2+M_{34}C_4+M_{35}C_5+M_{36}C_6),\\
0 &= (R_1+L_1p)C_1+(R_5+L_5p)C_5-(R_2+L_2p)C_2\\
&\quad +p(M_{12}C_2+M_{13}C_3+M_{14}C_4+M_{15}C_5+M_{16}C_6)\\
&\quad +p(M_{51}C_1+M_{52}C_2+M_{53}C_3+M_{54}C_4+M_{56}C_6)\\
&\quad -p(M_{21}C_1+M_{23}C_3+M_{24}C_4+M_{25}C_5+M_{26}C_6),\\
0 &= (R_3+L_3p)C_3-(R_4+L_4p)C_4-(R_5+L_5p)C_5\\
&\quad +p(M_{31}C_1+M_{32}C_2+M_{34}C_4+M_{35}C_5+M_{36}C_6)\\
&\quad -p(M_{41}C_1+M_{42}C_2+M_{43}C_3+M_{45}C_5+M_{46}C_6)\\
&\quad -p(M_{51}C_1+M_{52}C_2+M_{53}C_3+M_{54}C_4+M_{56}C_6).
\end{aligned}\right\} \quad \text{..........(65}c\text{)}$$

Now, eliminate C_1, C_2, C_6 by the continuity conditions

$$C_1=C_3+C_5, \qquad C_2=C_4-C_5, \qquad C_6=C_3+C_4, \quad \text{...... .(66}c\text{)}$$

giving us

$$\left.\begin{aligned}
e_6 &= X_{11}C_3+X_{12}C_4+X_{13}C_5,\\
0 &= X_{21}C_3+X_{22}C_4+X_{23}C_5,\\
0 &= X_{31}C_3+X_{32}C_4+X_{33}C_5,
\end{aligned}\right\} \quad \text{...................... (67}c\text{)}$$

where the X's are functions of p and constants. Solve for C_5. Then we see that

$$X_{21}X_{32}=X_{22}X_{31} \quad \text{..........................(68}c\text{)}$$

is the complex condition of conjugacy of branches 5 and 6. This could be more simply deduced by assuming $C_5=0$ at the beginning, but it may be as well to give the values of all the X's, although we want but four of them. Thus

$$\left.\begin{aligned}
X_{11} &= R_1+R_3+R_6+(L_1+L_3+L_6+2M_{61}+2M_{63}+2M_{31})p,\\
X_{12} &= R_6+(L_6+M_{62}+M_{64}+M_{12}+M_{14}+M_{16}+M_{32}+M_{34}+M_{36})p,\\
X_{13} &= R_1+(L_1+M_{61}-M_{62}+M_{65}-M_{12}+M_{15}+M_{31}-M_{32}+M_{35})p,\\
X_{21} &= R_1+(L_1+M_{13}+M_{15}+M_{16}+M_{53}+M_{56}-M_{21}-M_{23}-M_{26})p,\\
X_{22} &= -R_2+(-L_2+M_{12}+M_{14}+M_{16}+M_{52}+M_{54}+M_{56}-M_{24}-M_{26})p,\\
X_{23} &= R_1+R_2+R_5+(L_1+L_2+L_5+2M_{15}-2M_{25}-2M_{12})p,\\
X_{31} &= R_3+(L_3+M_{31}+M_{36}-M_{41}-M_{43}-M_{46}-M_{51}-M_{53}-M_{56})p,\\
X_{32} &= -R_4+(-L_4+M_{32}+M_{34}+M_{36}-M_{42}-M_{46}-M_{52}-M_{54}-M_{56})p,\\
X_{33} &= -R_5+(-L_5+M_{31}-M_{32}+M_{35}-M_{41}+M_{42}-M_{45}+M_{52}-M_{51})p.
\end{aligned}\right\} \text{...(69}c\text{)}$$

Now, using the required four of these in (68*c*), and arranging in powers of p, it becomes

$$0=A_0+A_1p+A_2p^2. \quad\text{.....................(70}c)$$

So $A_0=0$ gives the resistance-balance; $A_1=0$, in addition, makes the integral transient current vanish; and $A_2=0$, in addition, wipes out all trace of current.

There is also the periodic balance,

$$A_1=0, \qquad A_0=A_2n^2, \quad\text{.....................(71}c)$$

if the frequency is $n/2\pi$.

The values of A_0 and A_1 are

$$A_0=R_2R_3-R_1R_4, \quad\text{.........................(72}c)$$

$$\begin{aligned}A_1= \;& R_2L_3+R_3L_2-R_1L_4-R_4L_1\\ &+R_2(M_{31}+M_{36}-M_{41}-M_{43}-M_{46}-M_{51}-M_{53}-M_{56})\\ &+R_3(M_{24}+M_{26}-M_{12}-M_{14}-M_{16}-M_{52}-M_{54}-M_{56})\\ &+R_1(M_{32}+M_{34}+M_{36}-M_{42}-M_{46}-M_{52}-M_{54}-M_{56})\\ &+R_4(M_{21}+M_{23}+M_{26}-M_{13}-M_{16}-M_{15}-M_{53}-M_{56}). \quad\text{.........(73}c)\end{aligned}$$

In this last, let the coefficients of R_2, R_3, R_1, R_4 in the brackets be q_2, q_3, q_1, q_4. Then the value of A_2 is

$$A_2=L_2L_3-L_1L_4+L_2q_2+L_3q_3+L_1q_1+L_4q_4+q_2q_3-q_1q_4. \quad\text{...(74}c)$$

It is with the object of substituting one investigation for a large number of simpler ones that the above full expressions for A_1 and A_2 are written out.

Examination of Special Cases. Reduction of the Three Conditions of Balance to Two.

If we take all the M's as zero, we fall back upon the self-induction balance (25*c*) to (27*c*). Next, by taking all the M's as zero except one, we arrive at the fifteen sets of three conditions. Of these we may write out three sets, or, rather, the two conditions in each case besides the condition of resistance-balance, which is always the same.

All M's $=0$, except M_{36}.

$$\left.\begin{aligned}R_1R_4(x_1+x_4-x_2-x_3)&=(R_1+R_2)M_{36},\\ L_1L_4-L_2L_3&=(L_1+L_2)M_{36}.\end{aligned}\right\}\quad\text{................(75}c)$$

All M's $=0$, except M_{46}.

$$\left.\begin{aligned}R_1R_4(x_1+x_4-x_2-x_3)&=-(R_1+R_2)M_{46},\\ L_1L_4-L_2L_3&=-(L_1+L_4)M_{46}.\end{aligned}\right\}\quad\text{................(76}c)$$

As these only differ in the sign of the M, we may unite these two cases, allowing induction between 6 and 3, and 6 and 4. The two conditions will be got by writing $M_{36}-M_{46}$ for M_{36} in (75*c*).

All M's $=0$, except M_{56} (Prof. Hughes's method).

$$\left.\begin{aligned}0&=R_1R_4(x_1+x_4-x_2-x_3)+M_{56}(R_1+R_2+R_3+R_4),\\ 0&=L_1L_4-L_2L_3+M_{56}(L_1+L_2+L_3+L_4).\end{aligned}\right\}\quad\text{.........(77}c)$$

Now choose a ratio of equality, $R_1 = R_2$, $L_1 = L_2$, which is the really practical way of using induction-balances in general. In the M_{36} case the two conditions (75*c*) unite to form the single condition

$$L_4 - L_3 = 2M_{36}, \quad \text{.........................(78c)}$$

and in the M_{46} case (76*c*) unite to form the single condition

$$L_4 - L_3 = -2M_{46}. \quad \text{.........................(79c)}$$

We know already that the same occurs in the case of the simple Christie, as in (29*c*), making

$$L_4 = L_3; \quad \text{..............................(80c)}$$

so that we have three ways of uniting the second and third conditions. Now examine all the other M's, one at a time, on the same assumption, $R_1 = R_2$, $L_1 = L_2$. With M_{12} we obtain

$$(L_4 - L_3)(L_1 - M_{12}) = 0, \qquad \text{and} \qquad L_4 = L_3.$$

But $L_1 - M_{12}$ cannot vanish; so that

$$L_4 = L_3 \quad \text{...............................(81c)}$$

is the single condition. Similarly, in case of M_{43},

$$L_4 = L_3 \quad \text{...............................(82c)}$$

again. All these, (77*c*) to (82*c*), were given in the paper referred to; the last two mean that M_{12} and M_{34} have absolutely no influence on the balance of self-induction.

All the rest are double conditions. Thus, in A_1 and A_2 put $R_1 = R_2$, $R_3 = R_4$, and $L_1 = L_2$; then the two conditions are

$$\begin{aligned} 0 = L_4 - L_3 &+ (1 + R_4/R_1)(M_{14} - M_{23} + M_{51} + M_{52} + M_{53} + M_{54} + 2M_{56}) \\ &+ 2(M_{46} - M_{36}) + (1 - R_4/R_1)(M_{24} - M_{13}) + 2(R_4/R_1)(M_{16} - M_{26}); \quad (83c) \end{aligned}$$

$$\begin{aligned} 0 = L_1(L_4 - L_3) &+ L_3(M_{12} + M_{14} + M_{16} + M_{52} + M_{54} + M_{56} - M_{24} - M_{26}) \\ &+ L_4(M_{13} + M_{15} + M_{16} + M_{53} + M_{56} - M_{12} - M_{23} - M_{26}) \\ &+ L_1(M_{41} + M_{42} + M_{51} + M_{52} + M_{53} + M_{54} - M_{31} - M_{32} + 2M_{46} + 2M_{56} - 2M_{36}) \\ &+ (M_{13} + M_{15} + M_{16} + M_{53} + M_{56} - M_{21} - M_{23} - M_{26}) \\ &\qquad \times (M_{42} + M_{46} + M_{54} + M_{52} + M_{56} - M_{32} - M_{34} - M_{36}) \\ &+ (M_{41} + M_{43} + M_{46} + M_{51} + M_{53} + M_{56} - M_{31} - M_{36}) \\ &\qquad \times (M_{24} + M_{26} - M_{12} - M_{14} - M_{16} - M_{52} - M_{54} - M_{56}); \quad (84c) \end{aligned}$$

which are convenient for deriving the conditions when several M's are operative at the same time. Thus, one at a time, excepting the few already examined :—

$$M_{51} \ldots\ldots \begin{cases} 0 = L_4 - L_3 + M_{51}(1 + R_4/R_1) \\ 0 = L_4 - L_3 + M_{51}(1 + L_4/L_1) \end{cases}, \quad \text{.................(85c)}$$

$$M_{52} \ldots\ldots \begin{cases} 0 = L_4 - L_3 + M_{52}(1 + R_4/R_1) \\ 0 = L_4 - L_3 + M_{52}(1 + L_3/L_1) \end{cases}, \quad \text{.................(86c)}$$

$$M_{53} \ldots\ldots \begin{cases} 0 = L_4 - L_3 + M_{53}(1 + R_4/R_1) \\ 0 = L_4 - L_3 + M_{53}(1 + L_4/L_1) \end{cases}, \quad \text{.................(87c)}$$

$$M_{54}\dots\dots\left\{\begin{matrix}0=L_4-L_3+M_{54}(1+R_4/R_1)\\0=L_4-L_3+M_{54}(1+L_3/L_1)\end{matrix}\right\}, \quad\dots\dots\dots\dots(88c)$$

$$M_{56}\dots\dots\left\{\begin{matrix}0=L_4-L_3+2M_{56}(1+R_4/R_1)\\0=L_4-L_3+M_{56}\{2+(L_4+L_3)/L_1\}\end{matrix}\right\}, \quad\dots\dots\dots\dots(89c)$$

$$M_{16}\dots\dots\left\{\begin{matrix}0=L_4-L_3+2M_{16}R_4/R_1\\0=L_4-L_3+M_{16}(L_3+L_4)/L_1\end{matrix}\right\}, \quad\dots\dots\dots\dots(90c)$$

$$M_{26}\dots\dots\left\{\begin{matrix}0=L_4-L_3-2M_{26}R_4/R_1\\0=L_4-L_3-M_{26}(L_3+L_4)/L_1\end{matrix}\right\}, \quad\dots\dots\dots\dots(91c)$$

$$M_{13}\dots\dots\left\{\begin{matrix}0=L_4-L_3-M_{13}(1-R_4/R_1)\\0=L_4-L_3-M_{13}(1-L_4/L_1)\end{matrix}\right\}, \quad\dots\dots\dots\dots(92c)$$

$$M_{24}\dots\dots\left\{\begin{matrix}0=L_4-L_3+M_{24}(1-R_4/R_1)\\0=L_4-L_3+M_{24}(1-L_3/L_1)\end{matrix}\right\}, \quad\dots\dots\dots\dots(93c)$$

$$M_{14}\dots\dots\left\{\begin{matrix}0=L_4-L_3+M_{14}(1+R_4/R_1)\\0=L_4-L_3+M_{14}(1+L_3/L_1)-M_{14}^2/L_1\end{matrix}\right\}, \quad\dots\dots\dots\dots(94c)$$

$$M_{23}\dots\dots\left\{\begin{matrix}0=L_4-L_3-M_{23}(1+R_4/R_1)\\0=L_4-L_3-M_{23}(1+L_4/L_1)+M_{23}^2/L_1\end{matrix}\right\}. \quad\dots\dots\dots\dots(95c)$$

If we compare the two general conditions (83*c*), (84*c*), we shall see that whenever

$$q_1q_4-q_2q_3=0,$$

we may obtain the reduced forms of the conditions by adding together the values of L_3-L_4 given by every one of the M's concerned. We may therefore bracket together certain sets of the M's. To illustrate this, suppose that M_{13} and M_{24} are existent together, and all the other M's are zero. Then (92*c*) and (93*c*) give, by addition,

$$L_3-L_4=(M_{24}-M_{13})(1-R_4/R_1),$$
$$L_3-L_4=M_{24}-M_{13}+M_{13}L_4/L_1-M_{24}L_3/L_1,$$

which are the conditions required.

Similarly M_{12} and M_{34} may be bracketed. Also M_{61}, M_{62}, M_{63}, M_{64}, and M_{65}. Also M_{51}, M_{52}, M_{53}, M_{54}, and M_{56}. But M_{14} and M_{23} will *not* bracket.

Miscellaneous Arrangements. Effects of Mutual Induction between the Branches.

As already observed, the self-induction balance (28*c*), (29*c*) is independent of M_{12} and M_{34}, when these are the sole mutual inductances concerned; that is, when $R_1=R_2$, $L_1=L_2$, $R_3=R_4$, $L_3=L_4$. By (92*c*) and (93*c*) we see that independence of M_{13} and M_{24} is secured by making all four branches 1, 2, 3, 4 equal in resistance and inductance.

But it is unsafe to draw conclusions relating to independence when several coils mutually influence, from the conditions securing balance when only two of the coils at a time influence one another. Let us examine what (83*c*) and (84*c*) reduce to when there is induction between all the four branches 1, 2, 3, 4, but none between 5 and the rest or between 6 and the rest. Put all M's = 0 which have either $_5$ or $_6$ in their double suffixes, and put $L_4 = L_3$. Then we may write the conditions thus :—

$$0 = (1 + R_4/R_1)(M_{14} - M_{23}) + (1 - R_4/R_1)(M_{24} - M_{13}), \quad \text{.................(96c)}$$

$$0 = (L_1 + L_4)(M_{14} - M_{23}) + (L_1 - L_4)(M_{24} - M_{13}) + M_{23}^2 - M_{14}^2$$
$$+ (M_{24} - M_{13})(M_{34} - M_{12}) + (M_{14} - M_{23})(M_{24} + M_{13} - M_{12} - M_{34}). \quad \text{(97c)}$$

The simplest way of satisfying these is by making

$$M_{14} = M_{23} \quad \text{and} \quad M_{24} = M_{13}. \quad \text{...............(98c)}$$

If these equalities be satisfied, we have independence of M_{12} and M_{34}.

Now, if we make the four branches 1, 2, 3, 4 equal in resistance and inductance, so that in (96c) and (97c) we have $R_1 = R_4$ and $L_1 = L_4$, the first reduces to

$$0 = M_{14} - M_{23}, \quad \text{..........................(99c)}$$

so that it is first of all absolutely necessary that $M_{14} = M_{23}$, if the balance is to be preserved; whilst, subject to this, the second condition reduces to

$$0 = (M_{24} - M_{13})(M_{34} - M_{12}), \quad \text{.................(100c)}$$

so that either $M_{24} = M_{13}$, or else $M_{34} = M_{12}$. Thus there are two ways of preserving the balance when all four branches are equal, viz., $M_{14} = M_{23}$ and $M_{24} = M_{13}$, independent of the values of M_{12} and M_{34}; and $M_{14} = M_{23}$ and $M_{34} = M_{12}$, independent of the values of M_{24} and M_{13}.

The verification of these properties, (98c) and later, makes some very pretty experiments, especially when the four branches consist, not merely of one coil each, but of two or more. The meanings of some of the simpler balances are easily reasoned out without mathematical examination of the theory; but this is not the case when there is simultaneous induction between many coils, and their resultant action on the telephone-branch is required.

Returning to (96c) and (97c), the nearest approach we can possibly make to independence of the self-induction balance of the values of all the M's therein concerned, consistent with keeping wires 3 and 4 away from one another for experimental purposes, is by winding the equal wires 1 and 2 together. Then, whether they be joined up straight, which makes $M_{13} = M_{23}$ and $M_{14} = M_{24}$ identically, or reversed, making $M_{13} = -M_{23}$ and $M_{14} = -M_{24}$, we shall find that

$$M_{14} = M_{23}$$

is the necessary and sufficient condition of preservation of balance.

At first sight it looks as if M_{31} and M_{32} must cancel one another when wires 1 and 2 are reversed. But although 1 and 2 cancel on 3, yet 3 does not cancel on 1 and 2 as regards the telephone in 5. The effects are added. On the other hand, when wires 1 and 2 are straight, 3 cancels on them as regards the telephone, but 1 and 2 add their effects on 3. Similar remarks apply to the action between 4 and the equal wires 1 and 2 when straight or reversed; hence the necessity of the condition represented by the last equation.

On the other hand, M_{61} and M_{62} cancel when 1 and 2 are straight, and add their effects when they are reversed; whilst M_{51} and M_{52} cancel when 1 and 2 are reversed, and add their effects when they are straight, results which are immediately evident. But wires 1 and 2 must be thoroughly well twisted, before being wound into a coil, if it is desired to get rid of the influence of, say, M_{61} and M_{62}, when it is a coil that operates in 6, and this coil is brought near to 1 and 2.

This leads me to remark that a simple way of proving that the mutual induction between iron and copper (fine wires) is the same as between copper and copper, which is immensely more sensitive than the comparison of separate measurements of the induction in the two cases, is to take two fine wires of equal length, one of iron, the other of copper, twist them together carefully, wind into a coil, and connect up with a telephone differentially. On exposure of the double coil to the action of an external coil in which strong intermittent currents or reversals are passing, there will be hardly the slightest sound in the telephone, if the twisting be well done, with several twists in every turn. But if it be not well done, there will be a residual sound, which can be cancelled by allowing induction between the external or primary coil and a turn of wire in the telephone-circuit. A rather curious effect takes place when we exaggerate the differential action by winding the wires into a coil without twists, in a certain short part of its length. The now comparatively loud sound in the telephone may be cancelled by inserting a nonconducting iron core in the secondary coil, provided it be not pushed in too far, or go too near or into the primary coil. This paradoxical result appears to arise from the secondary coil being equivalent to two coils close together, so that insertion of the iron core does not increase the mutual inductance of the primary and secondary in the first place, but first decreases it to a minimum, which may be zero, and later increases it, when the core is further inserted. Reversing the secondary coil with respect to the primary makes no difference. Of course insertion of the core into the primary always increases the mutual inductance and multiplies the sound. The fact that one of the wires in the secondary happens to be iron has nothing to do with the effect.

Another way of getting unions of the two conditions of the induction-balance is by having branches 1 and 3 equal, instead of 1 and 2. Thus, if we take $R_1=R_3$, $L_1=L_3$, $R_2=R_4$ in A_1 and A_2, (73*c*) and (74*c*), we obtain fifteen sets of double conditions similar to those already given, out of which just four (as before) unite the two conditions. Thus, using M_{13} only, we have

$$L_2=L_4, \qquad \text{(101}c\text{)}$$

and the same if we use M_{24} only, and the same when both M_{13} and M_{24} are operative. That is, the self-induction balance is independent of M_{13} and M_{24}. This corresponds to (81*c*) and (82*c*).

The other two are M_{25} and M_{45}. With M_{25} we have

$$0=L_2-L_4-2M_{25}, \qquad \text{(102}c\text{)}$$

and with M_{45},

$$0=L_2-L_4-2M_{45}. \qquad \text{(103}c\text{)}$$

The remaining eleven double conditions corresponding to (85c) to (95c) need not be written down.

Several special balances of a comparatively simple kind can be obtained from the preceding by means of inductionless resistances, double-wound coils whose self-induction is negligible under certain circumstances, allowing us to put the L's of one, two, or three of the four branches 1, 2, 3, 4 equal to zero. We may then usefully remove the ratio-of-equality restriction if required. This vanishing of the L of a branch of course also makes the induction between it and any other branch vanish.

For instance, let $L_1 = L_2 = L_4 = 0$; then

$$0 = R_2 L_3 + M_{36}(R_1 + R_2) \quad \text{.....................(104}c\text{)}$$

gives the induction-balance when M_{36} is used, subject to $R_1 R_4 = R_2 R_3$. And

$$0 = R_2 L_3 - M_{35}(R_2 + R_4) \quad \text{.....................(105}c\text{)}$$

is the corresponding condition when M_{35} is used. But M_{56} will not give balance, except in the special case of S.H. currents, with a false resistance-balance. The method (104c) is one of Maxwell's. His other two have been already described.

In the general theory of reciprocity, it is a force at one place that produces the same flux at a second as the same force at the second place does at the first. That the reciprocity is between the force and the flux, it is sometimes useful to remember in induction-balances. Thus the above-mentioned second way of having a ratio of equality is merely equivalent to exchanging the places of the force and the vanishing flux. We must not, in making the exchange, transfer a coil that is operative. For example, in the M_{64} method (79c), there is induction between branches 6 and 4; M_{45} (equation (88c)), on the other hand, fails to give balance. But if we exchange the branches 5 and 6, it is the battery and telephone that have to be exchanged; so that we now use M_{54}, which gives silence, whilst M_{64} will not.

I have also employed the differential telephone sometimes, having had one made some five years ago. But it is not so adaptable as the quadrilateral to various circumstances. I need say nothing as to its theory, that having been, I understand, treated by Prof. Chrystal. Using a pair of equal coils, it is very similar to that of the equal-ratio quadrilateral.

PART VII.

Some Notes on Part VI. (1). *Condenser and Coil Balance.*

After my statement [p. 260, vol. II.] of the general condition of conjugacy of a pair of conductors, and the interpretation of the set of equations into which it breaks up, I stated that in cases where, by the presence of inverse powers of p, there could not be any steady current in either of the to-be conjugate conductors due to impressed voltage in the other, a true resistance-balance was still wanted to ensure conjugacy when the currents vary. I am unable to maintain this hasty generalisation. In the example I gave, equations (59*c*) to (61*c*), in which each side of the quadrilateral consists of a condenser and a coil in sequence, so that there can be no steady current in the bridge-wire, it is true that the obvious simple way of getting conjugacy is to have a true resistance-balance. The conditions may then be written

$$R_1R_4=R_2R_3, \qquad S_1S_4=S_2S_3, \qquad L_1L_4=L_2L_3; \quad \text{......(1}d\text{)}$$

and either

$$\left.\begin{matrix} x_1=x_2, \\ x_3=x_4, \end{matrix} \quad \text{and} \quad \begin{matrix} y_1=y_2, \\ y_3=y_4; \end{matrix} \quad \text{or else} \quad \begin{matrix} x_1=x_3, \\ x_2=x_4, \end{matrix} \quad \text{and} \quad \begin{matrix} y_1=y_3, \\ y_2=y_4; \end{matrix}\right\} \text{.....(2}d\text{)}$$

where R stands for the resistance and L for the inductance of a coil, S for the permittance of the corresponding condenser, x for the coil time-constant L/R, and y for the condenser time-constant RS; that is, we require either vertical or else horizontal equality of time-constants, electrostatic and magnetic, subject to certain exceptional peculiarities similar to those mentioned in connection with the self-induction balance. It is also the case that on first testing the power of evanescence of the other factor on the right of equation (61*c*), it seemed to always require negative values to be given to some of the necessarily positive quantities concerned. But a closer examination shows that this is not necessary. As an example, choose

$$\left.\begin{matrix} R_1=1, & R_2=2, & R_3=3, & R_4=10, \\ L_1=\frac{10}{7}, & L_2=5, & L_3=\frac{10}{21}, & L_4=\frac{5}{3}, \\ S_1=7, & S_2=5, & S_3=\frac{21}{29}, & S_4=\frac{15}{29}. \end{matrix}\right\} \text{......(3}d\text{)}$$

It will be found that these values satisfy the whole of equations (61*c*), and yet the resistance-balance is not established. No doubt simpler illustrations can be found. We must therefore remove the requirement of a resistance-balance when there can be no steady current, although the condition of a resistance-balance, when fulfilled, leads to the simple way of satisfying all the conditions.

(2). *Similar Systems.*

If $V=Z_1C$ be the characteristic equation of one system and $V=Z_2C$ that of a second, V being the voltage and C the current at the terminals, they are similar when

$$Z_1/Z_2=n, \quad \text{any numeric.} \quad \text{......................(4}d\text{)}$$

Here Z is the symbol of the generalised resistance of a system between its terminals, when it is, save for its terminal connexions, independent of all other systems; a condition which is necessary to allow of the form $V=ZC$ being the full expression of the relation between V and C, Z being a function of constants and of p, p^2, p^3, etc., and p being d/dt. To ensure the possession of the property (4*d*), we require first of all that one system should have the same arrangement as the other, as a coil for a coil, a condenser for a condenser, or equivalence (as, for instance, by two condensers in sequence being equivalent to one); and, next, that every resistance and inductance in the first system be n times the corresponding resistance and inductance in the second system, and every permittance in the second system be n times the corresponding one in the first.

Then, if the two systems be joined in parallel, and exposed to the same external impressed voltage at the terminals, the potentials and voltages will be equal in corresponding parts, whilst the current in any part of the second system will be n times that in the corresponding part of the first. Also the electric energy, the magnetic energy, the dissipativity, and the energy-current in any part of the second system are n times those in the corresponding part of the first.

The induction-balance got by joining together corresponding points through a telephone is, of course, far more general than the Christie balance, limited to four branches, each subject to $V=ZC$; at the same time, however, it is less general than the conditions which result when the full differential equation is worked out.*

By the above, any number of similar systems may be joined in parallel, having then equal voltages, and their currents in the ratio of the conductances. They will behave as a single similar system, the conductance of any part of which is the sum of the conductances of the corresponding parts in the real systems; and similarly for the permittances and for the reciprocals of the inductances. If, on the other hand, they be put in sequence, the resultant Z is the sum of the separate Z's, the current in all is the same, and the voltages are proportional to the resistances.

When the systems are not independent the above simplicity is lost; and I have not formulated the necessary conditions of similarity in an extended sense except in some simple cases, of which a very simple one will occur later in connexion with another matter.

(3). *The Christie Balance of Resistance, Self and Mutual Induction.*

The three general conditions of this are given in equations (72*c*) to (74*c*). If, now, we introduce the following abbreviations,

$$\left.\begin{aligned}
m_1 &= L_1+L_2+L_5+2(M_{15}-M_{12}-M_{25}),\\
m_3 &= L_3+L_4+L_5+2(M_{45}-M_{34}-M_{35}),\\
m_6 &= L_2+L_4+L_6+2(M_{62}+M_{64}+M_{34}),\\
m_{13} &= -L_5+M_{13}-M_{14}-M_{15}-M_{23}+M_{24}+M_{25}+M_{35}-M_{45},\\
m_{61} &= -L_2+M_{12}+M_{14}+M_{16}-M_{24}-M_{26}+M_{25}+M_{45}+M_{56},\\
m_{36} &= -L_4+M_{32}+M_{34}+M_{36}-M_{24}-M_{46}-M_{25}-M_{45}-M_{56},
\end{aligned}\right\} \quad (5d)$$

the conditions mentioned reduce simply to

$$\left.\begin{gathered}
R_1R_4=R_2R_3,\\
(m_1+m_{13}+m_{16})R_4-m_{36}R_1=(m_{31}+m_3+m_{36})R_2-m_{16}R_3,\\
(m_1+m_{13})m_{36}=(m_3+m_{13})m_{16}.
\end{gathered}\right\} \ldots\ldots (6d)$$

* This general property is, it will be seen, of great value in enabling us to avoid useless and lengthy mathematical investigations. In another place [p. 115, vol. II.], I have shown how to apply it to the at first sight impossible feat of balancing iron against copper.

The interpretation is, that as there are only three independent currents in the Christie arrangement, there can be only six independent inductances, viz., three self and three mutual; and these may be chosen to be the above m's, whose meanings are as follows. Let the three circuits be AB_1B_2A, CB_2B_1C, and AB_2CA in the figure, so that the currents in them are C_1, C_3, and C_6. Then m_1, m_3, and m_6 are the self, and m_{13}, m_{36}, m_{61} the mutual inductances of the three circuits.

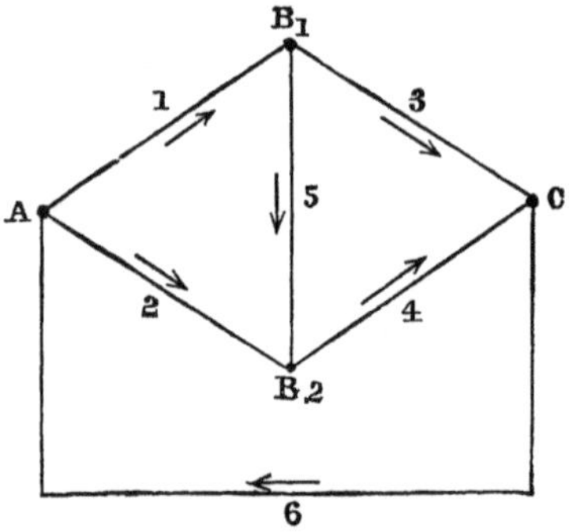

Now if the four sides of the quadrilateral consist merely of short pieces of wire, which are not bent into nearly closed curves, it is clear that (6d) are the true conditions, to which alone can definite meaning be attached; the inductance of a short wire being an indefinite quantity, depending upon the position of other wires. We may therefore start *ab initio* with only these six inductances, and immediately deduce [p. 107, vol. II.] the conditions (6d), saving a great deal of preliminary work. But, on coming to practical cases, in which the inductances do admit of being definitely localised in and between the six branches of the Christie, we have to expand the m's properly, using (5d) or as much of them as may be wanted, and so obtain the various results in Part VI. Therefore equations (6d) are only useful as a short registration of results, subject to (5d), and in the remarkably short way in which they may be got; a method which is, of course, applicable to any network, which can only have as many independent inductances as there are independent circuits, *plus* the number of pairs of the same.

(4). *Reduction of Coils in Parallel to a Single Coil.*

In Part VI. [p. 267, vol. II.], in speaking of the inductometer, I referred to the most useful property that a pair of equal coils in parallel behave as one coil to external voltage, whatever be the amount of mutual induction between them; a property which, excepting in the mention of mutual induction, I had pointed out in 1878 [p. 111, vol. I.]. But, although there appears to be no other case in which this property is true for *any* value of the mutual inductance, which is the property wanted, yet, if a special value be given to it, any two coils in parallel will be made equivalent to one.

The condition required is obviously that Z, the generalized resistance of the two coils in parallel, should reduce to the form $R+Lp$. Equation (30c) gives Z; to make the reduction possible, on dividing the denominator into the numerator, the second remainder must vanish. Performing this work, we find

$$Z=\frac{r_1r_2}{r_1+r_2}+\frac{l_1l_2-m^2}{l_1+l_2-2m}p\,; \qquad \text{.................}(7d)$$

which shows the effective resistance and inductance of the coils in parallel, r_1 and r_2 being their resistances, and l_1, l_2, m the inductances; subject to

$$\frac{r_1}{r_2}=\frac{l_1-m}{l_2-m}; \qquad \text{.............................(8}d\text{)}$$

giving a special value to m, which, if it be possible, will allow the coils to behave as one coil, so that, when put in one side of the Christie, the self-induction balance can be made. This equation (8d) is the expression of the making of coils 1 and 2 *similar*, in the extended sense, being the simple case to which I referred above. Let a unit current flow in the circuit of the two coils. Then l_1-m and l_2-m are the inductions through them, and these must be proportional to the resistances, making therefore the actual inductions through them always the same.

Similarly, if any number of coils be in parallel, exposed to the same impressed voltage V, with the equations

$$\left.\begin{aligned} V&=(r_1+l_1p)C_1+m_{12}pC_2+m_{13}pC_3+\dots,\\ V&=m_{21}pC_1+(r_2+l_2p)C_2+m_{23}pC_3+\dots,\end{aligned}\right\} \qquad \text{..............(9}d\text{)}$$

we have, by solution,

$$\left.\begin{aligned} DC_1/V&=N_{11}+N_{21}+N_{31}+\dots,\\ DC_2/V&=N_{21}+N_{22}+N_{23}+\dots,\end{aligned}\right\} \qquad \text{..................(10}d\text{)}$$

if D be the determinant of the coefficients of the C's in (9d), and N_{rs} the coefficient of m_{rs} in D. So, if $C=C_1+C_2+\dots$ be the total current, we have

$$C=V(\Sigma N)/D; \quad \text{therefore} \quad Z=D/(\Sigma N), \qquad \text{.........(11}d\text{)}$$

where the summation includes all the N's. To reduce Z to the single-coil form, we require the satisfaction of a set of conditions whose number is one less than the number of coils.

The simplest way to obtain these conditions is to take advantage of the fact that, if any number of coils in parallel behave as one, the currents in them must at any moment be in the ratio of their conductances. Then, since by (9d),

$$\left.\begin{aligned} V-r_1C_1&=p(l_1C_1+m_{12}C_2+m_{13}C_3+\dots),\\ V-r_2C_2&=p(m_{21}C_1+l_2C_2+m_{23}C_3+\dots),\\ V-r_3C_3&=p(m_{31}C_1+m_{32}C_2+l_3C_3+\dots),\\ &\dots\dots\dots\dots\dots\dots\end{aligned}\right\} \qquad \text{..............(12}d\text{)}$$

are the equations of voltage, when we introduce

$$r_1C_1=r_2C_2=r_3C_3=\dots \qquad \text{...........................(13}d\text{)}$$

into them, we obtain the required conditions:—

$$\frac{l_1}{r_1}+\frac{m_{12}}{r_2}+\frac{m_{13}}{r_3}+\dots=\frac{m_{21}}{r_1}+\frac{l_2}{r_2}+\frac{m_{23}}{r_3}+\dots=\frac{m_{31}}{r_1}+\frac{m_{32}}{r_2}+\frac{l_3}{r_3}+\dots. \qquad (14d)$$

The induction through every coil at any moment is the same in amount; also the voltage due to its variation, and the voltage supporting current, and the impressed voltage.

(5). *Impressed Voltage in the Quadrilateral. General Property of a Linear Network.*

In my remarks on [p. 271, vol. II.], relating to the behaviour of batteries when put in the quadrilateral, I, for brevity in an already long article, left out any reference to the theory. As is well known, in the usual Christie arrangement (see figure, above) the steady current in 5, due to an impressed voltage in any one of 1, 2, 3, 4, is the same whether 6 be open or closed, if a steady impressed voltage in 6 give no current in 5. But the distribution of current is not the same in the two cases; so that, when we change from one to the other, the current in 5 changes temporarily; as may be seen in making Mance's test of the resistance of a battery, or by simply measuring the resistance of the battery in the same way as if it had no E.M.F., using another battery in 6, but taking the galvanometer-zero differently. We, in either case, have not to observe the absence of a deflection; or, which is similar, the absence of any change in the deflection; but the equivalence of two deflections at different moments of time, between which the deflection changes. Hence Mance's method is not a true nul method, unless it be made one by having an induction-balance as well as one of resistance; in which case, if the battery behave as a mere coil or resistance, which is sometimes nearly true, especially if the battery be fresh, we may employ the telephone instead of the galvanometer.

The proof that the complete self-induction condition, $Z_1Z_4 = Z_2Z_3$, where the Z's stand for the generalised resistances of the four sides of the quadrilateral, when satisfied, makes the current in the bridge-wire due to impressed force in, for example, side 1, the same whether branch 6 be opened or closed, without any transient disturbance, is, formally, a mere reproduction of the proof in the problem relating to steady currents. Thus, suppose

$$C_5 = \frac{Ae_1}{B},$$

where e_1 is a steady impressed force in side 1, and A and B the proper functions of the resistances, in the case of the common Christie, but without the special condition $R_1R_4 = R_2R_3$ which makes a resistance-balance. Then we know that if we introduce this condition into A and B, the resistance R_6 can be altogether eliminated from the quotient A/B, making C_5 due to e_1 independent of R_6.

Now, in the extended problem, in which it is still possible to represent the equation of a branch by $V = ZC$, wherein Z is no longer a resistance, we have merely to write Z for R in the expansion of A/B to obtain the differential equation of C_5; and consequently, on making $Z_1Z_4 = Z_2Z_3$, we make A/B independent of Z_6. Hence, the current in the bridge-wire is independent of branch 6 altogether when the general condition of an induction-balance is satisfied, making branches 5 and 6 conjugate.

But, as is known to all who have had occasion to work out problems concerning the steady distribution of current in a network, there is a great deal of labour involved, which, when it is the special state involved in a resistance-balance, is wholly unnecessary. This remark applies with immensely greater force when the balance is to be a universal one, for transient as well as permanent currents; so that the proper course is either to assume the existence of the property required at the beginning, and so avoid the reductions from the complex general to the simple special state, or else to purposely arrange so that the reductions shall be of the simplest character. Thus, to show that C_5 is independent of branch 6, when there is an impressed voltage in (say) side 1, making no assumptions concerning the nature of branch 6, we may ask this question, Under what circumstances is C_5 independent of C_6? And, to answer it, solve for C_5 in terms of e_1 and C_6, and equate the coefficient of C_6 to zero.

Thus, writing down the equations of voltage in the circuits AB_1B_2A and $B_1CB_2B_1$ in the above figure, we have

$$\left.\begin{aligned} e_1 &= Z_1C_1 + Z_5C_5 - Z_2C_2, \\ 0 &= Z_3C_3 + Z_4C_4 - Z_5C_5, \end{aligned}\right\} \qquad (15d)$$

when there is no mutual induction between different branches, but not restricting Z to a particular form; and now putting

$$C_4 = C_6 - C_1 + C_5, \qquad C_3 = C_1 - C_5, \qquad C_2 = C_6 - C_1, \qquad (16d)$$

we obtain

$$\left.\begin{aligned} e_1 + Z_2C_6 &= (Z_1 + Z_2)C_1 + Z_5C_5, \\ Z_4C_6 &= (Z_3 + Z_4)C_1 - (Z_3 + Z_4 + Z_5)C_5; \end{aligned}\right\} \qquad (17d)$$

which give

$$C_5 = \frac{(Z_3 + Z_4)e_1 + (Z_2Z_3 - Z_1Z_4)C_6}{(Z_1 + Z_2)(Z_3 + Z_4) + Z_5(Z_1 + Z_2 + Z_3 + Z_4)}, \qquad (18d)$$

making C_5 independent of C_6 when the condition of conjugacy of branches 5 and 6 is satisfied.

If there are impressed voltages in all four sides of the quadrilateral, then (18d) obviously becomes

$$C_5 = \frac{(Z_3 + Z_4)(e_1 - e_2) - (Z_1 + Z_2)(e_3 - e_4) + (Z_2Z_3 - Z_1Z_4)C_6}{(Z_1 + Z_2)(Z_3 + Z_4) + (Z_1 + Z_2 + Z_3 + Z_4)Z_5}, \quad (19d)$$

which makes C_5 always zero if $e_1 = e_2$, $e_3 = e_4$, and $Z_1Z_4 = Z_2Z_3$. As an example, let $e_2 = 0$, $e_4 = 0$; then, if there is conjugacy of 5 and 6, and also

$$e_1/(Z_1 + Z_2) = e_3/(Z_3 + Z_4), \qquad (20d)$$

the impressed forces are also balanced. Putting, therefore, batteries in sides 1 and 3, and letting them work an intermitter in branch 6, we obtain a simultaneous balance of their resistances and voltages, and know the ratio of the latter. If self-induction be negligible, we may take Z as R, the resistance; if not negligible, it must be separately balanced.

But should there be mutual induction between different branches, this working-out of problems relating to transient states by merely turning R to Z partly fails. We may then proceed thus:—As before, write down the equations of voltage in the circuits AB_1B_2A and

CB_2B_1C, using the six independent inductances of these and of the circuit CAB_2C. Thus,

$$\left.\begin{aligned} e_1 &= R_1C_1 + R_5C_5 - R_2C_2 + p(m_1C_1 + m_{13}C_3 + m_{16}C_6), \\ 0 &= R_3C_3 - R_4C_4 - R_5C_5 + p(m_{31}C_1 + m_3C_3 + m_{36}C_6), \end{aligned}\right\} \quad \ldots\ldots (21d)$$

if there is an impressed voltage in side 1. As before, eliminate C_2, C_3, and C_4 by (16d), and we obtain

$$\left.\begin{aligned} e_1 + (R_2 - pm_{16})C_6 &= \{R_1 + R_2 + p(m_1 + m_{13})\}C_1 + (R_5 - pm_{13})C_5, \\ (R_4 - pm_{36})C_6 &= \{R_3 + R_4 + p(m_3 + m_{13})\}C_1 - (R_3 + R_4 + R_5 + pm_3)C_5, \end{aligned}\right\} \quad (22d)$$

which, by solution for C_5, gives its differential equation at once in terms of e_1 and C_6. To be independent of C_6, we require

$$(R_2 - pm_{16})\{R_3 + R_4 + p(m_3 + m_{13})\} = (R_4 - pm_{36})\{R_1 + R_2 + p(m_1 + m_{13})\}, \quad (23d)$$

which, expanded, gives us the three equations (6d) again, showing that C_5 depends upon e_1 and the nature of sides 1, 2, 3, and 4, subject to (23d), and of 5, but is independent of the nature of C_6 altogether, except in the fact that the mutual induction between branch 6 and other parts of the system must be of the proper amounts to satisfy (23d) or (6d).

The extension that is naturally suggested of this property to any network whose branches may be complex, and not independent, is briefly as follows. The equations of voltage of the branches will be of the form

$$\left.\begin{aligned} e_1 + V_1 &= Z_1C_1 + Z_{12}C_2 + Z_{13}C_3 + \ldots, \\ e_2 + V_2 &= Z_{21}C_1 + Z_2C_2 + Z_{23}C_3 + \ldots, \\ &\ldots\ldots\ldots\ldots\ldots\ldots \end{aligned}\right\} \quad \ldots\ldots\ldots\ldots (24d)$$

wherein the Z's are differentiation-operators.

Suppose branches m and n are to be conjugate, so that a voltage in m can cause no current in n. First exclude m's equation from (24d) altogether, and, with it, Z_{mm}. Then write down the equations of voltage in all the independent circuits of the remaining branches, by adding together equations (24d) in the proper order; this excludes the V's, and leaves us equations between the e's and all the independent C's, but one fewer in number than them. Put the C_m terms on the left side, then we can solve for all the currents (except C_m) in terms of C_m and the e's. That the coefficient of C_m in the C_n solution shall vanish is the condition of conjugacy, and when this happens, C_n is not merely independent of e_m but also of Z_{mm}, though not of Z_{m1}, Z_{m2}, etc.

I have dwelt somewhat upon this property, and how to prove it for transient states, because, although it is easy enough to understand how the current in one of the conjugate branches, say n, is independent of current arising from causes in the other conjugate branch, m, yet it is far less easy to understand how, when m is varied in its nature, and therefore wholly changes the distribution of current in all the branches (except one of the conjugate ones) due to impressed forces in them, it does not also change the current in the excepted branch n. Conscientious learners always need to work out the full results in a problem relating to the steady-flow of current before they can completely satisfy themselves that the property is true.

Note on Part III. Example of Treatment of Terminal Conditions. Induction-Coil and Condenser.

One of the side-matters left over for separate examination when giving the main investigation of Parts I. to IV. was the manner of treatment of terminal conditions when normal solutions are in question, especially with reference to the finding of the terms in the complete solution arising from an arbitrary initial state which are due to the terminal apparatus, concerning which I remarked in Part III. that the matter was best studied in the concrete application. There is also the question of finding the nature of the terminal arbitraries from the mere form of the terminal equation, without knowledge of the nature of the arrangement in detail, except what can be derived from the terminal equation.

Let, for example, in the figure, the thick line to the right be the beginning of the telegraph-line, and what is to the left of it the terminal apparatus, consisting of an induction-coil and a shunted condenser. The line is joined through the primary of the induction-coil, of resistance R_1, to the condenser of permittance S_0, whose shunt has the conductance K_0, and whose further side is connected to earth, as symbolised by the arrow-head.* Let R_2 be the resistance of the secondary coil, and L_1, L_2, M the inductances, self and mutual, of the primary and the secondary. At the distant end of the line, where $z=l$, we may have

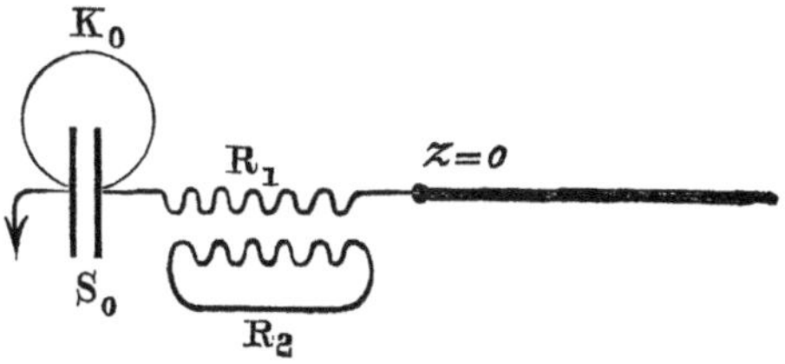

another arrangement of apparatus, also joined through to earth, though this is not necessary. The line and the two terminal arrangements form the complete system, supposed to be independent of all other systems.

Now suppose there to be no impressed voltage in any part of the system, so that its state at a given moment depends entirely upon its initial state at the time of removal of the impressed voltage; after which, owing to the existence of resistance, it must subside to a state of zero electric force and zero magnetic force everywhere (with some

* It is not altogether improbable that the arrangement shown in the figure, with the receiving instrument placed in the *secondary* circuit, would be of advantage. A preliminary examination of the form of the arrival-curve when this arrangement is used for receiving at the end of a long cable, with $K_0=0$, yields a favourable result. But the examination did not wholly include the influence of the resistances on the form of the curve.

exceptional cases in which there is ultimately electric force, though not magnetic force), the manner of the subsidence to the final state depending upon the connexions of the system. The course of events at any place depends upon the initial state of every part, including the terminal apparatus, which may be arbitrary, since any values may be given to the electrical variables which serve to fully specify the amount and distribution of the electric and magnetic energies.

Suppose that V, the transverse voltage, and C, the current in the line, are sufficient to define its state, *i.e.* as electrical variables, when the nature of the line is given, and that u and w are the normal functions of V and C in a normal system of subsidence. Then, at time t, we have

$$V = \Sigma\, Aue^{pt}, \qquad C = \Sigma\, Awe^{pt}, \qquad \text{.....................(1}e\text{)}$$

wherein the p's are known from the connexions of the whole system; each normal system having its own p, and also a constant A to fix its magnitude. The value of A is thus what depends upon the initial state, and is to be found by an integration extending over every part of the system. In one case, viz., when the initial state is what could be set up finally by any distribution of steadily acting impressed force, we do not need to perform this complex integration, since we may obtain what we want by solving the inverse problem of the setting up of the final state due to the impressed force, as done by one method in Part III., and by another in Part IV. If also the initial state of the apparatus be neutral, so that it is the state of the line only that determines the subsequent state, we can pretty easily represent matters, viz., by giving to A the value

$$A = \int_0^l \frac{(SUu - LWw)dz}{\Delta}, \qquad \text{........................(2}e\text{)}$$

wherein U and W are the initial V and C in the line, whose permittance and inductance per unit length are S and L; so that the numerator of A is the excess of the mutual electric over the mutual magnetic energy of the initial and a normal state, whilst the denominator Δ is twice the excess of the electric over the magnetic energy of the normal state itself, which quantity may be either expressed in the form of an integration extending over the whole system, or, more simply, and without any of the labour this involves, in the form of a differentiation with respect to p of the determinantal equation. For instance, when we assume $L = 0$, and we make the line-constants to be simply R and S, its resistance and permittance per unit length (constants), as we may approximately do in the case of a submarine cable that is worked sufficiently slowly to make the effects of inertia insensible, in which case we have

$$-\frac{dV}{dz} = RC, \qquad -\frac{dC}{dz} = S\dot{V}, \qquad \frac{d^2V}{dz^2} = RS\dot{V}; \qquad \text{............(3}e\text{)}$$

so that we may take

$$u = \sin(mz + \theta), \qquad w = -\frac{m}{R}\cos(mz + \theta), \qquad \text{................(4}e\text{)}$$

if $-m^2 = RSp$; then equation (2e) becomes

$$A = \frac{S\int_0^l Uu\,dz + Y_0 + Y_1}{\frac{1}{2}Sl\left\{1 - \cos^2 ml \, . \, \frac{d}{d(ml)}F(ml)\right\}}, \qquad \text{(5e)}$$

where the undefined terms Y_0 and Y_1 in the numerator depend upon the terminal apparatus, and F in the denominator is defined by

$$\tan ml = \frac{m}{Sp} \cdot \frac{Z_1 - Z_0}{(m/Sp)^2 + Z_1 Z_0} = F(ml), \qquad \text{(6e)}$$

which is the determinantal equation arising out of the terminal conditions

$$V = Z_0 C \quad \text{at} \quad z = 0, \qquad \text{and} \qquad V = Z_1 C \quad \text{at} \quad z = l. \qquad \text{(7e)}$$

(See equations (177) to (180), Part IV.) We have now to add on to the numerator of A the terms corresponding to the initial state of the terminal apparatus, when it is not neutral. As the process is the same at both ends of the line, we may confine ourselves to the $z=0$ apparatus, according to the figure. First we require the form of Z_0, the negative of the generalized resistance of the terminal apparatus. It consists of three parts, one due to the condenser, a second to the primary coil, and a third to the presence of the secondary; thus,

$$-Z_0 = (K_0 + S_0 p)^{-1} + (R_1 + L_1 p) - M^2 p^2 (R_2 + L_2 p)^{-1}, \qquad \text{(8e)}$$

showing the three parts in the order stated. Now as shown in Part III., dZ_0/dp expresses twice the excess of the electric over the magnetic energy in a normal system (when p becomes a constant), per unit square-of-current. Performing the differentiation, we have

$$\frac{dZ_0}{dp} = \frac{S_0}{(K_0 + S_0 p)^2} - L_1 + \frac{2M^2 p}{R_2 + L_2 p} - \frac{L_2 M^2 p^2}{(R_2 + L_2 p)^2}. \qquad \text{(9e)}$$

Here we may at once recognise that the first term represents twice the electric energy of the condenser per unit square-of-current, that the second term is the negative of twice the magnetic energy of the unit primary current, and that the fourth is similarly the negative of twice the magnetic energy of the secondary current per unit primary current; whilst the third, which at first sight appears anomalous, is the negative of twice the mutual magnetic energy of the unit primary and corresponding secondary current. Thus, if w_0 be the normal current-function, that is, by (4e), $w_0 = -(m/R)\cos\theta$, we have

$$\frac{w_0}{K_0 + S_0 p}, \qquad w_0, \qquad \text{and} \qquad -\frac{Mpw_0}{R_2 + L_2 p}, \qquad \text{(10e)}$$

as the expressions for the normal voltage of the condenser, for the primary current, and for the secondary current. If then V_0, C_1, and C_2 are the initial quite arbitrary values of the voltage of the condenser, and of the primary and secondary currents, their expansions must be

$$V_0 = \sum A \frac{w_0}{K_0 + S_0 p}, \qquad C_1 = \Sigma A w_0, \qquad C_2 = \sum A \frac{(-Mpw_0)}{R_2 + L_2 p}. \qquad \text{(11e)}$$

Also, the excess of the mutual electric over the mutual magnetic energy of the initial state V_0, C_1, C_2, and the normal state represented by (10*e*) is

$$Y_0 = w_0\left\{\frac{S_0V_0}{K_0+S_0p} - (L_1C_1+MC_2) + (L_2C_2+MC_1)\frac{Mp}{R_2+L_2p}\right\}, \quad (12e)$$

and this is what must be added to the numerator in (5*e*) to obtain the complete value of A, if we also add the corresponding expression Y_1 for the apparatus at the other end, if it be not initially neutral. Using this value of A in (1*e*) and in (11*e*) with the time-factor ϵ^{pt} attached, and in the corresponding expansions for the other end, we thus express the state of the whole system at any time.

Since, initially, V is U, and independent of the state of the terminal apparatus, it follows that in the expansion

$$U = \Sigma\, Au,$$

the parts of A depending on the apparatus contribute nothing to U, so that, by (5*e*) and (12*e*), we have the identities

$$0 = \sum A\frac{w_0 u}{K_0+S_0p}, \qquad 0 = \Sigma\, Aw_0 u, \qquad 0 = \sum A\frac{w_0 pu}{R_2+L_2p}, \quad \text{...(13}e)$$

for all the values of z from 0 to l.

It may have been observed in the above that the use of (9*e*) was quite unnecessary, owing to the forms of the normal functions in (10*e*) being independently obtainable from our *à-priori* knowledge of the terminal apparatus in detail, from which knowledge the form of Z_0 in (8*e*) was deduced; so that, without using (9*e*), we could form (11*e*) and (12*e*). I have, however, introduced (9*e*) in order to illustrate how we can find the complete solution, without knowing the detailed terminal connexions, from a given form of Z. We must either decompose dZ_0/dp into the sum of squares of admissible functions of p, multiplied by constants, say,

$$\frac{dZ_0}{dp} = a_0 f_1^2 + a_2 f_2^2 + a_3 f_3^2 + \dots, \quad \text{......................(14}e)$$

where a_1, a_2, etc., are the constants, and f_1, f_2, ... the functions of p; or else into the form of the sum of squares and products, thus

$$\frac{dZ_0}{dp} = a_1 f_1^2 + a_2 f_2^2 + a_3 f_3^2 + b_1 f_1 f_2 + b_2 f_1 f_3 + \dots. \quad \text{.......(15}e)$$

When this is done, we know that the terminal arbitraries are

$$F_1 = \Sigma\, Af_1 w_0, \qquad F_2 = \Sigma\, Af_2 w_0, \qquad F_3 = \Sigma\, Af_3 w_0, \quad \text{...(16}e)$$

and that

$$Y_0 = w_0\{a_1F_1f_1 + a_2F_2f_2 + a_3F_3f_3 + \dots\} \quad \text{............(17}e)$$

in the case (14*e*) of sums of squares, wherein the F's may have any values, assuming that we have satisfied ourselves that they are all independent; with the identities

$$0 = \Sigma\, Af_1 u, \qquad 0 = \Sigma\, Af_2 u, \quad \text{etc. ....(18}e)$$

Thus, in the case (9*e*), the first, second, and fourth terms are of the proper form for reduction to (14*e*), but the third is not. We are certain, therefore, that there cannot be more than three arbitraries, if there be so many. Now, if we do not recognise the connection between the third term and those which precede and follow it (as may easily happen in some other case), we should rearrange the terms to bring it to the form (14*e*); for instance, thus:—

$$\frac{dZ_0}{dp}=\frac{S_0}{(K_0+S_0p)^2}-\left(L_1-\frac{M^2}{L_2}\right)-\frac{M^2R_2^2/L_2}{(R_2+L_2p)^2}, \quad \ldots\ldots\ldots(19e)$$

which is what we require. We may then take

$$\left.\begin{aligned} f_1&=(K_0+S_0p)^{-1}, & f_2&=1, & f_3&=(R_2+L_2p)^{-1},\\ a_1&=S_0, & a_2&=-(L_1-M^2/L_2), & a_3&=-M^2R_2^2/L_2. \end{aligned}\right\}\ldots\ldots(20e)$$

Further, we can certainly conclude, provided a_1 is positive, and a_2 and a_3 are negative, that the first term on the right of (19*e*) stands for electric (or potential) energy, and the remainder for magnetic (or kinetic). It is clear that we may assume any form of Z that we please of an admissible kind (*e.g.*, there must be no such thing as $p^{\frac{1}{2}}$), find the arbitraries, and fully solve the problem that our data represent, whether it be or be not capable of a real physical interpretation on electrical principles. I have pursued this subject in some detail for the sake of verifications; it is an enormous and endless subject, admitting of infinite development. Owing, however, to the abstractly mathematical nature of the investigations—to say nothing of the length to which they expand, although when carried on upon electrical principles they are much simplified, and made to have meaning—I merely propose to give later one or two examples in which circular functions of p are taken to represent Z.

Although, however, the state of the line at any moment is fully determinable for any form of the terminal Z's, when they alone are given, from the initial state of the line, provided the initial values of the terminal arbitraries be taken to be zero, and although it is similarly determinable when particular values are given to the arbitraries, whose later values also are determinable by affixing the time-factor, it does not appear that this determinateness of the later values of the terminal arbitraries is always of a complete character, when the sole data relating to them are the form of Z and their initial values. For it is possible for a terminal arrangement to have a certain portion conjugate with respect to the line; and although the state of the line will not be affected by initial energy in that portion, yet it will influence the later values of the other terminal arbitraries. This might wholly escape notice in an investigation founded upon a given form of Z with undetailed connections, owing to the disappearance from Z of terms depending upon the conjugate portion. In such a case the reduced form of Z cannot give us the least information concerning the influence of the portion conjugate to the line. It is as if it were non-existent. If, however, Z be made more general, so as to contain terms depending upon the conjugate portion, although they be capable of immediate elimination from Z, it would seem that the indeterminateness must be removed.

Some Notes on Part IV. Looped Metallic Circuits. Interferences due to Inequalities, and consequent Limitations of Application.

It is scarcely necessary to remark that, in the investigation of Parts I. and II., the choice of a round wire or tube surrounded by a coaxial tube for return-conductor was practically necessitated in order to allow of the use of the well-known J_0 and J_1 functions and their complements, because it was not merely the total current in the wire with which we were concerned, but also with its distribution. Next, in order that it should be a question of self-induction, and not one of mutual induction also, with fearful complications, it was necessary to impose the condition that the wire, tubular dielectric, and outer tube should be a self-contained system, making the magnetic force zero at the outer boundary. It is true that no external inductive effect is observable when the double-tube circuit is of moderate length. But electrostatic induction is cumulative; and it is certain that, by sufficiently lengthening the double tube, we should ultimately obtain observable inductive interferences. Our investigation, then, only applies strictly when the double tube is surrounded on all sides, to an infinite distance, by a medium of infinite elastivity and resistivity.

(Maxwell termed $4\pi/c$, when c is the dielectric constant, the electric elasticity. I make this the elastivity: first, to have one word for two; next, to avoid confusion with mechanical elasticity; and, thirdly, to harmonise with the nomenclature I have used for some time past. Thus:—

Flux.			Force.
Conduction-Current .	Resistivity. Conductivity.	Resistance. Conductance.	Electric.
Induction	Inductivity.	Inductance.	Magnetic.
Displacement . . .	Elastivity. Permittivity.	Elastance. Permittance.	Electric.

The elastance of a condenser is the reciprocal of its permittance, and elastivity is the elastance of unit volume, as resistivity is the resistance of unit volume, and conductivity the conductance of unit volume. As for "permittivity" and "permittance," there are not wanting reasons for their use instead of "specific inductive capacity" (electric), and "electrostatic capacity." The word capacity alone is too general; it must be capacity for something, as electrostatic capacity. It is an essential part of my scheme to always use *single* and unmistakable words, because people will abbreviate. Again, capacity is an unadaptable word, and is altogether out of harmony with the rest of the scheme. Now the flux concerned is the electric displacement, involving elastic resistance to yielding from one point of view, and a capacity for permitting the yielding from the inverse; hence elastance and permittance, the latter being the electrostatic capacity of a condenser. There are now only two gaps left, viz. for the reciprocals of inductivity and inductance. "Resistance to lines of force" and "magnetic resistance" will obviously not do for permanent use.)

If this restriction be removed, we have self- and mutual-induction concerned, and interferences; or, even if there be no external conductors, we have still the electric current of elastic displacement, and with it electric and magnetic energy outside the double tube. But, ignoring these, we have the following striking peculiarities:—Putting on one side the question of the propagation of disturbances *into* the conductors, which is so interesting a one in itself, we find that the electrical constants are three in number—the resistance, permittance, and inductance of the double-tube per unit of its length; whilst the electrical variables are two—the current in each conductor, and the transverse voltage. The effective resistance per unit length is the sum of their resistances, which may be divided between the two conductors in any ratio; the permittance is that of the dielectric between them; and the inductance is the sum of that of the dielectric, inner, and outer conductors. Another remarkable peculiarity is, that equal impressed forces, similarly directed in the two conductors at corresponding places, can do nothing; from which it follows that the effective impressed force may, like the effective resistance, be divided between the conductors in any proportion we please.

In Part IV., having in view the rapidly extending use of metallic circuits of double wires looped, excluding the earth, consequent upon the development of telephonic communication in a manner to eliminate inductive interferences, I extended the above-described method to a looped circuit consisting of a pair of parallel wires. So far as propagation into the wires is concerned, it is merely necessary that they should not be too close to one another, to allow of the application of the J_0 and J_1 functions to them separately. Now suspended wires are usually of iron, and are not set too close, so that the application is justified. On the other hand, buried twin wires, though very near one another, are of copper, and also considerably smaller than the iron suspended wires; so that the diffusion-effect, though not so well representable by the above-named functions, is made insignificant. Dismissing, as before, this question of inward propagation, we have, just as in the tubular case, two electrical variables and three constants, viz. the transverse voltage, the current in each wire, and the effective resistance, permittance, and inductance.

First of all, let the wires be alone in an infinite dielectric. Then we have similar results to those concerning the double-tube. The effective resistance, which is the sum of the resistances of the wires, may be divided between them in any proportions; and so may be the effective impressed voltage. The effective permittance is that of the condenser consisting of the dielectric bounded by the two wires, the surface of one being the positive, and that of the other the negative coating. Or, in another form, the effective permittance is the reciprocal of the elastance from one wire to the other. In the standard medium, this elastance is, in electrostatic units, the same as the inductance of the dielectric in electromagnetic units. Thus,

$$L_0 = 2\mu \log \frac{r_{12}^2}{r_1 r_2}, \quad \text{.........................} (1f)$$

if r_1 and r_2 be the radii of the wires, and r_{12} their distance apart (between axes), and μ the inductivity of the dielectric. And

$$S = c\left(2 \log \frac{r_{12}^2}{r_1 r_2}\right)^{-1}. \quad \text{.........................(2f)}$$

Their product, when in the same units, is v^{-2}, the reciprocal of the square of the speed of undissipated waves through the dielectric. The two variables, transverse voltage and current, fully define the state of the wires, except as regards the diffusion-effect in them, of course, and an effect due to outward propagation into the unbounded dielectric from the seat of impressed force, which is made insignificant by the limitation of the magnetic field (in sensible intensity) due to the nearness of the wires as compared with their length. To L_0 has to be added a variable quantity, whose greatest value is $\frac{1}{2}\mu_1 + \frac{1}{2}\mu_2$, if μ_1 and μ_2 are the inductivities of the wires, to obtain the complete inductance per unit length.

So far, then, there is a perfect correspondence between the double-tube and the double-wire problem. But when we proceed to make allowance for the presence of neighbouring conductors, as, for instance, the earth, although there is a formal resemblance between the results in the two cases, when proper values are given to the constants concerned, yet the fact that in one case the outer conductor encloses the inner, whilst in the other this is not so, causes practical differences to exist. For example, there are two constants of permittance concerned in the coaxial tube case, that of the dielectric between them, and that of the dielectric outside the outer tube. But in the case of looped wires there are three, which may be chosen to be the permittance of each wire with respect to earth including the other wire, and a coefficient of mutual permittance. There are, similarly, three constants of inductance, and two of resistance, and at least two of leakage, viz. from each wire to earth, with a possible third direct from wire to wire. This is when the wires are treated in a quite general manner, and arbitrarily operated upon; so that there must be four electrical variables, viz., two currents and two potential-differences or voltages. I have somewhat developed this matter in my paper "On Induction between Parallel Wires" [p. 116, vol. I.]; and as regards the values of the constants of capacity concerned, in my paper "On the Electrostatic Capacity of Suspended Wires" [p. 42, vol. I.]. As may be expected, the solutions tend to become very complex, except in certain simple cases. If, then, we can abolish this complexity, and treat the double wire as if it were a single one, having special electrical constants, we make a very important improvement. I have at present to point out certain peculiarities connected with the looped-wire problem in addition to those described in Part IV., and to make the necessary limitations of application of the method and the results which are required by the presence of the earth.

First of all, even though the wires be not connected to earth, if they be charged and currented in the most arbitrary manner possible, we must employ the four electrical variables and the ten or eleven electrical constants as above mentioned. On the other hand, going back to the looped wires far removed from other conductors, there are but two electrical variables and four constants (counting one for leakage). Now bring these parallel wires to a distance above the earth which is a large multiple of their distance apart. The constant S of permittance is a little increased. The method of images gives

$$S=c\left(2\log\frac{s_1 s_2 r_{12}^2}{r_1 r_2 s_{12}^2}\right)^{-1}; \quad \ldots\ldots\ldots\ldots\ldots\ldots(3f)$$

where r_1, r_2 are the radii of the wires, r_{12} their distance apart, s_1, s_2 their distances from their images, and s_{12} the distance from either to the image of the other; but, owing to $s_1 s_2/s_{12}^2$ being nearly unity, the permittance S does not sensibly differ from the value in an infinite dielectric, or the earth has scarcely anything to do with the matter.* If, however, the wires be brought close to the earth, the increase of permittance will become considerable; this is also the case when the wires are buried. The extreme is reached when each wire is surrounded by dielectric to a certain distance, and the space between and surrounding the two dielectrics is wholly filled up with well-conducting matter. Then the permittance S becomes the reciprocal of the sum of the elastances of the two wires with respect to the enveloping conductive matter; in another form, the effective elastance is the sum of the elastances of the two dielectrics. Returning to the suspended wires, if the earth were infinitely conducting, the effective inductance would be the reciprocal of S in $(3f)$ with μ written for c, in electromagnetic units, with $\frac{1}{2}(\mu_1+\mu_2)$ added; whilst, allowing for the full extension of the magnetic field into the earth, we should have the formula $(1f)$, giving a slightly greater value. The effective resistance is of course the sum of the resistances, and the effective leakage-resistance would be the sum of the leakage-resistances of the two wires with respect to earth, if that were the only way of getting leakage between the wires, but it must be modified in its measure by leakage being mostly from wire to wire over the insulators, arms, and only a part of the poles.

But if there be any inequalities between the wires, differential effects will result, due to the presence of the earth, in spite of its little influence on the value of the effective permittance; whereby the current in one wire is made not of the same strength as in the other, and the charge on one wire not the negative of that on the other. The propagation of signals from end to end of the looped-circuit will not then take place exactly in the same manner as in a single wire. To allow for this, we may either bring in the full, comprehensive system of electrical constants and variables; or, perhaps better, exhibit the differential effects separately by taking for variables the sum of the

* On the other hand, Mr. W. H. Preece, F.R.S., assures us that the capacity is half that of either wire (Proc. Roy. Soc. March 3, 1887, and Journal S. T. E. and E., Jan. 27 and Febr. 10, 1887). This is simple, but inaccurate. It is, however, a mere trifle in comparison with Mr. Preece's other errors; he does not fairly appreciate the theory of the transmission of signals, even keeping to the quite special case of a long and slowly worked submarine cable, whose theory, or what he imagines it to be, he applies, in the most confident manner possible, universally. There is hardly any resemblance between the manner of transmission of currents of great frequency and slow signals. [See also p. 160, vol. II.]

potentials of the wires (taking earth at zero potential) and half the difference of the strength of current in them, in addition to the difference of potential of the wires and half the sum of the current-strengths, which last are the sole variables when the wires are in an infinite dielectric, or else are quite equal. By adopting the latter course our solutions will consist of two parts, one expressing very nearly the same results as if the differential effects did not exist, the other the differential effects by themselves.

Another result of inequalities is to produce inductive interferences from parallel wires which would not exist were the wires equal. As an example, let an iron and a parallel copper wire be looped, and telephones be placed at the ends of the circuit. Even if the wires be well twisted, there is current in the telephones caused by rapid reversals in a parallel wire whose circuit is completed through the earth. Again, if two precisely equal wires be twisted, and telephones placed at the ends as before, the insertion of a resistance into either wire intermediately will upset the induction-balance and cause current in the terminal telephones when exposed to interference from a parallel wire. This interference can be removed by the insertion of an equal resistance in the companion-wire at the same place. In the working of telephone metallic circuits with intermediate stations and apparatus, we not only introduce great impedance by the insertion of the intermediate apparatus, thus greatly shortening the length of line that can be worked through, but we produce inductive interferences from parallel wires, unless the intermediate apparatus be double, one part being in circuit with one wire, the other part (quite similar) in circuit with the other. In mentioning my brother's system of bridge-working of telephones (in Part V.), whereby the intermediate impedance is wholly removed, I mentioned, without explanation, the cancelling of inductive interferences. The present and preceding paragraphs supply the needed explanation of that remark. The intermediate apparatus, being in bridges across from one wire to the other, do not in the least disturb the induction-balance, so that transmission of speech is not interfered with by foreign sounds.

But theory goes much further than the above in predicting interferences than practice up to the present time verifies. For instance, if two perfectly equal wires be suspended at the same height above the ground and be looped at the ends, terminal telephones will not be interfered with by variations of current in a parallel wire equidistant from both wires of the loop-circuit, having its own circuit completed through the earth. But if the loop-circuit be in a vertical plane, so that one wire is at a greater height above the ground than the other, there must be terminal disturbance produced, even when the disturbing wire is equidistant. Similarly in the many other cases of inequality that can be mentioned.

The two matters, preservation of the induction-balance, and transmission of signals in the same manner as on a single wire, are intimately connected. If we have one, we also have the other. The limitations of application of the method of Part IV. may be summed up in saying that the loop-circuit must either be far removed from all conductors, in which case equivalence of the wires is quite needless; or else they must be equal in their electrical constants. In the latter case the effective resistance R is the double of that of either wire, and the effective permittance, inductance, and leakage are to be measured as before described, whilst the variables are the transverse voltage from wire to wire and the current in each. But the four electrical constants may vary in any (not too rapid) manner along the line. And the impressed force (in the investigations of Part IV.) may also be an arbitrary function of the distance, provided it be put, half in one wire, half in the other, oppositely directed in space. For, although equal, similarly directed impressed forces will cause no terminal disturbance (and none anywhere if other conductors be sufficiently distant), yet disturbances at intermediate parts of the line will result. It is true that the most practical case of impressed voltage is when it is situated at one end only of the circuit, when it is of course equally in both wires, or not in them at all; but there is such a great gain in the theoretical treatment of these problems by generalising, that it is worth while to point out the above restriction.

Besides this case of equality of wires, which is precisely the one that obtains in practice, there are other cases in which, by proper proportioning of the electrical constants of the two looped wires, the induction-balance is preserved; and, simultaneously, we obtain transmission of signals as on a single wire. [But this is not an invariable rule.] Their investigation is a matter of scientific interest, though scarcely of practical importance.

I have yet to add investigations by the method of waves (mentioned in Part IV.), by which I have reached interesting results in a simple manner.

PART VIII.

The Transmission of Electromagnetic Waves along Wires without Distortion.

One feature of solutions of physical problems by expansions in infinite series of normal solutions is the very artificial nature of the process. If it be a case of subsidence towards a state of equilibrium, then, if a sufficient time has elapsed since the commencement of the subsidence to allow the great mass of (singly) insignificant systems to nearly vanish, leaving only two or three important systems, which may be readily examined—or merely one, the most important—then the process is natural enough. It is the early stage of the subsidence that is so artificially represented, when the resultant of a very large number of normal solutions must be found before we come to what we want. Sometimes, too, the full investigation of the normal systems in detail is prevented by mathematical difficulties connected with the roots of transcendental equations. This goes very far to neutralise the advantage presented by the ease with which solutions in terms of normal functions may be obtained.

In some respects these difficulties are evaded by the consideration of the solution due to a sinusoidal impressed force. The method is very powerful; and, by considering the nature of the results through a sufficiently wide range of frequencies, we may indirectly gain, with comparatively little trouble, knowledge that is unattainable by the method of normal systems.

But the real desideratum, which, if it can be reached, is of paramount importance, is to get solutions which can be understood and appreciated at first sight, and followed into detail with ease, presenting to us, as nearly as possible, the effects as they really occur in the physical problem, disconnected from the often unavoidable complications due to the form of mathematical expression. To illustrate this, it is sufficient to refer to the elementary theory of the transmission of waves without dissipation along a stretched flexible cord. If we employ Fourier-series, we are doing mathematical exercises. But only use the other method, in which arbitrary disturbances are transferred bodily in either direction at constant speed, *e.g.*,

$$u = f(z - vt),$$

and we get rid of the mathematical complications, and can interpret results as we see their physical representatives in reality—for instance, when we agitate one end of a long cord.

Now there is one case, and, so far as I know at present, only one, in the many-sided question of the transmission of electromagnetic disturbances along wires, which admits of this simple and straightforward method of treatment. Singularly enough, it is not by the simplifying process of equating to zero certain constants, and so ignoring certain effects, that we reach this unique state of things, but rather the other way, generalising to some extent. It is usual to ignore the leakage of conductors, sometimes also the inductance, and sometimes the permittance. But we must take all the four properties into account which are symbolised by resistance, leakage-conductance, inductance, and permittance, to reach the much-desired result. Briefly stated, the effects are these, roughly speaking. If there be only resistance and permittance, there is, when disturbances of an irregular character are sent along a long circuit, both very great attenuation and very great distortion produced. The distortion at the end of an Atlantic cable is enormous. Now if we introduce leakage, we shall lessen the distortion considerably, but at the same time increase the attenuation. On the other hand, if we introduce inductance (instead of leakage) we shall lessen the attenuation as well as the distortion. And, finally, if we have both leakage and inductance, in addition to resistance and permittance, we may so adjust matters, by the effects of inductance and of leakage being opposite as regards distortion, as to annihilate the distortion altogether, leaving only attenuation. The solutions can now be followed into detail in various cases without any laborious and roundabout calculations. Besides this, they cast much light upon the more difficult problems which occur when not so many physical actions are in question.

In my usual notation, let R, L, S, and K be the resistance, inductance, permittance, and leakage-conductance of a circuit, per unit length, all to be treated, in the present theory, as constants; and let V and C be the transverse voltage and the current at distance z. The fundamental equations are

$$-\frac{dV}{dz}=(R+Lp)C, \qquad -\frac{dC}{dz}=(K+Sp)V, \quad \text{............(1g)}$$

p standing for d/dt. Here C is related to the space-variation of V in the same formal manner as is V to the space-variation of C. This property allows us to translate solutions in an obvious manner, and gives rise to the distortionless state of things. Let

$$LSv^2=1, \qquad \text{and} \qquad R/L=K/S=q. \quad \text{..............(2g)}$$

The equation of V is then

$$v^2\frac{d^2V}{dz^2}=(q+p)^2V, \quad \text{..........................(3g)}$$

and the complete solution consists of waves travelling at speed v with attenuation but without distortion. Thus, if the wave be positive, or travel in the direction of increasing z, we shall have, if $f_1(z)$ be the state of V initially,

$$V_1=\epsilon^{-qt}f_1(z-vt), \quad \text{..........................(4g)}$$

$$C_1=V_1/Lv. \quad \text{...................................(5g)}$$

If V_2, C_2 be a negative wave, travelling the other way,

$$V_2=\epsilon^{-qt}f_2(z+vt), \quad \text{..........................(6g)}$$

$$C_2=-V_2/Lv. \quad \text{..............................(7g)}$$

Thus, any initial state being the sum of V_1 and V_2 to make V, and of C_1 and C_2 to make C, the decomposition of an arbitrarily given initial state of V and C into the waves is effected by

$$V_1=\tfrac{1}{2}(V+LvC), \qquad V_2=\tfrac{1}{2}(V-LvC). \quad \text{..............(8g)}$$

We have now merely to move V_1 bodily to the right at speed v, and V_2 bodily to the left at speed v, and attenuate them to the extent ϵ^{-qt}, to obtain the state at time t later, provided no changes of conditions have occurred. The solution is therefore true for all future time in an infinitely long circuit. But when the end of a circuit is reached, a reflected wave usually results, which must be added on to obtain the real result.

In any portion of a solitary wave, positive or negative, the electric and magnetic energies are equal, thus

$$\tfrac{1}{2}LC_1^2=\tfrac{1}{2}SV_1^2. \quad \text{.............................(9g)}$$

The dissipation of energy is half in the wires and half without, thus

$$\tfrac{1}{2}RC_1^2=\tfrac{1}{2}KV_1^2. \quad \text{..........................(10g)}$$

When a positive and a negative wave coexist, and energies are added, cross-products disappear. Thus the total energy is always

$$S(V_1^2+V_2^2), \qquad \text{or} \qquad L(C_1^2+C_2^2)\,; \quad \text{..............(11g)}$$

the total dissipativity is always

$$R(C_1^2+C_2^2), \qquad \text{or} \qquad K(V_1^2+V_2^2); \quad \text{............(12}g\text{)}$$

and the total energy-flux is always

$$V_1C_1+V_2C_2. \quad \text{..........................(13}g\text{)}$$

The relation $V_1=LvC_1$ is equivalent to $C_1=SvV_1$; *i.e.*, a charge SV moving at speed v is the equivalent of a current C of strength equal to their product. But it is practically best to employ Lv, the ratio of the force V to the flux C being then at once expressible or measurable in ohms. For v is 30 ohms, and L is a convenient numeric, say from 2 up to 100, according to circumstances. $L=20$ is a convenient rough measure in the case of a pair of suspended copper wires. This makes our critical impedance 600 ohms. It must not be confounded with resistance, of course, though measurable in ohms. The electric and magnetic forces are perpendicular. It is the total flux of energy which is expressed by the product VC, not the dissipativity.

Regarding v, its possible greatest value is the speed of light *in vacuo*. When there is distortion also, making the apparent speed variable, it does not appear that under any circumstances the speed can exceed v. Now the classical experiments of Wheatstone indicated a speed half as great again as that of light. Would it not be of scientific interest to have these important experiments carefully repeated, on a straight circuit (as well as of other forms), to ascertain whether, on the straight circuit, the speed is not always less than, rather than greater than, that of light, and whether there was any difference made by curving the circuit?

The following remark may be useful. In treatises on electromagnetism by the German methods, a *current-element* and its properties of attraction, repulsion, etc., occupy an important place. It is, however, quite an abstraction, and devoid of physical significance when by itself. But the current-element in our theory above, say $V=V_1$ constant through unit distance, $C=V_1/Lv$ through the same unit distance, V and C zero everywhere else, is a physical reality (with limitations to be mentioned). It is a complete electromagnetic system of itself, with the electric currents closed. To fix ideas most simply, the two conductors may be a wire with an enveloping tube separated by a dielectric, and by our current-element we imply a definite electric field, magnetic field, and dissipation of energy, which can exist apart from all other current-elements. It is only an abstraction in this quite different sense, that we could not really terminate the element *quite* suddenly, and that in the process of travelling it must be distorted from causes not considered in our fundamental equations, one cause being the diffusion of current in the conductors in time, which alone serves to prevent the propagation of an abrupt wave-front, either in our distortionless system, or when there is marked distortion. Even assuming that Maxwell's representation of the electromagnetic field is not correct, there seems to me to be very marked advantage in assuming its correctness, even as a working hypothesis, from its exceeding physical explicitness in dynamical interpretation, without specifying a special mechanism to correspond. We have also the inimitable advantage of abolishing once for all the speculations about unclosed currents, and the insoluble problems they present. In Maxwell's scheme currents always close themselves, and cannot help it.

It will be seen that our waves, in the above, do not in any way differ from plane waves of light (in Maxwell's theory), save in being attenuated by dissipation of energy in the dielectric (when it is a tubular conducting dielectric bounded by a pair of conductors that is in question), and also in the bounding conductors, and in being practically of quite a different order of wave-length. The lines of energy-flux are parallel to the wires, (a wave simply carries its energy with it, less the amount dissipated); these are also the lines of *pressure*, for the electrostatic attraction equals and cancels the electromagnetic repulsion. The variation of the pressure constitutes a mechanical force, half derived from the electromagnetic force, half from the magneto-electric force. Here, however, I am bound to say I cannot follow readily. If this mechanical force exist, there must be corresponding acceleration of momentum; if it do not exist, or be balanced, the stress supposed is not the real stress, though it may be a part of it. Again, if it be the real stress, and there be the corresponding acceleration of momentum, this is equivalent to introducing an *impressed* force (mechanical), and it must be allowed for. The matter is difficult all round. Yet Maxwell's stresses, assumed to exist in the fluid dielectric between conductors, account perfectly for the forces between them, when the electric and magnetic fields are stationary. But when they vary, then the region of mechanical force due to stress-variation extends into the dielectric medium. As for Maxwell's stress in a magnetised medium, there are so many different arrangements of stress that will serve equally well, that I cannot have any faith whatever in the special form given by Maxwell.

It is also well to remember that we are not exactly representing Maxwell's scheme, but a working simplification thereof. The lines of energy-transfer are not quite parallel to the conductors, but converge upon them at a very acute angle on both sides of the dielectric. Only by having conductors to bound it of infinite conductivity can we make truly plane waves. Then they will be greatly distorted, unless we at the same time remove the leakage by making the dielectric a non-conductor instead of a feeble conductor; when we have undissipated waves without attentuation or distortion.

Properties of the Distortionless Circuit itself, and Effect of Terminal Reflection and Absorption.

Now to mention some properties of the distortionless circuit. A pair of equal disturbances, travelling opposite ways, on coincidence, double V and cancel C. But if the electrifications be opposite, V is annulled and C doubled on coincidence.

On arrival of a disturbance at the end of a circuit, what happens depends upon the connections there. One case is uniquely simple. Let there be a resistance inserted of amount Lv. It introduces the condition $V=LvC$ if at say B, the positive end of the circuit, and $V=-LvC$ if at the negative end A, or beginning. These are the characteristics of a positive and of a negative wave respectively; it follows that any disturbance arriving at the resistance is at once absorbed. Thus, if the circuit be given in any state whatever, without impressed force, it is wholly cleared of electrification and current in the time l/v at the most, if l be the length of the circuit, by the complete absorption of the two waves into which the initial state may be decomposed.

But let the resistance be of amount R_1 at say B; and let V_1 and V_2 be corresponding elements in the incident and reflected wave. Since we have

$$V_1=LvC_1, \qquad V_2=-LvC_2, \qquad V_1+V_2=R_1(C_1+C_2), \qquad (14g)$$

we have the reflected wave given by

$$\frac{V_2}{V_1}=\frac{R_1-Lv}{R_1+Lv}. \qquad (15g)$$

If R_1 be greater than the critical resistance of complete absorption, the current is negatived by reflection, whilst the electrification does not change sign. If it be less, the electrification is negatived, whilst the current does not reverse.

Two cases are specially notable. They are those in which there is no absorption of energy. If $R_1=0$, meaning a short-circuit, the reflected wave of V is a perverted and inverted copy of the incident. But if $R_1=\infty$, representing insulation, it is C that is inverted and perverted.

After reflection, of course, we have the original wave travelling to the absorber or absorbing reflector, or pure reflector, and the reflected wave coming from it. Let ρ_0 be the coefficient of attenuation at A, and ρ_1 at B, these being the values of the ratio of the reflected to the incident waves at A and at B, which may be + or −, due to terminal resistances (without self-induction or other cause to produce a modified reflected wave; some of these will come later); and let ρ be the attenuation from end to end of the circuit (A to B or B to A), viz.,

$$\rho=\epsilon^{-Rl/Lv}. \qquad (16g)$$

Then an elementary positive disturbance V_0 starting from A becomes attenuated to ρV_0 on reaching B; becomes $\rho_1\rho V_0$ by reflection at B; travels to A, when it becomes $\rho^2\rho_1 V_0$; is reflected, becoming $\rho_0\rho_1\rho^2 V_0$; and so on, over and over again, until it becomes infinitesimal, by the continuous dissipation of energy in the circuit, and the periodic losses on reflection. But if the circuit have no resistance and no leakage, and the terminal resistances be either zero or infinity, there is no subsidence, and the to-and-fro passages with the reversals at A and B continue for ever.

If an impressed force e be inserted anywhere, say at distance z_1, it causes a difference of potential of amount e there, which travels both ways ($+\frac{1}{2}e$ to the right, and $-\frac{1}{2}e$ to the left) at speed v, with the proper attenuation as the waves progress. That is, taking for simplicity the zero of z at the seat of impressed force, we set up a positive wave

$$V_1 = \tfrac{1}{2} e\, \epsilon^{-Rz/Lv}, \qquad \text{(17}g\text{)}$$

and a negative wave

$$V_2 = -\tfrac{1}{2} e\, \epsilon^{+Rz/Lv}; \qquad \text{(18}g\text{)}$$

these being true when z is less than vt in the first, and $-z$ is less than vt in the second. On arrival at A and B these waves are reflected in the manner before described. It will be understood that the original waves still keep pouring in, so long as e is kept on. By successive attenuations we at length arrive at a steady state, which is that calculable by Ohm's law, allowing for leakage.

If the impressed force be at A, and the circuit be short-circuited there, making $\rho_0 = -1$, the two initial waves are converted into one, thus,

$$V_1 = e\, \epsilon^{-Rz/Lv}, \qquad \text{(19}g\text{)}$$

true when z is not greater than vt. On arrival at B, if the resistance there be Lv, nothing more happens, *i.e.*, (19g) is the complete solution. This is something quite unique in its way. If e at A vary in any manner with the time, the current at B varies in the same manner at a time l/v later. Thus, if $e = f(t)$, the current at B is

$$C_B = \frac{f(t - l/v)}{Lv} \epsilon^{-Rl/Lv}. \qquad \text{(20}g\text{)}$$

But if we short-circuit at B, we superimpose first a negative wave

$$V_2 = -e\rho . \epsilon^{-R(l-z)/Lv} = -e\rho^2 . \epsilon^{Rz/Lv}, \qquad \text{(21}g\text{)}$$

beginning at time l/v and travelling towards A; then at time $2l/v$ add a positive wave

$$V_3 = e\rho^2 . \epsilon^{-Rz/Lv}, \qquad \text{(22}g\text{)}$$

and so on, *ad inf.*, settling down to the steady state.

The Fourier-series solution in this case (got by the method of Part IV.) is

$$V = e . \frac{\rho \epsilon^{\frac{Rz}{Lv}} - \rho^{-1} \epsilon^{-\frac{Rz}{Lv}}}{\rho - \rho^{-1}} - \frac{2e}{l} \sum v \sin mz\, \epsilon^{-qt} \frac{(q \sin + vm \cos) vmt}{q^2 + v^2 m^2}. \qquad \text{(23}g\text{)}$$

This includes the whole process of setting up the final state, but requires laborious examination to extract its real meaning, which we have already described. (ml goes from π, 2π, 3π, ..., up to ∞.) When the summation vanishes, we have left the term independent of t, of which the positive part is the sum of the positive waves V_1, V_3, etc., and the negative is the sum of the negative waves V_2, etc., above ((19g), (21g), (22g)).

The uniquely simple case of complete absorption at B of the first wave is much more troublesome by Fourier-series than is the really more complex (23g) case. In some other cases in which we can by the method of waves solve completely, and in a rational manner, the Fourier-series are difficult to interpret.

Let us construct the complete solution when the terminal resistances have any values; by (15g) we know ρ_0 and ρ_1, and by (16g) we express ρ. First of all we have the positive wave

$$V_1 = \tfrac{1}{2}e(1-\rho_0)\epsilon^{-qz/v}, \quad \text{..........................(24}g\text{)}$$

true when z is not greater than vt. When $t=l/v$ it is complete, and remains on. Then begins

$$V_2 = \tfrac{1}{2}e(1-\rho_0)\rho^2\rho_1 . \epsilon^{qz/v}, \quad \text{........................(25}g\text{)}$$

travelling towards A, when it is complete and remains on. The third wave then begins:—

$$V_3 = \tfrac{1}{2}e(1-\rho_0)\rho^2\rho_1\rho_0 . \epsilon^{-qz/v}, \quad \text{.....................(26}g\text{)}$$

which reaches B at time $t=3l/v$, and remains on. The fourth wave then starts:—

$$V_4 = \tfrac{1}{2}e(1-\rho_0)\rho^4\rho_1^2\rho_0 . \epsilon^{qz/v}, \quad \text{.....................(27}g\text{)}$$

reaching A at time $4l/v$; and so on. We thus follow the whole history of the establishment of the final state. The resultant positive wave is the sum of V_1, V_3, ..., and the resultant negative wave the sum of V_2, V_4, ..., which are in geometrical progression; so that finally we have

$$V = \frac{\tfrac{1}{2}e(1-\rho_0)}{1-\rho^2\rho_0\rho_1}(\epsilon^{-qz/v} + \rho^2\rho_1\epsilon^{qz/v}). \quad \text{.....................(28}g\text{)}$$

In the positive component-waves the current is got by dividing V by Lv, and in the negative waves by $-Lv$, so that we get the resultant final current by dividing V in (28g) by Lv and changing the sign of the second term, expressing the negative waves of V.

Should L and S have their values changed in any way, the final state (28g) will be unaltered, but the manner in which it is established will not be the same, of course. We can, however, form a very fair idea of the process from the above, when R/L is not greatly different from K/S, especially if the circuit be sufficiently short to make the attenuation ρ be not great.

The case of no resistance is peculiar. There is no steady state if there be no resistance to make the to-and-fro waves (which may be regarded as a single wave overlapping itself) attenuate. Thus, if there be short-circuits at A and B, and also $R=0$, $K=0$, the first wave due to e at $z=0$ is

$$V_1 = e \qquad \text{from } z=0 \text{ to } z=vt.$$

Then, when this is completed, we have to add on the reflected wave

$$V_2 = -e \qquad \text{from } z=l \text{ to } z=2l-vt,$$

so that when B is reached, there is no electrification left. This is a period, and the state of electrification repeats itself in the same way. But the current doubles itself the moment the first wave reaches B, and the region of doubled current then extends itself to A, where it is at once increased to a trebled value; and so on, *ad inf.*, every reflection adding e/Lv to the current. Thus the current in time mounts up infinitely, though never becoming permanently steady at any spot. The least resistance anywhere inserted will cause a settling down to (or mounting up to) a final steady current.

Effect of Resistances and Conducting Bridges Intermediately Inserted.

Let us now examine the effect of an intermediately inserted resistance r. (If the circuit be a double wire, then, in accordance with the Section on Interferences in Part VII., half the resistance should be put in one wire, and half in the other, just opposite.) Let a wave be going towards r, and let V_1, V_2, and V_3 be corresponding elements in the incident, reflected, and transmitted waves. As we have

$$\left.\begin{aligned} V_1 + V_2 &= V_3 + rC_3, \\ C_1 + C_2 &= C_3, \end{aligned}\right\} \qquad (29g)$$

and V_1 and V_3 are positive waves, whilst V_2 is a negative wave, therefore

$$V_3/V_1 = (1 + r/2Lv)^{-1}, \qquad (30g)$$

and

$$V_1 = V_2 + V_3. \qquad (31g)$$

From (31g) we see that an element of the original wave, on arriving at the resistance, is divided into two parts, both of the same sign as regards electrification, of which one goes forward, the other backward, increasing the electrification behind. The attenuation caused by the resistance is expressed by (30g). If there be n resistances r, such that $nr = Rz$, equidistantly arranged, the attenuation produced in the distance z will be the nth power of the right member of (30g), and in the limit, when the resistances are packed infinitely closely, each being infinitely small, the attenuation in distance z becomes

$$\epsilon^{-Rz/2Lv}. \qquad (32g)$$

This, it will be observed, is when there is *no* leakage. R is the resistance per unit length, uniformly distributed.

Now consider the effect of a bridge of conductance k, in the absence of resistance in the wires, or of uniform leakage. We now have

$$\left.\begin{aligned} V_1 + V_2 &= V_3, \\ C_1 + C_2 &= C_3 + kV_3, \end{aligned}\right\} \qquad (33g)$$

if V_1, V_2, V_3 be corresponding incident, reflected, and transmitted elements. Consequently

$$V_3/V_1 = (1 + k/2Sv)^{-1}, \qquad (34g)$$

and

$$C_1 = C_2 + C_3. \qquad (35g)$$

Compare with (30g), (31g). Observe the changes from voltage to current, inductance to permittance, and resistance to conductance. It is the current that now splits *without loss*, (like the charge before), so that the reflected electrification is negative, if the incident be positive.

The attenuation in distance z due to uniformly distributed leakage-conductance K per unit length is therefore

$$\epsilon^{-Kz/2Sv}$$

We may infer from this opposite behaviour of a resistance in the main circuit, and of a bridge across it, that if $r/L = k/S$, there will be no reflected wave. We must, however, see whether combining the resistance and bridge does not alter the nature of the result. When the resistance r and the bridge of conductance k coexist at the same spot, we shall have

$$\left.\begin{array}{l} V_1+V_2=(1+r/Lv)V_3, \\ V_1-V_2=V_3+(V_1+V_2)k/Sv, \end{array}\right\} \qquad (36g)$$

whence

$$\frac{V_2}{V_1}=\frac{r-(k/Sv)(r+Lv)}{r+2Lv+(k/Sv)(r+Lv)}. \qquad (37g)$$

So the reflected wave is annulled when

$$\frac{r}{Lv}=\frac{k}{Sv}+rk, \qquad (38g)$$

or by $r/L=k/S$ when r and k are infinitely small. When this happens, the attenuation is

$$V_3/V_1=(1+r/Lv)^{-1}, \qquad (39g)$$

and, therefore, when R and K are uniformly distributed,

$$\epsilon^{-Rz/Lv}$$

is the attenuation in distance z. We have thus a complete electrical explanation of the distortionless system; reflection due to conductance in the dielectric itself is annulled by reflection due to the boundary resistance (of the wires). If there be no leakage, any travelling isolated disturbance will cast a slender tail behind it, whose electrification is similarly signed to that of the nucleus, whilst the current in the tail points to its tip. On the other hand, if there be leakage, but no resistance in the wires, the travelling disturbance will cast off a tail of a different kind, viz., of the opposite electrification to the nucleus, and of the same current as in the nucleus. And when the resistances in the wires and in the dielectric are properly balanced, the formation of tails is prevented altogether.

From this manner of viewing the matter we can get hints as to the solution of other and more difficult partial differential equations than the one we are concerned with. Keeping to it, however, we may somewhat generalise it by making the attenuation-rate a function of the distance, and also the speed, but managing so that there shall be no tailing. Thus, it is clear that if L and S be constant, whilst R and K are functions of z such that their ratio is constant, the speed will be constant, and there will be no tailing, whilst the attenuation in distance $z-z_0$ will be

$$\exp\left(-\frac{1}{Lv}\int_{z}^{z} R\,dz\right). \qquad (40g)$$

Now if we make the speed also variable, we must inquire how to prevent tailing due to what is equivalent to a change of medium, as when light goes from air into glass perpendicularly. The condition that there be no reflected ray is $\mu_1 v_1=\mu_2 v_2$ in that case, μ_1 and μ_2 being the inductivities, and v_1 and v_2 the speeds. In our present case it is $L_1v_1=L_2v_2$ when the wires and the dielectric have no resistance and no conductance respectively; L_1, v_1 being the values on one side, L_2, v_2 those on the other side of the discontinuity. That is, the quantity Lv must not vary with z, if there is to be no tailing.

We should, however, make sure that this is the condition when we have simultaneously L, S, R, and K in operation. Let, then, r and k be the resistance in the main circuit, and the conductance of a bridge across it, at a place where the main circuit changes in inductance and permittance from L, S to L', S', the main circuit being supposed to have itself no resistance or leakage. Let V_1, V_2, and V_3 be corresponding elements of an incident, reflected and transmitted wave. We have, by common electrical principles, united with the properties $V = \pm LvC$,

$$\left.\begin{aligned} V_1 + V_2 &= (1 + r/L'v')V_3, \\ (V_1 - V_2)/Lv &= V_3/L'v' + k(V_1 + V_2); \end{aligned}\right\} \qquad (41g)$$

from which

$$\frac{V_2}{V_1} = \frac{1 - k/Sv - Lv/(r + L'v')}{1 + k/Sv + Lv/(r + L'v')}, \qquad (42g)$$

$$\frac{V_3}{V_1} = 2 \div \left(1 + \frac{Lv}{L'v'} + \frac{r}{L'v'} + \frac{k}{Sv} + \frac{rk}{L'v'Sv}\right). \qquad (43g)$$

There is no reflected wave when the numerator on the right of (42g) vanishes, or when

$$\frac{r}{L'v'} + 1 = \frac{k}{Sv} + \frac{Lv}{L'v'} + \frac{rk}{SvL'v'}; \qquad (44g)$$

and then

$$V_3/V_1 = (1 + r/L'v')^{-1}. \qquad (45g)$$

So, if we take $Lv = L'v'$, we secure the desired result, because the product rk ultimately vanishes when we distribute resistance and conductance continuously. That is to say, if Lv does not vary, and $R/L = K/S$ always, there will be no tailing, the speed will be a function of z, viz.:

$$v = (LS)^{-\frac{1}{2}},$$

and the attenuation-rate will be a function of z, as indicated by (40g).

To verify, observe that our fundamental equations (1) may be written, if $R/L = K/S$,

$$-v\frac{dV}{dz} = (q + p)LvC, \qquad -v\frac{dC}{dz} = (q + p)\frac{V}{Lv}; \qquad (46g)$$

hence, if Lv be constant, we have

$$-v\frac{dV}{dz} = (q + p)(LvC), \qquad -v\frac{d(LvC)}{dz} = (q + p)V, \qquad (47g)$$

which become *identical* if $V = \pm LvC$, indicating a complete satisfaction when q and v are functions of z. Then

$$\left(\pm v\frac{d}{dz} + \frac{d}{dt} + q\right)V = 0 \qquad (48g)$$

are the equations of positive or negative waves.

Approximate Method of following the Growth of Tails, and the Transmission of Distorted Waves.

The substitution of isolated resistances and conducting bridges for continuously distributed resistance and leakage leads to a very easy way of following the course of events when there is distortion by a want of the balance between the resistance in the main circuit and the leakage which is required to wholly remove the distortion. As may be expected, the results are only rough approximations, but the method is so easy to follow, and gives so much information of a rough kind, that it is worthy of attention. The subject is quite a large one in itself, and would need a large number of diagrams to fully illustrate. I shall therefore only briefly indicate the nature of the process.

Suppose there is no leakage whatever. Then, unless the resistance in the main circuit be low, there will usually be much distortion due to tailing, unless the waves be of great frequency, making R/Ln small. The smaller this quantity is, by either reducing R, or increasing L or the frequency, the nearer do we approximate to a state of little distortion, and to attenuation represented by

$$\epsilon^{-Rz/2Lv}$$

in the distance z. In fact, in long-distance telephony we do not need any excessive leakage to bring about an approximation to the state of things which prevails in our distortionless system (where, however, disturbances of any kind, not merely waves of very great frequency, are propagated without distortion), and the attenuation is of course less than when there is leakage. As this, however, would require us to examine the sinusoidal solutions of Parts II. and V., we may now keep to the question of tailing and its approximate representation.

Let it be required to find how a charge, initially given existent in a small portion of the circuit, and at rest, divides, when left to itself. We know that if there were no resistance, it would immediately separate into equal halves, which would travel with speed v in opposite directions without attenuation or distortion. And, if there be resistance, but accompanied by proper leakage to match, the same thing will happen, with attenuation. Now there is to be no leakage; this keeps the total charge unchanged. If then there were no tailing there would be no attenuation. But the charges, on separation, cast out slender tails behind them, so that they are joined by a band (the two tails superimposed). The heads, therefore, or nuclei, are attenuated, besides being distorted; the loss of charge from them is to be found in the tails. It is sufficient to consider the progress of one of the two halves of the initial disturbance, say that which moves to the right, and the tail it casts behind it.

Localise the resistance at points, between which there is no resistance, and let the attenuation in passing each resistance (equidistantly placed) be any convenient large proper fraction, say $\frac{9}{10}$; though this is scarcely large enough it is convenient, as all operations will consist in multiplications by 9 and simple additions. Let the initial charge, moving to the right, be 10,000, extending uniformly over a complete section between two resistances, and let a be the time taken to travel one section. Then first we have

$$\overrightarrow{10{,}000}\,;$$
$$\overleftarrow{1{,}000},\quad \overrightarrow{9{,}000}\,;$$
$$\overleftarrow{900},\quad \overrightarrow{100},\quad \overleftarrow{900},\quad \overrightarrow{8{,}100}\,;$$
$$\overleftarrow{810},\quad \overrightarrow{90},\quad \overleftarrow{820},\quad \overrightarrow{180},\quad \overleftarrow{810},\quad \overrightarrow{7{,}290}.$$

The figures in the successive lines show the distribution of the charge in the consecutive sections to right and left, initially and after intervals a, $2a$, $3a$, etc. First of all $\frac{9}{10}$ of the initial charge passes into the next section to the right, and the other $\frac{1}{10}$ is reflected back by the resistance to where it was at the beginning. Then these two charges similarly divide, $\frac{9}{10}$ of each going forward, the other $\frac{1}{10}$ backward. The arrows indicate the direction of motion of a charge. All subsequent operations consist in pairing the charges which are moving towards one another in the proportions $\frac{9}{10}$ and $\frac{1}{10}$. After seven operations we have this result:—

$$\overleftarrow{531},\ 59,\ 566,\ 120,\ 583,\ 184,\ 591,\ 245,\ 583,\ 302,\ 565,\ 371,\ 530,\ \overrightarrow{4773}\,;$$

so that more than half the original charge is in the tail. The directions of motion are alternately to left and to right, so that it is only necessary to know this, and not to continue drawing the arrow-heads. The currents are alternately + and −.

But we should, to approach reality, extend the original charge at least over two sections, instead of one only. To do this, we have merely to add each of the numbers to the one following it. After seven operations, therefore, an initial charge of 20,000 extending over two sections, and moving to the right, becomes distributed thus:—

531, 590, 625, 686, 703, 767, 775, 836, 828, 885, 867, 936, 901, 5303, 4773;

which is really something like its distribution when the resistances are uniformly spread. The corresponding current is not represented by these figures, of course, owing to the opposite direction of current in alternate segments when the original charge extended over only one segment. Allowing for this fact, the current, after seven operations, due to 20,000 over two sections initially, is represented by

$$\overleftarrow{531},472,507,446,463,399,407,346,338,281,263,\overleftarrow{194},169,\overrightarrow{4243},\overrightarrow{4773}.$$

In the head the current is positive. In the whole of the tail (represented by the small numbers) the current is negative. We see that the division of the initial charge over two sections has not been sufficient to remove the fluctuations wholly, though the reversals have disappeared.

In course of time, if the circuit be sufficiently long, the nucleus is so attenuated as to practically make the charge one long tail stretching out both ways, and tending to do so equally, so that the greatest V-disturbance is at or near the origin to the right of it. The current is then negative in the hinder part and also in a portion of the forward part, and positive in the rest. That is, the region of positive current extends gradually from the nucleus into the tail.

Now pass to the other kind of tail, due to reflection by leakage. If there be no resistance in the circuit, but uniform leakage instead, we have tailing and distortion of a distinct kind. It is the current-element that splits into two parts, one going forward, the other backward on passing a bridge, whilst the electrification in the reflected wave is the negative of that in the incident. If, then, the attenuation be $\frac{9}{10}$ as before (ratio of transmitted to incident wave), at every one of the isolated conducting bridges which we use to replace uniformly distributed leakage-conductance, we shall have the same results as above precisely, except that current takes the place of transverse voltage. Thus the first row of figures (after seven operations) shows the current distribution (everywhere positive) due to an initial charge 10,000 (with corresponding current as before) extending over one section; the second row that due to 20,000 over two sections; and the third row the corresponding distribution of electrification, positive in the head, and negative in all the rest. Observe that as, when there was no leakage, the line-integral of V remained constant, so now that there is leakage, the line integral of C remains constant. In one case it is really conservation or persistence of the electrification $\int SVdz$; in the other, of the momentum $\int LCdz$. In the one case the momentum-integral subsides, the time-factor being $\epsilon^{-Rt/L}$; in the other the electrification-integral subsides, the time-factor being $\epsilon^{-Kt/S}$. In both cases the energy subsides towards zero, in spite of the persistence of electrification or of momentum.

When we have both resistance in the conductors and leakage, the tail is positive or negative (referring to the electrification), according as R/L is greater or less than K/S. The latter case is quite out of ordinary practice, which aims at high insulation; the results are consequently very singular, when considered in more detail, which cannot be done now.

In a somewhat similar manner to that in which we have roughly followed the growth of tails, we may follow the progress of signals through a circuit, and obtain the arrival-curves of the current at the distant end, or rather, we may obtain curves resembling the real ones somewhat by drawing curves through the zigzags which result. The method has no recommendation whatever in point of accuracy: its real recommendation lies in the facility with which a general knowledge of the whole course of events may be obtained, and I daresay some people may think that of not insignificant moment.

To make the method intelligible, without going into detail elaborately, let the circuit be perfectly insulated, and in only seven sections, at each of the six junctions of which is concentrated one-sixth part of the resistance of the real circuit. The results will now depend materially upon the ratio Rl/Lv, whether it be a large number, or small. First, let it be small, say $Rl=\frac{4}{5}Lv$. The attenuation at each resistance $(Rl/6)$ is then $\frac{9}{10}$ as before. Let us also insert resistances of amount Lv at both ends, to stop reflections and complications. Then, starting with 10,000 in the first section, we proceed thus:—

A.	$\overrightarrow{10{,}000}$;							
	$\overleftarrow{1{,}000}$,	$\overrightarrow{9{,}000}$;						
	0,	$\overleftarrow{900}$,	$\overrightarrow{8{,}100}$;					
	$\overleftarrow{810}$,	$\overrightarrow{90}$,	$\overleftarrow{810}$	$\overrightarrow{7{,}290}$;				
	0,	$\overleftarrow{738}$,	$\overrightarrow{162}$,	$\overleftarrow{729}$,	$\overrightarrow{6{,}561}$;			
	$\overleftarrow{664}$,	$\overrightarrow{74}$,	$\overleftarrow{672}$,	$\overrightarrow{219}$,	$\overleftarrow{656}$,	$\overrightarrow{5{,}905}$;		
	0,	$\overleftarrow{612}$,	$\overrightarrow{134}$,	$\overleftarrow{612}$,	$\overrightarrow{262}$,	$\overleftarrow{590}$,	$\overrightarrow{5{,}314}$;	B.
	$\overleftarrow{551}$,	$\overrightarrow{61}$,	$\overleftarrow{564}$,	$\overrightarrow{181}$,	$\overleftarrow{557}$,	$\overrightarrow{295}$,	0;	
	0,	$\overleftarrow{514}$,	$\overrightarrow{112}$,	$\overleftarrow{520}$,	$\overrightarrow{219}$,	$\overleftarrow{29}$,	$\overrightarrow{266}$.	

If a = time of going one section, this gives the whole history of the circuit from the moment of putting on a steady impressed force at A up to $9a$, or $2a$ after commencement of arrival of the current at B. The calculation is precisely that by which we should calculate (by the previously described method) the progress of a charge 10,000 initially in the first section and moving to the right. In time a, 9,000 goes forward to the second section, 1,000 is reflected back. After another step the 1,000 is absorbed, whilst $\frac{9}{10}$ of the 9,000 goes forward, and $\frac{1}{10}$ is reflected back. This brings us to the third line. The first arrival at B is of 5,314, the second of 266, and so on (not carried further). The sum total of all the arrivals at B when carried further is 5,999, which really means 6,000. That is, $\frac{6}{10}$ of the charge would go out at B and $\frac{4}{10}$ at A. Now the same figures serve with the impressed force, which we have to imagine continuously sending into the first section the 10,000 wave. The real state of electrification of the line at any stage is to be found by summing up the columns, and the real state of current by summing up the columns with allowance made for the fact that all charges moving to the left mean negative currents. Thus the current at A falls to its final strength, whilst at B it rises to it. Of course the current would not really arrive at B in a perfectly sudden manner to $\frac{53}{60}$ of its final strength, though it would arrive far more suddenly than the current arrives at the end of an Atlantic cable. The final current is $(e/2Lv)\times\cdot 6$. If we increase the number of sections so greatly that the first arrival at B is *insensible*, then the arrival-curve will resemble that at the end of an Atlantic cable (or even much shorter cables). The value of $\epsilon^{-Rl/Lv}$ is exceedingly small in such a case.

Now if we short-circuit at A and B the process is essentially the same, although we must not absorb all reflected waves arriving at A, and all transmitted waves arriving at B, but reflect them properly. This causes there to be a sort of bore running to and fro, in addition to the regular action, so that the arriving current at B gives a sudden jump at regular intervals $2l/v$ apart; these jumps get smaller and smaller rapidly at each repetition, of course. But should the circuit be so long that the first increment of current at B is insensible, this jumping cannot occur. It is also to be remarked that the insertion of terminal resistances stops the oscillatory action.

It was my intention to have given the equations of the tails, positive or negative, or mixed, but as the investigation would unduly extend the length of the present communication, I propose to consider the tails in the next Part IX. At present I may remark that the equation is in the form of a series of rising powers of $(vt \mp z)$, true when $\pm z < vt$; this gives the results very simply in the early stages of development. But later on, it is desirable to transform first into powers of z multiplied into Bessel's functions of the time, and then into other forms, working down to inertialess solutions.

Conditions Regulating the Improvement of Transmission.

The general lines to be followed to improve the capabilities of telegraph or telephone circuits (long-distance) for getting signals through with the least distortion and least attenuation combined are these. First of all R/L is usually far greater than K/S. We should therefore reduce R/L and increase K/S. The former may be done by either reducing the resistance or by increasing the inductance, or by both together. This will lessen both the attenuation and the distortion. So remarkable is this effect, that without changing either the resistance or the permittance of an Atlantic cable, we could, by increasing the inductance (with sinusoidal currents), make the current-amplitude at B be nearly twice as great as the full strength of steady current (the doubling being due to absence of terminal resistance). It is scarcely necessary to remark that it is wholly impracticable to go anything like so far as this; the illustration serves however to show the extraordinary range of possibilities implied in a single theory. The other way is to increase K and reduce S, or both together. By increasing the leakage-conductance we lessen the distortion, but at the same time increase the attenuation. Thus, if the resistance and the permittance be fixed, we should increase the inductance as much as possible, and then increase the leakage-conductance until the attenuation goes as far as is permissible. We shall then have the least distortion possible with the given resistance and permittance. (It is, however, assumed that we are only approximating towards equalizing R/L and K/S, whilst R/L still remains the larger, as, for instance in the case of a very long cable.)

It seems very probable that the iron-sheathing of a submarine cable may be beneficial, though it is not at all easy to precisely state its full effect. But it is naturally suggested to increase the inductance by the use of an irony insulator. In Part VI. I described the use of nonconducting iron to demonstrate the strict proportionality of magnetic force to induction variations when the range is small. This was an insulator impregnated with iron dust, and it shows, with small range of magnetic force (with which alone we are concerned in signalling) no sign of increased resistance, which is to be avoided, of course, since we require the lowest possible resistance to reduce attenuation and distortion. It is possible, therefore, that such an insulator might be of great service in cables for telephony and telegraphy, especially as its insulation-resistance could not be so high as is ordinarily the case. The changed permittance must also be allowed for, though.

As regards open wires, if of copper, and of low resistance, good telephony is possible to ridiculously great distances, further than any one wants to speak, without troubling about getting the leakage to be large.

There is a value of L which gives the least attenuation. For since, in the distortionless system, the received current is

$$C_B = \frac{e}{Lv}\epsilon^{-Rl/Lv}, \qquad \text{......(50}g\text{)}$$

if short-circuited at A, but with resistance Lv at B; or one half this amount, if there be resistance Lv both at A and at B, we see that

$$Rl = Lv, \qquad \text{......(51}g\text{)}$$

makes C_B a maximum. But the attenuation is then so trifling that to carry this out (by increasing L) would be, if possible, quite unnecessary in the case of a long circuit.

Again, in the case of no leakage at all, it may be shown by an examination of the sinusoidal solution in Part V., that if R/Ln be small, we approximate towards the same formula but with the index $-Rl/2Lv$, so that

$$Rl = 2Lv \qquad \text{......(52}g\text{)}$$

gives the value of Lv which makes the current received at B a maximum to suit a given resistance of circuit. It may also be shown by the same formula that if the receiver have small inductance, the resistance it should have (when of a given size and shape) to make the magnetic force a maximum approximates to Lv, which is the critical resistance that absorbs all arriving disturbances.

May 7, 1887.

XLI. ON TELEGRAPH AND TELEPHONE CIRCUITS.*

[February, 1887; but now first published.]

App. A. *On the Measure of the Permittance and Retardation of Closed Metallic Circuits.*

Owing to the fact that most of the circuits of which mention is made in my brother's paper consist of or contain a considerable amount of

* [This article consists of the three appendices that I wrote to the paper of Mr. A. W. Heaviside and myself on "The Bridge System of Telephony," which paper

buried wires, and therefore possess considerable permittance, combined with the fact that these buried wires have very high resistance, as much as 45 ohms per mile, and with the further fact that the self-induction of these lines is small, we may, leaving on one side the question of the apparatus (which is no unimportant one in itself), regard the transmission of telephonic currents through the lines as being governed mainly by the three factors—resistance, permittance, and length of line. Take, therefore, for starting-point the now well-known theory of the submarine cable promulgated by Sir W. Thomson in 1855, which was so curiously foreshadowed by Ohm in 1827, in his celebrated memoir on the galvanic circuit, when guided by an analogy between the flow of electricity and the flow of heat, which is now known to be entirely erroneous.

A translation of Ohm's memoir is contained in vol. II. of Taylor's "Scientific Memoirs," and Sir W. Thomson's writings on the subject of the submarine cable are collected in vol. II. of his "Mathematical and Physical Papers."

Electromagnetic induction is wholly ignored. The line is a single wire, fully defined by the three data—its length, and its resistance and permittance per unit length. The circuit is completed through the "earth," supposed to have no resistance, and to extend right up to the dielectric material which envelops it, whose outer boundary is therefore taken to be permanently at potential zero. On these suppositions, a single quantity V, the potential of the wire, when given along it, fully expresses its state at a given moment, and we may exactly calculate the effect at the distant end of the line (or at any other part), due to arbitrarily varying the potential by a battery at the beginning; the periods of time concerned being, in lines of different lengths, governed by the important law of the squares. Thus if R be the resistance, and S the permittance per mile of a cable of length l, the retardation is proportional to RSl^2, a certain interval of time, which, if R be in ohms, and S in microfarads, is expressed in millionths of a second, owing to the ohm being 10^9 and the microfarad 10^{-15} c.g.s. electromagnetic units. If there be two cables, with constants R_1, S_1, l_1, and R_2, S_2, l_2, and we operate similarly upon them, the time required to set up a given state in the first will be to that required to set up the corresponding state in the second, as $R_1S_1l_1^2$ is to $R_2S_2l_2^2$. For instance, if it take 1 second to bring the current at the distant end to $\frac{9}{10}$ of its full strength due to a steady impressed voltage at the beginning of the first cable, and the

was intended for presentation to the Soc. Tel. Eng. and Electricians, but which never got so far, owing to the objections of the official censor. I have omitted the portion of Appendix C relating to the distortionless circuit, as the matter is more fully treated elsewhere in this volume. The portions of the obnoxious paper contributed by myself (about 20 pages) are also omitted, for a similar reason. I was given to understand that the official censor ordered it all to be left out, because he considered that the Society was saturated with self-induction, and should be given credit for knowing all about it. See, however, Art. XXXVIII., p. 160, in this volume for evidence to the contrary. The present article may now usefully serve as appendices to the preceding one "On the Self-Induction of Wires," since it consists mainly of practical applications of the theory contained therein.]

retardation $R_2S_2l_2^2$ of the second cable be 5 times that of the first, it will take 5 seconds to bring the current at the distant end of the second line to $\frac{9}{10}$ of its final strength. The final currents will not, of course, be equal, unless the impressed voltages are in proportion to the resistance of the lines. The way the current rises at the distant end due to suddenly raising the potential at the beginning to, and keeping it at, a constant amount, is precisely similar to the way a current of heat appears at the distant end of a metallic bar when its beginning receives a sudden accession of temperature, which is maintained constant there, provided the bar be prevented from losing heat laterally. This reservation is necessary, because it is usually the case that submarine cables are well-insulated; whilst, on the other hand, there is considerable lateral loss of heat from a bar through which a current of heat is sent. But if the amounts of loss be properly adjusted in the two cases, there will still be a perfect similarity, if the loss per unit length be proportional to temperature-difference in the one case, and to potential-difference in the other.

The effect of terminal resistances, as of the battery at the beginning and of the receiving instrument at the distant end of the line, is to increase the retardation considerably, whilst at the same time somewhat modifying the manner of rise of the current, so that a strict comparison of a cable with terminal resistances to one without them is not possible; although if both have terminal resistances, and they be properly adjusted in amount, we may render the systems similar, and allow strict comparison. The influence of resistance at either end, or at both ends of a line, on the nature of the arrival-curve, was given by me in my paper "On Signalling through Heterogeneous Conductors" [Art. xv., p. 61, vol. I.], the main object of which was to explain the very singular phenomenon of a marked difference in the speed of working through a submarine cable having land-lines of widely different lengths at its two ends, which was first observed by myself in October, 1869, when making trials of the speed of working, both by reversing key and by automatic transmitter, on the then newly-laid Anglo-Danish cable; when I also had the opportunity of being present at both ends of the line (not quite at the same time, however,) so as to be sure that the anomalous symptoms did not arise from some easily remediable local cause, but had their cause deep-seated in the electrical system.

The insertion of a condenser between line and earth at the receiving end, and more especially the insertion of condensers at both ends of the line, has, on the other hand, a remarkable accelerating power on the signalling, more than doubling the speed of working—a performance that contrasts with the effect of the most ingeniously arranged curbing keys, especially when the excessive simplicity of the means by which this result is attained is remembered. This remarkable power seems to have been found out by pure accident, the practice of signalling through condensers having arisen out of Mr. Willoughby Smith's system of testing cables during submersion. It is indeed true that Mr. C. F. Varley had previously patented the method in what Mr. W. Smith has called a fishing patent, but it does not appear that Mr. Varley or anyone else had foreseen the extraordinary merits of the condenser-method. The theory of the influence of terminal condensers I have given in my paper "On Telegraphic Signalling with Condensers" [Art. XIII., p. 47, vol. I.], and again, more completely, in my paper "On the Theory of Faults in Cables" [Art. XVI., p. 71, vol. I.], in which the theory of the almost equally remarkable accelerating effect on the speed of working due to a leakage-fault in the cable is considered, and it is shown how to take account of the influence of any terminal arrangements, with the solutions in several simple cases.

Suppose now we take for granted that we know precisely how signals are propagated through a single submarine cable, with given terminal arrangements; and next, take two equal but quite independent cables, with independent batteries and instruments, and operate upon them similarly and simultaneously, as is symbolically represented in

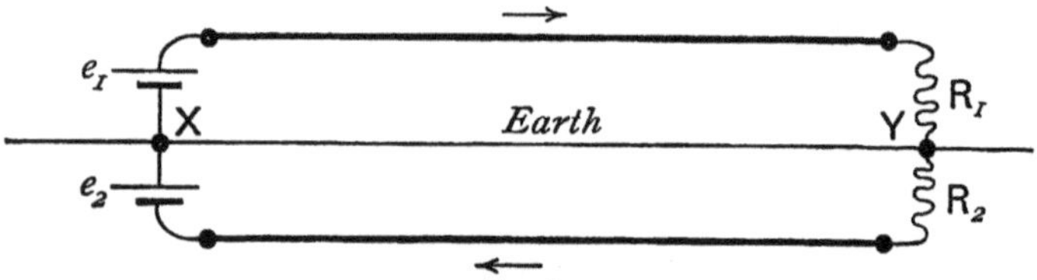

FIG. 1.

fig. 1. If the batteries be both with positive or both with negative poles to line, the phenomena produced in the two cables will be identically the same at the same time at corresponding places, owing to the equality of the cables and of the other circumstances. We could, therefore, by substituting for the two cables one of double the permittance and half the resistance of either of the old; and for the two batteries, one of the same E.M.F. and half the resistance of either; and for the two instruments, one of half the resistance and half the inductance; and, if there be terminal condensers, a single condenser for the two at either end, but of double the permittance of either; signal through the new line in precisely the same manner as through the former two, the new potential being the same as that in both the old cables, whilst the new current is the sum of the currents in the former case.

But if, on the other hand, as in fig. 1, the equal batteries have always opposite poles to line, the potentials at corresponding points will be equal and oppositely signed, and the currents will be equal and oppositely directed in space, or in the same direction in the circuit of

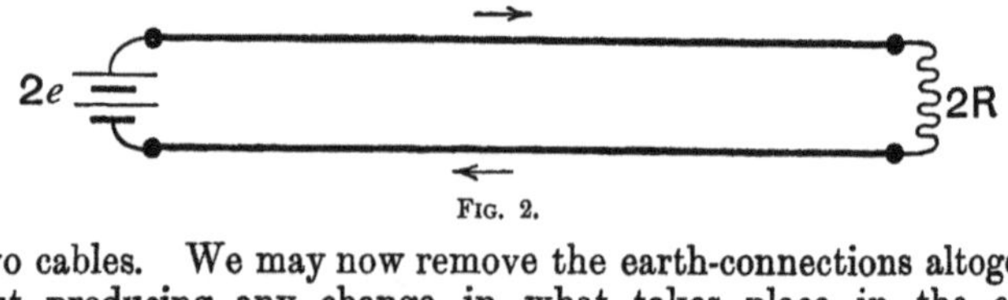

FIG. 2.

the two cables. We may now remove the earth-connections altogether, without producing any change in what takes place in the cables, thus making a closed metallic circuit, as in fig. 2. We see, therefore, that by the abolition of the earth as a return-conductor, and by the substitution of a return through an equal and independent cable, we vary the current in the same manner as before the change, provided we double the E.M.F. and the resistance of the battery, and double the resistance and inductance of the receiver, and if there be terminal condensers, halve their permittances.

So far, therefore, as signals from end to end are concerned, we may treat the new circuit as a single wire with earth-return, if instead of R and S being the constants of either wire, we take them to be $2R$ and $\frac{1}{2}S$ per mile of the new circuit; and, at the same time, take for V, not the potential of either wire, but their difference of potential at a given place; whilst C, the current in the single wire, becomes the current in either wire of the loop-circuit. The new resistance is the resistance per mile of line, and the new permittance is the effective permittance per mile of line. The electrostatic retardation of the line is unchanged. (But if we do not, in passing from single-wire to double, alter the terminal arrangements in proportion, we naturally accelerate signalling.)

The halving of the permittance is, in another form, a doubling of what might be called the electrostatic "resistance," if it were not desirable to refrain from multiplying applications of the term resistance; owing to the condensers, first wire to earth, and earth to second wire, being in sequence, whilst the earth itself counts for nothing except a perfect conductor, for our present purpose. We may, however, perhaps appropriately speak of the doubling of the "elastance" of a condenser, defining the elastance to be the reciprocal of the permittance; for this is at once in accord with Maxwell's "electric elasticity," the reciprocal of the specific inductive capacity, and with the general terminology that I have proposed, thus:—

Conduction Current	Resistance, Resistivity. Conductance, Conductivity.
Magnetic Induction	Inductance, Inductivity. [Reluctance, Reluctivity.]
Electric Displacement	Elastance, Elastivity. [Permittance, Permittivity.]

Resistance and conductance are reciprocal, as are resistivity and conductivity, which refer to the unit volume. Inductivity and elastivity also refer to the unit volume; whilst inductivity is to inductance as conductivity is to conductance; and elastivity is to elastance as resistivity is to resistance. In the cases of the fluxes induction and displacement, it may be observed that appropriate reciprocals are wanting. This system, I find, works well practically, except in this respect. Although elastance is supported by Maxwell's elasticity, yet it does not at all harmonize with displacement, which is, by itself, quite appropriate, though it does not lend itself to the variations that are wanted. Again, elasticity might be confounded with mechanical elasticity, unless we prefix the adjective electric, which prefixing of adjectives is just one of the things that we should try to avoid in a convenient terminology. This objection is, however, completely removed by the substitution of elastivity, which has also the advantage of more perfectly harmonising with conductivity and inductivity. As for going to the dead languages for more new words, which may be quite unaccommodative, I must regard that as a barbarous practice. A good and adaptable substitute for displacement is therefore wanted, and from it a pair of words which shall stand for the reciprocals of the above elastance and elastivity, which are convenient. Now capacity, the present term for the reciprocal of elastance, may mean anything; it is too general a term; we should rather have a word suggestive of *elastic yielding*; capacity seems to suggest the power of holding electricity, a notion which is thoroughly antagonistic to Maxwell's notion of the functions of a dielectric. Again, the reciprocals of inductivity and inductance are wanted. It is quite painful to read of "magnetic resistance" to "lines of force." [I have now inserted the additional words coined after writing the above, and have substituted permittance for capacity in the text.]

After this little digression upon a subject which is important to all who desire the improvement of electrical nomenclature in a systematic and convenient manner that will harmonize with Maxwell's theory of electricity and its later developments, we may return to the looped cables. The earth between them has, or rather has been assumed to have, merely the function of a conductor of negligible resistance; which, though not true, for there would be some small mutual action between the cables, is perhaps sufficiently true practically when cables are submerged. The above reasoning therefore applies to a pair of buried wires, provided they be each wholly surrounded by fairly well-conducting matter, either existent all the way between them, or at least in good conductive connection, if the matter does not extend from the outside of the insulator of one wire to that of the other and surround both. But if this be not the case, it is clear that the effective elastance will be increased by the substitution of dielectric for conducting matter, or the effective permittance will be reduced, thus reducing the retardation. Hence the greatest possible measure of the electrostatic retardation of a pair of equal buried wires in loop is that of either alone, when buried in the technical "earth," and it may be considerably less. Experiment on this point is wanting to see how wires buried in pipes behave as regards permittance. It is no use at all to measure the permittance of each wire by itself with respect to earth; the proper way is (as I have before pointed out) to measure the effective permittance as it really is, that from one wire to the other, modified in amount to an unknown extent (in the present case) by the amount of moisture present, and by the parallel conductors.

If the radius of a wire be r, and that of its (homogeneous) insulator s, its greatest permittance, viz., when earth comes close up to the outside of the insulator, is

$$S = c\left\{2 \log \frac{s}{r}\right\}^{-1}, \qquad \text{.........................(1)}$$

per unit length, where c is the permittivity of the dielectric, this being the well-known formula due to Sir W. Thomson. When the covering consists of concentric layers of different permittivities c_1, c_2, etc., of outer radii s_1, s_2, etc., we get the permittance at once by taking the reciprocal of the sum of the elastances; thus,

$$S=\left[\frac{2}{c_1}\log\frac{s_1}{r}+\frac{2}{c_2}\log\frac{s_2}{s_1}+\frac{2}{c_3}\log\frac{s_3}{s_2}+\dots\right]^{-1}. \quad \dots\dots\dots\dots(2)$$

To illustrate the way of getting this formula, let this wire be suspended in the air, and its permittance with respect to earth be wanted; we shall have to add on the elastance between the outside of the solid covering and the earth, to obtain the total elastance, when, of course, its reciprocal, the permittance, is greatly reduced.

If c vary continuously with the radius, then

$$S^{-1}=\int\frac{2}{cr}dr,$$

taken between the proper limits, is the elastance. Thus, if c vary inversely as r, the elastance is simply proportional to the thickness of the dielectric. If it vary as r, the elastance is proportional to the difference of the reciprocals of the radii, so that the permittance is finite when the outer radius is infinite, instead of zero, as is the case when c is constant, or varies inversely as r. The permittance of an infinitely thick cylindrical dielectric with finite internal radius, is zero or finite according as, if $c=c_0r^n$, n is negative (including zero) or positive, the general formula being

$$S^{-1}=\int_a^b\frac{2dr}{c_0r^{n+1}}=\frac{2}{c_0n}\left\{\frac{1}{a^n}-\frac{1}{b^n}\right\},$$

when the outer and inner radii are b and a.

Similarly, when the dielectric layers are spherical, since the elastance of a layer of thickness dr is $(4\pi/c)(dr/4\pi r^2)$, $4\pi/c$ being the elastivity, we have

$$S^{-1}=\int_a^b\frac{dr}{cr^2}$$

as the expression for the elastance between the proper limits. And if $c=c_0r^n$, we have only to change n to $n+1$ in the cylinder case to obtain the spherical results; *e.g.*, permittance inversely as thickness if $n=-2$.

The *strict* application of this method to magnetic induction problems is not possible on account of the circuital property, except in some peculiar cases of magnetic circuits. But its partial application is useful enough.

[It is, I believe, to Mr. F. C. Webb, in his work "Electrical Accumulation and Conduction," 1862, that we must give the credit of first recognising and employing in electrostatic problems the idea of the addition of elastances, rather than that of the compounding of permittances. It is, however, unfortunate that the application of the method is so limited.]

In the case of a pair of twin wires in pipes, we only safely know the greatest possible effective permittance, which is $\frac{1}{2}S$, where S is given by (1) or (2); whilst the effective resistance is double that of either wire; and that this measure of the permittance may be considerably reduced. But using the proper value, whatever it may be, we may apply the submarine-cable theory, as if a single wire were in question, but taking V to represent the difference of potential of the two wires.

Let us now pass to the other extreme, by removing all conducting matter from the neighbourhood of the wires to a very great distance; for instance, imagine the twin wires to go from the earth to the moon. If the wires be at the same distance apart as before, the permittance is brought to a minimum. (It is, of course, nonsense to talk of the permittance of the wires, strictly speaking, as it is really the permittance of the dielectric between them that is in question.) Let one be charged positively, the other equally negatively; the ratio of this charge to the difference of potential is the permittance required. Its value was given in my paper "On the electrostatic capacity of suspended wires" [Art. XII., vol. I., p. 42]. If r_1 and r_2 are the radii, and r_{12} their distance apart (between axes or centres),

$$S=\left\{2\log\frac{r_{12}^2}{r_1 r_2}\right\}^{-1} \quad \text{..............................(3)}$$

is the permittance per unit length (in electrostatic units), if the dielectric has the unit permittivity. But if the wires are covered with solid dielectrics in concentric layers, this formula (3), or rather the reciprocal, S^{-1}, will only represent the elastance between the external coverings supposed of radii r_1 and r_2; we must then add the elastances of the various concentric layers, as per equation (2), for each wire, to obtain the total elastance between the wires; and, lastly, its reciprocal is the required permittance.

But, keeping to (3), with a dielectric of unit permittivity all the way from wire to wire, the resistance to be coupled with S will be the sum of the resistances of the two wires per unit length. Observe that the radii of the wires need not be equal, nor their resistances. Quite independently of equality of the wires, the propagation of signals from end to end will take place according to the single-wire theory, with R and S as just defined, and V taken to be the fall of potential across the dielectric. (As to the permittance of either wire by itself in space, that is zero, or else meaningless, if it be infinitely long.) But whether magnetic induction will now be ignorable will depend upon the values of R, S, and the inductance, which last is not now in question.

If one conductor surround the other concentrically, and be far removed from other conductors, we of course use formula (1) for the permittance, whilst the effective resistance is the sum of the resistances of the wire and sheath, and V is their difference of potential. But if other conductors be brought close, their presence will necessitate the consideration of the external permittance of the sheath, and somewhat modify the propagation of signals according to the single-wire theory.

Returning to the previous case, let the wires be equal, and be not infinitely removed from other conductors, but still be at a distance from them which is a large multiple of their distance apart; for instance, let them be suspended above the ground in the usual manner. Clearly they will cancel one another to a great extent as regards their influence in charging the earth, when they are equally and oppositely charged by the battery.

Hence the formula (3), with $r_1 = r_2$, or

$$S = \left\{4 \log \frac{r_{12}}{r}\right\}^{-1}, \quad \text{.........................(4)}$$

will be approximately true. But this value of S will be rather less than the true value, which is a little increased by the presence of the earth. The value of the permittance between two unequal wires of radii r_1 and r_2, distant r_{12} between centres, at heights $\frac{1}{2}s_1$ and $\frac{1}{2}s_2$ above the ground, is, if s_{12} be the distance between either wire and the image of the other (the image being a parallel similar imaginary wire as much vertically under as the real wire is above the ground), given by

$$S = \left\{2 \log \frac{s_1 s_2 r_{12}^2}{r_1 r_2 s_{12}^2}\right\}^{-1}. \quad \text{.........................(5)}$$

(To get this and other formulæ, see the paper last referred to, and pair wires.) So, when the wires are of equal radii, and at equal heights, we shall have

$$S = \left\{4 \log \frac{s r_{12}}{r s_{12}}\right\}^{-1}; \quad \text{.........................(6)}$$

and, since s/s_{12} is nearly unity, (4) is nearly equivalent. On the other hand, the permittance between either wire and earth is

$$S = \left\{2 \log \frac{s}{r}\right\}^{-1}, \quad \text{.............................(7)}$$

and we see that one-half of this has no necessary equivalence whatever to the true S of (4) or (6). There may be an accidental equivalence. But, whilst (4) assumes the earth to be infinitely distant, and (6) allows for the increase due to the earth's nearness, there is still a further increase to be practically reckoned on account of the proximity of parallel wires (*i.e.*, when there are any, as is usual). The amount of this increase, which is not at all insignificant, I have calculated in the paper referred to, when the earth is the return-conductor. To get the results when wires are looped, we have merely to pair the wires properly.

It is necessary for the wires to be at the same height above the ground, and to be equal in other respects, for the looped circuit to behave strictly as a single wire in the propagation of signals from end to end. Otherwise, differential effects are produced, due to the currents not being quite equal in the two wires. The extension of the meaning of a "line" to include looped wires, generally to be equal, but sometimes with a complete removal of this restriction, leads to a great simplicity in the treatment of problems relating to the transmission of signals from end to end, doing away with a vast quantity of roundabout work that occurs when each wire is considered independently, with its own constants and potential and current. I have developed this in my paper "On the Self-Induction of Wires," [Art. XL., vol. II.]; a more elementary treatment is contained in "Electromagnetic Induction and its Propagation," Sections XXXII. to XXXV. [Art. XXXV., vol. II., p. 76].

In further illustration of this matter, go back to fig. 1, in which let the wires be, not equal, but have the same time-constants of retardation, or $R_1S_1l^2 = R_2S_2l^2$. Let the upper wire have N times the resistance of the lower, and the lower have N times the permittance of the upper, between wire and earth. The top wire should then have a battery of N times the resistance of that of the battery on the lower wire, and also an instrument of N times the resistance and inductance; whilst any condensers in the lower terminal arrangements should have N times the permittance of those in the upper. In short, the two systems are to be similar; one to be an enlarged copy of the other, the ratio being N.

If, now, the earth be kept on for return-conductor, and similar poles of batteries of equal voltage be to line, the potentials at corresponding points will be equal, though not the currents, so that the two wires behave like one, having the same time-constant. And, if the batteries be with opposite poles to line, with voltages in the ratio R_1/R_2, we have equal but oppositely signed charges and currents, and the earth-connections may be removed, leaving a metallic circuit, which, if V be taken as the fall of potential from wire to wire, is equivalent to a single wire with earth-return, of resistance equal to the sum of the resistances of the two wires, and elastance equal to the sum of the elastances, so that the electrostatic time-constant is unchanged.

This applies to all wires whose dielectric coverings are externally joined by matter of negligible resistance. On the other hand, when there is dielectric everywhere about the wires, we have the case of equation (3) again, if sufficiently distant from earth and other conductors. But if not sufficiently distant, we shall have differential effects produced, and the propagation of signals will not take place strictly according to the single-wire theory, but will have to be, if the differential effects are great enough to make it worth while to allow for them, calculated according to the methods appropriate to self *and mutual* induction of wires, electrostatic and magnetic, as developed in my paper "On Induction between Parallel Wires" [Art. XIX., vol. I., p. 116]. As an extreme case, let one wire be suspended, and the other, of equal resistance, be buried in the ground. Here the differential effects will be very large. But this is a mere curiosity, from the practical point of view. What is important is, that in the practical cases that have arisen of late years, principally owing to the extension of the use of the telephone, in which metallic circuits are employed, the wires are practically equal in all respects, so that the circuit may be treated as a single wire with very great accuracy in the manner I have exemplified here in some elementary cases and developed elsewhere, extended to include self-induction and leakage.

See my paper on "Induction between Parallel Wires" already referred to. For two parallel wires the equations are [vol. I., p. 140]

$$v_1'' = \left(k_1 + s_1\frac{d}{dt}\right)\left(\frac{v_1}{i_1} + c_1\dot{v}_1 + c_{12}\dot{v}_2\right) + s_{12}\frac{d}{dt}\left(\frac{v_2}{i_2} + c_2\dot{v}_2 + c_{12}\dot{v}_1\right),$$

$$v_2'' = \left(k_2 + s_2\frac{d}{dt}\right)\left(\frac{v_2}{i_2} + c_2\dot{v}_2 + c_{12}\dot{v}_1\right) + s_{12}\frac{d}{dt}\left(\frac{v_1}{i_1} + c_1\dot{v}_1 + c_{12}\dot{v}_2\right),$$

where v_1 and v_2 are the potentials of wires 1 and 2 at distance x; k_1 and k_2 the resistances; i_1 and i_2 the insulation-resistances; s_1, s_2, s_{12} the magnetic induction-coefficients; c_1, c_2, and c_{12} the electrostatic induction-coefficients; the dot standing for time-differentiation, and the accent for x-differentiation.

Now let the wires be equal, and loop them. Let

$$\begin{aligned} v_1 - v_2 = V &= \text{difference of potential,} \\ C &= \text{current,} \\ 2k = R &= \text{resistance of line per unit length,} \\ \tfrac{1}{2}(c - c_{12}) = S &= \text{permittance } ,, \quad ,, \quad ,, \\ 2(s - s_{12}) = L &= \text{inductance } ,, \quad ,, \quad ,, \\ (2i)^{-1} = K &= \text{leakage-conductance } ,, \quad ,, \end{aligned}$$

Then we shall have

$$-\frac{dV}{dx} = RC + L\dot{C}, \qquad -\frac{dC}{dx} = KV + S\dot{V};$$

and the potential equation is, by subtracting the equation of v_2 from that of v_1,

$$\frac{d^2V}{dx^2} = \left(R + L\frac{d}{dt}\right)\left(K + S\frac{d}{dt}\right)V.$$

These are the equations of a single wire with earth-return and constants R, L, S, and K, potential V, and current C, as in equation (25) of the same paper [p. 139, vol. I.]. There are several other cases in which a similar simplification results.

It would appear from the results given in my brother's paper, and from others of a similar nature, that the greatest value of the time-constant of a buried circuit with wires of high resistance which it is possible to work through practically with telephones is about

$$RSl^2 = \cdot 015 \text{ second.}$$

From the results obtained in the early days of the telephone I concluded that ·01 second was something like it. But it is really a quite indefinite quantity, depending upon so many circumstances, including not only the instruments, but also the absurdly-called personal "equation." One man might go on to ·015, and another declare that ·0075 was past bearing, a difference of 100 per cent. But on this point I wish it to be distinctly understood, so far as my own views are concerned, that, taking this ·015 second as expressing the practical utmost limit of what it claims to represent, it only applies when the line can be treated as a submarine cable. And, to emphasize this remark, I will add that if any one would pay the cost, which would be considerable, I would undertake to erect a line of such length and permittance that its electrostatic time-constant should be several times this ·015 second, and yet work the telephone beautifully through it. It would not be a submarine cable, that is all. The submarine cable would have no more to do with it than Mrs. Harris.

Apparatus is a matter of considerable importance. Nearly all the progress to efficiency described in my brother's paper was in getting rid of apparatus retardation, and allowing the lines to have the best chance. When, however, it comes to the complete removal of all intermediate apparatus (leaving only apparatus in bridge), and then to working through the longest distance possible, it is clear that, if the terminal apparatus is fairly good, the substitution of one telephone for another cannot (unless they are of widely different natures) be accompanied by any important change in the greatest working distance.

App. B. *On Telephone Lines (Metallic Circuits) considered as Induction-Balances.*

It is needless to say that a circuit consisting of a single wire with earth-return is not balanced against the inductive interference of parallel wires at all. But, as is remarked in my brother's paper, a double-wire telephone line is an induction-balance. More correctly speaking, it ought to be made one. The disturbances of balance referred to in the paper are, from the scientific point of view, of considerable interest. In the following the theory of these disturbances is illustrated by investigating some comparatively simple analogous cases.

Take two long wires and thoroughly twist them together; and join them up with a telephone so that any current in the circuit must go up one wire and down the other; and then try to induce currents in the circuit by means of intermittences or reversals in an external wire. If this be done as a laboratory experiment, there will be no sound in the telephone. It is true that we can easily detect the induction between the primary and a single loop (or half a complete twist) of the secondary, especially if we make a loop in the primary of about the same size; but there is practically not the least effect when it is not one loop, but hundreds in the secondary that are in question. In fact, the two wires of the secondary circuit change places so often that they may, in the mean, be regarded as identically situated, and have precisely equal E.M.F.'s induced in them by the primary current. There is, then, no observable current in the secondary; nor does it matter whether the wires have the same resistance or not, (though there might perhaps be an observable current if the wires were of widely different sizes, especially if the thicker one be iron), nor whether resistance is inserted in the circuit or not. It is simply a question of the resistance and inductance of the secondary circuit; and since there is no E.M.F. in it on the whole, there is no current.

But the case becomes different when we stretch out the double wire to many miles in length; for then electrostatic permittance comes sensibly into play, which allows current to *leave* the wires, and therefore permits current to exist *in* them. The difference between the long and the short line is, however, only one of degree in this respect. In fig. 1, let the two horizontal lines represent a pair of telephone wires in loop, which are to be imagined to be twisted (or not, as we please), and let

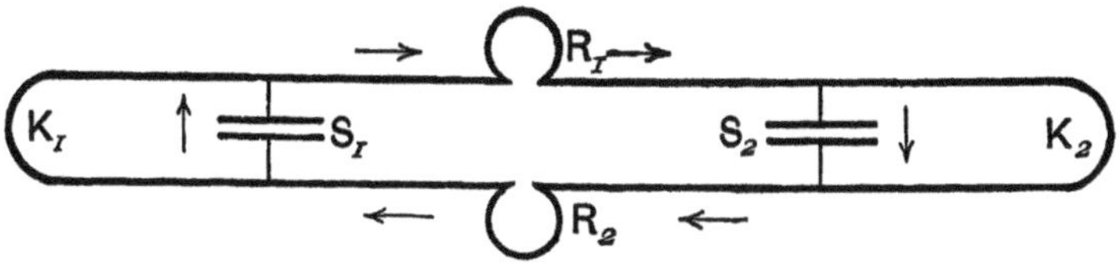

FIG. 1.

K_1, K_2 be the terminal apparatus. There is no interference from parallel wires to be observed at K_1 and K_2 in general, but if a resistance R_1 be inserted intermediately in one of the wires, there is. It can be abolished by inserting an equal resistance R_2 in the other wire at the same place. If unequal, there is still interference. If R_1 is a coil and R_2 a mere resistance, equal to that of the coil, there is still interference. We must make R_2 an equal coil to get rid of it. These interferences are weak, and are not observable when it is a telephone-wire that is the primary; but when the primary is a Wheatstone-transmitter wire, they disturb speech on the telephone circuit, and require removal. The way in which the Bridge-system absolutely cures the evil is one of the most interesting things about it, though not the most important, which is of course the entire removal of the impedance of intermediate apparatus.

Now, if electromagnetic induction were alone concerned, there could be no such interference, either at the terminals or anywhere else. The interference is therefore connected with the permittance of the wires. Imagine, first, the circuit to be so far removed from other conductors that the permittance is appreciably the reciprocal of the elastance from one wire to the other in an infinite dielectric. For illustration in a simple manner, concentrate the permittance at two places, represented by the condensers S_1 and S_2 in fig. 1. Then let the wires be cut by the lines of magnetic force of a primary current, causing equal and similarly directed E.M.F.'s in them between K_1 and S_1, also between S_1 and S_2, and between S_2 and K_2. We shall call these the impressed forces and ignore the external agency. It is easily seen that those between K_1 and S_1 can produce no current; neither can those between S_2 and K_2; as there is no permittance attached to those parts of the circuit. But between S_1 and S_2 the wires are not conductively connected. Yet the impressed forces can still produce no current, because any current there might be is constrained to be of the same strength in both wires, and to be oppositely directed. This conclusion is wholly independent of the resistances concerned, as well as of the permittances of the condensers; so that there could be, in this case also, no interference effect.

Thus, formally, let the arrows indicate the directions of positive current and E.M.F.; let R_1 and R_2 be the resistances of the upper and lower middle sections; L the inductance of the circuit $R_1S_2R_2S_1$; e_1 and e_2 the impressed forces in R_1 and R_2, and V_1, V_2 the falls of potential through the condensers. Then, if C is the current in R_1 and R_2,

$$e_1+e_2=V_1+V_2+(R_1+R_2+Lp)C \quad \text{.................(1)}$$

is the equation of E.M.F. in the circuit $R_1S_2R_2S_1$, where p stands for the time-differentiator. Also, the condenser-equations are

$$C=(K_1+S_1p)V_1=(K_2+S_2p)V_2, \quad \text{.................(2)}$$

if the currents in K_1 and K_2 in the figure be K_1V_1 and K_2V_2. Here K_1 and K_2 may be arbitrary, depending upon the nature of the line, etc., to the left of S_1 and to the right of S_2, if we take K_1V_1 and K_2V_2 to be the currents which shunt the condensers. From these data, if $e_1+e_2=0$, we have $C=0$, so far as impressed force in R_1 and R_2 is concerned.

If e_1+e_2 be not zero, V_1 and V_2 may be found by

$$\left.\begin{aligned} V_1&=\frac{(K_2+S_2p)(e_1+e_2)}{K_1+K_2+(S_1+S_2)p+(K_1+S_1p)(K_2+S_2p)(R_1+R_2+Lp)},\\ V_2&=\frac{(K_1+S_1p)(e_1+e_2)}{\text{same denominator}}.\end{aligned}\right\} \quad \text{.....(3)}$$

Thus, since the equal and similarly-directed impressed forces in the two wires between the condensers can produce no current, and since the same reasoning applies to any number of condensers with any resistances and inductances between them, we may conclude that there will be no current induced in any part of a circuit consisting of two wires twisted together, however unequal they may be, provided the effective permittance be the permittance in the sense above mentioned. This is most intimately connected with the fact that under these circumstances the propagation of signals from end to end of the line takes place in the same manner as on a single wire with earth-return.

As the interference is not due to the mutual permittance, we must refer it to the permittances of the wires with respect to external conductors, or rather, to inequalities therein. Let it be the earth that is the external conductor, and now modify fig. 1 thus, to make fig. 2. Here the pair of condensers S_1 and S_2 represent the permittances of the

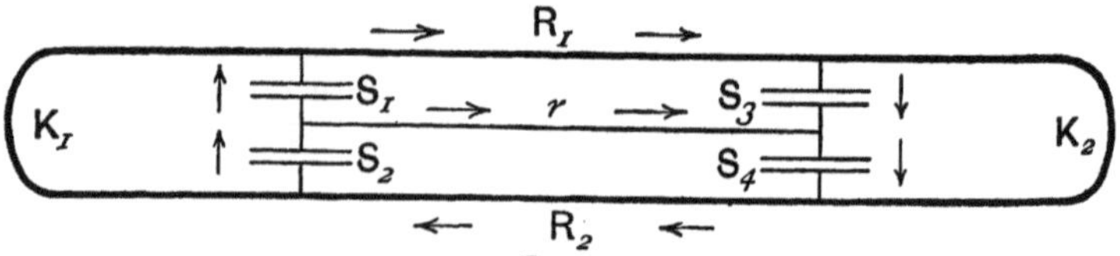

FIG. 2.

upper and lower wires with respect to earth at one place, and S_3 and S_4 do the same at another place, the earth being represented by a wire joining the condensers together as represented, to which wire we may attribute resistance r, which may be zero if we please. As before, the arrows indicate the direction of positive current and E.M.F. Let C_1, C_2, and c be the currents in R_1, R_2, and r; and let L_1, L_2, and M be the inductances, self and mutual, of the circuits $R_1S_3rS_1$ and $rS_4R_2S_2$. Then the equations of E.M.F. in these circuits are

$$\left.\begin{aligned} e_1 &= V_1 + V_3 + (R_1 + L_1p)C_1 - rc + MpC_2, \\ e_2 &= V_2 + V_4 + (R_2 + L_2p)C_2 + rc + MpC_1, \end{aligned}\right\} \quad \text{.................(4)}$$

where V_1, V_2, etc., are the falls of potential through the condensers. Also, the condenser-equations are

$$\left.\begin{aligned} C_1 &= S_3pV_3 + K_3(V_3 + V_4) = S_1pV_1 + K_1(V_1 + V_2), \\ C_2 &= S_4pV_4 + K_3(V_3 + V_4) = S_2pV_2 + K_1(V_1 + V_2), \end{aligned}\right\} \quad \text{...............(5)}$$

if K_1 and K_3 be the conductances (generalised) of the systems to the left between the upper side of S_1 and the lower of S_2, and to the right between the upper side of S_3 and the lower of S_4. Finally, to complete the relations, we have

$$C_1 + c = C_2. \quad \text{.................................(6)}$$

As the current is not now constrained to be of the same strength in the two wires, on account of the auxiliary conductor r, we shall usually have differential effects and interferences. Let us then enquire how to make the currents in K_1 and K_3 zero when $e_1 + e_2 = 0$. That is,

$$V_1 + V_2 = 0, \quad \text{and} \quad V_3 + V_4 = 0, \quad \text{when} \quad e_1 + e_2 = 0. \quad \text{.........(7)}$$

Introduce these into (4), and we get

$$\left.\begin{aligned} e_1 - V_1 - V_3 &= (R_1 + L_1p)C_1 - rc + MpC_2, \\ -(e_1 - V_1 - V_3) &= (R_2 + L_2p)C_2 + rc + MpC_1, \end{aligned}\right\} \quad \text{....................(8)}$$

by adding which there results

$$0 = (R_1 + L_1p + Mp)C_1 + (R_2 + L_2p + Mp)C_2. \quad \text{..............(9)}$$

Also, by (7) in (5), we have

$$C_1 = S_1pV_1 = S_3pV_3, \qquad C_2 = -S_2pV_1 = -S_4pV_3, \quad \text{.........(10)}$$

from which we see that $C_2/C_1 = -S_2/S_1 = -S_4/S_3$; which, used in (9), give

$$0 = (R_1 - R_2S_2/S_1)C_1 + \{(L_1 + M) - (L_2 + M)S_2/S_1\}pC_1, \quad \text{.........(11)}$$

which must be identically satisfied. Hence, finally,

$$\frac{R_1}{R_2} = \frac{S_2}{S_1} = \frac{S_4}{S_3} = \frac{L_1 + M}{L_2 + M}, \quad \text{..........................(12)}$$

are the complete conditions of the induction balance. Notice the independence of the auxiliary wire's resistance. From this we see that in the previous case (got by making r infinite here), the induction-balance was merely true because $C_1 = C_2$; then, as we saw before, R_1, R_2, etc., may have any values. Notice also that M is negative, and that $L_1 + M$ and $L_2 + M$ are the inductions through the circuits $R_1S_3rS_1$ and $R_2S_2rS_4$ due to unit current in the circuit $R_1S_3S_4R_2S_2S_1$.

If, as in the simplest case, the wires are equal, and $R_1 = R_2$, etc., we of course upset the induction-balance by putting a coil in sequence with one of the wires R_1, R_2, and restore it by putting an equal coil in sequence with the other. In this case of equality, undisturbed, we have, since $R_1 = R_2$, $L_1 = L_2$, $S_1 = S_2$, $S_3 = S_4$,

$$C_1 = -C_2 = -\tfrac{1}{2}c = S_1pV_1 = S_3pV_3 = S_2pV_2 = S_4pV_4;$$

and the equation of E.M.F. in the circuit of R_1, R_2 in parallel, and the condensers and return, is

$$e_1 = (\tfrac{1}{2}R_1 + r)2C_1 + \left(\frac{1}{2S_1p} + \frac{1}{2S_3p}\right)2C_1 + \tfrac{1}{2}(L_1 - M)p \,.\, 2C_1. \quad \text{....(13)}$$

We have now equal and similarly directed currents in R_1 and R_2, passing through the condensers and returning combined through the auxiliary wire. The equal wires may be replaced by one of half the resistance, and of inductance $\tfrac{1}{2}(L_1 - M)$; the terminal condensers S_1 and S_2 by one of double the permittance, and similarly for S_3 and S_4, when put in sequence with the substituted wire on the one hand and r on the other. Then $2C_1$ is the going and return current.

It may, perhaps, be worth while to give the full equations in the general case of disturbed balance. They are

$$e_1 = A_3V_3 + A_4V_4, \qquad e_2 = B_3V_3 + B_4V_4, \quad \text{............(14)}$$

in which the A's and B's have the expressions

$$\left.\begin{aligned}
A_3 &= 1 + \frac{K_1S_3 + K_3S_2 + S_2S_3p}{K_1(S_1 + S_2) + S_1S_2p} + (R_1 + L_1p)(K_3 + S_3p) + (rS_3 + MK_3)p,\\
A_4 &= \frac{K_3S_2 - K_1S_4}{\cdots\cdots\cdots} + (R_1 + L_1p)K_3 - rS_4p + Mp(K_3 + S_4p),\\
B_3 &= \frac{K_3S_1 - K_1S_3}{\cdots\cdots\cdots} + (R_2 + L_2p)K_3 - rS_3p + Mp(K_3 + S_3p),\\
B_4 &= 1 + \frac{K_1S_4 + K_3S_1 + S_1S_4p}{\cdots\cdots\cdots} + (R_2 + L_2p)(K_3 + S_4p) + rS_4p + MpK_3.
\end{aligned}\right\} \quad (15)$$

From these we may deduce (12) by taking

$$e_1 + e_2 = 0, \qquad V_3 + V_4 = 0.$$

When, instead of two pairs of condensers only, as in fig. 2, we have a large number of pairs, the earth-wire r must run on and join the middles of every pair. We see from this that the equal E.M.F.'s in R_1 and R_2 will cause currents in them similarly directed which will not return immediately by the wire r in the figure, but only partly there, the rest going further and getting to the auxiliary wire through other condensers. Supposing, then, we have the condensers, etc., uniformly distributed, if the impressed forces be also uniformly distributed along the two wires, there would be, by their mutual cancelling, little if any effect produced (not referring to the balance at the terminals, which is independent of uniformity of distribution of the equal E.M.F.'s). But, generally, the E.M.F.'s will not be thus uniformly distributed.

The general equations of self- and mutual-induction of parallel wires, given in "Induction between Parallel Wires" [vol. I., p. 116], show that if we start with a pair of equal wires looped, and then introduce some inequality, we cause the induction-balance to be a little upset, and simultaneously we cause the circuit to behave not quite the same as a single wire, as described in App. A. Thus, if the wires be equal in all respects, and be at the same height above the ground, they behave as one; and also, if exposed to the interference of a parallel wire equidistant from them, the balance will not be upset. But if the paired wires be in a vertical plane, and therefore at different heights above the ground, we cause a small departure from behaviour as a single wire, and also slightly upset the balance, even although the interfering wire be equidistant from the paired two. Both effects will be small, and it is questionable whether they would be observable. But I am informed by my brother that the interference arising from one wire being of iron and the other of copper has been observed in his district.

When the circuit is completed by a concentric tube, the external permittance of the tube will give rise to interference, if the circuit be long enough. This has not yet been observed.

Practical telephonists who keep their eyes open have unusual opportunities of observing very curious and interesting electrostatic and magnetic effects. Unfortunately, however, the demands of business, to say nothing of other reasons, usually prevent their careful examination, record, and explanation.

App. C. *On the Propagation of Signals along Wires of Low Resistance, especially in reference to Long-Distance Telephony.*

A wholly exaggerated importance has been attached by some writers to electrostatic retardation. I do not desire to underrate its importance in the least—its influence is sometimes paramount,—but the application of reasoning based solely upon electrostatic considerations should certainly be limited to such cases where the application is legitimate. Now some writers, without any justification, take Sir W. Thomson's theory of the submarine cable to be the theory for universal (or almost universal) application, supposing that magnetic induction is merely a disturbing cause, introducing additional retardation, but only to an extent which is practically negligible in copper circuits. This is very wide of the truth. What has yet to be distinctly recognised by practicians, is that the theory of the transmission of signals along wires is a many-sided one, and that the electrostatic theory shows only one side—a very important one, but having only a limited application in some of the more modern developments of commercial electricity, notably in telephony, especially through wires of low resistance. Sometimes magnetic inertia itself becomes a main controlling factor.

In my paper "On the Extra Current" [Art. XIV., vol. I., p. 53] I brought the consideration of magnetic induction into the theory of the propagation of disturbances along a wire, by the introduction of the E.M.F. of inertia, according to Maxwell's system, in accordance with which the inductance per unit length of wire is twice the magnetic energy of the unit current in the wire. Calling this L, the momentum is LC and the E.M.F. due to its variation is $-L\dot{C}$ per unit length.

In my paper "On Induction between Parallel Wires" [Art. XIX., vol. I., p. 116] I have further considered the question; and more recently, 1885-6-7, in the course of my articles "Electromagnetic Induction and its Propagation," and "The Self-Induction of Wires," I have given a tolerably comprehensive theory of the propagation of disturbances, and have worked out certain important parts of it in detailed solutions suitable for numerical calculation. In the present place I propose to give some practical applications of the formulæ, in addition to what I have already given, to be followed by an account of the principal properties of a distortionless circuit, which casts considerable light on the subject by reason of the simplicity of treatment it allows.

Roughly speaking, we may divide circuits into five classes:—

(1). Circuits of considerable permittance, to be regarded as submarine cables in general, according to the electrostatic theory, unless the wave-frequency be great or the resistance very low. Long overhead wires of comparatively small permittance may sometimes be included, especially if the resistance be high.

(2). Short lines which may be treated by disregarding the electrostatic permittance altogether, and considering only the resistance and inductance, provided the frequency be not too great. Ordinary *short* telephone-circuits usually come under this class.

(3). An intermediate class, in which both the electrostatic and magnetic sides have to be considered simultaneously. This class is rather troublesome to manage in general.

(4). Yet another class brought into existence by the late extensions of the telephone in America and on the Continent, and of rapidly increasing importance, in which wires of small resistance and small permittance are used combined with high frequencies, and in which the permittance (though small) must not be ignored, since, in combination with the inductance it produces an approximation towards the transmission of signals without distortion. The theory is then, even when the line is thousands of miles long, quite unlike the electrostatic theory.

(5). Distortionless circuits, now to be first described, in which, by means of a suitable amount of leakage, the distortion of waves is abolished. Though rather outside practice, except that extreme cases of the last class resemble it, this class is very important in the comprehensive theory, because it supplies a sort of royal road to the more difficult parts of the subject.

There may also be sub-classes derived from the above. For instance, a leaky submarine cable, in which resistance, permittance and leakage-conductance control matters, whilst inertia may be of insensible influence.

The peculiarity that is brought in by magnetic inertia (symbolised by the inductance) combined with electric displacement, is propagation by elastic waves (similar to the waves that may be sent along a flexible cord, or perhaps better, a common clothes-line, though even then there is not usually enough resistance), as distinguished from the waves of diffusion (as of heat in metals) which is the main characteristic of the slow signalling through an Atlantic cable. The two features are always both present, but sometimes one is paramount, as in class (1), and sometimes the other, as in classes (4) and (5). [The Americans who went in for wires of low resistance had, I think, no idea of the important theoretical significance of the step they took, but did it because they wanted long-distance telephony, and because wires of high resistance would not go—a characteristically American way of doing things. Yet their action led the way to a rapid recognition of the sound practical merits of Maxwell's theory of the dielectric.]

Let R, S, L, K be the resistance, permittance, inductance, and leakage-conductance respectively, per unit length of circuit, which may be a single wire with earth-return, or a pair of wires in loop, in which case the wires should generally be equal, to avoid the interferences which would remain in spite of the twisting by which the greater part of the interferences from other circuits may be eliminated. Also, let V and C be the potential-difference and current at distance z; then

$$-\nabla V=(R+Lp)C, \qquad -\nabla C=(K+Sp)V, \quad \text{............(1)}$$

where ∇ stands for d/dz and p for d/dt, are the fundamental equations.

Now suppose that an oscillatory impressed force acts at the beginning of the line. Let ρ denote the ratio of its amplitude to that of the current. At $z=0$, ρ is plainly the impedance of the circuit to the impressed force. If the line were perfectly insulated, and had no permittance, ρ would be a constant for the whole circuit, at a given frequency. But the range of the current is not everywhere the same (besides varying in phase), so that ρ is a function of z. The term impedance is strictly applicable only at the place of impressed force, therefore. But to avoid coining a new word, I shall extend its use, and term ρ anywhere the "equivalent impedance." It is with the equivalent impedance at the far end of the circuit, say $z=l$, that we are principally concerned. Call it I, this being the ratio of the amplitude of the impressed force at $z=0$ to that of the current at $z=l$. Let

$$LSv^2=1, \qquad R/Ln=f, \qquad K/Sn=g, \quad \text{................(2)}$$

where $n/2\pi$ is the frequency. Also let

$$P \text{ or } Q=\frac{n}{v}(\tfrac{1}{2})^{\frac{1}{2}}\{(1+f^2)^{\frac{1}{2}}(1+g^2)^{\frac{1}{2}}\pm(fg-1)\}^{\frac{1}{2}}. \quad \text{...........(3)}$$

On these understandings, the value of I is

$$I=\tfrac{1}{2}Lv\left(\frac{1+f^2}{1+g^2}\right)^{\frac{1}{2}}\{\epsilon^{2Pl}+\epsilon^{-2Pl}-2\cos 2Ql\}^{\frac{1}{2}}, \quad \text{...........(4)}$$

provided the line be short-circuited at both ends. Terminal apparatus will be considered later.

If $S=0$, $L=0$, $K=0$, then $I=Rl$, the steady resistance of the circuit. If only $S=0$, $K=0$, then $I=l(R^2+L^2n^2)^{\frac{1}{2}}$, the magnetic impedance. If $L=0$, $K=0$, then

$$I=\tfrac{1}{2}\left(\frac{R}{Sn}\right)^{\frac{1}{2}}\{\epsilon^{2Pl}+\epsilon^{-2Pl}-2\cos 2Pl\}^{\frac{1}{2}}, \quad \text{.................(5)}$$

in which

$$Pl=(\tfrac{1}{2}nRSl^2)^{\frac{1}{2}}. \quad \text{............................(6)}$$

Now the significance of (4) depends materially upon the values of the ratios f, g, and on the frequency. First as regards g. A leakage-resistance of 1 megohm per kilom. makes $K=10^{-20}$, and a permittance of 1 microf. per kilom. makes $S=10^{-20}$ also. Therefore on a land-line of 1 megohm per kilom. insulation-resistance and ·01 microf. per kilom. permittance, we have $g=100/n$. Thus g is important at low frequencies, and becomes a small fraction at high frequencies, even with this relatively low insulation. Thus, $n=1000$ makes $g=\cdot 1$, and $n=20{,}000$ makes $g=\cdot 005$. These correspond to frequencies of about 160 and 3200. We see that in telephony, even with poor insulation, g is always small. By bettering the insulation it is made smaller still. Therefore we may practically take $g=0$ in telephony through a fairly well-insulated line. Notice here that the effect of g in attenuating the current may be considerable when the frequency is low, and yet be small when the frequency is high.

Now the frequency is low on long submarine cables. Consequently g, if there is sensible leakage, has an important attenuating effect. But the above formula does not inform us what other effects leakage has, except by examination through a large range of frequencies. It has a remarkable effect in removing the distortion of the signals, by neutralising the effect of electrostatic retardation. This is marked when the frequency is low, and becomes less marked when it is high. But in the latter case, if the frequency be only high enough, there is little distortion even when the insulation is perfect, or $g=0$, provided the resistance be small. Thus g has a large attenuating and also a large rectifying effect when the frequency is low; when it is high, then it does not attenuate so much and does not rectify so much, nor is so much rectification wanted. But the full nature of this rectifying action will be seen later in the distortionless circuit.

Now consider f. This depends on the resistance, inductance, and frequency. Now 1 ohm per kilom. makes $R=10^4$; consequently, if r be the resistance in ohms per kilom.,

$$f=10^4 r/Ln. \qquad\qquad (7)$$

In a long submarine cable r is small, but n is also small, and L is small, or certainly not great; therefore f is big. So we may take its reciprocal to be zero; or, what will come to the same thing, take $L=0$. We have then the formula (5) for the equivalent impedance (unless leakage is important); and since we can work up to such frequencies that ϵ^{2Pl} is big, we may then write

$$I=\tfrac{1}{2}(R/Sn)^{\frac{1}{2}}\epsilon^{Pl}, \qquad\qquad (8)$$

or

$$\rho=I/Rl=\epsilon^{Pl}(8^{\frac{1}{2}}Pl)^{-1}, \qquad\qquad (9)$$

where Pl is as in (6). This Pl may be as big as 10 on an Atlantic cable. Equation (8) shows the extent to which the line's resistance appears to be multiplied, and is according to Sir W. Thomson's theory.

Now consider buried wires of 45 ohms per mile, such as are used in telephony by the Post Office. Being twin wires, L is small; so, when n is even as high as 10^4, f is made rather large. Consequently we may still apply the electrostatic theory, even in telephony, so far as the buried wires mentioned are concerned, although it will somewhat fail at the higher frequencies: and we see that it is by reason of their *high* resistance and *low* inductance that we can ignore the influence of inertia in them. But this does not apply to the suspended wires which are in circuit with the buried wires, as we shall see presently.

Consider a pair of open or suspended wires. Take 20 ohms per kilom. as the resistance, or 10 ohms each wire. This will, by (7), make $f=2$ if $L=10$ and $n=10{,}000$; and $f=\cdot 2$ if $L=100$. Now the last value of L is extreme. It could only be got with an iron wire, and its inductivity would need to be large even then; besides that, the frequency would need to be low in order to allow the large L to operate, on account of the increased resistance due to the tendency to skin-conduction at high frequencies. Such a large value of L may usually be put on one side, so far as practical work is concerned; but $L=50$ would be more reasonable, remembering that in L is included the part due to the dielectric surrounding the wire. The data regarding the inductivity of iron telegraph-wires are not copious; from my own observations, I believe that, with the weak magnetic forces concerned in telephony, $\mu=200$ is high, and it may be as low as 100. The point is, however, that f, from being large, may be made small by increasing the inductance without other changes. Still, however, with the assumed steady resistance of 20 ohms per kilom., we could not treat f as a small fraction, especially as the increased resistance due to the imperfect penetration of the magnetic induction into the wires will increase f, as will also the reduced inductance due to the same cause. Thus f must be kept in the formula for the equivalent impedance, though not to be treated as either very large or very small in general. That is, we have the form of theory of class (3) mentioned above. Similar remarks apply to long suspended copper wires if the resistance be several ohms per kilom., and they be at the usual distance apart; for although with high frequencies f will be small, yet it will not be small enough at the low frequencies to allow of its treatment as a small quantity. We should therefore use equation (4) with only $g=0$ in general.

But now come to a copper wire of only 1 ohm per kilom., in loop with a similar wire, making $R=20^4$ or $r=2$. Now $n=10^4$ makes

$$f=2/L; \qquad\qquad (10)$$

from which we see that f may be so small a fraction as to lead to a simplified form of theory. We now have the fourth class of circuits; well-insulated, of low resistance, and of fairly high inductance, making R/Ln small, and a tolerably close approach to distortionless transmission.

To estimate the value of L, go back to equation (2) defining v. Here v is a speed, always less than that of light, but of the same order of magnitude. If the wires are of iron, it is considerably less; but if of copper it is so little less that we may neglect the difference. Now

$v^2 = 90^{20}$ and 1 microf. $= 10^{-15}$, so that if s_0 is the permittance in microf. per kilom.,

$$L = (9s_0)^{-1}, \quad \text{.......................(11)}$$

which is useful in giving an immediate notion of the size of L in terms of the permittance, when that is known. Thus ·01 microf. per kilom. makes $L = 11$, so that $f = \frac{2}{11}$ when $n = 10{,}000$, when the resistance per kilom. is 2 ohms; and f is only $\frac{1}{11}$ at the higher frequency $20{,}000/2\pi$.

But this estimate (11) will always be too small a one, and sometimes much too small, if s_0 be the measured permittance per kilom. It was found by Professor Jenkin that the measured permittance was twice as great as that calculated on the assumption that the wire was solitary. The explanation (or a part of it) which I have before given [Art. XII., vol. I., p. 42, and XXXVII., vol. II., p. 159] is that the neighbouring wires themselves largely increase the permittance. Therefore, if s_0 be the measured permittance in presence of earthed wires, the real L must be considerably greater than by equation (11). On the other hand, there is a set-off by reason of L being reduced by the induction of currents in the neighbouring wires, though not so greatly as to counteract the preceding effect. Again, the magnetic field penetrates the earth, which increases L. But, to avoid these complexities, which require us to consider the various mutual effects of circuits, let our circuit be quite solitary. Then, if r = radius of each wire, and s = distance apart,

$$L = 1 + 4 \log (s/r) \quad \text{.......................(12)}$$

when $\mu = 1$, as with copper wires, the 1 standing for $\frac{1}{2}\mu_1 + \frac{1}{2}\mu_2$, if μ_1 and μ_2 are the inductivities of the two wires. These terms are important in the case of iron wires; but not with copper, unless the wires are very close, when they become relatively important on account of the smallness of the total inductance. The other part of L is the inductance of the dielectric, and it is this which, when multiplied by S, gives the reciprocal of the square of the speed of light, subject to the proper limitations. Now $L = 20$ requires $s/r = 148$; or if r be $\frac{1}{8}$ inch (which is about what is wanted to make the resistance 1 ohm per mile), s must be $18\frac{1}{2}$ inches. We therefore see that $L = 20$ is quite a reasonable value with copper loop-circuits. It gives $f = 1$ when $n = 1000$, and $\frac{1}{10}$ when $n = 10{,}000$. Thus f is less than unity throughout the whole range of telephonic frequencies, and becomes a small fraction even at practical frequencies.

Take, then, $g = 0$ and f small in (3) and (4). We get

$$P = \frac{1}{2}\frac{n}{v}\frac{R}{Ln} = \frac{R}{2Lv}, \qquad Q = \frac{n}{v}, \quad \text{.......................(13)}$$

and the equivalent impedance formula (4) reduces to

$$I = \tfrac{1}{2}Lv\left(\epsilon^{Rl/Lv} + \epsilon^{-Rl/Lv} - 2\cos\frac{2nl}{v}\right)^{\frac{1}{2}}, \quad \text{.......................(14)}$$

in the fourth class of circuits.

The further significance of this formula will depend materially upon the value of the ratio $Rl/2Lv$ (that is, the value of Pl), the ratio of the resistance of the circuit to $2Lv$, which is, in the present case, 1200 ohms. If the length of the circuit be a small fraction of 600 kiloms., the impedance depends upon the frequency in a fluctuating manner, going down nearly to $\frac{1}{2}Rl$ and then running up nearly to Lv, as the circular function goes from -1 to $+1$, on raising the frequency. Thus the least possible equivalent impedance at $z=l$ is one half the steady resistance of the line, and the greatest is Lv.

According to (14) this would go on indefinitely, as the frequency was raised continuously. But another effect would come into play, viz., the increased resistance due to skin-conduction, with a corresponding small change in L. As the result of this increased resistance the value of $Rl/2Lv$ will rise, and the range in the fluctuations of I decrease; and if the frequency be pushed high enough the fluctuations will tend to disappear. But this could not happen in telephony at any reasonable frequency, say $n=20,000$.

The physical cause of the low value $\frac{1}{2}Rl$ at certain frequencies is the timing together of the impressed force at the beginning of the circuit and the reflected waves. It is akin to resonance. Thus, if the line had no resistance at all we should have

$$I = Lv \sin(nl/v), \qquad \text{.........................(15)}$$

with the circular function taken always positive. When $nl/v=\pi$, $I=0$. Then $2\pi/n = 2l/v$, or the period of the impressed force coincides with the time of a double transit (to the end of the circuit and back again).

In connection with (15) I may mention that an approximate formula for the impedance, when nl/v is in the first quadrant, and especially in its early part, is

$$I = (R^2 + L^2n^2)^{\frac{1}{2}} l \times \frac{\sin(nl/v)}{nl/v}, \qquad \text{....................(16)}$$

which shows the beginning of the action of the permittance in reducing the impedance from its magnetic value as the frequency is raised.

But to use wires of such low resistance for comparatively short lines would be wastefully extravagant. Such wires admit of very long circuits being worked. Therefore increase the length of the line in equation (14); as we do this the range in the oscillation in I falls, until, when $Rl=2Lv$, I does not depend much upon the circular function. We may then, and at all higher frequencies, write simply

$$I = \tfrac{1}{2}Lv \,.\, \epsilon^{Rl/2Lv}, \qquad \text{.........................(17)}$$

or

$$\rho = \frac{I}{Rl} = \frac{Lv}{2Rl}\epsilon^{Rl/2Lv}. \qquad \text{.........................(18)}$$

Compare with (8), the corresponding cable-formula, and note the differences. The impedance is now nearly independent of the frequency, and there is nearly distortionless transmission of signals, provided R/Ln be small, and $Rl/Lv=2$ or 3 or more.

The following table gives the values of ρ calculated by (14), which only assumes that R/Ln is small, for a series of values of $Rl/Lv = y$.

y.	Min. ρ.	Mean ρ.	Max. ρ.	y.	ρ.	y.	ρ.
$\frac{1}{2}$	·505	1·500	2·063	6	1·678	12	16·81
1	·521	·878	1·128	7	2·365	14	39·3
2	·587	·686	·771	8	3·378	16	93·2
2·0653	·594	·685	·766	9	5·000	18	225
3	·710	·748	·784	10	7·420	20	550
4	·907	·924	·940				
5	1·210	1·218	1·226				

Here the "mean," "maximum," and "minimum" values of ρ mean the values when the cosine is 0, $+1$, and -1. The fluctuations are very large when y is small, going from $\frac{1}{2}Rl$ to Lv; but they are insensible when y is bigger. Remember that the line is short-circuited. The receiving apparatus, by absorbing energy, reduces the fluctuations, and we shall see later that they can be nearly abolished.

When $Rl/Lv = y$ is variable, the value of I/Rl is made a minimum by taking $Rl = 2{\cdot}06\,Lv$, say $2Lv$. This is a little over 1200 ohms in our example of $L = 20$; and makes the length of circuit be 600 kilom., when the resistance is 2 ohms per kilom. After $y = 3$ we may disregard the fluctuations.

Now this length of only 600 kilom is still far too short to make it necessary to employ so expensive a wire. One of much higher resistance would answer quite well enough for practical telephony, in which a considerable amount of distortion is permissible, because transmission would be nearly perfect over 600 kilom. according to the above data. The question arises, upon what principles can we compare one circuit with another, and is it possible to lay down the law from theory as to the limiting distance of telephony? The answer is plainly that it is not possible, because the types of telephonic circuits differ. A cable or other circuit with inertia ignored is radically different from one in which there is a marked approach to elastic wave-propagation. Even if we fix the type, and take, say, the above example of low resistance, 2 ohms per kilom. and $L = 20$ per centim., and the question be asked, How far can you telephone?—the answer is that there is no fixed limit, as it depends upon so many circumstances, some of which are unstated, and are hardly susceptible of measurement when stated.

Consider, first, the circuit without terminal influences. We may distinguish two connected, but yet entirely different, things in operation. We set up electromagnetic vibrations at A somehow, not regular vibrations of one frequency, but irregular, and of almost any type. Now, during transmission along the circuit, the vibrations are attenuated for one thing, and distorted, or changed in type, for another. With perfect transmission there would be neither attenuation nor distortion. This would require perfect conductors, which would not permit the waves to enter them from the dielectric and be dissipated, but would let them slip along like greased lightning. Then there is a kind of circuit which is distortionless, but in which there is considerable attenuation. Here, plainly, any distance can be worked through, provided the attenuation is not too great. Trial alone could settle how far it would be practicable with a given type. Coming to more practical cases, there is the approximately distortionless circuit above described. Here the attenuation is not nearly so great as in the distortionless circuit of the same type (that is, only differing in the leakage needed to remove the remaining distortion), so that the distance to be worked through is much greater with similarly sensitive instruments, or with instruments graduated to make the currents received and sounds produced be about equal in the different cases compared. Here, again, trial alone can settle how far we may work safely. Supposing, for instance, we had reached a practical limit with nearly distortionless transmission, it is clear that we could increase that limit by the simple expedient of increasing the current sent out or the sensitiveness of the receiver. So we cannot fix a limit at all on theoretical principles. But undoubtedly the distortion will increase as the circuit is lengthened (except in the ideal distortionless circuit); this will tend to fix a limit, though we cannot precisely define it, independently of the attenuation. Nor should interferences be forgotten, and their distorting effects. When thousands of miles are in question, many other things may come in to interfere, all tending to fix a limit. Independently of the line, too, there are the terminal arrangements to be considered. A practical limit in a given case might be fixed merely by the inadequate intensity of the received currents to work the receiver suitably. But apart from intensity of action, both the transmitter and the receiving telephone distort the proper "signals" themselves. The distortion due to the electrical part of the receiver may, however, be minimized by a suitable choice of its impedance, and especially by making its inductance the smallest possible consistent with the possession of the other necessary qualifications. The conditions as regards perfect silence in reception are also of importance. Finally, there is "personal equation." It is clear, then, that in such a mixed-up problem as this is, we cannot safely estimate what amount of distortion is permissible in transit along the circuit, and how much attenuated and distorted we may allow the vibrations to become before human speech ceases to be recognisable as such, and to be intelligibly guessable.

It is, however, surprising what a large amount of distortion is permissible, not merely on long lines, but on short ones. It is, indeed, customary, or certainly was on the first introduction of the telephone, and for long after, for people to enlarge upon the wonderful manner in which a receiving telephone exactly reproduces, in all details, the sounds that are communicated to the transmitter, and to be astonished at the power the disc possesses of doing it, and to explain it by harmonic analysis, and so forth. Well, the disc does not do it. If it did, as it would be in quite mechanical obedience to the forces acting upon it, there would be nothing to wonder at; or the reason for wonder would be shifted elsewhere. It would be really wonderful if we could get perfect reproduction of speech. The best telephony is bad to the critical ear, if a high standard be selected, and not one based upon mere intelligibility. (As a commentary upon the reports of "perfect articulation," etc., I may mention that we sometimes see the amusingly innocent remarks added that even whistling could be heard, and one voice distinguished from another.) Consider the difficulties in the way. We cannot even make the diaphragm of the transmitter precisely follow the vibrations set up by the vocal organs (which vibrations are, by the way, distorted between the larynx and the diaphragm, though this is not an important matter), because it is not a dead-beat arrangement, and responds differently to different tones. Here is one cause of distortion. A second occurs in trying to make the primary current variations copy the motion of the diaphragm. A third is in the transformation to the secondary circuit, though perhaps this and the last transformation may be taken together with advantage. So to begin with, we have considerably distorted our signals before getting them on to the telephone line. Then, there is the distortion in transit, which may be very little or very great, according to the nature of the line. Next, the received-current variations ought to be exactly copied by the magnetic stress between the disc and magnet of the receiver. But the inductance of the receiver prevents that, even if the resistance be suitably chosen to nearly stop the reaction of the instrument on the line. Then we should get the disc of the telephone to exactly copy the magnetic-force variations, which it cannot do at all well, on account of the want of dead-beatness, and the augmentation of certain tones and weakening of others. The remaining transformations, from the brain to the vocal organs at one end, and from the disc to the brain *viâ* the air and ear at the other end of the circuit, we need not consider. And yet, after all these transformations and distortions, practical telephony is possible. The real explanation is, I think, to be found in the human mind, which has been continuously trained during a lifetime (assisted by inherited capacity) to interpret the indistinct indications impressed upon the human ear; of which some remarkable examples may be found amongst partially deaf persons, who seem to hear very well even when all they have to go by (which practice makes sufficient) is as like articulate speech as a man's shadow is like the man.

In connection with these transformations, I may mention that one of them, viz., in the telephone receiver itself, was until recently unexplained. Writers have before now remarked upon the necessity of a permanent magnetic field, and speculated as to its cause, and recently Prof. Silvanus Thompson recalled attention to the matter, and candidly confessed his ignorance of the explanation, beyond what was furnished by M. Giltay, who had also considered the matter, and found that the permanent field was needed to eliminate the vibrations of doubled frequency that would result were there no permanent field. This is true in a sense; but it is not the really important part of what is, I think, the true explanation, because the vibrations of doubled frequency would be very feeble. What the permanent field does is to vastly magnify the effect of the weak telephonic currents, and make them workable. The disc is attracted by the magnet, and the stress between them varies as the square of the intensity of magnetic force in the intermediate space. We want the disc to vibrate sensibly by very weak variations of magnetic force. If the permanent magnet were not there, we should have insensible vibrations of doubled frequency. But the permanent field makes the stress-variations vary as the product of the intensity of the permanent field and that of the weak variation due to the current-variations; they are therefore proportional to the received current-variations, and are also greatly magnified, so that the telephone becomes efficient. [See Art. XXXVI., vol. II., p. 155.]

Returning to the telephone-circuit itself, the following would appear to be what should be aimed at (apart from improvements in terminal transmission and reception) in efficient long-distance telephony. Setting up an arbitrary train of disturbances at one end, causing the despatch of a continuously varying train of waves into the circuit, the waves should travel to the distant end of the line as little distorted as possible, and with as nearly equal attenuation as possible, which attenuation should not be too great; and, finally, on reaching the terminal telephone, the waves should be absorbed by it, as nearly as possible, without reflex action. This ideal may be illustrated by a long cord, along which we can, by forcibly agitating one end, despatch a train of waves, which travel along it only slightly distorted, and which should then be absorbed by some mechanical arrangement at the further end. Theoretically this only needs the further end to have its motion resisted by a force proportional to its velocity, the coefficient of resistance depending upon the mass and tension of the cord. At any intermediate point we may correctly register the disturbances passing it. It is evident that the reflected wave from the distant end should be done away with, in order that the disturbances passing (and reaching the distant end) may be a correct copy of those originally despatched. This ideal state of things is fairly-well reached in the fourth class of circuits above mentioned, and perfectly in the fifth class, whilst the low-resistance long-distance circuits introduced in America are somewhere between the third and the fourth classes.

In passing from the fourth class to the third, by increasing the resistance of the line from very low to more common values, the effect is to introduce a considerable amount of distortion which may be (somewhat imperfectly) ascribed to electrostatic retardation. The limiting distance of telephony will therefore now depend more upon the circuit itself (apart from terminal arrangements) than before. Still we cannot fix it. Only by passing to the extreme case of such high resistance of the line acting in conjunction with the permittance that the effect of inertia is really insensible, do we so magnify the effect of the distortion in transit as to make the limiting distance be determined approximately by the value of the electrostatic time-constant RSl^2. We now come to the first class we began with, and Sir W. Thomson's law of the squares may be applied in making comparisons. The distortion in transit is very great, if the line be long, and we therefore to some extent swamp the terminal apparatus as regards the total distortion.

But there is only a tendency to the electrostatic theory, not a complete fulfilment. In the case of a cable of the Atlantic type, used as a telephone-circuit (of course not across the Atlantic) the resistance is rather low, and this is quite sufficient, in conjunction with the inductance, to greatly improve matters from the electrostatic theory, in spite of the large permittance. In fact, a small amount of inductance is sufficient to render telephony possible under circumstances which would preclude possibility were it non-existent. To show this, consider the following table :—

n.	$L=0$.	$L=2{\cdot}5$.	$L=5$.	$L=10$.
1250	1·723	1·567	1·437	1·235
2500	3·431	2·649	2·251	1·510
5000	10·49	5·587	3·176	1·729
10,000	58·87	10·496	4·169	1·825
20,000	778	16·707	4·670	1·854

In the first column we have the frequency-constant $n=2\pi\times$ frequency, so that the frequency ranges through four octaves. It is supposed that the resistance is 4 ohms and the permittance $\frac{1}{4}$ microf. per kilom., being somewhat like what obtains in an Atlantic cable. The remaining columns show the values of the equivalent impedance ρ at the distant end according to the already-given formula (4), with the values of L given at the tops of the columns. (Take $g=0$ in (4).)

Thus in the second column we have the figures given by the electrostatic theory showing such an extremely rapid increase of attenuation with the frequency that telephony would I think be quite impossible.

But the third column shows that the small inductance of 2·5 per centim. immensely improves matters, especially with the great frequencies.

The fourth column, with $L=5$, shows a far greater improvement, and I should think good telephony would be possible.

The fifth column, with $L=10$, is very remarkable, as it shows an approach to distortionless transmission.

This remarkable result is wholly due to the inductance, in presence of the rather low resistance. Whereabouts the effective inductance really lies it is hard to say, but it must surely be greater than 2·5, though it may not be much more, as the iron sheathing does not make the effective L run up in the way that might be supposed at first sight.

With $L=0$, $n=10{,}000$ makes $\rho=58$, or the received current 1/58 of the steady current. To have the same result in our low-resistance circuit, we see by the first table that $Rl=15\,Lv$ about does it, giving $Rl=15\times600=9000$ ohms, and $l=4500$ kilom. Now is it possible to work a telephone fairly well through a mere resistance of 58×9000 or say 50,000 ohms (ignoring complications due to the telephone not being a mere resistance), remembering that our currents will be fairly uniformly attenuated? If so, then this circuit of 4500 kilom. will work with good articulation, under favourable conditions—freedom from interferences, etc. But I do not fix this limit, nor any, for reasons before given.

This difference should be noted. In the case of the cable of *no* inductance, the reduction to 1/58 part applies only to $n = 10,000$. If $n = 1250$, at the lower limit, the reduction is only to 10/17 of the steady current; thus there is plenty of sound, but very inarticulate. This is the reverse of what occurs in our other case, in which there is little sound, but with good articulation, and therefore usefully admitting of magnification.

If, on the other hand, we take the electrostatic time-constant as ·02 second, the attenuation at $n = 10,000$ is, by the second table, to 1/778 of the steady current; and this value, by the first table, gives $Rl/Lv =$ say 20, and $Rl = 12,000$ and $l = 6000$ kilom., and the equivalent impedance $= 778 \times 12,000$ ohms. Of course this is excessively large. If component vibrations on a cable really suffer attenuation to 1/778 part, such vibrations might as well be altogether omitted, leaving only the lower tones. On the other hand, a sufficient magnification in the 6000 kilom. case would render telephony possible. But the probable fact is that ·01 second with $L = 0$ is not possible, far less ·02 second. When it is said to be done, the reason is that L is not zero. In the north of England examples there are usually buried wires and overhead wires in sequence, so that it is still more true that self-induction comes in to help, although the theory of such composite circuits cannot be easily brought down to numerical calculation.

But, returning to the 4500 kilom. example, it appears reasonable that the circuit might be worked under favourable circumstances. Let us see what its electrostatic time-constant is. We get, by (11), $s_0 = \frac{1}{180}$ microf. per kilom. Hence

$$RSl^2 = 2 \times \frac{1}{180} \times \frac{(4500)^2}{10^6} = \frac{2}{9} \text{ second,}$$

which is no less than 22 times the supposed maximum of ·01 second. Even if we make a large allowance, and suppose that an attenuation to $\frac{1}{10}$ part only of the steady current, instead of $\frac{1}{58}$ part, is the utmost allowable, we shall see by the table that this makes $Rl = 11\,Lv$ (instead of the previous 15), so that the electrostatic time-constant is still a large multiple of the value ·01 obtained by observation of wires of high resistance.

Again, to contrast the two theories, let us inquire what length of line makes ·01 sec. the electrostatic time-constant. The result is $300\sqrt{10}$ or say 900 kilom., of resistance 1800 ohms, which is only three times Lv; so that there is nearly perfect transmission on the line of low resistance, whilst there is extreme distortion on the circuit having the same electrostatic time-constant if destitute of inductance.

Since there is a minimum value of the attenuation-ratio I/Rl when the ratio Rl/Lv is variable, let it be merely L that is variable, without change of length or resistance. This may be done by simply varying the distance between the two wires in the circuit. The minimum attenuation at the distant end comes about (by first table) when

$$L = \frac{Rl}{2{\cdot}065v}, \qquad \text{or,} \qquad L = \frac{Rl}{2v} = \frac{Rl \text{ ohms}}{60 \text{ ohms}}, \quad \text{nearly.}$$

When $l = 600$ kilom. we have $L = 20$, as we saw before. If $l = 300$ kilom., then $L = 10$, which change is easily made by bringing the wires closer. But if $l = 1200$ kilom., we require $L = 40$, and a wide separation is necessary, according to equation (12). But there is another thing to be remembered. The distance between the wires should continue to be a small fraction of the height above the ground, in order that the property $LSv^2 = 1$ should remain fairly true. Although the permittance does not appear explicitly in formula (14), it is implicitly present in v, and in such a way that a doubling of S and halving of L are equivalent. (But this does not apply to the table, where L and S may vary independently.) Now, if we separate wires very widely without raising them any higher, S tends to become simply the reciprocal of the sum of the elastances from the first wire to earth and from the earth to the second wire; that is, half the permittance of either. It therefore tends to constancy instead of varying inversely as L, which goes on increasing slowly as the wires are further separated. Hence the necessity of raising the wires, as well as of separating them, if the full advantage of L is to be secured when it is large.

In passing, I may add that if the earth were perfectly conducting, so as to shut out the magnetic field from itself, the product LSv^2, where L is the inductance of the dielectric and S its permittance, calculated so as to suit the propagation of plane-waves, would remain unity always, however the wires were shifted, provided parallelism were maintained.

It seems at first sight anomalous that when the permittance is so small that we might expect the common magnetic formula to apply, we should increase the amplitude of current of any (not too low) frequency by increasing the inductance. It seems to show how careful we should be not to extend too widely the application of professedly approximate formulæ. Equation (4) has quite different significations under varied circumstances; and, general as it is, it is yet not general enough to meet extreme cases, even when, as in my original statement of it (*The Electrician*, July 23, 1886) [vol. II., p. 61], the increased resistance and reduced inductance due to the tendency towards skin-conduction are allowed for. Besides the propagation of disturbances through the dielectric following the wires, after the manner of plane-waves, there is an outward propagation from the source of energy, which seems to me, however, to be quite a secondary matter, and insignificant, especially when the circuit is a metallic loop, which concentrates the electromagnetic field considerably. But when there is an earth-return, there is a wide extension of the magnetic field, and distances *from* the line should be compared with its *length*, in making estimates of the range of disturbances of appreciable magnitude, appreciable by cumulative action on a distant wire. There are also the modifications due to the presence of neighbouring wires, which may be calculated by the equations of a system of parallel wires. But perhaps the most important modifying influence of all is that of the terminal apparatus.

I have considered the effect of any terminal apparatus in my paper, "On the Self-Induction of Wires," Part V., [vol. II., p. 247]. It is very complex in general. But so far as relates to a long circuit of low resistance, we do not want the full formulæ. Take (17) as the formula when the wires are short-circuited at the sending and receiving ends. Then, when we put on terminal apparatus containing no impressed force except the one sinusoidally varying force at the beginning of the circuit, (which may be in any part of the main circuit of the terminal apparatus there), the result is to alter the attenuation-ratio from the former ρ to ρ_1, given by

$$\rho_1 = \rho \times G_0^{\frac{1}{2}} \times G_1^{\frac{1}{2}}, \quad \text{.........................(19)}$$

where $G_0^{\frac{1}{2}}$ and $G_1^{\frac{1}{2}}$ are the terminal factors for the sending and receiving ends, to be calculated in the following manner. Let R_1 and L_1 be the "effective" resistance and inductance of the apparatus at the receiving end, then

$$G_1 = 1 + (R^2 + L^2n^2)^{-1}$$
$$\times \Big((P^2+Q^2)(R_1^2 + L_1^2n^2) + 2P(RR_1 + LL_1n^2) + 2Q(LnR_1 - RL_1n)\Big), \quad (20)$$

without assumptions regarding the size of f and g. Now take $g=0$, and f a small fraction, and we reduce (20), when the fraction fL_1n/Lv is small, to

$$G_1^{\frac{1}{2}} = 1 + R_1/Lv. \quad \text{.........................(21)}$$

Therefore (19) becomes

$$\rho_1 = \tfrac{1}{2}Lv \,.\, \epsilon^{Rl/2Lv}(1 + R_0/Lv)(1 + R_1/Lv). \quad \text{..............(22)}$$

Note that the full expression for G_0 is obtainable from (20) by changing R_1 and L_1 to R_0 and L_0. But if we only assume f to be small and g zero, then, instead of (21), we have

$$G_1 = (1 + R_1/Lv)^2 + (L_1n/Lv)(L_1n/Lv - f) + f^2(R_1/Lv). \quad \text{........(23)}$$

Now let it be merely a telephone that is the receiving apparatus, of resistance and inductance R_1 and L_1, or something equivalent to a mere coil. If it be a mere coil, and also, though less easily, if a telephone, we may vary L_1 independently by changing the form of the coil or by inserting non-conducting iron. We see, then, that the terminal factor is made a minimum, with L_1 alone variable, when

$$2L_1n^2 = Rv,$$

which, with $R = 20^4$ and $n = 10^4$ makes $2L_1 = 60^6$, quite a reasonable value for a *small* telephone. But if $n = 20^3$, the result is 150^7, twenty-five times as large.

Next let it be, not the current, but the magnetic force of the coil that is a maximum, on the assumption that L_1/R_1, the time-constant of the coil, is fixed. This is nearly true when the size of the wire is varied, if it be a mere coil that is concerned, and is an approach to the truth when there is iron. It is now G_1/R_1 that has to be a minimum, subject to R_1/L_1 = constant. This happens when

$$(R_1^2 + L_1^2 n^2)^{\frac{1}{2}} = Lv, \qquad \text{.........................(24)}$$

or when the impedance of the coil equals the critical Lv.

I showed in my paper "On Electromagnets," etc. [Art. XVII., vol. I., p. 99], that in the magnetic theory the condition of maximum magnetic force of the coil is that its impedance should equal that of the rest of the circuit, which contains the impressed force. We may easily verify that Lv is the impedance in the present case (with f small). Now $Lv = 600$ ohms when $L = 20$; this is the extreme value of the resistance of the coil, which should really be less on account of the term $L_1 n$. For instance, if the time-constant be ·0002 second, and $n = 10^4$, we require $2{\cdot}24\,R_1 = Lv$. We see further that this does make $fL_1 n/Lv$ small, because f is small, and $L_1 n/Lv < 1$. Therefore, using (23), we have

$$G_1^{\frac{1}{2}} = 2^{\frac{1}{2}}(1 + R_1/Lv)^{\frac{1}{2}} = 2^{\frac{1}{2}}\{1 + (1 + n^2 a^2)^{-\frac{1}{2}}\}^{\frac{1}{2}}, \qquad \text{..........(25)}$$

which, with $n = 10^4$ and the time-constant $a =$ ·0002, becomes $G_1^{\frac{1}{2}} = 1{\cdot}7$. This is, of course, a far larger value of the terminal factor than need be. In fact, the conditions of maximum magnetic force of the coil and of maximum received current are not usually identical, and may be quite antagonistic. For instance, if we should make the terminal factor nearly unity, we should have the biggest current, but with the least power.

But a remarkable property should be mentioned, which may be proved by the general formula from which (19) is derived. It is that if the receiver be a mere resistance, the choice of its resistance to equal Lv will, when R/Ln is small, nearly annihilate the reflected wave, and so do away with the fluctuations and the distortion due to them, whether the circuit be a long or a short one. Under these circumstances we have practically perfect reception of signals.

The general condition making G_1/R a minimum on a long circuit, subject to constancy of a, is by (19) and (20),

$$1 = (R^2 + L^2 n^2)^{-1}(P^2 + Q^2)(R_1^2 + L_1^2 n^2),$$

or

$$R_1^2 + L_1^2 n^2 = \left(\frac{R^2 + L^2 n^2}{K^2 + S^2 n^2}\right)^{\frac{1}{2}} = \left(\frac{1 + f^2}{1 + g^2}\right)^{\frac{1}{2}} L^2 v^2. \qquad \text{...........(26)}$$

The right member expresses the square of the impedance of the circuit to a S.H. impressed force at its end. When f and g are small we obtain the former result. The property of equal impedances is, however, a general one, so that all we do in verifying it is to see that no glaring error has crept in. If a coil connect two points of any arrangement in which a S.H. state is kept up by impressed force, and we vary the size of wire without varying the size and shape of the coil, we bring the magnetic force of the coil to a maximum by making its impedance equal to that external to it, if the thickness of covering vary similarly to that of the wire.

XLII. ON RESISTANCE AND CONDUCTANCE OPERATORS, AND THEIR DERIVATIVES, INDUCTANCE AND PERMITTANCE, ESPECIALLY IN CONNECTION WITH ELECTRIC AND MAGNETIC ENERGY.

[*Phil. Mag.*, December, 1887, p. 479.]

General Nature of the Operators.

1. IF we regard for a moment Ohm's law merely from a mathematical point of view, we see that the quantity R, which expresses the resistance, in the equation $V=RC$, when the current is steady, is the operator that turns the current C into the voltage V. It seems, therefore, appropriate that the operator which takes the place of R when the current varies should be termed the resistance-operator. To formally define it, let any self-contained electrostatic and magnetic combination be imagined to be cut anywhere, producing two electrodes or terminals. Let the current entering at one and leaving at the other terminal be C, and let the voltage be V, this being the fall of potential from where the current enters to where it leaves. Then, if $V=ZC$ be the differential equation (ordinary, linear) connecting V and C, the resistance-operator is Z.

All that is required to constitute a self-contained system is the absence of impressed force within it, so that no energy can enter or leave it (except in the latter case by the irreversible dissipation concerned in Joule's law) until we introduce an impressed force; for instance, one producing the above voltage V at a certain place, when the product VC expresses the energy-current, or flux of energy into the system per second.

The resistance-operator Z is a function of the electrical constants of the combination and of d/dt, the operator of time-differentiation, which will in the following be denoted by p simply. As I have made extensive use of resistance-operators and connected quantities in previous papers,* it will be sufficient here, as regards their origin and manipulation, to say that resistance-operators combine in the same way as if they represented mere resistances. It is this fact that makes them of so much importance, especially to practical men, by whom they will be much employed in the future. I do not refer to practical men in the very limited sense of anti- or extra-theoretical, but to theoretical men who desire to make theory practically workable by the simplification and systematisation of methods which the employment of resistance-operators and their derivatives allows, and the substitution of simple for more complex ideas. In this paper I propose to give a connected account of most of their important properties, including some new ones, especially in connection with energy, and some illustrations of extreme cases, which are found, on examination, to "prove the rule."

2. If we put $p=0$ in the resistance-operator of any system as above defined, we obtain the steady resistance, which we may write Z_0. If all the operations concerned in Z involve only differentiations, it is

* Especially Part III., and after, "On the Self-Induction of Wires," [vol. II., pp. 201 to 361 generally. Also vol. I., p. 415].

clear that when C is given completely, V is known completely. But if inverse operations (integrations) have to be performed, we cannot find V immediately from C completely; but this does not interfere with the use of the resistance-operator for other purposes.

It is sometimes more convenient to make use of the converse method. Thus, let Y be the reciprocal of Z, so that $C=YV$. If we make p vanish in Y, the result, say Y_0, is the conductance of the combination. Therefore Y is the conductance-operator.

The fundamental forms of Y and Z are

$$Z=R+Lp, \qquad (1)$$

$$Y=K+Sp. \qquad (2)$$

In the first case, it is a coil of resistance R and inductance L that is in question, with the momentum LC and magnetic energy $\frac{1}{2}LC^2$ In the second case, it is a condenser of conductance K and permittance S, with the charge SV and electric energy $\frac{1}{2}SV^2$; or its equivalent, a perfectly nonconducting condenser having a shunt of conductance K.

In a number of magnetic problems (no electric energy) the resistance-operator of a combination, even a complex one, reduces to the simple form (1). The system then behaves precisely like a simple coil, so far as externally impressed force is concerned, and is indistinguishable from a coil, provided we do not inquire into the internal details. I have previously given some examples.* Substituting condensers for coils, permittances for inductances, we see that corresponding reductions to the simple form (2) occur in electrostatic combinations (no magnetic energy).

But such cases are exceptional; and, should a combination store both electric and magnetic energy, it is not possible to effect the above simplifications except in some very extreme circumstances. There are, however, two classes of problems which are important practically, in which we can produce simplicity by a certain sacrifice of generality. In the first class the state of the whole combination is a sinusoidal or simple-harmonic function of the time. In the second class we ignore altogether the manner of variation of the current, and consider only the integral effects in passing from one steady state to another, which are due to the storage of electric and magnetic energy.

S.H. Vibrations, and the effective R′, L′, K′, *and* S′.

3. If the voltage at the terminals be made sinusoidal, the current will eventually become sinusoidal in every part of the system, unless it be infinitely extended, when consequences of a singular nature result. At present we are concerned with a finite combination. Then, if $n/2\pi$ be the periodic frequency, we have the well-known property $p^2=-n^2$; which substitution, made in Z and Y, reduces them to the forms

$$Z=R'+L'p, \qquad (3)$$

$$Y=K'+S'p; \qquad (4)$$

* "On the Self-Induction of Wires," Parts VI. and VII. [vol. II., pp. 268 and 292.]

where R', L', K', S' are functions of the electrical constants and of n^2, and are therefore constants at a given frequency.

In the first case we compare the combination to a coil whose resistance is R' and inductance L', so that R' and L' are the effective resistance and inductance of the combination, originally introduced by Lord Rayleigh* for magnetic combinations. In my papers, however, there is no limitation to cases of magnetic energy only,† and it would be highly inconvenient to make a distinction.

In a similar way, in the second case we compare the combination to a condenser, and we may then call K' the effective conductance and S' the effective permittance at the given frequency. R' reduces to Z_0, and K' to Y_0 at zero frequency. But it is important to remember that the two comparisons are of widely different natures: and that the effective resistance [in the coil-comparison] is not the reciprocal of the effective conductance [in the condenser-comparison].

Y and Z in (3) and (4) are reciprocal, or $YZ=1$, just as the general Y and Z of (1) and (2) are reciprocal.

If (V) and (C) denote the amplitudes of V and C, we have, by (3) and (4),

$$(V)/(C)=(R'^2+L'^2n^2)^{\frac{1}{2}}=I, \quad \text{say}, \qquad \text{......(5)}$$

$$(C)/(V)=(K'^2+S'^2n^2)^{\frac{1}{2}}=J, \quad \text{say}. \qquad \text{......(6)}$$

I and J are also reciprocal. The former, I, being the ratio of the force to the flux (amplitudes), is the impedance of the combination. It is naturally suggested to call J the "admittance" of the combination. But it is not to be anticipated that this will meet with so favourable a reception as impedance, which term is now considerably used, because the methods of representation (1), (3), and (5) are more useful in practice than (2), (4), and (6); although theoretically the two sets are of equal importance.‡

To obtain the relations between R' and K', and L' and S', we have

$$Y=(R'+L'p)^{-1}=(R'-L'p)I^{-2}, \qquad \text{......(7)}$$

$$Z=(K'+S'p)^{-1}=(K'-S'p)J^{-2}; \qquad \text{......(8)}$$

from which we derive

$$\left.\begin{array}{ll} I^2K'=R', & J^2R'=K', \\ -I^2S'=L', & -J^2L'=S', \\ L'/R'=-S'/K', & R'/K'=I^2=-L'/S', \end{array}\right\} \qquad \text{......(9)}$$

all of which are useful relations.

* *Phil. Mag.*, May, 1886.

† In Part V. of "On the Self-Induction of Wires" I have given a few examples of mixed cases of an elementary nature, in connexion with the problem of finding the effect of an impressed force in a telegraph circuit.

‡ The necessity of the term impedance (or some equivalent) to take the place of the various utterly misleading expressions that have been used, has come about through the wonderful popularisation of electromagnetic knowledge due to the dynamo, and its adoption to Sir W. Thomson's approval of it and of one or two other terms.

4. By (3) and (4) we have the equations of activity

$$VC = R'C^2 + p(\tfrac{1}{2}L'C^2), \quad \text{.......................(10)}$$

$$VC = K'V^2 + p(\tfrac{1}{2}S'V^2), \quad \text{.......................(11)}$$

in general. Now, if we take the mean values, the differentiated terms go out, leaving

$$\overline{VC} = R'\overline{C^2} = K'\overline{V^2}, \quad \text{..........................(12)}$$

the bars denoting mean values. The three expressions in (12) each represent the mean dissipativity, or heat per second. R' and K' are therefore necessarily positive. It should be noted that $R'C^2$ or $K'V^2$ do not represent the dissipativity at any moment. The dissipativity fluctuates, of course, because the square of the current fluctuates; but besides that, there is usually a fluctuation in the resistance, because the distribution of current varies, and it is only by taking mean values that we can have a definite resistance at a given frequency.

If the combination be magnetic, and T denote the magnetic energy, its mean value is given by

$$\overline{T} = \tfrac{1}{2}L'\overline{C^2}, \quad \text{............................(13)}$$

so that L' is necessarily positive and S' negative. But $\tfrac{1}{2}L'C^2$ is not usually the magnetic energy at any moment.

If the combination be electrostatic, and U denote the electric energy, its mean value is

$$\overline{U} = \tfrac{1}{2}S'\overline{V^2}, \quad \text{...........................(14)}$$

so that S' is positive and L' negative. The electric energy at any moment is not usually $\tfrac{1}{2}S'V^2$.

But, in the general case of both energies being stored, we have

$$\overline{T} - \overline{U} = \tfrac{1}{2}L'\overline{C^2} = -\tfrac{1}{2}S'\overline{V^2}. \quad \text{.......................(15)}$$

If the mean magnetic energy preponderates, the effective inductance is positive, and the permittance negative; and conversely if the electric energy preponderates. If there be no condensers, the comparison with a coil is obviously most suitable, and if there be no magnetic energy we should naturally use the comparison with a condenser; but when both energies coexist, which method of representation to adopt is purely a matter of convenience in the special application concerned.

If the mean energies, electric and magnetic, be equal, then

$$L' = 0 = S', \quad R'K' = 1, \quad I = R', \quad J = K'. \quad \text{..........(16)}$$

That is, by equalising the mean energies we bring the current and voltage into the same phase, annihilate the effective inductance (and also permittance), and make the effective conductance the reciprocal of the effective resistance, which now equals the impedance itself. It should be noted that the vanishing of the energy-difference only refers to the mean value. The two energies are not equal and do not vanish simultaneously. Sometimes, however, their sum is constant at every moment, but this is exceptional. (Example, a coil and a condenser in sequence.)

Impulsive Inductance and Permittance. General Theorem relating to the Electric and Magnetic Energies.

5. Passing now to the second class referred to in § 2, imagine, first, the combination to be magnetic, and that V is steady, producing a steady C, dividing in the system in a manner solely settled by the distribution of conductivity. Although we cannot treat the combination as a coil as regards the way the current varies when the impressed force is put on, we may do so as regards the integral effect at the terminals produced by the magnetic energy. The last is the well-known quadratic function of the currents in different parts of the system,

$$T = \tfrac{1}{2}L_1C_1^2 + MC_1C_2 + \tfrac{1}{2}L_2C_2^2 + \dots \quad \text{.................(17)}$$

Now put every one of these C's in terms of *the* C, the total current at the terminals, which may be done by Ohm's law. This reduces T to

$$T = \tfrac{1}{2}L_0C^2, \quad \text{..............................(18)}$$

where L_0 is a function of the real inductances, self and mutual, of the parts of the system, and of their resistances. This L_0 may be called the impulsive inductance of the system. For although it is, in a sense, the effective steady inductance, taking the current C at the terminals as a basis, being, in fact, the value of the sinusoidal inductance L' at zero frequency; yet, as it is only true for impulses that the combination behaves as a coil of inductance L_0, it is better to signify this fact in the name, to avoid confusion. This will be specially useful in the more general case in which both energies are concerned.

Secondly, let the system be electrostatic. Then, in a similar way, we may write the electric energy in the form

$$U = \tfrac{1}{2}S_0V^2, \quad \text{..............................(19)}$$

in terms of the V at the terminals, where S_0 is a function of the real permittances and of the resistances. S_0 is the impulsive permittance of the combination. It is also the sinusoidal S' at zero frequency.

In (18) L_0 is positive, and in (19) S_0 is positive. The momentum or electromotive impulse [or the voltaic impulse, if we use the modern "voltage" to signify the old "electromotive force"] at the terminals in the former case is L_0C, and in the latter case is $-S_0RV$, where R is the steady resistance. The true analogue of momentum, however, is charge, or time-integral of current, and this, at the terminals, is $-S_0V$, corresponding to L_0C.

6. Passing to the general case, and connecting with the resistance-operator, let Γ be the current at the terminals at time t when varying, so that

$$V = Z\Gamma = (Z_0 + pZ_0' + \tfrac{1}{2}p^2Z_0'' + \dots)\Gamma, \quad \text{................(20)}$$

where the accents denote differentiations to p, and the zero suffixes indicate that the values when $p = 0$ are taken. The coefficients of the powers of p are therefore constants. Integrating to the time,

$$\int Vdt = \int Z_0\Gamma dt + Z_0'[\Gamma] + \tfrac{1}{2}Z_0''[\dot{\Gamma}] + \dots \quad \text{.............(21)}$$

If the current be steady at beginning and at end,

$$\int (V - Z_0\Gamma)dt = Z'[\Gamma], \quad \text{......................(22)}$$

and if the initial current be zero, and the final value be C,

$$\int (V - Z_0\Gamma)dt = Z_0'C; \quad \text{......................(23)}$$

so that $Z_0'C$ is the voltaic impulse employed in setting up the magnetic and the electric energy of the steady state due to steady V at the terminals. Thus

$$L_0 = Z_0' \quad \text{..............................(24)}$$

finds the impulsive inductance from the resistance-operator. Or,

$$L_0 = (Z - Z_0)p^{-1} \quad \text{with} \quad p = 0. \quad \text{..................(25)}$$

In a similar manner, we may show that

$$S_0 = Y_0' = -Z_0^{-2}Z_0'. \quad \text{.......................(26)}$$

finds the impulsive permittance from the conductance-operator. L_0C and $-S_0Z_0V$ are equivalent expressions for the voltaic impulse.

If Z_0 should be infinite, then use Y. For instance, the insertion of a nonconducting condenser of permittance S_1 in the main circuit of the current makes Z_0 infinite, since the resistance-operator of the condenser is $(S_1p)^{-1}$. There is no final steady current, and L_0 is infinite. We should then use (26) instead of (24), especially as the energy is wholly electric in the steady state.

7. To connect with the energy, multiply (23) by C, the final current, and, for simplicity, let V be steady; giving

$$\int (V - R\Gamma)Cdt = Z_0'C^2 = \int V(C - \Gamma)dt. \quad \text{.............(27)}$$

It may be anticipated from the preceding that these equated quantities express twice the excess of the magnetic over the electric energy.

In connexion with this I may quote from Maxwell, vol. ii., art. 580. A purely electromagnetic system is in question. "If the currents are maintained constant by a battery during a displacement in which a quantity of work, W, is done by electromotive force, the electrokinetic energy of the system will be at the same time increased by W. Hence the battery will be drawn upon for a double quantity of energy, or $2W$, in addition to that which is spent in generating heat in the circuit. This was first pointed out by Sir W. Thomson. Compare this result with the electrostatic property in art. 93." The electrostatic property referred to relates to conductors charged by batteries. If "their potentials are maintained constant, they tend to move so that the energy of the system is increased, and the work done by the electrical forces during the displacement is equal to the increment of the energy of the system. The energy spent by the batteries is equal to double of either of these quantities, and is spent half in mechanical, half in electrical work."

Although of a somewhat similar nature, these properties are not what is at present required, which is contained in the following general theorem given by me* :—Let any steady impressed electric forces be suddenly started and continued in a medium permitting linear relations between the two forces, electric and magnetic, and the three fluxes—conduction current, electric displacement, and magnetic induction (but with no rotational property allowed, even for conduction current); the whole work done by the impressed forces during the establishment of the steady state exceeds what would have been done had this state been instantly established (but then without any electric or magnetic energy) by twice the excess of the electric over the magnetic energy. That is,

$$\int_0^\infty dt \sum \mathbf{e}(\Gamma - \Gamma_0)dt = 2(U - T), \qquad (28)$$

where $\mathbf{e}$ stands for an element of impressed force, Γ the current-density at time t, Γ_0 the final value, and Σ the space-integration to include all the impressed forces. (Black letters for vectors.) The theorem (28) seems the most explicit and general representation of what has been long recognised in a general way, that permitting electric displacement increases the activity of a battery, whilst permitting magnetisation decreases it. The one process is equivalent to allowing elastic yielding, and the other to putting on a load (not to increasing the resistance, as is sometimes supposed).

Applying (28) to our present case of one impressed voltage V, producing the final current C, we obtain

$$\int dt V(\Gamma - C)dt = 2(U - T), \qquad (29)$$

comparing which with (27), we see that

$$T - U = \tfrac{1}{2}Z_0'C^2 = \tfrac{1}{2}L_0C^2 = -\tfrac{1}{2}S_0V^2, \qquad (30)$$

confirming the generality of our results.

General Theorem of Dependence of Disturbances solely on the Curl of the Impressed Forcive.

8. It is scarcely necessary to remark that the properties of Z and Z' previously discussed do not apply merely to combinations consisting of coils of fine wire and condensers; the currents may be free to flow in conducting masses or dielectric masses. Solid cores, for example, may be inserted in coils within the combination. The only effect is to make the resultant resistance-operator at a given place more complex.

But a further very remarkable property we do not recognise by regarding only common combinations of coils and condensers. If we, in the complex medium above defined, select any unclosed surface, or surface bounded by a closed line, and make it a shell of impressed voltage (analogous to a simple magnetic shell), thereby producing a potential-difference V between its two faces, and C be the current through the shell in the direction of the impressed voltage, there must be a definite resistance-operator Z connecting them, depending upon

* *Electrician*, April 25, 1885, p. 490, [vol. I., p. 464.]

the distribution of conductivity, permittivity, and inductivity through all space, and determinable by a sufficiently exhaustive analysis. The remarkable property is that the resistance-operator is the same for any surfaces having the same bounding-edge. For a closed shell of impressed voltage of uniform strength can produce no flux whatever. This is instructively shown by the equation of activity,

$$\Sigma \mathbf{e}\Gamma = Q + \dot{U} + \dot{T}, \quad \text{.........................(31)}$$

indicating that the sum of the activities of the impressed forces, or the energy added to the system per second, equals the total dissipativity Q, *plus* the rate of increase of the stored energies, electric and magnetic, throughout the system. Now here Γ is circuital; if, therefore, the distribution of $\mathbf{e}$ be polar, or $\mathbf{e}$ be the vector space-variation of a single-valued scalar potential, of which a simple closed shell of impressed force is an example, the left member of (31) vanishes, so that the dissipation, if any, is derived entirely from the stored energy. Start, then, with no electric or magnetic energy in the system; then the positivity of Q, U, and T ensures that there never can be any, under the influence of polar impressed force. Hence two shells of impressed force of equal uniform strength produce the same fluxes if their edges be the same; not merely the steady fluxes possible, but the variable fluxes anywhere at corresponding moments after commencing action. The only difference made when one shell is substituted for the other is in the manner of the transfer of energy at the places of impressed force; for we have to remember that the effective force producing a flux, or the "force of the flux," equals the sum of the impressed force and the "force of the field"; whereas the transfer of energy is determined by the vector product of the two forces of the field, electric and magnetic respectively. In (31) no count is taken of energy transferred from one seat of impressed force to another, reversibly, all such actions being eliminated by the summation.

It is well to bear in mind, when considering the consequences of this transferability of impressed force, especially in cases of electrolysis or the Volta-force, not only that the three physical properties of conductivity, permittivity, and inductivity, though sufficient for the statement of the main facts of electromagnetism, are yet not comprehensive, but also that they have no reference to molecules and molecular actions; for the equations of the electromagnetic field are constructed on the hypothesis of the ultimate homogeneity of matter, or, in another form, only relate to elements of volume large enough to allow us to get rid of the heterogeneity.

As the three fluxes are determined solely by the vorticity (to borrow from liquid motion) of the vector impressed force, we cannot know the distribution of the latter from that of the former, but have to find where energy transformations are going on; for the denial of the law that $\mathbf{e}\Gamma$ not only measures the activity of an impressed electric force $\mathbf{e}$ on the current Γ, but represents energy received by the electromagnetic system at the very same place, lands us in great difficulties.

Again, as regards the "electric force of induction." We cannot find the distribution through space of this vector from the Faraday-law that its line-integral in a closed circuit equals the rate of decrease of induction through the circuit. We may add to any distribution satisfying this law any polar distribution without altering matters, except that a different potential function arises. In this case we do not even alter the transfer of energy. The electric force of the field is always definite; but when we divide it into two distinct distributions, and call one of them the electric force of induction, and the other the force derived from electric potential, it is then quite an indeterminate problem how to effect the division, unless we choose to make the quite arbitrary assumption that the electric force of induction has nothing of the polar character about it (or has no divergence anywhere), when of course it is the other part that possesses the whole of the divergence. This fact renders a large part of some mathematical work on the electromagnetic field that I have seen redundant, as we may write down the final results at the beginning. In the course of some investigations concerning normal electromagnetic distributions in space I have been forcibly struck with the utter inutility of dividing the electric field into two fields, and by the simplicity that arises by not doing so, but confining oneself to the actual forces and fluxes, which describe the real state of the medium and have the least amount of artificiality about them. Similar remarks apply to Maxwell's vector-potential **A**. Has it divergence or not? It does not matter in the least, on account of the auxiliary polar force. When the electric force itself is made the subject of investigation, the question of divergence of the vector-potential does not present itself at all.

The lines of vorticity, or vortex-lines of the vector impressed force, are of the utmost importance, because they are the originating places of all disturbances. This is totally at variance with preconceived notions founded upon the fluid analogy, which is, though so useful in the investigation of steady states, utterly misleading when variable states are in question, owing to the momentum and energy belonging to the magnetic field, not to the electric current. Every solution involving impressed forces consists of waves emanating from the vortex-lines of impressed force (electric or magnetic as the case may be, but only the electric are here considered), together with the various reflected waves produced by change of media and other causes. At the first moment of starting an impressed force the only disturbance is at the vortex-lines, which are the first lines of magnetic induction.

Examples of the Forced Vibrations of Electromagnetic Systems.

(*a*). Thus a uniform field of impressed force suddenly started over all space can produce no effect. For, either there are no vortex-lines at all, or they are at an infinite distance, so that an infinite time must elapse to produce any effect at a finite distance from the origin.

(*b*). Copper and zinc put in contact. Whether the Volta-force be at the contact or over the air-surfaces away from and terminating at the contact (if perfectly metallic), the vortex-line is the common meeting-place of air, zinc, and copper; the first line of magnetic force is there, and from it the disturbance proceeds into the metals and out into the air, which ends in the steady electric field.*

Since the vortex-lines or tubes are closed, we need only consider one at present—say, that due to a simple shell of impressed force. If it be wholly within a conductor, the initial wave emanating from it is so rapidly attenuated by the conductivity (the process being akin to repeated internal reflexions, say reflexion of 9 parts and transmission of 1 part, repeated at short intervals) that the transmission to a distance through the conductor (if good) becomes a very slow process, that of diffusion. Consequently, when the impressed force is rapidly alternated, there is no sensible disturbance except at and near the vortex-line.

But if there be a dielectric outside the conductor, the moment disturbances reach it, and therefore instantly if the vortex-line be on the boundary, waves travel through the dielectric at the speed of light unimpeded, and without the attenuating process within the conductor, which therefore becomes exposed to electric force all over its boundary in a very short time; hence diffusion inward from the boundary. The electric telegraph would be impossible without the dielectric. It would take ages if the wire itself had to be the seat of transfer of energy.

(*c*). In the magnetic theory of the rise of current in a wire we have, at first sight, an exception to the law that at the first moment there is no disturbance except at the vortex-lines of impressed force. But it is that theory which is incorrect, in assuming that there is no displacement. This is equivalent to making the speed of propagation through the dielectric infinitely great; so that we have results mathematically equivalent to distributing the impressed force throughout the whole circuit, and therefore its vortex-lines over the whole boundary.† In reality, with finite speed, the disturbances come from the real vortex-lines in time.

There is still a limitation of the disturbances to the neighbourhood of the vortex-lines when they are on the boundary of the conductor, and the periodic frequency is sufficiently great, the impressed force being within the conductor. [The attenuation by resistance is referred to.]

But in a nonconducting dielectric this effect does not occur, at least in any case I have examined. On the contrary, as the frequency is raised, there is a tendency to constancy of amplitude of the waves sent out from the edge of a simple sheet of impressed force, or from a shell of vortex-lines of the same, in a dielectric. Very remarkable results follow from the coexistence of the primary and reflected waves. Thus:

(*d*). If a spherical portion of an infinitely extended dielectric have a uniform field of alternating impressed force within it, and the radius a, the wave-frequency $n/2\pi$, and the speed v be so related that

$$\tan\frac{na}{v}=\frac{na}{v},$$

* "Some Remarks on the Volta Force," *Journal S. T. E. & E.*, 1885 [vol. I., p. 425].

† *The Electrician*, June 25, 1886, p. 129 [vol. II., p. 60].

there is no disturbance outside the sphere. There are numerous similar cases; but this is a striking one, because, from the distribution of the impressed force, it looks as if there *must* be external displacement produced by it. There is not, because the above relation makes the primary wave outward from the surface of the sphere, which is a shell of vorticity, be exactly neutralised by the reflexion, from the centre, of the primary wave inward from the surface.

(*e*). If, instead of alternating, the uniform field of impressed force in (*d*) be steady, the final steady electric field due to it takes the time $(r+a)/v$ to be established at distance r from the centre. The moment the primary wave inward reaches the centre, the steady state is set up there; and as the reflected wave travels out, its front marks the boundary between the steady field (final) and a spherical shell of depth $2a$, within which is the uncancelled first portion of the primary wave outward from the surface; which carries out to an infinite distance an amount of energy equal to that of the final steady electric field. This is the loss by radiation. (The magnetic energy in this shell equals half the final electric energy on the whole journey; the electric energy in the shell is greater, but ultimately becomes the same.) In practical cases this energy would be mostly, perhaps wholly dissipated in conductors.

(*f*). If a uniformly distributed impressed force act alternatingly longitudinally within an infinitely long circular cylindrical portion of a dielectric, the axis is the place of reflexion of the primary wave inward, and the reflected wave cancels the outward primary wave when

$$J_1(na/v)=0;$$

so that there is no external disturbance, except at first. Here $a=$ radius of cylinder.

(*g*). There is a similar result when the vorticity of impressed force takes the place of impressed force in (*f*).

(*h*). If the alternating impressed force act uniformly and longitudinally in a thin conducting-tube of radius a, with air within and without, then

$$J_0(na/v)=0$$

destroys the external field and makes the conduction-current depend upon the impressed force only. And if we put a barrier at distance x to serve as a perfect reflector, that is, a tube of infinite conductivity,

$$J_0(nx/v)=0$$

makes the electric force of the field in the inner tube be the exact negative of the impressed force; so that there is no conduction-current. The electromagnetic field is in stationary vibration. If the inner tube be situated at one of the nodal surfaces of electric force, the vibrations mount up infinitely.

(*i*). If, in case (*h*), the impressed force act circularly about the axis of the inner tube (which may be replaced by a solenoid of small depth),

$$J_1(na/v)=0$$

destroys the external field, and

$$J_1(nx/v)=0$$

makes the electric force of the field the negative of the impressed force, and so destroys the conduction-current.

(j). We can also destroy the longitudinal force of the field in a conductor without destroying the external field. Let it be a wire of steady resistance in a dielectric, and the impressed force in it be

$$e = e_0 \cos mx \cos nt$$

per unit length. Then $m = n/v$ makes e be the force of the flux, in the wire; so that the current is Ke, if K be the conductance of unit length.

These examples are mostly selected from a paper I am now writing on the subject of electromagnetic waves, which I hope to be permitted to publish in this Journal.

If the electric and magnetic energies, and the dissipation of energy, in a given system be bounded in their distribution, it is clear that the resistance operator is a rational function of p. But should the field be boundless, as when conductors are contained in an infinitely extended dielectric, then just as complete solutions in infinite series of normal solutions may become definite integrals by the infinite extension, so may the resistance-operator become irrational. We may also have to modify the meaning of the sinusoidal R' from representing mean resistance only, on account of the never-ceasing outward transfer of energy so long as the impressed force continues.

Induction-Balances—General, Sinusoidal, and Impulsive.

9. Returning to a finite combination represented by $V = ZC$, there are at least three kinds of induction-balances possible. First, true balances of similar systems, where we balance one combination against another which either copies it identically or upon a reduced scale, without any reference to the manner of variation of the impressed force. Along with these we may naturally include all cases in which the Z of a combination, in virtue of peculiar internal relations, reduces to a simpler form representing another combination, equivalent so far as V and C are concerned. The telephone may be employed with great advantage, and is, in fact, the only proper thing to use, especially for the observation of phenomena.

There are, next, the sinusoidal-current balances. These are also true, in being independent of the time, so that the telephone may be used; but are of course of a very special character otherwise. Here any combination is made equivalent to a mere coil if L' be positive, or to a condenser if S' be positive (§§ 3 and 4), and so may be balanced by one or the other. But intermittences of current cannot be safely taken to represent sinusoidality, and large errors may result from an assumed equivalence.

In the third kind of balances it is the impulsive inductance that is balanced against some other impulsive inductance, positive or negative as the case may be; or perhaps the impulsive inductance of a combination is made to vanish, by equating the electric and magnetic energies in it when its state is steady. The rule that the impulsive balance in a Christie arrangement without mutual induction between the four sides is given by equating to zero the coefficient of p in the expansion of $Z_1Z_4 - Z_2Z_3$ in powers of p, where Z_1, etc. are the resistance-operators of the four sides,* is in agreement with the rule derived from (24) or (25) above, to make the impulsive inductance of one combination vanish. Impulsive, or "kick" balances, naturally require a galvanometer. Even then, however, the method is sometimes unsatisfactory, when the opposing influences which make up the impulse are not sufficiently simultaneous, as has been pointed out by Lord Rayleigh.†

There is also the striking method of cumulation of impulses employed by Ayrton and Perry,‡ employing false resistance-balances. It seems complex, and of rather difficult theory; but, just as a watch is a complex piece of mechanism, and is yet thoroughly practical, so perhaps the secohmmeter may have a brilliant career before it.

Several interesting papers relating to the comparison of inductances and permittances have appeared lately. It is usually impulsive balances that are in question, probably because it is not the observation of phenomena that is required, but a direct, even if rough, measurement of the inductance or permittance concerned, often under circumstances that do not well admit of the use of the telephone. Only one of these papers, however, contains anything really novel, scientifically, viz., that of Mr. W. H. Preece, F.R.S.,§ who concludes, from his latest researches, that the "coefficient of self-induction" of copper telegraph-circuits is nearly zero, the results he gives being several hundred times smaller than the formula derived from electromagnetic principles asserts it to be. Here is work for the physicist.

10. To equate the expressions for the electric and magnetic energies of a combination is, I find, in simple cases, the easiest and most direct way of furnishing the condition that the impulsive inductance shall vanish. Thus, if there be but one condenser and one coil, $SV^2 = LC^2$ is the condition, S and L being the permittance and the inductance respectively, V the voltage of the condenser, and C the current in the coil. The relation between V and C will be, of course, dependent upon the resistances concerned.‖ But in complex cases, and to obtain the value of the impulsive inductance when it is not zero, equation (24) is most useful.

The Resistance Operator of a Telegraph Circuit.

The following illustration of the properties of Z and Z_0' is a complex one, but I choose it because of its comprehensive character, and because it leads to some singular extreme cases, interesting both mathematically

* "On the Self-Induction of Wires," Part VI., *Phil. Mag.*, Feb. 1887 [vol. II., p. 263].

† *Electrical Measurements*, p. 65.

‡ *Journ. Soc. Tel. Engineers and Electricians*, 1887.

§ B.A. Meeting, 1887: "On the Coefficient of Self-Induction of Iron and Copper Wires."

‖ If the condenser shunts the coil, making $V = RC$, we get the case brought before the S.T.E. & E. by Mr. Sumpner, with developments.

and in the physical interpretation of the apparent anomalies. Let the combination be a telegraph-circuit, say a pair of parallel copper wires, of length l; resistance R, permittance S, inductance L, and leakage-conductance K, all per unit length, and here to be considered strictly constants, or independent of p. Let the two wires be joined through an arrangement whose resistance-operator is Z_1 at the distant end B; then the resistance-operator at the beginning A of the circuit is given by*

$$Z = \frac{(R+Lp)l\{(\tan ml)/ml\} + Z_1}{1+(K+Sp)lZ_1\{(\tan ml)/(ml)\}}, \qquad (32)$$

if

$$-m^2 = (R+Lp)(K+Sp). \qquad (33)$$

Take $Z_1 = 0$ for the present, or short-circuit at B. This makes

$$Z = (R+Lp)l(\tan ml)/ml, \qquad (34)$$

and the steady resistance at A is therefore

$$Z_0 = Rl\,(\tan m_0 l)/m_0 l, \qquad (35)$$

if $-m_0^2 = RK$. Also, differentiating (34) to p, and then making $p=0$, we find

$$Z_0' = L_0 = \tfrac{1}{2}l\frac{\tan m_0 l}{m_0 l}\left(L - \frac{RS}{K}\right) + \tfrac{1}{2}l\sec^2 m_0 l\left(L + \frac{RS}{K}\right) \qquad (36)$$

represents the impulsive inductance.

If we put $S=0$ in (36) we make the arrangement magnetic, and then L_0 is positive. If we put $L=0$, we make it electrostatic, and L_0 is negative, or S_0, the impulsive permittance, is positive. It is to be noticed that there is no confusion when both energies are present; that is, there are no terms in Z_0' containing products of real permittances and inductances, which is clearly a general property of resistance-operators, otherwise the two energies would not be independent.

We may make L_0 vanish by special relations. Thus, if there be no leakage, or $K=0$, (36) is

$$L_0 = Ll - \tfrac{1}{3}Rl\,.\,RSl^2; \qquad (37)$$

so that the magnetic must be one third of the electrostatic time-constant to make the "extra-current" and the static charge balance. (The length of the circuit required for this result may be roughly stated as about 60 kilometres if it be a single copper wire of 6 ohms per kilometre, 4 metres high, with return through the ground; but it varies considerably, of course.)

But if leakage be now added, it will increase the relative importance of the magnetic energy, so that the length of the circuit requires to be increased to produce a balance. This goes on until K reaches the value RS/L, when, as an examination of (36) will show, the length of the circuit needs to be infinitely great. The same formula also shows that if K be still greater, L_0 cannot be made to vanish at all, being then always positive.

* "On the Self-Induction of Wires," Part IV., *Phil. Mag.*, Nov., 1886 [vol. II., p. 232; also p. 247 and p. 105.]

11. Now let the circuit be infinitely long. Equation (35) reduces to the irrational form

$$Z = \pm(R+Lp)^{\frac{1}{2}}(K+Sp)^{-\frac{1}{2}}, \quad \text{.................... (38)}$$

with ambiguity of sign. Of course the positive sign must be taken. The negative appears to refer to disturbances coming from an infinite distance, which are out of the question in our problem, as there can be no reflexion from an infinite distance. But equation (38) may be obtained directly in a way which is very instructive as regards the structure of resistance-operators. Since the circuit is infinitely long, Z cannot be altered by cutting-off from the beginning, or joining on, any length. Now first add a coil of resistance R_1 and inductance L_1 in sequence, and a condenser of conductance K_1 and permittance S_1, in bridge, at A, the beginning of the circuit. The effect is to increase Z to Z_2, where

$$Z_2 = \{K_1 + S_1 p + (R_1 + L_1 p + Z)^{-1}\}^{-1}; \quad \text{............. (39)}$$

i.e., the reciprocal of the new Z_2, or the new conductance-operator, equals the sum of the conductance-operators of the two branches in parallel, one the conducting condenser, the other the coil and circuit in sequence. (39) gives the quadratic

$$Z_2^2 + (R_1 + L_1 p)Z_2 = (R_1 + L_1 p)(K_1 + S_1 p)^{-1}. \quad \text{.......... (40)}$$

Now choose R_1, L_1, K_1, S_1, in exact proportion to R, L, K, and S, and then make the former set infinitely small. The result is that we have added to the original circuit a small piece of the same type, so that Z_2 and Z are identical, and that the coefficient of the first power of Z_2 in (40) vanishes. Therefore (40) becomes

$$Z = (R+Lp)^{\frac{1}{2}}(K+Sp)^{-\frac{1}{2}}. \quad \text{.................... (41)}$$

This fully serves to find the sinusoidal solution. Differentiating it, we find

$$L_0 = \frac{1}{2}\frac{L}{(RK)^{\frac{1}{2}}}\left(1 - \frac{RS}{KL}\right), \quad \text{...................... (42)}$$

corroborating the previous result as to the vanishing of L_0 when the circuit is infinitely long by equality of RS and KL, and the positivity of L_0 when $KL > RS$.

The Distortionless Telegraph Circuit.

12. Now, in the singular case of $R/L = K/S$, we have, by (41) and (42),

$$Z = Lv, \qquad L_0 = 0, \quad \text{......................... (43)}$$

if $v = (LS)^{-\frac{1}{2}}$, the speed of transmission of disturbances along the circuit. The resistance-operator has reduced to an absolute constant, and the current and transverse voltage are in the same phase, altogether independent of the frequency of wave-period, or indeed of the manner of variation. The quantity Lv, or $L \times 30$ ohms, approximately, if the dielectric be air, is strictly, and without any reservation, the impedance of the circuit at A, but it is only exceptionally the resistance.

Make $V=f(t)$, at A, an arbitrary function of the time; then, if V_x and C_x are the transverse voltage and the current at distance x from A at time t, we shall have

$$V_x=f(t-x/v)\epsilon^{-Rx/Lv}, \qquad C_x=V_x/Lv, \quad \text{................}(44)$$

or all disturbances originating at A are transmitted undistorted along the circuit at the speed v, attenuating at a rate indicated by the exponential function. (I have elsewhere* fully developed the properties of this distortionless circuit, and only mention such as are necessary to understand the peculiarities connected with the present subject-matter.) The electric and magnetic energies are always equal, not only on the whole, but in any part of the circuit; this accounts for the disappearance of L_0, and the bringing of V_x and C_x to the same phase, as we should expect from § 4. But in the present case Z_0, or Lv, or R', for they are all equal, is only the resistance when the steady state due to the steady V at A is arrived at (asymptotically), or the effective resistance at a given frequency when V is sinusoidal, and sufficient time has elapsed to have allowed V_x and C_x to become sinusoidal to such a distance from A that we can neglect the remainder of the circuit into which greatly attenuated disturbances are still being transmitted.

13. Now, since the impedance is unaltered by joining on at A any length of circuit of the same type, and is a constant, it follows that the impedance at A of a distortionless circuit as above described, but of *finite* length, stopping at B, where $x=l$, with a *resistance* of amount Lv inserted at B, is also a constant, viz. the same Lv. To corroborate, take $RS=KL$ and $Z_1=Lv$ in the full formula (32). The result is $Z=Lv$. The interpretation in this case is that all disturbances sent from A are absorbed completely by the resistance at B immediately on arrival, so that the finite circuit behaves as if it were infinitely long. The permanent state due to a steady V at A is arrived at in the time l/v. The impedance and the resistance then become identical.

14. If, in the case of § 12, we further specialize by taking $R=0$, $K=0$, producing a perfectly insulated circuit of no resistance, the impedance is, as before, Lv; but no part of it is resistance, or ever can be, in spite of the identity of phase of V and C. However long we may keep on a steady V at A, we keep the impressed force working at the same rate, the energy being entirely employed in increasing the electric and magnetic energies at the front of the wave, which is unattenuated, and cannot return.

But if we cut the circuit at B, at a finite distance l, and there insert a resistance Lv, the effect is that, as soon as the front of the wave reaches B, the inserted resistance immediately becomes the resistance of the whole combination; or the impedance instantly becomes the resistance, without change of value.

15. As a last example of singularity, substitute a short-circuit for the terminal resistance Lv just mentioned. Since there is now no resistance in any part of the system, if we make the state sinusoidal everywhere,

* "Electromagnetic Induction and its Propagation," Sections XL. to L., *Electrician*, 1887 [vol. II., pp. 119 to 155].

by V sinusoidal at A, R' must vanish, or V and C be in perpendicular phases, due to the infinite series of to-and-fro reflexions. We now have, by (32),

$$Z' = Lpl\frac{\tan(pli/v)}{pli/v} = Llp\frac{\tan(nl/v)}{nl/v}, \quad \text{..............(45)}$$

if $n/2\pi =$ frequency, and R' has disappeared.

If, on the other hand, V be steady at A, the current increases without limit, every reflexion increasing it by the amount V/Lv at A or at B (according to which end the reflexion takes place at), which increase then extends itself to B or A at speed v. The magnetic energy mounts up infinitely. On the other hand, the electric energy does not, fluctuating perpetually between 0 when the circuit is uncharged, and $\frac{1}{2}SlV^2$ when fully charged. The impedance of the circuit to the impressed force at A is Lv for the time $2l/v$ after starting it; then $\frac{1}{3}Lv$ for a second period $2l/v$; then $\frac{1}{5}Lv$ for a third period, and so on.

It will have been observed that I have, in the last four paragraphs, used the term impedance in a wider sense than in § 3, where it is the ratio of the amplitude of the impressed force to the amplitude of the flux produced at the place of impressed force when sufficient time has elapsed to allow the sinusoidal state to be reached, when that is possible. The justification for the extension of meaning is that, since in the distortionless circuit of infinite length, or of finite length with a terminal resistance to take the place of the infinite extension, we have nothing to do with the periodic frequency, or with waiting to allow a special state to be established, it is quite superfluous to adhere to the definition of the last sentence; and we may enlarge it by saying that the impedance of a combination is simply the ratio of the force to the flux, when it happens to be a constant, which is very exceptional indeed. I may add that R, L, K, and S need not be constants, as in the above, to produce the propagation of waves without tailing. All that is required is $R/L = K/S$, and $Lv =$ constant; so that R and L may be functions of x. The speed of the current, and the rate of attenuation, now vary from one part of the circuit to another.

The Use of the Resistance-Operator in Normal Solutions.

16. In conclusion, consider the application of the resistance-operator to normal solutions. If we leave a combination to itself without impressed force, it will subside to equilibrium (when there is resistance) in a manner determined by the normal distributions of electric and magnetic force, or of charges of condensers and currents in coils; a normal system being, in the most extended sense, a system that, in subsiding, remains similar to itself, the subsidence being represented by the time-factor ϵ^{pt}, where p is a root of the equation $Z = 0$. It is true that each part of the combination will usually have a distinct resistance-operator; but the resistance-operators of all parts involve, and are contained in, the same characteristic function, which is merely the Z of any part cleared of fractions. It is sometimes useful to remember that we should clear of fractions, for the omission to do so

may lead to the neglect of a whole series of roots; but such cases are exceptional and may be foreseen; whilst the employment of a resistance-operator rather than the characteristic function is of far greater general utility, both for ease of manipulation and for physical interpretation.

Given a combination containing energy and left to itself, it is upon the distribution of the energy that the manner of subsidence depends, or upon the distribution of the electric and magnetic forces in those parts of the system where the permittivity and the inductivity are finite, or are reckoned finite for the purpose of calculation. Thus conductors, if they be not also dielectrics, have only to be considered as regards the magnetic force, whilst in a dielectric we must consider both the electric and the magnetic force. (The failure of Maxwell's general equations of propagation arises from the impossibility of expressing the electric energy in terms of his potential function. The variables should always be capable of expressing the energy.) Now the internal connexions of a system determine what ratios the variables chosen should bear to one another in passing from place to place in order that the resultant system should be normal; and a constant multiplier will fix the size of the normal system. Thus, supposing u and w are the normal functions of voltage and current, which are in most problems the most practical variables, the state of the whole system at time t will be represented by

$$V=\Sigma A u \epsilon^{pt}, \qquad C=\Sigma A w \epsilon^{pt}; \qquad \text{.................(46)}$$

V being the real voltage at a place where the corresponding normal voltage is u, and C the real current where the normal current is w, the summation extending over all the p-roots of the characteristic equation. The size of the systems, settled by the A's (one for each p) are to be found by the conjugate property of the vanishing of the mutual energy-difference of any pair of p-systems, applied to the initial distributions of V and C.

17. To find the effect of impressed force is a frequently recurring problem in practical applications; and here the resistance-operator is specially useful, giving a general solution of great simplicity. Thus, suppose we insert a steady impressed force e at a place where the resistance-operator is Z, producing $e=ZC$ thereafter. Find C in terms of e and Z. The following demonstration appears quite comprehensive. Convert the problem into a case of subsidence first, by substituting a condenser of permittance S, and initial charge Se, for the impressed force. By making S infinite later we arrive at the effect of the steady e. In getting the subsidence solution we have only to deal with the energy of the condenser, so that a knowledge of the internal connexions of the system is quite superfluous.

The resistance-operator of the condenser being $(Sp)^{-1}$, that of the combination, when we use the condenser, is Z_1, where

$$Z_1=(Sp)^{-1}+Z. \qquad \text{.........................(47)}$$

Let V and C be the voltage and the current respectively, at time t after insertion of the condenser, and due entirely to its initial charge.

Equations (46) above express them, if u and w have the special ratio proper at the condenser, given by

$$w = -Spu, \quad \text{.............................(48)}$$

because the current equals the rate of decrease of its charge. Initially, we have $e = \Sigma Au$ and $\Sigma Aw = 0$. So, making use of the conjugate property,* we have

$$Seu = 2(U_p - T_p)A, \quad \text{.........................(49)}$$

if U_p be the electric and T_p the magnetic energy in the normal system. But the following property of the resistance-operator is also true,*

$$2(T_p - U_p) = \frac{dZ_1}{dp}w^2\,; \quad \text{.........................(50)}$$

that is, dZ_1/dp is the impulsive inductance in the p system at a place where the resistance-operator is Z_1, p being a root of $Z_1 = 0$; just as dZ_1/dp with $p = 0$ is the impulsive inductance (complete) at the same place. Using (50) in (49) gives

$$A = -(Seu) \div \left(w^2\frac{dZ_1}{dp}\right). \quad \text{.......................(51)}$$

Now use (48) in (51) and insert the resulting A in the second of (46), and there results

$$C = \sum \frac{e}{pZ_1'}\epsilon^{pt}, \quad \text{..............................(52)}$$

where the accent means differentiation to p. This is the complete subsidence solution. Now increase S infinitely, keeping e constant Z_1 ultimately becomes Z; but, in doing so, one root of $Z_1 = 0$ becomes zero. We have, by (47), and remembering that $Z_1 = 0$,

$$pZ_1' = -(Sp)^{-1} + pZ' = Z + pZ'\,; \quad \text{...................(53)}$$

so, when $S = \infty$ and $Z = 0$, we have $pZ_1' = pZ'$ for all roots except the one just mentioned, in which case p tends to zero and Z' is finite, making in the limit $pZ_1' = Z_0$, by (53), where Z_0 is the $p = 0$ value of Z, or the steady resistance. Therefore, finally,

$$C = \frac{e}{Z_0} + \sum \frac{e}{pZ'}\epsilon^{pt}, \quad \text{..........................(54)}$$

where the summation extends over the roots of $Z = 0$, shows the manner of establishment of the current by the impressed force e. The use of this equation (54), even in comparatively elementary problems, leads to a considerable saving of labour, whilst in cases involving partial differential equations it is invaluable.† To extend it to show the rise of the current at any other part of the system than where the impressed

* "On the Self-Induction of Wires," *Phil. Mag.*, Oct. 1886 [vol. II., pp. 202 to 206].

† In Part III. of "On the Self-Induction of Wires," I employed the Condenser Method, with application to a special kind of combination; but, as we have seen from the above proof, (54) is true for any electrostatic and electromagnetic combination provided it be finite.

force is, it is necessary to know the connections, so that we may know the ratio of the current in a normal system at the new place to that at the old; inserting this ratio in the summation, and modifying the external Z_0 to suit the new place, furnishes the complete solution there. Or, use the more general resistance-operator Z_{xy}, such that $e_x = Z_{xy} C_y$, connecting the impressed force at any place x with the current at another place y.

18. When the initial current is zero, as happens when there is self-induction without permittance at the place of e, and in other cases, (54) gives

$$\frac{1}{Z_0} = \sum \frac{1}{-pZ'}; \qquad (55)$$

showing that the normal systems may be imagined to be arranged in parallel, the resistance of any one being $(-pZ')$.

To express the impulsive inductance Z_0' in terms of the normal Z's, multiply (54) by e and take the complete time-integral. We obtain

$$\int e\left(C - \frac{e}{Z_0}\right)dt = 2(U - T) = -\sum \frac{e^2}{p^2 Z'}, \qquad (56)$$

remembering (29). Or, using (26),

$$Y_0' = \sum \frac{1}{p^2 Z'}. \qquad (57)$$

In electrostatic problems the roots of $Z = 0$ are real and negative, as is also the case in magnetic problems. There are never any oscillatory results in either case, and the vanishing of Z' is then accompanied by vanishing of the corresponding normal functions, to prevent the oscillations which seem on the verge of occurring by the repetition of a root which $Z' = 0$ implies.* When both energies are present, the real parts of the imaginary roots are always compelled to be negative by the positivity of U, T, and of Q the dissipativity.

When Z is irrational, it is probable that the complete solution corresponding to (54) might be immediately derived from Z. In the case of (41),† however, the application is not obvious, although there is no difficulty in passing from the (54) solution to the corresponding definite integrals which arise when the length of the circuit is infinitely increased.

* [See p. 529, vol. I. Also Thomson and Tait, Part I., § 343c and after, relating to Routh's Theorem, given in his Adam's Prize Essay, "Stability of Motion."]

† [Done in "El. Mag. Waves," 1888. Arts. XLIII. and XLIV. later.]

XLIII. ON ELECTROMAGNETIC WAVES, ESPECIALLY IN RELATION TO THE VORTICITY OF THE IMPRESSED FORCES; AND THE FORCED VIBRATIONS OF ELECTROMAGNETIC SYSTEMS.

[*Phil. Mag.*, 1888; Part I., February, p. 130; Part II., March, p. 202; Part III., May, p. 379; Part IV., October, p. 360; Part V., November, p. 434; Part VI., December, 1888, p. 488.]

PART I.

Summary of Electromagnetic Connections.

1. To avoid indistinctness, I start with a short summary of Maxwell's scheme, so far as its essentials are concerned, in the form given by me in January, 1885.*

Two forces, electric and magnetic, **E** and **H**, connected linearly with the three fluxes, electric displacement **D**, conduction-current **C**, and magnetic induction **B**; thus

$$\mathbf{B}=\mu\mathbf{H}, \qquad \mathbf{C}=k\mathbf{E}, \qquad \mathbf{D}=(c/4\pi)\mathbf{E}. \qquad (1)$$

Two currents, electric and magnetic, Γ and **G**, each of which is proportional to the curl or vorticity of the *other* force, not counting impressed; thus,

$$\operatorname{curl}(\mathbf{H}-\mathbf{h})=4\pi\Gamma, \qquad (2)$$

$$\operatorname{curl}(\mathbf{e}-\mathbf{E})=4\pi\mathbf{G}; \qquad (3)$$

where **e** and **h** are the impressed parts of **E** and **H**. These currents are also directly connected with the corresponding forces through

$$\Gamma=\mathbf{C}+\dot{\mathbf{D}}, \qquad \mathbf{G}=\dot{\mathbf{B}}/4\pi. \qquad (4)$$

An auxiliary equation to exclude unipolar magnets, viz.

$$\operatorname{div}\mathbf{B}=0, \qquad (5)$$

expressing that **B** has no divergence. The most important feature of this scheme is the equation (3), as a fundamental equation, the natural companion to (2).

The derived energy-relations are not necessary, but are infinitely too useful to be ignored. The electric energy U, the magnetic energy T, and the dissipativity Q, all per unit volume, are given by

$$U=\tfrac{1}{2}\mathbf{ED}, \qquad T=\tfrac{1}{2}\mathbf{HB}/4\pi, \qquad Q=\mathbf{EC}. \qquad (6)$$

The transfer of energy **W** per unit area is expressed by a vector product,

$$\mathbf{W}=\mathrm{V}(\mathbf{E}-\mathbf{e})(\mathbf{H}-\mathbf{h})/4\pi, \qquad (7)$$

and the equation of activity per unit volume is

$$\mathbf{e}\Gamma+\mathbf{hG}=Q+\dot{U}+\dot{T}+\operatorname{div}\mathbf{W}, \qquad (8)$$

from which **W** disappears by integration over *all* space.

The equations of propagation are obtained by eliminating either **E** or

* See the opening sections of "Electromagnetic Induction and its Propagation," *Electrician*, Jan. 3, 1885, and after [Art. xxx., vol. I., p. 429].

H between (2) and (3), and of course take different forms according to the geometrical coordinates selected.

In a recent paper I gave some examples* illustrating the extreme importance of the lines of vorticity of the impressed forces, as the sources of electromagnetic disturbances. Those examples were mostly selected from the extended developments which follow. Although, being special investigations, involving special coordinates, vector methods will not be used, it will still be convenient occasionally to use the black letters when referring to the actual forces or fluxes, and to refer to the above equations. The German or Gothic letters employed by Maxwell I could never tolerate, from inability to distinguish one from another in certain cases without looking very hard. As regards the notation **EC** for the scalar product of **E** and **C** (instead of the quaternionic – S**EC**) it is the obvious practical extension of EC, the product of the tensors, what **EC** reduces to when **E** and **C** are parallel.†

Plane Sheets of Impressed Force in a Nonconducting Dielectric.

2. We need only refer to impressed electric force **e**, as solutions relating to **h** are quite similar. Let an infinitely extended nonconducting dielectric be divided into two regions by an infinitely extended plane (x, y), on one side of which, say the left, or that of $-z$, is a field of **e** of uniform intensity e, but varying with the time. If it be perpendicular to the boundary, it produces no flux. Only the tangential component can be operative. Hence we may suppose that **e** is parallel to the plane, and choose it parallel to **x**. Then **E**, the force of the flux, is parallel to **x**, of intensity E say, and the magnetic force, of intensity H, is parallel to **y**. Let $e=f(t)$; the complete solutions due to the impressed force are then

$$E = \mu vH = -\tfrac{1}{2}f(t-z/v) \quad\text{.........................(9)}$$

on the right side of the plane, where z is $+$, and

$$-E = \mu vH = -\tfrac{1}{2}f(t+z/v) \quad\text{.....................(10)}$$

on the left side of the plane, where z is $-$. In the latter case we must deduct the impressed force from E to obtain the force of the field, say F, which is therefore

$$F = -f(t) + \tfrac{1}{2}f(t+z/v). \quad\text{........................(11)}$$

* *Phil. Mag.* Dec. 1887, "On Resistance and Conductance Operators," § 8, p. 487 [Art. XLII., vol. II., p. 363].

† In the early part of my paper "On the Electromagnetic Wave-Surface," *Phil. Mag.*, June, 1885 [Art. XXXI., vol. II., p. 1] I have given a short introduction to the Algebra of vectors (not quaternions) in a practical manner, *i.e.*, without metaphysics. The result is a thoroughly practical working system. The matter is not an insignificant one, because the extensive use of vectors in mathematical physics is bound to come (the sooner the better), and my method furnishes a way of bringing them in without any study of Quaternions (which are scarcely wanted in Electromagnetism, though they may be added on), and allows us to work without change of notation, especially when the vectors are in special type, as they should be, being entities of widely different nature from scalars. I denote a vector by (say) **E**, its tensor by E, and its x, y, z components, when wanted, by E_1, E_2, E_3. The perpetually occurring scalar product of two vectors requires no prefix. The prefix V of a vector product should be a special symbol.

The results are most easily followed thus:—At the plane itself, where the vortex-lines of e are situated, we, by varying e, produce simultaneous changes in H, thus,

$$-H = e/2\mu v, \qquad \qquad (12)$$

at the plane. This disturbance is then propagated both ways undistorted at the speed $v = (\mu c)^{-\frac{1}{2}}$.

On the other hand, the corresponding electric displacements are oppositely directed on the two sides of the plane.

Since the line-integral of H is electric current, and the line-integral of e is electromotive force, the ratio of e to H is the resistance-operator of an infinitely long tube of unit area; a constant, measurable in ohms, being 60 ohms in vacuum, or 30 ohms on each side. Why it is a constant is simply because the waves cannot return, as there is no reflecting barrier in the infinite dielectric.

3. If the impressed force be confined to the region between two parallel planes distant $2a$ from one another, there are now two sources of disturbances, which are of opposite natures, because the vorticity of e is oppositely directed on the two planes, so that the left plane sends out both ways disturbances which are the negatives of those simultaneously emitted by the right plane. Thus, if the origin of z be midway between the planes, we shall have

$$E = \mu v H = -\tfrac{1}{2} f\left(t - \frac{z-a}{v}\right) + \tfrac{1}{2} f\left(t - \frac{z+a}{v}\right) \qquad (13)$$

on the right side of the stratum of e, and

$$-E = \mu v H = -\tfrac{1}{2} f\left(t + \frac{z-a}{v}\right) + \tfrac{1}{2} f\left(t + \frac{z+a}{v}\right) \qquad (14)$$

on the left side. If therefore e vary periodically in such a way that

$$f(t) = f(t + 2a/v), \qquad (15)$$

there is no disturbance outside the stratum, after the initial waves have gone off, the disturbance being then confined to the stratum of impressed force.

Decreasing the thickness of the stratum indefinitely leads to the result that the effect due to $e = f(t)$ in a layer of thickness dz at $z = 0$ is, on the right side,

$$H = -\frac{1}{2\mu v}\left\{ f\left(t - \frac{z}{v}\right) - f\left(t - \frac{z + dz}{v}\right)\right\} = -\frac{c\,dz}{2} f'\left(t - \frac{z}{v}\right), \qquad (16)$$

since $\mu c v^2 = 1$; on the left side the + sign is required.

We can now, by integration, express the effect due to $e = f(z, t)$, viz.,

$$H = -\frac{c}{2}\int_{-\infty}^{z} \frac{d}{dt} f\left(t - \frac{z-z'}{v}, z'\right) dz' + \frac{c}{2}\int_{z}^{\infty} \frac{d}{dt} f\left(t + \frac{z-z'}{v}, z'\right) dz', \qquad (17)$$

$$E = e - \frac{1}{2v}\int_{-\infty}^{z} \frac{d}{dt} f\left(t - \frac{z-z'}{v}, z'\right) dz' - \frac{1}{2v}\int_{z}^{\infty} \frac{d}{dt} f\left(t + \frac{z-z'}{v}, z'\right) dz'. \qquad (18)$$

In these, however, a certain assumption is involved, viz. that e vanishes at ∞ both ways, because we base the formulæ upon (16), which concerns a layer of e on both sides of which e is zero. Now the disturbances really depend upon de/dz, for there can be none if this be zero. By (12) the elementary de/dz through distance dz instantly produces

$$H=\frac{1}{2\mu v}\frac{de}{dz}dz \qquad \text{.........................(19)}$$

at the place. If, therefore, $e=f(z,t)$, the H-solution at any point consists of the positive waves coming from planes of de/dz on the left, producing say, H_1, and of H_2, due to the negative waves from the planes of de/dz on the right side, making the complete solution

$$H=H_1+H_2, \qquad E=\mu v(H_1-H_2); \qquad \text{..............(20)}$$

where

$$H_1=\frac{1}{2\mu v}\int_{-\infty}^{z}\frac{d}{dz'}f\left(t-\frac{z-z'}{v}, z'\right)dz', \qquad \text{................(21)}$$

$$H_2=\frac{1}{2\mu v}\int_{z}^{\infty}\frac{d}{dz'}f\left(t+\frac{z-z'}{v}, z'\right)dz'. \qquad \text{................(22)}$$

This is the most rational form of solution, and includes the case of $e=f(t)$ only. The former may be derived from it by effecting the integrations in (21) and (22); remembering in doing so that the differential coefficient under the sign of integration is not the complete one with respect to z', as it occurs twice, but only to the second z', and further assuming that $e=0$ at infinity.

Waves in a Conducting Dielectric. How to remove the Distortion due to the Conductivity.

4. Let us introduce a new physical property into the conducting medium, namely that it cannot support magnetic force without dissipation of energy at a rate proportional to the square of the force, a property which is the magnetic analogue of electric conductivity. We make the equations (2) and (3) become, if $p=d/dt$,

$$\text{curl}\,\mathbf{H}=(4\pi k+cp)\mathbf{E}, \qquad \text{.........................(23)}$$

$$-\text{curl}\,\mathbf{E}=(4\pi g+\mu p)\mathbf{H}; \qquad \text{.........................(24)}$$

if there be no impressed force at the spot, where g is the new coefficient of magnetic conductivity, analogous to k.

Let

$$\left.\begin{array}{lll} 4\pi k/2c=q_1, & q_1+q_2=q, & \mathbf{E}=\epsilon^{-qt}\mathbf{E}_1, \\ 4\pi g/2\mu=q_2, & q_1-q_2=s, & \mathbf{H}=\epsilon^{-qt}\mathbf{H}_1. \end{array}\right\} \text{.........(25)}$$

Substitution in (23), (24) leads to

$$\text{curl}\,\mathbf{H}_1=c(s+p)\mathbf{E}_1, \qquad \text{.........................(26)}$$

$$-\text{curl}\,\mathbf{E}_1=\mu(-s+p)\mathbf{H}_1. \qquad \text{.......................(27)}$$

If $s=0$, these are the equations of electric and magnetic force in a nonconducting dielectric. If therefore the new g be of such magnitude as to make $s=0$, we cause disturbances to be propagated in the conducting dielectric in identically the same manner as if it were nonconducting, but with a uniform attenuation at a rate indicated by the time-factor ϵ^{-gt}.

Undistorted Plane Waves in a Conducting Dielectric.

5. Taking **z** perpendicular to the plane of the waves, we now have, as special forms of (23), (24),

$$-dH/dz = (4\pi k + cp)E, \quad \text{......................(28)}$$

$$-dE/dz = (4\pi g + \mu p)H, \quad \text{......................(29)}$$

E being the tensor of **E**, parallel to **x**, and H the tensor of **H**, parallel to **y**, and both being functions of z and t.

Given $E = E_0$ and $H = H_0$ at time $t = 0$, functions of z only, decompose them thus,

$$2f_1 = E_0 + \mu v H_0, \quad \text{..........................(30)}$$

$$2f_2 = E_0 - \mu v H_0. \quad \text{..........................(31)}$$

Here f_1 makes the positive and f_2 the negative wave, and at time t the solutions are, due to the initial state, when $s = 0$,

$$E = \epsilon^{-gt}\{f_1(z - vt) + f_2(z + vt)\}, \quad \text{..................(32)}$$

$$\mu v H = \epsilon^{-gt}\{f_1(z - vt) - f_2(z + vt)\}. \quad \text{..................(33)}$$

The only difference from plane waves in a nonconducting dielectric is in the uniform attenuation that goes on, due to the dissipation of energy, which is so balanced on the electric and magnetic sides as to annihilate the distortion the waves would undergo were s finite, whether positive or negative.

Practical Application. Imitation of this Effect.

6. When I introduced * the new property of matter symbolized by the coefficient g, it was merely to complete the analogy between the electric and magnetic sides of electromagnetism. The property is non-existent, so far as I know. But I have more recently found how to precisely imitate its effect in another electromagnetic problem, also relating to plane waves, making use of electric conductivity to effect the functions of both k and g in §§ 4 and 5. In the case of § 5, first remove both conductivities, so that we have plane waves unattenuated and undistorted. Next put a pair of parallel wires of no resistance in the dielectric, parallel to **z**, and let the lines of electric force terminate upon them, whilst those of magnetic force go round the wires. We shall still have these plane electromagnetic waves with curved lines of force propagated undistorted and unattenuated, at the same speed v. If V be the line-integral of **E** across the dielectric from one wire to the other, and $4\pi C$ be the line-integral of **H** round either wire, we shall have

$$-dV/dz = LpC, \quad \text{..........................(34)}$$

$$-dC/dz = SpV, \quad \text{..........................(35)}$$

* See first footnote [p. 375].

(34) taking the place of (29), and (35) of (28), with k and g both zero Here L and S are the inductance and permittance of unit length of the circuit of the parallel wires, and $v=(LS)^{-\frac{1}{2}}$.

Next let the wires have constant resistance R per unit length to current in them, and let the medium between them be conducting (to a very low degree), making K the conductance per unit length across from one wire to the other. We then turn the last equations into

$$-dV/dz=(R+Lp)C, \qquad (36)$$

$$-dC/dz=(K+Sp)V, \qquad (37)$$

and have a complete imitation of the previous unreal problem. The two dissipations of energy are now due to R in the wires, and to K in the dielectric, it being that in the wires which takes the place of the unreal magnetic dissipation. The relation $R/L=K/S$, which does not require excessive leakage when the wires are of copper of low resistance, removes the distortion otherwise suffered by the waves. I have, however, found that when the alternations of current are very rapid, as in telephony, there is very little distortion produced by copper wires, even without the leakage required to wholly remove it, owing to R/Ln becoming small, $n/2\pi$ being the frequency; an effect which is greatly assisted by increasing the inductance (see Note A, [p. 392]). Of course there is little resemblance between this problem and that of the long and slowly-worked submarine cable, whether looked at from the physical side or merely from the numerical point of view, the results being then of different orders of magnitude. A remarkable misconception on this point seems to be somewhat generally held. It seems to be imagined that self-induction is harmful* to long-distance telephony. The precise contrary is the case. It is the very life and soul of it, as is proved both by practical experience in America and on the Continent on very long copper circuits, and by examining the theory of the matter. I have proved this in considerable detail;† but they will not believe it. So far does the misconception extend that it has perhaps contributed to leading Mr. W. H Preece to conclude that the coefficient of self-induction in copper circuits is negligible (several hundred times smaller than it can possibly be), on the basis of his recent remarkable experimental researches.

The following formula, derived from my general formulæ‡, will show the *rôle* played by self-induction Let R and L be the resistance and inductance per unit length of a perfectly insulated circuit of length l, short-circuited at both ends. Let a rapidly sinusoidal impressed force of amplitude e_0 act at one end, and let C_0 be the amplitude of the

* W. H. Preece, F.R.S., "On the Coefficient of Self-Induction of Copper Wires," B. A. Meeting, 1887.

† "El. Mag. Ind. and its Propagation," *Electrician*, Sections XL. to L. (1887) [vol. II., pp. 119 to 155].

‡ See the sinusoidal solutions in Part II. and Part. V. of "On the Self-Induction of Wires," *Phil. Mag.*, Sept. 1886 and Jan. 1887 [vol. II., pp. 194 and 247. Also p. 62].

current at the distant end. Then, if the circuit be very long,

$$C_0 = \frac{2e_0}{Lv}\epsilon^{-Rl/2Lv}, \quad \text{.........................(38)}$$

where v is the speed $(LS)^{-\frac{1}{2}} = (\mu c)^{-\frac{1}{2}}$, provided R/Ln be small, say $\frac{1}{4}$. It may be considerably greater, and yet allow (38) to be nearly true. We can include nearly the whole range of telephonic frequencies by using suspended copper wires of low resistance.*

It is resistance that is so harmful, not self-induction; as, in combination with the electrostatic permittance, it causes immense distortion of waves, unless counteracted by increasing the inductance, which is not often practicable (see Note B, [p. 393]).

Distorted Plane Waves in a Conducting Dielectric.

7. Owing to the fact that, as above shown, we can fully utilize solutions involving the unreal g, by changing the meaning of the symbols, whilst still keeping to plane electromagnetic waves, we may preserve g in our equations (28) and (29), remembering that H has to become C, E become V, $4\pi k$ become K, c become S, $4\pi g$ become R, and μ become L, when making the application to the possible problem; whilst, when dealing with a real conducting dielectric, g has to be zero.

Required the solutions of (28) and (29) due to any initial states E_0 and H_0, when s is not zero. Using the notation and transformations of (25), (or direct from (26), (27)), we produce

$$-dH_1/dz = c(s+p)E_1, \quad \text{......................(39)}$$

$$-dE_1/dz = \mu(-s+p)H_1; \quad \text{..................(40)}$$

from which

$$v^2(d^2H_1/dz^2) = (p^2 - s^2)H_1, \quad \text{......................(41)}$$

with the same equation for E_1.

The complete solution may be thus described. Let, at time $t=0$, there be $H = H_0$ through the small distance a at the origin. This immediately splits into two plane waves of half the amplitude, which travel to right and left respectively at speed v, attenuating as they progress, so that at time t later, when they are at distances $\pm vt$ from the origin, their amplitudes equal

$$\tfrac{1}{2}H_0\epsilon^{-qt}, \quad \text{..............................(42)}$$

with corresponding E's, viz.,

$$\tfrac{1}{2}\mu v H_0\epsilon^{-qt} \quad \text{and} \quad -\tfrac{1}{2}\mu v H_0\epsilon^{-qt}, \quad \text{...............(43)}$$

on the right and left sides respectively. These extend through the

* The explanation of the $\frac{1}{2}Lv$ dividing e_0 in (38), instead of the Lv we might expect from the μv resistance-operator of a tube of unit section infinitely long one way only, is that, on arrival at the distant end of the line, the current is immediately doubled in amplitude by the reflected wave. The second and following reflected waves are negligible, on account of the length of the line.

distance a. Between them is a diffused disturbance, given by

$$H = \epsilon^{-qt}\frac{H_0 a}{2v}\left(s + \frac{d}{dt}\right)J_0\left\{\frac{s}{v}(z^2 - v^2t^2)^{\frac{1}{2}}\right\}, \quad \text{..............(44)}$$

$$E = \epsilon^{-qt}\frac{H_0 a}{2cv}\left(- \frac{d}{dz}\right)J_0\left\{\frac{s}{v}(z^2 - v^2t^2)^{\frac{1}{2}}\right\}, \quad \text{..............(45)}$$

in which $v^2t^2 > z^2$.

In a similar manner, suppose initially $E = E_0$ through distance a at the origin. Then, at time t later, we have two plane strata of depth a at distance vt to right and left respectively, in which

$$E = \tfrac{1}{2}E_0\epsilon^{-qt} = \pm \mu vH, \quad \text{......................(46)}$$

the + sign to be used in the right-hand stratum, the − in the left. And, between them, the diffused disturbance given by

$$E = \epsilon^{-qt}\frac{E_0 a}{2v}\left(- s + \frac{d}{dt}\right)J_0\left\{\frac{s}{v}(z^2 - v^2t^2)^{\frac{1}{2}}\right\}, \quad \text{..............(47)}$$

$$H = \epsilon^{-qt}\frac{E_0 a}{2\mu v}\left(- \frac{d}{dz}\right)J_0\left\{\frac{s}{v}(z^2 - v^2t^2)^{\frac{1}{2}}\right\}. \quad \text{..............(48)}$$

Knowing thus the effects due to initial elements of E_0 and H_0, we have only to integrate with respect to z to find the solutions due to any arbitrary initial distributions. I forbear from giving a detailed demonstration, leaving the satisfaction of the proper conditions to be the proof of (42) to (48); since, although they were very laboriously worked out by myself, yet, as mathematical solutions, are more likely to have been given before in some other physical problem than to be new.

Another way of viewing the matter is to start with $s = 0$, and then examine the effect of introducing s, either + or − Let an isolated plane disturbance of small depth be travelling along in the positive direction undistorted at speed v. We have $E = \mu vH$ in it. Now suddenly increase k, making s positive. The disturbance still keeps moving on at the same speed, but is attenuated with greater rapidity. At the same time it leaves a tail behind it, the tip of which travels out the other way at speed v, so that at time t after commencement of the tailing, the whole disturbance extends through the distance $2vt$. In this tail H is of the same sign as in the head, and its integral amount is such that it exactly accounts for the extra-attenuation suffered by H in the head. On the other hand, E in the tail is of the opposite sign to E in the head; so that the integral amount of E in head and tail decreases faster. As a special case, let, in the first place, there be no conductivity, $k = 0$ and $g = 0$. Then, keeping g still zero, the effect of introducing k is to cause the above-described effect, except that as there was no attenuation at first, the attenuation later is entirely due to k, whilst the line-integral of H along the tail, or

$$\int H dz,$$

including H in the head, remains constant. This is the persistence of momentum.

If, on the other hand, we introduce g, the statements made regarding H are now true as regards E, and conversely The tail is of a different nature, E being of same sign in the tail as in the head, and H of the opposite sign. Hence, of course, when we have both k and g of the right amounts, there is no tailing. This subject is, however, far better studied in the telegraphic application, owing to the physical reality then existent, than in the present problem, and also then by elementary methods.*

8. Owing to the presence of d/dz in (45) and (47) we are enabled to give some integral solutions in a finite form. Thus, let $H = H_0$ (constant) and $E = 0$ initially on the whole of the negative side of the origin, with no E or H on the positive side. The E at time t later is got by integrating (45), giving

$$E = \frac{H_0}{2cv} J_0 \left\{ \frac{s}{v}(z^2 - v^2t^2)^{\frac{1}{2}} \right\} \epsilon^{-qt}, \quad \text{......................(49)}$$

which holds between the limits $z = \pm vt$, there being no disturbance beyond, except the H_0 on the left side. When $g = 0$ and z/vt is small, it reduces to

$$E = \frac{H_0}{4\pi}\left(\frac{\mu}{kt}\right)^{\frac{1}{2}} \epsilon^{-\pi k \mu z^2/t}. \quad \text{..............................(50)}$$

This is the pure-diffusion solution, suitable for good conductors.

If initially $E = E_0$, constant, on the left side of the origin, and zero on the right side, then at time t the H due to it is, by (48),

$$H = \frac{E_0}{2\mu v} J_0 \left\{ \frac{s}{v}(z^2 - v^2t^2)^{\frac{1}{2}} \right\} \epsilon^{-qt}. \quad \text{......................(51)}$$

The result of taking $c = 0$, $g = 0$, in this formula is zero, as we may see by observing that c in (49) becomes μ in (51). It is of course obvious that as the given initial electric field has no energy if $c = 0$, it can produce no effect later.

The H-solution corresponding to (49) cannot be finitely expressed. It is

$$H = \tfrac{1}{2} H_0 \epsilon^{-qt} \left[1 + \int_z^{vt} \frac{dz}{v} \left(s + \frac{d}{dt} \right) J_0 \left\{ \frac{s}{v}(z^2 - v^2t^2)^{\frac{1}{2}} \right\} \right],$$

which, integrated, gives

$$H = \tfrac{1}{2} H_0 \epsilon^{-qt} \left[\epsilon^{st} - \frac{sz}{v}(J_0 - iJ_1) + \frac{1}{\lfloor 3}\left(\frac{sz}{v}\right)^3 \frac{1}{st}(-iJ_1 - J_2) \right.$$
$$\left. - \frac{1.3}{\lfloor 5}\left(\frac{sz}{v}\right)^5 \frac{1}{s^2t^2}(-J_2 + iJ_3) + \frac{1.3.5}{\lfloor 7}\left(\frac{sz}{v}\right)^7 \frac{1}{s^3t^3}(iJ_3 + J_4) + \dots \right], \quad (52)$$

where all the J's operate on $st\sqrt{-1}$; thus, *e.g.* (Bessel's),

$$J_3 = J_3(st\sqrt{-1}).$$

* "Electromagnetic Induction and its Propagation," *Electrician*, Sections XLIII. to L. (1887) [vol. II., pp. 132 to 155].

But a much better form than (52), suitable for calculating the shape of the wave speedily, especially at its start, may be got by arranging in powers of $z-vt$, thus

$$H=\tfrac{1}{2}H_0\epsilon^{-qt}\left\{1+stf_1\left(1-\frac{z}{vt}\right)+\frac{s^2t^2}{\lfloor 2}f_2\left(1-\frac{z}{vt}\right)^2+\frac{s^3t^3}{\lfloor 3}f_3\left(1-\frac{z}{vt}\right)^3+\dots\right\}, \quad (53)$$

true when $z<vt$, where f_1, f_2, etc., are functions of t only, of which the first five are given by

$$f_1=1+\frac{st}{2},\qquad f_2=\frac{st}{2}\left(1+\frac{st}{4}\right),\qquad f_3=-\tfrac{1}{2}\left(1+\frac{st}{4}\right)+\frac{s^2t^2}{2.4}\left(1+\frac{st}{6}\right),$$

$$f_4=-\tfrac{3}{4}\frac{st}{2}\left(1+\frac{st}{6}\right)+\frac{s^3t^3}{2.4.6}\left(1+\frac{st}{8}\right),$$

$$f_5=\tfrac{3}{8}\left(1+\frac{st}{6}\right)-\frac{s^2t^2}{2.4}\left(1+\frac{st}{8}\right)+\frac{s^4t^4}{2.4.6.8}\left(1+\frac{st}{10}\right).$$

At the origin, H is given by

$$H=\tfrac{1}{2}H_0\epsilon^{-2q_2t}, \quad \dots\dots(54)$$

and is therefore permanently $\tfrac{1}{2}H_0$ when $g=0$. At the front of the wave, where $z=vt$,

$$H=\tfrac{1}{2}H_0\epsilon^{-qt}. \quad \dots\dots(55)$$

Now, to represent the E-solution corresponding to (51), we have only to turn H to E and H_0 to E_0 in (53), and change the sign of s throughout, *i.e.* explicit, and in the f's. Similarly in (52). Thus, at the origin,

$$E=\tfrac{1}{2}E_0\epsilon^{-2q_1t}, \quad \dots\dots(56)$$

and at the front of the wave

$$E=\tfrac{1}{2}E_0\epsilon^{-qt}. \quad \dots\dots(57)$$

9. Again, let $H=\tfrac{1}{2}H_0$ on the left side, and $H=-\tfrac{1}{2}H_0$ on the right side of the origin, initially. The E that results from each of them is the same, and is half that of (49); so that (49) still expresses the E-solution. This case corresponds to an initial electric current of surface-density $H_0/4\pi$ on the $z=0$ plane, with the full magnetic field to correspond, and from it immediately follows the E-solution due to any initial distribution of electric current in plane layers.

Owing to H being permanently $\tfrac{1}{2}H_0$ at the origin in the case (49), (54), when $g=0$, we may state the problem thus:—An infinite conducting dielectric with a plane boundary is initially free from magnetic induction, and its boundary suddenly receives the magnetic force $\tfrac{1}{2}H_0=$ constant. At time t later (49) and (52) or (53) give the state of the conductor at distance $z<vt$ from the boundary. In a good conductor the attenuation at the front of the wave is so enormous that the diffusion-solution (50) applies practically. It is only in bad conductors that the more complete form is required.

Effect of Impressed Force.

10. We can show that the initial effect of impressed force is the same as if the dielectric were nonconducting. In equations (23), (24),

let $p=ni$, where $n/2\pi$ = periodic frequency, supposing e to alternate rapidly. By increasing n we can make the second terms on the right sides be as great multiples of the first terms as we please, so that in the limit we have results independent of k and g, in this respect, that as the frequency is raised infinitely, the true solutions tend to be infinitely nearly represented by simplified forms, in which k and g play the part of small quantities. An inspection of the sinusoidal solution for plane waves shows that **E** and **H** get into the same phase, and that k and g merely present themselves in the exponents of factors representing attenuation of amplitude as the waves pass away from the seat of vorticity of impressed force.

Consequently, in the plane problem, the initial effect of an abrupt discontinuity in e, say e = constant on the left, and zero on the right side of the plane through the origin, is to produce

$$H = -e/2\mu v \qquad (58)$$

all over the plane of vorticity; and

$$E = \mp \tfrac{1}{2}e \qquad (59)$$

on its right and left sides respectively. We may regard the plane as continuously emitting these disturbances to right and left at speed v so long as the impressed force is in operation, but their subsequent history can only be fully represented by the tail-formulæ already given.

Irrespective of the finite curvature of a surface, any element thereof may be regarded as plane. Therefore every element of a sheet of vortex-lines of impressed force acts in the way just described as being true of the elements of an infinite plane sheet. But it is only in comparatively simple cases, of which I shall give examples later, that the subsequent course of events does not so greatly complicate matters as to render it impossible to go into details after the first moment. On first starting the sheet, it becomes a sheet of magnetic induction, whose lines coincide with the vortex-lines of impressed force. If f be the measure of the vorticity per unit area, $f/2\mu v$ is the intensity of the magnetic force. In the imaginary good conductor of *no* permittivity, this is zero, owing to v being then assumed to be infinite.

Notice that whilst the vorticity of e produces magnetic induction, that of **h** produces electric displacement, and whilst in the former case **E** is made discontinuous at a plane of finite vorticity, in the latter case it is **H** that is initially discontinuous.

True Nature of Diffusion in Conductors.

11. The process of diffusion of magnetic induction in conductors appears to be fundamentally one of repeated internal reflexions with partial transmission. Thus, let a plane wave $E_1 = \mu v H_1$ moving in a nonconducting dielectric strike flush an exceedingly thin sheet of metal. Let $E_2 = \mu v H_2$ be the transmitted wave in the dielectric on the other side, and $E_3 = -\mu v H_3$ be the reflected wave. At the sheet we have

$$E_1 + E_3 = E_2, \qquad (60)$$

$$H_1 + H_3 = H_2 + 4\pi k_1 z E_2, \qquad (61)$$

if k_1 be the conductivity of the sheet of thickness z. Therefore

$$\frac{E_2}{E_1}=\frac{H_2}{H_1}=\frac{E_1+E_3}{E_1}=\frac{1}{1+2\pi\mu k_1 zv}. \qquad (62)$$

H is reflected positively and E negatively. A perfectly conducting barrier is a perfect reflector; it doubles the magnetic force and destroys the electric force on the side containing the incident wave, and transmits nothing.

Take $k_1=(1600)^{-1}$ for copper, and $\mu v=3\times 10^{10}$ centim. per sec.

Then we see that to attenuate the incident wave H_1 to $\frac{1}{2}H_1$ by transmission through the plate, requires

$$z=(2\pi\mu k_1 v)^{-1}=\frac{8}{3\pi}\times 10^{-8}\text{ centim.}, \qquad (63)$$

which is a very small fraction of the wave-length of visible light. The H-disturbance is made $\frac{3}{2}H_1$, the E reduced to $\frac{1}{2}E_1$, on the transmission side. There is, however, persistence of H, although there is dissipation of E. To produce dissipation of H with persistence of E requires the plate to be a magnetic, not an electric conductor.

Now, imagine an immense number of such plates to be packed closely together, with dielectric between them, forming a composite dielectric conductor, and let the outermost sheet be struck flush by a plane wave as above. The first sheet transmits $\frac{1}{2}H_1$, the second $\frac{1}{4}H_1$, the third $\frac{1}{8}H_1$, and so on. This refers to the front of the wave, going into the composite conductor at speed v. It is only necessary to go a very short distance to attenuate the front of the wave to nothing; the immense speed of propagation does not result in producing any sensible immediate effect at a distance, which comes on quite slowly as the complex result of all the internal reflexions and transmissions between and at the sheets. Observe that there is an initial accumulation of H, so to speak, at the boundary of the conductor, due to the reflexion. (Example: the current-density may be greater at the outermost layer * of a round wire when the current is started in it than the final value, and the total current in the wire increases faster than if it were constrained to be uniformly distributed.)

Thus a good conductor may have very considerable permittivity, much greater than that of air, and yet show no signs of it, on account of the extraordinary attenuation produced by the conductivity. Now this is rather important from the theoretical point of view. It is commonly assumed that good conductors, *e.g.*, metals, are not dielectrics at all. This makes the speed of propagation of disturbances through them infinitely great. Such a hypothesis, however, should have no place in a rational theory, professing to represent transmission in time by stresses in a medium occupying the space between molecules of gross matter. But by admitting that not only bad conductors, but all conductors, are also dielectrics, we do away with the absurdity of infinitely rapid action through infinite distances in no time at all, and make the method of propagation, although it practically differs so greatly from

* "On the S.I. of Wires," Part I., *Phil. Mag.*, August 1886 [vol. II. p. 181].

that in a nonconducting dielectric, be yet fundamentally the same, with its characteristic features masked by repeated internal reflexions with loss of energy. We need not take any account of the electric displacement in actual reckonings of the magnitude of the effects which can be observed in the case of good conductors, but it is surely a mistake to overlook it when it is the nature of the actions involved that is in question. (See Note C, [p. 153.])

Why conductors act as reflectors is quite another question, which can only be answered speculatively. If molecules are perfect conductors, they are perfect reflectors, and if they were packed quite closely, we should nearly have a perfect conductor in bulk, impenetrable by magnetic induction; and we know that cooling a metal and packing the molecules closer does increase its conductivity. But as they do not form a compact mass in any substance, they must always allow a partial transmission of electromagnetic waves in the intervening dielectric medium, and this would lead to the diffusion method of propagation. We do not, however, account in this way for the dissipation of energy, which requires some special hypothesis.

The diffusion of heat, too, which is, in Fourier's theory, done by instantaneous action to infinite distances, cannot be physically true, however insignificant may be the numerical departures from the truth. What can it be but a process of radiation, profoundly modified by the molecules of the body, but still only transmissible at a finite speed? The very remarkable fact that the more easily penetrable a body is to magnetic induction the less easily it conducts heat, in general, is at present a great difficulty in the way, though it may perhaps turn out to be an illustration of electromagnetic principles eventually.

Infinite Series of Reflected Waves. Remarkable Identities. Realized Example.

12. When, in a plane-wave problem, we confine ourselves to the region between two parallel planes, we can express our solutions in Fourier series, constructed so as to harmonize with the boundary conditions which represent the effect of the whole of the ignored regions beyond the boundaries in modifying the phenomena occurring within the limited region. Now the effect of the boundaries is usually to produce reflected waves. Hence a solution in Fourier series must usually be decomposable into an infinite series of separate solutions, coming into existence one after the other in time if the speed v be finite, or all in operation at once from the first moment if the speed be made infinite (as in pure diffusion). If the boundary conditions be of a simple nature, this decomposition can sometimes be easily explicitly represented, indicating remarkable identities, of which the following investigation leads to one. We may either take the case of plane-waves in a conducting dielectric bounded by infinitely conductive planes, making $E=0$ the boundary condition; or, similarly, by perfect magnetically conductive planes producing $H=0$. But the most practical way, and the most easily followed, is to put a pair of parallel wires in the dielectric, and produce a real problem relating to a telegraph-circuit.

Let A and B be its terminations at $z=0$ and $z=l$ respectively. Let them be short-circuited, producing the terminal conditions $V=0$ at A and B in the absence of impressed force at either place. Now, the circuit being free from charge and current initially, insert a steady impressed force e_0 at A. Required the effect, both in Fourier series and in detail, showing the whole history of the phenomena that result.

Equations (36) and (37) are the fundamental connections of V and C at any distance z from A. Let R, L, K, S be the resistance, inductance, leakage-conductance, and permittance per unit length of circuit, and

$$s_1=R/2L, \qquad s_2=K/2S, \qquad q=s_1+s_2, \qquad s_0=s_1-s_2, \quad \text{....(64)}$$

$$\lambda=(m^2v^2-s_0^2)^{\frac{1}{2}}. \quad \text{..........................(65)}$$

It may be easily shown, by the use of the resistance-operator, or by testing satisfaction of conditions, that the required solutions are

$$V=V_0-\frac{2e_0}{l}\sum\frac{m\sin mz}{RK+m^2}\epsilon^{-qt}\left[\cos\lambda t+\frac{q}{\lambda}\sin\lambda t\right], \quad \text{........(66)}$$

$$C=C_0-\frac{e_0}{Rl}\epsilon^{-2s_1t}-\frac{2e_0}{Rl}\sum\frac{\cos mz}{m^2+RK}\epsilon^{-qt}\left[RK\left(\cos-\frac{s_0}{\lambda}\sin\right)\lambda t-2s_1m^2\frac{\sin\lambda t}{\lambda}\right], \text{(67)}$$

where $m=j\pi/l$, and j includes all integers from 1 to ∞; whilst V_0 and C_0 represent the final steady V and C, which are

$$V_0=e_0\left(\cos m_0z-\frac{\sin m_0z}{\tan m_0l}\right), \quad \text{.....................(68)}$$

$$C_0=\frac{m_0e_0}{R}\left(\sin m_0z+\frac{\cos m_0z}{\tan m_0l}\right), \quad \text{.................(69)}$$

where $m_0^2=-RK$.

Now if the circuit were infinitely long both ways and were charged initially to potential-difference $2e_0$ on the whole of the negative side of A, with no charge on the positive side, and no current anywhere, the resulting current at time t later at distance z from A would be

$$C_1=\frac{e_0}{Lv}\epsilon^{-qt}J_0\left\{\frac{s_0}{v}(z^2-v^2t^2)^{\frac{1}{2}}\right\}, \quad \text{.....................(70)}$$

by §§ 7 and 8; and if, further, $K=0$, V at A would be permanently e_0, which is what it is in (66). Hence the C-solution (67) can be finitely decomposed into separate solutions of the form (70) in the case of perfect insulation, when (67) takes the form

$$C=\frac{e_0}{Rl}(1-\epsilon^{-2qt})+\frac{2e_0}{Ll}\epsilon^{-qt}\sum\cos mz\frac{2q}{\lambda}\sin\lambda t, \quad \text{..........(71)}$$

where $q=s_1=s_0$, by the vanishing of s_2 in (64).

Therefore (70) represents the real meaning of (71) from $t=0$ to l/v, provided $vt>z$. But on arrival of the wave C_1 at B, V becomes zero, and C doubled by the reflected wave that then commences to travel from B to A This wave may be imagined to start when $t=0$ from a point distant l beyond B, and be the precise negative of the first wave as regards V but the same as regards C. Thus

$$C_2 = \frac{e_0}{Lv}\epsilon^{-qt}J_0\left\{\frac{q}{v}[(2l-z)^2 - v^2t^2]^{\frac{1}{2}}\right\} \quad \text{.................(72)}$$

expresses the second wave, starting from B when $t = l/v$, and reaching A when $t = 2l/v$. The sum of C_1 and C_2 now expresses (71) where the waves coexist, and C_1 alone expresses (71) in the remainder of the circuit.

The reflected wave arising when this second wave reaches A may be imagined to start when $t = 0$ from a point distant $2l$ from A on its negative side, and be a precise copy of the first wave. Thus

$$C_3 = \frac{e_0}{Lv}\epsilon^{-qt}J_0\left\{\frac{q}{v}[(2l+z)^2 - v^2t^2]^{\frac{1}{2}}\right\} \quad \text{...................(73)}$$

expresses the third wave; and now (71) means $C_1 + C_2 + C_3$ in those parts of the circuit reached by C_3, and $C_1 + C_2$ in the remainder.

The fourth wave is, similarly,

$$C_4 = \frac{e_0}{Lv}\epsilon^{-qt}J_0\left\{\frac{q}{v}[(4l-z)^2 - v^2t^2]^{\frac{1}{2}}\right\}, \quad \text{.................(74)}$$

starting from B when $t = 3l/v$, and reaching A when $t = 4l/v$. And so on, *ad inf.**

If we take $L = 0$ in this problem, we make $v = \infty$, and bring the whole of the waves into operation immediately. (70) becomes

$$C_1 = e_0\left(\frac{S}{\pi Rt}\right)^{\frac{1}{2}}\epsilon^{-RSz^2/4t}; \quad \text{.......................(75)}$$

and similarly for C_2, C_3, etc. In this simplified form the identity is that obtained by Sir W. Thomson † in connexion with his theory of the submarine cable; also discussed by A. Cayley ‡ and J. W. L. Glaisher. [See also vol. I., p. 88.]

In order to similarly represent the history of the establishment of V_0, we require to use the series for E due to E_0, corresponding to (53), or some equivalent. In other respects there is no difference.

Whilst it is impossible not to admire the capacity possessed by solutions in Fourier series to compactly sum up the effect of an infinite series of successive solutions, it is greatly to be regretted that the Fourier solutions themselves should be of such difficult interpretation.

* It is not to be expected that in a real telegraph-circuit the successive waves have abrupt fronts, as in the text. There are causes in operation to prevent this, and round off the abruptness. The equations connecting V and C express the first approximation to a complete theory. Thus the wires are assumed to be instantaneously penetrated by the magnetic induction as a wave passes over their surfaces, as if the conductors were infinitely thin sheets of the same resistance. It is only a very partial remedy to divide a wire into several thinner wires, unless we at the same time widely separate them. If kept quite close it would, with copper, be no remedy at all.

† *Math. and Physical Papers*, vol. ii., art. lxxii.; with Note by A. Cayley.

‡ *Phil. Mag.*, June 1874.

Perhaps there will be discovered some practical way of analysing them into easily interpretable forms.

Some special cases of (66), (67) are worthy of notice. Thus V is established in the same way when $R=0$ as when $K=0$, provided the value of K/S in the first case be the same as that of R/L in the second. Calling this value $2q$, we have in both cases

$$V=e_0\left(1-\frac{z}{l}\right)-\frac{2e_0}{l}\epsilon^{-qt}\sum\frac{\sin mz}{m}\left(\cos\lambda t+\frac{q}{\lambda}\sin\lambda t\right). \quad\text{........(76)}$$

But the current is established in quite different manners. When it is K that is zero, (71) is the solution; but if R vanish instead, then (67) gives

$$C=\frac{e_0t}{Ll}+\frac{e_0Kl}{2}\left(1-\frac{z}{l}\right)^2-\frac{2e_0K}{l}\epsilon^{-qt}\sum\frac{\cos mz}{m^2}\left\{\cos\lambda t-\left(\frac{m^2v^2}{2q}-q\right)\frac{\sin\lambda t}{\lambda}\right\}. \quad (77)$$

C now mounts up infinitely. But the leakage-current, which is KV, becomes steady, as (76) shows.

In connexion with this subject I should remark that the distortionless circuit produced by taking $R/L=K/S$ is of immense assistance, as its properties can be investigated in full detail by elementary methods, and are most instructive in respect to the distortional circuits in question above.*

Modifications made by Terminal Apparatus. Certain Cases easily brought to Full Realization.

13. Suppose that the terminal conditions in the preceding are $V=-Z_0C$ and $V=Z_1C$, Z_0 and Z_1 being the "resistance-operators" of terminal apparatus at A and B respectively. In a certain class of cases the determinantal equation so simplifies as to render full realization possible in an elementary manner. Thus, the resistance-operator of the circuit, reckoned at A, is†

$$\phi=Z_0+\frac{(R+Lp)l(\tan ml)/ml+Z_1}{1+(K+Sp)lZ_1(\tan ml)/ml}, \quad\text{................(78)}$$

where

$$m^2=-(R+Lp)(K+Sp). \quad\text{......................(79)}$$

That is, $e=\phi C$ is the linear differential equation of the current at A. Now, to illustrate the reductions obviously possible, let $Z_0=0$, and

$$Z_1=n_1l(R+Lp). \quad\text{.........................(80)}$$

This makes the apparatus at B a coil whose time-constant is L/R, and reduces ϕ to

$$\phi=(R+Lp)l\left(\frac{\tan ml}{ml}+n_1\right)\left\{1-m^2n_1l^2\frac{\tan ml}{ml}\right\}^{-1}, \quad\text{.......(81)}$$

so that the roots of $\phi=0$ are given by

$$R+Lp=0. \quad\text{..........................(82)}$$

$$\tan ml+mln_1=0; \quad\text{..........................(83)}$$

* "Electromagnetic Induction and its Propagation," Arts. XL. to L. [vol. II., p. 119].

† "On the Self-Induction of Wires," Part IV. [vol. II., p. 232].

i.e., a solitary root $p = -R/L$, and the roots of (83), which is an elementary well-known form of determinantal equation.

The complete solution due to the insertion of the steady impressed force e_0 at A will be given by*

$$V = V_0 + \Sigma\, e_0 u \epsilon^{pt} \div \left(p\frac{d\phi}{dp}\right), \quad \text{......................(84)}$$

$$C = C_0 + \Sigma\, e_0 w \epsilon^{pt} \div \left(p\frac{d\phi}{dp}\right), \quad \text{......................(85)}$$

where the summations range over all the p roots of $\phi = 0$, subject to (79); whilst u and w are the V and C functions in a normal system, expressed by

$$w = \cos mz, \qquad u = m \sin mz \div (K + Sp)\,; \quad \text{...........(86)}$$

and V_0, C_0 are the final steady V and C. In the case of the solitary root (82) we shall find

$$-p\frac{d\phi}{dp} = Rl(1 + n_1), \quad \text{........................(87)}$$

but for all the rest

$$-p\frac{d\phi}{dp} = \frac{l}{2(K+Sp)}\,\frac{dm^2}{dp}(1 + n_1\cos^2 ml). \quad \text{..............(88)}$$

Realizing (84), (85) by pairing terms belonging to the two p's associated with one m^2 through (79), we shall find that (66), (67) express the solutions, provided we make these simple changes :—Divide the general term in both the summations by $(1 + n_1\cos^2 ml)$, and the term following C_0 outside the summation in (67) by $(1 + n_1)$. Of course the m's have now different values, as per (83), and V_0, C_0 are different.

14. There are several other cases in which similar reductions are possible. Thus, we may have

$$Z_0 = n_0(R + Lp) + n_0'(K + Sp)^{-1},$$
$$Z_1 = n_1(R + Lp) + n_1'(K + Sp)^{-1},$$

simultaneously, n_0, n_0', n_1, n_1' being any lengths. That is, apparatus at either end consisting of a coil and a condenser in sequence, the time-constant of the coil being L/R and that of the condenser S/K. Or, the condenser may be in parallel with the coil. In general we have, as an alternative form of $\phi = 0$, equation (78),

$$\frac{\tan ml}{ml} = -\frac{(Z_0 + Z_1)\{(R + Lp)l\}^{-1}}{1 - m^2l^2 Z_0 Z_1\{(R + Lp)l\}^{-2}}\,; \quad \text{.............(89)}$$

from which we see that when

$$\frac{Z_0}{(R + Lp)l} \quad \text{and} \quad \frac{Z_1}{(R + Lp)l}$$

are functions of ml, equation (89) finds the value of m^2 immediately, *i.e.* not indirectly as functions of p. In all such cases, therefore, we may

* *Ib.* Parts III. and IV. *Phil. Mag.*, Oct. and Nov. 1886; or "On Resistance and Conductance Operators," *Phil. Mag.*, Dec. 1887, § 17, p. 500 [vol. II., p. 373].

advantageously have the general solutions (84), (85) put into the realized form. They are

$$V=V_0-\frac{2e_0}{l}\sum\frac{(\sin mz+\tan\theta\cos mz)m\epsilon^{-qt}(\cos+q\lambda^{-1}\sin)\lambda t}{\sec^2\theta(m^2+RK)\left(1-\cos^2 ml\frac{d}{d(ml)}\tan ml\right)}, \quad\text{.............(90)}$$

$$C=C_0-\frac{2e_0}{l}\sum\frac{(\cos mz-\tan\theta\sin mz)\epsilon^{-qt}K\{\cos-(2s_2\lambda)^{-1}(\lambda^2+qs_0)\sin\}\lambda t}{\text{same denominator}}, \quad (91)$$

where q, λ, s_0, s_2 are as in (64), (65). The differentiation shown in the denominator is to be performed upon the function of ml to which $\tan ml$ is equated in (89), after reduction to the form of such a function in the way explained; and θ depends upon Z_0 thus,

$$\tan\theta=-m^{-1}(K+Sp)Z_0, \qquad \sec^2\theta=1+m^{-2}Z_0^2(K+Sp)^2, \quad (92)$$

which are also functions of ml. It should be remarked that the terms depending upon solitary roots, occurring in the case $m^2=0$, are not represented in (90), (91). They must be carefully attended to when they occur.

NOTE A. *The Electromagnetic Theory of Light.*

An electromagnetic theory of light becomes a necessity, the moment one realizes that it is the same medium that transmits electromagnetic disturbances and those concerned in common radiation. Hence *the* electromagnetic theory of Maxwell, the essential part of which is that the vibrations of light are really electromagnetic vibrations (whatever they may be), and which is an undulatory theory, seems to possess far greater intrinsic probability than *the* undulatory theory, because that is not an electromagnetic theory. Adopting, then, Maxwell's notion, we see that the only difference between the waves in telephony (apart from the distortion and dissipation due to resistance) and light-waves is in the wave-length; and the fact that the speed, as calculated by electromagnetic data, is the same as that of light, furnishes a powerful argument in favour of the extreme relative simplicity of constitution of the ether, as compared with common matter in bulk. There is observational reason to believe that the sun sometimes causes magnetic disturbances here of the ordinary kind. It is impossible to attribute this to any amount of increased activity of emission of the sun so long as we only think of common radiation. But, bearing in mind the long waves of electromagnetism, and the constant speed, we see that disturbances from the sun may be hundreds or thousands of miles long of one kind (*i.e.* without alternation), and such waves, in passing the earth, would cause magnetic "storms," by inducing currents in the earth's crust and in telegraph-wires. Since common radiation is ascribed to molecules, we must ascribe the great disturbances to movements of large masses of matter.

There is nothing in the abstract electromagnetic theory to indicate whether the electric or the magnetic force is in the plane of polarization, or rather, surface of polarization. But by taking a concrete example, as the reflexion of light at the boundary of transparent dielectrics, we get Fresnel's formula for the ratio of reflected to incident wave, on the assumption that his "displacement" coincides with the electric displacement; and so prove that it is the magnetic flux that is in the plane of polarization.

NOTE B. *The Beneficial Effect of Self-Induction.*

I give these numerical examples :—

Take a circuit 100 kilom. long, of 4 ohms and $\frac{1}{4}$ microf. per kilom. and *no* inductance in the first place, and also no leakage in any case. Short-circuit at beginning A and end B. Introduce at A a sinusoidal impressed force, and calculate the amplitude of the current at B by the electrostatic theory. Let the ratio of the full steady current to the amplitude of the sinusoidal current be ρ, and let the frequency range through 4 octaves, from $n=1250$ to $n=20{,}000$; the frequency being $n/2\pi$. The values of ρ are

1·723, 3·431, 10·49, 58·87, 778.

It is barely credible that any kind of speaking would be possible, owing to the extraordinarily rapid increase of attenuation with the frequency. Little more than murmuring would be the result.

Now let $L=2\frac{1}{2}$ (very low indeed), L being inductance per centim. Calculate by the combined electrostatic and magnetic formula. The corresponding figures are

1·567, 2·649, 5·587, 10·496, 16·607.

The change is marvellous. It is only by the preservation of the currents of great frequency that good articulation is possible, and we see that even a very little self-induction immensely improves matters. There is no "dominant" frequency in telephony. What should be aimed at is to get currents of any frequency reproduced at B in their proper proportions, attenuated to the same extent.

Change L to 5. Results :—

1·437, 2·251, 3·176, 4·169, 4·670.

Good telephony is now possible, though much distortion remains.

Increase L to 10. Results :—

1·235, 1·510, 1·729, 1·825, 1·854.

This is first class, showing approximation towards a distortionless circuit. Now this is all done by the self-induction carrying forward the waves undistorted (relatively) and also with much less attenuation.

I should add that I attach no importance to the above figures in point of exactness. The theory is only a first approximation. In order to emphasize the part played by self-induction, I have stated that by sufficiently increasing it (without other change, if this could be possible) we could make the amplitude of current at the end of an Atlantic cable greater than the steady current (by the *quasi*-resonance).

NOTE C. *The Velocity of Electricity.*

In Sir W. Thomson's article on the "Velocity of Electricity" (Nichols's *Cyclopædia*, 2nd edition, 1860, and Art. lxxxi. of 'Mathematical and Physical Papers,' vol. ii.) is an account of the chief results published up to that date relating to the "velocity" of transmission of electricity, and a very explicit statement, except in some respects as regards inertia, of the theoretical meaning to be attached to this velocity under different circumstances. This article is also strikingly illustrative of the remarkable contrast between Sir W. Thomson's way of looking at things electrical (at least at that time) and Maxwell's views; or perhaps I should say Maxwell's plainly evident views combined with the views which his followers have extracted from that mine of wealth 'Maxwell,' but which do not lie on the surface. (As charity begins at home, I may perhaps illustrate by a personal example the difference between the patent and the latent, in

Maxwell. If I should claim (which I do) to have discovered the true method of establishment of current in a wire—that is, the current starting on its boundary, as the result of the initial dielectric wave outside it, followed by diffusion inwards,—I might be told that it was all "in Maxwell." So it is; but entirely latent. And there are many more things in Maxwell which are not yet discovered.) This difference has been the subject of a most moving appeal from Prof. G. F. Fitzgerald, in *Nature*, about three years since. There really seemed to be substance in that appeal. For it is only a master-mind that can adequately attack the great constructional problem of the ether, and its true relation to matter; and should there be reason to believe that the master is on the wrong track, the result must be, as Prof. Fitzgerald observed (in effect) disastrous to progress. Now Maxwell's theory and methods have stood the test of time, and shown themselves to be eminently rational and developable.

It is not, however, with the general question that we are here concerned, but with the different kinds of "velocity of electricity." As Sir W. Thomson points out, his electrostatic theory, by ignoring magnetic induction, leads to infinite speed of electricity through the wire. Interpreted in terms of Maxwell's theory, this speed is not that of electricity through the wire at all, but of the waves through the dielectric, guided by the wire. It results, then, from the assumption $\mu=0$, destroying inertia (not of the electric current, but of the magnetic field), and leaving only forces of elasticity and resistance.

But he also points out another way of getting an infinite speed, when we, in the case of a suspended wire, not of great length, ignore the static charge. This is illustrated by the pushing of incompressible water through an unyielding pipe, constraining the current to be the same in all parts of the circuit. This, in Maxwell's theory, amounts to stopping the elastic displacement in the dielectric, and so making the speed of the wave through it infinite. As, however, the physical actions must be the same, whether a wire be long or short, the assumption being only warrantable for purposes of calculation, I have explained the matter thus. The electromagnetic waves are sent to and fro with such great frequency (owing to the shortness of the line) that only the mean value of the oscillatory V at any part can be perceived, and this is the final value; at the same time, by reason of current in the negative waves being of the same sign as in the positive, the current C mounts up by little jumps, which are, however, packed so closely together as to make a practically continuous rise of current in a smooth curve, which is that given by the magnetic theory. This curve is of course practically the same all over the circuit, because of the little jumps being imperceptible.

But in any case this speed is not the speed of electricity through the wire, but through the dielectric outside it. Maxwell remarked that we know nothing of the speed of electricity in a wire supporting current; it may be an inch in an hour, or immensely great. This is on the assumption, apparently, that the electric current in a wire really consists in the transfer of electricity through the wire. I have been forced, to make Maxwell's scheme intelligible to myself, to go further, and add that the electricity may be standing still, which is as much as to say that there is no current, in a literal sense, inside a conductor. (The slipping of electrification over the surface of a wire is quite another thing. That is merely the movement of the wave through the dielectric, guided by the wire. It occurs in a distortionless circuit, owing to the absence of tailing, in the most plainly evident manner.) In other words, take Maxwell's definition of electric current in terms of magnetic force as a basis, and ignore the imaginary fluid behind it as being a positive hindrance to progress, as soon as one leaves the elementary field of *steady* currents and has to deal with variable states.

The remarks in the text on the subject of the speed of waves in conductors relates to a speed that is not considered in Sir W. Thomson's article. It is the speed of transmission of magnetic disturbances into the wire, in cylindrical waves, which begins at any part of a wire as soon as the primary wave through the dielectric reaches that part. It would be no use trying to make signals through a wire if we had not the outer dielectric to carry the magnetizing and electrizing force to its boundary. The slowness of diffusion in large masses is surprising. Thus a sheet of copper covering the earth, only 1 centim. in thickness, supporting a current whose external field imitates that of the earth, has a time-constant of about a fortnight. If the copper extended to the centre of the earth, the time-constant of the most slowly subsiding normal system would be millions of years.

In the article referred to, Sir W. Thomson mentions that Kirchhoff's investigation, introducing magnetic induction, led to a velocity of electricity considerably greater than* that of light, which is so far in accordance with Wheatstone's observation. Now it seems to me that we have here a suggestion of a probable explanation of why Sir W. Thomson did not introduce self-induction into his theory. There were presumably more ways than one of doing it, as regards the measure of the electric force of induction. When we follow Maxwell's equations, there is but one way of doing it, which is quite definite, and leads to a speed which cannot possibly exceed that of light, since it is the speed $(\mu c)^{-\frac{1}{2}}$ through the dielectric, and cannot be sensibly greater than 3×10^{10} centim., though it may be less. Kirchhoff's result is therefore in conflict with Maxwell's statement that the German methods lead to the same results as his. Besides that, Wheatstone's classical result has not been supported by any later results, which are always less than the speed of light, as is to be expected (even in a distortionless circuit). But a reference to Wheat-

* (*Note by* Sir William Thomson.) In this statement I inadvertently did injustice to Kirchhoff. In the unpublished investigation referred to in the article *Electricity, Velocity of* [Nichols's *Cyclopædia*, second edition, 1860; or my 'Collected Papers,' vol. ii. page 135 (3)], I had found that the ultimate velocity of propagation of electricity in a long insulated wire in air is equal to the number of electrostatic units in the electromagnetic unit; and I had correctly assumed that Kirchhoff's investigation led to the same result. But, owing to the misunderstanding of two electricities or one, referred to in § 317 of my 'Electrostatics and Magnetism,' I imagined Weber's measurement of the number of electrostatic units in the electromagnetic to be $2 \times 3{\cdot}1 \times 10^{10}$ centimetres per second, which would give for the ultimate velocity of electricity through a long wire in air twice the velocity of light. In my own investigation, for the submarine cable, I had found the ultimate velocity of electricity to be equal to the number of electrostatic units in the electromagnetic unit divided by $\sqrt{k}$; k denoting the specific inductive capacity of the gutta-percha. But at that time no one in Germany (scarcely any one out of England) believed in Faraday's "specific inductive capacity of a dielectric."

Kirchhoff himself was perfectly clear on the velocity of electricity in a long insulated wire in air. In his original paper, "Ueber die Bewegung der Electricität in Drähten" (Pogg. *Ann.* Bd. c. 1857; see pages 146 and 147 of Kirchhoff's Volume of Collected Papers, Leipzig, 1882), he gives it as $c/\sqrt{2}$, which is what I then called the number of electrostatic units in the electromagnetic unit; and immediately after this he says, "ihr Werth ist der von 41950 Meilen in einer Sekunde, also sehr nahe gleich der Geschwindigkeit des Lichtes im leeren Raume."

Thus clearly to Kirchhoff belongs the priority of the discovery that the velocity of electricity in a wire insulated in air is very approximately equal to the velocity of light.

[*Note by* The Author. In Maxwell's theory, however, as I understand it, we are not at all concerned with the velocity of electricity in a wire (except the transverse velocity of lateral propagation). The velocity is that of the waves in the dielectric outside the wire.]

stone's paper on the subject will show, first, that there was confessedly a good deal of guesswork; and, next, that the repeated doubling of the wire on itself made the experiment, from a modern point of view, of too complex a theory to be examined in detail, and unsuitable as a test.

PART II.

NOTE ON PART I. *The Function of Self-Induction in the Propagation of Waves along Wires.**

An editorial query, the purport of which I did not at first understand, has directed my attention to Prof. J. J. Thomson's paper "On Electrical Oscillations in Cylindrical Conductors" (*Proc. Math. Soc.*, vol. xvii., Nos. 272, 273), a copy of which the author has been so good as to send me. His results, for example, that an iron wire of $\frac{1}{6}$ centim. radius, of inductivity 500, carries a wave of frequency 100 per second about 100,000 miles before attenuating it from 1 to ϵ^{-1}, and similar results, summed up in his conclusion that the carrying-power of an iron-wire cable is very much greater than that of a copper one of similar dimensions, are so surprisingly different from my own, deduced from my developed sinusoidal solutions, in the accuracy of which I have perfect confidence (having had occasion last winter to make numerous practical applications of them in connexion with a paper which was to have been read at the S. T. E. and E.) [see Art XLI., vol. II., p. 323], that I felt sure there must be some serious error of a fundamental nature running through his investigations. On examination I find this is the case, being the use of an erroneous boundary condition in the beginning, which wholly vitiates the subsequent results [relating to the effect of magnetisation]. It is equivalent to assuming that the tangential component of the flux magnetic induction is continuous at the surface of separation of the wire and dielectric, where the inductivity changes value, from a large value to unity, when the wire is of iron. The true conditions are continuity of tangential *force* and of normal *flux*.

As regards my own results, and how increasing the inductance is favourable, the matter really lies almost in a nutshell; thus. In order to reduce the full expression of Maxwell's connexions to a practical working form I make two assumptions. First, that the longitudinal component of current (parallel to the wires) in the dielectric is negligible, in comparison with the total current in the conductors, which makes C one of the variables, C being the current in either conductor; and next, what is equivalent to supposing that the wave-length of disturbances transmitted along the wires is a large multiple of their distance apart. The result is that the equations connecting V and C become

$$-dV/dz = R''C, \qquad -dC/dz = KV + S\dot{V};$$

S being the permittance and K the conductance of the dielectric per unit length of circuit, whilst R'' is a "resistance-operator," depending

* This note may be regarded as a continuation of Note B [p. 393, vol. II.].

upon the conductors, and their mutual position, which, in the sinusoidal state of variation, reduces to

$$R'' = R' + L'(d/dt),$$

where R' and L' are the effective resistance and inductance of the circuit respectively, per unit length, to be calculated entirely upon magnetic principles. It follows that the fully developed sinusoidal solution is of precisely the same form as if the resistance and inductance were constants. Disregarding the effect of reflexions, we have

$$V = V_0 \epsilon^{-Pz} \sin (nt - Qz),$$

due to $V_0 \sin nt$ impressed at $z=0$; where P and Q are functions of R', L', S, K, and n.

Now if $R'/L'n$ is large, and leakage is negligible (a well-insulated slowly-worked submarine cable, and other cases), we have

$$P = Q = (\tfrac{1}{2}R'Sn)^{\frac{1}{2}},$$

as in the electrostatic theory of Sir W. Thomson. There is at once great attenuation in transit, and also great distortion of arbitrary waves, owing to P and Q varying with n.

But in telephony, n being large, P and Q may have widely different values, because $R'/L'n$ may be quite small, even a fraction. In such case we have no resemblance to the former results. If $R'/L'n$ is small, P and Q approximate to

$$P = R'/2L'v' + K/2Sv', \qquad Q = n/v',$$

where $v' = (L'S)^{-\frac{1}{2}}$. This also requires K/Sn to be small. But it is always very small in telephony.

Now take the case of copper wires of low resistance. L' is practically L_0, the inductance of the dielectric, and v' is practically v, the speed of undissipated waves, or of all elementary disturbances, through the dielectric, whilst R' may be taken to be R, the steady resistance, except in extreme cases. Hence, with perfect insulation,

$$P = R/2L_0v, \qquad Q = n/v,$$

or the speed of the waves is v, and the attenuating coefficient P is practically independent of the frequency, and is made smaller by reducing the resistance, and by *increasing* the inductance *of the dielectric.*

The corresponding current is

$$C = V/L_0v$$

very nearly, or V and C are nearly in the same phase, like undissipated plane waves. There is very little distortion in transit.

How to increase L_0 is to separate the conductors, if twin wires, or raise the wire higher from the ground, if a single wire with earth-return. It is not, however, to be concluded that L_0 could be increased indefinitely with advantage. If l is the length of the circuit,

$$Rl = 2L_0v$$

shows the value of L_0 which makes the received current greatest. It is then far greater than is practically wanted, so that the difficulty of increasing L_0 sufficiently is counterbalanced by the non-necessity. The best value of L_0 is, in the case of a long line, out of reach; so that we may say, generally, that increasing the inductance is always of advantage to reduce the attenuation and the distortion.

Now if we introduce leakage, such that $R/L_0 = K/S$, we entirely remove the distortion, not merely when R/L_0n is small, but of any sort of waves. It is, however, at the expense of increased attenuation. The condition of greatest received current, L_0 being variable, is now

$$Rl = L_0v.$$

We have thus two ways of securing good transmission of electromagnetic waves: one very perfect, for any kind of signals; the other less perfect, and limited to the case of R/L_0n small, but quite practical. The next step is to secure that the receiving-instrument shall not introduce further distortion by the quasi-resonance that occurs. In the truly distortionless circuit this can be done by making the resistance of the receiver be L_0v (whatever the length of the line); this causes complete absorption of the arriving waves. In the other case, of R/L_0n small, with good insulation, we require the resistance of the receiver to be also L_0v, to secure this result approximately. I have also found that this value of the receiver's resistance is exactly the one that (when size of wire in receiver is variable) makes the magnetic force, and therefore the strength of signal, a maximum. Some correction is required on account of the self-induction of the receiver; but in really good telephones of the best kind, with very small time-constants, it is not great. We see therefore that telephony, so far as the electrical part of the matter is concerned, can be made as nearly perfect as possible on lines of thousands of miles in length. But the distortion that is left, due to imperfect translation of sound waves into electromagnetic waves at the sending-end, and the reproduction of sound-waves at the receiving-end, is still very great; though, practically, any fairly good telephonic speech is a sufficiently good imitation of the human voice.

There is one other way of increasing the inductance which I have described, viz., in the case of covered wires to use a dielectric impregnated with iron dust. I have proved experimentally that L_0 can be multiplied several times in this way without any increase in resistance; and the figures I have given above (in Note B) prove what a wonderful difference the self-induction makes, even in a cable, if the frequency is great. Hence, if this method could be made practical, it would greatly increase the distance of telephony through cables.

Now, passing to iron wires, the case is entirely different, on account of the great increase in resistance that the substitution of iron for copper of the same size causes, which increases P and the attenuation. Taking for simplicity the very extreme case of such an excessive frequency as to make the formula

$$R' = (\tfrac{1}{2}R\mu n)$$

nearly true, R being the steady and R' the actual resistance, we see that increasing either R or μ increases R' and therefore P, because $L'v'$ tends to the value L_0v. Thus the carrying power of iron is not greatly above, but greatly below that of copper of the same size.

I have, however, pointed out a possible way of utilizing iron (other than that above mentioned), viz., to cover a bundle of fine iron wires with a copper sheath. The sheath is to secure plenty of conductance;

the division of the iron to facilitate the penetration of current, and so lower the resistance still more, to the greatest extent, whilst at the same time increasing the inductance. But the theory is difficult, and it is doubtful whether this method is even theoretically legitimate. First class results were obtained by Van Rysselberghe on a 1000-mile circuit in America (2000 miles of wire), using copper-covered steel wire. Here the resistance was very low, on account of the copper, and the inductance considerable, on account of the dielectric alone; so that there is no certain evidence that the iron did any good except by lowering the resistance. But about the advantage of increasing the inductance of the dielectric there can, I think, be no question. It imparts momentum to the waves, and that carries them on.

In Note B to the first part of this paper [p. 393 *ante*], I gave four sets of numerical results showing the influence of increasing the inductance, selecting a cable of large permittance (constant) in order to render the illustrations more forcible. The formula used was equation (82), Part II. of my paper "On the Self-Induction of Wires" [p. 195 *ante*], which is

$$C_0 = 2V_0\frac{(Sn)^{\frac{1}{2}}}{(R'^2+L'^2n^2)^{\frac{1}{4}}}(\epsilon^{2Pl}+\epsilon^{-2Pl}-2\cos 2Ql)^{-\frac{1}{2}};$$

where P or $Q = (\frac{1}{2}Sn)^{\frac{1}{2}}\{(R'^2+L'^2n^2)^{\frac{1}{2}} \mp L'n\}^{\frac{1}{2}}$.

Here C_0 is the amplitude of current at $z=l$ due to impressed force $V_0 \sin nt$ at $z=0$, with terminal short-circuits. When the circuit is long enough to make ϵ^{-Pl} small, we obtain

$$\rho = \frac{(R'^2+L'^2n^2)^{\frac{1}{4}}}{2Rl(Sn)^{\frac{1}{2}}}\epsilon^{Pl}$$

as the expression for the ratio ρ of the steady current to the amplitude of the sinusoidal current.

The following table is constructed to show the fluctuating manner of variation of the amplitude with the frequency. Drop the accents, and let R/Ln be small. Then, approximately,

$$\rho = \frac{1}{2y}\left(\epsilon^{y}+\epsilon^{-y}-2\cos 2\frac{nl}{v}\right)^{\frac{1}{2}},$$

where $y = Rl/Lv$,

under no restriction as regards the length of the circuit. Now give y a succession of values, and calculate ρ with the cosine taken as -1, 0, and $+1$. Call the results the maximum, mean, and minimum values of ρ.

y.	Min. ρ.	Mean ρ.	Max. ρ.	y.	ρ.	y.	ρ.
$\frac{1}{2}$	·505	1·500	2·063	6	1·678	12	16·81
1	·521	·878	1·128	7	2·365	14	39·3
2	·587	·686	·771	8	3·378	16	93·2
2·065	·594	·685	·766	9	5·000	18	225
3	·710	·748	·784	10	7·420	20	550
4	·907	·924	·940				
5	1·210	1·218	1·226				

It will be seen that when the resistance of the circuit varies from a small fraction to about the same magnitude as Lv (which may be from 300 to 600 ohms in the case of a suspended copper wire), the variation in the value of ρ as the frequency changes through a sufficiently wide range, is great, merely by reason of the reflexions causing reinforcement or reduction of the strength of the received current. The theoretical least value of ρ is $\frac{1}{2}$, when R/Ln is vanishingly small, indicating a doubling of the amplitude of current. But, as y increases, the range of ρ gets smaller and smaller. After $y=5$ it is negligible.

It is, however, the mean ρ that is of most importance, because the influence of terminal resistances is to lower the range in ρ, and to a variable extent. The value $y=2{\cdot}065$, or, practically, $Rl=2Lv$, makes the mean ρ a minimum. As I pointed out in the paper before referred to, these fluctuations can only be prejudicial to telephony. In the present Note I have described how to almost entirely destroy them. The principle may be understood thus. Let the circuit be infinitely long first. Then its impedance to an intermediate impressed force alternating with sufficient frequency to make R/Ln small will be $2Lv$, viz., Lv each way. The current and transverse voltage produced will be in the same phase, and in moving away from the source of energy they will be similarly attenuated according to the time-factor $\epsilon^{-Rt/2L}$. In order that the circuit, when of finite length, shall still behave as if of infinite length, the constancy of the impedance suggests to us that we should make the terminal apparatus a mere resistance, of amount Lv, by which the waves will be absorbed without reflexion.

That this is correct we may prove by my formula for the amplitude of received current when there is terminal apparatus, equation (19*b*), Part V. "On the Self-Induction of Wires" (*Phil. Mag.*, Jan. 1887). It is

$$C_0=2V_0\left[\frac{K^2+S^2n^2}{R'^2+L'^2n^2}\right]^{\frac{1}{2}}[G_0G_1\epsilon^{2Pl}+H_0H_1\epsilon^{-2Pl}-2(G_0G_1H_0H_1)^{\frac{1}{2}}\cos 2(Ql+\theta)]^{-\frac{1}{2}}.$$

Here C_0 is the amplitude of received current at $z=l$ due to $V_0\sin nt$ impressed force at $z=0$, R' and L' the effective resistance and inductance per unit length of circuit; K and S the leakage-conductance and permittance per unit length,

$$P \text{ or } Q=(\tfrac{1}{2})^{\frac{1}{2}}\Big((R'^2+L'^2n^2)^{\frac{1}{2}}(K^2+S^2n^2)^{\frac{1}{2}}\pm(KR'-L'Sn^2)\Big)^{\frac{1}{2}};$$

G_0, H_0, are terminal functions depending upon the apparatus at $z=0$; G_1, H_1, upon that at $z=l$; the apparatus being of any kind, specified by resistance-operators, making R_0', L_0' the effective resistance and inductance of apparatus at $z=0$, and R_1', L_1', at $z=l$. G_0 is given by

$$G_0=1+(R'^2+L'^2n^2)^{-1}\Big((P^2+Q^2)(R_0'^2+L_0'^2n^2)$$
$$+2P(R'R_0'+L'L_0'n^2)+2Qn(R_0'L'-R'L_0')\Big),$$

from which H_0 is derived by changing the signs of P and Q; whilst

G_1 and H_1 are the same functions of R_1', L_1' as G_0 and H_0 are of R_0', L_0'.

Now drop the accents, since we have only copper wires of low resistance (but not very thick) in question, and the terminal apparatus are to be of the simplest character. K/Sn will be vanishingly small practically, so take $K=0$. Next let R/Ln be small, and let the apparatus at $z=l$ be a mere coil, R_1, of negligible inductance first. We shall now have

$$P=R/2Lv, \qquad Q=n/v,$$

and these make $\quad G_1^{\frac{1}{2}}=(1+R_1/Lv), \qquad H_1^{\frac{1}{2}}=(1-R_1/Lv).$

Thus $R_1=Lv$ makes H_1 vanish, whatever the length of lines, and the terms due to reflexions disappear.

We now have

$$C_0=(V_0/Lv).\epsilon^{-Rl/2Lv}\times G_0^{-\frac{1}{2}},$$

where $G_0^{-\frac{1}{2}}$ expresses the effect of the apparatus at $z=0$ in reducing the potential-difference there, V_0 being the impressed force, and the value of G_0 being unity when there is a short-circuit.

Now, to show that $R_1=Lv$ makes the magnetic force of the receiver the greatest, go back to the general formula, let ϵ^{-Pl} be small, and let the size of the wire vary, whilst the size of the receiving-coil is fixed. It will be easily found, from the expression for G_1, that the magnetic force of the coil is a maximum when

$$R_1^2+L_1^2n^2=\left(\frac{R^2+L^2n^2}{K^2+S^2n^2}\right)^{\frac{1}{2}},$$

where we keep in L_1, the inductance of the receiver. Or, when R/Ln and K/Sn are both small,

$$(R_1^2+L_1^2n^2)^{\frac{1}{2}}=Lv,$$

or, as described, $R_1=Lv$ when the receiver has a sufficiently small time-constant. The rule is, equality of impedances.

We may operate in a similar manner upon the terminal function at the sending end. Suppose the apparatus to be representable as a resistance containing an electromotive force, and that by varying the resistance we cause the electromotive force to vary as its square root. Then, according to a well-known law, the arrangement producing the maximum external current is given by $R_0=Lv$, equality of impedances again. This brings us to

$$C_0=(V_0/2Lv)\epsilon^{-Rl/2Lv},$$

as if the circuit were infinitely long both ways, with maximum efficiency secured at both ends.

Lastly, the choice of L such that $Rl=2Lv$ makes the circuit, of given resistance, most efficient.

In long-distance telephony using wires of low resistance, the waves are sent along the circuit in a manner closely resembling the transmission of waves along a stretched elastic cord, subject to a small amount of friction. In order to similarly imitate the electrostatic theory, we must so reduce the mass of the cord, or else so exaggerate the friction, that there cannot be free vibrations. We may suppose that the displacement of the cord represents the transverse voltage in both cases. But the current will be in the same phase as the transverse voltage in one case, and proportional to its variation along the circuit in the other.

We may conveniently divide circuits, so far as their signalling peculiarities are concerned, into five classes. (1). Circuits of such short length, or so operated upon, that any effects due to electric displacement are insensible. The theory is then entirely magnetic, at least so far as numerical results are concerned. (2). Circuits of such great length that they can only be worked so slowly as to render electromagnetic inertia numerically insignificant in its effects. Also some telephonic circuits in which R/Ln is large. Then, at least so far as the reception of signals is concerned, we may apply the electrostatic theory. (3). The exceedingly large intermediate class in which both the electrostatic and magnetic sides have to be considered, not separately, but conjointly. (4). The simplified form of the last to which we are led when the signals are very rapid and the wires of low resistance. (5). The distortionless circuit, in which, by a proper amount of uniform leakage, distortion of signals is abolished, whether fast or slow. Regarded from the point of view of practical application, this class lies on one side. But from the theoretical point of view, the distortionless circuit lies in the very focus of the general theory, reducing it to simple algebra. I was led to it by an examination of the effect of telephones bridged across a common circuit (the proper place for intermediate apparatus, removing their impedance) on waves transmitted along the circuit. The current is reflected positively, the charge negatively, at a bridge. This is opposite to what occurs when a resistance is put in the main circuit, which causes positive reflexion of the charge, and negative of the current. Unite the two effects and the reflexion of the wave is destroyed, approximately when the resistance in the main circuit and the bridge-conductance are finite, perfectly when they are infinitely small, as in a uniform distortionless circuit.

PART III.

SPHERICAL ELECTROMAGNETIC WAVES.

15. Leaving the subject of plane waves, those next in order of simplicity are the spherical. Here, at the very beginning, the question presents itself whether there can be anything resembling condensational waves?

Sir W. Thomson ("Baltimore Lectures", as reported by Forbes in *Nature*, 1884) suggested that a conductor charged rapidly alternately + and − would cause condensational waves in the ether. But there is no other way of charging it than by a current from somewhere else, so he suggested two conducting spheres to be connected with the poles of an alternating dynamo. The idea seems to be here that electricity would be forced out of one sphere and into the other to and fro with great rapidity, and that between the spheres there might be condensational waves.

But in this case, according to the Faraday law of induction, the result would be the setting up of alternating electromagnetic disturbances in the dielectric, exposing the bounding surfaces of the two spheres to rapidly alternating magnetizing and electrizing force, causing waves, approximately spherical at least, to be transmitted into the spheres, in the diffusion manner, greatly attenuating as they progressed inward.

Perhaps, however, there can be condensational waves if we admit that a certain quite hypothetical something called electricity is compressible, instead of being incompressible, as it must be if we in Maxwell's scheme make the unnecessary assumption that an electric current is the motion through space of the something. In fact, Prof. J. J. Thomson has calculated* the speed of condensational waves supposed to arise by allowing the electric current to have convergence. But a careful examination of his equations will show that the condensational waves there investigated do not exist, *i.e.*, the function determining them has the value zero.†

16. To construct a perfectly general spherical wave we may proceed thus. The characteristic equation of **H**, the magnetic force, in a homogeneous medium free from impressed force is, by (2) and (3),

$$\nabla^2 \mathbf{H} = (4\pi\mu k p + \mu c p^2)\mathbf{H}. \qquad (93)$$

Now, let **r** be the vector distance from the origin, and Q any *scalar* function satisfying this equation. Let

$$\mathbf{H} = \text{curl}\,(\mathbf{r}Q). \qquad (94)$$

Then this derived vector will satisfy (93), and have no convergence, and have no radial component, or will be arranged in spherical sheets. From it derive the other electromagnetic quantities. Change **H** to **E** to obtain spherical sheets of electric force.

This method leads to the spherical sheets depending upon any kind of spherical harmonic. They are, however, too general to be really useful except as mathematical exercises. For the examination of the manner of origin and propagation of waves, zonal harmonics are more useful, besides leading to the solution of more practical problems. It is then not difficult to generalize results to suit any kind of spherical harmonics.

The Simplest Spherical Waves.

17. Let the lines of **H** be circles, centred upon the axis from which θ is measured, and let r be the distance from the origin. We have no concern with ϕ (longitude) as regards **H**, so that the simple specification

* B.A. Report on Electrical Theories.

† I ought to qualify this by adding that the investigation seems very obscure, so that, although I cannot make the system work, yet others may.

of its intensity H fully defines it. Under these circumstances the equation (93) becomes

$$\left.\begin{aligned}(rH)'' + \frac{\nu}{r^2}(\nu r H)^{\grave{}\grave{}} &= (4\pi\mu_0 kp + \mu_0 cp^2)rH, \\ &= q^2 rH, \quad \text{say,}\end{aligned}\right\} \quad \text{.............(95)}$$

where the acute accent denotes differentiation to r, and the grave accent to $\cos\theta$ or μ, whilst ν stands for $\sin\theta$. The inductivity will be now μ_0, to avoid confusing with the μ of zonal harmonics. Equation (95) also defines q in the three forms it can assume in a conductor, dielectric, and conducting dielectric.

Now try to make $r\mathbf{H}$ be an undistorted spherical wave, *i.e.* H varying inversely as the distance, and travelling inward or outward at speed v. Let

$$rH = Af(r - vt), \quad \text{............................(96)}$$

where A is independent of r and t. Of course we must have $k = 0$, making $q = p/v$. Now (96) makes

$$v^2(rH)'' = rp^2 H; \quad \text{................................(97)}$$

which, substituted in (95), gives

$$\nu(\nu H)^{\grave{}\grave{}} = 0; \quad \text{......................................(98)}$$

therefore

$$A\nu = A_1\mu + B_1. \quad \text{.............................(99)}$$

From these we find the required solutions to be

$$H = E/\mu_0 v = \frac{A_1\mu + B_1}{r\nu} F_0'(r - vt), \quad \text{..................(100)}$$

$$F = \mu_0 v \frac{A_1}{r^2} F_0(r - vt); \quad \text{..............................(101)}$$

where F_0 is any function, A_1 and B_1 constants, E and F the two components of the electric force, F being the radial component out, and E the other component coinciding with a line of longitude, the positive direction being that of increasing θ, or from the pole. Similarly, if the lines of $\mathbf{E}$ be circular about the axis, we have the solutions

$$E = -\mu_0 v H_\theta = -\mu_0 v \frac{A_1\mu + B_1}{r\nu} F_0'(r - vt), \quad \text{............(102)}$$

$$H_r = \frac{A_1}{r^2} F_0(r - vt), \quad \text{.......................(103)}$$

where H_r and H_θ are the radial and tangential components of $\mathbf{H}$.

But both these systems involve infinite values at the axis. We must therefore exclude the axis somehow to make use of them. Here is one way. Describe a conical surface of any angle θ_1, and outside it another of angle θ_2, and let the dielectric lie between them. Make the tangential component of $\mathbf{E}$ at the conical surfaces vanish, requiring infinite conductivity there, and we make F vanish in (101), and produce the solution

$$E = \mu_0 v H = \frac{B}{r\nu} f(r - vt), \quad \text{......................(104)}$$

exactly resembling plane waves as regards rvE. Here B is the same as $\mu_0 vB_1$, and f the same as F_0', in equation (100).*

18. Now bring in zonal harmonics. Split equation (95) into the two

$$(rH)'' = \left\{q^2 + \frac{m(m+1)}{r^2}\right\} rH, \quad \text{.................(105)}$$

$$\frac{\nu}{r^2}(\nu H)'' = -\frac{m(m+1)}{r^2} H. \quad \text{.....................(106)}$$

The equation (106) has for solution

$$H = A\nu Q_m',$$

where A is independent of θ, and is to be found from (105).

The most practical way of getting the r functions is that followed by Professor Rowland in his paper† wherein he treats of the waves emitted when the state is sinusoidal with respect to the time. We shall come across the same waves in some problems.

Let $$H = P_m \frac{\epsilon^{qr}}{r} \nu Q_m'. \quad \text{...........................(107)}$$

Then the equation of P_m is, by insertion of (107) in (105),

$$P'' + 2qP' = \frac{m(m+1)}{r^2} P; \quad \text{....................(108)}$$

* In order to render this arrangement (104) intelligible in terms of more every-day quantities, let the angles θ_1 and θ_2 be small, for simplicity of representation; then we have two infinitely conducting tubes of gradually increasing diameter enclosing between them a non-conducting dielectric. Now change the variables. Let V be the line-integral of E across the dielectric, following the direction of the force; it is the transverse voltage of the conductors. Let $4\pi C$ be the line-integral of H round the inner tube; it is the same for a given value of r, independent of θ; C is therefore what is commonly called the current in the conductor. We shall have

$$V = LvC, \qquad C = SvV, \qquad LSv^2 = 1;$$

where L is the inductance and S the permittance, per unit length of the circuit. The value of L is

$$L = 2\mu_0 \log\left[(\tan\tfrac{1}{2}\theta_2) \div (\tan\tfrac{1}{2}\theta_1)\right];$$

so that the circuit has uniform inductance and permittance. The value of C in terms of (104) is

$$C = \frac{B}{2\mu_0 v} f(r - vt).$$

When the tubes have constant radii a_1 and a_2, the value of L reduces to the well known

$$L = 2\mu_0 \log(a_2/a_1),$$

of concentric cylinders. The wave may go either way, though only the positive wave is mentioned.

† *Phil. Mag.*, June 1884, "On the Propagation of an Arbitrary Electromagnetic Disturbance, Spherical Waves of Light, and the Dynamical Theory of Refraction." Prof. J. J. Thomson has also considered spherical waves in a dielectric in his paper "On Electrical Oscillations and Effects produced by the Motion of an Electrified Sphere," *Proc. London Math. Soc.* vol. xv., April 3, 1884. [See also Stoke's *Mathematical and Physical Papers*, and Rayleigh's *Sound* on the subject of these functions.]

and the solution, for practical purposes with complete harmonics, is

$$P=1-\frac{m(m+1)}{2qr}+\frac{m(m^2-1^2)(m+2)}{2.4q^2r^2}-\frac{m(m^2-1)(m^2-2^2)(m+3)}{2.4.6q^3r^3}+\dots \quad (109)$$

We shall find the first few useful, thus :—

$$\left.\begin{aligned} P_1&=1-(qr)^{-1},\\ P_2&=1-3(qr)^{-1}+3(qr)^{-2},\\ P_3&=1-6(qr)^{-1}+15(qr)^{-2}-15(qr)^{-3}. \end{aligned}\right\} \quad (110)$$

Now let $U=\epsilon^{qr}P$, so that U is the r function in Hr. If we change the sign of q in U, producing, say, W, it is the required second solution of (105). Thus

$$U_1=\epsilon^{qr}\left(1-\frac{1}{qr}\right), \qquad W_1=\epsilon^{-qr}\left(1+\frac{1}{qr}\right), \quad (111)$$

in the very important case of Q_1, when $m=1$.

The conjugate property of U and W is

$$UW'-U'W=-2q, \quad (112)$$

which is continually useful.

We have next to combine U and W so as to produce functions suitable for use inside spheres, right up to the centre, and finite there. Let

$$u=\tfrac{1}{2}(U+W), \qquad w=\tfrac{1}{2}(U-W), \quad (113)$$

It will be found that when m is even, w/r is zero and u/r infinite at the origin; but that when m is odd, it is u/r that is zero at the origin and w infinite.

The conjugate property of u and w is

$$uw'-u'w=q, \quad (114)$$

corresponding to (112).

Construction of the Differential Equations connected with a Spherical Sheet of Vorticity of Impressed Force.

19. Now let there be two media—one extending from $r=0$ to $r=a$, in which we must therefore use the u-function or w-function, according as m is odd or even, and an outer medium, or at least one in which q has a different form in general. Then, within the sphere of radius a, we have

$$H=Ar^{-1}u, \quad (115)$$

$$-k_1E=Ar^{-1}u', \quad (116)$$

where $k_1=4\pi k+cp$, and we suppose m odd. It follows that

$$\frac{E}{H}=-\frac{1}{k_1}\frac{u'}{u}. \quad (117)$$

In the outer medium use W, if the medium extends to infinity, or both U and W if there be barriers or change of medium. First, let it be an infinitely extended medium. Then, in it,

$$H=Br^{-1}(u-w), \quad (118)$$

$$-k_2E=Br^{-1}(u'-w'), \quad (119)$$

where $k_2 = 4\pi k + cp$ in the outer medium. From these

$$\frac{E}{H} = -\frac{1}{k_2}\frac{u' - w'}{u - w}. \qquad (120)$$

(117) and (120) show the forms of the resistance-operators on the two sides.*

Now, at the surface of separation, $r = a$, H is continuous (unless we choose to make it a sheet of electric current, which we do not); so that the H in (117) and (120) are the same. We only require a relation between the E's to complete the differential equation.

Let there be vorticity of impressed force on the surface $r = a$, and nowhere else (the latter being already assumed). Then

$$\text{curl}\,\mathbf{e} = \text{curl}\,\mathbf{E} \qquad (121)$$

is the surface-condition which follows; or, if f be the measure of the curl of **e**,

$$f = E_2 - E_1, \qquad (122)$$

E_2 meaning the outer and E_1 the inner E. Therefore

$$f = H_a\left(\frac{E_2}{H_2} - \frac{E_1}{H_1}\right), \qquad (123)$$

H_a denoting the surface H. So, by (117) and (120), used in (123),

$$f = \left(\frac{1}{k_1}\frac{u_1'}{u_1} - \frac{1}{k_2}\frac{u_2' - w_2'}{u_2 - w_2}\right)H_a, \quad (r = a), \qquad (124)$$

the required differential equation. Observe that u_1 only differs from u_2 and w_1 from w_2 in the different values of q inside and outside (when different), and that $r = a$ in all.

* Some rather important considerations are presented here. On what principles should we settle which functions to use internally and externally, seeing that these functions U and W are not quantities, but differential operators? First, as regards the space outside the surface of origin of disturbances. The operator ϵ^{qr} turns $f(t)$ into $f(t + r/v)$, and can therefore only be possible with a negative wave, coming to the origin. But there cannot be such a wave without a barrier or change of medium to produce it. Hence the operator ϵ^{-qr} alone can be involved in the external solution when the medium is unbounded, and we must use W. Next, go inside the sphere $r = a$. It is clear that both U and W are now needed, because disturbances come to any point from the further as well as from the nearer side of the surface, thus coming from and going to the centre. Two questions remain: Why take U and W in equal ratio; and why their sum or their difference, according as m is odd or even? The first is answered by stating the facts that, although it is convenient to assume the origin to be a place of reflection, yet it is really only a place where disturbances cross, and that the H produced at any point of the surface is (initially) equal on both sides of it. The second question is answered by stating the property of the Q_m^1 function, that it is an even function of μ when m is odd, and conversely; so that when m is odd the H disturbances arriving at any point on a diameter from its two ends are of the same sign, requiring $U + W$; and when m is even, of opposite sign, requiring $U - W$.

Similar reasoning applies to the operators concerned in other than spherical waves. Cases of simple diffusion are brought under the same rules by generalizing the problem so as to produce wave-propagation with finite speed. On the other hand, when there are barriers, or changes of media, there is no difficulty, because the boundary conditions tell us in what ratio U and W must be taken.

Equation (124) applies to any odd m. When m is even, exchange u and w, also u' and w'. In the mth system we may write

$$f_m = \phi_m H_a, \quad \text{.............................(125)}$$

the form of ϕ being given in (124). The vorticity of the impressed force is of course restricted to be of the proper kind to suit the mth zonal harmonic. Thus, any distribution of vorticity whose lines are the lines of latitude on the spherical surface may be expanded in the form

$$\Sigma f_m \nu Q'_m, \quad \text{.............................(126)}$$

and it is the mth of these distributions which is involved in the preceding.

20. Both media being supposed to be identical, ϕ reduces to

$$\phi = \frac{1}{k_1} \frac{q}{u_a(u_a - w_a)}, \quad \text{.........................(127)}$$

by using (114) in (124). This is with m odd; if even, we shall get

$$\phi = \frac{1}{k_1} \frac{q}{w_a(u_a - w_a)}. \quad \text{.........................(128)}$$

In a non-dielectric conductor, $k_1 = 4\pi k$, and $q^2 = 4\pi\mu kp$; so that, keeping to m odd,

$$\phi = \left(\frac{\mu p}{4\pi k}\right)^{\frac{1}{2}} \frac{1}{u_a(u_a - w_a)}. \quad \text{....................(129)}$$

In a non-conducting dielectric, $k_1 = cp$, and $q = p/v$; so

$$\phi = \frac{\mu_0 v}{u_a(u_a - w_a)}. \quad \text{.............................(130)}$$

In this case the complete differential equation is

$$H_a = \sum \frac{\nu Q'_m}{\mu_0 v} u_a(u_a - w_a) f_m, \quad \text{....................(131)}$$

when there is any distribution of impressed force in space whose vorticity is represented by (126).

Outside the sphere, consequently,

$$(\text{out}) \begin{cases} H = \sum \dfrac{\nu Q'_m}{\mu_0 v} \dfrac{a}{r} u_a(u - w) f_m, & \text{................(132)} \\ -cpE = \sum \dfrac{\nu Q'_m}{\mu_0 v} \dfrac{a}{r} u_a(u' - w') f_m, & \text{...............(133)} \end{cases}$$

understanding that when no letter is affixed to u or w, the value at distance r is meant. We see at once that $u_a = 0$ makes the external field vanish, *i.e.*, the field of the particular f concerned. This happens when f is a sinusoidal function of the time, at definite frequencies. Also, inside the sphere,

$$(\text{in}) \begin{cases} H = \sum \dfrac{\nu Q'_m}{\mu_0 v} \dfrac{a}{r} u(u_a - w_a) f_m, & \text{.................(134)} \\ -cpE = \sum \dfrac{\nu Q'_m}{\mu_0 v} \dfrac{a}{r} u'(u_a - w_a) f_m. & \text{................(135)} \end{cases}$$

As for the radial component F, it is not often wanted. It is got thus from H:—

$$-cpF = r^{-1}(\nu H)^{\backprime}, \qquad (136)$$

where for cp write $4\pi k + cp$ in the general case. Thus, the internal F corresponding to (135) is

$$\text{(in)} \qquad cpF = \sum \frac{m(m+1)}{\mu_0 v} \frac{a}{r^2} u(u_a - w_a) f_m Q_m. \qquad (137)$$

Practical Problem. Uniform Impressed Force in the Sphere.

21. If there be a uniform field of impressed force in the sphere, parallel to the axis, of intensity f_1, its vorticity is represented by $f_1 \sin\theta$ on the surface of the sphere. It is therefore the case $m = 1$ in the above. Let this impressed force be suddenly started. Find the effect produced. We have, by (132),*

$$\text{(out)} \qquad H = u_a(u - w) \frac{f_1 \nu a}{\mu_0 v r}; \qquad (138)$$

or, in full, referring to the forms of u and w, equations (110) to (113),

$$H = \frac{\nu a}{2\mu_0 v r}\left\{\epsilon^{-q(r-a)}\left(1 - \frac{1}{qa}\right)\left(1 + \frac{1}{qr}\right) + \epsilon^{-q(r+a)}\left(1 + \frac{1}{qa}\right)\left(1 + \frac{1}{qr}\right)\right\} f_1. \qquad (139)$$

Effect the integrations indicated by the inverse powers of q or p/v; thus

$$\frac{f_1}{q^n} = f_1 \frac{(vt)^n}{\underline{|n}}, \qquad (140)$$

if f_1 be zero before and constant after $t = 0$. As for the exponentials, use Taylor's theorem, as only differentiations are involved. We get, after the process (140) has been applied to (139), and then Taylor's theorem carried out,

$$H = \frac{f_1 \nu a}{2\mu_0 v r}\left\{\left(1 - \frac{vt_1}{a} + \frac{vt_1}{r} - \frac{v^2 t_1^2}{2ar}\right) + \left(1 + \frac{vt_2}{a} + \frac{vt_2}{r} + \frac{v^2 t_2^2}{2ar}\right)\right\}, \qquad (141)$$

where $\qquad vt_1 = vt - r + a, \qquad vt_2 = vt - r - a.$

* It will be observed that the operator connecting f_1 and H is of such a nature that the process of expansion of H in a series of normal functions fails. I have examined several cases of this kind. The invariable rule seems to be that when there is a surface of vorticity of e, leading to an equation of the form $f = \phi H$, and there is a change of medium somewhere, or else barriers, causing reflected waves, the form of ϕ is such that we can, when f is constant, starting at $t = 0$, solve thus [p. 373, vol. II.]

$$H = \frac{f}{\phi_0} + \sum \frac{f\epsilon^{pt}}{p(d\phi/dp)},$$

extending over all the (algebraical) p-roots of $\phi = 0$, which is the determinantal equation. But should there be no change of medium, the conjugate property of the functions concerned comes into play. It causes a great simplification in the form of ϕ, and makes the last method fail completely, all trace of the roots having disappeared. But if we pass continuously from one case to the other, then the last formula becomes a definite integral. On the other hand, we can immediately integrate $f = \phi H$ in its simplified form, and obtain an interpretable equivalent for the definite integral, the latter being more ornamental than useful. In the simplified form, ϕ may be either rational or irrational. The integration of the irrational forms will be given in some later problems.

It is particularly to be noticed that the t_1 part of (141) only comes into operation when t_1 reaches zero, and similarly as regards the t_2 part. Thus, the first part expresses the primary wave out from the surface; the second, arriving at any point $2a/v$ later than the first, is the reflected wave from the centre, arising from the primary wave inward from the surface.

The primary wave outward may be written

$$H=\left(\frac{f_1\nu}{2\mu_0 v}\right)\frac{1}{2}\left(1+\frac{a^2-v^2t^2}{r^2}\right), \qquad \text{(142)}$$

where $vt>(r-a)$, and the second wave by its exact negative, with $vt>(r+a)$. Now, by comparing (132) with (134), we see that the internal solution is got from the external by exchanging a and r in the {}'s in (139) and (141), including also in t_1 and t_2. The result is that (142) represents the internal H in the primary inward wave, vt having to be $>(a-r)$; whilst its negative represents the reflected wave, provided $vt>(a+r)$.

The whole may be summed up thus. First, vt is $<a$. Then (142) represents H everywhere between $r=a+vt$ and $r=a-vt$. But when vt is $>a$, H is given by the same formula between the limits $r=vt-a$ and $vt+a$. In both cases H is zero outside the limits named.

The reflected wave, superimposed on the primary, annuls the H disturbance, which is therefore, after the reflexion, confined to a spherical shell of depth $2a$ containing the uncancelled part of the primary wave outward.

The amplitude of H at the front of the two primary waves, in and out, before the former reaches the centre, is

$$(f_1\nu a)\div(2\mu_0 vr).$$

After the inward wave has reached the centre, however, the amplitude of H on the front of the reflected wave is the negative of that of the primary wave at the same distance, which is itself negative.

The process of reflexion is a very remarkable one, and difficult to fully understand. At the moment $t=a/v$ that the disturbance reaches the centre, we have $H=(f_1\nu)\div(4\mu_0 v)$, constant, all the way from $r=0$ to $2a$, which is just half the initial value of H on leaving the surface of the sphere. But just before reaching the centre, H runs up infinitely for an infinitely short time, infinitely near the centre; and just after the centre is reached we have $H=-\infty$ infinitely near the centre, where the H-disturbance is always zero, except in this singular case when it is seemingly finite for an infinitely short time, though, of course, ν is indeterminate.

With respect to this running-up of the value of H in the inward primary wave, it is to be observed that whilst H is increasing so fast at and near its front, it is falling elsewhere, viz., between near the front and the surface of the sphere; so that just before the centre is reached H has only half the initial value, except close to the centre, where it is enormously great.

After reflexion has commenced, the H-disturbance is negative in the hinder part of the shell of depth $2a$ which goes out to infinity, positive of course still in the forward part. At a great distance these portions become of equal depth a; at the front of the shell $H=(f_1va)(2\mu_0vr)^{-1}$, at its back $H=$ – ditto; using of course a different value of r.

22. As regards the electric field, we have, by (133),

$$\text{(out)} \qquad E=-\frac{1}{cp}\frac{\nu}{\mu_0 v}\frac{a}{r}u_a(u'-w')f_1; \qquad (143)$$

which, expanded, is

$$E=\frac{\nu a}{2r}\left\{\epsilon^{q(a-r)}\left(1-\frac{1}{qa}\right)\left(1+\frac{1}{qr}+\frac{1}{q^2r^2}\right)+\epsilon^{-q(a+r)}\left(1+\frac{1}{qa}\right)\left(1+\frac{1}{qr}+\frac{1}{q^2r^2}\right)\right\}f_1; \quad (144)$$

comparing which with (139), we see that

$$E=\mu vH+\frac{\nu a}{2r}\left\{\epsilon^{-q(r-a)}\frac{1}{q^2r^2}\left(1-\frac{1}{qa}\right)+\epsilon^{-q(r+a)}\frac{1}{q^2r^2}\left(1+\frac{1}{qa}\right)\right\}f_1. \qquad (145)$$

We have, therefore, only to develop the second part, which is not in the same phase with H. It is, in the same manner as before,

$$\frac{f_1\nu a}{2r}\left(\frac{v^2t_1^2}{2r^2}-\frac{v^3t_1^3}{6r^2a}\right)+\frac{f_1\nu a}{2r}\left(\frac{v^2t_2^2}{2r^2}+\frac{v^3t_2^3}{6r^2a}\right), \qquad (146)$$

only operating when $vt_1=vt-r+a$, and $vt_2=vt-r-a$ are positive. Or,

$$\frac{f_1\nu a}{4r^3}\left\{\left(\frac{2a^2}{3}+a(vt-r)-\frac{(vt-r)^3}{3a}\right)_1+\left(\frac{2a^2}{3}-a(vt-r)+\frac{(vt-r)^3}{3a}\right)_2\right\}, \quad (146a)$$

1 and 2 referring to the two waves. So, when $vt>(r+a)$, and the two are coincident, we have the sum

$$E=\frac{f_1\nu a^3}{3r^3}, \qquad (147)$$

which is the tangential component of the steady electric field left behind.

The radial component F is, by (137),

$$\text{(out)} \quad F=\frac{a\cos\theta}{r}\left\{\epsilon^{-q(r-a)}\left(\frac{1}{qr}+\frac{1}{q^2r^2}-\frac{1}{q^2ra}-\frac{1}{q^3r^2a}\right)+\dots\right\}f_1, \quad (148)$$

where the unwritten term ... may be obtained from the preceding by changing the sign of a. Or

$$F=\frac{f_1a\cos\theta}{r^2}\left\{\left(vt_1+\frac{v^2t_1^2}{2r}-\frac{v^2t_1^2}{2a}-\frac{v^3t_1^3}{6ra}\right)+\dots\right\}, \qquad (149)$$

where $vt_1=vt+a-r$. Or,

$$F=\frac{f_1a\cos\theta}{r^2}\left\{\frac{a^2}{3r}+\frac{a}{2}+\frac{a}{2r}(vt-r)-\frac{1}{2a}(vt-r)^2-\frac{1}{6ra}(vt-r)^3+\dots\right\}; \quad (150)$$

so that, when both waves coincide, we have their sum,

$$F=\frac{2f_1a^3\cos\theta}{3r^3}, \qquad (151)$$

which is the radial component of the steady field left behind by the part of the primary wave whose magnetic field is wholly cancelled.

To verify; the uniform field of impressed force of intensity f_1, by elementary principles, produces the external electric potential

$$\Omega = f_1 \cos\theta \frac{a^3}{3r^2},$$

whose derivatives, radial and tangential, taken negatively, are (151) and (147). The corresponding internal potential is

$$\Omega = \tfrac{1}{3} f_1 r \cos\theta.$$

But its slope does not give the force **E** left behind within the sphere, because this **E** is the force of the flux. Any other distribution of impressed force, with the same vorticity, will lead to the same **E**. Our equation (135) and its companion for F, derived from (134) by using (136), lead to the steady field (residual)

$$E = -\tfrac{2}{3} f_1 \sin\theta, \qquad F = \tfrac{2}{3} f_1 \cos\theta, \quad \text{..............(152)}$$

the components of the true force of the flux. Add **e** to the slope of Ω to produce **E**.*

F is always zero at the front of the primary wave outward, and $E = \mu_0 vH$. At the front of the primary wave inward F is also zero, and $E = -\mu_0 vH$. After reflection, F at the front of the reflected wave is still zero, but now $E = \mu_0 vH$.

The electric energy U_1 set up is the volume-integral of the scalar product $\frac{1}{2}$**eD**. That is,

$$U_1 = \tfrac{1}{2} f_1 \times \tfrac{2}{3}\frac{cf_1}{4\pi} \times \frac{4\pi a^3}{3} = \frac{ca^3 f_1^2}{9}. \quad \text{..................(153)}$$

But the total work done by **e** is $2U_1$, by the general law that the whole work done by impressed forces suddenly started exceeds the amount representing the waste by Joule-heating at the final rate (when there is any), supposed to start at once, by twice the excess of the electric over the magnetic energy of the steady field set up. It is clear, then, that when the travelling shell has gone a good way out, and it has become nearly equivalent to a plane wave, its electric and magnetic energies are nearly equal, and each nearly $\frac{1}{2}U_1$ in value. I did not, however, anticipate that the magnetic energy in the travelling shell would turn out to be constant, viz., $\frac{1}{2}U_1$ during the whole journey, from $t = a/v$ to $t = \infty$, so that it is the electric energy in the shell which gradually decreases to $\frac{1}{2}U_1$. Integrate the square of H according to (142) to verify.

23. The most convenient way of reckoning the work done, and also the most appropriate in this class of problems, is by the integral of the

* Sometimes the flux is apparently wrongly directed. For example, a uniform field of impressed force from left to right in all space except a spherical portion produces a flux from right to left in that portion. This is made intelligible by the above. Let the impressed force act in the space between $r=a$ and $r=b$, a being small and b great. In the inner sphere the first effects are those due to the $r=a$ vorticity, and the flux left behind is against the force. But after a time comes the wave from the $r=b$ vorticity, which sets matters right. The same applies in the case of conductors, when, in fact, a long time might have to elapse before the second and real permanent state conquered the first one.

scalar product of the curl of the impressed force and the magnetic force. Thus, in our problem

$$2U_1 = \int dt\, \Sigma\, \mathbf{e}\Gamma = \int dt\, \Sigma\, \mathbf{H}\, \mathrm{curl}\, \mathbf{e}/4\pi = \iint \frac{f_1 v dS}{4\pi}\int H_a dt, \quad \text{......(154)}$$

where dS is an element of the surface $r = a$. So we have to calculate the time-integral of the magnetic force at the place of vorticity of $\mathbf{e}$, the limits being 0 and $2a/v$. This can be easily done without solving the full problem, not only in the case of $m = 1$, but $m =$ any integer. The result is, if U_m be the electric energy of the steady field due to f_m,

$$\int H_a dt = \frac{caf_m v Q_m^{\backprime}}{2m+1}, \quad \text{........................(155)}$$

and, therefore, by surface-integration according to (154),

$$2U_m = a^3 c f_m^2 \frac{m(m+1)}{(2m+1)^2}. \quad \text{......................(156)}$$

$\frac{1}{2}U_m$ is the magnetic energy in the mth travelling shell. I have entered into detail in the case of $m = 1$, because of its relative importance, and to avoid repetition. In every case the magnetic field of the primary wave outward is cancelled by that of the reflection of the primary wave inward, producing a travelling shell of depth $2a$, within which is the final steady field. There, are, however, some differences in other respects, according as m is even or odd.

Thus, in the case $m = 2$, we have, by (110) to (113),

$$\tfrac{1}{2}(U_a - W_a)W$$
$$= \tfrac{1}{2}\left\{\epsilon^{q(a-r)}\left(1 - \frac{3}{qa} + \frac{3}{q^2a^2}\right) - \epsilon^{-q(r+a)}\left(1 + \frac{3}{qa} + \frac{3}{q^2a^2}\right)\right\} \times \left(1 + \frac{3}{qr} + \frac{3}{q^2r^2}\right). \quad (157)$$

Making this operate upon f_2, zero before and constant after $t = 0$, we obtain, by (132), (140), and Taylor's theorem,

$$\text{(out)} \quad H = \frac{f_2 av Q_2^{\backprime}}{2\mu_0 vr}\left\{\frac{1}{4} + \frac{3}{8}\left(\frac{a^2}{r^2} + \frac{r^2}{a^2}\right) - v^2t^2\frac{3}{4}\left(\frac{1}{r^2} + \frac{1}{a^2}\right) + \frac{3}{8}\frac{v^4t^4}{a^2r^2} - \dots\right\}. (158)$$

In the wave represented, $vt > (r - a)$, it being the primary wave out. The unrepresented part, to be obtained by changing the sign of a within the { }, is the reflected wave, in which $vt > (r + a)$.

To obtain the internal H exchange a and r within the { } in (158). The result is that

$$H = \frac{f_2 av Q_2^{\backprime}}{2\mu_0 vr}\left\{-\tfrac{1}{2} + \frac{3}{8a^2r^2}(v^2t^2 - a^2 - r^2)^2\right\} \quad \text{...........(159)}$$

expresses the H-solution always, provided that when $vt < a$ the limits for r are $a - vt$ and $a + vt$; but when $vt > a$, they are $vt - a$ and $vt + a$.

At the surface of the sphere,

$$H_a = \frac{f_2 v Q_2^{\backprime}}{2\mu_0 v}\left\{1 - \frac{3}{2}\left(\frac{vt}{a}\right)^2 + \frac{3}{8}\left(\frac{vt}{a}\right)^4\right\}, \quad \text{.................(160)}$$

from $t = 0$ to $2a/v$. It vanishes twice, instead of only once, intermediately, finishing at the *same* value that it commenced at, instead of at the opposite, as in the $m = 1$ case.

The radial component F of $\mathbf{E}$ is always zero at the front of either of the primary waves or of the reflected wave, and $E = \pm \mu_0 v H$, according as the wave is going out or in. In the travelling shell H changes sign m times, thus making $m+1$ smaller shells of oppositely directed magnetic force. At its outer boundary

$$E = \mu_0 v H = \tfrac{1}{2} f_m \nu Q'_m (a/r), \qquad \text{.......................(161)}$$

and at the inner boundary the same formula holds, with $\pm$ prefixed according as m is even or odd.

In case $m=3$, the magnetic force at the spherical surface is

$$H_a = \frac{f_3 \nu Q'_3}{2\mu_0 v}\left\{1 - \frac{3v^2t^2}{a^2} + \frac{15}{8}\frac{v^4t^4}{a^4} - \frac{5}{16}\frac{v^6t^6}{a^6}\right\} \qquad \text{............(162)}$$

from $t=0$ to $2a/v$; after which, zero.

Spherical Sheet of Radial Impressed Force.

24. If the surface $r=a$ be a sheet of radial impressed force, it is clear that the vorticity is wholly on the surface. Let the intensity be independent of ϕ, so that

$$e = \Sigma e_m Q_m. \qquad \text{...............................(163)}$$

The steady potential produced is

$$\text{(in)} \qquad V_1 = -\sum e_m Q_m \frac{m+1}{2m+1}\left(\frac{r}{a}\right)^m, \qquad \text{...............(164)}$$

$$\text{(out)} \qquad V_2 = +\sum e_m Q_m \frac{m}{2m+1}\left(\frac{a}{r}\right)^{m+1}, \qquad \text{...........(165)}$$

because, at $r=a$, these make

$$V_2 - V_1 = e, \qquad \text{and} \qquad dV_1/dr = dV_2/dr; \qquad \text{............(166)}$$

i.e., potential-difference e, and continuity of displacement. The normal component of displacement is

$$-\frac{c}{4\pi}\frac{dV_1}{dr} = \frac{c}{4\pi a}\sum e_m Q_m \frac{m(m+1)}{2m+1}; \qquad \text{.............(167)}$$

therefore, integrating over the sphere, the total work done by e is

$$2U = \sum cae_m^2 \frac{m(m+1)}{(2m+1)^2}, \qquad \text{.....................(168)}$$

which agrees with the estimate (156), because

$$f = -\frac{de}{ad\theta} = \frac{\nu}{a}\frac{de}{d\mu} \qquad \text{..........................(169)}$$

finds the vorticity, f, from the radial impressed force e; or, taking

$$e = e_m Q_m, \qquad e_m \nu Q'_m a^{-1} = \text{vorticity},$$

so that the old $f_m = e_m/a$.

Single Circular Vortex Line.

25. There are some advantages connected with transferring the impressed force to the surface of the sphere, as it makes the force of the

flux and the force of the field identical both outside and inside. At the boundary F is continuous, E discontinuous.

Let the impressed force be a simple circular shell of radius a, and strength e. Let it be the equatorial plane, so that the equator is the one line of vorticity. Substitute for this shell a spherical shell of strength $\frac{1}{2}e$ on the positive hemisphere, $-\frac{1}{2}e$ on the negative, the impressed force acting radially. Expand this distribution in zonal harmonics. The result is

$$\Sigma e_m Q_m = \frac{e}{2}\left\{\frac{3}{2}Q_1 - \frac{7}{2.4}Q_3 + \frac{11.3}{2.4.6}Q_5 - \frac{15.1.3.5}{2.4.6.8}Q_7 + \dots\right\}, \quad \dots\dots(170)$$

so that we are only concerned with the odd m's. This equation settling the value of e_m, the vorticity is

$$\Sigma e_m a^{-1}\nu Q'_m = \Sigma f_m \nu Q'_m. \quad \dots\dots(171)$$

We know therefore, by the preceding, the complete solution due to sudden starting of the single vortex-line. That is, we know the individual waves in detail produced by e_1, e_3, etc. The resultant travelling disturbance is therefore confined between two spherical surfaces of radii $vt-a$ and $vt+a$, after the centre has been reached, or of radii $a-vt$ and $a+vt$ before the centre is reached. But it cannot occupy the whole of either of the regions mentioned.

The actual shape of the boundaries, however, may be easily found. It is sufficient to consider a plane section through the axis of the sphere. Let A and B be the points on this plane cut by the vortex-line. Describe circles of radius vt with A and B as centres. If $vt<a$, the circles do not intersect; the disturbance is therefore wholly within them. But when vt is $> a$, the intersecting part contains no **H**, and only the **E** of the steady field due to the vortex-line, which we know by § 24.

That within the part common to both circles there is no **H** we may prove thus. The vortex-line in question may be imagined to be a line of latitude on any spherical surface passing through A and B, and centred upon the axis. Let a_1 be the radius of any sphere of this kind. Then, at a time making $vt>a$, the disturbance must lie between the surfaces of spheres of radii $vt-a_1$ and $vt+a_1$, whose centre is that of the sphere a_1. Now this excludes a portion of the space between the $vt-a$ and $vt+a$ circles, referring to the plane section; and by varying the radius a_1 we can find the whole space excluded. Thus, find the locus of intersections of circles of radius

$$vt-(a^2+z^2)^{\frac{1}{2}},$$

with centre at distance z from the origin, upon the axis. The equation of the circle is

$$(x-z)^2+y^2=\{vt-(a^2+z^2)^{\frac{1}{2}}\}^2,$$

or

$$x^2+y^2-2xz=v^2t^2+a^2-2vt(a^2+z^2)^{\frac{1}{2}}. \quad \dots\dots(172)$$

Differentiate with respect to z, giving

$$z(v^2t^2-x^2)=ax, \quad \dots\dots(173)$$

and eliminate z between (173) and (172). After reductions, the result is

$$x^2+(y\pm a)^2=v^2t^2, \qquad\text{(174)}$$

indicating two circles, both of radius vt, whose centres are at A and B. Within the common space, therefore, the steady electric field has been established.

If this case be taken literally, then, since it involves an infinite concentration in a geometrical line of a finite amount of vorticity of **e**, the result for the steady field is infinite close up to that line, and the energy is infinite. But imagine, instead, the vorticity to be spread over a zone at the equator of the sphere $r=a$, half on each side of it, and its surface-density to be $f_1\nu$, where f_1 is finite. Consider the effect produced at a point in the equatorial plane. From time $t=0$ to $t_1=(r-a)/v$ (if the point be external) there is no disturbance. But from time t_1 to $t_2=b/v$, where b is the distance from the point to the edges of the zone, the disturbance must be identically the same as if the harmonic distribution $f_1\nu$ were complete, viz. by (142),

$$H=\left(\frac{f_1\nu}{4\mu_0 v}\right)\left(1+\frac{a^2-v^2t^2}{r^2}\right). \qquad\text{(175)}$$

After this moment t_2, the formula of course fails. Now narrow the band to width $ad\theta$ at the equator and simultaneously increase f_1, so as to make $f_1 ad\theta=e$, the strength of the shell of impressed force when there is but one. The formula (175) will now be true only for a very short time, and in the limit it will be true only momentarily, at the front of the wave, viz.,

$$f_1a/2\mu_0vr=H=e/2\mu_0vr\,d\theta, \qquad\text{(176)}$$

going up infinitely as $d\theta$ is reduced. To avoid infinities in the electric and magnetic forces we must seemingly keep either to finite volume or finite surface-density of vorticity of **e**, just as in electrostatics with respect to electrification.

Instead of a simple shell of impressed electric force, it may be one of magnetic force, with similar results. As a verification, calculate the displacement through circle ν on the sphere $r=a$ due to a vortex-circle at ν_1 on the same surface, the latter being of unit strength. It is

$$\sum\frac{a\nu}{2}\frac{ce_m\nu Q_m^1}{2m+1}, \qquad\text{(177)}$$

due to Σe_mQ_m, through the circle ν. Take then

$$e_m=\frac{(2m+1)\nu_1^2Q_{1m}^1}{2m(m+1)}, \qquad\text{(178)}$$

which represents e_m due to vortex-line of unit strength at ν_1. Use this in the preceding equation (177), and we obtain

$$D=\sum\frac{ca}{4}\frac{\nu^2Q_m^1\nu_1^2Q_{1m}^1}{m(m+1)}, \qquad\text{(179)}$$

as the displacement through ν due to unit vortex-line at ν_1. Applying this result to a circular electric current, $\mathbf{B}=\mu_0\mathbf{H}$ takes the place of

$\mathbf{D} = (c/4\pi)\mathbf{E}$, as the flux concerned, whilst if h be the strength of the shell of impressed magnetic force, $h/4\pi$ is the equivalent bounding electric current. The induction through the circle ν due to unit electric current in the circle ν_1 is therefore obtainable from (179) by turning c to μ_0 and multiplying by $(4\pi)^2$. The result agrees with Maxwell's formula for the coefficient of mutual induction of two circles (vol. II., art. 697).

It must be noted that in the magnetic-shell application there must be no conductivity, if the wave-formulæ are to apply.

An Electromotive Impulse. m = 1.

26. Returning to the case of impressed electric force, let in a spherical portion of an infinite dielectric a uniform field of impressed force act momentarily. We know the result of the continued application of the force. We have, then, to imagine it cancelled by an oppositely directed force, starting a little later. Let t_1 be the time of application of the real force, and let it be a small fraction of $2a/v$, the time the travelling shell takes to traverse any point. The result is evidently a shell of depth vt_1 at $r = vt + a$, in which the electromagnetic field is the same as in the case of continued application of the force, and a similar shell situated at $r = vt - a$, in which H is negative. Within this inner shell there is no $\mathbf{E}$ or $\mathbf{H}$. But between the two thin shells just mentioned there is a diffused disturbance, of weak intensity, which is due to the sphericity of the waves, and would be non-existent were they plane waves. In fact, at time $t = t_1$, when the initial disturbance $H = f_1\nu/2\mu_0 v$ has extended itself a small distance vt_1 on each side of the surface of the sphere, there is a radial component F at the surface itself, since, by (150),

$$F_a = f_1 \cos\theta\left(\frac{vt}{a} - \frac{v^3t^3}{6a^3}\right), \qquad \text{(180)}$$

so that the sudden removal of f_1 leaves two waves which do not satisfy the condition $E = \mu_0 vH$ at their common surface of contact. On separation, therefore, there must be a residual disturbance between them. The discontinuity in E at the moment of removing f_1 is abolished by instantaneous assumption of the mean value, but it is impossible to destroy the radial displacement which joins the two shells at the moment they separate. Put on f_1 when $t = 0$, then $-f_1$ at time t_1 later. The H at time t due to both is, by (142),

$$H = \frac{f_1\nu}{4\mu_0 vr^2}v^2(t_1^2 - 2tt_1)\,; \qquad \text{(181)}$$

which, when t_1 is infinitely small, becomes

$$H = -\frac{f_1\nu tt_1 v^2}{2\mu_0 vr^2}. \qquad \text{(182)}$$

First of all, at a point distant r from the centre, comes the primary disturbance or head,

$$H = \frac{f_1\nu a}{2\mu_0 vr}, \qquad \text{(183)}$$

when $vt = r - a$, lasting for the time t_1. It is followed by the diffused negative disturbance, or tail, represented by (182), lasting for the time

$2a/v$. At its end comes the companion to (183), its negative, when $vt=r+a$, lasting for time t_1, after which it is all over. This description applies when $r>a$. If $r<a$, the interval between the beginning and end of the H-disturbance is only $2r/v$. From the above follows the integral solution expressing the effect of f_1 varying in any manner with the time.

Alternating Impressed Forces.

27. If the impressed force in the sphere, or wherever it may be, be a sinusoidal function of the time, making $p^2=-n^2$, if $n=2\pi\times$ frequency, the complete solutions arise from (132) to (135) so immediately that we can almost call them the complete solutions. Of course in any case in which we have developed the connection between the impressed force and the flux, say $e=ZC$, or $C=Z^{-1}e$, where Z is the resistance-operator, we may call this equation *the* solution in the sinusoidal case, if we state that p^2 is to mean $-n^2$. But there is usually a lot of work needed to bring the solution to a practical form. In the present instance, however, there is scarcely any required, because u and w are simple functions of qr, and q^2 is real. The substitution $p^2=-n^2$ in u results in a real function of nr/v, and in w in a real function $\times(-1)^{\frac{1}{2}}$. Thus:—

$$\left.\begin{aligned} u_1&=\cos\frac{nr}{v}-\frac{v}{nr}\sin\frac{nr}{v},\\ w_1&=i\left(\sin\frac{nr}{v}+\frac{v}{nr}\cos\frac{nr}{v}\right);\end{aligned}\right\}\quad\text{(184)}$$

$$\left.\begin{aligned} u_2&=\left(1-\frac{3v^2}{n^2r^2}\right)\cos\frac{nr}{v}-\frac{3v}{nr}\sin\frac{nr}{v},\\ w_2&=i\left\{\left(1-\frac{3v^2}{n^2r^2}\right)\sin\frac{nr}{v}+\frac{3v}{nr}\cos\frac{nr}{v}\right\}.\end{aligned}\right\}\quad\text{(185)}$$

In the case $m=1$, if $(f_1)\cos nt$ is the form of f_1, so that (f_1) represents the amplitude, we find, writing this case fully because it is the most important:—

$$\left.\begin{aligned} &\text{(out)}\ H=\frac{(f_1)va}{\mu_0vr}\left(\cos-\frac{v}{na}\sin\right)\frac{na}{v}.\left(\cos-\frac{v}{nr}\sin\right)\left(\frac{nr}{v}-nt\right),\\ &\text{(in)}\ \ H=\frac{(f_1)va}{\mu_0vr}\left(\cos-\frac{v}{nr}\sin\right)\frac{nr}{v}.\left(\cos-\frac{v}{na}\sin\right)\left(\frac{na}{v}-nt\right),\end{aligned}\right\}\quad\text{(185}a\text{)}$$

$$\left.\begin{aligned} &\text{(out)}\ F=-\frac{2(f_1)va\mu}{nr^2}\left(\cos-\frac{v}{na}\sin\right)\frac{na}{v}.\left(\sin+\frac{v}{nr}\cos\right)\left(\frac{nr}{v}-nt\right),\\ &\text{(in)}\ \ F=-\frac{2(f_1)va\mu}{nr^2}\left(\cos-\frac{v}{nr}\sin\right)\frac{nr}{v}.\left(\sin+\frac{v}{na}\cos\right)\left(\frac{na}{v}-nt\right),\end{aligned}\right\}\quad\text{(185}b\text{)}$$

$$\left.\begin{aligned} &\text{(out)}\ E=\frac{(f_1)av}{r}\left(\cos-\frac{v}{na}\sin\right)\frac{na}{v}.\left\{\left(1-\frac{v^2}{n^2r^2}\right)\cos-\frac{v}{nr}\sin\right\}\left(\frac{nr}{v}-nt\right),\\ &\text{(in)}\ \ E=-\frac{(f_1)av}{r}\left\{\left(1-\frac{v^2}{n^2r^2}\right)\sin+\frac{v}{nr}\cos\right\}\frac{nr}{v}.\left(\sin+\frac{v}{na}\cos\right)\left(\frac{na}{v}-nt\right).\end{aligned}\right\}\text{(185}c\text{)}$$

It is very remarkable, on first acquaintance, that the impressed force produces no external effect at all when

$$u_a = 0, \qquad \text{or} \qquad \tan\frac{na}{v} = \frac{na}{v}.$$

For the impressed force may be most simply taken to be a uniform field of inteusity $(f_1)\cos nt$ in the sphere of radius a acting parallel to the axis, and it looks as if external displacement must be produced. Of course, on acquaintance with the reason, the fact that the solution is made up of two sets of waves, those outward from the lines of vorticity and those going inward, and then reflected out, the mystery disappears.

To show the positive and negative waves explicitly, we may write the first of (185a) in the form

$$\text{(out)}\ H = \frac{(f_1)av}{2\mu_0 vr}\Bigg[\left\{\left(1 - \frac{v^2}{n^2ar}\right)\cos + \left(\frac{v}{na} + \frac{v}{nr}\right)\sin\right\}\left(nt - \frac{n(a+r)}{v}\right)$$
$$+ \left\{\left(1 + \frac{v^2}{n^2ar}\right)\cos + \left(\frac{v}{nr} - \frac{v}{na}\right)\sin\right\}\left(nt + \frac{n(a-r)}{v}\right)\Bigg], \quad (185d)$$

the second line showing the primary wave out, the first the reflected wave.* Exchange a and r within the [ ] to obtain the internal H. The disturbance, at the surface, of the primary wave going both ways is, from $t = 0$ to $2a/v$,

$$\frac{(f_1)\nu}{2\mu_0 v}\left\{\cos nt + \frac{v^2}{n^2ar}(\cos nt - 1)\right\}. \quad \ldots\ldots\ldots\ldots\ldots\ldots(185e)$$

The amplitude due to both waves is

$$\frac{(f_1)\nu}{\mu_0 v}u_a\left(1 + \frac{v^2}{n^2a^2}\right)^{\frac{1}{2}}. \quad \ldots\ldots\ldots\ldots\ldots\ldots\ldots\ldots(185f)$$

The time-rate of outward transfer of energy per unit area at any distance r is $EH/4\pi$. In the mth system this is

$$\frac{EH}{4\pi} = -\frac{(f_m)^2a^2(\nu Q_m^1)^2}{4\pi(\mu_0 v)^2 r}\cdot\frac{u_a^2}{cnr}.\{u'\sin - (-iw')\cos\}nt.\{u\cos + (-iw)\sin\}nt, \quad (186)$$

where m is supposed odd, whilst u and $-iw$ are the real functions of

* In reference to this formula (185d), and the corresponding ones for other values of m, it is not without importance to know that a very slight change suffices to make (185d) represent the solution from the first moment of starting the impressed force. Thus, let it start when $t = 0$, and let the f_1 in equation (139) be $(f_1)\cos nt$. Effect the two integrations thus,

$$\frac{f_1}{q} = (f_1)\frac{v}{n}\sin nt, \qquad \frac{f_1}{q^2} = (f_1)\frac{v^2}{n^2}(1 - \cos nt),$$

vanishing when $t = 0$, and then operate with the exponentials, and we shall obtain (185d) thus modified:—To the first line must be added

$$\frac{(f_1)\nu a}{2\mu_0 vr}\,\frac{v^2}{n^2ar},$$

and to the second line its negative. Thus modified, (185d) is true from $t = 0$, understanding that the second line begins when $t = (r - a)/v$, and the first when $t = (r + a)/v$. The first of (185a) is therefore true up to distance $r = vt - a$, when this is positive. In the shell of depth $2a$ beyond, it fails.

nr/v obtained in the same way as (184). The mean value of the t function is, by the conjugate property of u and w, equation (114),

$$= -n/2v.$$

Using this, and integrating (186) over the complete surface of radius r, giving

$$\iint (\nu Q_m^1)^2 dS = \frac{4\pi r^2 m(m+1)}{2m+1}, \quad \text{.....................(187)}$$

we find the mean transfer of energy outward per second through any surface enclosing the sphere to be

$$\frac{m(m+1)}{2(2m+1)} \frac{(f_m)^2 u_a^2 a^2}{\mu_0 v}, \quad \text{........................(188)}$$

if $(f_m)\nu Q_m^1 \cos nt$ is the vorticity of the impressed force. [When m is even substitute $-w_a^2$.]

In the case $m=1$, the waste of energy per second is

$$\frac{(f_1)^2 a^2 u_a^2}{3\mu_0 v}, \quad \text{............................(189)}$$

due to the uniform alternating field of impressed force of intensity $(f_1) \cos nt$ within the sphere.

In reality, the impressed force must have been an infinitely long time in operation to make the above solutions true to an infinite distance, and have therefore already wasted an infinite amount of energy. If the impressed force has been in operation any finite time t, however great, the disturbance has only reached the distance $r=vt+a$. Of course the solutions are true, provided we do not go further than $r=vt-a$. We see, therefore, that the real function of the never-ceasing waste of energy is to set up the sinusoidal state of **E** and **H** in the boundless regions of space which the disturbances have not yet reached. The above outward waves are the same as in Rowland's solutions.* Here, however, they are explicitly expressed in terms of the impressed forces causing them.

$u_a=0$ makes the external field vanish when m is odd; and $w_a=0$ when m is even; that is, when the sinusoidal state has been assumed. It takes only the time $2a/v$ to do this, as regards the sphere $r=a$; the initial external disturbance goes out to infinity and is lost. This vanishing of the external field happens whatever may be the nature of the external medium away from the sphere, except that the initial external disturbance will behave differently, being variously reflected or absorbed according to circumstances.

Conducting Medium. m = 1.

28. Now consider the same problem in an infinitely extended conductor of conductivity k. We may remark at once that, unless the conductivity is low, the solution is but little different from what it would be were the conductor not greatly larger than the spherical

*In paper referred to in § 18.

portion within it on whose surface lie the vortex-lines of the impressed force, owing to the great attenuation suffered by the disturbances as they progress from the surface. In a similar manner, if the sphere be large, or the periodic frequency great, or both, we may remove the greater part of the interior of the sphere without much altering matters.

We have now

$$q=(4\pi\mu_0 kp)^{\frac{1}{2}}=(1+i)x, \qquad \text{if} \qquad x=(2\pi\mu_0 kn)^{\frac{1}{2}}. \quad \text{..........(190)}$$

The realization is a little troublesome on account of this $p^{\frac{1}{2}}$. The result is that the uniform alternating field of impressed force of intensity $(f_1)\cos nt$, gives rise to the internal solution

$$\left[H=\frac{(f_1)va}{r}\left(\frac{4\pi k}{\mu_0 p}\right)^{\frac{1}{2}}uW_a; \quad \text{see (129), § 20;}\right]$$

$$\text{(in)} \qquad H=\left(\frac{\pi k}{2\mu_0 n}\right)^{\frac{1}{2}}\frac{(f_1)va}{r}\{(A+B)\cos nt+(A-B)\sin nt\}, \quad \text{.....(191)}$$

where A and B are the functions of r expressed by

$$A=\epsilon^{x(r-a)}\left[\left(1+\frac{1}{2xa}-\frac{1}{2xr}\right)\cos+\left(\frac{1}{2xr}-\frac{1}{2xa}+\frac{2}{2xr.2xa}\right)\sin\right]x(a-r)$$

$$+\epsilon^{-x(r+a)}\left[\left(1+\frac{1}{2xa}+\frac{1}{2xr}\right)\cos-\left(\frac{1}{2xa}+\frac{1}{2xr}+\frac{2}{2xr.2xa}\right)\sin\right]x(a+r); \quad (192)$$

$$B=\epsilon^{x(r-a)}\left[\left(\frac{1}{2xr}-\frac{1}{2xa}+\frac{2}{2xr.2xa}\right)\cos-\left(1-\frac{1}{2xr}+\frac{1}{2xa}\right)\sin\right]x(a-r)$$

$$-\epsilon^{-x(r+a)}\left[\left(\frac{1}{2xr}+\frac{1}{2xa}+\frac{2}{2xr.2xa}\right)\cos+\left(1+\frac{1}{2xr}+\frac{1}{2xa}\right)\sin\right]x(a+r). \quad (193)$$

Equation (191) showing the internal H, the external is got by exchanging a and r in the functions A and B.

Now xa is easily made large, in a good conductor; then, anywhere near the boundary, $(r=a)$, we have

$$A=\epsilon^{-x(a-r)}\cos x(a-r), \qquad -B=\epsilon^{-x(a-r)}\sin x(a-r), \quad \text{.....(194)}$$

and (191) becomes

$$\text{(in)} \qquad H=\left(\frac{\pi k}{\mu_0 n}\right)^{\frac{1}{2}}\frac{(f_1)av}{r}\epsilon^{-x(a-r)}\cos\left\{nt-x(a-r)-\frac{\pi}{4}\right\}. \quad \text{.......(195)}$$

The wave-length λ is

$$\lambda=\left(\frac{2\pi}{\mu_0 kn}\right)^{\frac{1}{2}}. \quad \text{..........................(196)}$$

Thus, in copper, a frequency of 1600 to 1700 makes $\lambda=1$ centim. Both λ and the attenuation-rate depend inversely on the square roots of the inductivity, conductivity, and frequency, whereas the amplitude varies directly as the square root of the conductivity, and inversely as the square roots of the others.

[The attenuation in distance λ is $\epsilon^{-x\lambda}=\epsilon^{-2\pi}$; therefore we may say it is nearly insensible further on. If we introduce an auxiliary impressed force to keep the current straight, we shall, when xa is large, just double the external H and the activity.]

To verify that very great frequency ultimately limits the disturbance to the vortex-line of e when there is but one, we may use the last solution to construct that due to a sheet of impressed force

$$\cos nt \, \Sigma e_m Q_m,$$

acting radially on the surface of the sphere. Thus,

$$\text{(in)} \qquad H = \left(\frac{\pi k}{\mu_0 n}\right)^{\frac{1}{2}} \sum \frac{e_m \nu Q'_m}{r} \epsilon^{-x(a-r)} \cos\left\{nt - x(a-r) - \frac{\pi}{4}\right\}, \qquad (197)$$

when xa is very great. When the vorticity is confined to one line of latitude, H in (197) vanishes everywhere except at the vortex-line. But a further approximation is required, or a different form of solution, to show the disturbance round the vortex-line explicitly, *i.e.*, when n is great, though not infinitely great.

A Conducting Dielectric. m = 1.

29. Here, if k is the conductivity, c the permittivity, and μ_0 the inductivity, let

$$q = (4\pi\mu_0 kp + \mu_0 cp^2)^{\frac{1}{2}} = n_1 + n_2 i, \qquad (198)$$

when $p = ni$. Then n_1 and n_2 will be given by

$$n_1 \text{ or } n_2^2 = \frac{n^2}{2v^2}\left[\left\{1 + \left(\frac{4\pi k}{cn}\right)^2\right\}^{\frac{1}{2}} \mp 1\right]. \qquad (199)$$

Using this q in the general external H-solution, but ignoring the explicit connexion with the impressed force, we shall arrive at

$$\text{(out)} \qquad H = \frac{C_0 \nu}{r}\epsilon^{-n_1 r}\left[\left(1 + \frac{n_1}{r(n_1^2 + n_2^2)}\right)\cos - \frac{n_2}{r(n_1^2 + n_2^2)}\sin\right](n_2 r - nt), \qquad (200)$$

where C_0 is an undetermined constant, depending upon the magnitude of the disturbance at $r = a$. So far as the external solution goes, however, the internal connexions are quite arbitrary save in the periodicity and confinement to producing magnetic force proportional in intensity to the cosine of the latitude. The solution (200) may be continued unchanged as near to the centre as we please. Stopping it anywhere, there are various ways of constructing complementary distributions in the rest of space, from which (200) is excluded.

n_1 is zero when $k = 0$. We then have the dielectric solution, with $n_2 = n/v$. On the other hand, $c = 0$ makes

$$n_1 = n_2 = (2\pi\mu_0 kn)^{\frac{1}{2}} = x,$$

as in § 28. The value of $n_1^2 + n_2^2$ is

$$\frac{n^2}{v^2}\left(1 + \left(\frac{4\pi k}{cn}\right)^2\right)^{\frac{1}{2}} = \frac{n^2}{v^2}\left(1 + \left(\frac{4\pi k\mu_0 v^2}{n}\right)^2\right)^{\frac{1}{2}}. \qquad (201)$$

Enormously great frequency brings us to the formulæ of the non-conducting dielectric, with a difference, thus: n_1 and n_2 become

$$n_1 = 2\pi k\mu_0 v, \qquad n_2 = n/v, \qquad (202)$$

when $4\pi k/cn$ is a small fraction. The attenuation due to conductivity still exists, but is independent of the frequency. We have now

$$\text{(out)} \qquad H = \frac{C_0 v}{r} \epsilon^{-n_1 r} \left(\cos - \frac{v}{nr} \sin \right) \left(\frac{nr}{v} - nt \right), \quad \text{.........(203)}$$

differing from the case of no conductivity only in the presence of the exponential factor.

It is, however, easily seen by the form of n_1 in (202) that in a good conductor the attenuation in a short distance is very great, so that the disturbances are practically confined to the vortex-lines of the impressed force, where the H-disturbance is nearly the same as if the conductivity were zero, as before concluded. It follows that the initial effect of the sudden introduction of a steady impressed force in the conducting dielectric is the emission from the seat of its vorticity of waves in the same manner as if there were no conductivity, but attenuated at their front to an extent represented by the factor $\epsilon^{-n_1 r}$, with the (202) value of n_1, in addition to the attenuation by spreading which would occur were the medium nonconducting. This estimate of the attenuation applies at the front only.

Current in Sphere constrained to be uniform.

30. Let us complete the solution (200) of § 29 by means of a current of uniform density parallel to the axis within the sphere of radius a, beyond which (200) is to be the solution. This will require a special distribution of impressed force, which we shall find. Equation (200) gives us the normal component of electric current at $r = a$, by differentiation. Let this be $\Gamma \cos\theta$. Then Γ is the density of the internal current. The corresponding magnetic field must have the boundary-value according to (200), and vary in intensity as the distance from the axis, its lines being circles centred upon it, and in planes perpendicular to it. Thus the internal **H** is also known. The internal **E** is fully known too, being $k^{-1}\Gamma$ in intensity and parallel to the axis. It only remains to find **e** to satisfy

$$\operatorname{curl}(\mathbf{e} - \mathbf{E}) = \mu \dot{\mathbf{H}}, \quad \text{........................(3) } bis$$

within the sphere, and at its boundary (with the suitable surface interpretation), as it is already satisfied outside the sphere. The simplest way appears to be to first introduce a uniform field of **e** parallel to the axis, of such intensity e_1 as to neutralize the difference between the tangential components of the internal and external **E** at the boundary, and so make continuity there in the force of the field; and next, to find an auxiliary distribution $\mathbf{e}_2$, such that

$$\operatorname{curl} \mathbf{e}_2 = \mu \dot{\mathbf{H}},$$

and having no tangential component on the boundary. This may be done by having $\mathbf{e}_2$ parallel to the axis, of intensity proportional to

$$(a^2 - r^2) \sin\theta.$$

The result is that the internal H is got from the external by putting $r = a$ in (200) and then multiplying by r/a; Γ from the internal H by multiplying by $(2\pi r \sin\theta)^{-1}$; e_1 from the difference of the tangential components E outside and inside is given by

$$e_1 = \frac{C_0}{4\pi a^2}\epsilon^{-n_1 a}\left\{k^2+\left(\frac{cn}{4\pi}\right)^2\right\}^{-1}\left[\left\{k\left(3+n_1a+\frac{3n_1}{a(n_1^2+n_2^2)}\right)+\frac{cn}{4\pi}\left(n_2a-\frac{3n_2}{a(n_1^2+n_2^2)}\right)\right\}\cos(n_2a-nt)\right.$$
$$\left.-\left\{k\left(\frac{3n_2}{a(n_1^2+n_2^2)}-n_2a\right)+\frac{cn}{4\pi}\left(3+n_1a+\frac{3n_1}{a(n_1^2+n_2^2)}\right)\right\}\sin(n_2a-nt)\right]. \quad\text{.........(204)}$$

Finally, the auxiliary force has its intensity given by

$$e_2 = \mu_0 n C_0 \frac{a^2-r^2}{2a^2}\epsilon^{-n_1 a}\left\{\left(1+\frac{n_1}{a(n_1^2+n_2^2)}\right)\sin+\frac{n_2}{a(n_1^2+n_2^2)}\cos\right\}(n_2a-nt). \quad\text{......(205)}$$

A remarkable property of this auxiliary force, which (or an equivalent) is absolutely required to keep the current straight, is that it does no work on the current, on the average; the mean activity and waste of energy being therefore settled by e_1.

Nov. 27, 1887.

PART IV.

Spherical Waves (with Diffusion) in a Conducting Dielectric.

31. In an infinitely extended homogeneous isotropic conducting dielectric, let the surface $r=a$ be a sheet of vorticity of impressed electric force; for simplicity, let it be of the first order, so that the surface-density is represented by $f\nu$. By (127), § 20, the differential equation of H, the intensity of magnetic force is, at distance r from the origin, outside the surface of f, (ν meaning $\sin\theta$),

$$H = \frac{k_1}{q}\left(\frac{\nu a}{r}\right)\epsilon^{-qr}\left(1+\frac{1}{qr}\right)\left\{\cosh qa - \frac{\sinh qa}{qa}\right\}f, \quad\text{.........(206)}$$

where f may be any function of the time. Here, in the general case, including the unreal "magnetic conductivity" g,* we have

$$\left.\begin{aligned} q &= [(4\pi k + cp)(4\pi g + \mu p)]^{\frac{1}{2}} = v^{-1}[(p+\rho)^2-\sigma^2]^{\frac{1}{2}}, \\ k_1 &= 4\pi k + cp; \end{aligned}\right\} \quad\text{...........(207)}$$

if, for subsequent convenience,

$$\left.\begin{aligned} \rho_1 &= 4\pi k/2c, & \rho &= \rho_1+\rho_2, & \\ \rho_2 &= 4\pi g/2\mu, & \sigma &= \rho_1-\rho_2, & v=(\mu c)^{-\frac{1}{2}}. \end{aligned}\right\} \quad\text{.........(208)}$$

The speed is v, and ρ_1, ρ_2 are the coefficients of attenuation of the parts transmitted of elementary disturbances due to the real electric conductivity k and the unreal g; that is, $\epsilon^{-\rho t}$ is the factor of attenuation due to conductivity. On the other hand, the distortion produced by conductivity depends on σ, and vanishes with it. There is some utility

* Owing to the lapse of time, I should mention that the physical and other meanings of the coefficient g are explained in Part I. of this paper. Also k = electric conductivity; μ = magnetic inductivity; and $c/4\pi$ = electric permittivity. All the problems in this paper, except in § 43, relate to spherical waves; the geometrical coordinates are r and θ. Unless otherwise mentioned, p always signifies the operator d/dt, t being the time.

in keeping in g, because it sometimes happens that the vanishing of k, making $\rho = -\sigma$, leads to a solvable case. We can then produce a real problem by changing the meaning of the symbols, turning the magnetic into an electric field, with other changes to correspond.

The Steady Magnetic Field due to f *Constant.*

32. Let f be zero before, and constant after $t=0$, the whole medium having been previously free from electric and magnetic force. All subsequent disturbances are entirely due to f. The steady field which finally results is expressed by (206); by taking $p=0$; that is, k_1 has to mean $4\pi k$, and $q=4\pi(kg)^{\frac{1}{2}}$, by (207). To obtain the corresponding internal field, exchange a and r in (206), except in the first a/r. The same values of k_1 and q used in the corresponding equations of E and F give the final electric field. The steady magnetic field here considered depends upon g, and vanishes with it.

Variable State when $\rho_1 = \rho_2$. *First Case. Subsiding* f.

33. There are cases in which we already know how the final state is reached, viz., the already given case of a nonconducting dielectric (§§ 21, 22), and the case $\sigma=0$ in (208), which is an example of the theory of § 4. In the latter case the impressed force must subside at the same rate as do the disturbances it sends out from the surface of f. Thus, given $f=f_0\epsilon^{-\rho t}$, starting when $t=0$, with f_0 constant, the resulting electric and magnetic fields are represented by those in the corresponding case in a nonconducting dielectric, when multiplied by $\epsilon^{-\rho t}$. The final state is zero because f subsides to zero; the travelling shell also loses all its energy. But there are, in a sense, two final states; the first commencing at any place as soon as the rear of the travelling shell reaches it, and which is entirely an electric field; the second is zero, produced by the subsidence of this electric field. There is no magnetic force to correspond, and therefore no "true" electric current, in Maxwell's sense of the term, except in the shell.

Second Case. f *Constant.*

34. But let the impressed f be constant. Then, by effecting the integrations in (206), we are immediately led to the full solution

$$H=\frac{fva}{2\mu vr}\Big[\epsilon^{-\frac{\rho}{v}(r-a)}\Big(1+\frac{v}{\rho r}\Big)\Big(1-\frac{v}{\rho a}\Big)+\epsilon^{-\rho t}(1+\rho t)\frac{v^2}{ra\rho^2}$$

$$+\text{same function of } -a\Big], \quad \text{...(209)}$$

where the fully-represented part expresses the primary wave out from the surface of f, reaching r at time $(r-a)/v$; whilst the rest expresses the second wave, reaching r when $t=(r+a)/v$. After that, the actual H is their sum, viz.,

$$H=\frac{fva}{\mu vr}\epsilon^{-\rho r/v}\Big(1+\frac{v}{\rho r}\Big)\Big[\cosh-\frac{v}{\rho a}\sinh\Big]\frac{\rho a}{v}, \quad \text{.........(210)}$$

agreeing with (206), when we give q therein the special value ρ/v at present concerned, and $k_1 = 4\pi k$.

At the front of the first wave we have

$$H = \epsilon^{-\rho t} fva/2\mu vr, \quad \text{.........................(211)}$$

so that the energy in the travelling shell still subsides to zero. Equation (211) also expresses H at the front of the inward wave, both before and after reaching the centre of the sphere. The exchange of a and r in the [] in (209) produces the corresponding internal solution.

Unequal ρ_1 and ρ_2. General Case.

35. If we put $d/dr = \nabla$, we may write (206) thus,

$$H = \frac{va}{2r}\frac{k_1}{q^3}\left[\left(\nabla - \frac{1}{r}\right)\left(\nabla + \frac{1}{a}\right)\epsilon^{-q(r-a)} + \left(\nabla - \frac{1}{r}\right)\left(\nabla - \frac{1}{a}\right)\epsilon^{-q(r+a)}\right]f. \quad (212)$$

It is, therefore, sufficient to find

$$\epsilon^{-q(r-a)}q^{-3}f, \quad \text{.............................(213)}$$

to obtain the complete solution of (212); namely, by performing upon the solution of (213) the differentiations ∇ and the operation k_1. This refers to the first half of (212); the second half only requires the changed sign of a in the [] to be attended to.

Now (213) is the same as

$$v^3\epsilon^{-\rho t}\epsilon^{-\frac{r-a}{v}(p^2-\sigma^2)^{\frac{1}{2}}}(p^2-\sigma^2)^{-\frac{3}{2}}(f\epsilon^{\rho t}). \quad \text{..............(214)}$$

Expand the two functions of p in descending powers of p, thus,

$$(p^2-\sigma^2)^{-\frac{3}{2}} = p^{-3}\left[1 + \frac{3}{2}\frac{\sigma^2}{p^2} + \frac{3\,.\,5}{2^2\lfloor 2}\frac{\sigma^4}{p^4} + \frac{3\,.\,5\,.\,7}{2^3\lfloor 3}\frac{\sigma^6}{p^6} + \dots\right], \quad \text{...(215)}$$

$$\epsilon^{-\frac{r-a}{v}(p^2-\sigma^2)^{\frac{1}{2}}} = \epsilon^{-\frac{p}{v}(r-a)}\left[1 + \frac{\sigma}{p}h_1 + \frac{\sigma^2}{p^2}h_2 + \dots\right], \quad \text{........(216)}$$

where the h's are functions of r, but not of p. Multiplying these together, we convert (213) or (214) to

$$v^3\epsilon^{-\rho t}\epsilon^{-\frac{p}{v}(r-a)}\frac{1}{p^3}\left[1 + \frac{\sigma}{p}i_1 + \frac{\sigma^2}{p^2}i_2 + \dots\right](f\epsilon^{\rho t}), \quad \text{.........(217)}$$

where the i's are functions of r, but not of p. The integrations can now be effected. Let f be constant, first. Then, f starting when $t = 0$, we have

$$p^{-3}(f\epsilon^{\rho t}) = f\rho^{-3}(\epsilon^{\rho t} - 1 - \rho t - \tfrac{1}{2}\rho^2 t^2) = \rho^{-3}f(\epsilon^{\rho t})_3 \quad \text{say; .....(218)}$$

etc., etc. Next, operating with the exponential containing p in (217) turns t to $t - (r-a)/v$, and gives the required solution in the form

$$H = \frac{fvav^2}{2\mu vr}\epsilon^{-\rho t}(\sigma + p)\left[\left(\nabla - \frac{1}{r}\right)\left(\nabla + \frac{1}{a}\right)\left\{\frac{(\epsilon^{\rho t_1})_3}{\rho^3} + \sigma i_1\frac{(\epsilon^{\rho t_1})_4}{\rho^4} + \dots\right\}\right.$$
$$\left. + \text{ same function of } -a\right], \quad \text{..........(219)}$$

where $t_1 = t - (r-a)/v$; the represented part beginning when t_1 reaches zero, and the rest when $t-(r+a)/v$ reaches zero.

Fuller Development in a Special Case. Theorems involving Irrational Operators.

36. As this process is very complex, and (219) does not admit of being brought to a readily interpretable form, we should seek for special cases which are, when fully developed, of a comparatively simple nature. Write the first half of (212) thus,

$$H = \frac{cv^3 va}{2r}\epsilon^{-\rho t}\left(\nabla - \frac{1}{r}\right)\left(\nabla + \frac{1}{a}\right)\frac{1}{p^2-\sigma^2}\left[\left(\frac{p+\sigma}{p-\sigma}\right)^{\frac{1}{2}}\epsilon^{-\frac{r-a}{v}(p^2-\sigma^2)^{\frac{1}{2}}}(f\epsilon^{\rho t})\right]. \quad (220)$$

Now the part in the square brackets can be finitely integrated when $f\epsilon^{\rho t}$ subsides in a certain way. We can show that

$$\left(\frac{p+\sigma}{p-\sigma}\right)^{\frac{1}{2}}\epsilon^{-\frac{r-a}{v}(p^2-\sigma^2)^{\frac{1}{2}}}(\epsilon^{-\sigma t}) = J_0\left\{\frac{\sigma}{v}[(r-a)^2 - v^2t^2]^{\frac{1}{2}}\right\}, \quad \text{......(221)}$$

in which, observe, the sign of σ may be changed, making no difference on the right side (the result), but a great deal on the left side.

The simplest proof of (221) is perhaps this. First let $r = a$. Then

$$\left(\frac{p+\sigma}{p-\sigma}\right)^{\frac{1}{2}}(\epsilon^{-\sigma t}) = \epsilon^{-\sigma t}\left(1 - \frac{2\sigma}{p}\right)^{-\frac{1}{2}}(1), \quad \text{..............(222)}$$

by getting the exponential to the left side, so as to operate on unity. Next, by the binomial theorem,

$$= \epsilon^{-\sigma t}\left[1 + \frac{1}{2}\frac{2\sigma}{p} + \frac{1\,.\,3}{2^2\lfloor 2}\left(\frac{2\sigma}{p}\right)^2 + \dots\right](1). \quad \text{...........(223)}$$

Now integrate, and we have (f commencing when $t=0$),

$$\left.\begin{aligned} &= \epsilon^{-\sigma t}\left(1 + \sigma t + \frac{1\,.\,3}{\lfloor 2 \lfloor 2}\sigma^2 t^2 + \frac{1\,.\,3\,.\,5}{\lfloor 3 \lfloor 3}\sigma^3 t^3 + \dots\right), \\ &= \epsilon^{-\sigma t}\epsilon^{\sigma t}J_0(\sigma t i); \end{aligned}\right\} \text{.........} \quad (224)$$

so that, finally,

$$\left(\frac{p+\sigma}{p-\sigma}\right)^{\frac{1}{2}}(\epsilon^{-\sigma t}) = J_0(\sigma t i). \quad \text{.....................(225)}$$

It is also worth notice that, integrating in a similar manner,

$$\left(1 - \frac{\sigma^2}{p^2}\right)^{-\frac{1}{2}}(1) = 1 + \tfrac{1}{2}\frac{\sigma^2 t^2}{\lfloor 2} + \frac{1\,.\,3}{2^2\lfloor 2}\frac{\sigma^4 t^4}{\lfloor 4} + \dots = J_0(\sigma t i). \quad \text{.......(226)}$$

These theorems present themselves naturally in problems relating to a telegraph-circuit, when treated by the method of resistance-operators. A special case of (225) is

$$p^{\frac{1}{2}}(1) = (\pi t)^{-\frac{1}{2}}, \quad \text{..........................(227)}$$

which presents itself in the electrostatic theory of a submarine cable.*

We have now to generalize (225) to meet the case (221). The left member of (221) satisfies the partial differential equation

$$v^2\nabla^2 = p^2 - \sigma^2, \qquad\qquad (228)$$

so we have to find the solution of (228) which becomes $J_0(\sigma t i)$ when $r=a$. Physical considerations show that it must be an even function of $(r-a)$, so that it is suggested that the t in $J_0(\sigma t i)$ has to become, not $t-(r-a)/v$ or $t+(r-a)/v$, but that t^2 has to become their product. In any case, the right member of (221) does satisfy (228) and the further prescribed condition, so that (221) is correct.

If a direct proof be required, expand the exponential operator in (221) containing r in the way indicated in (216), and let the result operate upon $J_0(\sigma t i)$. The integrated result can be simplified down to (221).

37. Now use (221) in (220). Let $f\epsilon^{\rho t} = f_0\epsilon^{-\sigma t}$, where f_0 is constant; and the square bracket in (220) becomes known, being in fact the right member of (221) multiplied by f_0. So, making use also of (228), we bring (220) to

$$H = \frac{vaf_0}{2\mu vr}\epsilon^{-\rho t}\left[1 + \left\{\left(\frac{1}{a} - \frac{1}{r}\right)v^2\frac{d}{dr} - \frac{v^2}{ar}\right\}\frac{1}{p^2-\sigma^2}\right]J_0\left\{\frac{\sigma}{v}\left[(r-a)^2 - v^2t^2\right]^{\frac{1}{2}}\right\}; \qquad (229)$$

to which must be added the other part, beginning $2a/v$ later, got by negativing a, except the first one. The operation $(p^2-\sigma^2)^{-1}$ may be replaced by two integrations with respect to r.

Let r and a be infinitely great, thus abolishing the curvature. Let $r-a=z$, and $f_0 va/r$, which is now constant, be called e_0. Then we have simply

$$H = \frac{e_0}{2\mu v}\epsilon^{-\rho t}J_0\left\{\frac{\sigma}{v}(z^2 - v^2t^2)^{\frac{1}{2}}\right\}, \qquad\qquad (230)$$

showing the H produced in an infinite homogeneous conducting dielectric medium at time t after the introduction of a plane sheet (at $z=0$), of vorticity of impressed electric force, the surface-density of

* Thus, let an infinitely long circuit, with constants R, S, K, L, be operated upon by impressed force at the place $z=0$, producing the potential-difference V_0 there, which may be any function of the time. Let C be the current and V the potential-difference at time t at distance z. Then

$$C = \left(\frac{K+Sp}{R+Lp}\right)\epsilon^{-qz}V_0,$$

where $q=(R+Lp)^{\frac{1}{2}}(K+Sp)^{\frac{1}{2}}$. Take $K=0$, and $L=0$; then, if V_0 be zero before and constant after $t=0$, the current at $z=0$ is given by

$$C_0 = V_0(S/R)^{\frac{1}{2}}p^{\frac{1}{2}}(1),$$

and (227) gives the solution. Prove thus: let b be any constant, to be finally made infinite; then

$$p^{\frac{1}{2}}(1) = b^{\frac{1}{2}}(1+bp^{-1})^{-\frac{1}{2}} = b^{\frac{1}{2}}J_0(\tfrac{1}{2}bti)\epsilon^{-bt/2},$$

by the investigation in the text. Now put $b=\infty$, and (227) results.

In the similar treatment of cylindrical waves in a conductor, $p^{\frac{1}{4}}$, $p^{\frac{3}{4}}$, etc., occur. We may express the results in terms of Gamma-functions.

vorticity being $e_0\epsilon^{-2\rho_1 t}$. This corroborates the solution in § 8, equation (51) [Part I., p. 383], whilst somewhat extending its meaning.

The condition to which f is subject may be written, by (208),

$$f = f_0\epsilon^{-2\rho_1 t}, \qquad \text{.......}(231)$$

where f_0 is constant. If, then, we desire f to be constant, ρ_1 must vanish, which, by (208), requires $k=0$, whilst g may be finite.

But we can make the problem real thus. In (229) change H to E and μv to cv; we have now the solution of the problem of finding the electric field produced by suddenly magnetizing uniformly a spherical portion of a conducting dielectric; *i.e.*, the vorticity of the impressed magnetic force is to be on the surface of the sphere $r=a$, parallel to its lines of latitude, and of surface-density fv, such that $fv\epsilon^{2\rho_2 t}$ is constant. This makes f constant when $g=0$ and k finite, representing a real conducting dielectric.

The Electric Force at the Origin due to fν *at* r $=$ a.

38. Returning to the case of impressed electric force, the differential equation of F, the radial component of electric force inside the sphere on whose surface $r=a$ the vorticity of **e** is situated, is, by § 20, equations (136), (137),

$$F = \frac{2a\cos\theta}{qr^2}\epsilon^{-qa}\left(1+\frac{1}{qa}\right)\left(\cosh qr - \frac{\sinh qr}{qr}\right)f. \qquad \text{.......}(232)$$

At the centre, therefore, the intensity of the full force, which call F_0, whose direction is parallel to the axis, is

$$F_0 = \tfrac{2}{3}(1+qa)\epsilon^{-qa}f = \tfrac{2}{3}\left(1 - a\frac{d}{da}\right)\epsilon^{-qa}f. \qquad \text{.......}(233)$$

Unless otherwise specified, I may repeat that the forces referred to are always those of the fluxes, thus doing away with any consideration of the distribution of the impressed force, and of scalar potential, of varying form, which it involves. (233) is equivalent to

$$F_0 = \tfrac{2}{3}\epsilon^{-\rho t}\left(1 + av^{-1}(p^2-\sigma^2)^{\frac{1}{2}}\right)\epsilon^{-(a/v)(p^2-\sigma^2)^{\frac{1}{2}}}(f\epsilon^{\rho t}). \qquad \text{.......}(234)$$

Let f be constant, and $\rho=\sigma$, or $g=0$. Then (234) becomes

$$F_0 = \tfrac{2}{3}f\epsilon^{-\sigma t}\left[\frac{a}{v}(p+\sigma)\left(\frac{p-\sigma}{p+\sigma}\right)^{\frac{1}{2}} + 1\right]\epsilon^{-(a/v)(p^2-\sigma^2)^{\frac{1}{2}}}(\epsilon^{\sigma t}), \qquad \text{......}(235)$$

of which the complete solution is, by (221),

$$F_0 = (\tfrac{2}{3}f)\left(\epsilon^{-\sigma t}av^{-1}(p+\sigma)J_0\{\sigma v^{-1}(a^2-v^2t^2)^{\frac{1}{2}}\} + X_a\right), \qquad \text{......}(236)$$

where, subject to $g=0$, $\qquad \epsilon^{-qa}(1) = X_a; \qquad \text{.......}(236a)$

or, solved,

$$X_a = 1 - \epsilon^{-\sigma t}\left[\frac{\sigma a}{v}\left(J_0 + \frac{J_1}{i}\right) - \frac{1}{\underline{|3}}\left(\frac{\sigma a}{v}\right)^3\frac{1}{\sigma t}\left(\frac{J_1}{i}+\frac{J_2}{i^2}\right) + \frac{1.3}{\underline{|5}}\left(\frac{\sigma a}{v}\right)^5\frac{1}{\sigma^2t^2}\left(\frac{J_2}{i^2}+\frac{J_3}{i^3}\right) - \dots\right], \qquad \text{.....}(237)$$

in which $i=(-1)^{\frac{1}{2}}$, and all the J's operate upon σti. This solution (236) begins when $t=a/v$. The value of σ is $4\pi k/2c$.

In a good conductor σ is immense. Then assume $c=0$, or do away with the elastic displacement, and reduce (236) to the pure-diffusion formula, which is

$$F_0=(\tfrac{2}{3}f)\left[\left(\frac{2}{\pi}\right)^{\frac{1}{2}}y\epsilon^{-\frac{1}{2}y^2}+1-\left(\frac{2}{\pi}\right)^{\frac{1}{2}}\left\{y-\frac{y^3}{\underline{|3}}+\frac{1.3}{\underline{|5}}y^5-\dots\right\}\right], \tag{238}$$

where $y=(4\pi\mu ka^2/2t)^{\frac{1}{2}}$. The relation of X_a in (236) to the preceding terms is explained by equations (233) or (235).

Effect of uniformly magnetizing a Conducting Sphere surrounded by a Nonconducting Dielectric.

39. Here, of course, it is the lines of **E** that are circles centred upon the axis, both inside and outside. Let **h** be the impressed magnetic force, and $h\nu$ the surface-density of its vorticity, at $r=a$, outside which the medium is nonconducting, and inside a conducting dielectric. The differential equation of E_a, the surface-value of the tensor of **E** at $r=a$, is (compare (124), § 19)

$$\frac{h\nu}{E_a}=\left(\frac{1}{\mu p}\frac{W'}{W}\right)_{\text{out}}-\left(\frac{1}{\mu p}\frac{U'+W'}{U+W}\right)_{\text{in}}; \qquad \text{.............. (239)}$$

in which $r=a$, and μ and q are to have the proper values on the two sides of the surface.

Now, by (111),

$$W'/W=-q\left\{1+(qr)^{-1}(1+qr)^{-1}\right\} \qquad \text{.............. (240)}$$

in the case of $m=1$, (first order), here considered. This refers to the external dielectric, in which $q=p/v$. Let $v=\infty$, making

$$W'/W=-a^{-1}. \qquad \text{.......................... (241)}$$

This assumption is justifiable when the sphere has sensible conductivity, on account of the slowness of action it creates in comparison with the rapidity of propagation in the dielectric outside. Then (239) becomes

$$-\frac{h\nu}{E_a}=\frac{1}{\mu_1 pa}\cdot\frac{q_1 a\sinh q_1 a}{\cosh q_1 a-(q_1 a)^{-1}\sinh q_1 a}+\frac{1}{pa}\left(\frac{1}{\mu_0}-\frac{1}{\mu_1}\right), \qquad \text{...... (242)}$$

if μ_0 is the external and μ_1 the internal inductivity, and q_1 the internal q. When the inductivities are equal, there is a material simplification, leading to

$$E_a=-\mu pa\frac{\cosh q_1 a-(q_1 a)^{-1}\sinh q_1 a}{q_1 a\sinh q_1 a}h\nu, \qquad \text{.............. (243)}$$

where $q_1=\{(4\pi k_1+c_1 p)\mu_1 p\}^{\frac{1}{2}}$. First let $c_1=0$, in the conductor, making $q_1^2=4\pi\mu_1 k_1 p=-s^2$, say. Then

$$E_a=-\frac{1}{4\pi k_1 a}\cdot\frac{\cos sa-(sa)^{-1}\sin sa}{(sa)^{-1}\sin sa}h\nu. \qquad \text{.............. (244)}$$

From this we see that $\sin sa = 0$ is the determinantal equation of normal systems. The slowest is

$$sa = \pi, \qquad \text{or} \qquad -p^{-1} = 4\mu_1 k_1 a^2/\pi. \qquad \text{.................(245)}$$

This time-constant is about $(1250)^{-1}$ second if the sphere be of copper of 1 centim. radius; about 8 seconds if of 1 metre radius, and about 10 million years if of the size of the earth.

At distance r from the centre of the sphere, within it, at time t after starting h, we have

$$E = -\frac{hv}{4\pi k_1 r}\sum \frac{\cos sr - (sr)^{-1}\sin sr}{p(d/dp)\{(sa)^{-1}\sin sa\}}\cdot \epsilon^{pt}, \qquad \text{...............(246)}$$

subject to the determinantal equation, over whose roots the summation extends, p being now algebraic. Effecting the differentiation indicated, we obtain

$$E = -\frac{2hv}{4\pi k_1 r}\sum \frac{\cos sr - (sr)^{-1}\sin sr}{\cos sa}\epsilon^{pt}. \qquad \text{...............(247)}$$

The corresponding solution for the radial component of the magnetic force, say H_r, is

$$H_r = (\tfrac{2}{3}h\cos\theta) - 4h\cos\theta \sum \frac{\cos sr - (sr)^{-1}\sin sr}{s^2 r^2 \cos sa}\epsilon^{pt}. \qquad \text{.........(248)}$$

At the centre of the sphere, let H_0 be the intensity of the actual magnetic force. It is, by (248),

$$H_0 = \tfrac{2}{3}h\Big(1 + 2\,\Sigma\,(\cos sa)^{-1}\epsilon^{pt}\Big). \qquad \text{...................(249)}$$

Thus the magnetic force arrives at the centre of the sphere in identically the same manner as current arrives at the distant end of an Atlantic cable according to the electrostatic theory, when a steady impressed force is applied at the beginning, with terminal short-circuits. In the case of the cable the first time-constant is

$$-p^{-1} = RSl^2/\pi^2,$$

where Rl is the total resistance and Sl the total permittance. It is not greatly different from 1 second, so that, by (245), the sphere should be about a foot in radius to imitate, at its centre, the arrival-curve of the cable.

To be precise we should not speak of magnetizing the sphere, because (ignoring the minute diamagnetism) it does not become magnetized. The principle, however, is the same. We set up the flux magnetic induction. But the magnetic terminology is defective. Perhaps it would be not objected to if we say we inductize* the sphere, whether we magnetize it or not. This is, at any rate, better than extending the meaning of the word magnetize, which is already precise in the mathematical theory, though of uncertain application in practice, from the variable behaviour of iron.

* Accent the first syllable, like magnetize. Practical men sometimes speak of energizing a core, etc. But energize is too general; by using inductize we specify what flux is set up.

40. The following is the alternative form of solution showing the waves, when c_1 is finite. With the same assumption as before that $v=\infty$ outside the sphere, the equation of H_r, the radial component of **H**, is

$$H_r=\frac{2\cos\theta}{q^2r^2}\cdot\frac{\cosh qr-(qr)^{-1}\sinh qr}{(qa)^{-1}\sinh qa}h, \quad \ldots\ldots\ldots\ldots(250)$$

which, at $r=0$, becomes

$$H_0=\tfrac{2}{3}qa\,(\sinh qa)^{-1}h. \quad \ldots\ldots\ldots\ldots\ldots\ldots\ldots(251)$$

Expand the circular function, giving

$$H_0=\tfrac{4}{3}qa\,\epsilon^{-qa}(1+\epsilon^{-2qa}+\epsilon^{-4qa}+\ldots)h\,; \quad \ldots\ldots\ldots\ldots(252)$$

or, since here $q=v^{-1}\{(p+\sigma)^2-\sigma^2\}^{\frac{1}{2}}$, where $\sigma=4\pi k/2c$,

$$H_0=\frac{4}{3}\frac{a}{v}\epsilon^{-\sigma t}(p+\sigma)\left(\frac{p-\sigma}{p+\sigma}\right)^{\frac{1}{2}}\left[\epsilon^{-\frac{a}{v}(p^2-\sigma^2)^{\frac{1}{2}}}+\epsilon^{-\frac{3a}{v}(p^2-\sigma^2)^{\frac{1}{2}}}+\ldots\right](h\epsilon^{\sigma t}), \quad (253)$$

so, using (221), we get finally

$$H_0=\frac{4}{3}h\frac{a}{v}\epsilon^{-\sigma t}(p+\sigma)\left[J_0\left\{\frac{\sigma}{v}(a^2-v^2t^2)^{\frac{1}{2}}\right\}+J_0\left\{\frac{\sigma}{v}(9a^2-v^2t^2)^{\frac{1}{2}}\right\}+\ldots\right]. \quad (254)$$

The J_0 functions commence when $vt=a$, $3a$, $5a$, etc., in succession, and the successive terms express the arrival of the fi t wave and of the reflexions from the surface which follow. In the cas of pure diffusion, this reduces to

$$H_0=(\tfrac{2}{3}h)2a(4\pi k_1\mu_1/\pi t)^{\frac{1}{2}}[\epsilon^{-\pi k_1\mu_1a^2/t}+\epsilon^{-9\pi\mu_1k_1a^2/4t}+\ldots], \quad \ldots..(255)$$

which is the alternative form of (249), involving instantaneous action at a distance. The theorem (in diffusion)

$$\epsilon^{-xp^{\frac{1}{2}}}.\,p^{\frac{1}{2}}(\dot{1})=(\pi t)^{-\frac{1}{2}}\epsilon^{-x^2/4t} \quad \ldots\ldots\ldots\ldots\ldots\ldots\ldots(256)$$

becomes generalized to

$$\epsilon^{-xq}q(1)=v^{-1}\epsilon^{-\sigma t}(p+\sigma)J_0\{\sigma v^{-1}(x^2-v^2t^2)\}, \quad \ldots\ldots\ldots(257)$$

if

$$q=v^{-1}(p^2+2\sigma p)^{\frac{1}{2}}.$$

On the right side of (257), the p means, as usual, differentiation to t. The two quantities σ and v may have any positive values; to reduce to (256), make v infinite whilst keeping σ/v^2 finite.

Diffusion of Waves from a Centre of Impressed Force in a Conducting Medium.

41. In equation (206) let a be infinitely small. It then becomes

$$H=\tfrac{1}{3}a^3vr^{-2}(4\pi k+cp)(1+qr)\epsilon^{-qr}f, \quad \ldots\ldots\ldots\ldots\ldots(258)$$

the equation of H at distance r from an element of impressed electric force at the origin. Comparing with (233), we see that the solution of (258) may be derived, when f is constant, starting when $t=0$. Take $g=0$, making $\rho=\sigma=4\pi k/2c$. Then

$$H=\frac{H_0}{4\pi k}(4\pi k+cp)\left\{\epsilon^{-\sigma t}\frac{r}{v}(p+\sigma)J_0\left[\frac{\sigma}{v}(r^2-v^2t^2)^{\frac{1}{2}}\right]+X_r\right\}, \quad (259)$$

where X_r is what the X_a of (237) becomes on changing a to r; and

$$H_0 = \nu k r^{-2} \times \text{vol. integral of } f, \quad \text{.................(260)}$$

supposing the impressed force to be confined to the infinitely small sphere, so that its volume-integral is the "electric moment," by analogy with magnetism. The solution (259) begins at r as soon as $t = r/v$. It is true from infinitely near the origin to infinitely near the front; but no account is given of the state of things at the front itself. H_0 is the final value of H. We may also write X_r thus,

$$\epsilon^{\sigma t} X_r = 1 + \int_r^{vt} \frac{dr}{v}(\sigma + p) J_0 \left\{ \frac{\sigma}{v}(r^2 - v^2 t^2)^{\frac{1}{2}} \right\}; \quad \text{..........(261)}$$

and (259) may also be written

$$H = \frac{H_0}{4\pi k}(4\pi k + cp)\left(1 - r\frac{d}{dr}\right) X_r. \quad \text{................(262)}$$

When $c = 0$, (259) or (262) reduce to

$$H = H_0 \left[\left(\frac{2}{\pi}\right)^{\frac{1}{2}} y \epsilon^{-\frac{1}{2}y^2} + 1 - \left(\frac{2}{\pi}\right)^{\frac{1}{2}} \left\{ y - \frac{1}{\underline{|3}} y^3 + \frac{1.3}{\underline{|5}} y^5 - \dots \right\} \right], \quad (263)$$

where $\qquad y = (2\pi\mu k r^2 / t)^{\frac{1}{2}}.$

Conducting Sphere in a Nonconducting Dielectric. Circular Vorticity of **e**. *Complex Reflexion. Special very Simple Case.*

42. At distance r from the origin, outside the sphere of radius a, which is the seat of vorticity of **e**, represented by $f\nu$, we have

$$H = \phi^{-1}(W/W_a) f \nu a / r. \quad \text{......................(264)}$$

The operator ϕ will vary according to the nature of things on both sides of $r = a$. When it is a uniform conducting medium inside, and nonconducting outside, to infinity, we shall have

$$\phi = \phi_1 + \phi_2,$$

when ϕ_1, depending upon the inner medium, is given by

$$\phi_1 = \frac{q_1}{4\pi k_1 + c_1 p} \cdot \frac{\{1 + (q_1 a)^{-2}\} \sinh q_1 a - (q_1 a)^{-1} \cosh q_1 a}{\cosh q_1 a - (q_1 a)^{-1} \sinh q_1 a}, \quad (265)$$

and ϕ_2, depending upon the outer medium, is given by

$$\phi_2 = \mu v \frac{1 + (qa)^{-1} + (qa)^{-2}}{1 + (qa)^{-1}}. \quad \text{...................(266)}$$

The solution arising from the sudden starting of f constant is therefore

$$H = \frac{f\nu a}{r} \sum \frac{W}{W_a} \frac{\epsilon^{pt}}{p\frac{d\phi}{dp}}, \quad \text{.......................(267)}$$

where p is now algebraical, and the summation ranges over the roots of $\phi = 0$. There is no final H in this case, if we assume $g = 0$ all over.

But the determinantal equation is very complex, so that this (267) solution is not capable of easy interpretation. The wave-method is also impracticable, for a similar reason.

In accordance, however, with Maxwell's theory of the impermeability of a "perfect" conductor to magnetic induction from external causes, the assumption $k_1=\infty$ makes the solution depend only upon the dielectric, modified by the action of the boundary, and an extraordinary simplification results. ϕ_1 vanishes, and the determinantal equation becomes $\phi_2=0$, which has just two roots,

$$qa=pa/v=-\tfrac{1}{2}\pm i(\tfrac{3}{4})^{\frac{1}{2}}; \quad \text{......................(268)}$$

and these, used in (267), give us the solution

$$H=(fva/3\mu vr)\epsilon^{-z}\Big(3\cos-3^{\frac{1}{2}}(1-2a/r)\sin\Big)z\sqrt{3}, \quad \text{......(269)}$$

where $$z=\{vt-(r-a)\}/2a.$$

Correspondingly, the tangential and radial components of $\mathbf{E}$ are

$$E=\mu vH+fva^3r^{-3}\Big(1-\tfrac{1}{3}\epsilon^{-z}(3\cos+3^{\frac{1}{2}}\sin)z\sqrt{3}\Big), \quad \text{............(270)}$$

$$F=\frac{2a^3}{r^3}f\cos\theta\left[1-\frac{1}{3}\frac{r}{a}\epsilon^{-z}\left\{\frac{3a}{r}\cos-\sqrt{3}\Big(2-\frac{a}{r}\Big)\sin\right\}z\sqrt{3}\right]. \quad (271)$$

This remarkably simple solution, considering that there is reflexion, corroborates Prof. J. J. Thomson's investigation* of the oscillatory discharge of an infinitely conducting spherical shell initially charged to surface-density proportional to the sine of the latitude, for, of course, it does not matter how thin or thick the shell may be when infinitely conducting, so that it may be a solid sphere. (269) to (271) show the establishment of the permanent state. Take off the impressed force, and the oscillatory discharge follows. But the impressed force keeping up the charge on the sphere need not be an external cause, as supposed in the paper referred to. There seems no other way of doing it than by having impressed force with vorticity fv on the surface, but in other respects it is immaterial whether it is internal or external, or superficial.

It may perhaps be questioned whether the sphere does reflect, seeing that its surface is the seat of f. But we have only to shift the seat of f to an outer spherical surface in the dielectric, to see at once that the surface of the conductor is the place of continuous reflexion of the wave incident upon it coming from the surface of f. The reflexion is not, however, of the same simple character that occurs when a plane wave strikes a plane boundary ($k=\infty$) flush, which consists merely in sending back again every element of $\mathbf{H}$ unchanged, but with its $\mathbf{E}$ reversed; the curvature makes it much more complex. When we bring the surface of f right up to the conducting sphere, we make the reflexion instantaneous. At the front of the wave we have $z=0$ and

$$H=fva/\mu vr=E/\mu v,$$

* "On Electrical Oscillations and the Effects produced by the Motion of an Electrified Sphere," *Proc. Math. Soc.*, vol. xv., p. 210.

by (269) and (270). This is exactly double what it would be were the conductor replaced by dielectric of the same kind as outside, the doubling being due to the instantaneous reflexion of the inward-going wave by the conductor.

The other method of solution may also be applied, but is rather more difficult. We have

$$H=\frac{av}{\mu vr}\epsilon^{-q(r-a)}\left(1+\frac{1}{qr}\right)\left(1-\frac{1}{qa}\right)\left(1-\frac{1}{q^3a^3}\right)^{-1}f. \quad \text{.......(272)}$$

Expand the last factor in descending powers of $(qa)^3$, and integrate. The result may be written

$$H=\frac{afv}{\mu vr}\left[\frac{d^2}{dx^2}+\left(\frac{a}{r}-1\right)\frac{d}{dx}-\frac{a}{r}\right]\left(\frac{x^2}{\underline{|2}}+\frac{x^5}{\underline{|5}}+\frac{x^8}{\underline{|8}}+\dots\right), \quad \text{.....(273)}$$

where $x=a^{-1}(vt-r+a)$. Conversion to circular functions reproduces (269).

Same Case with Finite Conductivity. Sinusoidal Solution.

42A. It is to be expected that with finite conductivity, even with the greatest at command, or $k=(1600)^{-1}$, the solution will be considerably altered, being controlled by what now happens in the conducting sphere. To examine this point, consider only the value of H at the boundary. We have, by (264),

$$H_a=\phi^{-1}fv=(\phi_1+\phi_2)^{-1}fv. \quad \text{..................(274)}$$

Let f vary sinusoidally with the time, and observe the behaviour of ϕ_1 and ϕ_2 as the frequency changes. The full development which I have worked out is very complex. But it is sufficient to consider the case in which k is big enough, in concert with the radius a and frequency $n/2\pi$, to make the disturbances in the sphere be practically confined to a spherical shell whose depth is a small part of the radius. Let $s=(2\pi\mu_1k_1na^2)^{\frac{1}{2}}$; then our assumption requires ϵ^{-s} to be small. This makes

$$\phi_1=-\frac{1}{4\pi k_1a}\left(1-2s^2\frac{s+i(s-1)}{1-2s+2s^2}\right), \quad \text{..............(275)}$$

and, if further, s itself be a large number, this reduces to

$$\phi_1=(1+i)(\mu_1n/8\pi k_1)^{\frac{1}{2}}. \quad \text{.........................(276)}$$

Adding on the other part of ϕ, similarly transformed by $p^2=-n^2$, we obtain

$$\phi=\left[\mu v\frac{(na/v)^2}{1+(na/v)^2}+\left(\frac{\mu_1n}{8\pi k_1}\right)^{\frac{1}{2}}\right]-i\left[\frac{\mu v}{(na/v)+(na/v)^3}-\left(\frac{\mu_1n}{8\pi k_1}\right)^{\frac{1}{2}}\right], \quad (277)$$

where the terms containing k_1 show the difference made by its not being infinite. The real part is very materially affected. Thus, copper, let

$$k_1=(1600)^{-1}, \quad \mu_1=1, \quad 2\pi n=1600, \quad a=10, \quad \therefore \quad s=10.$$

These make s large enough. Now na/v is very small, but, on the other hand,

$$(\mu_1n/8\pi k_1)^{\frac{1}{2}}=130,$$

so that the real part of ϕ depends almost entirely on the sphere, whilst the other part is little affected.

Now make n extremely great, say $na/v = 1$; else the same. Then

$$\phi = (\tfrac{3}{2} \times 10^{10} + 44 \times 10^4) - i(\tfrac{3}{2} \times 10^{10} - 44 \times 10^4),$$

from which we see that the dissipation in space has become *relatively* important. The ultimate form, at infinite frequency, is

$$\phi = \mu v + (\mu_1 n / 8\pi k_1)^{\frac{1}{2}}(1+i)\,; \quad \text{.....................(278)}$$

so that we come to a third state, in which the conductor puts a stop to all disturbance. This is, however, because it has been assumed not to be a dielectric also, so that inertia ultimately controls matters. But if, as is infinitely more probable, it is a dielectric, the case is quite changed. We shall have

$$\phi_1 = (4\pi g_1 + \mu_1 p)^{\frac{1}{2}}(4\pi k_1 + c_1 p)^{-\frac{1}{2}}, \quad \text{............(279)}$$

when the frequency is great enough, and this tends to $\mu_1 v_1$, μ_1 being the inductivity and v_1 the speed in the conductor, whatever g and k may be, provided they are finite. Thus, finally,

$$\phi = \mu_1 v_1 + \mu v \quad \text{...........................(280)}$$

represents the impedance, or ratio of $f\nu$ to H_a, which are now in the same phase.

At any distance outside we know the result by the dielectric-solution for an outward wave. But there is only superficial disturbance in the conducting sphere.

Resistance at the front of a Wave sent along a Wire.

43. In its entirety this question is one of considerable difficulty, for two reasons, if not three. First, although we may, for practical purposes, when we send a wave along a telegraph-circuit, regard it as a plane wave, in the dielectric, on account of the great length of even the short waves of telephony, and the great speed, causing the lateral distribution (out from the circuit) of the electric and magnetic fields to be, to a great distance, almost rigidly connected with the current in the wires and the charges upon them; yet this method of representation must to some extent fail at the very front of the wave. Secondly, we have the fact that the penetration of the electromagnetic field into the wires is not instantaneous; this becomes of importance at the front of the wave, even in the case of a thin wire, on account of the great speed with which it travels over the wire.* The resistance per unit length must vary rapidly at the front, being much greater there than in the body of the wave; thus causing a throwing back, equivalent to electrostatic or "jar" retardation.

* The distance within which, reckoned from the front of the wave backward, there is materially increased resistance, we may get a rough idea of by the distance travelled by the wave in the time reckoned to bring the current-density at the axis of the wire to, say, nine-tenths of the final value. It has all sorts of values. It may be 1 or 1000 kilometres, according to the size of wire and material. At the front, on the assumption of constant resistance, the attenuation is according to $\epsilon^{-Rt/2L}$, R being the resistance, and L the inductance of the circuit per unit length. Hence the importance of the increased resistance in the present question.

Now, according to the magnetic theory, the resistance must be infinitely great at the front. Thus, alternate the current sufficiently slowly, and the resistance is practically the steady resistance. Do it more rapidly, and produce appreciable departure from uniformity of distribution of current in the wire, and we increase the resistance to an amount calculable by a rather complex formula. But do it very rapidly, and cause the current to be practically confined to near the boundary, and we have a simplified state of things in which the resistance varies inversely as the area of the boundary, which may, in fact, be regarded as plane. The resistance now increases as the square root of the frequency, and must therefore, as said, be infinitely great at the front of a wave, which is also clear from the fact that penetration is only just commencing.

But for many reasons, some already mentioned, it is far more probable that the wire is a dielectric. If, as all physicists believe, the ether permeates all solids, it is *certain* that it is a dielectric. Now this becomes of importance in the very case now in question, though of scarcely any moment otherwise. Instead of running up infinitely, the resistance per unit area of surface of a wire tends to the finite value $4\pi\mu_1 v_1$. This is great, but far from infinity, so that the attenuation and change of shape of wave at its front produced by the throwing back cannot be so great as might otherwise be expected.

Thus, in general, at such a great frequency that conduction is nearly superficial, we have, if μ, c, k, and g belong to the wire,

$$E/H = (4\pi g + \mu p)^{\frac{1}{2}}(4\pi k + cp)^{-\frac{1}{2}}, \quad \ldots\ldots\ldots\ldots\ldots\ (281)$$

if E is the tangential electric force and H the magnetic force, also tangential, at the boundary of a wire. Now let R' and L' be the resistance and inductance of the wire per unit of its length. We must divide H by 4π to get the corresponding current in the wire, as ordinarily reckoned. So $4\pi A^{-1}$ times the right member of (281) is the resistance-operator of unit length, if A is the surface per unit length; so, expanding (281), we get

$$R' \text{ or } L'n = \frac{4\pi}{A}\frac{\mu v}{\sqrt{2}}\left\{\left(\frac{4\rho_2^2+n^2}{4\rho_1^2+n^2}\right)^{\frac{1}{2}} \pm \frac{4\rho_1\rho_2+n^2}{4\rho_1^2+n^2}\right\}^{\frac{1}{2}}, \quad \ldots\ldots(282)$$

where ρ_1, ρ_2 are as before, in (208). Here $n/2\pi =$ frequency.

Disregarding g, and therefore ρ_2, we have

$$R' \text{ or } L'n = (\tfrac{1}{2})^{\frac{1}{2}} 4\pi\mu v A^{-1}\{B \pm B^2\}^{\frac{1}{2}}, \quad \ldots\ldots\ldots\ldots(283)$$

where

$$B = n(4\rho_1^2+n^2)^{-\frac{1}{2}} = nc\{(4\pi k)^2 + n^2c^2\}^{-\frac{1}{2}}.$$

When c is zero, R' and $L'n$ tend to equality, as shown by Lord Rayleigh. But when c is finite, $L'n$ tends to *zero*, and R' to $4\pi\mu v A^{-1}$, as indeed we can see from (281) at once, by the relative evanescence of k and g, when finite.

But the frequency needed to bring about an approximation towards the constant resistance is excessive; in copper we require trillions per second. This brings us to the third reason mentioned; we have no knowledge of the properties of matter under such circumstances, or of ether either. The net result is that although it is infinitely more probable that the resistance should tend to constancy than to infinity, yet the real value is quite speculative.* Similar remarks apply to sudden discharges, as of lightning along a conductor. The above R', it should be remarked, is real resistance, in spite of its ultimate form, suggestive of impedance without resistance.† The present results are corroborative of those in Part I., and, in fact, only amount to a special application of the same.

Reflecting Barriers.

44. Let the medium be homogeneous between $r = a_0$ and $r = a_1$, where there is a change of some kind, yet unstated. Let between them the surface $r = a$ be a sheet of vorticity of e of the first order. We already know what will happen when fv is started, for a certain time, until in fact the inward wave reaches the inner boundary, and, on the other side, until the outward wave reaches the outward boundary; though, when the surface of f is not midway between the boundaries, the reflected wave from the nearest barrier may reach into a portion of the region beyond f, by the time the further barrier is reached by the primary wave. The subsequent history depends upon the constitution of the media beyond the boundaries, which can be summarized in two boundary conditions. The expression for E/H is, in general,

$$\frac{E}{H} = -(4\pi k + cp)^{-1}\frac{u' - yw'}{u - yw}, \qquad (284)$$

by (120), extended, the extension being the introduction of y, which is a differential operator of unstated form, depending upon the boundary

* The above was written before the publication of Professor Lodge's highly interesting lectures before the Society of Arts. Some of the experiments described in his second lecture are seemingly quite at variance with the magnetic theory. I refer to the smaller impedance of a short circuit of fine iron wire than of thick copper, as reckoned by the potential-difference at its beginning needed to spark across the circuit between knobs. Should this be thoroughly verified, it has occurred to me as a possible explanation that things may be sometimes so nicely balanced that the occurrence of a discharge may be determined by the state of the skin of the wire. A wire cannot be homogeneous right up to its boundary, with then a perfectly abrupt transition to air; and the electrical properties of the transition-layer are unknown. In particular, the skin of an iron wire may be nearly unmagnetisable, μ varying from 1 to its full value, in the transition-layer. Consequently, in the above formula, resistance $4\pi\mu v$ per unit surface, we may have to take $\mu = 1$ in the extreme, in the case of an iron wire. But even then, the explanation of Professor Lodge's results is capable of considerable elucidation. Perhaps resonance will do it. [Professor Lodge has since examined the theory of the apparently anomalous behaviour; and concludes that it was due to the great effective resistance of iron producing very rapid attenuation of the oscillations.]

† There is a tendency at present amongst some writers to greatly extend the meaning of resistance in electromagnetism; to make it signify cause/effect. This seems a pity, owing to the meaning of resistance having been thoroughly specialized in electromagnetism already, in strict relationship to "frictional" dissipation of energy. What the popular meaning of "resistance" may be is beside the point. I would suggest that what is now called the magnetic resistance be called the magnetic reluctance; and per unit volume, the reluctancy [or reluctivity].

conditions. Let y_0 and y_1 be the y's on the inner and outer side of the surface of f. The differential equation of H_a, the magnetic force there, is then

$$f\nu = \{(E/H)_{(\text{out})} - (E/H)_{(\text{in})}\} H_a, \quad \dots\dots\dots\dots\dots\dots (285)$$

as in § 19. Applying (284) and the conjugate property (114) of the functions u and w (since there is no change of medium at the surface of f), this becomes

$$H_a = \frac{4\pi k + cp}{q} \frac{(u_a - y_0 w_a)(u_a - y_1 w_a)}{y_1 - y_0} f\nu; \quad \dots\dots\dots\dots (286)$$

from which the differential equation of H at any point between a_0 and a is obtained by changing $u_a - y_0 w_a$ to $(a/r)(u - y_0 w)$; and at any point between a and a_1 by changing $u_a - y_1 w_a$ to $(a/r)(u - y_1 w)$.

Unless, therefore, there are singularities causing failure, the determinantal equation is

$$y_1 - y_0 = 0, \quad \dots\dots\dots\dots\dots\dots\dots\dots\dots\dots (287)$$

and the complete solution between a_0 and a_1 due to f constant may be written down at once. Thus, at a point outside the surface of f we have

$$\text{(out)} \qquad H = \frac{4\pi k + cp}{q} \cdot \frac{a}{r} \cdot \frac{(u_a - y_0 w_a)(u - y_1 w)}{y_1 - y_0} f\nu = \phi^{-1} f; \quad \dots\dots (288)$$

and therefore, if f starts when $t = 0$,

$$H = \frac{f}{\phi_0} + \frac{fa\nu}{r} \sum \frac{(u_a - y w_a)(u - y w)}{p(d/dp)(y_1 - y_0)} \cdot \frac{4\pi k + cp}{q} \epsilon^{pt}, \quad \dots\dots\dots (289)$$

p being now algebraic, given by (287); ϕ_0 the steady ϕ, from (288); and y the common value of the (now) equal y's; which identity makes (289) applicable on both sides of the surface of f.

Construction of the Operators y_1 *and* y_0.

45. In order that y_1 and y_0 should be determinable in such a way as to render (286) true, the media beyond the boundaries must be made up of any number of concentric shells, each being homogeneous, and having special values of c, k, μ, and g. For the spherical functions would not be suitable otherwise, except during the passage of the primary waves to the boundaries, or until they reached places where the departure from the assumed constitution commenced. Assuming the constitution in homogeneous spherical layers, there is no difficulty in building up the forms of y_0 and y_1 in a very simple and systematic manner, wholly free from obscurities and redundancies. In any layer the form of E/H is as in (284), containing one y. Now at the boundary of two layers E is continuous, and also H (provided the physical constants are not infinite), so E/H is continuous. Equating, therefore, the expressions for E/H in two contiguous media expresses the y of one in terms of the y of the other. Carrying out this process from the origin up to the medium between a_0 and a, expresses y_0 in terms of the y of the medium containing the origin; this is zero, so that y_0 is found as an explicit function of the values of u, w, u', w' at all the boundaries

between the origin and a_0. In a similar manner, since the y of the outermost region, extending to infinity, is 1, we express y_1, belonging to the region between a and a_1, in terms of the values of u, etc., at all the boundaries between a and ∞. Each of these four functions will occur twice for each boundary, having different values of the physical constants with the same value of r. I mention this method of equation of E/H operators because it is a far simpler process than what we are led to if we use the vector and scalar potentials; for then the force of the flux has three component vectors—the impressed force, the slope of the scalar potential, and the time-rate of decrease of the vector potential. The work is then so complex that a most accomplished mathematician may easily go wrong over the boundary conditions. These remarks are not confined in application to spherical waves.

If an infinite value be given to a physical constant, special forms of boundary condition arise, usually greatly simplified; *e.g.*, infinite conductivity in one of the layers prevents electromagnetic disturbances from penetrating into it from without; so that they are reflected without loss of energy.

Knowing y_1 and y_0 in (288), we virtually possess the sinusoidal solution for forced vibrations, though the initial effects, which may or may not subside or be dissipated, will require further investigation for their determination; also the solution in the form of an infinite series showing the effect of suddenly starting f constant; also the solution arising from any initial distribution of **E** and **H** of the kind appropriate to the functions, viz., such as may be produced by vorticity of **e** in spherical layers, proportional to ν (or $\nu Q_m^\backslash$ in general). But it is scarcely necessary to say that these solutions in infinite series, of so very general a character, are more ornamental than useful. On the other hand, the immediate integration of the differential equations to show the development of waves becomes excessively difficult, from the great complexity, when there is a change of medium to produce reflexion.

Thin Metal Screens.

46. This case is sufficiently simple to be useful. Let there be at $r=a_1$ a thin metal sheet interposed between the inner and outer non-conducting dielectrics, the latter extending to infinity. If made infinitely thin, E is continuous, and H discontinuous to an amount equal to 4π times the conduction-current (tangential) in the sheet. Let K_1 be the conductance of the sheet (tangential) per unit area; then

$$(H/E)_{\text{in}} - (H/E)_{\text{out}} = 4\pi K_1 \qquad \text{at} \qquad r=a_1.$$

Therefore by (284), when the dielectric is the same on both sides,

$$cp\left(\frac{u_1-w_1}{u_1'-w_1'} - \frac{u_1-y_1w_1}{u_1'-y_1w_1'}\right) = 4\pi K_1,$$

where the functions u_1, etc., have the $r=a_1$ values. From this,

$$y_1 = \frac{1-(4\pi K_1/cpq)u_1'(u_1'-w_1')}{1-(4\pi K_1/cpq)w_1'(u_1'-w_1')} \qquad \text{.................(290)}$$

expresses y_1 for an outer thin conducting metal screen, to be used in (286). If of no conductivity, it has no effect at all, passing disturbances freely, and $y_1 = 1$. At the other extreme we have infinite conductivity, making $y_1 = u_1'/w_1'$, with complete stoppage of outward-going waves, and reflexion without absorption, destroying the tangential electric disturbance.

When the screen, on the other hand, is within the surface of f, say at $r = a_0$, of conductance K_0 per unit area, we shall find

$$y_0 = \frac{(4\pi K_0/cpq)u_0'^2}{1 + (4\pi K_0/cpq)u_0'w_0'}, \qquad \text{.....................(291)}$$

where u_0, etc., have the $r = a_0$ values. The difference of form from y_1 arises from the different nature of the r functions in the region including the origin. As before, no conductivity gives transparency ($y_0 = 0$), and infinite conductivity total reflexion ($y_0 = u_0'/w_0'$). When the inner screen is shifted up to the origin, we make $y_0 = 0$, and so remove it.

Solution with Outer Screen; $K_1 = \infty$; f constant.

47. Let there be no inner screen, and let the outer be perfectly conducting. As J. J. Thomson has considered these screens,* I will be very brief, regarding them here only in relation to the sheet of f and to former solutions. The determinantal equation is

$$u_1' = 0, \qquad \text{or} \qquad \tan x = x(1 - x^2)^{-1}, \qquad \text{.............(292)}$$

if $x = ipa_1/v$. Roots nearly π, 2π, 3π, etc.; except the first, which is considerably less. The solution due to starting f constant, by (289), is therefore

$$H = \frac{fva}{\mu vr} \sum \frac{uu_a w_1'}{a_1 u_1''} \epsilon^{pt}; \qquad \text{.......................(293)}$$

which, developed by pairing terms, leads to

$$H = \frac{fva}{\mu vr} \sum \frac{x^4 - x^2 + 1}{x^3(x^2 - 2)} 2 \sin \frac{vtx}{a_1} \left(\cos - \frac{a_1}{xr} \sin\right) \frac{xr}{a_1} \left(\cos - \frac{a_1}{xa} \sin\right) \frac{xa}{a_1}, \quad (294)$$

which of course includes the effects of the infinite series of reflexions at the barrier. By making $a_1 = \infty$, however, the result should be the same as if the screen were non-existent, because an infinite time must elapse before the first reflexion can begin, and we are concerned only with finite intervals. The result is

$$H = \frac{fva}{\mu vr} \cdot \frac{2}{\pi} \int_0^\infty dx_1 \frac{\sin x_1 vt}{x_1} \left(\cos - \frac{1}{x_1 r} \sin\right) x_1 r \left(\cos - \frac{1}{x_1 a} \sin\right) x_1 a, \quad (295)$$

which must be the equivalent of the simple solution (142) of § 21, showing the origin and progress of the wave.

Now reduce it to a plane wave. We must make a infinite, and $r - a = z$ finite. Also take $fv = e$, constant. We then have

$$H = \frac{e}{4\mu v} \frac{2}{\pi} \int_0^\infty dx_1 \left(\frac{\sin x_1(vt - z)}{x_1} + \frac{\sin x_1(vt + z)}{x_1} \right), \qquad \text{.........(296)}$$

* In the paper before referred to.

showing the H at z due to a plane sheet of vorticity of e situated at $z=0$. This is the equivalent of the solution (12) of § 2, indicating the continuous uniform emission of $H=e/2\mu v$ both ways from the plane $z=0$. [But the sign of e is changed from that of § 2.]

Returning to (294), it is clear that from $t=0$ to $t=(a_1-a)/v$, the solution is the same as if there were no screen. Also if a is a very small fraction of a_1, the electromagnetic wave of depth $2a$ will, when it strikes the screen, be reflected nearly as from a plane boundary. It would therefore seem that this wave would run to-and-fro between the origin and boundary unceasingly. This is to a great extent true; and therefore there is no truly permanent state (the electric flux, namely, alone); but examination shows that the reflexion is not clean, on account of the electrification of the boundary, so that there is a spreading of the magnetic field all over the region within the screen.

Alternating f *with Reflecting Barriers. Forced Vibrations.*

48. Let the medium be nonconducting between the boundaries a_0 and a_1. Equation (288) then becomes

$$H=\frac{\nu a}{\mu vr}\frac{(u_a-y_0w_a)(u-y_1w)}{y_1-y_0}f, \quad\text{.....................(297)}$$

giving H outside the surface of f. We see that $y_0=0$ and $u_a=0$ make $H=0$. That is, the forced vibrations are confined to the inside of the surface of f only, at the frequencies given by $u_a=0$, provided there is no internal screen to disturb, but independently of the structure of the external medium (since y_1 is undetermined so far), with possible exceptions due to the vanishing of y_1 simultaneously. But (297), sinusoidally realized by $p^2=-n^2$, does not represent the full final solution, unless the nature of y_0 and y_1 is such as to allow the initial departure from this solution to be dissipated in space or killed by resistance. Ignoring the free vibrations, let $y_0=0$, and $y_1=u_1'/w_1'$, meaning no internal, and an infinitely conducting external screen. Then

$$\left.\begin{aligned}&\text{(out)}\quad H=(\nu a/\mu vr)u_a\{uw_1'/u_1'-w\}f,\\&\text{(in)}\quad H=(\nu a/\mu vr)u\,\{u_aw_1'/u_1'-w_a\}f.\end{aligned}\right\}\text{...............(298)}$$

If $w_1'=0$, or in full,

$$(v/na_1)\tan(na_1/v)=1-(v/na_1)^2,$$

we obtain a simplification, viz.

$$H_{(\text{in or out})}=-(\nu a/\mu vr)(uw_a\quad\text{or}\quad u_aw)f; \quad\text{..............(299)}$$

and the corresponding tangential components of electric force are

$$E_{(\text{in or out})}=(\nu a/\mu vr)(u'w_a\quad\text{or}\quad u_aw')(cp)^{-1}f. \quad\text{..........(300)}$$

But if $u_1'=0$, the result is infinite. This condition indicates that the frequency coincides with that of one of the free vibrations possible within the sphere $r=a_1$ without impressed force. But, considering that we may confine our impressed force to as small a space as we please round the origin, the infinite result is not easily understood, as regards its development.

But the development of infinitely great magnetic force by a *plane* sheet of f is very easily followed in full detail, not merely with sinusoidal f, but with f constant. Considering the latter case, the emission of H is continuous, as before described, from the surface of f. Now place a plane infinitely-conducting barrier parallel to f, say on the left side. We at once stop the disturbances going to the left and send them back again, unchanged as regards **H**, reversed as regards **E**. The H-disturbance on the left side of f therefore commences to be doubled after the time a/v has elapsed, a being the distance of the reflecting barrier from the plane of f, and on the right side after the interval $2a/v$. Next, put a second infinitely-conducting barrier on the right side of f. It also doubles the H-disturbances as they arrive; so that, by the inclusion of the plane of f between impermeable barriers, combined with the continuous emission of **H**, the magnetic disturbance mounts up infinitely, in a manner which may be graphically followed with ease. Similarly with f alternating, at particular frequencies depending upon the distances of the two barriers from f.

Returning to the spherical case, an infinitely-conducting internal screen, with no external, produces

$$H_a = \frac{(u_a w_0' - w_a u_0')(u_a - w_a)}{\mu v(w_0' - u_0')} f\nu. \qquad \text{(301)}$$

We cannot produce infinite H in this case, because the absence of an external barrier will not let it accumulate. Shifting the surface of f right up to the screen, or conversely, simplifies matters greatly, reducing to the case of § 42.

May 8, 1888.

PART V.

CYLINDRICAL ELECTROMAGNETIC WAVES.

49. In concluding this paper I propose to give some cases of cylindrical waves. They are selected with a view to the avoidance of mere mathematical developments and unintelligible solutions, which may be multiplied to any extent; and for the illustration of peculiarities of a striking character. The case of vibratory impressed E.M.F. in a thin tube is very rich in this respect, as will be seen later. At present I may remark that the results of this paper have little application in telegraphy or telephony, when we are only concerned with long waves. Short waves are, or may be, now in question, demanding a somewhat different treatment.* We do, however, have very short waves in the

* The waves here to be considered are essentially of the same nature as those considered by J. J. Thomson, "On Electrical Oscillations in a Cylindrical Conductor," *Proc. Math. Soc.* vol. XVII., and in Parts I. and II. of my paper, "On the Self-Induction of Wires," *Phil. Mag.*, August and September, 1886; viz. a mixture of the plane and cylindrical. But the peculiarities of the telegraphic problem make it practically a case of plane waves as regards the dielectric, and cylindrical in the wires. The "resonance" effects described in my just-mentioned paper arise from the to-and-fro reflexion of the plane waves in the dielectric, moving parallel

discharge of condensers, and in vacuum-tube experiments, so that we are not so wholly removed from practice as at first appears. But independently of considerations of practical realization, I am strongly of opinion that the study of very unrealizable problems may be of use in forwarding the supply of one of the pressing wants of the present time or near future, a practicable ether—mechanically, electromagnetically, and perhaps also gravitationally comprehensive.

Mathematical Preliminary.

50. On account of some peculiarities in Bessel's functions, which require us to change the form of our equations to suit circumstances, it is desirable to exhibit separately the purely mathematical part. This will also considerably shorten and clarify what follows it.

Let the axis of z be the axis of symmetry, and let r be the distance of any point from it. Either the lines of **E**, electric force, or of **H**, magnetic force, may be circular, centred on the axis. For definiteness, choose **H** here. Then the lines of **E** are either longitudinal, or parallel to the axis; or there is, in addition, a radial component of **E**, parallel to **r**. Thus the tensor H of **H**, and the two components of **E**, say E longitudinal and F radial, fully specify the field. Their connexions are these special forms of equations (2) and (3):—

$$\frac{1}{r}\frac{d}{dr}rH=(4\pi k+cp)E, \qquad -\frac{dH}{dz}=(4\pi k+cp)F, \qquad \frac{dE}{dr}-\frac{dF}{dz}=\mu pH, \qquad (302)$$

where (and always later) p stands for d/dt. This is in space where neither the impressed electric nor the impressed magnetic force has curl, it being understood that **E** and **H** are the forces of the fluxes, so as to include impressed. From (302) we obtain

$$\left.\begin{aligned}\frac{1}{r}\frac{d}{dr}r\frac{dE}{dr}+\frac{d^2E}{dz^2}&=(4\pi k+cp)\mu pE,\\ \frac{d}{dr}\frac{1}{r}\frac{d}{dr}rH+\frac{d^2H}{dz^2}&=(4\pi k+cp)\mu pH,\end{aligned}\right\}\qquad (303)$$

the characteristics of E and H. Let now

$$q^2=-s^2=(4\pi k+cp)\mu p-d^2/dz^2; \qquad (304)$$

then the first of (303) becomes the equation of $J_0(sr)$ and its companion, whilst the second becomes that of $J_1(sr)$ and its companion. Thus E is associated with J_0 and H with J_1, when **H** is circular; conversely when **E** is circular.

to the wire. This is also practically true in Prof. Lodge's recent experiments, discharging a Leyden jar into a miniature telegraph-circuit. On the other hand, most of such effects in the present paper depend upon the cylindrical waves in the dielectric; and in order to allow the dielectric fair play for their development, the contaminating influence of diffusion is done away with by using tubes only, when there are conductors. In Hertz's recent experiments the waves are of a very mixed character indeed.

We have first Fourier's cylinder function

$$J_{0r}=J_0(sr)=1-\frac{(sr)^2}{2^2}+\frac{(sr)^4}{2^24^2}-\dots; \quad \text{.................(305)}$$

and its companion,* which call G_0, is

$$\left.\begin{aligned} G_{0r}&=G_0(sr)=(2/\pi)[J_{0r}(\log sr-\beta)+L_{0r}], \\ \text{where}\qquad L_{0r}&=\frac{(sr)^2}{2^2}-(1+\tfrac{1}{2})\frac{(sr)^4}{2^24^2}+(1+\tfrac{1}{2}+\tfrac{1}{3})\frac{(sr)^6}{2^24^26^2}-\dots .\end{aligned}\right\}\text{.........(306)}$$

The coefficient $2/\pi$ is introduced to simplify the solutions. The function $J_1(sr)$ or J_{1r} is the negative of the first derivative of J_{0r} with respect to sr. Let $G_1(sr)$ or G_{1r} be the function similarly derived from G_{0r}. The conjugate property, to be repeatedly used, is

$$(J_0G_1-J_1G_0)_r=-2/\pi sr. \quad \text{.......................(307)}$$

We have also Stokes's formula for J_{0r}, useful when sr is real and not too small, viz.

$$J_{0r}=\left(\frac{1}{\pi sr}\right)^{\frac{1}{2}}\Big(R(\cos+\sin)sr+Si(\sin-\cos)sr\Big), \quad \text{.........(308)}$$

where R and Si are functions of sr to be presently given. The corresponding formula for G_{0r} is obtained by changing cos to sin and sin to $-$cos in (308).

Besides these two sets of solutions, we sometimes require to use a third set. A pair of solutions of the J_0 equation is

$$\left.\begin{aligned} U&=r^{-\frac{1}{2}}\epsilon^{qr}(R+S), \qquad W=r^{-\frac{1}{2}}\epsilon^{-qr}(R-S), \\ \text{where}\qquad R\pm S&=1\pm\frac{1}{8qr}+\frac{1^23^2}{\lfloor 2(8qr)^2}\pm\frac{1^23^25^2}{\lfloor 3(8qr)^3}+\dots .\end{aligned}\right\}\text{..........(309)}$$

The last also defines the R and Si in (308). R is real whether q^2 be +

* [In investigations where we are concerned with the complementary function to $J_0(sr)$ between boundaries, the constant β (which I now introduce) may be omitted *ab initio*, being superfluous. If retained, it will go out later, by the β's of one boundary cancelling those of the other. This is true in the resultant differential equations as well as in solutions. For this reason β has been omitted in the previous investigations in this work. But in the following investigations we are often concerned with the G_0 function when the outer boundary is removed to infinity, that is, when there is no outer boundary. We should then standardize G_0 so as to vanish at infinity. This requirement is satisfied by the form

$$G_0(sr)=(\pi sr)^{-\frac{1}{2}}\Big(R(\sin-\cos)sr-Si(\cos+\sin)sr\Big), \quad \text{..............(308a)}$$

derived from (308) in the manner described above. But the form (306) requires β to be retained, for evanescence at infinity. Its value is

$$\beta=\log 2-\gamma=\log 2-\cdot 5772=\cdot 11593, \quad \text{.......................(308b)}$$

where γ is Euler's constant

$$\gamma=1+\tfrac{1}{2}+\tfrac{1}{3}+\dots+\frac{1}{\infty}-\log\infty=\cdot 5772. \quad \text{....................(308c)}$$

An evaluation of this β will be found in Lord Rayleigh's Sound, vol. II. The process is not free from difficulty, and a different estimate has been given, but I have corroborated the above estimate by two other independent methods. Note that (306) with β and (308a) are equivalent.]

or $-$, whilst S is unreal when q^2 is $-$, or Si is then real, s^2 being $+$. [Take $q=si$ in (309), then we have

$$R=1-\frac{1^2 3^2}{\lfloor 2(8sr)^2}+\frac{1^2 3^2 5^2 7^2}{\lfloor 4(8sr)^4}-\dots,\qquad Si=\frac{1}{8sr}-\frac{1^2 3^2 5^2}{\lfloor 3(8sr)^3}+\frac{1^2 3^2 5^2 7^2 9^2}{\lfloor 5(8sr)^5}-\dots,$$

to be used in (308).]

When qr is a $+$ numeric, the solution U is meaningless, as its value is infinity. But in our investigations q^2 is a differential operator, so that the objection to U on that score is groundless. We shall use it to calculate the shape of an inward progressing wave, whilst W goes to find an outward wave. The results are fully convergent within certain limits of r and t. From this alone we see that a comprehensive theory of ordinary linear differential equations [by themselves] is sometimes impossible. They must be generalized into partial differential equations before they can be understood.*

The conjugate property of U and W is

$$UW'-U'W=-2q/r, \qquad \text{(310)}$$

if the $'=d/dr$. An important transformation sometimes required is

$$J_{0r}-iG_{0r}=2iW(2\pi q)^{-\frac{1}{2}}; \qquad \text{(311)}$$

or, which means the same,

$$W=-\left(\frac{2q}{\pi}\right)^{\frac{1}{2}}[J_{0r}(\log qr-\beta)+L_{0r}]. \qquad \text{(312)}$$

* [We may, however, use U to calculate the numerical value of $J_0(sri)$ or $I_0(qr)$ when qr is not too small, namely, by wholly rejecting the infinite divergent part of the series. Thus

$$I_0(qr)=1+\frac{(qr)^2}{2^2}+\frac{(qr)^4}{2^2 4^2}+\dots=\frac{\epsilon^{qr}}{(2\pi qr)^{\frac{1}{2}}}\left\{1+\frac{1}{8qr}+\frac{1^2 3^2}{\lfloor 2(8qr)^2}+\dots\right\} \qquad \text{(309a)}$$

expresses the equivalence, the convergent series being suitable for small, and the divergent for large values of the argument. But the convergent series admits of exact calculation, whilst the divergent series does not, though by stopping at the smallest term we obtain the nearest approach to the true value of $I_0(qr)$. This contrasts with the behaviour of U as a complex differentiator, when the whole series is operative.

It is difficult to imagine a direct transformation from the convergent to the divergent series by ordinary mathematics, for, owing to the terms in the latter being all positive, it makes nonsense. The following transformation is the only one I have been able to make up. Let t be the variable, and p the differentiator d/dt. Then, q being a constant,

$$I_0(qt)=1+\frac{(qt)^2}{2^2}+\frac{(qt)^4}{2^2 4^2}+\dots=1+\frac{q^2}{p^2}\frac{2}{2^2}+\frac{q^4}{p^4}\frac{\lfloor 4}{2^2 4^2}+\dots, \qquad \text{(309b)}$$

by applying $p^{-n}=t^n/\lfloor n$, understanding here and later that when no operand is expressed, the operand is 1, that is, zero before and 1 after $t=0$. Therefore, by the binomial theorem,

$$I_0(qt)=\left(1-\frac{q^2}{p^2}\right)^{-\frac{1}{2}}=\frac{p}{(p^2-q^2)^{\frac{1}{2}}}. \qquad \text{(309c)}$$

Now we also have $\epsilon^{qt}=\dfrac{p}{p-q}$, or $p=(p-q)\epsilon^{qt}$. (309d)

Substituting this for the numerator in the last form we get

$$I_0(qt)=\frac{p-q}{(p^2-q^2)^{\frac{1}{2}}}\epsilon^{qt}=\left(\frac{p-q}{p+q}\right)^{\frac{1}{2}}\epsilon^{qt}. \qquad \text{(309e)}$$

When we have obtained the differential equation in any problem, the assumption s^2 = a constant * converts it into the solution due to impressed force sinusoidal with respect to t and z; this requires $d^2/dz^2 = -m^2$, and $d^2/dt^2 = -n^2$, where m and n are positive constants, being 2π times the wave-shortness along z and 2π times the frequency of vibration respectively.

After (309) we became less exclusively mathematical. To go further in this direction, and come to electromagnetic waves, observe that we need not concern ourselves at all with F the radial component, in seeking for the proper differential equation connected with a surface of curl of impressed force; it is E and H only that we need consider, as the boundary conditions concern them. The second of (302) derives F from H.

When **H** is circular, the operator E/H is given by

$$\frac{E}{H}=\frac{s}{4\pi k+cp}\cdot\frac{J_{0r}-yG_{0r}}{J_{1r}-yG_{1r}}, \qquad (313)$$

where y is undetermined. When **E** is circular, the operator E/H is given by

$$\frac{E}{H}=\frac{s}{4\pi k+cp}\cdot\frac{J_{1r}-yG_{1r}}{J_{0r}-yG_{0r}}. \qquad (314)$$

The use of these operators greatly facilitates and systematizes investigation. The meaning is that (313) or (314) is the characteristic equation connecting E and H.

Longitudinal Impressed E.M.F. in a Thin Conducting Tube.

51. Let an infinitely long thin conducting tube of radius a have conductance K per unit of its surface to longitudinal current, and be bounded by a dielectric on both sides. Strictly speaking, the tube should be infinitely thin, in order to obtain instantaneous magnetic penetration, and yet be of finite conductance without possessing infinite

Now shift the new operand ϵ^{qt} to the left (or make 1 the operand again) and we change p to $p+q$, giving

$$I_0(qt)=\epsilon^{qt}\left(\frac{p}{p+2q}\right)^{\frac{1}{2}}. \qquad (309f)$$

So far is equivalent to the work on p. 427, vol. II. But now use the result $p^{\frac{1}{2}}=(\pi t)^{-\frac{1}{2}}$, make it the operand, and expand the radical denominator in *rising* powers of p. Then (309f) gives

$$I_0(qt)=\epsilon^{qt}\left(\frac{p}{2q}+1\right)^{-\frac{1}{2}}\left(\frac{p}{2q}\right)^{\frac{1}{2}}=\epsilon^{qt}\left(\frac{p}{2q}+1\right)^{-\frac{1}{2}}\frac{1}{(2\pi qt)^{\frac{1}{2}}}$$

$$=\epsilon^{qt}\left\{1-\frac{p}{4q}+\frac{1.3}{\underline{|2}}\left(\frac{p}{4q}\right)^2-\dots\right\}\frac{1}{(2\pi qt)^{\frac{1}{2}}}. \qquad (309g)$$

Lastly, perform the differentiations, and we get

$$I_0(qt)=\frac{\epsilon^{qt}}{(2\pi qt)^{\frac{1}{2}}}\left\{1+\frac{1}{8qt}+\frac{1^2 3^2}{\underline{|2}(8qt)^2}+\dots\right\}, \qquad (309h)$$

which is the required result.]

* [When $k=0$, then $p=ni$ and $d^2/dz^2=-m^2$ makes s^2 constant, either + or −. In a conducting dielectric s^2 is complex. We have $p=ni$, $q=si$, in the rest.]

conductivity, because that would produce opacity. In this tube let impressed electric force, of intensity e per unit length, act longitudinally, e being any function of t and z. We have to connect e with E and H internally and externally.

The magnetic force being circular, (313) is the resistance-operator required. Within the tube take $y=0$ if the axis is to be included; else find y by some internal boundary-condition. Outside the tube take $y=i$ when the medium is homogeneous and boundless, because that is the only way to prevent waves from coming from infinity; else find y by some outer boundary-condition. There is no difficulty in forming the y to suit any number of coaxial cylinders possessing different electrical constants, by the continuity of E and H at each boundary, which equalizes the E/H's of its two sides, and so expresses the y on one side in terms of that on the other; but this is useless for our purpose. For the present take $y=0$ inside, and leave it unstated outside.

At $r=a$, E_a has the same value on both sides of the tube, on account of its thinness. In the substance of the tube $e+E_a$ is the force of the flux. On the other hand H is discontinuous at the tube, thus

$$4\pi K(e+E)=H_{(\text{out})}-H_{(\text{in})}=\left(\frac{H}{E}(\text{out})-\frac{H}{E}(\text{in})\right)E_a. \quad \text{......(315)}$$

In this use (313), and the conjugate property (307), and we at once obtain

$$e=\left[-1+\frac{4\pi k+cp}{4\pi Ks}\frac{2y}{\pi sa}\frac{1}{J_{0a}(J_{0a}-yG_{0a})}\right]E_a, \quad \text{...... .....(316)}$$

from which all the rest follows. Merely remarking concerning k that the realization of (316) when k is finite requires the splitting up of the Bessel functions into real and imaginary parts, that the results are complex, and that there are no striking peculiarities readily deducible; let us take $k=0$ at once, and keep to nonconducting dielectrics. Then, from (316), follow the equations of E and H, in and out; thus

$$E_{(\text{in})} \text{ or }_{(\text{out})}=\frac{J_{0r}(J_{0a}-yG_{0a}) \text{ or } J_{0a}(J_{0r}-yG_{0r})}{\frac{cp}{4\pi Ks}\cdot\frac{2y}{\pi sa}-J_{0a}(J_{0a}-yG_{0a})}e, \quad \text{......(317)}$$

$$H_{(\text{in})} \text{ or }_{(\text{out})}=\frac{cp}{s}\cdot\frac{J_{1r}(J_{0a}-yG_{0a}) \text{ or } J_{0a}(J_{1r}-yG_{1r})}{\text{same denominator}}e, \quad \text{.....(318)}$$

which we can now examine in detail.

Vanishing of External Field. $J_{0a}=0$.

52. The very first thing to be observed is that $J_{0a}=0$ makes E and H and therefore also F vanish outside the tube, and that this property is independent of y, or of the nature of the external medium. We require the impressed force to be sinusoidal or simply periodic with respect to z and t, thus

$$e=e_0\sin(mz+\alpha)\sin(nt+\beta), \quad \text{....................(319)}$$

so that, ultimately, $s^2=n^2/v^2-m^2$; ....(320)

and any one of the values of s given by $J_{0a}=0$ causes the evanescence of the external field. The solutions just given reduce to

$$\text{(in)}\qquad \left.\begin{aligned} H &= -4\pi K(J_{1r}/J_{1a})e, \\ E &= (s/cn)4\pi K(J_{0r}/J_{1a})ie, \\ F &= -(cn)^{-1}4\pi K(J_{1r}/J_{1a})i(de/dz), \end{aligned}\right\} \qquad (321)$$

which are fully realized, because i signifies p/n, or involves merely a time-differentiation performed on the e of (319).

The electrification is solely upon the inner surface of the tube. In its substance H falls from $-4\pi Ke$ inside to zero outside, and E_a being zero, the current in the tube is Ke per unit surface.

The independence of y raises suspicion at first that (321) may not represent the state which is tended to after e is started. But since the resistance of the tube itself is sufficient to cause initial irregularities to subside to zero, even were there a perfectly reflecting barrier outside the tube to prevent dissipation of these irregularities in space, there seems no reason to doubt that (321) do represent the state asymptotically tended to. Changing the form of y will only change the manner of the settling down. We may commence to change the nature of the medium immediately at the outer boundary of the tube. We cannot, however, have those abrupt assumptions of the steady or simply periodic state which characterize spherical waves, owing to the geometrical conditions of a cylinder.

Case of Two Coaxial Tubes.

53. If there be a conducting tube anywhere outside the first tube, there is no current in it, except initially. From this we may conclude that if we transfer the impressed force to the outer tube, there will be no current in the inner. Thus, let there be an outer tube at $r=x$, of conductance K_1 per unit area, containing the impressed force e_1. We have

$$E_x = \frac{4\pi K_1 e_1}{Y_3 - Y_2 - 4\pi K_1}, \qquad (322)$$

where Y_3 and Y_2 are the H/E operators just outside and inside the tube, whilst E_x is the E at x, on either side of the tube, resulting from e_1. We have

$$Y_3 = \frac{cp}{s}\frac{J_{1x} - y_1 G_{1x}}{J_{0x} - y_1 G_{0x}}, \qquad Y_2 = \frac{cp}{s}\frac{J_{1x} - yG_{1x}}{J_{0x} - yG_{0x}}, \qquad (323)$$

where y_1 is settled by some external and y by some internal condition. In the present case the inner tube at $r=a$, if it contains no impressed force, produces the condition

$$Y_2 - Y_1 = 4\pi K \quad \text{at} \quad r=a, \qquad (324)$$

where Y_1 is the internal H/E operator. Or

$$4\pi K = \frac{cp}{s}\left(\frac{J_{1a} - yG_{1a}}{J_{0a} - yG_{0a}} - \frac{J_{1a}}{J_{0a}}\right),$$

giving

$$y = \frac{4\pi K J_{0a}^2}{\dfrac{2}{\pi s a}\dfrac{cp}{s} + 4\pi K J_{0a} G_{0a}}. \qquad (325)$$

Now, using (323) in (322) brings it to

$$E_x = \frac{(J_{0x} - yG_{0x})(J_{0x} - y_1 G_{0x})4\pi K_1 e_1}{\frac{cp}{s}(y_1 - y)\frac{2}{\pi s x} - 4\pi K_1 (J_{0x} - yG_{0x})(J_{0x} - y_1 G_{0x})}, \quad \text{......(326)}$$

in which y is given by (325), and from (326) the whole state due to e_1 follows, as modified by the inner tube.

Now $J_{0a} = 0$ makes $y = 0$; this reduces (326) to

$$E_x = \frac{J_{0x}(J_{0x} - y_1 G_{0x})4\pi K_1 e_1}{\frac{cp}{s} y_1 \frac{2}{\pi s x} - 4\pi K_1 J_{0x}(J_{0x} - y_1 G_{0x})}; \quad \text{..............(327)}$$

and, by comparison with (317), we see that it is now the same as if the inner tube were non-existent. That is, when it is situated at a nodal surface of E due to impressed force in the outer tube, and there is therefore no current in it (except transversely, to which the dissipation of energy is infinitely small), its presence does nothing, or it is perfectly transparent.

It is clearly unnecessary that the external impressed force should be in a tube. Let it only be in tubular layers, without specification of actual distribution or of the nature of the medium, except that it is in layers so that c, k, and μ are functions of r only; then if the axial portion be nonconducting dielectric, the J_{0r} function specifies E and allows there to be nodal surfaces, for instance $J_{0a} = 0$, where a conducting tube may be placed without disturbing the field. Admitting this property *ab initio*, we can conversely conclude that e in the tube at $r = a$ will, when $J_{0a} = 0$, make *every* external cylindrical surface a nodal surface, and therefore produce no external disturbance at all.

54. Now go back to § 51, equations (317), (318). There are no *external* nodal surfaces of E in general (exception later). We cannot therefore find a place to put a tube so as not to disturb the existing field due to e in the tube at $r = a$. But we may now make use of a more general property. To illustrate simply, consider first the magnetic theory of induction between linear circuits. Let there be any number of circuits, all containing impressed forces, producing a determinate varying electromagnetic field. In this field put an additional circuit of infinite resistance. The E.M.F. in it, due to the other circuits, will cause no current in it of course, so that no change in the field takes place. Now, lastly, close the circuit or make its resistance finite, and simultaneously put in it impressed force which is at every moment the negative of the E.M.F. due to the other circuits. Since no current is produced there will still be no change, or everything will go on as if the additional circuit were non-existent.

Applying this to our tubes, we may easily verify by the previous equations that when there are two coaxial tubes, both containing impressed forces, we can reduce the resultant electromagnetic field everywhere to that due to the impressed force in one tube, provided we suitably choose the impressed force in the second to be the negative of the electric force of field due to e in the first tube when the second is non-existent. That is, we virtually abolish the conductance of the second tube and make it perfectly transparent.

Perfectly Reflecting Barrier. Its Effects. Vanishing of Conduction Current.

55. To produce nodal surfaces of E outside the tube containing the vibrating impressed force, we require an external barrier, which shall prevent the passage of energy or its absorption, by wholly reflecting all disturbances which reach it. Thus, let there be a perfect conductor at $r=x$. This makes $E=0$ there. This requires that the y in (317), (318) shall have the value J_{0x}/G_{0x}, whereas without any bound to the dielectric it would be i. We can now choose m and n so as to make $J_{0x}=0$. This reduces those equations to

$$\text{(in and out)} \qquad E=-\frac{J_{0r}}{J_{0a}}e, \qquad \left.\begin{aligned} F&=+\frac{1}{s}\frac{J_{1r}}{J_{0a}}\frac{de}{dz},\\ H&=-\frac{1}{s}\frac{J_{1r}}{J_{0a}}cpe. \end{aligned}\right\} \qquad (328)$$

This solution is now the same inside and outside the tube containing the impressed force, and there is no current in the tube, that is, no longitudinal current.

To understand this case, take away the impressed force and the tube. Then (328) represents a conservative system in stationary vibration. Now, by the preceding, we may introduce the tube at a nodal surface of E without disturbing matters, provided there be no impressed force in the tube. But if we introduce the tube anywhere else, where E is not zero, we require, by the preceding, an impressed force which is at every moment the negative of the undisturbed force of the field, in order that no change shall occur. Now this is precisely what the solution (328) represents, e in the tube being cancelled by the force of the field, so that there is no conduction-current. The remarkable thing is that it is the impressed force in the tube itself that sets up the vibrating field, and gradually ceases to work, so that in the end it and the tube may be removed without altering the field. That a perfect conductor as reflector is required is a detail of no moment in its theoretical aspect.

Shifting the tube, with a finite impressed force in it, towards a nodal surface of E, sends up the amplitude of the vibrations to any extent.

$K=0$ *and* $K=\infty$.

56. If the tube have no conductance, e produces no effect. This is because the two surfaces of curl of **e** are infinitely close together, and therefore cancel, not having any conductance between them to produce a discontinuity in the magnetic force.

But if the tube have infinite conductance, we produce complete mutual independence of the internal and external fields, except in the quite unessential particular that the two surfaces of curl e are of opposite kind and time together. Equations (317), (318) reduce to

$$\text{(in)} \qquad E = -\frac{J_{0r}}{J_{0a}}e, \qquad F = +\frac{1}{s}\frac{J_{1r}}{J_{0a}}\frac{de}{dz}, \qquad H = -\frac{1}{s}\frac{J_{1r}}{J_{0a}}cpe. \quad \text{....(329)}$$

$$\text{(out)} \qquad \left\{ \begin{array}{l} E = -\dfrac{J_{0r} - yG_{0r}}{J_{0a} - yG_{0a}}e, \qquad F = \dfrac{1}{s}\dfrac{J_{1r} - yG_{1r}}{J_{0a} - yG_{0a}}\dfrac{de}{dz}, \\ H = -\dfrac{1}{s}\dfrac{J_{1r} - yG_{1r}}{J_{0a} - yG_{0a}}cpe. \end{array} \right\} \text{..........(330)}$$

Observe that (329) is the same as (328). The external solution (330) requires y to be stated. When $y = i$, for a boundless dielectric, the realization is immediate.

s = 0. *Vanishing of* E *all over, and of* F *and* H *also internally.*

57 This is a singularity of quite a different kind. When $n = mv$, we make $s = 0$. Of course there is just one solution with a given wave-length along z; a great frequency with small wave-length, and conversely.

E vanishes all over, that is, both inside and outside the tube containing e, provided s/y is zero. The internal H and therefore also F vanish. Thus within the tube is no disturbance, and outside, (317) (318) reduce to

$$\text{(out)} \qquad H = \frac{a}{r}4\pi Ke, \qquad F = \frac{1}{cn}\frac{a}{r}4\pi Ki\frac{de}{dz}. \quad \text{..............(331)}$$

Observe that H and F do not fluctuate or alternate along r, but that H has the same distribution (out from the tube) as if e were steady and did not vary along z.

A special case is $m = 0$. Then also $n = 0$, or e is steady and independent of z. F vanishes, and the first of (331) expresses the steady state.

Without this restriction, the current in the tube is Ke per unit surface, owing to the vanishing of the opposing longitudinal E of the field. This property was, by inadvertence, attributed by me in a former paper* to a wire instead of a tube. The wave-length must be great in order to render it applicable to a wire, because instantaneous penetration is assumed.

I mentioned that s/y must vanish. This occurs when $y = i$, or the external dielectric is boundless. But it also occurs when $E = 0$ at $r = x$, produced by a perfectly conductive screen. This is plainly allowable because it does not interfere with the $E = 0$ all-over property. What the screen does is simply to terminate the field abruptly. Of course it is electrified.

s = 0 *and* $H_x = 0$.

58. But with other boundary conditions, we do not have the solutions (331). Thus, let $H_x = 0$, instead of $E_x = 0$. This makes $y = J_{1x}/G_{1x}$ in

* "On Resistance and Conductance Operators," *Phil. Mag.*, Dec. 1887, p. 492, Ex. . [vol. II., p. 366].

(317), (318). There are at least two ways (theoretical) of producing this boundary condition. First, there may be at $r=x$ a screen made of a perfect magnetic conductor ($g=\infty$). Or, secondly, the whole medium beyond $r=x$ may be infinitely elastive and resistive ($c=0$, $k=0$) to an infinite distance.

Now choose $s=0$ in addition, and reduce (317), (318). The results are

$$\left.\begin{aligned} &E=-\frac{e}{1+\frac{1}{2}x^2cp/4\pi Ka}; \qquad F=-\frac{1}{cp}\frac{dH}{dz}, \\ &\text{(in) or (out)} \quad H=-\frac{cpe}{1+\frac{1}{2}x^2cp/4\pi Ka}\left(\frac{r}{2} \text{ or } \frac{r}{2}-\frac{x^2}{2r}\right), \end{aligned}\right\} \quad \ldots\ldots (332)$$

which are at once realized by removing p from the denominator to the numerator.

Although E is not now zero, it is independent of r, only varying with t and z.

When s^2 is negative, or $n < m/v$, the solutions (317), (318) require transforming in part because some of the Bessel functions are unreal. Use (312), because q is now real. There are no alternations in E or H along r. They only commence when $n > mv$.

Separate Actions of the Two Surfaces of curl **e**.

59. Since all the fluxes depend solely upon the curl of **e**, and not upon its distribution, and there are two surfaces of curl **e** in the tube problem, their actions, which are independent, may be separately calculated. The inner surface may arise from e in the − direction in the inner dielectric, or by the same in the + direction in the tube and beyond it. The outer may be due to **e** in the − direction beyond the tube, or in the + direction in the tube and inner dielectric.

We shall easily find that the inner surface of curl of **e**, say of surface-density f_1, produces

$$\left.\begin{aligned} &\text{(in)} \quad E=J_{0r}\frac{(J_{1a}-yG_{1a})-(J_{0a}-yG_{0a})4\pi Ks/cp}{2y/\pi sa-J_{0a}(J_{0a}-yG_{0a})4\pi Ks/cp}f_1, \\ &\text{(out)} \quad E=\frac{J_{1a}(J_{0r}-yG_{0r})}{\text{same denominator}}f_1, \end{aligned}\right\} \quad \ldots\ldots\ldots (333)$$

from which H may be got by the E/H operator.

The external sheet, say f_2, produces

$$\left.\begin{aligned} &\text{(in)} \quad E=\frac{J_{0r}(J_{1a}-yG_{1a})}{\cdots\cdots\cdots\cdots\cdots}f_2, \\ &\text{(out)} \quad E=(J_{0r}-yG_{0r})\frac{J_{1a}+J_{0a}4\pi Ks/cp}{\cdots\cdots\cdots\cdots\cdots\cdots}f_2, \end{aligned}\right\} \quad \ldots\ldots\ldots (334)$$

where the unwritten denominators are as in the first of (333). Observe that when $J_{1a}=0$, f_1 produces no external field (in tube or beyond it). It is then only f_2 that operates in the tube and beyond.

Now take $f_2 = e$ and $f_1 = -e$ in (333) and (334) and add the results. We then obtain (317), (318); and it is now $J_{0a} = 0$ that makes the external field vanish, instead of $J_{1a} = 0$ when f_1 alone is operative.

Having treated this problem of a tube in some detail, the other examples may be very briefly considered, although they too admit of numerous singularities.

Circular Impressed Force in Conducting-Tube.

60. The tube being as before, let the impressed force **e** (per unit length) act circularly in it instead of longitudinally, and let **e** be a function of t only, so that we have an inner and an outer cylindrical surface of longitudinally directed curl of **e**. **H** is evidently longitudinal and **E** circular, so that we now require to use the (314) operator.

At the tube E_a is continuous, this being the tensor of the force of the flux on either side, and H is discontinuous thus,

$$H_{(\text{in})} - H_{(\text{out})} = 4\pi K(e + E_a),$$

or

$$e = -\left\{1 + \frac{1}{4\pi K}\left(\frac{H}{E}(\text{out}) - \frac{H}{E}(\text{in})\right)\right\}E_a. \quad \text{..........(335)}$$

Substituting the (314) operator, with $y = 0$ inside, and y undetermined outside, and using the conjugate property (307), we obtain

$$H_{(\text{in})} \text{ or }_{(\text{out})} = -i\frac{(J_{1a} - yG_{1a})J_{0r} \text{ or } J_{1a}(J_{0r} - yG_{0r})}{\mu v J_{1a}(J_{1a} - yG_{1a}) + \dfrac{y}{4\pi K}\dfrac{2v}{\pi a p}}e, \quad \text{.....(336)}$$

$$E_{(\text{in})} \text{ or }_{(\text{out})} = -\mu v\frac{(J_{1a} - yG_{1a})J_{1r} \text{ or } J_{1a}(J_{1r} - yG_{1r})}{\text{same denominator}}e. \quad \text{....(337)}$$

When e is simply periodic, $J_{1a} = 0$ makes the external E and H vanish independent of the nature of y. The complete solution is then

$$H_{(\text{in})} = 4\pi K\frac{J_{0r}}{J_{0a}}e, \qquad E_{(\text{in})} = -4\pi K\mu v\frac{J_{1r}}{J_{0a}}ie. \quad \text{..........(338)}$$

The conduction-current in the tube is Ke per unit area of surface.

To make the conduction-current vanish by balancing the impressed force against the electric force of the field that it sets up, put an infinitely-conducting screen at $r = x$ outside the tube, and choose the frequency to make $J_{1x} = 0$, since we now have $y = J_{1x}/G_{1x}$. We shall then have the same solution inside and outside, viz.

$$H = -\frac{1}{\mu v}\frac{J_{0r}}{J_{1a}}ie, \qquad E = -\frac{J_{1r}}{J_{1a}}e; \quad \text{..................(339)}$$

so that at the tube itself, $E = -e$. This case may be interpreted as in § 55, the tube being at a nodal surface of E.

A special case of (338) is when $n = 0$, or **e** is steady. Then there is merely the longitudinal **H** inside the tube, given by $H = 4\pi Ke$.

Cylinder of Longitudinal curl of **e** *in a Dielectric.*

61. In a nonconductive dielectric let the impressed electric force be such that its curl is confined to a cylinder of radius a, in which it is uniformly distributed, and is longitudinal. Let f be the tensor of curl **e**, and let it be a function of t only. Since **E** is circular and **H** longitudinal, we have (314) as operator, in which k is to be zero. This is outside the cylinder. Inside, on the other hand, on account of the existence of curl **e**, the equation corresponding to (314) is

$$\frac{E}{H - f/\mu p} = \frac{s}{cp}\frac{J_{1r}}{J_{0r}}. \qquad \text{.........................(340)}$$

At the boundary $r = a$ both E and H are continuous; so, by taking $r = a$ in (340) and in the corresponding (314) with $k = 0$, and eliminating E_a or H_a between them, we obtain the equation of the other. We obtain

$$(\text{out}) \left\{ \begin{aligned} E &= \tfrac{1}{2}\pi a y^{-1} J_{1a}(J_{1r} - yG_{1r})f, \\ H &= \tfrac{1}{2}\pi a y^{-1} J_{1a}(J_{0r} - yG_{0r})(\mu v)^{-1} if, \end{aligned} \right\} \qquad \text{..............(341)}$$

in which y, as usual, is to be fixed by an external boundary condition, or, if the medium be boundless, $y = i$.

We see at once that $J_{1a} = 0$, with f simply-periodic, makes the external fluxes vanish. We should not now say that it makes the external field vanish, though the statement is true as regards H, because the electric force of the field does not vanish; it cancels the impressed force, so that there is no flux. This property is apparently independent of y. But, since there is no resistance concerned, except such as may be expressed in y, it is clear that (341), sinusoidally realized, cannot represent the state which is tended to after starting f, unless there be either no barrier, so that initial disturbances can escape, or else there be resistance somewhere, to be embodied in y, so that they can be absorbed, though only through an infinite series of passages between the boundary and the axis of the initial wave and its consequences.

Thus, with a conservative barrier producing $E = 0$ at $r = x$, and $y = J_{1x}/G_{1x}$, there is no escape for the initial effects, which remain in the form of free vibrations, whilst only the forced vibrations are got by taking $s^2 = +$ constant in (341). The other part of the solution must be separately calculated. If $J_{1x} = 0$, E and H run up infinitely. If $J_{1a} = 0$ also, the result is ambiguous.

With no barrier at all, or $y = i$, we have

$$(\text{out}) \left\{ \begin{aligned} E &= -(2a)^{-1} J_{1a}(G_{1r} + iJ_{1r}) f_0, \\ H &= (2a\mu v)^{-1} J_{1a}(J_{0r} - iG_{0r}) f_0, \end{aligned} \right\} \qquad \text{.................(342)}$$

which are fully realized. Here $f_0 = f\pi a^2$, which may be called the strength of the filament. We may most simply take the impressed force to be circular, its intensity varying as r within, and inversely as r outside the cylinder. Then $f = 2e_a/a$, if e_a is the intensity at $r = a$.

When nr/v is large, (342) becomes, by (308), writing $f_0 \sin nt$ for f_0,

$$(\text{out})\ E = \mu v H = \frac{f_0 n}{4v}\left(\frac{2v}{\pi nr}\right)^{\frac{1}{2}} \sin\left(nt - \frac{nr}{v} + \frac{\pi}{4}\right) \qquad \text{..........(343)}$$

approximately. $2\pi r$ should be a large multiple, and $2\pi a$ a small fraction of the wave-length along r.

Filament of curl **e**. *Calculation of Wave.*

62. In the last, let f_0 be constant, whilst a is made infinitely small. It is then a mere filament of curl of **e** at the axis that is in operation. We now have, by the second of (342), with $J_{1a} = \frac{1}{2}na/v$,

$$\frac{cn}{4}(J_{0r} - iG_{0r})f_0 = H = -(cp/4)(iJ_{0r} + G_{0r})f_0, \quad \text{............(344)}$$

which may be regarded as the simply-periodic solution or as the differential equation of H. In the latter case, put in terms of W by (311), then

$$H = (2\mu v)^{-1}(q/2\pi)^{\frac{1}{2}} W f_0 \,; \quad \text{........................(345)}$$

or, expanding by (309),

$$H = \frac{1}{2\mu v}\frac{1}{(2\pi r)^{\frac{1}{2}}}\epsilon^{-qr}\left(1 - \frac{1}{8qr} + \frac{1^2 3^2}{\underline{|2}(8qr)^2} - \dots\right)q^{\frac{1}{2}}f_0, \quad \text{........(346)}$$

in which f_0 may be any function of the time. Let it be zero before, and constant after $t = 0$. Then, first,

$$q^{\frac{1}{2}}f_0 = f_0(\pi vt)^{-\frac{1}{2}}. \quad \text{..........................(347)}$$

Next effect the integrations of this function indicated by the inverse powers of q or p/v, thus

$$\left(1 - \frac{1}{8qr} + \dots\right)(\pi vt)^{-\frac{1}{2}} = \left(1 - \tfrac{1}{2}\left(\frac{vt}{2r}\right) + \frac{1.3}{2^2\underline{|2}}\left(\frac{vt}{2r}\right)^2 - \dots\right)(\pi vt)^{-\frac{1}{2}}$$

$$= (1 + vt/2r)^{-\frac{1}{2}}(\pi vt)^{-\frac{1}{2}} = (2r/\pi)^{\frac{1}{2}}[vt(vt + 2r)]^{-\frac{1}{2}}. \quad \text{........(348)}$$

Lastly, operating on this by ϵ^{-qr} turns vt to $vt - r$, and brings (346) to

$$H = (f_0/2\pi\mu v)(v^2t^2 - r^2)^{-\frac{1}{2}}, \quad \text{....................(349)}$$

which is ridiculously simple. Let Z be the time-integral of H, then

$$Z = \frac{cf_0}{2\pi}\log\left[\frac{vt}{r} + \left(\frac{v^2t^2}{r^2} - 1\right)^{\frac{1}{2}}\right], \quad \text{................(350)}$$

from which we may derive E; thus

$$\text{curl}\,\mathbf{Z} = c\mathbf{E}, \quad\text{or}\quad E = -\frac{1}{c}\frac{dZ}{dr} = \frac{vtf_0}{2\pi r(v^2t^2 - r^2)^{\frac{1}{2}}}. \quad (351)$$

The other vector-potential **A**, such that $\mathbf{E} = -p\mathbf{A}$, is obviously

$$A = -\frac{1}{2\pi v}\left(\frac{v^2t^2}{r^2} - 1\right)^{\frac{1}{2}}f_0. \quad \text{....................(352)}$$

All these formulæ of course only commence when vt reaches r. The infinite values of E and H at the wave-front arise from the infinite concentration of the curl of **e** at the axis.

Notice that

$$E = Ht/rc \quad \text{............................(353)}$$

everywhere. It follows from this connexion between E and H (or from their full expressions) that

$$cE^2 - \mu H^2 = ce^2 = c(f_0/2\pi r)^2\,; \qquad \text{(354)}$$

where e denotes the intensity of impressed force at distance r, when it is of the simplest type, above described. That is, the excess of the electric over the magnetic energy at any point is independent of the time. Both decrease at an equal rate; the magnetic energy to zero, the electric energy to that of the final steady displacement $c\mathbf{e}/4\pi$.

62 A. The above E and H solutions are fundamental, because all electromagnetic disturbances due to impressed force depend solely upon, and come from, the lines of curl of the impressed force. From them, by integration, we can find the disturbances due to any collection of rectilinear filaments of $\mathbf{f}$. Thus, to find the H due to a plane sheet of parallel uniformly distributed filaments, of surface-density f, we have, by (349), at distance a from the plane, on either side,

$$H = \int \frac{f\,dy}{2\pi\mu v(v^2t^2a - {}^2 - y^2)^{\frac{1}{2}}} = \frac{f}{2\pi\mu v}\left[\sin^{-1}\frac{y}{(v^2t^2 - a^2)^{\frac{1}{2}}}\right],$$

where the limits are $\pm(v^2t^2 - a^2)^{\frac{1}{2}}$. Therefore

$$H = f/2\mu v$$

after the time $t = a/v$; before then, H is zero. [Compare with § 2, equation (12).]

62 B. Similarly, a cylindrical sheet of longitudinal $\mathbf{f}$ produces

$$H = \frac{fa}{2\pi\mu v}\int\frac{d\theta}{(v^2t^2 - b^2)^{\frac{1}{2}}}\,;$$

where b is the distance of the point where H is reckoned from the element $a\,d\theta$ of the circular section of the sheet, a being its radius. The limits have to be so chosen as to include all elements of f which have had time to produce any effect at the point in question. When the point is external and vt exceeds $a + r$ the limits are complete, viz. to include the whole circle. The result is then, at distance r from the axis of the cylinder,

$$H = \frac{fa/\mu v}{(v^2t^2 - a^2 - r^2)^{\frac{1}{2}}}\left[1 + \frac{1.3}{2^2\,\underline{|2}}\frac{x}{2} + \frac{1.3.5.7}{2^4\,\underline{|4}}\cdot\frac{x^2}{2^3}\cdot\frac{4.3}{1.2} + \frac{1.3.5.7.9.11}{2^6\,\underline{|6}}\cdot\frac{x^3}{2^5}\cdot\frac{6.5.4}{1.2.3} + \dots\right], \qquad \text{(355)}$$

where $$x = (2ar)^2(v^2t^2 - a^2 - r^2)^{-2}.$$

This formula begins to operate when $x = 1$, or $vt = a + r$. As time goes on, x falls to zero, leaving only the first term.

PART VI.

Cylindrical Surface of Circular curl $\mathbf{e}$ *in a Dielectric.*

63. Let the curl of the impressed electric force be wholly situated on the surface of a cylinder, of radius a, in a nonconducting dielectric. The

impressed force e to correspond may then be most conveniently imagined to be either longitudinal, within or without the cylinder, uniformly distributed in either case (though oppositely directed), and the density of curl e will be e; or, the impressed force may be transferred to the surface of the cylinder, by making e radial, but confined to an infinitely thin layer. The measure of the surface-density of curl e will now be

$$f=\frac{de}{dz}=E_{(\text{in})}-E_{(\text{out})}, \qquad (356)$$

where e is the total impressed force (its line-integral through the layer). The second form of this equation shows the effect produced on the electric force **E** of the flux, outside and inside the surface. This **E** is, as it happens, also the force of the field; but in the other case, when e is uniformly distributed within the cylinder, producing $f=e$, we have the same discontinuity produced by f.

H being circular, we use the operator (313). Applying it to (356), we obtain

$$f=\frac{s}{cp}\left(\frac{J_{0a}}{J_{1a}}-\frac{J_{0a}-yG_{0a}}{J_{1a}-yG_{1a}}\right)H_a; \qquad (357)$$

from which, by the conjugate property (307), and the operator (313), we derive

$$E_{(\text{in})} \text{ or }_{(\text{out})}=\frac{\pi as}{2y}\Big(J_{0r}(J_{1a}-yG_{1a}) \quad \text{or} \quad J_{1a}(J_{0r}-yG_{0r})\Big)f, \qquad (358)$$

$$H_{(\text{in})} \text{ or }_{(\text{out})}=\frac{\pi acp}{2y}\Big(J_{1r}(J_{1a}-yG_{1a}) \quad \text{or} \quad J_{1a}(J_{1r}-yG_{1r})\Big)f, \qquad (359)$$

in which f is a function of t, and it may be also of z. If so, then we have the radial component F of electric force given by

$$F_{(\text{in})} \text{ or }_{(\text{out})}=-\frac{\pi a}{2y}\Big(J_{1r}(J_{1a}-yG_{1a}) \quad \text{or} \quad J_{1a}(J_{1r}-yG_{1r})\Big)\frac{df}{dz}. \qquad (360)$$

From these, by the use of Fourier's theorem, we can build up the complete solutions for any distribution of f with respect to z; for instance, the case of a single circular line of curl e.

$J_{1a}=0$. *Vanishing of External Field.*

64. Let f be simply-periodic with respect to t and z; then $J_{1a}=0$, or

$$J_1\{a\sqrt{n^2/v^2-m^2}\}=0, \qquad (361)$$

produces evanescence of E and H outside the cylinder. The independence of this property of y really requires an unbounded external medium, or else boundary-resistance, to let the initial effects escape or be dissipated, because no resistance appears in our equations except in y. The case $s=0$ or $n=mv$ is to be excepted from (361); it is treated later.

y = i. *Unbounded Medium.*

65. When $n/v > m$, s is real, and our equations give at once the fully realized solutions in the case of no boundary, by taking $y = i$,

$$\left.\begin{aligned} H_{(\text{in})} \text{ or }_{(\text{out})} &= \tfrac{1}{2}\pi acn\Big(J_{1r}(J_{1a} - iG_{1a}) \quad \text{or} \quad J_{1a}(J_{1r} - iG_{1r})\Big)f, \\ E_{(\text{in})} \text{ or }_{(\text{out})} &= -\tfrac{1}{2}\pi as\Big(J_{0r}(G_{1a} + iJ_{1a}) \quad \text{or} \quad J_{1a}(G_{0r} + iJ_{0r})\Big)f, \\ F_{(\text{in})} \text{ or }_{(\text{out})} &= \tfrac{1}{2}\pi a\Big(J_{1r}(G_{1a} + iJ_{1a}) \quad \text{or} \quad J_{1a}(G_{1r} + iJ_{1r})\Big)(df/dz), \end{aligned}\right\} \quad (362)$$

in which i means p/n.

The instantaneous outward transfer of energy per unit length of cylinder is (by Poynting's formula)

$$-EH/4\pi \times 2\pi r,$$

and the mean value with respect to the time comes to

$$(cn/8\pi)(f_0 \pi a \cos mz\, J_{1a})^2, \quad \text{.....................}(363)$$

if f_0 is the maximum value of f, [thus, $f = f_0 \cos mz \sin nt$]. This may of course be again averaged to get rid of the cosine.

s = 0. *Vanishing of External* E.

66. When $n = mv$, we make $s = 0$, and then (362) reduce to the singular solution

$$\left.\begin{aligned} H_{(\text{in})} &= \tfrac{1}{2}rcpf, & E_{(\text{in})} &= f, & F_{(\text{in})} &= -\tfrac{1}{2}r\frac{df}{dz}, \\ H_{(\text{out})} &= \tfrac{1}{2}\cdot\frac{a^2}{r}\cdot cpf, & E_{(\text{out})} &= 0, & F_{(\text{out})} &= -\tfrac{1}{2}\cdot\frac{a^2}{r}\cdot\frac{df}{dz}. \end{aligned}\right\} \text{....}(364)$$

Observe that the internal longitudinal displacement is produced entirely by the impressed force (*if* it be internal), though there is radial displacement also, on account of the divergence of e (if internal). Outside the cylinder, the displacement is entirely perpendicular to it.

H and F do not alternate along r. This is also true when s^2 is negative, or n lies between 0 and mv. Then, q^2 being positive, we have

$$E_{(\text{out})} = \tfrac{1}{2}a^2q^2\Big(\frac{2}{sa}J_{1a}\Big)\Big(J_{0r}(\log qr - \beta) + L_{0r}\Big)f, \quad \text{............}(365)$$

as the rational form of the equation of the external E when the frequency is too low to produce fluctuations along r.

The system (364) may be obtained directly from (358) to (360) on the assumption that s/y is zero when s is zero. But (364) appears to require an unbounded medium. Even in the case of the boundary condition $E = 0$ at $r = x$, which harmonizes with the vanishing of E externally in (364), there will be the undissipated initial effects continuing.

If, on the other hand, $H_x = 0$, making $y = J_{1x}/G_{1x}$, we shall not only have the undissipated initial effects, but a different form of solution for the forced vibrations. Thus, using this expression for y, and also $s=0$, in (358) to (360), we obtain

$$\left.\begin{aligned} H_{(\text{in})} &= \frac{a}{2}\left(1-\frac{a^2}{x^2}\right)\frac{r}{a}cpf, & E_{(\text{in})} &= \left(1-\frac{a^2}{x^2}\right)f; \\ H_{(\text{out})} &= \frac{a}{2}\left(1-\frac{r^2}{x^2}\right)\frac{a}{r}cpf, & E_{(\text{out})} &= -\frac{a^2}{x^2}f; \end{aligned}\right\}\qquad(366)$$

representing the forced vibrations.

Effect of suddenly Starting a Filament of **e**.

67. The vibratory effects due to a vibrating filament we find by taking a infinitely small in (362), that is $J_{1a}=\frac{1}{2}sa$. To find the wave produced by suddenly starting such a filament, transform equations (358), (359) by means of (311). We get [e being intensity of longitudinal **e**]

$$\left.\begin{aligned} E_{(\text{in})} &= -(\pi/2q)^{\frac{1}{2}}aJ_{0r}W'_a e, & H_{(\text{in})} &= -(\pi q/2)^{\frac{1}{2}}\frac{ar}{2\mu v}\left(\frac{2}{sr}J_{1r}\right)W'_a e, \\ E_{(\text{out})} &= -\tfrac{1}{2}(\pi q^3/2)^{\frac{1}{2}}a^2\left(\frac{2}{sa}J_{1a}\right)We, & H_{(\text{out})} &= -(\pi q/2)^{\frac{1}{2}}\frac{a^2}{2\mu v}\left(\frac{2}{sa}J_{1a}\right)W'e; \end{aligned}\right\}\quad(367)$$

where W is given by (309); the accent means differentiation to r, and the suffix a means the value at $r=a$.

In these, let $e_0=\pi a^2 e$, which we may call the strength of the filament, and let a be infinitely small. We then obtain

$$(\text{out})\quad H=-(q/2\pi)^{\frac{1}{2}}(2\mu v)^{-1}W'e_0,\qquad E=-\tfrac{1}{2}(q^3/2\pi)^{\frac{1}{2}}We_0.\qquad(368)$$

Now if e_0 is a function of t only, it is clear that there is no scalar electric potential involved. We may therefore advantageously employ (and for a reason to be presently seen) the vector-potential **A**, such that

$$E=-pA,\qquad\text{or}\qquad A=-p^{-1}E;\qquad\text{and}\qquad \mu H=-\frac{dA}{dr}\qquad(369)$$

The equation of A is obviously, by the first of (369) applied to second of (368),

$$A=\tfrac{1}{2}(p/2\pi v^3)^{\frac{1}{2}}We_0.\quad\ldots\ldots(370)$$

Comparing this equation with that of H in (345) (problem of a filament of curl of **e**), we see that f_0 there becomes e_0 here, and μH there becomes A here. The solution of (370) may therefore be got at once from the solution of (345), viz. (349). Thus

$$A=\frac{e_0}{2\pi v(v^2t^2-r^2)^{\frac{1}{2}}};\quad\ldots\ldots(371)$$

from which, by (369),

$$E=\frac{e_0vt}{2\pi(v^2t^2-r^2)^{3/2}},\qquad H=-\frac{e_0r}{2\pi\mu v(v^2t^2-r^2)^{\frac{3}{2}}},\quad\ldots\ldots(372)$$

the complete solution. It will be seen that

$$A=Et+r\mu H,\quad\ldots\ldots(373)$$

whilst the curious relation (353) in the problem of a filament of curl e is now replaced by

$$A = r\mu Z/t, \qquad \text{.............................(374)}$$

where Z is the time-integral of the magnetic force; so that

$$H = pZ, \qquad \text{and} \qquad \text{curl}\,\mathbf{Z} = c\mathbf{E}, \qquad \text{...............(375)}$$

Z being merely the vectorised Z. It is the vector-potential of the magnetic current.

The following reciprocal relation is easily seen by comparing the differential equations of an infinitely fine filament e_0 and a finite filament. The electric current-density at the axis due to a longitudinal cylinder of e (uniform) of radius a is numerically identical with the total current through the circle of radius a due to the same total impressed force (that is, $\pi a^2\mathbf{e}$) concentrated in a filament at the axis, at corresponding moments.

68. Having got the solutions (372) for a filament e_0, it might appear that we could employ them to build up the solutions in the case of, for instance, a cylinder of longitudinal impressed force of finite radius a. But, according to (372), E would be positive and H negative everywhere and at every moment, in the case of the cylinder, because the elementary parts are all positive or all negative. This is clearly a wrong result. For it is certain that, at the first moment of starting the longitudinal impressed force of intensity e in the cylinder, E just outside it is negative; thus

$$E = \pm \tfrac{1}{2}e, \quad \text{in or out,} \quad \text{at} \quad r = a, \quad t = 0;$$

and that H is positive; viz.

$$H = e/2\mu v \quad \text{at} \quad r = a, \quad t = 0.$$

We know further that, as E starts negatively just outside the cylinder, E will be always negative at the front of the outward wave, and H positive; thus

$$-E = \mu v H = \tfrac{1}{2}e \times (a/r)^{\frac{1}{2}}, \qquad \text{......................(376)}$$

the variation in intensity inversely as the square root of the distance from the axis being necessitated in order to keep the energy constant at the wave-front. The same formula with $+E$ instead of $-E$ will express the state at the front of the wave running in to the axis. There is thus a momentary infinity of E at the axis, viz., when $t = a/v$.

So far we can certainly go. Less securely, we may conclude that during the recoil, E will be settling down to its steady value e within the cylinder, and therefore the force of the field there will be positive, and, by continuity, also positive outside the cylinder. Similarly, H must be negative at any distance within which E is decreasing. We conclude, therefore, that the filament-solutions (372) only express the settling down to the final state, and are not comprehensive enough to be employed as fundamental solutions.

Sudden Starting of e *longitudinal in a Cylinder.*

69. In order to fully clear up what is left doubtful in the last paragraph, I have investigated the case of a cylinder of e comprehensively.

The following contains the leading points. We have to make four independent investigations: viz., to find (1), the initial inward wave; (2), the initial outward wave; (3), the inside solution after the recoil; (4), the outside solution ditto. We may indeed express the whole by a definite integral, but there does not seem to be much use in doing so, as there will be all the labour of finding out *its* solutions, and they are what we now obtain from the differential equations.

Let E_1 and E_2 be the E's of the inward and outward waves. Their equations are

$$E_1 = -(a/2q)W_a' Ue, \quad \text{.....................(377)}$$

$$E_2 = -(a/2q)WU_a'e; \quad \text{.....................(378)}$$

where U and W are given by (309), the accent means differentiation to r, and the suffix indicates the value at $r=a$. To prove these, it is sufficient to observe that U and W involve ϵ^{qr} and ϵ^{-qr} respectively, so that (377) expresses an inward and (378) an outward wave; and further that, by (310), we have

$$E_1 - E_2 = e \quad \text{at} \quad r=a, \quad \text{always}; \quad \text{..............(379)}$$

which is the sole boundary condition at the surface of curl of e.

Expanding (377), we get

$$E_1 = \tfrac{1}{2}\left(\frac{a}{r}\right)^{\frac{1}{2}}\epsilon^{q(r-a)}(R+S)\left[1+\frac{3}{y}-\frac{3.5}{\underline{|2}y^2}+\frac{3^2.5.7}{\underline{|3}y^3}-\frac{3^2.5^2.7.9}{\underline{|4}y^4}+\dots\right]e, \quad (380)$$

where $R+S$ is given by (309), and $y=8qa$. Now, e being zero before and constant after $t=0$, effect the integrations indicated by the inverse powers of p, and then turn t to t_1, where

$$vt_1 = vt + r - a.$$

The result is

$$E_1 = \tfrac{1}{2}e\left(\frac{a}{r}\right)^{\frac{1}{2}}\left[1+3z_1-\frac{3.5}{\underline{|2}\,\underline{|2}}z_1^2+\frac{1^2.3^2.5.7}{\underline{|3}\,\underline{|3}}z_1^3-\frac{1^2.3^2.5^2.7.9}{\underline{|4}\,\underline{|4}}z_1^4+\dots\right.$$
$$+\frac{a}{r}z_1\left(1+\frac{3}{\underline{|2}}z_1-\frac{3.5}{\underline{|2}\,\underline{|3}}z_1^2+\frac{1^2.3^2.5.7}{\underline{|3}\,\underline{|4}}z_1^3-\dots\right)$$
$$\left.+\frac{1^2 3^2}{\underline{|2}}\frac{a^2}{r^2}z_1^2\left(\frac{1}{\underline{|2}}+\frac{3}{\underline{|3}}z_1-\frac{3.5}{\underline{|2}\,\underline{|4}}z_1^2+\frac{1^2.3^2.5.7}{\underline{|3}\,\underline{|5}}z_1^3-\dots\right)+\dots\right], \quad \text{.......(381)}$$

the structure of which is sufficiently clear. Here $z_1 = vt_1/8a$.

This formula, when $vt < a$, holds between $r=a$ and $r=a-vt$. But when $vt > a$ though $< 2a$, it holds between $r=a$ and $vt-a$. Except within the limits named, it is only a partial solution.

70. As regards E_2, it may be obtained from (381) by the following changes. Change E_1 to $-E_2$ on the left, and on the right change z_1 to $-z_2$, where

$$z_2 = (vt + a - r)/8a.$$

It is therefore unnecessary to write out E_2. This E_2 formula will hold from $r=a$ to $r=vt+a$, when $vt < 2a$; but after that, when the front of the return wave has passed $r=a$, it will only hold between $r=vt-a$ and $vt+a$.

71. Next to find E_3, the E in the cylinder when $vt > a$ and the solution is made up of two oppositely going waves, and E_4 the external E after $vt = 2a$, when it is made up of two outward going waves. I have utterly failed to obtain intelligible results by uniting the primary waves with a reflected wave. But there is another method which is easier, and free from the obscurity which attends the simultaneous use of U and W Thus, the equations of E_3 and E_4 are

$$E_3 = -(\pi/2q)^{\frac{1}{2}} a J_{0r} W'_a e, \qquad (382)$$

$$E_4 = -(\pi q^3/2)^{\frac{1}{2}} \tfrac{1}{2} a^2 \left(\frac{2}{sa} J_{1a}\right) W.e, \qquad (383)$$

by (367); and a necessity of their validity is the presence of two waves inside the cylinder, because of the use of J_0 and J_1; it is quite inadmissible to use J_0 when only one wave is in question, because $J_{0r} = 1$ when $r = 0$, and being a differential operator in rising powers of p, the meaning of (382) is that we find E_3 at r by differentiations from E_3 at $r = 0$; thus (382) only begins to be valid when $vt = a$.

To integrate (382), (383), it saves a little trouble to calculate the time-integrals of E_3 and E_4, say

$$A_3 = -p^{-1} E_3, \qquad A_4 = -p^{-1} E_4. \qquad (384)$$

The results are

$$-A_3 = J_{0r} \cdot \frac{e}{v} (v^2t^2 - a^2)^{\frac{1}{2}}, \qquad (385)$$

$$A_4 = \left(\frac{2}{sa} J_{1a}\right) \frac{a^2 e}{2v} (v^2t^2 - r^2)^{-\frac{1}{2}}. \qquad (386)$$

From these derive E_3 and E_4 by time-differentiation, and H_3, H_4 by space-differentiation, according to

$$\operatorname{curl} \mathbf{A} = \mu \mathbf{H}, \qquad \text{or} \qquad H = -\frac{1}{\mu}\frac{dA}{dr}. \qquad (387)$$

We see that the value of E_3 at the axis, say E_0, is

$$E_0 = evt(v^2t^2 - a^2)^{-\frac{1}{2}}; \qquad (388)$$

and by performing the operation J_{0r} in (385) we produce, if $u = (v^2t^2 - a^2)^{\frac{1}{2}}$,

$$-A_3 = \frac{e}{v}\Big[u + \frac{r^2}{2^2}\left(\frac{1}{u} - \frac{v^2t^2}{u^3}\right) + \frac{3r^4}{2^2 4^2}\left(-\frac{1}{u^3} + \frac{6v^2t^2}{u^5} - \frac{5v^4t^4}{u^7}\right)$$
$$+ \frac{45r^6}{2^2 4^2 6^2}\left(\frac{1}{u^5} - \frac{15v^2t^2}{u^7} + \frac{35v^4t^4}{u^9} - \frac{21v^6t^6}{u^{11}}\right) + \dots\Big]; \qquad (389)$$

from which we derive

$$E_3 = \frac{evt}{u}\Big[1 + \frac{3a^2r^2}{4u^4} + \frac{15a^2r^4}{2^2 4^2 u^8}(3a^2 + 4v^2t^2)$$
$$+ \frac{5.7.9a^2r^6}{2^2.4^2.6^2u^{12}}(5a^4 + 20v^2t^2a^2 + 8v^4t^4) + \dots\Big]. \qquad (390)$$

These formulæ commence to operate when $vt = a$ at the axis, and when $vt = a + r$ at any point $r < a$, and continue in operation for ever after.

72. Lastly, perform the operation $(2/sa)J_{1a}$ in (386), and we obtain

$$A_4=\frac{a^2e}{2v}\Big[\frac{1}{u}+\frac{a^2}{8}\Big(-\frac{1}{u^3}+\frac{3v^2t^2}{u^5}\Big)+\frac{a^4}{64}\Big(\frac{3}{u^5}-\frac{30v^2t^2}{u^7}+\frac{35v^4t^4}{u^9}\Big)$$
$$+\frac{45a^6}{4.36.64}\Big(-\frac{5}{u^7}+\frac{135v^2t^2}{u^9}-\frac{315v^4t^4}{u^{11}}+\frac{231v^6t^6}{u^{13}}\Big)+\dots\Big], \quad (391)$$

from which we derive

$$E_4=\frac{a^2evt}{2u^3}\Big[1+\frac{3a^2}{8u^4}(2v^2t^2+3r^2)+\frac{5a^4}{64u^8}(8v^4t^4+40v^2t^2r^2+15r^4)$$
$$+\frac{45a^6}{4.36.64u^{12}}(112v^6t^6+1176v^4t^4r^2+1470v^2t^2r^4+245r^6)+\dots\Big]. \quad (392)$$

These begin to operate at $r=a$ when $vt=2a$; and later, the range is from $r=a$ to $r=vt-a$.

This completes the mathematical work. As a check upon the accuracy, we may test satisfaction of differential equations, and of the initial condition, and that the four solutions join together with the proper discontinuities.

73. The following is a general description of the manner of establishing the steady flux. We put on e in the cylinder when $t=0$. The first effect inside is $E_1=\frac{1}{2}e$ at the surface, and $H_1=E_1/\mu v$. This primary disturbance runs in to the axis at speed v, varying at its front inversely as the square root of the distance from the axis, thus producing a momentary infinity there. At this moment $t=a/v$, E_1 is also very great near the axis. In the meantime, E_1 has been increasing generally all over the cylinder, so that, from being $\frac{1}{2}e$ initially at the boundary, it has risen to $\cdot 77\,e$, whilst the simultaneous value at $r=\frac{1}{2}a$ is about $\cdot 95\,e$.

Now consider E_3 within the cylinder, it being the natural continuation of E_1. The large values of E_1 near the axis subside with immense rapidity. But near the boundary E_1 still goes on increasing. The result is that when $vt=2a$, and the front of the return-wave reaches the boundary, E_3 has fallen from ∞ to $1\cdot 154\,e$ at the axis; at $r=\frac{1}{2}a$ the value is $1\cdot 183\,e$; at $r=\frac{3}{4}a$ it is $1\cdot 237\,e$; and at the boundary the value has risen to $1\cdot 71\,e$, which is made up thus, $1\cdot 21\,e+\frac{1}{2}e$; the first of these being the value just before the front of the return-wave arrives, the second part the sudden increase due to the wave-front. E_3 is now a minimum at the axis and rises towards the wave-front, the greater part of the rise being near the wave-front.

Thirdly, go back to $t=0$ and consider the outward wave. First, $E_2=-\frac{1}{2}e$ at $r=a$. This runs out at speed v, varying at the front inversely as $r^{\frac{1}{2}}$. As it does so, the E_2 that succeeds rises, that is, is less negative. Thus when $vt=a$, and the front has got to $r=2a$, the values of E_2 are $-\cdot 232\,e$ at $r=a$ and $-\cdot 353\,e$ at $r=2a$. Still later, as this wave forms fully, its hinder part becomes positive. Thus, when fully formed, with front at $r=3a$, we have $E_2=-\cdot 288\,e$ at $r=3a$; $-\cdot 145\,e$ at $r=2a$; and $\cdot 21\,e$ at $r=a$. This is at the moment when the return-wave reaches the boundary, as already described.

The subsequent history is that the wave E_2 moves out to infinity, being negative at its front and positive at its back, where there is a sudden rise due to the return-wave E_4, behind which there is a rapid fall in E_4, not a discontinuity, but the continuation of the before-mentioned rapid fall in E_3 near its front. The subsidence to the steady state in the cylinder and outside is very rapid when the front of E_4 has moved well out. Thus, when $vt=5a$, we have $E_3=1{\cdot}022\,e$ at $r=a$, and of course, just outside, we have $E_4={\cdot}022\,e$; and when $vt=10a$, we have $E_3=1{\cdot}005\,e$, $E_4={\cdot}005\,e$, at $r=a$.

As regards H, starting when $t=0$ with the value $e/2\mu v$ at $r=a$ only, at the front of the inward or outward wave it is $E=\pm\mu vH$, as usual. It is positive in the cylinder at first, and then changes to negative. Outside, it is first positive for a short time, and then negative for ever after.

74. We can now see fully why the solution for a filament e_0 of $\mathbf{e}$ can *not* be employed to build up more complex solutions in general, whilst that for a filament f_0 of curl $\mathbf{e}$ *can* be so employed. For, in the latter case, the disturbances come, *ab initio*, from the axis, because the lines of curl $\mathbf{e}$ are the sources of disturbance, and they become a single line at the axis. But in the former case it is not the body of the filament, but its surface only, that is the real source, however small the filament may be, producing first $\mathbf{E}$ negative (or against $\mathbf{e}$) just outside the filament, and, immediately after, $\mathbf{E}$ positive. Now when the diameter of the filament is indefinitely reduced, we lose sight altogether of the preliminary negative electric and positive magnetic force, because their duration becomes infinitely small, and our solutions (372) show only the subsequent state of positive electric and negative magnetic force during the settling down to the final state, but not its real commencement, viz., at the front of the wave.

75. The occurrence of momentary infinite values of $\mathbf{E}$ or of $\mathbf{H}$, in problems concerning spherical and cylindrical electromagnetic waves, is physically suggestive. By means of a proper convergence to a point or an axis, we should be able to disrupt the strongest dielectric, starting with a weak field, and then discharging it. Although it is impossible to realize the particular arrangements of our solutions, yet it might be practicable to obtain similar results in other ways.*

It may be remarked that the solution worked out for an infinitely

* If we wish the solution for an infinitely long cylinder to be quite unaltered, when of finite length l, let at $z=0$ and $z=l$ infinitely conducting barriers be placed. Owing to the displacement terminating upon them perpendicularly, and the magnetic force being tangential, no alteration is required. Then, on taking off the impressed force, we obtain the result of the discharge of a condenser consisting of two parallel plates of no resistance, charged in a certain portion only; or, by integration, charged in any manner.

To abolish the momentary infinity at the axis, in the text, substitute for the surface distribution of curl of $\mathbf{e}$ a distribution in a thin layer. The infinity will be replaced by a large finite value, without other material change. Of course the theory above assumes that the dielectric does not break down. If it does, we change the problem, and have a conducting (or resisting) path, possibly with oscillations of great frequency if the resistance be not too great, as Prof. Lodge believes to be the case in a lightning discharge.

long cylinder of longitudinal **e** is also, to a certain extent, the solution for a cylinder of finite length. If, for instance, the length is $2l$, and the radius a, disturbances from the extreme terminal lines of **f** (or curl **e**) only reach the centre of the axis after the time $(a^2+l^2)^{\frac{1}{2}}/v$, whilst from the equatorial line of **f** the time taken is a/v, which may be only a little less, or very greatly less, according as l/a is small or large. If large, it is clear that the solutions for **E** and **H** in the central parts of the cylinder are not only identical with those for an infinitely long cylinder until disturbances arrive from its ends, but are not much different afterwards.

Cylindrical Surface of Longitudinal **f**, *a Function of* θ *and* t.

76. When there is no variation with θ, the only Bessel functions concerned are J_0 and J_1. The extension of the vibratory solutions to include variation of the impressed force or its curl as $\cos\theta$, $\cos 2\theta$, etc., is so easily made that it would be inexcusable to overlook it. Two leading cases will be very briefly considered. Let the curl of the impressed force be wholly upon the surface of a cylinder of radius a, longitudinally directed, and be a function of t and θ, its tensor being f, the measure of the surface-density. **H** is also longitudinal, of course, whilst **E** has two components, circular E and radial F. The connections are

$$-\frac{dH}{dr}=cpE, \qquad \frac{1}{r}\frac{dH}{d\theta}=cpF, \qquad \frac{1}{r}\frac{d}{dr}rE-\frac{1}{r}\frac{dF}{d\theta}=-\mu pH, \qquad (393)$$

from which the characteristic of H is

$$\frac{1}{r}\frac{d}{dr}r\frac{dH}{dr}+\left(s^2-\frac{m^2}{r^2}\right)H=0, \qquad \text{.....................(394)}$$

if $s^2=-p^2/v^2$ and $m^2=-d^2/d\theta^2$. Consequently

$$H=(J_{mr}-yG_{mr})\cos m\theta\times\text{function of }t \quad \text{.............(395)}$$

when m^2 is constant, and the E/H operator is

$$\frac{E}{H}=-\frac{1}{cp}\frac{J'_{mr}-yG'_{mr}}{J_{mr}-yG_{mr}}, \qquad \text{.........................(396)}$$

if J_{mr} or $J_m(sr)$ is the mth Bessel-function, and G_{mr} its companion, whilst the $'$ means d/dr.

The boundary condition is

$$E_1=E_2-f \qquad \text{at} \qquad r=a, \qquad \text{...................(397)}$$

E_1 being the inside, E_2 the outside value of the force of the flux. Therefore, using (396) with $y=0$ inside, we obtain

$$H_a=\frac{J_{ma}(J_{ma}-yG_{ma})}{y(J_{ma}G'_{ma}-J'_{ma}G_{ma})}cpf=\frac{axcp}{y}J_{ma}(J_{ma}-yG_{ma})f, \quad \text{........(398)}$$

where x is a constant, being $\pi/2$ when $m=0$, according to (307), and always $\pi/2$ if G_m has the proper numerical factor to fix its size.

We see that if

$$f=f_0\cos m\theta\cos nt,$$

where f_0 is constant, the boundary H, and with it the whole external field, electric and magnetic, vanishes when

$$J_{ma} = 0.$$

If $m = 0$, or there is no variation with θ, the impressed force may be circular, outside the cylinder, and varying as r^{-1}.

If $m = 1$, the impressed force may be transverse, within the cylinder, and of uniform intensity.

Conducting Tube. **e** *Circular, a Function of θ and* **t**.

77. This is merely chosen as the easiest extension of the last case. In it let there be two cylindrical surfaces of **f**, infinitely close together. They will cancel one another if equal and opposite, but if we fill up the space between them with a tube of conductance K per unit area, we get the case of **e** circular in the tube, **e** varying with θ and t, and produce a discontinuity in **H** (which is still longitudinal, of course). Let E_a be the common value of E just outside and inside the tube; $e + E_a$ is then the force of the flux in the substance of the tube, and

$$H_1 - H_2 = 4\pi K(e + E_a), \qquad \text{.......................(399)}$$

the discontinuity equation, leads, by the use of (396) and the conjugate property of J_m and G_m as standardized* in the last paragraph, through

$$\left(\frac{H_1}{E_1} - \frac{H_2}{E_2} - 4\pi K\right) E_a = 4\pi K e,$$

to the equation of E_a, viz.,

$$E_a = \frac{4\pi K e}{-4\pi K + \frac{2cpy}{\pi a}[J'_{ma}(J'_{ma} - yG'_{ma})]^{-1}}, \qquad \text{..............(400)}$$

from which we see that it is $J'_{ma} = 0$ that now makes the external field vanish.

78. This concludes my treatment of electromagnetic waves in relation to their sources, so far as a systematic arrangement and uniform method is concerned. Some cases of a more mixed character must be reserved. It is scarcely necessary to remark that all the dielectric solutions may be turned into others, by employing impressed magnetic instead of electric force. The hypothetical magnetic conductor is required to obtain full analogues of problems in which electric conductors occur.

August 10, 1888.

* [If we take Stokes's formula for J_m, thus

$$J_m(z) = \left(\frac{2}{\pi z}\right)^{\frac{1}{2}} \left[\left\{ 1 - \frac{(1^2 - 4m^2)(3^2 - 4m^2)}{\underline{|2}(8z)^2} + \dots \right\} \cos + \left\{ \frac{1^2 - 4m^2}{\underline{|1}(8z)} - \dots \right\} \sin \right] \left(z - \frac{\pi}{4} - \frac{n\pi}{\cdot 2} \right),$$

then the substitution of sin for cos and − cos for sin will give the G_m function standardized as in the text. Also note that the infiniteness of G_0 when β is omitted, referred to in footnote p. 445, arises when q^2 is +].

XLIV. THE GENERAL SOLUTION OF MAXWELL'S ELECTROMAGNETIC EQUATIONS IN A HOMOGENEOUS ISOTROPIC MEDIUM, ESPECIALLY IN REGARD TO THE DERIVATION OF SPECIAL SOLUTIONS, AND THE FORMULÆ FOR PLANE WAVES.

[*Phil. Mag.*, Jan. 1889, p. 30.]

Equations of the Field.

1. Although, from the difficulty of applying them to practical problems, general solutions frequently possess little practical value, yet they may be of sufficient importance to render their investigation desirable, and to let their applications be examined as far as may be practicable. The first question here to be answered is this. Given the state of the whole electromagnetic field at a certain moment, in a homogeneous isotropic conducting dielectric medium, to deduce the state at any later time, arising from the initial state alone, without impressed forces.

The equations of the field are, if p stand for d/dt,

$$\text{curl}\,\mathbf{H} = (4\pi k + cp)\mathbf{E}, \qquad (1)$$

$$-\,\text{curl}\,\mathbf{E} = (4\pi g + \mu p)\mathbf{H}\,; \qquad (2)$$

the first being Maxwell's well-known equation defining electric current in terms of the magnetic force $\mathbf{H}$, k being the electric conductivity and $c/4\pi$ the electric permittivity (or permittance of a unit cube condenser), and $\mathbf{E}$ the electric force; whilst the second is the equation introduced by me* as the proper companion to the former to make a complete system suitable for practical working, g being the magnetic conductivity and μ the magnetic inductivity. This second equation takes the place of the two equations

$$\mathbf{E} = -\dot{\mathbf{A}} - \nabla\Psi, \qquad \text{curl}\,\mathbf{A} = \mu\mathbf{H}, \qquad (3)$$

of Maxwell, where $\mathbf{A}$ is the electromagnetic momentum at a point, and Ψ the scalar electric potential. Thus Ψ and $\mathbf{A}$ are murdered, so to speak, with a great gain in definiteness and conciseness. As regards g, however, standing for a physically non-existent quality, such that the medium cannot support magnetic force without a dissipation of energy at the rate $g\mathbf{H}^2$ per unit volume, it is only retained for the sake of mathematical completeness, and on account of the singular telegraphic application in which electric conductivity is made to perform the functions of both the real k and the unreal g.

Let

$$\left.\begin{matrix} \rho_1 = 4\pi k/2c, & \rho = \rho_1 + \rho_2, & \\ \rho_2 = 4\pi g/2\mu, & \sigma = \rho_1 - \rho_2, & v = (\mu c)^{-\frac{1}{2}}. \end{matrix}\right\} \qquad (4)$$

The speed of propagation of all disturbances is v, and the attenuating effects due to the two conductivities depend upon ρ_1 and ρ_2, whilst σ determines the distortion due to conductivity.

* "Electromagnetic Induction and its Propagation," *The Electrician*, January 3, 1885, and later [vol. I., p. 449.]

General Solutions.

2. Let q^2 denote the operator

$$q^2 = -(v\,\text{curl})^2 + \sigma^2; \qquad \text{.........................(5)}$$

or, in full, when operating upon **E** for example,

$$q^2\mathbf{E} = v^2\nabla^2\mathbf{E} - v^2\nabla\,\text{div}\,\mathbf{E} + \sigma^2\mathbf{E}. \qquad \text{...................(6)}$$

Now it may be easily found by ordinary "symbolical" work which it is not necessary to give, that, given $\mathbf{E}_0$, $\mathbf{H}_0$, the values of **E** and **H** when $t=0$, and satisfying (1) and (2), those at time t later are given by

$$\left.\begin{aligned} \mathbf{E} &= \epsilon^{-\rho t}\left[\left(\cosh qt - \frac{\sigma}{q}\sinh qt\right)\mathbf{E}_0 + \frac{\sinh qt}{q}\cdot\frac{\text{curl}\,\mathbf{H}_0}{c}\right], \\ \mathbf{H} &= \epsilon^{-\rho t}\left[\left(\cosh qt + \frac{\sigma}{q}\sinh qt\right)\mathbf{H}_0 - \frac{\sinh qt}{q}\cdot\frac{\text{curl}\,\mathbf{E}_0}{\mu}\right]. \end{aligned}\right\} \text{..........(7)}$$

A sufficient proof is the satisfaction of the equations (1), (2), and of the two initial conditions.

An alternative form of (7) is

$$\left.\begin{aligned} \mathbf{E} &= \epsilon^{-\rho t}\left(\cosh qt + \frac{\sinh qt}{q}(p+\rho)\right)\mathbf{E}_0, \\ \mathbf{H} &= \epsilon^{-\rho t}\left(\cosh qt + \frac{\sinh qt}{q}(p+\rho)\right)\mathbf{H}_0, \end{aligned}\right\} \text{..................(7a)}$$

showing the derivation of **E** from $\mathbf{E}_0$ and $p\mathbf{E}_0$ in precisely the same way as **H** from $\mathbf{H}_0$ and $p\mathbf{H}_0$. In this form of solution the initial values of $p\mathbf{E}_0$ and $p\mathbf{H}_0$ occur. But they are not arbitrary, being connected by equations (1), (2). The form (7) is much more convenient, involving only $\mathbf{E}_0$ and $\mathbf{H}_0$ as functions of position, although (7*a*) looks simpler. The form (7) is also the more useful for interpretations and derivations.

If, then, $\mathbf{E}_0$ and $\mathbf{H}_0$ be given as continuous functions admitting of the performance of the differentiations involved in the functions of q^2, (7) will give the required solutions. The original field should therefore be a real one, not involving discontinuities. We shall now consider special cases.

Persistence or Subsidence of Polar Fields.

3. We see immediately by (7) that the **E** resulting from $\mathbf{H}_0$ depends solely upon its curl, or on the initial electric current, and, similarly, that the **H** due to $\mathbf{E}_0$ depends solely upon its curl, or on the magnetic current. Notice also that the displacement due to $\mathbf{H}_0$ is related to $\mathbf{H}_0$ in the same way as the induction $\div -4\pi$ due to $\mathbf{E}_0$ is related to $\mathbf{E}_0$. Or, if it be the electric and magnetic currents that are considered, the displacement due to electric current is related to it in the same way as the induction $\div 4\pi$ due to magnetic current is related to it.

Observe, also, that in passing from the **E** due to $\mathbf{E}_0$ to the **H** due to $\mathbf{H}_0$, the sign of σ is changed.

By (7), a distribution of $\mathbf{H}_0$ which has no curl, or a polar magnetic field, does not, in subsiding, generate electric force; and, similarly, a polar electric field does not, in subsiding, generate magnetic force. Let then $\mathbf{E}_0$ and $\mathbf{H}_0$ be polar fields, in the first place. Then, by (5),

$$q^2 = \sigma^2,$$

that is, a constant; and, using this in (7), we reduce the general solutions to

$$\mathbf{E} = \mathbf{E}_0 \epsilon^{-2\rho_1 t}, \qquad \mathbf{H} = \mathbf{H}_0 \epsilon^{-2\rho_2 t}. \quad \text{..................(8)}$$

The subsidence of the electric field requires electric conductivity, that of the magnetic field requires magnetic conductivity; but the two phenomena are wholly independent. The first of (8) is equivalent to Maxwell's solution.* The second is its magnetic analogue.

As, in the first case, there must be initial electrification, so in the second, there should be "magnetification," its volume-density to be measured by the divergence of the induction $\div 4\pi$. Now the induction can have no divergence. But it might have, if g existed.

There is no true electric current during the subsidence of $\mathbf{E}_0$, and there would be no true magnetic current during the subsidence of $\mathbf{H}_0$. In both cases the energy is frictionally dissipated on the spot, or there is no transfer of energy.† The application of (8) will be extended later.

Circuital Distributions.

4. By a circuital ‡ distribution, I mean one which has no divergence anywhere. Any field of force vanishing at infinity may be uniquely divided into two fields, one of which is polar, the other circuital; the proof thereof resting upon Sir W. Thomson's well-known theorem of Determinancy. Now we know exactly what happens to the polar fields. Therefore dismiss them, and let $\mathbf{E}_0$ and $\mathbf{H}_0$ be circuital. Then

$$q^2 = v^2\nabla^2 + \sigma^2, \quad \text{............................(9)}$$

where ∇^2 is the usual Laplacean operator. Of course $\cosh qt$ and $q^{-1}\sinh qt$ are rational functions of q^2, so that if the differentiations are possible we shall obtain the solutions out of (7).

Distortionless Cases.

5. Let the subsidence-rates of the polar electric and magnetic fields be equal. We then have

$$\sigma = 0, \qquad q^2 = -(v\,\mathrm{curl})^2, \qquad \rho = 4\pi k/c = 4\pi g/\mu, \quad \text{.....(10)}$$

in the solutions (7). The fields change in precisely the same manner as if the medium were nonconducting, as regards the relative values at different places; that is, there is no distortion due to the conductivities;

* Vol. I. chap. X., art. 325, equation (4).

† This is of course obvious without any reference to Poynting's formula. The only other simple case of no transfer of energy, which had been noticed before that formula, is that of conduction-current kept up by impressed force so distributed as to require no polar force to supplement it.

‡ [Lord Kelvin's word "circuital" is here substituted for "purely solenoidal."]

but there is a uniform subsidence all over brought in by them,* expressed by the factor $\epsilon^{-\rho t}$. This property I have explained by showing the opposite nature of the tails left behind by a travelling plane-wave according as σ is + or −

The above applies to a homogeneous medium. But if, in

$$\text{curl}\,(\mathbf{H}-\mathbf{h}) = (4\pi k + cp)\mathbf{E}, \qquad \text{(1}a\text{)}$$

$$\text{curl}\,(\mathbf{e}-\mathbf{E}) = (4\pi g + \mu p)\mathbf{H}, \qquad \text{(2}a\text{)}$$

differing from (1), (2) only in the introduction of impressed forces e and h, we write

$$(\mathbf{H}, \mathbf{h}, \mathbf{E}, \mathbf{e}) = (\mathbf{H}_1, \mathbf{h}_1, \mathbf{E}_1, \mathbf{e}_1)\epsilon^{-\rho t},$$

we reduce them to

$$\left.\begin{aligned} \text{curl}\,(\mathbf{H}_1-\mathbf{h}_1) &= c(\sigma+p)\mathbf{E}_1, \\ \text{curl}\,(\mathbf{e}_1-\mathbf{E}_1) &= \mu(-\sigma+p)\mathbf{H}_1, \end{aligned}\right\} \qquad \text{(11)}$$

and these, if $\sigma = 0$, are the equations of a nonconducting dielectric. That is,

$$\rho = 4\pi k/c = 4\pi g/\mu = \text{constant}$$

is the required condition. Therefore c and μ may vary anyhow, independently, provided k and g vary similarly.† The impressed forces should subside according to $\epsilon^{-\rho t}$, in order to preserve similarity to the phenomena in a nonconducting dielectric.

Observe that there will be tailing now, on account of the variability of $(\mu/c)^{\frac{1}{2}}$ or μv. That is, there are reflexions and refractions due to change of medium. The peculiarity is that they are of the same nature with as without conductivity.

First Special Case.

6. A special case of (11) is given by taking $\mu = 0$ and $g = 0$; that is, a real conducting dielectric possessing no magnetic inductivity, in which k/c is constant. If the initial field be polar, then

$$\mathbf{E} = \mathbf{E}_0\epsilon^{-\rho t}, \qquad \mathbf{H} = 0. \qquad \text{(12)}$$

This extension of Maxwell's before-mentioned solution I have given before, and also the extension to any initial field, and the inclusion of impressed forces.‡ The theory of the result has considerable light now thrown upon it.

If the initial field be arbitrary, the circuital part of the flux displacement disappears instantly, therefore (12) is the solution, provided $\mathbf{E}_0$ means the polar part of the initial field; that is, $\mathbf{E}_0$ must have no curl, and the flux $c\mathbf{E}_0/4\pi$ must have the same divergence as the arbitrarily given displacement.

Now an impressed force e produces a circuital flux only. Therefore it produces its full effect and sets up the appropriate steady flux instantaneously; and all variations of e in time and in space are kept

* "Electromagnetic Waves," Part I., § 7 [p. 381, vol. II.].

† In § 4 of the article referred to in the last footnote the property was described only in reference to a homogeneous medium.

‡ "Electromagnetic Induction" [vol. I., p. 534].

time to without lag by the conduction-current in spite of the electric displacement.

This property is seemingly completely at variance with ideas founded upon the retardation usually associated with combinations of resistances and condensers. But, being a special case of the distortionless theory, we can now understand it. For suppose we start with a nonconducting dielectric, and put on **e** uniform within a spherical portion thereof, and send out an electromagnetic wave to infinity and set up the steady flux. On now removing **e**, we send out another wave to infinity, and the flux vanishes. Now make the medium conducting, with both conductivities balanced, as in (10). Starting with the same steady flux, its vanishing will take place in the same manner precisely, but with an attenuation-factor $\epsilon^{-\rho t}$. Now gradually reduce g and μ at the same time, in the same ratio. The vanishing of the flux will take place faster and faster, and in the limit, when both μ and g are zero, will take place instantly, not by subsidence, but by instantaneous transference to an infinite distance when the impressed force is removed, owing to v being made infinite.

Second Special Case.

7. There is clearly a similar property when $k=0$ and $c=0$; that is, in a medium possessing magnetic inductivity and conductivity, but deprived of the electric correspondences. Thus, when g/μ is constant, the solution due to any polar field $\mathbf{H}_0$ is

$$\mathbf{H}=\mathbf{H}_0\epsilon^{-\rho t}, \qquad \mathbf{E}=0; \quad \text{......................(13)}$$

wherein $\rho=4\pi g/\mu$. But a circuital state of $\mu\mathbf{H}$ disappears at once, by instantaneous transference to infinity. Thus any varying impressed force **h** is accompanied without delay by the corresponding steady flux, the magnetic induction.

When the inertia associated with μ is considered, the result is rather striking and difficult to understand. It appears, however, to belong to the same class of (theoretical) phenomena as the following. If a coil in which there is an electric current be instantaneously shunted on to a second coil in which there is no current, then, according to Maxwell, the first coil instantly loses current and the second gains it, in such a way as to keep the momentum unchanged. Now we cannot set up a current in a coil instantly, so that we have a contradiction. But the disagreement admits of easy reconciliation. We cannot set up current instantly with a finite impressed force, but if it be infinite we can. In the case of the coils there is an electromotive impulse, or infinite electromotive force acting for an infinitely short time, when the coils are connected, with corresponding instantaneous changes in their momenta. A loss of energy is involved.

It is scarcely necessary to remark that the true physical theory involves other considerations, on account of the dielectric not being infinitely elastive, and on account of diffusion in the wires; so that we have sparking and very rapid vibrations in the dielectric. The energy which is not wasted in the spark, and which would go out to infinity were there no conducting obstacles, is probably all wasted practically in the heat of conduction-currents in them.

Impressed Forces.

8. Given initially $\mathbf{E}_0$ and $\mathbf{H}_0$, we know that the diverging parts must either remain constant or subside, and are, in a manner, self-contained; but the circuital parts, which would give rise to waves, may be kept from changing by means of impressed forces $\mathbf{e}_0$ and $\mathbf{h}_0$. Thus, let $\mathbf{E}_0$ and $\mathbf{H}_0$ be circuital. To keep them steady we have, in equations (1), (2), to get rid of $p\mathbf{E}$ and $p\mathbf{H}$. Thus

$$\left.\begin{aligned} \text{curl}\,(\mathbf{H}_0 - \mathbf{h}_0) &= 4\pi k\mathbf{E}_0, \\ \text{curl}\,(\mathbf{e}_0 - \mathbf{E}_0) &= 4\pi g\mathbf{H}_0, \end{aligned}\right\} \qquad (14)$$

are the equations of steady fields $\mathbf{E}_0$ and $\mathbf{H}_0$, these being the forces of the fluxes. Or

$$\left.\begin{aligned} \text{curl}\,\mathbf{h}_0 &= \text{curl}\,\mathbf{H}_0 - 4\pi k\mathbf{E}_0, \\ \text{curl}\,\mathbf{e}_0 &= \text{curl}\,\mathbf{E}_0 + 4\pi g\mathbf{H}_0, \end{aligned}\right\} \qquad (14a)$$

give the curls of the required impressed forces in terms of the given fluxes, and any impressed forces having these curls will suffice.

Now, on the sudden removal of $\mathbf{e}_0$, $\mathbf{h}_0$, the forces $\mathbf{E}_0$, $\mathbf{H}_0$, which had hitherto been the forces of the fluxes, become, instantaneously, the forces of the field as well. That is, the fluxes themselves do not change suddenly, except in such a case as a tangential discontinuity in a flux produced at a surface of curl of impressed force, when, at the surface itself, the mean value will be immediately assumed on removal of the impressed force. We know, therefore, the effects due to certain distributions of impressed force when we know the result of leaving the corresponding fluxes to themselves without impressed force. It is, however, the converse of this that is practically useful, viz., to find the result of leaving the fluxes without impressed force by solving the problem of the establishment of the steady fluxes when the impressed forces are suddenly started; because this problem can often be attacked in a comparatively simple manner, requiring only investigation of the appropriate functions to suit the surfaces of curl of the impressed forces. The remarks in this paragraph are not limited to homogeneity and isotropy.

Primitive Solutions for Plane Waves.

9. If we take z normal to the plane of the waves, we may suppose that both $\mathbf{E}$ and $\mathbf{H}$ have x and y components. This is, however, a wholly unnecessary mathematical complication, and it is sufficient to suppose that $\mathbf{E}$ is everywhere parallel to the x-axis, and $\mathbf{H}$ to the y-axis. The specification of an initial state is therefore E_0, H_0, the tensors of $\mathbf{E}$ and $\mathbf{H}$, given as functions of z; and the circuital equations (1), (2) become

$$-dH/dz = (4\pi k + cp)E, \qquad -dE/dz = (4\pi g + \mu p)H. \quad \ldots\ldots(15)$$

Now the operator q^2 in (5) becomes

$$q^2 = v^2\nabla^2 + \sigma^2; \quad \ldots\ldots(16)$$

where by ∇ we may now understand d/dz simply. Therefore, by (7), the solutions of (15) are

$$\left.\begin{aligned} E&=\epsilon^{-\rho t}\left[\left(\cosh qt-\frac{\sigma}{q}\sinh qt\right)E_0-\frac{\sinh qt}{q}\frac{\nabla}{c}H_0\right],\\ H&=\epsilon^{-\rho t}\left[\left(\cosh qt+\frac{\sigma}{q}\sinh qt\right)H_0-\frac{\sinh qt}{q}\frac{\nabla}{\mu}E_0\right].\end{aligned}\right\}\qquad(17)$$

When the initial states are such as $a\epsilon^{bz}$, or $a\cos bz$, the realization is immediate, requiring only a special meaning to be given to q in (17). But with more useful functions, as $a\epsilon^{-bz^2}$, etc., etc., there is much work to be performed in effecting the differentiations, whilst the method fails altogether if the initial distribution is discontinuous.

But we may notice usefully that when E_0 and H_0 are constants the solutions are

$$E=\epsilon^{-2\rho_1 t}E_0,\qquad H=\epsilon^{-2\rho_2}H_0,\qquad(18)$$

which are quite independent of one another. Further, since disturbances travel at speed v, (18) represents the solutions in any region in which E_0 and H_0 are constant, from $t=0$ up to the later time when a disturbance arrives from the nearest plane at which E_0 or H_0 varies.

Fourier-*Integrals*.

10. Now transform (17) to Fourier-integrals. We have Fourier's theorem,

$$f(z)=\frac{1}{\pi}\int_0^\infty\int_{-\infty}^\infty f(a)\cos m(z-a)\,dm\,da,\qquad(19)$$

and therefore $\phi(\nabla^2)f(z)=\frac{1}{\pi}\int_0^\infty\int_{-\infty}^\infty f(a)\phi(-m^2)\cos m(z-a)\,dm\,da\,;$ (20)

applying which to (17) we obtain

$$\left.\begin{aligned} E&=\frac{\epsilon^{-\rho t}}{\pi}\int_0^\infty\int_{-\infty}^\infty dm\,da\left[E_0\cos m(z-a)\left(\cosh-\frac{\sigma}{q}\sinh\right)qt\right.\\ &\qquad\qquad\left.+\frac{H_0}{c}m\sin m(z-a)\frac{\sinh qt}{q}\right],\\ H&=\frac{\epsilon^{-\rho t}}{\pi}\int_0^\infty\int_{-\infty}^\infty dm\,da\left[H_0\cos m(z-a)\left(\cosh+\frac{\sigma}{q}\sinh\right)qt\right.\\ &\qquad\qquad\left.+\frac{E_0}{\mu}m\sin m(z-a)\frac{\sinh qt}{q}\right],\end{aligned}\right\}\qquad(21)$$

in which, by (16), $$q^2=\sigma^2-m^2v^2,\qquad(22)$$

and E_0, H_0 are to be expressed as functions of a, whilst E and H belong to z. Discontinuities are now attackable.

The integrations with respect to m may be effected. In fact, I have done it in three different ways. First by finding the effect produced by impressed force. Secondly, by an analogous method applied to (17), transforming the differentiations to integrations. Thirdly, by direct integration of (21); this is the most difficult of all. The first method was given in a recent paper*; a short statement of the other two methods follows.

Transformation of the Primitive Solutions (17).

11. In (17) we naturally consider the functions of qt to be expanded in rising powers of q^2, and therefore of ∇^2, leading to differentiations to be performed upon the initial states. But if we expand them in descending powers of ∇, we substitute integrations, and can apply them to a discontinuous initial distribution.

The following are the expansions required :—

$$\left.\begin{aligned}\frac{\epsilon^{qt}}{q} &= \frac{1}{v\nabla}\left[U_0 + U_1\left(\frac{\sigma^2 t}{2v\nabla}\right) + \frac{U_2}{\lfloor 2}\left(\frac{\sigma^2 t}{2v\nabla}\right)^2 + \frac{U_3}{\lfloor 3}\left(\frac{\sigma^2 t}{2v\nabla}\right)^3 + \dots\right],\\ \epsilon^{qt} &= U_0 + U_0\left(\frac{\sigma^2 t}{2v\nabla}\right) + \frac{U_1}{\lfloor 2}\left(\frac{\sigma^2 t}{2v\nabla}\right)^2 + \frac{U_2}{\lfloor 3}\left(\frac{\sigma^2 t}{2v\nabla}\right)^3 + \dots,\end{aligned}\right\} \quad\dots(23)$$

where the U's are functions of $(v\nabla t)^{-1}$ given by

$$\left.\begin{aligned}&U_0 = \epsilon^{vt\nabla}, \qquad U_1 = \epsilon^{vt\nabla}\left(1 - \frac{1}{vt\nabla}\right), \qquad U_2 = \epsilon^{vt\nabla}\left(1 - \frac{3}{vt\nabla} + \frac{3}{(vt\nabla)^2}\right),\\ &U_r = \epsilon^{vt\nabla}\left[1 - \frac{r(r+1)}{2vt\nabla} + \frac{r(r^2-1^2)(r+2)}{2.4.(vt\nabla)^2} - \frac{r(r^2-1^2)(r^2-2^2)(r+3)}{2.4.6(vt\nabla)^3} + \dots\right];\end{aligned}\right\} \quad (24)$$

being in fact identically the same functions of $vt\nabla$ as those of r which occur in the investigation of spherical waves. [See p. 406, vol. II.]

Arranged in powers of $s = \sigma/v\nabla$, we have

$$\left.\begin{aligned}\frac{\epsilon^{qt}}{q} &= \frac{\epsilon^{vt\nabla}}{v\nabla}(1 + sg_1 + s^2 g_2 + \dots),\\ \epsilon^{qt} &= \epsilon^{vt\nabla}(1 + sh_1 + s^2 h_2 + \dots),\end{aligned}\right\} \quad\dots\dots(25)$$

where

$$\left.\begin{aligned}&g_1 = \frac{\sigma t}{2}, \qquad g_2 = -\tfrac{1}{2} + \frac{(\sigma t)^2}{2.4}, \qquad g_3 = -\tfrac{3}{4}\frac{\sigma t}{2} + \frac{(\sigma t)^3}{2.4.6},\\ &g_4 = \tfrac{3}{8} - \frac{(\sigma t)^2}{2.4} + \frac{(\sigma t)^4}{2.4.6.8}, \qquad g_5 = \tfrac{5}{8}\frac{\sigma t}{2} - \tfrac{5}{4}\frac{(\sigma t)^3}{2.4.6} + \frac{(\sigma t)^5}{2.4.6.8.10},\\ &g_6 = -\tfrac{5}{16} + \tfrac{15}{16}\frac{(\sigma t)^2}{2.4} - \tfrac{3}{2}\frac{(\sigma t)^4}{2.4.6.8} + \frac{(\sigma t)^6}{2.4\dots12};\end{aligned}\right\} \quad\dots\dots(26)$$

$$\left.\begin{aligned}&h_1 = \frac{\sigma t}{2}, \; h_2 = \frac{(\sigma t)^2}{2^2\lfloor 2}, \; h_3 = \frac{1}{2^2\lfloor 2}\left(-\sigma t + \frac{(\sigma t)^3}{2.3}\right), \; h_4 = \frac{1}{2^3\lfloor 3}\left(-3(\sigma t)^2 + \frac{(\sigma t)^4}{2.4}\right),\\ &h_5 = \frac{1}{2^3\lfloor 3}\left(3\sigma t - \frac{6(\sigma t)^3}{2.4} + \frac{(\sigma t)^5}{2^2.4.5}\right), \; h_6 = \frac{1}{2^4\lfloor 4}\left(15(\sigma t)^2 - \frac{10(\sigma t)^4}{2.5} + \frac{(\sigma t)^6}{2^2.5.6}\right),\\ &h_7 = \frac{1}{2^4\lfloor 4}\left(-15\sigma t + \frac{45(\sigma t)^3}{2.5} - \frac{15(\sigma t)^5}{2^2.5.6} + \frac{(\sigma t)^7}{2^3.5.6.7}\right).\end{aligned}\right\} \quad (27)$$

* "Electromagnetic Waves," Part IV. [p. 428, vol. II.].

The following properties of the g's and h's are useful. Understanding that g_0 and h_0 are unity, we have

$$\left.\begin{aligned} &g_r+\sigma t g_{r+1}+\frac{(\sigma t)^2}{\lfloor 2}g_{r+2}+\ldots=0, && \text{when } r \text{ is odd,}\\ &\text{and when } r \text{ is even,} \quad =1.3.5\ldots(r-1)(-1)^{\frac{1}{2}r}\frac{J_{\frac{1}{2}r}(\sigma t i)}{(\sigma t i)^{\frac{1}{2}r}}, \\ &\text{except } r=0, \text{ when} \quad =J_0(\sigma t i). \end{aligned}\right\}\ldots(28)$$

[Also, $h_r+\sigma t h_{r+1}+\frac{(\sigma t)^2}{\lfloor 2}h_{r+2}+\ldots=i(-1)^{\frac{1}{2}(r+1)}1.3.5\ldots(r-2)\frac{J_{\frac{1}{2}(r+1)}(\sigma t i)}{(\sigma t i)^{\frac{1}{2}(r-1)}}$, (29)

when r is odd, but is zero when r is even (except $r=0$, which case is not wanted), and $=-iJ_1(\sigma t i)$ when $r=1$.] Now if

$$\epsilon^{\sigma t}(1+\sigma/q)=\epsilon^{vt\nabla}(1+sf_1+s^2f_2+\ldots), \quad \ldots\ldots\ldots\ldots\ldots(30)$$

the f's * will be given by (25), viz.,

$$f_0=1, \qquad f_1=g_0+h_1, \qquad f_2=g_1+h_2, \qquad \text{etc.}; \quad \ldots\ldots\ldots(31)$$

and the properties of the f's corresponding to (28), (29) are

$$\left.\begin{aligned} f_r+\sigma t f_{r+1}+\frac{(\sigma t)^2}{\lfloor 2}f_{r+2}+\ldots &=\epsilon^{\sigma t} && \text{when } r=0,\\ &=0 && \text{when } r \text{ is even, except } 0; \end{aligned}\right\}(32)$$

and $\qquad = \pm 1.3.5\ldots(r-2)\frac{J_{\frac{1}{2}(r-1)}(\sigma t i)-iJ_{\frac{1}{2}(r+1)}(\sigma t i)}{(\sigma t i)^{\frac{1}{2}(r-1)}}; \quad \ldots\ldots\ldots(33)$

when r is odd, with the $+$ sign for $r=1, 5, 9, \ldots$, and the $-$ sign for the rest. The first case in (32), of $r=0$, is very important. But in case $r=1$, the coefficient in (33) is $+1$; thus,

$$=(J_0-iJ_1)(\sigma t i).$$

Special Initial States.

12. Now let there be an initial distribution of H_0 only, so that, by (17),

$$\left.\begin{aligned} H&=\epsilon^{-\rho t}\left(\cosh+\frac{\sigma}{q}\sinh\right)qt.H_0,\\ E&=-\epsilon^{-\rho t}\frac{\sinh qt}{q}\frac{\nabla}{c}H_0, \end{aligned}\right\}\ldots\ldots\ldots\ldots\ldots\ldots(34)$$

by (17). Let H_0 be zero on the right side and constant on the left side of the origin, and let us find H and E at a point on the right side. The operator $\epsilon^{vt\nabla}$ is inoperative, so that, by (30),

$$\left.\begin{aligned} H&=\tfrac{1}{2}\epsilon^{-\rho t}\epsilon^{-vt\nabla}(1-sf_1+s^2f_2-s^3f_3+\ldots)H_0,\\ E&=\tfrac{1}{2}\epsilon^{-\rho t}\epsilon^{-vt\nabla}(1-sg_1+s^2g_2-s^3g_3+\ldots)H_0\times\mu v, \end{aligned}\right\}\ldots\ldots\ldots(35)$$

* These f's are the same as in my paper "On Electromagnetic Waves," § 8 [vol. II., p. 384]; but s there is σ here.

the immediate integration of which gives

$$\left.\begin{aligned} H &= \tfrac{1}{2}H_0\epsilon^{-\rho t} \times \left\{1 + \sigma t f_1\left(1 - \frac{z}{vt}\right) + \frac{(\sigma t)^2}{\underline{|2}} f_2\left(1 - \frac{z}{vt}\right)^2 + \ldots\right\}, \\ E &= \tfrac{1}{2}\mu v H_0\epsilon^{-\rho t}\left\{1 + \sigma t g_1\left(1 - \frac{z}{vt}\right) + \frac{(\sigma t)^2}{\underline{|2}} g_2\left(1 - \frac{z}{vt}\right)^2 + \ldots\right\}. \end{aligned}\right\} \ldots(36)$$

To obtain the E due to E_0 constant from $z = -\infty$ to 0, use the first of (36); change H to E, H_0 to E_0, and change the sign of σ, not forgetting it in the f's. To obtain the corresponding H due to E_0, use the second of (36); change E to H, H_0 to E_0, and μ to c. So

$$\left.\begin{aligned} E &= \tfrac{1}{2}E_0\epsilon^{-\rho t} \times \left\{1 - \sigma t f_1'\left(1 - \frac{z}{vt}\right) + \frac{(\sigma t)^2}{\underline{|2}} f_2'\left(1 - \frac{z}{vt}\right)^2 - \ldots\right\}, \\ H &= \tfrac{1}{2}cvE_0\epsilon^{-\rho t}\left\{1 + \sigma t g_1\left(1 - \frac{z}{vt}\right) + \frac{(\sigma t)^2}{\underline{|2}} g_2\left(1 - \frac{z}{vt}\right)^2 + \ldots\right\}, \end{aligned}\right\} \ldots(37)$$

where the accent means that the sign of σ is changed in the f's.

From these, without going any further, we can obtain a general idea of the growth of the waves to the right and left of the origin, because the series are suitable for small values of σt. But, reserving a description till later, notice that E in (36) and H in (37) must be true on both sides of the origin; on expanding them in powers of z we consequently find that the coefficients of the odd powers of z vanish, by the first of (28), and what is left may be seen to be the expansion of

$$E = \tfrac{1}{2}\mu v H_0\epsilon^{-\rho t} J_0\left[\frac{\sigma}{v}(z^2 - v^2t^2)^{\frac{1}{2}}\right], \quad \ldots\ldots\ldots\ldots\ldots\ldots(38)$$

the complete solution for E due to H_0. Similarly,

$$H = \tfrac{1}{2}cvE_0\epsilon^{-\rho t} J_0\left[\frac{\sigma}{v}(z^2 - v^2t^2)^{\frac{1}{2}}\right] \quad \ldots\ldots\ldots\ldots\ldots\ldots(39)$$

is the complete solution for H due to E_0. In both cases the initial distribution was on the left side of the origin; but, if its sign be reversed, it may be put on the right side, without altering these solutions.

Similarly, by expanding the first of (36) and first of (37) in powers of z we get rid of the even powers of z, and produce the solutions given by me in a previous paper,* which, however, it is needless to write out here, owing to the complexity.

Arbitrary Initial States.

13. Knowing the solutions due to the above distributions, we find those due to initial $E_0 da$ at the origin, or $H_0 da$, by differentiation to z;

* "Electromagnetic Waves," § 8 [vol. II., p. 383].

and for this we do not need the firsts of (36) and (37), but only the seconds. The results bring the Fourier-integrals (21) to

$$\left.\begin{aligned} E &= \epsilon^{-\rho t}\Big[\tfrac{1}{2}(E_0+\mu vH_0)_{z-vt}+\tfrac{1}{2}(E_0-\mu vH_0)_{z+vt} \\ &\qquad + \tfrac{1}{2}\int_{z-vt}^{z+vt}\Big\{E_0\frac{-\sigma+p}{v}-\frac{H_0}{cv}\nabla\Big\}J_0(y)da\Big], \\ H &= \epsilon^{-\rho t}\Big[\tfrac{1}{2}(H_0+cvE_0)_{z-vt}+\tfrac{1}{2}(H_0-cvE_0)_{z+vt} \\ &\qquad + \tfrac{1}{2}\int_{z-vt}^{z+vt}\Big\{H_0\frac{\sigma+p}{v}-\frac{E_0}{\mu v}\nabla\Big\}J_0(y)da\Big]; \end{aligned}\right\} \qquad (40)$$

where $\qquad p=d/dt, \qquad \nabla=d/dz, \qquad y=\frac{\sigma}{v}\{(z-a)^2-v^2t^2\}$.

Another interesting form is got by the changes of variables

$$\left.\begin{aligned} U &= \tfrac{1}{2}\epsilon^{\rho t}(E-\mu vH), & u &= z-vt, \\ W &= \tfrac{1}{2}\epsilon^{\rho t}(E+\mu vH), & w &= z+vt. \end{aligned}\right\} \qquad (41)$$

These lead to

$$\left.\begin{aligned} U_{z,t} &= U_{w,0}+\int_u^w\Big(U_0\frac{d}{dw}-\frac{\sigma}{2v}W_0\Big)J_0\Big\{\frac{\sigma}{v}(u-a)^{\frac{1}{2}}(w-a)^{\frac{1}{2}}\Big\}da, \\ W_{z,t} &= W_{u,0}-\int_u^w\Big(W_0\frac{d}{du}+\frac{\sigma}{2v}U_0\Big)J_0\Big\{\frac{\sigma}{v}(u-a)^{\frac{1}{2}}(w-a)^{\frac{1}{2}}\Big\}da. \end{aligned}\right\} \qquad (42)$$

The connexions and partial characteristic of U or W are

$$\frac{dW}{dw}=-\frac{\sigma}{2v}U, \qquad \frac{dU}{du}=+\frac{\sigma}{2v}W, \qquad \frac{d^2U}{du\,dw}=-\Big(\frac{\sigma}{2v}\Big)^2U; \qquad (43)$$

and this characteristic has a solution

$$U=\Big(\frac{z+vt}{z-vt}\Big)^{\frac{m}{2}}J_m\Big[\frac{\sigma}{v}(z^2-v^2t^2)^{\frac{1}{2}}\Big], \qquad (44)$$

where m is any + integer, and in which the sign of the exponent may be reversed. We have utilized the case $m=0$ only.

Evaluation of Fourier-*Integrals.*

14. The effectuation of the integration (direct) of the original Fourier-integrals will be found to ultimately depend upon

$$\frac{2}{\pi}\int_0^\infty \cos mz\,\frac{\sinh qt}{q}\,dm=\frac{1}{v}J_0\Big[\frac{\sigma}{v}(z^2-v^2t^2)^{\frac{1}{2}}\Big], \qquad (45)$$

provided $vt>z$, where, as before,

$$q^2=\sigma^2-m^2v^2.$$

By equating coefficients of powers of z^2 in (45) we get

$$\frac{2}{\pi}\int\frac{\sinh qt}{q}m^{2r}dm=\sigma^{2r}\frac{1.3.5.(2r-1)}{v^{2r+1}}\,\frac{J_r(\sigma ti)}{(\sigma ti)^r}, \qquad (46)$$

except with $r=0$; then $\qquad =v^{-1}J_0(\sigma ti)$.

To prove (45), expand the q-function in powers of σ^2. Thus, symbolically written,

$$\frac{\sinh qt}{q} = \epsilon^{\frac{1}{2}\sigma^2 p^{-1} t}\left(\frac{\sin mvt}{mv}\right), \quad\text{.....................(47)}$$

the operand being in the brackets, and p^{-1} meaning integration from 0 to t with respect to t. Thus, in full,

$$\frac{2}{\pi}\int \cos mz \frac{\sinh qt}{q} dm = \frac{2}{\pi}\int_0^\infty \cos mz\left[\frac{\sin mvt}{mv} + \frac{\sigma^2}{2}\int_0^t t\frac{\sin mvt}{mv}dt + \frac{\sigma^2}{2}\frac{\sigma^2}{4}\int_0^t t\,dt\int_0^t t\,dt\frac{\sin mvt}{mv} + \dots\right]dm. \quad\text{....(48)}$$

Now the value of the first term on the right is

$$v^{-1}, \quad\text{or}\quad 0, \qquad\text{when } z \text{ is}\quad <, \quad\text{or}\quad > vt.$$

Thus, in (48), if $z > vt$, since the first term vanishes, so do all the rest, because their values are deduced from that of the first by integrations to t, which during the integrations is always $< z/v$. Therefore the value of the left member of (45) is zero when $z > vt$. In another form, disturbances cannot travel faster than at speed v.

But when $z < vt$ in (48), it is clear that whilst t' goes from 0 to t or from 0 to z/v, and then from z/v to t, the first integral is zero from 0 to z/v, so that the part z/v to t only counts. Therefore the second term is

$$\frac{2}{\pi}\frac{\sigma^2}{2}\int \cos mz\left[\int_{\frac{z}{v}}^t \frac{t\sin mvt}{mv}dt\right]dm = \frac{2}{\pi}\frac{\sigma^2}{2}\int_{\frac{z}{v}}^t t\,dt\int_0^\infty \cos mz\frac{\sin mvt}{mv}dm$$

$$= \frac{\sigma^2}{2}\cdot\frac{1}{v}\int_{\frac{z}{v}}^t t\,dt = \frac{1}{v}\frac{\sigma^2}{2^2}\left(t^2 - \frac{z^2}{v^2}\right).$$

The third is, similarly,

$$\frac{1}{v}\frac{\sigma^2}{2^2}\frac{\sigma^2}{4}\int_{\frac{z}{v}}^t t\left(t^2 - \frac{z^2}{v^2}\right)dt = \frac{1}{v}\frac{\sigma^4}{2^2 4^2}\left(t^2 - \frac{z^2}{v^2}\right)^2;$$

and so on, in a uniform manner, thus proving that the successive terms of (48) are the successive terms of the expansion of (45) (right member) in powers of σ^2; and therefore proving (45).

The following formulæ occur when the front of the wave is in question, where caution is needed in evaluations :—

$$\cosh \sigma t - \tfrac{1}{2} = \frac{2}{\pi}\int_0^\infty \frac{\sin mvt}{m}\cosh qt\,dm, \quad\text{.................(49)}$$

$$\frac{\sinh \sigma t}{\sigma} = \frac{2}{\pi}\int_0^\infty \frac{\sin mvt}{m}\frac{\sinh qt}{q}dm. \quad\text{.................(50)}$$

Interpretation of Results.

15. Having now given a condensation of the mathematical work, we may consider, in conclusion, the meaning and application of the formulæ.

In doing so, we shall be greatly assisted by the elementary theory of a telegraph circuit. It is not merely a mathematically analogous theory, but is, in all respects save one, essentially the same theory, physically, and the one exception is of a remarkable character. Let the circuit consist of a pair of equal parallel wires, or of a wire with a coaxial tube for the return, and let the medium between the wires be slightly conducting. Then, if the wires had no resistance, the problem of the transmission of waves would be the above problem of plane waves in a real dielectric, that is, with constants μ, c, and k, but without the magnetic conductivity; *i.e.* $g=0$ in the above.

The fact that the lines of magnetic and electric force are no longer straight is an unessential point. But it is, for convenience, best to take as variables, not the forces, but their line-integrals. Thus, if V be the line-integral of E across the dielectric between the wires, V takes the place of E. Then kE, the density of the conduction-current, is replaced by KV, where K is the conductance of the dielectric per unit length of circuit; and $cE/4\pi$, the displacement, becomes SV, where S is the permittance per unit length of circuit. The density of electric current $cpE/4\pi$ is then replaced by SpV. Also SV is the charge per unit length of circuit.

Next, take the line-integral of $H/4\pi$ round either conductor for magnetic variable. It is C, usually called the current in the wires. Then μH, the induction, becomes LC; where LC is the momentum per unit length of circuit, L being the inductance, such that $LSv^2=\mu cv^2=1$.

A more convenient transformation (to minimize the trouble with 4π's) is

$$E \text{ to } V, \qquad H \text{ to } C, \qquad \mu \text{ to } L, \qquad c \text{ to } S, \qquad 4\pi k \text{ to } K.$$

Now, lastly, the wires have resistance, and this is without any representation whatever in a real dielectric. But, as I have before shown, the effect of the resistance of the wires in attenuating and distorting waves is, to a first approximation (ignoring the effects of imperfect penetration of the magnetic field into the wires), representable in the same manner exactly as the corresponding effects due to g, the hypothetical magnetic conductivity of a dielectric.* Thus, in addition to the above,

$$4\pi g \quad \text{becomes} \quad R,$$

R being the resistance of the circuit per unit length.

16. In the circuit, if infinitely long and perfectly insulated, the total charge is constant. This property is independent of the resistance of the wires. If there be leakage, the charge Q at time t is expressed in terms of the initial charge Q_0 by

$$Q=Q_0\epsilon^{-Kt/S},$$

independently of the way the charge redistributes itself.

In the general medium, the corresponding property is persistence of

* "Electromagnetic Waves," § 6 [p. 379, vol. II.].

displacement, no matter how it redistributes itself, provided k be zero, whatever g may be. And, if there be electric conductivity,

$$\int_{-\infty}^{\infty} Ddz = \left(\int_{-\infty}^{\infty} D_0 dz\right)\epsilon^{-4\pi kt/c},$$

where D_0 is the initial displacement, and D that at time t, functions of z.

In the circuit, if the wires have no resistance, the total momentum remains constant, however it may redistribute itself. This is an extension of Maxwell's well-known theory of a linear circuit of no resistance. The conductivity of the dielectric makes no difference in this property, though it causes a loss of energy. When the wires have resistance, then

$$\int_{-\infty}^{\infty} LCdz = \left(\int_{-\infty}^{\infty} LC_0 dz\right)\epsilon^{-Rt/L}$$

expresses the subsidence of total momentum; and this is independent of the manner of redistribution of the magnetic force, and of the leakage.

In the general medium, when real, the corresponding property is persistence of the induction (or momentum); and when g is finite,

$$\int_{-\infty}^{\infty} \mu Hdz = \left(\int_{-\infty}^{\infty} \mu H_0 dz\right)\epsilon^{-4\pi gt/\mu}.$$

In passing, I may remark that, in my interpretation of Maxwell's views, it is not his vector-potential **A**, the so-called electrokinetic momentum, that should have the physical idea of momentum associated with it, but the magnetic induction **B**. To illustrate, consider Maxwell's theory of a linear circuit of no resistance, the simplest case of persistence of momentum. We may express the fact by saying that the induction through the circuit remains constant, or that the line-integral of **A** along or in the circuit remains constant. These are perfectly equivalent. Now, if we pass to an infinitely small closed circuit, the line-integral of **A** becomes **B** itself (per unit area). But if we consider an element of length only, we get lost at once.

Again, the magnetic energy being associated with **B**, (and **H**), so should be the momentum.

Suppose also we take the property that the line-integral of $-\dot{\mathbf{A}}$ is the E.M.F. in a circuit, and then consider $-\dot{\mathbf{A}}$ as the electric force of induction at a point. Its time-integral is **A**. But this is an electromotive impulse, not momentum.

Lastly, whilst **B** (or **H**) defines a physical property at a point, **A** does not, but depends upon the state of the whole field, to an infinite distance. In fact, it sums up, in a certain way, the effect which would arise at a point from disturbances coming to it from all parts of the field. It is therefore, like the scalar electric potential, a mathematical concept merely, not indicative in any way of the actual state of the medium anywhere.

The time-integral of **H**, whose curl is proportional to the displacement, has equal claims to notice as a mathematical function which is of occasional use for facilitating calculations, but which should not, in my opinion, be elevated to the rank of a fundamental quantity, as was done by Maxwell with respect to **A**.

Independently of these considerations, the fact that **A** has often a scalar potential parasite (and also the other function), sometimes causes great mathematical complexity and indistinctness; and it is, for practical reasons, best to murder the whole lot, or, at any rate, merely employ them as subsidiary functions.

17. Returning to the telegraph-circuit, let the initial state be one of uniform V on the whole of the left side of the origin, $V=0$ on the right side, and $C=0$ everywhere. The diagram will serve to show roughly what happens in the three principal cases.

First of all we have ABCD to represent the curve of V_0, the origin being at C. When the disturbance has reached Z, that is when $t=\mathrm{CZ}/v$, the curve is A 1 1 1 1 Z, if there be no leakage, when R and L are such that $\epsilon^{-Rt/2L}=\frac{1}{2}$. At the origin, $V=\frac{1}{2}V_0$; at the front, $V=\frac{1}{4}V_0$; and at the back, $V=\frac{3}{4}V_0$.

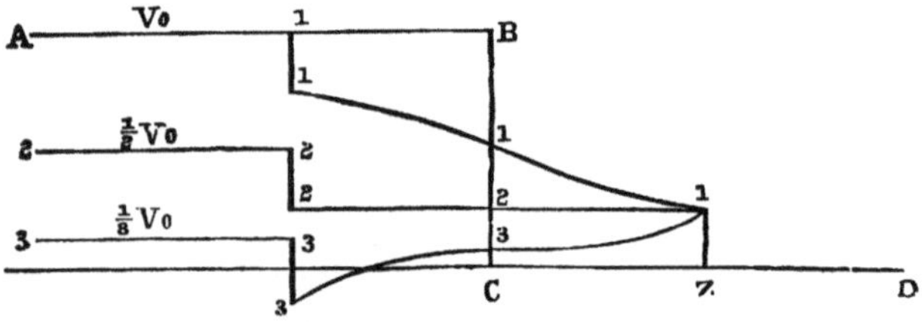

Now introduce leakage to make $R/L=K/S$. Then 2 2 2 2 1 Z shows the curve of V, provided $\epsilon^{-Kt/S}=\frac{1}{2}$. We have $V=\frac{1}{2}V_0$ on the left, and $V=\frac{1}{4}V_0$ in the rest.

Thirdly, let the leakage be in excess. Then, when V_0 has fallen, by leakage only, to $\frac{1}{8}V_0$ on the left, the curve 3 3 3 3 1 Z shows V; it is $\frac{1}{16}V_0$ at the origin, $-\frac{1}{8}V_0$ at the back, and $\frac{1}{4}V_0$ at the front.

[The third case is numerically wrong. Thus, at the front we have $V/V_0=\frac{1}{2}\epsilon^{-(\rho_1+\rho_2)t}$, at origin $\frac{1}{2}\epsilon^{-2\rho_1 t}$, and behind $\epsilon^{-2\rho_1 t}$. Now take $\rho_2=0$. Then, when $\epsilon^{-\rho_1 t}=\frac{1}{2}$, we have $V/V_0=\frac{1}{4}$ at front, $\frac{1}{8}$ at origin, 0 at back, and $\frac{1}{4}$ behind. It is later on that V becomes negative at the back. Thus, when $\epsilon^{-\rho_1 t}=\frac{1}{4}$, we have $V/V_0=\frac{1}{8}$ at front, $\frac{1}{32}$ at origin, $-\frac{1}{16}$ at back, and $\frac{1}{16}$ behind. And when $\epsilon^{-\rho_1 t}=\frac{1}{8}$, we have $V/V_0=\frac{1}{16}$ at front, $\frac{1}{128}$ at origin, $-\frac{3}{64}$ at back, and $\frac{1}{64}$ behind.]

Of course there has to be an adjustment of constants to make $\epsilon^{-(R/2L+K/2S)t}$ be the same $\frac{1}{2}$ in all cases, viz., the attenuation at the front.

18. Precisely the same applies when it is C_0 that is initially given instead of V_0, provided we change the sign of σ. That is, we have the curve 1 when the leakage is in excess, and the curve 3 when the leakage is smaller than that required to produce distortionless transmission.

19. Now transferring attention to the general medium, if we make the substitution of magnetic conductivity for the resistance of the wires, the curve 1 would apply when it is E_0 that is the initial state and g in excess, and 3 when it is deficient; whilst if H_0 is the initial state, 1

displacement, no matter how it redistributes itself, provided k be zero, whatever g may be. And, if there be electric conductivity,

$$\int_{-\infty}^{\infty} D dz = \left(\int_{-\infty}^{\infty} D_0 dz\right)\epsilon^{-4\pi kt/c},$$

where D_0 is the initial displacement, and D that at time t, functions of z.

In the circuit, if the wires have no resistance, the total momentum remains constant, however it may redistribute itself. This is an extension of Maxwell's well-known theory of a linear circuit of no resistance. The conductivity of the dielectric makes no difference in this property, though it causes a loss of energy. When the wires have resistance, then

$$\int_{-\infty}^{\infty} LC dz = \left(\int_{-\infty}^{\infty} LC_0 dz\right)\epsilon^{-Rt/L}$$

expresses the subsidence of total momentum; and this is independent of the manner of redistribution of the magnetic force, and of the leakage.

In the general medium, when real, the corresponding property is persistence of the induction (or momentum); and when g is finite,

$$\int_{-\infty}^{\infty} \mu H dz = \left(\int_{-\infty}^{\infty} \mu H_0 dz\right)\epsilon^{-4\pi gt/\mu}.$$

In passing, I may remark that, in my interpretation of Maxwell's views, it is not his vector-potential **A**, the so-called electrokinetic momentum, that should have the physical idea of momentum associated with it, but the magnetic induction **B**. To illustrate, consider Maxwell's theory of a linear circuit of no resistance, the simplest case of persistence of momentum. We may express the fact by saying that the induction through the circuit remains constant, or that the line-integral of **A** along or in the circuit remains constant. These are perfectly equivalent. Now, if we pass to an infinitely small closed circuit, the line-integral of **A** becomes **B** itself (per unit area). But if we consider an element of length only, we get lost at once.

Again, the magnetic energy being associated with **B**, (and **H**), so should be the momentum.

Suppose also we take the property that the line-integral of $-\dot{\mathbf{A}}$ is the E.M.F. in a circuit, and then consider $-\dot{\mathbf{A}}$ as the electric force of induction at a point. Its time-integral is **A**. But this is an electromotive impulse, not momentum.

Lastly, whilst **B** (or **H**) defines a physical property at a point, **A** does not, but depends upon the state of the whole field, to an infinite distance. In fact, it sums up, in a certain way, the effect which would arise at a point from disturbances coming to it from all parts of the field. It is therefore, like the scalar electric potential, a mathematical concept merely, not indicative in any way of the actual state of the medium anywhere.

The time-integral of **H**, whose curl is proportional to the displacement, has equal claims to notice as a mathematical function which is of occasional use for facilitating calculations, but which should not, in my of balanced exchanges). When there is propagation, and **H** is involved, we have

$$\mathbf{E} = -\nabla\Psi - \dot{\mathbf{A}}.$$

Now this is not an electromagnetic law specially, but strictly a truism, or mathematical identity. It becomes electromagnetic by the definition of **A**,

$$\text{curl}\,\mathbf{A} = \mu\mathbf{H},$$

leaving **A** indeterminate as regards a diverging part, which, however, we may merge in $-\nabla\Psi$. Supposing, then, **A** and Ψ to become fixed in this or some other way, the next question in connection with propagation is, Can we, instead of the propagation of **E** and **H**, substitute that of Ψ and **A**, and obtain the same knowledge, irrespective of the artificiality of Ψ and **A**? The answer is perfectly plain—we cannot do so. We could only do it if Ψ, **A**, given everywhere, found **E** and **H**. But they cannot. **A** finds **H**, irrespective of Ψ, but both together will not find **E**. We require to know a third vector, $\dot{\mathbf{A}}$. Thus we have Ψ, **A**, and $\dot{\mathbf{A}}$, required, involving *seven* scalar specifications to find the *six* in **E** and **H**. Of these three quantities, the utility of **A** is simply to find **H**, so that we are brought to a highly complex way of representing the propagation of **E** in terms of Ψ and $\dot{\mathbf{A}}$, giving no information about **H**, which is, it seems to me, as complex and artificial as it is useless and indefinite.

Again, merely to emphasize the preceding, the variables chosen should be capable of representing the energy stored. Now the magnetic energy may be expressed in terms of **A**, though with entirely erroneous localization; but the electric energy cannot be expressed in terms of Ψ. Maxwell (chap. XI. vol. II.) did it, but the application is strictly limited to electrostatics; in fact, Maxwell did not consider electric energy comprehensively. The full representation in terms of potentials requires Ψ and **Z**, the vector-potential of the magnetic current. (This is developed in my work "On Electromagnetic Induction and its Propagation" [vol. I., p. 507].) This inadequacy alone is sufficient to murder Ψ and **A**, considered as subjects of propagation.

Now take a concrete example, leaving the abstract mathematical reasoning. Let there be first no **E** or **H** anywhere. To produce any, impressed force is absolutely needed. Let it be impressed **e**, and of the simplest type, viz., an infinitely extended plane sheet of **e** of uniform intensity, acting normally to the plane. What happens? Nothing at all. Yet the potential on one side of the plane is made greater by the amount e (tensor of **e**) than on the other side. Say $\Psi = \frac{1}{2}e$ and $-\frac{1}{2}e$. Thus we have *instantaneous* propagation of Ψ to infinity. I prefer, however, to say that this is only a mathematical fiction, that nothing is propagated at all, that the electromagnetic mechanism is of such a nature that the applied forces are balanced *on the spot*, that is, in the sheet, by the reactions.

To emphasize this again, let the sheet be not infinite, but have a circular boundary. Let the medium be of uniform inductivity μ, and permittivity c. Then, irrespective of its conductivity, disturbances are propagated at speed $v = (\mu c)^{-\frac{1}{2}}$, and their source is the vortex-line of $\mathbf{e}$, on the edge of the disk. At any time t less than a/v, where a is the radius of the disk, the disturbance is confined within a ring whose axis is the vortex-line. Everywhere else, $\mathbf{E}=0$ and $\mathbf{H}=0$. On the surface of the ring, $E=\mu v H$, and $\mathbf{E}$ and $\mathbf{H}$ are perpendicular; there can be no normal component of either.

Now, we can naturally explain the absence of any flux in the central portion of the disk, by the applied forces being balanced by the reactions on the spot, until the wave arrives from the vortex-line. But how can we explain it in terms of Ψ, seeing that Ψ has now to change by the amount e at the disk, and yet be continuous everywhere else outside the ring? We cannot do it, so the propagation of Ψ fails altogether. Yet the actions involved must be the same whether the disk be small or infinitely great. We must therefore give up the idea altogether of the propagation of a Ψ to balance impressed force. In the ring itself, however, we may regard the propagation of Ψ (a different one), $\mathbf{A}$, and $\dot{\mathbf{A}}$; or, more simply, of $\mathbf{E}$ and $\mathbf{H}$.

If there be no conductivity, the steady electric field is assumed anywhere the moment the two waves from opposite ends of a diameter of the disk coexist; that is, as soon as the wave arrives from the more distant end.* But this simplicity is quite exceptional, and seems to be confined to plane and spherical waves. In general there is a subsidence to the steady state after the initial phenomena.

If it be remarked that incompressibility (or something equivalent or resembling it) is needed in order that the medium may behave as described (*i.e.*, no flux except at the vortex-line initially), and that if the medium be compressible we shall have other results (a pressural wave, for example, from the disk generally), the answer is that this is a wholly independent matter, not involved in Maxwell's dielectric theory, though perhaps needing consideration in some other theory. But the moment we let the electric current have divergence (the absence of which makes the vortex-lines of $\mathbf{e}$ to be the sources of disturbances), we at once (in my experience) get lost in an almost impenetrable fog of potentials. Maxwell's theory unamended, on the other hand, works perfectly and without a trace of indefiniteness, provided we regard $\mathbf{E}$ and $\mathbf{H}$ as the variables, and discard his "equations of propagation" containing the two potentials.†

October 22, 1888.

* "Electromagnetic Waves," § 25 [p. 415, vol. II.].

† [*March* 20, 1889.—Referring to the example given above of a circular disk, I strangely overlooked the fact that the absence of flux initially can be expressed by infinitely rapid propagation of both a Ψ and an $\mathbf{A}$. In the disk itself we must have $-\nabla\Psi - \dot{\mathbf{A}} = -$ impressed force, so that there is no flux *there*, and outside we must have $-\nabla\Psi - \dot{\mathbf{A}} = 0$. This makes it go. But as regards propagation, it only makes matters worse. It is a *reductio ad absurdum* to have an electrostatic field propagated infinitely rapidly, and, simultaneously, the electric force of induction, its exact negative, merely to cancel the former, itself quite hypothetical.

In my paper "On the Electromagnetic Effects due to Moving Electrification," *Phil. Mag.*, April, 1889 (vol. II., Art. L.), is an explicit example showing the absurdity of the thing.]

XLV. LIGHTNING DISCHARGES, ETC.

[*The Electrician*, Aug. 17, 1888, p. 479.]

THE gap between the electrical phenomena of common practice and those concerned in the transmission of light and heat, a gap that it once seemed almost impossible to bridge, is being gradually filled up, both from the theoretical and the experimental side; both from above, by the observation of dark heat and in other ways; and from below, by electrical means, as condenser-discharges, vacuum tube experiments, etc. Dr. Lodge's recent work on lightning discharges, especially the experiments described in his second lecture, deserves the most careful attention, as a substantial addition to our knowledge of the subject, and also because it is, so far as I know, the first serious attempt to treat the subject electromagnetically.

The fluids are played out; they are fast evaporating into nothingness. The whole field of electrostatics must be studied from the electromagnetic point of view to obtain an adequately comprehensive notion of the facts of the case; and it is here that Dr. Lodge's experiments are also useful.

Independently of this, I should not be surprised to find that a new fact is contained in some of the experiments. Now a new fact is a serious matter, and its existence can only be granted upon the most conclusive evidence, of varied nature. There is already some independent evidence, viz., in Kundt's recent paper on the speed of light in metals. But it is scarcely sufficient.

There is the plainest possible evidence that with waves of telephonic frequency the magnetic force and the flux induction are proportionate, and that their ratio is a large number in iron. I have observed, and I read that Ayrton and Perry have also observed, decrease of the inductivity with increased wave-frequency. But, at least with me, it went only a little way, and I had not the opportunity to extend the experiments.

Now a conducting wire at the first moment of receiving a wave (in the dielectric, of course) performs the important function of guiding it and preventing its dissipation in space; and besides that, the nature of the conductor partly determines what impedance the wave suffers, causing a reflection back, with heaping up behind, so to speak, of the electric disturbance. But at first the conduction-current is purely superficial. It is clear then that at the very front of a wave, where conduction is just commencing on the surface, the conductor cannot be treated as if it had the same properties (conductivity, inductivity, permittivity) as if it were material in bulk, for only a thin layer of molecules is concerned. We therefore do not know what the true boundary condition is when pushed to the extreme. And yet it may be that this unknown condition may sometimes serve to determine a choice of paths.

Thus, iron may behave, superficially, as if it were non-magnetic. (This does not mean that the inductivity of an iron wire is unity.) In

Kundt's experiments, electromagnetically interpreted, the inductivity of iron is nowhere; the conductivity, too, must, in other cases as well, be less than the steady value. This corroborates Maxwell's remarks concerning gold-leaf. Of course the application of electromagnetic principles to the passage of light through material substances is at present in a very tentative state; so that too much importance should not be attached to the speculations one may be led to make in these matters.

(If a conductor could be treated as homogeneous right up to its surface, the initial resistance of unit of surface I calculate to be $4\pi\mu v$, where μ is the inductivity and v the speed of transmission in the conductor. But neither μ nor v can be considered to be known in the case of iron.) [See p. 437, vol. II.]

Another matter I wish to direct attention to is this. Dr. Lodge has described some experiments relating to the reflection of waves sent along a circuit. It will also be in the knowledge of some readers that Sections XL. to XLVI. of my "Electromagnetic Induction and its Propagation," *Electrician*, June to September, 1887 (and a straggler, XLVII., December 31, 1887), deal with the subject of the transmission of waves along wires, their reflection, absorption, etc., by a new method.

Now I find that there is an idea prevalent that it is only possible for very advanced mathematicians to understand this subject. It is true that when it is comprehensively considered it is by no means easy. But I desire to call attention to the fact (as I did in one or more of the articles referred to) that all the main features of the transmission, reflection, absorption, etc., of waves can be worked out (as done there by me) by elementary algebra.

I was informed (substantially) that no one read my articles. Possibly some few may do so now, with Dr. Lodge's experiments in practical illustration of some of the matters considered.

My next communication, I may add (written in September, 1887), is on the important subject of the measure of the inductance of circuits, and its true effects, in amplification of preceding matter. It has also special reference to some experimental observations. It has also some valuable annotations by an eminent authority. [Art. XXXVIII., vol. II., p. 160.]

P.S.—In connection with lightning discharges, I may remark that it is usual, and seems very natural, to assume that the discharge is initiated at the place of the visible spark—the crack, so to speak. But my recent investigations lead me to conclude that this is by no means necessary, and that the strongest dielectric can be disrupted by a suitable convergence of a wave to a centre *or an axis*, starting with any steady field.

For instance, if in a cylindrical portion of a dielectric the displacement be uniform, and parallel to the axis, and it be allowed to discharge, the convergence of the resulting wave to the axis causes the electric force to mount up infinitely there, momentarily; hence disruption.

But I do not pretend to give a complete theory of the thundercloud. It is only a detail.

P.P.S.—In Dr. Fleming's recent articles on the theory of alternating currents, I observe that he calls the component Ln of the impedance $(R^2+L^2n^2)^{\frac{1}{2}}$ the "inductive resistance."

I should myself have scarcely thought that it deserved a name, for of course we must draw the line somewhere. But the fact that Dr. Fleming has given it a name is evidence that he found it convenient to do so. Taking it, then, for granted that it should have a special name, I can only object to the one chosen that it creates two kinds of resistance. I desire to recognise but one—*the* resistance. I might, for instance, call Ln the hindrance. Thus, in the case of a coil, R is the *electric* resistance, Ln the *magnetic* hindrance, and their resultant the impedance. But in any case it would not be a term for popular use.

August 13, 1888.

XLVI. PRACTICE *VERSUS* THEORY.—ELECTROMAGNETIC WAVES.

[*The Electrician*, Oct. 19, 1888, p. 772.]

THE remarkable leader in *The Electrician* for Oct. 12, 1888, states very lucidly some of the ways in which theory and practice seem to become antagonistic. There is, however, one point which does not, I think, receive the attention it deserves, which is, that it is the duty of the theorist to try to keep the engineer who has to make the practical applications straight, if the engineer should plainly show that he is behind the age, and has got shunted on to a siding. The engineer should be amenable to criticism.

Another point is this. It might appear from the concluding paragraph of the article to which I have referred that the points at issue between Mr. Preece's views and my own were mere matters of complicated corrections, not affecting the main argument much. But the case is far different. A complete change of type is involved.

Now, I shall have great pleasure, when opportunity offers, in endeavouring to demonstrate that such is the case, and that the despised self-induction is the great moving agent; that although Mr. Preece, in the presence of some distinguished mathematicians, recently boasted * that *he* made mathematics his *slave*, yet it is not wholly improbable that he is a very striking and remarkable example of the opposite procedure; that although Mr. Preece, who, as a practical engineer, knows all about electromagnetic inertia and throttling, does not see the use of inductance, impedance, and all that sort of thing, yet there is not wanting evidence to make it not wholly unbelievable that Mr. Preece is not quite fully acquainted with the subject as generally

* [The Discussion on Lightning Conductors at the Bath meeting of the B.A., reported at length in *The Electrician*, Sept. 21 and 28, 1888, is interesting reading, and is made quite amusing by Mr. Preece's attack upon mathematicians to his own exaltation, and the rejoinders thereto.]

understood; that, for example, his coefficient of self-induction is of very different size, and has very different properties, from the theoretical one; and that Mr. Preece's knowledge of the manner of transmission of signals, though it may not be "extensive," is certainly "peculiar."

I may take the opportunity of adding that on account of a certain peculiar concurrence and concatenation of circumstances last year rendering it impossible for me to communicate the practical applications of my theory (based upon Maxwell's views, so far as the higher developments are concerned), either *viâ* the S. T.-E. and E. or four other channels, the resultant effect of which was to screen Mr. Preece from criticism, combined with the fact that Mr. Preece, in his papers to the Royal Society, British Association, and S. T.-E. and E. has taken his stand upon Sir W. Thomson's celebrated theory of the submarine cable, I have been forced, with great reluctance, to assume what may have appeared to be, superficially, an apparently unnecessarily aggressive attitude towards the said theory. But those who are acquainted with the subject will know that there is no antagonism whatever between the electrostatic theory and the wider theory; and those, further, who may be acquainted with the peculiar concurrence I have mentioned will understand the meaning of the apparent aggressiveness.

In addition, it seems to me to be almost mathematically certain that Sir W. Thomson would emphatically repudiate the very notion of applying his theory of the diffusion of potential to cases to which it does not apply, and to which it was never meant to apply; and I cannot find any evidence in his writings that he ever would have made such a misapplication.

P.S.—Is self-induction played out? I think not. What *is* played out is what we may call (uniting the expressions of Ayrton, Preece, Thomson, and Lodge) the British engineer's self-induction, which stands still, and won't go. But the other self-induction, in spite of strenuous efforts to stop it, goes on moving; nay, more, it is accumulating momentum rapidly, and will, I imagine, never be stopped again. It is, as Sir W. Thomson is reported to have remarked, with a happy union of epigrammatic force and scientific precision, "in the air." Then there are the electromagnetic waves. Not so long ago they were nowhere; now they are everywhere, even in the Post Office. Mr. Preece has been advising Prof. Lodge to read Prof. Poynting's paper on the transfer of energy. This is progress, indeed! Now these waves are also in the air, and it is the "great bug" self-induction that keeps them going.

On this question of waves I take the opportunity of referring to a point mentioned at the Bath meeting by Prof. Fitzgerald. That physicist, in directing attention to Hertz's recent experiments, considered that they demonstrated the truth of the propagation of waves in time through the ether; but that, on the other hand, the waves sent along a circuit did not do so, because they might be explained by action at a distance.

It seems to me, however, that the more closely we look at the matter the less distinction there is between the two cases, and that to an unbiassed mind the experiments of Prof. Lodge, sending waves of short length into a miniature telegraph circuit, with consequent "resonance" effects, are equally conclusive to those of Hertz on the point named; in one respect, perhaps, more so, because their theory is simpler, and can be more closely followed.

But, after all, has it been demonstrated that we cannot explain the propagation of electromagnetic waves in time by action at a distance, pure and simple? I suggest the following as evidence to the contrary. Take the case of Maxwell's non-conducting dielectric. Let the electric-current element *cause* magnetic force at a distance according to Ampère's law, and let the magnetic current element *cause* electric force at a distance according to the same law with sign reversed. Then

$$\text{curl}\,\mathbf{H} = c\dot{\mathbf{E}}, \qquad \text{and} \qquad -\text{curl}\,\mathbf{E} = \mu\dot{\mathbf{H}}$$

follow, and propagation of waves in time follows. That is, by instantaneous mutual action at a distance between electric-current elements, and also between magnetic-current elements, we get propagation in time. Of course the currents may be oppositely moving electric or magnetic fluids or particles.

Whether there is any flaw here or not, it is scarcely necessary for me to remark that I do not believe in action at a distance. Not even gravitational.

XLVII. ELECTROMAGNETIC WAVES, THE PROPAGATION OF POTENTIAL, AND THE ELECTROMAGNETIC EFFECTS OF A MOVING CHARGE.

[*The Electrician*; Part I., Nov. 9, 1888, p. 23; Part II., Nov. 23, 1888, p. 83; Part. III., Dec. 7, 1888, p. 147; Part IV., Sept. 6, 1889, p. 458.]

PART I.

IN connection with the letters of Profs. Poynting and Lodge in *The Electrician*, Nov. 2, 1888, I believe that the following extract from a letter from Sir William Thomson (which I have permission to publish) will be of interest [see Postscript, p. 483, vol. II., to elucidate]:—

"I don't agree that velocity of propagation of electric potential is a merely metaphysical question. Consider an electrified globe, A, moved to and fro, with simple harmonic motion, if you please, to fix the ideas. Consider very quickly-acting electroscopes B, B′, at different distances from A. If the indications of B, B′ were exactly in the same phase, however their places are changed, the velocity of propagation of electric potential would be infinite; but if they showed differences of phase, they would demonstrate a velocity of propagation of electric potential.

"Neither is velocity of propagation of 'vector-potential' metaphysical. It is simply the velocity of propagation of electromagnetic force—the velocity of 'electromagnetic waves,' in fact."

Taking the second point first, it is, I think, clear that if by the propagation of vector-potential is to be *understood* that of electric and magnetic disturbances, it is merely the mode of expression that is in question. I am myself accustomed to mentally picture the electric and magnetic forces or fluxes, and their propagation, which takes place at the speed of light or thereabouts, because they give the most direct representation of the state of the medium, which, I think, must be agreed is the real physical subject of propagation. But if we regard the vector-potential directly, then we can only get at the state of the medium by complex operations, and we really require to know the vector-potential both as a function of position and of time, for its space-variation has to furnish the magnetic force, and its time-variation the electric force; besides which, there is sometimes the space-variation of a scalar potential in addition to be regarded, before we can tell what the electric force is. Besides this roundaboutness, it implies a knowledge of the full solution, and if we do not possess it, it is much simpler to think of the propagation of the electric and magnetic disturbances, and I find that this method works out much more easily in the solution of problems.

The other question will, I believe, be found to be ultimately of precisely the same nature. Start with the sphere A at rest, and the field steady, and consider two external points, P and P′, at different distances. The electric force at them has different values, and the whole field has a potential. But now give the sphere a displacement, and bring it to rest again in a new position. Is the readjustment of potential instantaneous? I should say, Certainly not, and describe what happens thus. When the sphere is moved, magnetic force is generated at its boundary (lines circles of latitude, if the axis be the line of motion), and with it there is necessarily disturbance of electric force. The two together make an electromagnetic wave, which goes out from the sphere at the speed of light, and at the front of the wave we have $E=\mu vH$, where E is the electric and H the magnetic force intensity. Before the front reaches P or P′ we have the electric field represented by the potential function, but after that it cannot be so represented until the magnetic force has wholly disappeared, when again we have a steady field representable by a potential function. It is difficult to see how to plainly differentiate any propagation of potential *per se*.

If the motion is simple-harmonic, there is a train of outward waves and no potential. I imagine that an electroscope, if infinitely sensitive and without reactions, would register the actual state of the electric field, irrespective of its steadiness. By an electroscope, as this is a purely theoretical question, I understand the very simplest one, a very small charge at a point; or, say, the unit charge, the force on which is the electric force of the field.

When these things are closely examined into, if the facts as regards the propagation of disturbances (electric and magnetic) are agreed on, the only subject of question is the best mode of expressing them, which I believe to be in terms of the forces, not potentials.

But there really is infinite speed of propagation of potential sometimes;

on examination, however, it is found to be nothing more than a mathematical fiction, nothing else being propagated at the infinite speed.

It will be understood that I preach the gospel according to my interpretation of Maxwell, and that any modification his theory of the dielectric may receive may involve a fresh kind of propagation at present not in question.

Nov. 5, 1888.

PART II.

The question raised by Prof. S. P. Thompson (in *The Electrician*, Nov. 16, 1888, p. 54) as to whether the motion of an uncharged dielectric through a field of electric force produces magnetic effects must, I think, be undoubtedly answered in the affirmative. As the distribution of displacement varies, its time-variation is the electric current, with determinable magnetic force to match. When the speed of motion is a small fraction of that of light, we may regard the displacement as having at every moment its proper steady distribution, so that there is no difficulty in estimating the magnetic effects, except, it may be, of a merely mathematical character. For instance, the case of a sphere moving in a field which would be uniform were the sphere absent, may be readily attacked, and does perfectly well to illustrate the general nature of the action.

But if the moved dielectric have the same electric permittivity as the surrounding medium, so that there is no difference made in the steady distribution, the question which may be now raised as to the possible production of transient disturbances is one to which the above theory does not present any immediate answer. I believe that the body will be magnetized transversely to the electric displacement and the velocity. [The motional magnetic force is referred to.]

Another question, somewhat connected, is contained in Prof. Poynting's suggestion (in letter to Prof. Lodge, *The Electrician*, p. 829, vol. XXI.) that electric displacement may possibly be produced without magnetic force by the agency of pyroelectricity. But, whatever the agency, it would, I conceive, be a new fact—quite outside Maxwell's theory legitimately developed. We may have subsidence of electric displacement without magnetic force; but I cannot see any way to produce it.

But the main subject of this communication is the electromagnetic effect of a moving charge. That a moving charge is equivalent to an electric current-element is undoubted, and to call it a convection-current, as Prof. S. P. Thompson does, seems reasonable. The true current has three components, thus,

$$\operatorname{curl} \mathbf{H} = 4\pi(\mathbf{C} + \dot{\mathbf{D}} + \rho\mathbf{u}),$$

where $\mathbf{H}$ is the magnetic force, $\mathbf{C}$ the conduction-current, $\mathbf{D}$ the displacement, and ρ the volume-density of electrification moving with velocity $\mathbf{u}$. The addition of the term $\rho\mathbf{u}$ is, I presume, the extension made by Prof. Fitzgerald to which Prof. S. P. Thompson refers. At any rate, I can at present see no other.

There are several ways of arriving at the conclusion that a moving charge must be regarded as an electric current; but, when that is admitted, we are very far from knowing what its magnetic effect is. No cut-and-dried statement of it can be made, because it varies according to circumstances. The magnetic field, whatever it be in a given case, is not that of a current-element (supposing the charge to be at a point), for that is anti-Maxwellian, but is that of the actual system of electric current, which is variable.

Thus, in the case of motion at a speed which is a small fraction of that of light, the magnetic field (as found by Prof. J. J. Thomson) is the same as that of Ampère's current-element represented by $\rho\mathbf{u}$; that is, a current-element whose direction is that of $\mathbf{u}$ and whose moment is ρu, if u is the tensor of $\mathbf{u}$ (understanding by "moment," current-density × volume); but the true current to correspond bears the same relation to the current-element as the induction of an elementary magnet bears to its magnetic moment. The magnetic energy due to the motion of a charge q upon a sphere of radius a in a medium of inductivity μ, at a speed u which is only a very small fraction of that of light, is expressed by $\frac{1}{3}\mu q^2u^2/a$. But if the speed be not a small fraction of that of light, the result is very different. Increasing the speed of the charge causes not merely greater magnetic force but changes its distribution altogether, and with it that of the electric field. It is no use discussing the potential. There is not one. The magnetic field tends to concentrate itself towards the equatorial plane, or plane through the charge perpendicular to the line of motion. When the speed equals that of light itself this process is complete, and the result is simply a plane wave (electromagnetic).

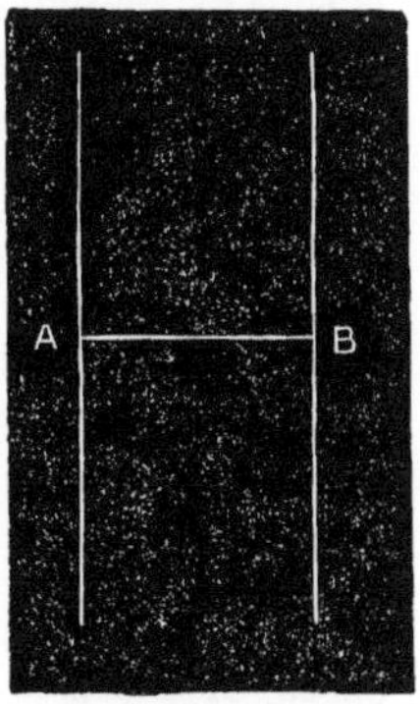

Since a charge at a point gives infinite values, it is more convenient to distribute it. Let it be, first, of linear density q along a straight line AB, moving in its own line at the speed of light. Then the field is contained between the parallel planes through A and B perpendicular to AB, and is completely given by

$$E/\mu v = H = 2qu/r,$$

where E and H are the intensities of the electric and magnetic forces at distance r from AB. The lines of $\mathbf{E}$ radiate uniformly from AB in all directions parallel to the planes; those of $\mathbf{H}$ are everywhere perpendicular to those of $\mathbf{E}$, or are circles centred upon AB. Outside this electromagnetic wave there is no disturbance. I should remark that the above is a description of the *exact* solution. It is, of course, nothing like the supposed field of a current-element AB.

To still further realize, we may substitute a cylindrical distribution for the linear, and then, again, terminate the lines of $\mathbf{E}$ on another cylindrical surface between the bounding planes. To find the resulting distributions of $\mathbf{E}$ and $\mathbf{H}$ (always perpendicular) may be done by superimposition of the elementary solutions, or by solving a bidimensional problem in a well-known manner.

Those who are acquainted with my papers in this journal will recognise that what we have arrived at is simply the elementary plane wave travelling along a distortionless circuit. All roads lead to Rome!

Returning to the case of a charge q at a point moving through a dielectric, if the speed of motion exceeds that of light, the disturbances are wholly left behind the charge, and are confined within a cone, AqB. The charge is at the apex, moving from left to right along Cq. The semi-angle, θ, of the cone, or the angle AqC, is given by

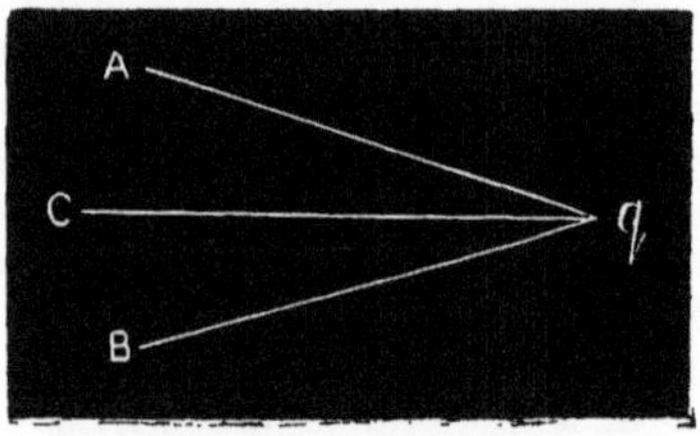

$$\sin\theta = v/u,$$

where v is the speed of light, and u that of the charge. The magnetic lines are circles round the axis, or line of motion. The displacement is away from q, of course, and of total amount q, but not uniformly distributed within the cone. The electric current is towards q in the inner part of the cone, and away from q in the outer.

It will be seen that the electric stress tends to pull the charge back. Therefore, applied force on q in direction Cq is required to keep up the motion. Its activity is accounted for by the continuous addition at a uniform rate which is being made to the electric and magnetic energies at q. For the motion at the wave-front, at any point on Aq or Bq, is perpendicularly outward, not towards q. Whilst the cone is thus expanding all over, the forward motion of q continually renews the apex, and keeps the shape unchanged.

Steady motion alone is assumed.

To avoid misconception I should remark that this is not in any way an account of what would happen if a charge were impelled to move through the ether at a speed several times that of light, about which I know nothing; but an account of what would happen if Maxwell's theory of the dielectric kept true under the circumstances, and if I have not misinterpreted it. [See footnote on p. 516, later.]

Nov. 18, 1888.

Part III.

All disturbances being propagated through the dielectric ether at the speed of light, when, therefore, a charge is in motion through the medium, the discussion of the effects produced naturally involves the consideration of three cases, those in which the speed u of the charge is less than, or equal to, or greater than v, that of light.

In a previous communication [Part II. above], I gave the complete and very simple solution of the intermediate case of equality of speeds. A formal demonstration is unnecessary, as the satisfaction of the necessary conditions may be immediately tested.

But I was not then aware that the case $u<v$ admitted of being presented in a nearly equally-simple form. That such is the fact is rather surprising, for it is very exceptional to arrive at simple results, and these now in question are sufficiently free from complexity to take a place in text-books of electricity.

Let the axis of z be the line of motion of the charge q at speed u. Everything is symmetrical with respect to this axis. The lines of electric force are radial out from the charge. Those of magnetic force are circles about the axis. The two forces are perpendicular. Having thus settled the directions, it only remains to specify their intensities at any point P distant r from the charge, the line r making an angle θ with the axis. Let E be the intensity of the electric, and H of the magnetic force. Then, if c is the permittivity and μ the inductivity, such that $\mu c v^2 = 1$, we have

$$(u<v) \quad \left\{ \begin{array}{ll} cE = \dfrac{\dfrac{q}{r^2}\left(1-\dfrac{u^2}{v^2}\right)}{\left(1-\dfrac{u^2}{v^2}\sin^2\theta\right)^{\frac{3}{2}}}, & \text{.....................(A)} \\ H = cEu\sin\theta. & \text{.........................(B)} \end{array} \right.$$

That (A), (B) represent the complete solution may be proved by subjecting them to the proper tests. Premising that the whole system is in steady motion at speed u, we have to satisfy the two fundamental laws of electromagnetism :—

(1). (Faraday's law). The electromotive force of the field [or voltage] in any circuit equals the rate of decrease of the induction through the circuit (or the magnetic current $\times -4\pi$).

(2). (Maxwell's law). The magnetomotive force of the field [or gaussage] in any circuit equals the electric current $\times 4\pi$ through the circuit.

Besides these, there is continuity of the displacement to be attended to. Thus :—

(3). (Maxwell) The displacement outward through any surface equals the enclosed charge.

Since (A) and (B) satisfy these tests, they are correct. And since no unrealities are involved, there is no room for misinterpretation.

When u/v is very small, we have, approximately,

$$cE = \frac{q}{r^2}, \qquad H = \frac{qu}{r^2}\sin\theta,$$

representing Prof. J. J. Thomson's solution—that is, the lines of displacement radiate uniformly from the charge, and the magnetic force is that of the corresponding displacement-currents together with the moving charge regarded as a current-element of moment qu. Instantaneous action through the medium is involved—that is, to make the solution quite correct.

That the lines of electric force should remain straight as the speed of the charge is increased is itself a rather remarkable result. Examining

(A), we see that the effect of increasing u is to concentrate the displacement about the equatorial plane $\theta=\frac{1}{2}\pi$. Self-induction does it. In the limit, when $u=v$, the numerator vanishes, making $E=0$, $H=0$ everywhere except at the plane mentioned, where, by reason of the denominator becoming infinitely small in comparison with the numerator, the displacement is all concentrated in a sheet, and with it the induction, forming a plane electromagnetic wave, as described (and realized) in my previous communication.

If we terminate the field described in (A) and (B) on a spherical surface of radius a, instead of continuing it up to the charge q at the origin, we have the case of a perfectly conducting sphere of radius a possessing a total charge q, moving steadily at speed u through the dielectric ether. As the speed is increased to v, the charge all accumulates at the equator of the sphere. [See footnote on p. 514, later.]

But after that? This brings us to the third case of $u>v$, and here I have so-far failed to find any solution which will satisfy all the necessary conditions without unreality. The description at the close of Part II. must therefore be received as a suggestion, at present unconfirmed. I hope to consider the matter in a future communication.

P.S.—In a recent number Mr. W. P. Granville raised the question of action through a medium being only action at a short distance instead of a long one, and asked for instruction. His inquiry has elicited no response. This is not, however, because there is nothing to be said about it. The matter did not escape the notice of the "anti-distance-action sage." My own opinion is that the question involved is, if not metaphysical, dangerously near to being so; consequently, whole books might be devoted to it. At present, however, I think it is more useful to try to find out what happens, and to construct a medium to make it happen; after that, perhaps, the matter referred to may be more advantageously discussed. The well of truth is bottomless.

PART IV.

In previous communications [above] I have discussed this matter. Referring to the case of steady rectilinear motion, I gave a description of the result when the speed of the charge exceeds that of light, obtained mainly by general reasoning, and stated my inability to find a solution to represent it. The displacement cannot be outside a certain cone of semi-vertical angle whose sine equals the ratio v/u of the speed of light to that of the charge, which is at the apex.

In the *Phil. Mag.* for July, 1889, Prof. J. J. Thomson has examined this question. Like myself, he fails to find a solution within the cone; but concludes that the displacement is confined to its surface. If so, it must form, along with the magnetic induction, an electromagnetic wave. But it may be readily seen that such a wave is impossible, having no stability.

For as the charge moves from A to B, a given surface-element, C, would move to D. In doing so its area would vary directly as its distance from the apex, and the energy in the element would therefore vary inversely as its distance from the apex, and the forces, electric and magnetic, would therefore vary inversely as the square root of the distance from the apex, instead of inversely as the distance, which is obviously necessary in order that the displacement may be confined to the surface. This conflict of conditions constitutes instability. In the *Phil. Mag.* for April, 1889, I suggested that whilst there must be a solution of some kind, one representing a *steady* state was impossible. This conclusion is confirmed by the failure of Prof. Thomson's proposed surface-wave to keep itself going.

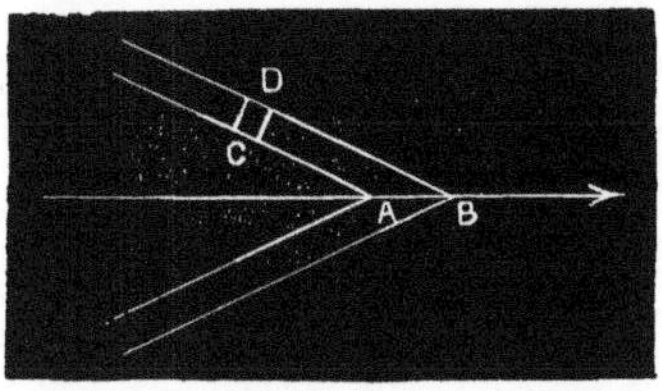

Prof. Thomson, who otherwise confirms my results, has also extended the matter by supposing that the medium itself is set in motion, as well as the electrification. This is somewhat beyond me. I do not yet know certainly that the ether can move, or its laws of motion if it can. Fresnel thought the earth could move through the ether without disturbing it; Stokes, that it carried the ether along with it, by giving irrotational motion to it. Perhaps the truth is between the two. Then there is the possibility of holes in the ether, as suggested by a German philosopher. When we get into one of these holes, we go out of existence. It is a splendid idea, but experimental evidence is much wanting.

But if we consider that the medium supporting the electric and magnetic fluxes is really set moving when a body moves, and assume a particular kind of motion, it is certainly an interesting scientific question to ask what influence the motion exerts on the electromagnetic phenomena. I do not, however, think that any new principles are involved.

The general connections of **E** and **H**, referred to fixed space without conductivity, being

$$\mathrm{curl}(\mathbf{e}-\mathbf{E}) = \mu p\mathbf{H}, \qquad (1)$$

$$\mathrm{curl}(\mathbf{H}-\mathbf{h}) = cp\mathbf{E}, \qquad (2)$$

where p stands for d/dt and **e** and **h** are the impressed parts of **E** and **H**; if there is also motion of electrification, we have to consider it to constitute a convection-current, a part of the true current, and so make (2) become

$$\mathrm{curl}(\mathbf{H}-\mathbf{h}) = cp\mathbf{E} + 4\pi\rho\mathbf{u}, \qquad (3)$$

where ρ is the density of electrification, whose velocity is **u**. [See Part II.] It now remains to specify **e** and **h**. They are zero when the medium supporting the fluxes is at rest. But if it moves, and its velocity is **w**, there is, first, the electric force due to motion in a magnetic field,

$$\mathbf{e} = \mu \mathrm{V}\mathbf{w}\mathbf{H}, \qquad (4)$$

which is well known; and next the magnetic force due to motion in an electric field,

$$\mathbf{h} = c\mathrm{V}\mathbf{E}\mathbf{w}, \qquad (5)$$

which is not so well known. (First, I believe, given by me in the third Section of "Electromagnetic Induction and its Propagation," *The Electrician*, January 24, 1885 [vol. I., p. 446]; again, obtained in a different way in Section XXII., January 15, 1886 [vol. I., p. 546]; see also *Phil. Mag.*, August, 1886 [vol. II., Art. L.], and an example of the use of (4) and (5) in *The Electrician*, April 12, 1889, p. 683 [vol. II., Art. LI.].)

The mechanical force called by Maxwell the "electromagnetic force" is $\mathbf{VCB}$, where $\mathbf{C}$ is the true current and $\mathbf{B}$ the induction. It is the force on the matter supporting electric current. Let it move. If $\mathbf{w}$ is its velocity, the activity of the force is

$$\mathbf{wVCB} = \mathbf{CVBw} = -\mathbf{eC}. \qquad (6)$$

Similarly, as I obtained in Section XXII. above referred to, there is a mechanical force (the magneto-electric) on matter supporting magnetic current $\mathbf{G} = \mu p\mathbf{H}/4\pi$, expressed by $4\pi\mathbf{VDG}$, and its activity is

$$4\pi\mathbf{wVDG} = 4\pi\mathbf{GVwD} = -\mathbf{hG}. \qquad (7)$$

Of course $\mathbf{e}$ and $\mathbf{h}$ are reckoned as impressed forces, which is the reason of the change of sign. *Their* activities are $\mathbf{eC}$ and $\mathbf{hG}$.

It should be remarked further, that the above expressions for $\mathbf{e}$ and $\mathbf{h}$ are not *certain*. For I have shown that the sources of all disturbances are the lines of curl of the impressed forces (*Phil. Mag.*, Dec., 1887) [vol. II., p. 362], and that the fluxes produced depend solely upon the curls of $\mathbf{e}$ and $\mathbf{h}$, both as regards the steady fluxes and the variable ones leading to them. We may, therefore, use any other expressions for $\mathbf{e}$ and $\mathbf{h}$ which have the same curls as the above. And, in fact, we see that equations (1) and (2) only contain their curls.

Equations (1) and (3), with $\mathbf{e}$ and $\mathbf{h}$ defined by (4) and (5), therefore enable us to determine the effect of the moving medium. Prof. Thomson also arrives at (4) and (5), and at the "magneto-electric force," in his paper to which I have referred, by an entirely different method. And to show how well things fit together, he concludes, from the consideration of the moving medium, that a moving electrified surface is a current-sheet, which is another way of saying that a convection current is a part of the true current, as expressed in (3). I must, however, disagree with Prof. Thomson's assumption that the motion must be irrotational. It would appear, by the above, that this limitation is unnecessary.

As an example, and to introduce a new point, take the case of a charge q moving at speed u along the axis of z. It will come to the same thing if we keep the charge at rest, and move the medium the other way. We then use the equations (1) and (2), and in them use (4) and (5) with $\mathbf{w} = -\mathbf{u}$. Now when the steady state is arrived at, we have $p = 0$, so (1) and (2) become

$$\text{curl}(\mu\mathbf{VHu} - \mathbf{E}) = 0, \qquad (8)$$

$$\text{curl}(\mathbf{H} - c\mathbf{VuE}) = 0. \qquad (9)$$

In addition, the divergence of **D** must be q at the origin, and the divergence of **B** must be zero. The latter gives, applied to (9),

$$\mathbf{H}=c\mathrm{V}\mathbf{u}\mathbf{E}, \qquad (10)$$

which gives **H** fully in terms of **E**. Eliminate **H** from (8) by means of (10), and we get

$$\mathrm{curl}\left(\mu c\mathrm{V}\mathbf{u}\mathrm{V}\mathbf{E}\mathbf{u}-\mathbf{E}\right)=0, \qquad (11)$$

or

$$\mathrm{curl}\left[\frac{u^2}{v^2}\left(\mathbf{E}-E_3\mathbf{k}\right)-\mathbf{E}\right]=0, \qquad (12)$$

where E_3 is the z-component of **E** and **k** a unit vector along **z**; or, integrating, and writing the three components,

$$E_1=-\frac{dP}{dx}, \qquad E_2=-\frac{dP}{dy}, \qquad E_3=-\left(1-\frac{u^2}{v^2}\right)\frac{dP}{dz}, \qquad (13)$$

where P is a scalar potential. Here is the new point. There *is* a potential, of a peculiar kind. The displacement due to the moving charge is distributed in precisely the same way as if it were at rest in an eolotropic medium, whose permittivity is c in all directions transverse to the line of motion, but is smaller, viz., $c(1-u^2/v^2)$, along that line and parallel to it. The potential P is given by

$$P=\frac{q}{c[(x^2+y^2)(1-u^2/v^2)+z^2]^{\frac{1}{2}}}. \qquad (14)$$

It is a particular case of eolotropy. In general, c_1, c_2, c_3, the principal permittivities, are all unequal. Then, with q at the origin, the potential is

$$P=\frac{q}{(c_1c_2c_3)^{\frac{1}{2}}\left(\frac{x^2}{c_1}+\frac{y^2}{c_2}+\frac{z^2}{c_3}\right)^{\frac{1}{2}}}. \qquad (15)$$

Observe that although the electric force in the substituted problem of a charge at rest in an eolotropic medium is the slope of a potential; yet it is not so when the medium is isotropic, and moves past the fixed charge, or *vice versâ*, although the distributions of displacement are the same.

When $u=v$, we abolish the permittivity along the z-axis in the substituted case, so that the displacement must be wholly transverse. We then have the plane electromagnetic wave. When u is greater than v it makes the permittivity negative along z; this is an impossible electrical problem, and furnishes another reason for supposing that there can be no steady state in the corresponding electromagnetic problem.

It now remains to find what *would* happen if electrification were conveyed through a medium faster than the natural speed of propagation of disturbances. There is the cone; but what takes place within it?

Aug. 25, 1889.

XLVIII. THE MUTUAL ACTION OF A PAIR OF RATIONAL CURRENT-ELEMENTS.

[*The Electrician*, Dec. 28, 1888, p. 229.]

STRICTLY speaking, there is no such thing, from the Maxwellian point of view, as mutual action between current-elements. Suppose, however, we have the well-known Ampèrian field of magnetic force usually ascribed to a current-element at one place, and a similar one centred at another place, it is clear that the forces concerned are quite definite, according to Maxwell's theory. The electric current of such an arrangement is closed. It is related to the nominal current, viz., in the element, in the same way as the induction of an elementary magnet is related to its magnetic moment, as regards the space-distribution. We may term the arrangement a rational current-element. If we take any number of equal rational current-elements and put them in line, with opposite poles in contact, only the terminal poles are left free, so that the current consists of a straight or curved line or tube of current, joining two points, A and B, with external continuity produced by means of an equal current diverging from the positive pole B in all directions uniformly, and converging to the negative pole A in a similar manner. Of course the tubes of current from B join on to those at A, and are curved; but it would only confuse matters to superimpose the two systems of polar current, which are much better kept separate. The rational current-element itself is to be regarded as an infinitely small volume with a uniform current distributed in it, *and* of the complementary currents from and to the poles. The moment is current-density multiplied by volume, ignoring the complementary currents altogether for the moment. What the actual current in the element may be does not matter much. It depends on the shape of the element. Thus, if spherical, the nominal strength of current, reckoned by its moment, is half as great again as the real, owing to the back action of the polar current. We need only consider the moment, which is fully representative of the external magnetic field, which, it should be remembered, is that due to the moment, according to Ampère's rule. To further illustrate, take the case of a charge, q, moving at speed u, small compared with that of light [p. 495, vol. II.], through a dielectric. The moment is qu; the magnetic force is qu/r^2 at distance r in the equatorial plane, and elsewhere proportional to the cosine of the latitude. The actual state of things in the element may require very complex calculations to discover, but is of little importance.

The mutual action of two German or irrational current-elements is indeterminate, and so we get a large number of so-called theories of electrodynamics. But the mutual action of a pair of rational current-elements is a legitimate subject of inquiry, is determinate, and does not involve any action at a distance. The quantity from which, by dynamical methods, we derive the forces (mechanical) on the elements, is the mutual magnetic energy (leaving out of consideration the electrostatic force, if any), that part of the magnetic energy due to both rational current-elements. If I have correctly calculated it, the mutual energy M of elements whose distance apart is r, in the medium of inductivity μ, is expressed by

$$M = \frac{\mu}{2r}\left(2u_1v_1 + u_2v_2 + u_3v_3\right),$$

where u_1, u_2, u_3 are the components of $\mathbf{C}_1$, the moment of the first element, and v_1, v_2, v_3 those of the second, $\mathbf{C}_2$, on the understanding that the axis of x is the line joining the elements, whilst the y and z axes are, as usual, perpendicular to it and to each other. In another form,

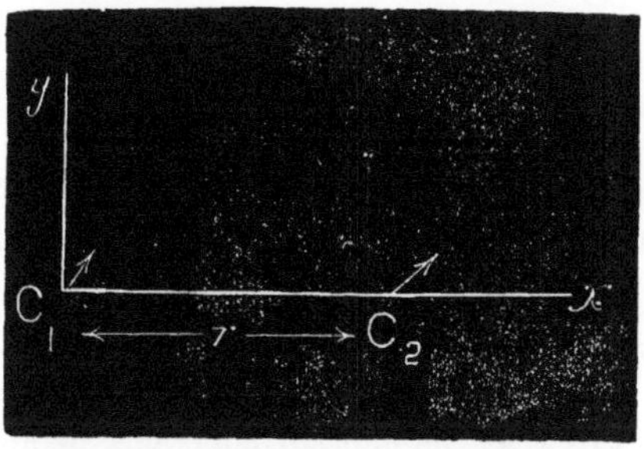

$$M = \mu\left(\frac{\cos\epsilon}{r} + \tfrac{1}{2}\frac{d^2r}{ds_1ds_2}\right)C_1C_2,$$

where ϵ is the inclination of the elements $\mathbf{C}_1$ and $\mathbf{C}_2$, parallel to $\mathbf{s}_1$ and $\mathbf{s}_2$.

If we substitute for r, in the differential coefficient, an arbitrary function R, we obtain the most general formula which will lead to Neumann's result for closed circuits. It is this R that is, by German methods, indeterminate. Rationalize the elements, and we fix it to be r. Clausius took $R=0$, I believe. It does not matter at all, so far as closed circuits are concerned, what formula we use, provided Neumann's result is complied with; but it is interesting to observe that the problem as stated by me has no uncertainty about it (except any possible working errors) and makes M definite, whilst it is not a mere mathematical abstraction (*i.e.*, the problem), but representative of (under certain circumstances) a reality. It is for these reasons that I mention the matter. For, as a matter of fact, I believe the whole method is fundamentally wrong, and of little practical service in the investigation of electromagnetism from the physical side, *i.e.*, with propagation in time through a medium. What does it matter about the current-elements? They are *not in it.* Still, such formulæ are sometimes of service, as, for instance, in the calculation of inductances.

It has been stated, on no less authority than that of the great Maxwell, that Ampère's law of force between a pair of current-elements is the cardinal formula of electrodynamics. If so, should we not be always using it? Do we *ever* use it? Did Maxwell, in his treatise? Surely there is some mistake. I do not in the least mean to rob Ampère of the credit of being the father of electrodynamics; I would only transfer the name of cardinal formula to another due to him, expressing the mechanical force on an element of a conductor supporting current in any magnetic field; the vector product of current and induction. There is something real about it; it is not like his force between a pair of unclosed elements; it is fundamental; and, as everybody knows, it is in continual use, either actually or virtually (through electromotive force) both by theorists and practicians.

Nov. 25, 1888.

XLIX. THE INDUCTANCE OF UNCLOSED CONDUCTIVE CIRCUITS.

In my communication on "The Mutual Action of Rational Current-Elements" [the last Art. XLVIII.] I described the meaning of, and gave the formula for, the mutual energy M of a pair of rational current-elements.

Thus, let C_1 and C_2 be their moments, r their distance apart, ϵ the angle between their directions $\mathbf{s}_1$ and $\mathbf{s}_2$, μ the magnetic inductivity of the medium (uniform), and M the mutual energy. Then,

$$M=\mu C_1 C_2\left(\frac{\cos\epsilon}{r}+\tfrac{1}{2}\frac{d^2 r}{ds_1 ds_2}\right). \qquad \text{(1)}$$

It follows immediately from this that the mutual inductance of any two linear circuits is

$$M=\mu\iint\left(\frac{\cos\epsilon}{r}+\tfrac{1}{2}\frac{d^2 r}{ds_1 ds_2}\right)ds_1 ds_2, \qquad \text{(2)}$$

M being now the mutual inductance. If the circuits are closed the second part contributes nothing, and we have

$$M_0=\mu\iint\frac{\cos\epsilon}{r}ds_1 ds_2, \qquad \text{(3)}$$

the common form of Neumann's equation, with the μ prefixed to adapt it to Maxwell's theory.

But if the lines are unclosed, then, according to my description of the nature of a rational current-element, the linear currents become closed by means of currents uniformly diverging from their positive ends, and uniformly converging to their negative ends. The second part of (2) is now finite. Let P_1 and P_2 be the positive poles, N_1 and N_2 the negative poles of the linear currents, and let the value of the second part of (2) be M_1. It is given by

$$M_1=\frac{\mu}{2}(P_1P_2+N_1N_2-P_1N_2-N_1P_2), \qquad \text{(4)}$$

where P_1N_2 means the length of the straight line joining P_1 to N_2, and similarly for the rest. We may, therefore, calculate M by Neumann's formula, applied to the linear circuits, and then add the correction (4) to obtain the complete expression.

A practical application is to the theory of a Hertzian oscillator, at least of a certain kind. Let a straight wire join two conducting spheres, or discs, etc. Imagine an impressed force to act in the wire, and to vary in any not too rapid manner. The current will leak out (or in) from (or to) the wire as well as the terminal conductors, but if they are relatively large nearly all the current will go across the air from one terminal conductor to the other, and we may ignore the wire-leakage. The permittance S is then that of the dielectric between the two spheres (say), and is quite definite. Also, if the changes of current are not too rapid, as mentioned, the current in the air will follow the lines or tubes of displacement. The inductance L is therefore also quite definite, in accordance with Maxwellian principles, so that the natural frequency of oscillation of the condenser-conductor circuit can be calculated with considerable precision from the dimensions.

If, as an illustrative approximation, we suppose the current to come from the centre of one sphere and go to that of the other, and then diverge or converge uniformly, we have to find the inductance L of a straight wire or tube of length l and radius a, with terminal continuations as before specified. In the *Phil. Mag.*, July, 1888, Prof. Lodge calculates L without any allowance for the current in the dielectric, viz., by Neumann's formula (3). We have therefore only to examine what the correction (4) amounts to.

In the case of two very close parallel lines, we may put

$$P_1P_2 = 0 = N_1N_2, \qquad \text{and} \qquad P_1N_2 = P_2N_1 = l,$$

so that the correction is simply $-l\mu$. That is, if the dielectric current is ignored, (3) overestimates M by the amount μl. The same applies when it is the inductance of a straight tube or solid wire that is in question. Deduct its length in centimetres from the uncorrected to obtain the true value, in c.g.s. electromagnetic units, *i.e.*, centimetres.

Prof. Lodge (*loc. cit.*) also gives the formula which Hertz says Maxwell's theory gives. On making the comparison, I find it is equivalent to adding, instead of deducting l, from the result of Neumann's formula.

It should be remarked, as an essential condition of the validity of the process described above, when practically applied, that the changes of current must not be too rapid. When the changes are slow the immense speed of propagation of disturbances through the air causes the electric displacement at any moment in the neighbourhood of the vibrator to be very nearly that which would obtain according to electrostatic principles, and the current to follow the tubes of displacement. But go to the other extreme, and imagine the changes to be so rapid that waves, whose length is a fractional part of the length of the vibrator, are produced. It is then clear that the theory would not apply at all, either as regards the inductance or the permittance. Now Hertz, in that series of brilliant experiments which have gone far towards practically establishing the truth of Maxwell's inimitable theory of the ether considered as a dielectric, sometimes employs waves which are not very much longer than the vibrator itself. Only close to the vibrator, therefore, do we have the electrostatic field (approximately) predominant, and we may expect a sensible error in applying the electrostatic theory. It is, however, quite easy—in fact, easier—to use longer waves. But in any case, the *exact* calculation of the permittance and inductance of a vibrator involves a good deal of mathematics to find relatively small corrections.

July 21, 1889.

L. ON THE ELECTROMAGNETIC EFFECTS DUE TO THE MOTION OF ELECTRIFICATION THROUGH A DIELECTRIC.

[*Phil. Mag.*, April, 1889, p. 324.]

Theory of the Slow Motion of a Charge.

1. THE following paper consists of, First, a short discussion of the theory of the *slow* motion of an electric charge through a dielectric, having for object the possible correction of previously published results. Secondly, a discussion of the theory of the electromagnetic effects due to motion of a charge at any speed, with the development of the complete solution in finite form when the motion is steady and rectilinear. Thirdly, a few simple illustrations of the last when the charge is distributed.

Given a steady electric field in a dielectric, due to electrification. It is sufficient to consider a charge q at a point, as we may readily extend results later. If this charge be shifted from one position to another, the displacement varies. In accordance, therefore, with Maxwell's inimitable theory of a dielectric, there is electric current produced. Its time-integral, which is the total change in the displacement, admits of no question; but it is by no means an elementary matter to settle its rate of change in general, or the electric current. But should the speed of the moving charge be only a very small fraction of that of the propagation of disturbances, or that of light, it is clear that the accommodation of the displacement to the new positions which are assumed by the charge during its motion is practically instantaneous in its neighbourhood, so that we may imagine the charge to carry about its stationary field of force rigidly attached to it. This fixation of the displacement at any moment definitely fixes the displacement-current. We at once find, however, that to close the current requires us to regard the moving charge itself as a current-element, of moment equal to the charge multiplied by its velocity; understanding by moment, in the case of a distributed current, the product of current-density and volume. The necessity of regarding the moving charge as an element of the "true current" may be also concluded by simply considering that when a charge q is conveyed *into* any region, an equal displacement simultaneously leaves it through its boundary.

Knowing the electric current, the magnetic force to correspond becomes definitely known if the distribution of inductivity be given;

and when this is constant everywhere, as we shall suppose now and later, the magnetic force is simply the circuital vector whose curl is 4π times the electric current; or the vector-potential of the curl of the current; or the curl of the vector-potential of the current, etc., etc. Thus, as found by J. J. Thomson,* the magnetic field of a charge moving at a speed which is a small fraction of that of light is that which is commonly ascribed to a current-element itself. I think it, however, preferable to regard the magnetic field as the primary object of attention; or else to regard the complete system of closed current derived from it by taking its curl as the unit, forming what we may term a rational current-element, inasmuch as it is not a mere mathematical abstraction, but is a complete dynamical system involving definite forces and energy.

2. Let the axis of z be the line of motion of the charge q at the speed u; then the lines of magnetic force **H** are circles centred upon the axis, in planes perpendicular to it, and its tensor H at distance r from the charge, the line **r** making an angle θ with the axis, is given by

$$H=\frac{q}{r^2}u\sin\theta=cEuv, \quad \text{.........................(1)}$$

where $\nu=\sin\theta$, E the intensity of the radial electric force, c the permittivity such that $\mu_0 cv^2=1$, if μ_0 is the other specific quality of the medium, its inductivity, and v is the speed of propagation.

Since, under the circumstance supposed of u/v being very small, the alteration in the electric field is insensible, and the lines of **E** are radial, we may terminate the fields represented by (1) at any distance $r=a$ from the origin. We then obtain the solution in the case of a charge q upon the surface of a conducting sphere of radius a, moving at speed u. This realization of the problem makes the electric and magnetic energies finite. Whilst, however, agreeing with J. J. Thomson in the fundamentals, I have been unable to corroborate some of his details; and since some of his results have been recently repeated by him in another place,† it may be desirable to state the changes I propose, before proceeding to the case of a charge moving at any speed.

The Energy and Forces in the Case of Slow Motion.

3. First, as regards the magnetic energy, say T. This is the space-summation $\Sigma\,\mu_0 H^2/8\pi$; or, by (1)‡,

$$T=\frac{\mu_0 q^2u^2}{8\pi}\iiint\frac{\nu^2}{r^2}dr\,d\mu\,d\phi=\frac{\mu_0 q^2u^2}{3a}. \quad \text{..................(2)}$$

The limits are such as include all space outside the sphere $r=a$. The coefficient $\frac{1}{3}$ replaces $\frac{2}{15}$.

4. Next, as regards the mutual magnetic energy M of the moving charge and any external magnetic field. This is the space-summation

* *Phil. Mag.*, April, 1881.

† "Applications of Dynamics to Physics and Chemistry," chap. iv., pp. 31 to 37.

‡ *The Electrician*, Jan. 24, 1885, p. 220 [vol. I., p. 446].

$\Sigma\mu_0\mathbf{H}_0\mathbf{H}/4\pi$, if $\mathbf{H}_0$ is the external field; and, by a well-known transformation, it is equivalent to $\Sigma\mathbf{A}_0\Gamma$, if $\mathbf{A}_0$ is any vector whose curl is $\mu_0\mathbf{H}_0$, whilst Γ is the current-density of the moving system. Further, if we choose $\mathbf{A}_0$ to be circuital, the polar part of Γ will contribute nothing to the summation, so that we are reduced to the volume-integral of the scalar product of the circuital $\mathbf{A}_0$ of the one system and the density of the convection-current in the other. Or, in the present case, with a single moving charge at a point, we have simply the scalar product $\mathbf{A}_0\mathbf{u}q$ to represent the mutual magnetic energy; or

$$M = \mathbf{A}_0\mathbf{u}q, \qquad \text{.................................(3)}$$

which is double J. J. Thomson's result.

5. When, therefore, we derive from (3) the mechanical force on the moving charge due to the external magnetic field, we obtain simply Maxwell's "electromagnetic force" on a current-element, the vector product of the moment of the current and the induction of the external field; or if $\mathbf{F}$ is this mechanical force,

$$\mathbf{F} = \mu_0 q\mathrm{V}\mathbf{u}\mathbf{H}_0, \qquad \text{.............................(4)}$$

which is also double J. J. Thomson's result. Notice that in the application of the "electromagnetic force" formula, it is the moment of the convection-current that occurs. This is not the same as the moment of the true current, which varies according to circumstances; for instance, in the case of a small dielectric sphere uniformly electrified throughout its volume, the moment of the true current would be only $\frac{2}{3}$ of that of the convection-current.

The application of Lagrange's equation of motion to (3) also gives the force on q due to the electric field so far as it can depend on M; that is, a force

$$-q\dot{\mathbf{A}}_0,$$

where the time-variation due to all causes must be reckoned, except that due to the motion of q itself, which is allowed for in (4). And besides this, there may be electric force not derivable from $\mathbf{A}_0$, viz.

$$-q\nabla\Psi_0,$$

where Ψ_0 is the scalar potential companion to $\mathbf{A}_0$.

6. Now if the external field be that of another moving charge, we shall obtain the mutual magnetic energy from (3) by letting $\mathbf{A}_0$ be the vector-potential of the current in the second moving system, constructed so as to be circuital. Now the vector-potential of the convection-current $q\mathbf{u}$ is simply $q\mathbf{u}/r$; this is sufficient to obtain the magnetic force by curling; but if used to calculate the mutual energy, the space-summation would have to include every element of current in the other system. To make the vector-potential circuital, and so be able to abolish this work, we must add on to $q\mathbf{u}/r$ the vector-potential of the displacement current to correspond. Now the complete current may be considered to consist of a linear element $q\mathbf{u}$ having two poles; a radial current outward from the + pole in which the current-density is $qu/4\pi r_1^2$; and a radial current inward to the − pole, in which the current-density is $-qu/4\pi r_2^2$; where r_1 and r_2 are the distances of any point from the

poles. The vector-potentials of these currents are also radial, and their tensors are $\frac{1}{2}qu$ and $-\frac{1}{2}qu$. We have now merely to find their resultant when the linear element is indefinitely shortened, add on to the former $q\mathbf{u}/r$, and multiply by μ_0, to obtain the complete circuital vector-potential of $q\mathbf{u}$, viz. :—

$$\mathbf{A} = \mu_0 q\left(\frac{\mathbf{u}}{r} - \tfrac{1}{2}u\nabla\frac{dr}{ds}\right), \qquad (5)$$

where r is the distance from q to the point P when $\mathbf{A}$ is reckoned, and the differentiation is to s, the axis of the convection-current. Both it and the space-variation are taken at P. The tensor of $\mathbf{u}$ is u. Though different and simpler in form (apart from the use of vectors) this vector-potential is, I believe, really the same as the one used by J. J. Thomson. From it we at once find, by the method described in § 4, the mutual energy of a pair of point-charges q_1 and q_2, moving at velocities $\mathbf{u}_1$ and $\mathbf{u}_2$, to be

$$M = \mu_0 q_1 q_2\left(\frac{\mathbf{u}_1\mathbf{u}_2}{r} - \tfrac{1}{2}u_1u_2\frac{d^2r}{ds_1ds_2}\right), \qquad (6)$$

when at distance r apart. Both axial differentiations are to be effected at one end of the line $\mathbf{r}$.

As an alternative form, let ϵ be the angle between $\mathbf{u}_1$ and $\mathbf{u}_2$, and let the differentiation to s_1 be at ds_1, that to s_2 at ds_2, as in the German investigations relating to current-elements; then *

$$M = \mu_0 q_1 q_2 u_1 u_2\left(\frac{\cos\epsilon}{r} + \tfrac{1}{2}\frac{d^2r}{ds_1ds_2}\right). \qquad (7)$$

Another form, to render its meaning plainer. Let λ_1, μ_1, ν_1 and λ_2, μ_2, ν_2 be the direction-cosines of the elements referred to rectangular axes, with the x-axis, to which λ_1 and λ_2 refer, chosen as the line joining the elements. Then†

$$M = \frac{\mu_0 q_1 q_2 u_1 u_2}{2r}(2\lambda_1\lambda_2 + \mu_1\mu_2 + \nu_1\nu_2). \qquad (8)$$

J. J. Thomson's estimate is ‡

$$M = \tfrac{1}{3}\mu_0 q_1 q_2 u_1 u_2\frac{\cos\epsilon}{r}. \qquad (9)$$

Comparing this with (8), we see that there is a notable difference.

7. The mutual energy being different, the forces on the charges, as derived by J. J. Thomson by the use of Lagrange's equations, will be different. When the speeds are constant, we shall have simply the before-described vector product (4) for the "electromagnetic force"; or

$$\mathbf{F}_1 = \mu_0 q_1 \mathrm{V}\mathbf{u}_1\mathbf{H}_2, \qquad \mathbf{F}_2 = \mu_0 q_2 \mathrm{V}\mathbf{u}_2\mathbf{H}_1, \qquad (10)$$

if $\mathbf{F}_1$ is the electromagnetic force on the first, and $\mathbf{F}_2$ that on the second element, whilst $\mathbf{H}_1$ and $\mathbf{H}_2$ are the magnetic forces. Similar changes are needed in the other parts of the complete mechanical forces.

* *The Electrician*, Dec. 28, 1888, p. 230 [p. 501, vol. II.].

† *The Electrician*, Jan. 24, 1885, p. 221 [vol. I., p. 446].

‡ "Applications of Dynamics to Physics and Chemistry," chap. iv.; and *Phil. Mag.*, April, 1881.

It may be remarked that (if my calculations are correct) equation (7) or its equivalents expresses the mutual energy of any two rational current-elements (see § 1) in a medium of uniform inductivity, of moments q_1u_1 and q_2u_2, whether the currents be of displacement, or conduction, or convection, or all mixed, it being in fact the mutual energy of a pair of definite magnetic fields. But, since the hypothesis of instantaneous action is expressly involved in the above, the application of (7) is of a limited nature.

General Theory of Convection Currents.

8. Now leaving behind altogether the subject of current-elements, in the investigation of which one is liable to be led away from physical considerations and become involved in mere exercises in differential coefficients, and coming to the question of the electromagnetic effects of a charge moving in any way, I have been agreeably surprised to find that my solution in the case of steady rectilinear motion, originally an infinite series of corrections, easily reduces to a very simple and interesting finite form, provided u be not greater than v. Only when $u > v$ is there any difficulty. We must first settle upon what basis to work. First the Faraday-law (p standing for d/dt),

$$-\operatorname{curl}\mathbf{E} = \mu_0 p\mathbf{H}, \qquad \text{.........................(11)}$$

requires no change when there is moving electrification. But the analogous law of Maxwell, which I understand to be really a *definition* of electric current in terms of magnetic force, (or a doctrine), requires modification if the true current is to be

$$\mathbf{C} + p\mathbf{D} + \rho\mathbf{u}; \qquad \text{.............................(12)}$$

viz., the sum of conduction-current, displacement-current, and convection-current $\rho\mathbf{u}$, where ρ is the volume-density of electrification. The addition of the term $\rho\mathbf{u}$ was, I believe, proposed by G. F. Fitzgerald.*

(This was not meant exactly for a new proposal, being in fact after Rowland's experiments; besides which, Maxwell was well acquainted with the idea of a convection-current. But what is very strange is that Maxwell, who insisted so strongly upon his doctrine of the *quasi*-incompressibility of electricity, never formulated the convection-current in his treatise. Now Prof. Fitzgerald pointed out that if Maxwell, in his equation of mechanical force,

$$\mathbf{F} = \mathrm{V}\mathbf{CB} - e\nabla\Psi - m\nabla\Omega,$$

had written $\mathbf{E}$ for $-\nabla\Psi$, as it is obvious he should have done, then the inclusion of convection-current in the true current would have followed naturally. (Here $\mathbf{C}$ is the true current, $\mathbf{B}$ the induction, e the density of electrification, m that of imaginary magnetic matter, Ψ the electrostatic and Ω the magnetic potential, and $\mathbf{E}$ the real electric force.)

Now to this remark I have to add that it is as unjustifiable to derive $\mathbf{H}$ from Ω as $\mathbf{E}$ from Ψ; that is, in general, the magnetic force is not the slope of a scalar potential; so, for $-\nabla\Omega$ we should write $\mathbf{H}$, the real magnetic force.

* Brit. Assoc., Southport, 1883.

But this is not all. There is possibly a fourth term in **F**, expressed by 4π**VDG**, where **D** is the displacement and **G** the magnetic current; I have termed this force the "magneto-electric force," because it is the analogue of Maxwell's "electromagnetic force," **VCB**. Perhaps the simplest way of deriving it is from Maxwell's electric stress, which was the method I followed.*

Thus, in a homogeneous nonconducting dielectric free from electrification and magnetization, the mechanical force is the sum of the "electromagnetic" and the "magnetoelectric," and is given by

$$\mathbf{F}=\frac{1}{v^2}\frac{d\mathbf{W}}{dt},$$

where $\mathbf{W}=\mathrm{V}\mathbf{EH}/4\pi$ is the transfer-of-energy vector.

It must, however, be confessed that the real distribution of the stresses, and therefore of the forces, is open to question. And when ether is the medium, the mechanical force in it, as for instance in a light-wave, or in a wave sent along a telegraph-circuit, is not easily to be interpreted.)

The companion to (11) in a nonconducting dielectric is now

$$\mathrm{curl}\,\mathbf{H}=cp\mathbf{E}+4\pi\rho\mathbf{u}. \qquad \text{.........................(13)}$$

Eliminate **E** between (11) and (13), remembering that **H** is circuital, because μ_0 is constant, and we get

$$(p^2/v^2-\nabla^2)\mathbf{H}=\mathrm{curl}\,4\pi\rho\mathbf{u}, \qquad \text{.........................(14)}$$

the characteristic of **H**. Here $\nabla^2=d^2/dx^2+\ldots$, as usual.

Comparing (14) with the characteristic of **H** when there is impressed force e instead of electrification ρ, which is

$$(p^2/v^2-\nabla^2)\mathbf{H}=\mathrm{curl}\,cp\mathbf{e},$$

we see that $\rho\mathbf{u}$ becomes $cp\mathbf{e}/4\pi$. We may therefore regard convection-current as *impressed* electric current. From this comparison also, we may see that an infinite plane sheet of electrification of uniform density cannot produce magnetic force by motion perpendicular to its plane. Also, we see that the sources of disturbances when ρ is moved are the places where $\rho\mathbf{u}$ has curl; for example, a dielectric sphere uniformly filled with electrification (which is imaginable), when moved, starts the magnetic force solely upon its boundary.

The presence of "curl" on the right side tells us, as a matter of mathematical simplicity, to make **H**/curl the variable. Let

$$\mathbf{H}=\mathrm{curl}\,\mathbf{A}, \qquad \text{.................................(15)}$$

and calculate **A**, which may be any vector satisfying (15). Its characteristic is

$$(p^2/v^2-\nabla^2)\mathbf{A}=4\pi\rho\mathbf{u}. \qquad \text{.........................(16)}$$

The divergence of **A** is of no moment, and it is only vexatious complication to introduce Ψ. The time-rate of decrease of **A** is not the real

* "El. Mag. Ind. and its Prop." XXII. *The Electrician*, Jan. 15, 1886, p. 187 [vol. I., p. 545].

distribution of electric force, which has to be found by the additional datum

$$\operatorname{div} c\mathbf{E} = 4\pi\rho, \qquad \text{.........................(17)}$$

where **E** is the real force.

9. "Symbolically" expressed, the solution of (16) is

$$\mathbf{A} = \frac{4\pi\rho\mathbf{u}}{p^2/v^2 - \nabla^2} = \frac{-4\pi\rho\mathbf{u}/\nabla^2}{1 - p^2/v^2\nabla^2}. \qquad \text{.....................(18)}$$

Here the numerator of the fraction to the right is the vector-potential of the convection-current. Calling it $\mathbf{A}_0$, we have

$$\mathbf{A}_0 = \frac{4\pi\rho\mathbf{u}}{-\nabla^2} = \sum \frac{\rho\mathbf{u}}{r}. \qquad \text{.........................(19)}$$

Inserting in (18) and expanding, we have

$$\mathbf{A} = \left\{1 + (p/v\nabla)^2 + (p/v\nabla)^4 + \dots\right\}\mathbf{A}_0. \qquad \text{................(20)}$$

Given then $\rho\mathbf{u}$ as a function of position and time, $\mathbf{A}_0$ is known by (19), and (20) finds **A**, whilst (15) finds **H**.

Complete Solution in the Case of Steady Rectilinear Motion. Physical Inanity of Ψ.

10. When the motion of the electrification is all in one direction, say parallel to the z-axis, **u**, $\mathbf{A}_0$, and **A** are all parallel to this axis, so that we need only consider their tensors. When there is simply one charge q at a point, we have

$$A_0 = qu/r,$$

and (20) becomes

$$A = q\left\{1 + (p/v\nabla)^2 + (p/v\nabla)^4 + \dots\right\}(u/r) \qquad \text{..............(21)}$$

at distance r from q. When the motion is steady, and the whole electromagnetic field is ultimately steady with respect to the moving charge, we shall have, taking it as origin,

$$p = -u(d/dz) = -uD,$$

for brevity; so that

$$A = qu\left\{1 + (uD/v\nabla)^2 + (uD/v\nabla)^4 + \dots\right\}r^{-1}. \qquad \text{...........(22)}$$

Now the property $\quad \nabla^2 r^{n+2} = (n+2)(n+3)r^n \quad$ ....(23)

brings (22) to $\quad A = qu\left\{\frac{1}{r} + \frac{u^2}{v^2}D^2\frac{r}{2!} + \frac{u^4}{v^4}D^4\frac{r^3}{4!} + \dots\right\}; \quad$ ....(24)

and the property $\quad D^{2n}r^{2n-1} = 1^2.3^2.5^2\dots(2n-1)^2\nu^{2n}/r, \quad$ ....(25)

where $\nu = \sin\theta$, θ being the angle between **r** and the axis, brings (24) to

$$A = \frac{qu}{r}\left\{1 + \frac{u^2}{v^2}\frac{\nu^2}{2}\left(1 + \frac{u^2}{v^2}\frac{3}{4}\nu^2\left(1 + \frac{u^2}{v^2}\frac{5}{6}\nu^2\left(1 + \dots\right\}; \qquad \text{.......(26)}$$

which, by the Binomial Theorem, is the same as

$$A=(qu/r)\{1-u^2\nu^2/v^2\}^{-\frac{1}{2}}, \qquad (27)$$

the required solution.

11. To derive H, the tensor of the circular **H**, let $r\nu=h$, the distance from the axis. Then, by (15),

$$H=-\frac{dA}{dh}=-\nu\frac{dA}{dr}+\frac{\mu\nu}{r}\frac{dA}{d\mu}=\frac{qu\nu}{r^2}\left(1+\mu\frac{d}{d\mu}\right)\left(1-\frac{u^2}{v^2}\nu^2\right)^{-\frac{1}{2}}, \quad (28)$$

by (27), if $\mu=\cos\theta$. Performing the differentiation, and also getting out E, the tensor of the electric force, we have the final result that the electromagnetic field is fully given by *

$$cE=\frac{q}{r^2}\cdot\frac{1-u^2/v^2}{(1-u^2\nu^2/v^2)^{\frac{3}{2}}}, \qquad H=cEu\nu, \qquad (29)$$

with the additional information that **E** is radial and **H** circular.

Now, as regards Ψ, if we bring it in, we have only got to take it out again. When the speed is very slow we may regard the electric field as given by $-\nabla\Psi$ *plus* a small correcting vector, which we may call the electric force of inertia. But to show the *physical* inanity of Ψ, go to the other extreme, and let u nearly equal v. It is now the electric force of inertia (supposed) that equals $+\nabla\Psi$ nearly (except about the equatorial plane), and its sole utility or function is to cancel the other $-\nabla\Psi$ of the (supposed) electrostatic field. It is surely impossible to attach any physical meaning to Ψ and to propagate it, for we require two Ψ's, one to cancel the other, and both propagated infinitely rapidly.

As the speed increases, the electromagnetic field concentrates itself more and more about the equatorial plane, $\theta=\frac{1}{2}\pi$. To give an idea of the accumulation, let $u^2/v^2=\cdot 99$. Then cE is $\cdot 01$ of the normal value q/r^2 at the pole, and 10 times the normal value at the equator. The latitude where the value is normal is given by

$$\nu=(v/u)\left[1-(1-u^2/v^2)^{\frac{2}{3}}\right]^{\frac{1}{2}}. \qquad (30)$$

Limiting Case of Motion at the Speed of Light. Application to a Telegraph Circuit.

12. When $u=v$, the solution (29) becomes a plane electromagnetic wave, E and H being zero everywhere except in the equatorial plane. As, however, the values of E and H are infinite, distribute the charge along a straight line moving in its own line, and let the linear-density be q. The solution is then †

$$H=Ecv=2qv/r \qquad (31)$$

at distance r from the line, between the two planes through the ends of the line perpendicular to it, and zero elsewhere.

To further realize, let the field terminate internally at $r=a$, giving a cylindrical-surface distribution of electrification, and terminate the tubes

* *The Electrician*, Dec. 7, 1888, p. 148 [p. 495, vol. II.].

† Ibid., Nov. 23, 1888, p. 84 [p. 493, vol. II.].

of displacement externally upon a coaxial cylindrical surface; we then produce a real electromagnetic plane wave with electrification, and of finite energy. We have supposed the electrification to be carried through the dielectric at speed v, to keep up with the wave, which would of course break up if the charge were stopped. But if perfectly-conducting surfaces be given on which to terminate the displacement, the natural motion of the wave will itself carry the electrification along them. In fact, we now have the rudimentary telegraph-circuit, with no allowance made for absorption of energy in the wires, and the consequent distortion. If the conductors be not coaxial, we only alter the distribution of the displacement and induction, without affecting the propagation without distortion.*

If we now make the medium conduct electrically, and likewise magnetically, with equal rates of subsidence, we shall have the same solutions, with a time-factor $\epsilon^{-\rho t}$ producing ultimate subsidence to zero; and, with only the real electric conductivity in the medium the wave is running through, it will approximately cancel the distortion produced by the resistance of the wires the wave is passing over when this resistance has a certain value.† We should notice, however, that it could not do so perfectly, even if the magnetic retardation in the wires due to diffusion were zero; because in the case of the unreal magnetic conductivity its correcting influence is where it is wanted to be, in the body of the wave; whereas in the case of the wires, their resistance, correcting the distortion due to the external conductivity, is outside the wave; so that we virtually assume instantaneous propagation laterally from the wires of *their* correcting influence, in the elementary theory of propagation along a telegraph-circuit which is symbolized by the equations

$$-dV/dz=(R+Lp)C, \qquad -dC/dz=(K+Sp)V, \quad \text{.........(32)}$$

where R, L, K, and S are the resistance, inductance, leakage-conductance, and permittance per unit length of circuit, C the current, and V what I, for convenience, term the potential-difference, but which I have expressly disclaimed‡ to represent the electrostatic difference of potential, and have shown to represent the transverse voltage or line-integral of the electric force across the circuit from wire to wire, including the electric force of inertia. Now in case of great distortion, as in a long submarine cable, this V approximates towards the electrostatic potential-difference, which it is in Sir W. Thomson's diffusion theory; but in case of little distortion, as in telephony through circuits of low resistance and large inductance, there may be a wide difference between my V and that of the electrostatic force. Consider, for instance, the extreme case of an isolated plane-wave disturbance with no spreading-out of the tubes of displacement. At the boundaries of the

* *The Electrician*, Jan. 10, 1885 [p. 440, vol. I.]. Also "Self-Induction of Wires," Part IV. *Phil. Mag.*, Nov. 1886 [p. 221, vol. II.].

† "Electromagnetic Waves," § 6, *Phil. Mag.*, Feb. 1888 [p. 379, vol. II.]. *The Electrician*, June, 1887 [p. 123, vol. II.].

‡ "Self-Induction of Wires," Part. II., *Phil. Mag.*, Sept. 1886 [vol. II., p. 189].

disturbance the difference between V and the electrostatic difference of potential is great.

But it is worth noticing, as a rather remarkable circumstance, that when we derive the system (32) by elementary considerations, viz., by extending the diffusion-system by the addition of the E.M.F. of inertia and leakage-current, we apparently as a matter of course take V to mean the same as in the diffusion-system. The resulting equations are correct, and yet the assumption is certainly wrong. The true way appears to be that given by me in the paper last referred to, by considering the line-integral of electric force in a closed curve [vol. II., p. 187. Also p. 87]. We cannot, indeed, make a separation of the electric force of inertia from $-\nabla\Psi$ without some assumption, though the former is quite definite when the latter is suitably defined, But, and this is the really important matter, it would be in the highest degree inconvenient, and lead to much complication and some confusion, to split V into two components, in other words, to bring in Ψ and $\mathbf{A}$.

In thus running down Ψ, I am by no means forgetful of its utility in other cases. But it has perhaps been greatly misused. The clearest course to pursue appears to me to invariably make $\mathbf{E}$ and $\mathbf{H}$ the primary objects of attention, and only use potentials when they naturally suggest themselves as labour-saving appliances.

Special Tests. The Connecting Equations.

13. Returning to the solutions (29), the following are the special tests of their accuracy. Let E_1 and E_2 be the z and h components of $\mathbf{E}$. Then, by (11) and (13), with the special meaning assumed by p, we have

$$\left.\begin{aligned} \frac{1}{h}\frac{d}{dh}hH &= -cu\frac{dE_1}{dz}, \\ -\frac{dH}{dz} &= -cu\frac{dE_2}{dz}, \qquad \text{or} \qquad H = cuE_2, \\ \frac{dE_1}{dh}-\frac{dE_2}{dz} &= -\mu_0 u\frac{dH}{dz}, \qquad \text{or} \qquad \frac{dE_1}{dh} = \left(1-\frac{u^2}{v^2}\right)\frac{dE_2}{dz}. \end{aligned}\right\} \qquad (33)$$

In addition to satisfying these equations, the displacement outward through any spherical surface centred at the charge may be verified to be q; this completes the test of the accuracy of (29).

But (33) are not limited to the case of a single point-charge, being true outside the electrification when there is symmetry with respect to the z-axis, and the electrification is all moving parallel to it at speed u.

When $u=v$, $E_1=0$, and $E_2=E=\mu vH$, so that we reduce to

$$\frac{1}{h}\frac{d}{dh}hH = 0, \qquad (34)$$

outside the electrification. Thus, if the electrification is on the axis of z, we have

$$E/\mu v = H = 2qv/r, \qquad (35)$$

differing from (31) only in that q, the linear density, may be any function of z.

The Motion of a Charged Sphere. The Condition at a Surface of Equilibrium (Footnote).

14. If, in the solutions (29), we terminate the fields internally at $r=a$, the perpendicularity of **E** and the tangentiality of **H** to the surface show that (29) represents the solutions in the case of a perfectly conducting sphere of radius a, moving steadily along the z-axis at the speed u, and possessing a total charge q. The energy is now finite. Let U be the total electric and T the total magnetic energy. By space-integration of the squares of E and H we find that they are given by

$$U=\frac{q^2}{2ca}\cdot\frac{1-u^2/v^2}{4}\left[1+\frac{\frac{3}{2}}{1-u^2/v^2}+\frac{\frac{3}{2}\tan^{-1}\dfrac{u/v}{(1-u^2/v^2)^{\frac{1}{2}}}}{(u/v)(1-u^2/v^2)^{\frac{3}{2}}}\right], \quad\text{.............(36)}$$

$$T=\frac{q^2}{2ca}\cdot\frac{1-u^2/v^2}{4}\left[1+\frac{2u^2/v^2-\frac{1}{2}}{1-u^2/v^2}+\frac{(2u^2/v^2-\frac{1}{2})\tan^{-1}\dfrac{u/v}{(1-u^2/v^2)^{\frac{1}{2}}}}{(u/v)(1-u^2/v^2)^{\frac{3}{2}}}\right], \quad (37)$$

in which $u<v$. When $u=v$, with accumulation of the charge at the equator of the sphere, we have infinite values, and it appears to be only possible to have finite values by making a zone at the equator cylindrical instead of spherical. The expression for T in (37) looks quite wrong; but it correctly reduces to that of equation (2) when u/v is infinitely small.*

* [I am indebted to Mr. G. F. C. Searle, of Cambridge, for the opportunity of making a somewhat important correction before going to press. In a private communication (August 19, 1892) he informed me that he had verified the accuracy of the solution for a point-charge, which he had also obtained in another way, from equations equivalent to (33), without the use of the function **A** of §§ 8 to 10; but he cast doubt upon the validity of the extension made in § 14, from a point-charge to a charged conducting sphere, and asked the plain question (in effect), What justification is there for terminating the displacement perpendicularly, to make a surface of equilibrium?

On examination, I find that there is no justification whatever, exceptions excepted. The true boundary condition may, however, be found without a fresh investigation. On p. 499 the problem of uniform motion of electrification through a dielectric medium, or conversely, of the uniform motion of the whole medium past stationary electrification, is reduced to a case of eolotropy in electrostatics. The effect of the motion of the isotropic medium on the displacement emanating from stationary electrification is there shown to be identical with the effect of keeping the medium stationary and reducing its permittivity in lines parallel to the (abolished) motion from c to $c(1-u^2/v^2)$, whilst keeping the transverse permittivity the same. The transverse concentration of the displacement is obvious. Now the function P (equation (14), p. 499) is the electrostatic potential in the stationary eolotropic problem, so that its slope $-\nabla P$, which call **F**, is the electric force, and the displacement **D** is a linear function thereof, say $\mathbf{D}=\lambda\mathbf{F}$, where λ is the permittivity operator. The condition of equilibrium is that **F** is perpendicular to the surface where it terminates, this being required to make curl $\mathbf{F}=0$, or the voltage zero in every circuit. Now, in the corresponding problem of the same electrification in a moving isotropic medium, we have the same function P (no longer the electrostatic potential) and the same derived vector **F**, whilst the displacement **D** is also derived from **F** in the same way. But whilst the meaning of **D** is the same in both cases, that of **F** is not. In the eolotropic case, **F** is the

The State when the Speed of Light is exceeded.

15. The question now suggests itself, What is the state of things when $u > v$? It is clear, in the first place, that there can be no disturbance at all in front of the moving charge (at a point, for simplicity). Next, considering that the spherical waves emitted by the charge in its motion along the z-axis travel at speed v, the locus of their fronts is a conical surface whose apex is at the charge itself, whose axis is that of z, and whose semiangle θ is given by

$$\sin \theta = v/u. \quad \text{......................(38)}$$

The whole displacement, of amount q, should therefore lie within this cone. And since the moving charge is a convection-current qu, the displacement-current should be towards the apex in the axial portion of the cone, and change sign at some unknown distance, so as to be away from the apex either in the outer part of the cone or else upon its boundary. The pulling back of the charge by the electric stress would require the continued application of impressed force to keep up the motion, and its activity would be accounted for by the continuous addition made to the energy in the cone; for the transfer of energy on its boundary is perpendicularly outward, and the field at the apex is being continuously renewed.

The above general reasoning seems plausible enough, but I cannot find any solution to correspond that will satisfy all the necessary conditions. It is clear that (29) will not do when $u > v$. Nor is it of any use to change the sign of the quantity under the radical, when needed, to make real. It is suggested that whilst there should be a definite solution, there cannot be one representing a *steady* condition of **E** and **H** with respect to the moving charge. As regards physical

electric force, and is not parallel to **D**. In the moving isotropic medium, on the other hand, **F** is not the electric force, which is **E**, parallel to **D**. Nevertheless, the same condition formally obtains, for we have curl **F** = 0 in the moving medium, requiring that **F** shall be perpendicular to a surface of equilibrium, not the electric force or displacement. P = constant is therefore the equation to a surface of equilibrium. That is, in the case of a point-charge, the surfaces of equilibrium are not spheres, but are concentric oblate spheroids, whose principal axes are proportional to the square roots of c, c, and $c(1-u^2/v^2)$, the principal permittivities in the eolotropic problem. In the extreme case of $u=v$, the spheroid reduces to a flat circular disc, with a single circular line of electrification on its edge. It would seem, however, to be a matter of indifference, in this extreme case, whether the conductor be a disc or a solid sphere. Bearing in mind the conditions assumed to prevail in the problem of motion of sources of displacement in a uniform medium, we see that if we introduce conductors, say by filling up spaces void of electric force with conducting matter, this should not interfere with the assumed motions. (See also "Electromagnetic Theory," § 164.)

Equations (36), (37) express the electric and magnetic energy outside a sphere of radius a, within which is either a point-source at the origin, or any equivalent spheroidal electrified surface.

In the corresponding bidimensional problem of § 17 in the text, with the solution (43), it is clear from the above that the surface of equilibrium is an elliptic cylinder, the shorter axis being in the direction of motion, and the axes themselves in the ratio 1 to $(1-u^2/v^2)^{\frac{1}{2}}$. This surface degenerates to a flat strip when $u=v$.]

possibility, in connexion with the structure of the ether, that is not in question.*

A Charged Straight Line moving in its own Line.

16. Let us now derive from (29), or from (27), the results in some cases of distributed electrification, in steady rectilinear motion. The integrations to be effected being all of an elementary character, it is not necessary to give the working.

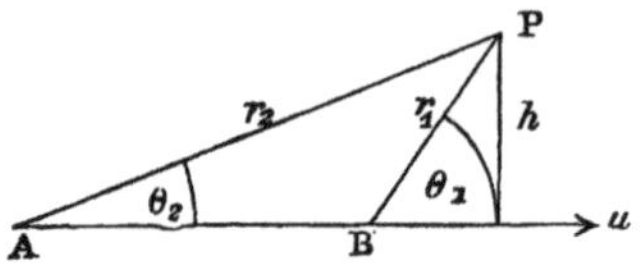

First, let a straight line AB be charged to linear density q, and be in motion at speed u in its own line from left to right. Then at P we shall have

$$A = qu \log \left(\frac{r_1}{r_2} \cdot \frac{\mu_1 + (1 - \nu_1^2 u^2/v^2)^{\frac{1}{2}}}{\mu_2 + (1 - \nu_2^2 u^2/v^2)^{\frac{1}{2}}} \right), \qquad (39)$$

from which $H = -dA/dh$ gives

$$H = qu\left(1 - \frac{u^2}{v^2}\right)\left[\frac{\nu_1}{r_1(1 - \nu_1^2 u^2/v^2) + r_1\mu_1(1 - \nu_1^2 u^2/v^2)^{\frac{1}{2}}} - \text{same f}^{\text{n}} \text{ of } r_2, \mu_2, \nu_2\right], \quad (40)$$

where $\mu = \cos\theta$, $\nu = \sin\theta$.

When P is vertically over B, and A is at an infinite distance, we shall find

$$H = qu/h, \qquad (41)$$

which is one half the value due to an infinitely long (both ways) straight current of strength qu. The notable thing is the independence of the ratio u/v.

* [The difficulty about the above method and solution (29) is that it is not explicit enough when $u > v$, and does not indicate the limits of application. It gives a real solution for the hinder cone, a real solution for the forward cone, and an unreal solution in the rest of space, but we have no instruction to reject the part for the forward cone and the unreal part, nor have we any means of testing that the remainder, confined to the hinder cone, is the proper solution, viz., by the test of divergence, to give the right amount of electrification. The integral displacement comes to $-\infty$. Now this may require to be supplemented by $+\infty + q$ on the boundary of the cone, but we have no way of testing it.

But certain considerations led me to the conclusion that the problem of $u > v$ was really quite as definite a one as that of $u < v$, and that a correct method of a general character (independent of the magnitude of u) would show this explicitly. I therefore (in 1890) attacked the problem from a different point of view, employing the method of resistance-operators (or an equivalent method). Form the complete differential equation $\mathbf{D} = \phi\mathbf{u}$, connecting the displacement $\mathbf{D}$ associated with a moving point-charge with its velocity $\mathbf{u}$, which is any function of the time t. Here ϕ is a differential operator, a function of p or d/dt. The solution of this equation gives $\mathbf{D}$ explicitly in terms of $\mathbf{u}$, whether steady or variable, and its structure indicates the limits of application.

Taking $\mathbf{u}$ = constant, we obtain the result (29) when $u < v$. But when $u > v$, the formula tells us to exclude all space except the hinder cone, and that in it, the solution is *not* (29), but *double as much*. That is, double the right member of the first of (29) when $u > v$. The boundary of the cone is also a displacement sheet. The displacement is to the charge in the cone, and from the charge on its surface. Being so near the end of the second volume, I regret that there is no space here for the mathematical investigation, which cannot be given in a few words, and must be reserved.]

But if $u=v$ in (40), the result is zero, unless $\nu_1=1$, when we have the result (41). But if P be still further to the left, we shall have to add to (41) the solution due to the electrification which is ahead of P. So when the line is infinitely long both ways, we have double the result in (41), with independence of u/v again.

But should q be a function of z, we do not have independence of u/v except in the already-considered case of $u=v$, with plane waves, and no component of electric force parallel to the line of motion.

A Charged Straight Line moving Transversely.

17. Next, let the electrified line be in steady motion perpendicularly to its length. Let q be the linear density (constant), the z-axis that of the motion, the x-axis coincident with the electrified line, and that of y upward on the paper. Then the A at P will be

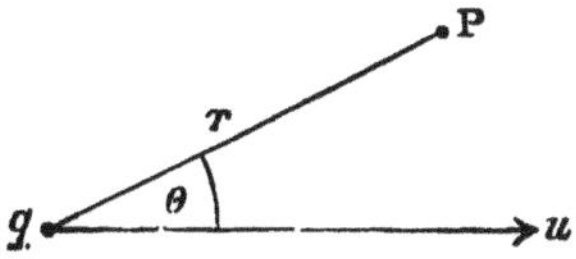

$$A=\frac{qu}{(1-u^2/v^2)}\log\frac{x_1+\{x_1^2+y^2+z^2(1-u^2/v^2)^{-1}\}^{\frac{1}{2}}}{x_2+\{x_2^2+y^2+z^2(1-u^2/v^2)^{-1}\}^{\frac{1}{2}}}; \quad \text{.......(42)}$$

where y and z belong to P, and x_1, x_2 are the limiting values of x in the charged line. From this derive the solution in the case of an infinitely long line. It is

$$cE=\frac{2q}{r}\cdot\frac{(1-u^2/v^2)^{\frac{1}{2}}}{1-\nu^2u^2/v^2}, \qquad H=cEu\nu, \quad \text{..............(43)}$$

where $\nu=\sin\theta$; understanding that **E** is radial, or along qP in the figure, and **H** rectilinear, parallel to the charged line.

Terminating the fields internally at $r=a$, we have the case of a perfectly conducting cylinder of radius a, charged with q per unit of length, moving transversely. When $u=v$ there is disappearance of E and H everywhere except in the plane $\theta=\frac{1}{2}\pi$, as in the case of the sphere, with consequent infinite values. It is the curvature that permits this to occur, *i.e.* producing infinite values; of course it is the self-induction that is the cause of the conversion to a plane wave, here and in the other cases. There is some similarity between (43) and (29). In fact, (43) is the bidimensional equivalent of (29).

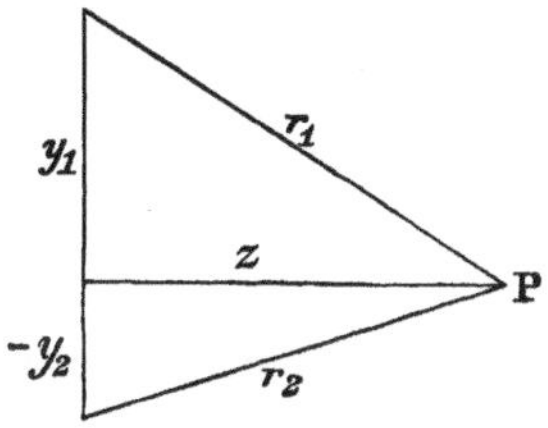

A Charged Plane moving Transversely.

18. Coming next to a plane distribution of electrification, let q be the surface-density, and the plane be moving perpendicularly to itself. Let it be of finite breadth and of infinite length, so that we may calculate H from (43). The result at P is

$$H=\frac{qu}{(1-u^2/v^2)^{\frac{1}{2}}}\log\frac{r_1^2-y_1^2u^2/v^2}{r_2^2-y_2^2u^2/v^2}. \quad \text{..................(44)}$$

When P is equidistant from the edges, H is zero. There is therefore no H anywhere due to the motion of an infinitely large uniformly charged plane perpendicularly to itself. The displacement-current is the negative of the convection-current and at the same place, viz. the moving plane, so there is no true current.

Calculating E_1, the z-component of $\mathbf{E}$, z being measured from left to right, we find

$$cE_1 = 2q\left\{\tan^{-1}\frac{y_1}{z}\left(1-\frac{u^2}{v^2}\right)^{\frac{1}{2}} - \tan^{-1}\frac{y_2}{z}\left(1-\frac{u^2}{v^2}\right)^{\frac{1}{2}}\right\}. \quad \text{..........(45)}$$

The component parallel to the plane is H/cu. Thus, when the plane is infinite, this component vanishes with H, and we are left with

$$cE_1 = cE = 2\pi q, \quad \text{............................(46)}$$

the same as if the plane were at rest.

A Charged Plane moving in its own Plane.

19. Lastly, let the charged plane be moving in its own plane. Refer to the first figure, in which let AB now be the trace of the plane when of finite breadth. We shall find that

$$H = 2qu\left[\tan^{-1}\frac{z}{h(1-u^2/v^2)^{\frac{1}{2}}}\right]_{z_1}^{z_2}, \quad \text{..............(47)}$$

z_1 and z_2 being the extreme values of z, which is measured parallel to the breadth of the plane.

Therefore, when the plane extends infinitely both ways, we have

$$H = 2\pi qu \quad \text{..............................(48)}$$

above the plane, and its negative below it. This differs from the previous case of vanishing displacement-current. There is H, and the convection-current is not now cancelled by coexistent displacement-current.

The existence of displacement-current, or changing displacement, was the basis of the conclusion that moving electrification constitutes a part of the true current. Now in the problem (48) the displacement-current has gone, so that the existence of H appears to rest merely upon the assumption that moving electrification is true current. But if the plane be not infinite, though large, we shall have (48) nearly true near it, and away from the edges; whilst the displacement-current will be strong near the edges, and almost nil where (48) is nearly true.

But in some cases of rotating electrification, there need be no displacement anywhere, except during the setting up of the final state. This brings us to the rather curious question whether there is any difference between the magnetic field of a convection-current produced by the rotation of electrification upon a good nonconductor and upon a good conductor respectively, other than that due to diffusion in the conductor. For in the case of a perfect conductor, it is easy to imagine that the electrification could be at rest, and the moved conductor merely slip past it. Perhaps Professor Rowland's forthcoming experiments on convection-currents may cast some light upon this matter.

December 27, 1888.

LI. DEFLECTION OF AN ELECTROMAGNETIC WAVE BY MOTION OF THE MEDIUM.

[*The Electrician*, April 12, 1889, p. 663.]

THIS subject is of interest in connection with theories of Aberration, which requires to be explained electromagnetically. A plane wave in a nonconducting dielectric is carried on at speed $v=(\mu c)^{-\frac{1}{2}}$, where μ is the inductivity and c the permittivity, and is not altered in any way, according to the rudimentary theory, that is to say, which overlooks dispersion. But if the medium be moving through the ether, it is altered in a manner depending upon the speed of motion and the angle it makes with the undisturbed direction of propagation.

Thus, let $E_0=\mu v H_0$ specify a plane wave in a medium at rest, E_0 being the tensor of the electric and H_0 of the magnetic force. Next set the medium in motion with velocity **u**, changing $\mathbf{E}_0$ to **E** and $\mathbf{H}_0$ to **H**, thus

$$\mathbf{E}=\mathbf{e}+\mathbf{E}_0, \qquad \mathbf{H}=\mathbf{h}+\mathbf{H}_0, \quad \text{.....................(A)}$$

where **e** and **h** are the auxiliary electric and magnetic forces due to the motion. To find them, we have, first, the electric force due to motion of matter in a magnetic field, or

$$\mathbf{e}=\mu\mathrm{V}\mathbf{uH}, \quad \text{................................(B)}$$

which formula is well known, and is included in Maxwell's treatise. Next, the magnetic force due to motion in an electric field, or

$$\mathbf{h}=c\mathrm{V}\mathbf{Eu}. \quad \text{................................(C)}$$

This equation, which is as necessary as (B), was, so far as I am at present aware, first given by me in Section III. of "Electromagnetic Induction and its Propagation," January 24, 1885 [vol. I., p. 446], and was again considered later on in connection with the "magneto-electric force," which is as necessary as Maxwell's "electromagnetic force."

We require one more relation, viz., between $\mathbf{E}_0$ and $\mathbf{H}_0$, viz.,

$$\mathbf{H}_0=c\mathrm{V}\mathbf{v}\mathbf{E}_0, \quad \text{.............................(D)}$$

the property of a plane wave, due to Maxwell; and we can now fully find the auxiliaries **e** and **h** in terms of the originals $\mathbf{E}_0$ and $\mathbf{H}_0$. Here **v** is the vectorized v of the wave when undisturbed.

In the above V is the symbol of vector product. Thus V**uH** is the vector perpendicular to **u** and to **H**, whose tensor equals the product of *their* tensors, u and H, into the sine of the angle between their directions. But this is merely used to state the general relations in a compact and intelligible form, instead of with Cartesian circumlocutions.

Instead of taking the general case, it is convenient to divide into three, viz., (1), **u** parallel to **v**; (2), **u** parallel to $\mathbf{E}_0$; (3), **u** parallel to $\mathbf{H}_0$. By putting the results together we shall obtain the mixed-up general case.

(1). **u** parallel to **v**. Here the medium is moving in the same direction as that of undisturbed propagation, and there is no alteration of direction of either $\mathbf{E}_0$ or $\mathbf{H}_0$, so that it is only necessary to specify the tensors of the auxiliaries $\mathbf{e}$ and $\mathbf{h}$. Thus:—

$$e = -\frac{u}{u+v}E_0, \qquad h = -\frac{u}{u+v}H_0. \quad \text{.................(1)}$$

If, for example, the medium be moving at half the speed v, and *with* it, the displacement and induction in a given length are spread over a space half as great again as if the medium were at rest, so that their intensities are reduced to two-thirds of the undisturbed values. There is no discontinuity when u is equal to or greater than v.

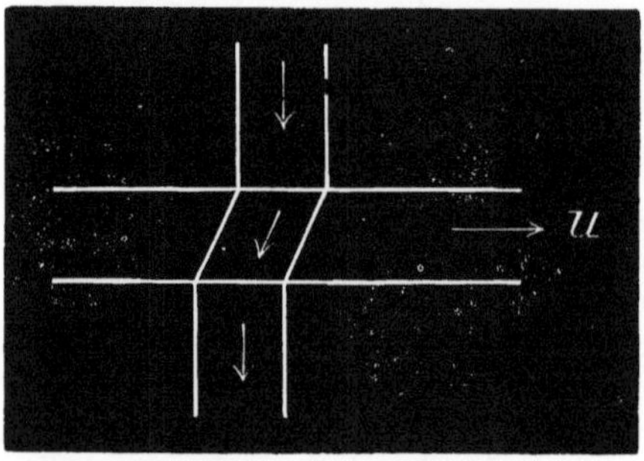

But if the medium move the other way there is compression into half the space, so that the intensities are doubled. As it is increased up to v, the compression increases infinitely. After that, with $u > v$, there is reversal of sign of $\mathbf{E}$ and $\mathbf{H}$ as compared with $\mathbf{E}_0$ and $\mathbf{H}_0$.

(2). $\mathbf{u}$ and $\mathbf{E}_0$ parallel. Here $\mathbf{h}_0$ is parallel to $\mathbf{H}_0$, but $\mathbf{e}_0$ is parallel to $\mathbf{v}$. Their tensors are given by

$$e = \frac{uv}{v^2 - u^2}E_0, \qquad h = \frac{u^2}{v^2 - u^2}H_0. \quad \text{.................(2)}$$

(3). Lastly, $\mathbf{u}$ and $\mathbf{H}_0$ parallel. Now $\mathbf{e}$ is parallel to $\mathbf{E}_0$, whilst $\mathbf{h}$ is parallel to $\mathbf{v}$. Their tensors are

$$e = \frac{u^2}{v^2 - u^2}E_0, \qquad h = \frac{uv}{v^2 - u^2}H_0. \quad \text{.................(3)}$$

In either case, (2) or (3), the angle of deflection θ is given by

$$\tan\theta = \frac{uv}{v^2 - u^2}; \quad \text{..........................(4)}$$

consequently the deflection is wholly independent of the plane of polarization.

Thus, let a slab of (say) glass move in its own plane at speed u, and a plane-wave from the upper medium strike the glass flush. The transmitted rays are deflected as shown in Fig. 1, the deflection being given by the above formula, where, observe, v is the speed in the *glass* when at rest, and u the speed *of* the glass with respect to the external medium.

The above working out of the effect of moving matter on a plane electromagnetic wave is (if done properly) strictly in accordance with electromagnetic principles. But it will be observed that Fresnel's result, relating to the alteration in the speed of light produced by moving a transparent medium through which it is passing, is not accounted for. It is said to have been thoroughly confirmed by Michelson. I should like to direct the attention of electromagneticians to this question, with a view to the discovery of a modification of the above data, or correction of the working, in order to explain Fresnel and Michelson, which *must* be done electromagnetically. Mr. Glazebrook has made Sir W. Thomson's extraordinary contractile ether do it by an auxiliary hypothesis; surely, then, Maxwell's ether equations could be appropriately modified.

LII. ON THE FORCES, STRESSES, AND FLUXES OF ENERGY IN THE ELECTROMAGNETIC FIELD.

[Royal Society. Received June 9, Read June 18, 1891.* Abstract in *Proceedings*, vol. 50, 1891; Paper in *Transactions*, A. 1892.]

(ABSTRACT.)

THE abstract nature of this paper renders its adequate abstraction difficult. The principle of conservation of energy, when applied to a theory such as Maxwell's, which postulates the definite localization of energy, takes a more special form, viz., that of the continuity of energy. Its general nature is discussed. The relativity of motion forbids us to go so far as to assume the objectivity of energy, and to identify energy, like matter; hence the expression of the principle is less precise than that of the continuity of matter (as in hydrodynamics), for all we can say in general is that the convergence of the flux of energy equals the rate of increase of the density of the energy; the flux of the energy being made up partly of the mere convection of energy by motion of the matter (or other medium) with which it is associated localizably, and partly of energy which is transferred through the medium in other ways, as by the activity of a stress, for example, not obviously (if at all) representable as the convection of energy. Gravitational energy is the chief difficulty in the way of the carrying out of the principle. It must come from the ether (for where else can it come from?), when it goes to matter; but we are entirely ignorant of the manner of its distribution and transference. But, whenever energy can be localized, the principle of continuity of energy is (in spite of certain drawbacks connected with the circuital flux of energy) a valuable principle which should be utilized to the uttermost. Practical forms are considered. In the electromagnetic application the flux of energy has a four-fold make-up, viz., the Poynting flux of energy, which occurs whether the medium be stationary or moving; the flux of energy due to the activity of the electromagnetic stress when the medium is moving; the convection of electric and magnetic energy; and the convection of other energy associated with the working of the translational force due to the stress.

As Electromagnetism swarms with vectors, the proper language for its expression and investigation is the Algebra of Vectors. An account

* Typographical troubles have delayed the publication of this paper. The footnotes are of date May 11, 1892.

is therefore given of the method employed by the author for some years past. The quaternionic basis is rejected, and the algebra is based upon a few definitions of notation merely. It may be regarded as Quaternions without quaternions, and simplified to the uttermost; or else as being merely a conveniently condensed expression of the Cartesian mathematics, understandable by all who are acquainted with Cartesian methods, and with which the vectorial algebra is made to harmonize. It is confidently recommended as a practical working system.

In continuation thereof, and preliminary to the examination of electromagnetic stresses, the theory of stresses of the general type, that is, rotational, is considered; and also the stress activity, and flux of energy, and its convergence and division into translational, rotational, and distortional parts; all of which, it is pointed out, may be associated with stored potential, kinetic, and wasted energy, at least so far as the mathematics is concerned.

The electromagnetic equations are then introduced, using them in the author's general forms, *i.e.*, an extended form of Maxwell's circuital law, defining electric current in terms of magnetic force, and a companion equation expressing the second circuital law; this method replacing Maxwell's in terms of the vector-potential and the electrostatic potential, Maxwell's equations of propagation being found impossible to work and not sufficiently general. The equation of activity is then derived in as general a form as possible, including the effects of impressed forces and intrinsic magnetization, for a stationary medium which may be eolotropic or not. Application of the principle of continuity of energy then immediately indicates that the flux of energy in the field is represented by the formula first discovered by Poynting. Next, the equation of activity for a moving medium is considered. It does not immediately indicate the flux of energy, and, in fact, several transformations are required before it is brought to a fully significant form, indicating (1), the Poynting flux, the form of which is settled; (2), the convection of electric and magnetic energy; (3), a flux of energy which, from the form in which the velocity of the medium enters, represents the flux of energy due to a working stress. Like the Poynting flux, it contains vector products. From this flux the stress itself is derived, and the form of translational force, previously tentatively developed, is verified. It is assumed that the medium in its motion carries its properties with it unchanged.

A side matter which is discussed is the proper measure of "true" electric current, in accordance with the continuity of energy. It has a four-fold make-up, viz., the conduction-current, displacement-current, convection-current (or moving electrification), and the curl of the motional magnetic force.

The stress is divisible into an electric and a magnetic stress. These are of the rotational type in eolotropic media. They do not agree with Maxwell's general stresses, though they work down to them in an isotropic homogeneous stationary medium not intrinsically magnetized or electrized, being then the well-known tensions in certain lines with equal lateral pressures.

Another and shorter derivation of the stress is then given, guided by the previous, without developing the expression for the flux of energy. Variations of the properties permittivity and inductivity with the strain can be allowed for. An investigation by Professor H. Hertz is referred to. His stress is not agreed with, and it is pointed out that the assumption by which it is obtained is equivalent to the existence of isotropy, so that its generality is destroyed. The obvious validity of the assumption on which the distortional activity of the stress is calculated is also questioned.

Another form of the stress-vector is examined, showing its relation to the fictitious electrification and magnetic current, magnetification and electric current, produced on the boundary of a region by terminating the stress thereupon; and its relation to the theory of action at a distance between the respective matters and currents.

The stress-subject is then considered statically. The problem is now perfectly indeterminate, in the absence of a complete experimental knowledge of the strains set up in bodies under electric and magnetic influence. Only the stress in the air outside magnets and conductors can be considered known. Any stress within them may be superadded, without any difference being made in the resultant forces and torques. Several stress-formulæ are given, showing a transition from one extreme form to another. A simple example is worked out to illustrate the different ways in which Maxwell's stress and others explain the mechanical actions. Maxwell's stress, which involves a translational force on magnetized matter (even when only inductively magnetized), merely because it is magnetized, leads to a very complicated and unnatural way of explanation. It is argued, independently, that no stress-formula should be allowed which indicates a translational force of the kind just mentioned.

Still the matter is left indeterminate from the statical standpoint. From the dynamical standpoint, however, we are led to a certain definite stress-distribution, which is also, fortunately, free from the above objection, and is harmonized with the flux of energy. A peculiarity is the way the force on an intrinsic magnet is represented. It is not by force on its poles, nor on its interior, but on its sides, referring to a simple case of uniform longitudinal magnetization; *i.e.*, it is done by a *quasi*-electromagnetic force on the fictitious electric current which would produce the same distribution of induction as the magnet does. There is also a force where the inductivity varies. This force on fictitious current harmonizes with the conclusion previously arrived at by the author, that when impressed forces set up disturbances, such disturbances are determined by the curl of the impressed forces, and proceed from their localities.

In conclusion it is pointed out that the determinateness of the stress rests upon the assumed localization of the energy and the two laws of circuitation, so that with other distributions of the energy (of the same proper total amounts) other results would follow; but the author has been unable to produce full harmony in any other way than that followed.

General Remarks, especially on the Flux of Energy.

§ 1. The remarkable experimental work of late years has inaugurated a new era in the development of the Faraday-Maxwellian theory of the ether, considered as the primary medium concerned in electrical phenomena—electric, magnetic, and electromagnetic. Maxwell's theory is no longer entirely a paper theory, bristling with unproved possibilities. The reality of electromagnetic waves has been thoroughly demonstrated by the experiments of Hertz and Lodge, Fitzgerald and Trouton, J. J. Thomson, and others; and it appears to follow that, although Maxwell's theory may not be fully correct, even as regards the ether (as it is certainly not fully comprehensive as regards material bodies), yet the true theory must be one of the same type, and may probably be merely an extended form of Maxwell's.

No excuse is therefore now needed for investigations tending to exhibit and elucidate this theory, or to extend it, even though they be of a very abstract nature. Every part of so important a theory deserves to be thoroughly examined, if only to see what is in it, and to take note of its unintelligible parts, with a view to their future explanation or elimination.

§ 2. Perhaps the simplest view to take of the medium which plays such a necessary part, as the recipient of energy, in this theory, is to regard it as continuously filling all space, and possessing the mobility of a fluid rather than the rigidity of a solid. If whatever possess the property of inertia be matter, then the medium is a form of matter. But away from ordinary matter it is, for obvious reasons, best to call it as usual by a separate name, the ether. Now, a really difficult and highly speculative question, at present, is the connection between matter (in the ordinary sense) and ether. When the medium transmitting the electrical disturbances consists of ether and matter, do they move together, or does the matter only partially carry forward the ether which immediately surrounds it? Optical reasons may lead us to conclude, though only tentatively, that the latter may be the case; but at present, for the purpose of fixing the data, and in the pursuit of investigations not having specially optical bearing, it is convenient to assume that the matter and the ether in contact with it move together. This is the working hypothesis made by H. Hertz in his recent treatment of the electrodynamics of moving bodies; it is, in fact, what we tacitly assume in a straightforward and consistent working out of Maxwell's principles without any plainly-expressed statement on the question of the relative motion of matter and ether; for the part played in Maxwell's theory by matter is merely (and, of course, roughly) formularized by supposing that it causes the etherial constants to take different values, whilst introducing new properties, that of dissipating energy being the most prominent and important. We may, therefore, think of merely one medium, the most of which is uniform (the ether), whilst certain portions (matter as well) have different powers of supporting electric displacement and magnetic induction from the rest, as well as a host of additional properties; and of these we can include the power of supporting conduction-current with dissipation of energy according to Joule's law, the change from isotropy to eolotropy in respect to the distribution of the several fluxes, the presence of intrinsic sources of energy, etc.*

§ 3. We do not in any way form the equations of motion of such a medium, even as regards the uniform simple ether, away from gross matter; we have only to discuss it as regards the electric and magnetic fluxes it supports, and the stresses and fluxes of energy thereby necessitated. First, we suppose the medium to be stationary, and examine the flux of electromagnetic energy. This is the Poynting flux of energy. Next we set the medium into motion of an unrestricted kind. We have now necessarily a convection of the electric and magnetic energy, as well as the Poynting flux. Thirdly, there must be a similar convection of the kinetic energy, etc., of the translational motion; and fourthly, since the motion of the medium involves the working of ordinary (Newtonian) force, there is associated with the previous a flux of energy due to the activity of the corresponding stress. The question is therefore a complex one, for we have to properly fit together these various fluxes of energy in harmony with the electromagnetic equations. A side issue is the determination of the proper measure of the activity of intrinsic forces, when the medium moves; in another form, it is the determination of the proper meaning of "true current" in Maxwell's sense.

§ 4. The only general principle that we can bring to our assistance in interpreting electromagnetic results relating to activity and flux of energy, is that of the persistence of energy. But it would be quite inadequate in its older sense referring to integral amounts; the definite localization by Maxwell, of electric and magnetic energy, and of its waste, necessitates the similar localization of sources of energy; and in the consideration of the supply of energy at certain places, combined with the continuous transmission of electrical disturbances, and therefore of the associated energy, the idea of a flux of energy through space, and therefore of the continuity of energy in space and in time, becomes forced upon us as a simple, useful, and necessary principle, which cannot be avoided.

When energy goes from place to place, it traverses the intermediate space. Only by the use of this principle can we safely derive the electromagnetic stress from the equations of the field expressing the two laws of circuitation of the electric and magnetic forces; and this

* Perhaps it is best to say as little as possible at present about the connection between matter and ether, but to take the electromagnetic equations in an abstract manner. This will leave us greater freedom for future modifications without contradiction. There are, also, cases in which it is obviously impossible to suppose that matter in bulk carries on with it the ether in bulk which permeates it. Either, then, the mathematical machinery must work between the molecules; or else, we must make such alterations in the equations referring to bulk as will be practically equivalent in effect. For example, the motional magnetic force **VDq** of equations (88), (92), (93) may be modified either in **q** or in **D**, by use of a smaller effective velocity **q**, or by the substitution in **D** or c**E** of a modified reckoning of c for the effective permittivity.

again becomes permissible only by the postulation of the definite localization of the electric and magnetic energies. But we need not go so far as to assume the objectivity of energy. This is an exceedingly difficult notion, and seems to be rendered inadmissible by the mere fact of the relativity of motion, on which kinetic energy depends. We cannot, therefore, definitely individualize energy in the same way as is done with matter.

If ρ be the density of a quantity whose total amount is invariable, and which can change its distribution continuously, by actual motion from place to place, its equation of continuity is

$$\text{conv}\, \mathbf{q}\rho = \dot{\rho}, \qquad \text{.............................(1)}$$

where $\mathbf{q}$ is its velocity, and $\mathbf{q}\rho$ the flux of ρ. That is, the convergence of the flux of ρ equals the rate of increase of its density. Here ρ may be the density of matter. But it does not appear that we can apply the same method of representation to the flux of energy. We may, indeed, write

$$\text{conv}\, \mathbf{X} = \dot{T}, \qquad \text{.............................(2)}$$

if $\mathbf{X}$ be the flux of energy from all causes, and T the density of localizable energy. But the assumption $\mathbf{X} = T\mathbf{q}$ would involve the assumption that T moved about like matter, with a definite velocity. A part of T may, indeed, do this, viz., when it is confined to, and is carried by matter (or ether); thus we may write

$$\text{conv}\, (\mathbf{q}T + \mathbf{X}) = \dot{T}, \qquad \text{..........................(3)}$$

where T is energy which is simply carried, whilst $\mathbf{X}$ is the total flux of energy from other sources, and which we cannot symbolize in the form $T\mathbf{q}$; the energy which comes to us from the Sun, for example, or radiated energy. It is, again, often impossible to carry out the principle in this form, from a want of knowledge of how energy gets to a certain place. This is, for example, particularly evident in the case of gravitational energy, the distribution of which, before it is communicated to matter, increasing its kinetic energy, is highly speculative. If it come from the ether (and where else *can* it come from?), it should be possible to symbolize this in $\mathbf{X}$, if not in $\mathbf{q}T$; but in default of a knowledge of its distribution in the ether, we cannot do so, and must therefore turn the equation of continuity into

$$S + \text{conv}\, (\mathbf{q}T + \mathbf{X}) = \dot{T}, \qquad \text{.......................(4)}$$

where S indicates the rate of supply of energy per unit volume from the gravitational source, whatever that may be. A similar form is convenient in the case of intrinsic stores of energy, which we have reason to believe are positioned within the element of volume concerned, as when heat gives rise to thermoelectric force. Then S is the activity of the intrinsic sources. Then again, in special applications, T is conveniently divisible into different kinds of energy, potential and kinetic. Energy which is dissipated or wasted comes under the same category, because it may either be regarded as stored, though irrecoverably, or passed out of existence, so far as any immediate useful purpose is concerned. Thus we have as a standard practical form of the equation of continuity of energy referred to the unit volume,

$$S + \text{conv}\,\{\mathbf{X} + \mathfrak{q}(U+T)\} = Q + \dot{U} + \dot{T}, \quad \ldots\ldots\ldots\ldots\ldots(5)$$

where S is the energy supply from intrinsic sources, U potential energy and T kinetic energy of localizable kinds, $\mathfrak{q}(U+T)$ its convective flux, Q the rate of waste of energy, and $\mathbf{X}$ the flux of energy other than convective, *e.g.*, that due to stresses in the medium and representing their activity. In the electromagnetic application we shall see that U and T must split into two kinds, and so must $\mathbf{X}$, because there is a flux of energy even when the medium is at rest.

§ 5. Sometimes we meet with cases in which the flux of energy is either wholly or partly of a circuital character. There is nothing essentially peculiar to electromagnetic problems in this strange and apparently useless result. The electromagnetic instances are paralleled by similar instances in ordinary mechanical science, when a body is in motion and is also strained, especially if it be in rotation. This result is a necessary consequence of our ways of reckoning the activity of forces and of stresses, and serves to still further cast doubt upon the "thinginess" of energy. At the same time, the flux of energy is going on all around us, just as certainly as the flux of matter, and it is impossible to avoid the idea; we should, therefore, make use of it and formularize it whenever and as long as it is found to be useful, in spite of the occasional failure to obtain readily understandable results.

The idea of the flux of energy, apart from the conservation of energy, is by no means a new one. Had gravitational energy been less obscure than it is, it might have found explicit statement long ago. Professor Poynting* brought the principle into prominence in 1884, by making use of it to determine the electromagnetic flux of energy. Professor Lodge† gave very distinct and emphatic expression of the principle generally, apart from its electromagnetic aspect, in 1885, and pointed out how much more simple and satisfactory it makes the principle of the conservation of energy become. So it would, indeed, could we only understand gravitational energy; but in that, and similar respects, it is a matter of faith only. But Professor Lodge attached, I think, too much importance to the identity of energy, as well as to another principle he enunciated, that energy cannot be transferred without being transformed, and conversely; the transformation being from potential to kinetic energy or conversely. This obviously cannot apply to the convection of energy, which is a true flux of energy; nor does it seem to apply to cases of wave-motion in which the energy, potential and kinetic, of the disturbance, is transferred through a medium unchanged in relative distribution, simply because the disturbance itself travels without change of type; though it may be that in the unexpressed internal actions associated with the wave-propagation there might be found a better application.

* Poynting, *Phil. Trans.*, 1884.

† Lodge, *Phil. Mag.*, June, 1885, "On the Identity of Energy."

It is impossible that the ether can be fully represented, even merely in its transmissive functions, by the electromagnetic equations. Gravity is left out in the cold; and although it is convenient to ignore this fact, it may be sometimes usefully remembered, even in special electromagnetic work; for, if a medium have to contain and transmit gravitational energy as well as electromagnetic, the proper system of equations should show this, and, therefore, include the electromagnetic. It seems, therefore, not unlikely that in discussing purely electromagnetic speculations, one may be within a stone's throw of the explanation of gravitation all the time. The consummation would be a really substantial advance in scientific knowledge.

On the Algebra and Analysis of Vectors without Quaternions. Outline of Author's System.

§ 6. The proper language of vectors is the algebra of vectors. It is, therefore, quite certain that an extensive use of vector-analysis in mathematical physics generally, and in electromagnetism, which is swarming with vectors, in particular, is coming and may be near at hand. It has, in my opinion, been retarded by the want of special treatises on vector-analysis adapted for use in mathematical physics, Professor Tait's well-known profound treatise being, as its name indicates, a treatise on Quaternions. I have not found the Hamilton-Tait notation of vector-operations convenient, and have employed, for some years past, a simpler system. It is not, however, entirely a question of notation that is concerned. I reject the quaternionic basis of vector-analysis. The anti-quaternionic argument has been recently ably stated by Professor Willard Gibbs.*. He distinctly separates this from the question of notation, and this may be considered fortunate, for whilst I can fully appreciate and (from practical experience) endorse the anti-quaternionic argument, I am unable to appreciate his notation, and think that of Hamilton and Tait is, in some respects, preferable, though very inconvenient in others.

In Hamilton's system the quaternion is the fundamental idea, and everything revolves round it. This is exceedingly unfortunate, as it renders the establishment of the algebra of vectors without metaphysics a very difficult matter, and in its application to mathematical analysis there is a tendency for the algebra to get more and more complex as the ideas concerned get simpler, and the quaternionic basis forms a real difficulty of a substantial kind in attempting to work in harmony with ordinary Cartesian methods.

Now, I can confidently recommend, as a really practical working system, the modification I have made. It has many advantages, and not the least amongst them is the fact that the quaternion does not appear in it at all (though it may, without much advantage, be brought

* Professor Gibbs's letters will be found in *Nature*, vol. 43, p. 511, and vol. 44, p. 79; and Professor Tait's in vol. 43, pp. 535, 608. This rather one-sided discussion arose out of Professor Tait stigmatizing Professor Gibbs as "a retarder of quaternionic progress." This may be very true; but Professor Gibbs is anything but a retarder of progress in vector analysis and its application to physics.

in sometimes), and also that the notation is arranged so as to harmonize with Cartesian mathematics. It rests entirely upon a few definitions, and may be regarded (from one point of view) as a systematically abbreviated Cartesian method of investigation, and be understood and practically used by any one accustomed to Cartesians, without any study of the difficult science of Quaternions. It is simply the elements of Quaternions without the quaternions, with the notation simplified to the uttermost, and with the very inconvenient *minus* sign before scalar products done away with.*

§ 7. Quantities being divided into scalars and vectors, I denote the scalars, as usual, by ordinary letters, and put the vectors in the plain black type, known, I believe, as Clarendon type, rejecting Maxwell's German letters on account of their being hard to read. A special type is certainly not essential, but it facilitates the reading of printed complex vector investigations to be able to see at a glance which quantities are scalars and which are vectors, and eases the strain on the memory. But in MS. work there is no occasion for specially formed letters.

Thus **A** stands for a vector. The tensor of a vector may be denoted by the same letter plain; thus A is the tensor of **A**. (In MS. the tensor is A_0.) Its rectangular scalar components are A_1, A_2, A_3. A unit vector parallel to **A** may be denoted by $\mathbf{A}_1$, so that $\mathbf{A} = A\mathbf{A}_1$. But little things of this sort are very much matters of taste. What is important is to avoid as far as possible the use of letter prefixes, which, when they come two (or even three) together, as in Quaternions, are very confusing.

The scalar product of a pair of vectors **A** and **B** is denoted by **AB**, and is defined to be

$$\mathbf{AB} = A_1B_1 + A_2B_2 + A_3B_3 = AB\cos\widehat{\mathbf{AB}} = \mathbf{BA}. \quad \text{.........(6)}$$

* §§ 7, 8, 9 contain an introduction to vector-analysis (without the quaternion), which is sufficient for the purposes of the present paper, and, I may add, for general use in mathematical physics. It is an expansion of that given in my paper "On the Electromagnetic Wave Surface," *Phil. Mag.*, June, 1885, (vol. II., pp. 4 to 8). The algebra and notation are substantially those employed in all my papers, especially in "Electromagnetic Induction and its Propagation," *The Electrician*, 1885.

Professor Gibbs's vectorial work is scarcely known, and deserves to be well known. In June, 1888, I received from him a little book of 85 pages, bearing the singular imprint NOT PUBLISHED, Newhaven, 1881-4. It is indeed odd that the author should not have published what he had been at the trouble of having printed. His treatment of the linear vector-operator is specially deserving of notice. Although "For the use of students in physics," I am bound to say that I think the work much too condensed for a first introduction to the subject.

In *The Electrician* for Nov. 13, 1891, p. 27, I commenced a few articles on elementary vector-algebra and analysis, specially meant to explain to readers of my papers how to work vectors. I am given to understand that the earlier ones, on the algebra, were much appreciated; the later ones, however, are found difficult. But the vector-algebra is identically the same in both, and is of quite a rudimentary kind. The difference is, that the later ones are concerned with analysis, with varying vectors; it is the same as the difference between common algebra and differential calculus. The difficulty, whether real or not, does not indicate any difficulty in the vector-algebra. I mention this on account of the great prejudice which exists against vector-algebra.

The addition of vectors being as in the polygon of displacements, or velocities, or forces; *i.e.*, such that the vector length of any closed circuit is zero; either of the vectors **A** and **B** may be split into the sum of any number of others, and the multiplication of the two sums to form **AB** is done as in common algebra; thus

$$(\mathbf{a}+\mathbf{b})(\mathbf{c}+\mathbf{d})=\mathbf{ac}+\mathbf{ad}+\mathbf{bc}+\mathbf{bd}=\mathbf{ca}+\mathbf{da}+\mathbf{cb}+\mathbf{db}. \quad \text{......(7)}$$

If **N** be a unit vector, **NN** or $\mathbf{N}^2=1$; similarly, $\mathbf{A}^2=A^2$ for any vector.

The reciprocal of a vector **A** has the same direction; its tensor is the reciprocal of the tensor of **A**. Thus

$$\mathbf{AA}^{-1}=\frac{\mathbf{A}}{\mathbf{A}}=1; \quad \text{and} \quad \mathbf{AB}^{-1}=\mathbf{B}^{-1}\mathbf{A}=\frac{\mathbf{A}}{\mathbf{B}}=\frac{A}{B}\cos\widehat{\mathbf{AB}}. \quad \text{.....(8)}$$

The vector product of a pair of vectors is denoted by **VAB**, and is defined to be the vector whose tensor is $AB\sin\widehat{\mathbf{AB}}$, and whose direction is perpendicular to the plane of **A** and **B**, thus

$$\mathbf{VAB}=\mathbf{i}(A_2B_3-A_3B_2)+\mathbf{j}(A_3B_1-A_1B_3)+\mathbf{k}(A_1B_2-A_2B_1)=-\mathbf{VBA}, \quad (9)$$

where **i**, **j**, **k**, are any three mutually rectangular unit vectors. The tensor of **VAB** is $\mathrm{V}_0\mathbf{AB}$; or

$$\mathrm{V}_0\mathbf{AB}=AB\sin\widehat{\mathbf{AB}}. \quad \text{...........................(10)}$$

Its components are **iVAB**, **jVAB**, **kVAB**.

In accordance with the definitions of the scalar and vector products, we have

$$\left.\begin{array}{lll} \mathbf{i}^2=1, & \mathbf{j}^2=1, & \mathbf{k}^2=1; \\ \mathbf{ij}=0, & \mathbf{jk}=0, & \mathbf{ki}=0; \\ \mathbf{Vij}=\mathbf{k}, & \mathbf{Vjk}=\mathbf{i}, & \mathbf{Vki}=\mathbf{j}; \end{array}\right\} \quad \text{..................(11)}$$

and from these we prove at once that

$$\mathbf{V}(\mathbf{a}+\mathbf{b})(\mathbf{c}+\mathbf{d})=\mathbf{Vac}+\mathbf{Vad}+\mathbf{Vbc}+\mathbf{Vbd},$$

and so on, for any number of component vectors. The order of the letters in each product has to be preserved, since $\mathbf{Vab}=-\mathbf{Vba}$.

Two very useful formulæ of transformation are

$$\mathbf{AVBC}=\mathbf{BVCA}=\mathbf{CVAB}$$

$$=A_1(B_2C_3-B_3C_2)+A_2(B_3C_1-B_1C_3)+A_3(B_1C_2-B_2C_1); \quad \text{....(12)}$$

and $\quad \mathbf{VAVBC}=\mathbf{B.CA}-\mathbf{C.AB}, \quad \text{or} \quad =\mathbf{B}(\mathbf{CA})-\mathbf{C}(\mathbf{AB}). \quad \text{......(13)}$

Here the dots, or the brackets in the alternative notation, merely act as separators, separating the scalar products **CA** and **AB** from the vectors they multiply. A space would be equivalent, but would be obviously unpractical.

As $\dfrac{\mathbf{A}}{\mathbf{B}}$ is a scalar product, so in harmony therewith, there is the vector product $\mathrm{V}\dfrac{\mathbf{A}}{\mathbf{B}}$. Since $\mathbf{VAB}=-\mathbf{VBA}$, it is now necessary to make a convention as to whether the denominator comes first or last in $\mathrm{V}\frac{\mathbf{A}}{\mathbf{B}}$. Say therefore, $\mathrm{V}\mathbf{A}\mathbf{B}^{-1}$. Its tensor is

$$\mathrm{V}_0\frac{\mathbf{A}}{\mathbf{B}}=\frac{A}{B}\sin\widehat{\mathbf{AB}}. \qquad (14)$$

§ 8. Differentiation of vectors, and of scalar and vector functions of vectors with respect to scalar variables is done as usual. Thus,

$$\left.\begin{aligned}\dot{\mathbf{A}}&=\mathbf{i}\dot{A}_1+\mathbf{j}\dot{A}_2+\mathbf{k}\dot{A}_3.\\ \frac{d}{dt}\mathbf{AB}&=\mathbf{A}\dot{\mathbf{B}}+\mathbf{B}\dot{\mathbf{A}}.\\ \frac{d}{dt}\mathbf{A}\mathrm{V}\mathbf{BC}&=\dot{\mathbf{A}}\mathrm{V}\mathbf{BC}+\mathbf{A}\mathrm{V}\dot{\mathbf{B}}\mathbf{C}+\mathbf{A}\mathrm{V}\mathbf{B}\dot{\mathbf{C}}.\end{aligned}\right\} \qquad (15)$$

The same applies with complex scalar differentiators, *e.g.*, with the differentiator

$$\frac{\partial}{\partial t}=\frac{d}{dt}+\mathbf{q}\nabla,$$

used when a moving particle is followed, q being its velocity. Thus,

$$\frac{\partial}{\partial t}\mathbf{AB}=\mathbf{A}\frac{\partial\mathbf{B}}{\partial t}+\mathbf{B}\frac{\partial\mathbf{A}}{\partial t}=\mathbf{A}\dot{\mathbf{B}}+\mathbf{B}\dot{\mathbf{A}}+\mathbf{A}.\mathbf{q}\nabla.\mathbf{B}+\mathbf{B}.\mathbf{q}\nabla.\mathbf{A}. \qquad (16)$$

Here $\mathbf{q}\nabla$ is a scalar differentiator given by

$$\mathbf{q}\nabla=q_1\frac{d}{dx}+q_2\frac{d}{dy}+q_3\frac{d}{dz}, \qquad (17)$$

so that $\mathbf{A}.\mathbf{q}\nabla.\mathbf{B}$ is the scalar product of $\mathbf{A}$ and the vector $\mathbf{q}\nabla.\mathbf{B}$; the dots here again act essentially as separators. Otherwise, we may write it $\mathbf{A}(\mathbf{q}\nabla)\mathbf{B}$.

The fictitious vector ∇ given by

$$\nabla=\mathbf{i}\nabla_1+\mathbf{j}\nabla_2+\mathbf{k}\nabla_3=\mathbf{i}\frac{d}{dx}+\mathbf{j}\frac{d}{dy}+\mathbf{k}\frac{d}{dz} \qquad (18)$$

is *very* important. Physical mathematics is very largely the mathematics of ∇. The name Nabla seems, therefore, ludicrously inefficient. In virtue of i, j, k, the operator ∇ behaves as a vector. It also, of course, differentiates what follows it.

Acting on a scalar P, the result is the vector

$$\nabla P=\mathbf{i}\nabla_1P+\mathbf{j}\nabla_2P+\mathbf{k}\nabla_3P, \qquad (19)$$

the vector rate of increase of P with length.

If it act on a vector $\mathbf{A}$, there is first the scalar product

$$\nabla\mathbf{A}=\nabla_1A_1+\nabla_2A_2+\nabla_3A_3=\operatorname{div}\mathbf{A}, \qquad (20)$$

or the divergence of $\mathbf{A}$. Regarding a vector as a flux, the divergence of a vector is the amount leaving the unit volume.

The vector product $\mathrm{V}\nabla\mathbf{A}$ is

$$\mathrm{V}\nabla\mathbf{A}=\mathbf{i}(\nabla_2A_3-\nabla_3A_2)+\mathbf{j}(\nabla_3A_1-\nabla_1A_3)+\mathbf{k}(\nabla_1A_2-\nabla_2A_1)=\operatorname{curl}\mathbf{A}. \quad (21)$$

The line-integral of **A** round a unit area equals the component of the curl of **A** perpendicular to the area.

We may also have the scalar and vector products **N**∇ and V**N**∇, where the vector **N** is not differentiated. These operators, of course, require a function to follow them on which to operate; the previous q∇.**A** of (16) illustrates.

The Laplacean operator is the scalar product ∇^2 or ∇∇; or

$$\nabla^2 = \nabla_1^2 + \nabla_2^2 + \nabla_3^2; \qquad (22)$$

and an example of (13) is

$$\mathrm{V}\nabla\mathrm{V}\nabla\mathbf{A} = \nabla.\nabla\mathbf{A} - \nabla^2\mathbf{A}, \quad \text{or} \quad \mathrm{curl}^2\mathbf{A} = \nabla\,\mathrm{div}\,\mathbf{A} - \nabla^2\mathbf{A}, \qquad (23)$$

which is an important formula.

Other important formulæ are the next three.

$$\mathrm{div}\,P\mathbf{A} = P\,\mathrm{div}\,\mathbf{A} + \mathbf{A}\nabla.P, \qquad (24)$$

P being scalar. Here note that $\mathbf{A}\nabla.P$ and $\mathbf{A}\nabla P$ (the latter being the scalar product of **A** and ∇P) are identical. This is not true when for P we substitute a vector. Also

$$\mathrm{div}\,\mathrm{V}\mathbf{AB} = \mathbf{B}\,\mathrm{curl}\,\mathbf{A} - \mathbf{A}\,\mathrm{curl}\,\mathbf{B}; \qquad (25)$$

which is an example of (12), noting that both **A** and **B** have to be differentiated. And

$$\mathrm{curl}\,\mathrm{V}\mathbf{AB} = \mathbf{B}\nabla.\mathbf{A} + \mathbf{A}\,\mathrm{div}\,\mathbf{B} - \mathbf{A}\nabla.\mathbf{B} - \mathbf{B}\,\mathrm{div}\,\mathbf{A}. \qquad (26)$$

This is an example of (13).

§ 9. When one vector **D** is a *linear* function of another vector **E**, that is, connected by equations of the form

$$\left.\begin{aligned} D_1 &= c_{11}E_1 + c_{12}E_2 + c_{13}E_3, \\ D_2 &= c_{21}E_1 + c_{22}E_2 + c_{23}E_3, \\ D_3 &= c_{31}E_1 + c_{32}E_2 + c_{33}E_3, \end{aligned}\right\} \qquad (27)$$

in terms of the rectangular components, we denote this simply by

$$\mathbf{D} = c\mathbf{E}, \qquad (28)$$

where c is the linear operator. The conjugate function is given by

$$\mathbf{D}' = c'\mathbf{E}, \qquad (29)$$

where $\mathbf{D}'$ is got from **D** by exchanging c_{12} and c_{21}, etc. Should the nine coefficients reduce to six by $c_{12} = c_{21}$, etc., **D** and $\mathbf{D}'$ are identical, or **D** is a self-conjugate or symmetrical linear function of **E**.

But, in general, it is the sum of **D** and $\mathbf{D}'$ which is a symmetrical function of **E**, and the difference is a simple vector-product. Thus

$$\left.\begin{aligned} \mathbf{D} &= c_0\mathbf{E} + \mathrm{V}\epsilon\mathbf{E}, \\ \mathbf{D}' &= c_0\mathbf{E} - \mathrm{V}\epsilon\mathbf{E}, \end{aligned}\right\} \qquad (30)$$

where c_0 is a self-conjugate operator, and ϵ is the vector given by

$$\epsilon = \mathbf{i}\frac{c_{32} - c_{23}}{2} + \mathbf{j}\frac{c_{13} - c_{31}}{2} + \mathbf{k}\frac{c_{21} - c_{12}}{2}. \qquad (31)$$

The important characteristic of a self-conjugate operator is

$$\mathbf{E}_1\mathbf{D}_2 = \mathbf{E}_2\mathbf{D}_1, \qquad \text{or} \qquad \mathbf{E}_1 c_0 \mathbf{E}_2 = \mathbf{E}_2 c_0 \mathbf{E}_1, \quad \text{............(32)}$$

where $\mathbf{E}_1$ and $\mathbf{E}_2$ are any two $\mathbf{E}$'s, and $\mathbf{D}_1$, $\mathbf{D}_2$, the corresponding $\mathbf{D}$'s. But when there is not symmetry, the corresponding property is

$$\mathbf{E}_1\mathbf{D}_2 = \mathbf{E}_2\mathbf{D}_1', \qquad \text{or} \qquad \mathbf{E}_1 c \mathbf{E}_2 = \mathbf{E}_2 c' \mathbf{E}_1. \quad \text{.............(33)}$$

Of these operators we have three or four in electromagnetism connecting forces and fluxes, and three more connected with the stresses and strains concerned. As it seems impossible to avoid the consideration of rotational stresses in electromagnetism, and these are not usually considered in works on elasticity, it will be desirable to briefly note their peculiarities here, rather than later on.

On Stresses, irrotational and rotational, and their Activities.

§ 10. Let $\mathbf{P}_N$ be the vector stress on the $\mathbf{N}$-plane, or the plane whose unit normal is $\mathbf{N}$. It is a linear function of $\mathbf{N}$. This will fully specify the stress on any plane. Thus, if $\mathbf{P}_1$, $\mathbf{P}_2$, $\mathbf{P}_3$ are the stresses on the $\mathbf{i}$, $\mathbf{j}$, $\mathbf{k}$ planes, we shall have

$$\left.\begin{aligned} \mathbf{P}_1 &= \mathbf{i}P_{11} + \mathbf{j}P_{12} + \mathbf{k}P_{13}, \\ \mathbf{P}_2 &= \mathbf{i}P_{21} + \mathbf{j}P_{22} + \mathbf{k}P_{23}, \\ \mathbf{P}_3 &= \mathbf{i}P_{31} + \mathbf{j}P_{32} + \mathbf{k}P_{33}. \end{aligned}\right\} \quad \text{........................(34)}$$

Let, also, $\mathbf{Q}_N$ be the conjugate stress; then, similarly,

$$\left.\begin{aligned} \mathbf{Q}_1 &= \mathbf{i}P_{11} + \mathbf{j}P_{21} + \mathbf{k}P_{31}, \\ \mathbf{Q}_2 &= \mathbf{i}P_{12} + \mathbf{j}P_{22} + \mathbf{k}P_{32}, \\ \mathbf{Q}_3 &= \mathbf{i}\mathrm{P}_{13} + \mathbf{j}P_{23} + \mathbf{k}P_{33}. \end{aligned}\right\} \quad \text{........................(35)}$$

Half the sum of the stresses $\mathbf{P}_N$ and $\mathbf{Q}_N$ is an ordinary irrotational stress; so that

$$\mathbf{P}_N = \phi_0\mathbf{N} + \mathrm{V}\epsilon\mathbf{N}, \qquad \mathbf{Q}_N = \phi_0\mathbf{N} - \mathrm{V}\epsilon\mathbf{N}, \quad \text{...........(36)}$$

where ϕ_0 is self-conjugate, and

$$2\epsilon = \mathbf{i}(P_{23} - P_{32}) + \mathbf{j}(P_{31} - P_{13}) + \mathbf{k}(P_{12} - P_{21}). \quad \text{.........(37)}$$

Here 2ϵ is the torque per unit volume arising from the stress $\mathbf{P}$.

The translational force, $\mathbf{F}$, per unit volume is (by inspection of a unit cube)

$$\mathbf{F} = \nabla_1\mathbf{P}_1 + \nabla_2\mathbf{P}_2 + \nabla_3\mathbf{P}_3 \quad \text{.......................(38)}$$

$$= \mathbf{i}\,\mathrm{div}\,\mathbf{Q}_1 + \mathbf{j}\,\mathrm{div}\,\mathbf{Q}_2 + \mathbf{k}\,\mathrm{div}\,\mathbf{Q}_3; \quad \text{...........(39)}$$

or, in terms of the self-conjugate stress and the torque,

$$\mathbf{F} = (\mathbf{i}\,\mathrm{div}\,\phi_0\mathbf{i} + \mathbf{j}\,\mathrm{div}\,\phi_0\mathbf{j} + \mathbf{k}\,\mathrm{div}\,\phi_0\mathbf{k}) - \mathrm{curl}\,\epsilon, \quad \text{.........(40)}$$

where $-\mathrm{curl}\,\epsilon$ is the translational force due to the rotational stress alone, as in Sir W. Thomson's latest theory of the mechanics of an "ether." *

* *Mathematical and Physical Papers*, vol. 3, Art. 99, p. 436.

Next, let **N** be the unit-normal drawn outward from any closed surface. Then

$$\Sigma \mathbf{P}_N = \Sigma \mathbf{F}, \quad \text{.............................(41)}$$

where the left summation extends over the surface and the right summation throughout the enclosed region. For

$$\mathbf{P}_N = N_1\mathbf{P}_1 + N_2\mathbf{P}_2 + N_3\mathbf{P}_3 = \mathbf{i}.\mathbf{N}\mathbf{Q}_1 + \mathbf{j}.\mathbf{N}\mathbf{Q}_2 + \mathbf{k}.\mathbf{N}\mathbf{Q}_3; \quad \text{......(42)}$$

so the well-known theorem of divergence gives immediately, by (39),

$$\Sigma \mathbf{P}_N = \Sigma(\mathbf{i} \operatorname{div} \mathbf{Q}_1 + \mathbf{j} \operatorname{div} \mathbf{Q}_2 + \mathbf{k} \operatorname{div} \mathbf{Q}_3) = \Sigma \mathbf{F}. \quad \text{..........(43)}$$

Next, as regards the equivalence of rotational effect of the surface-stress to that of the internal forces and torques. Let **r** be the vector distance from any fixed origin. Then **VrF** is the vector moment of a force, **F**, at the end of the arm **r**. Another (not so immediate) application of the divergence theorem gives

$$\Sigma \mathbf{V}\mathbf{r}\mathbf{P}_N = \Sigma \mathbf{V}\mathbf{r}\mathbf{F} + \Sigma 2\epsilon. \quad \text{.......................(44)}$$

Thus, any distribution of stress, whether rotational or irrotational, may be regarded as in equilibrium. Given any stress in a body, terminating at its boundary, the body will be in equilibrium both as regards translation and rotation. Of course, the boundary discontinuity in the stress has to be reckoned as the equivalent of internal divergence in the appropriate manner. Or, more simply, let the stress fall off continuously from the finite internal stress to zero through a thin surface-layer. We then have a distribution of forces and torques in the surface-layer which equilibrate the internal forces and torques.

To illustrate; we know that Maxwell arrived at a peculiar stress, compounded of a tension parallel to a certain direction, and an equal lateral pressure, which would account for the mechanical actions apparent between electrified bodies; and endeavoured similarly to determine the stress in the interior of a magnetized body to harmonize with the similar external magnetic stress of the simple type mentioned. This stress in a magnetized body I believe to be thoroughly erroneous; nevertheless, so far as accounting for the forcive on a magnetized body is concerned, it will, when properly carried out with due attention to surface-discontinuity, answer perfectly well, not because it *is* the stress, but because *any* stress would do the same, the only essential feature concerned being the external stress in the air.

Here we may also note the very powerful nature of the stress-function, considered merely as a mathematical engine, apart from physical reality. For example, we may account for the forcive on a magnet in many ways, of which the two most prominent are by means of forces on imaginary magnetic matter, and by forces on imaginary electric currents, in the magnet and on its surface. To prove the equivalence of these two methods (and the many others) involves very complex surface- and volume-integrations and transformations in the general case, which may be all avoided by the use of the stress-function instead of the forces.

§ 11. Next as regards the activity of the stress $\mathbf{P}_N$ and the equivalent translational, distortional, and rotational activities. The activity of $\mathbf{P}_N$ is $\mathbf{P}_N\mathbf{q}$ per unit area, if $\mathbf{q}$ be the velocity. Here

$$\mathbf{P}_N\mathbf{q} = q_1.\mathbf{NQ}_1 + q_2.\mathbf{NQ}_2 + q_3.\mathbf{NQ}_3, \qquad (45)$$

by (42); or, re-arranging,

$$\mathbf{P}_N\mathbf{q} = \mathbf{N}(q_1\mathbf{Q}_1 + q_2\mathbf{Q}_2 + q_3\mathbf{Q}_3) = \mathbf{N}\,\Sigma\, q\mathbf{Q} = \mathbf{N}q\mathbf{Q}_q, \qquad (46)$$

where $\mathbf{Q}_q$ is the conjugate stress on the $\mathbf{q}$-plane. That is, $q\mathbf{Q}_q$ or $\Sigma\,\mathbf{Q}q$ is the negative of the vector flux of energy expressing the stress-activity. For we choose P_{NN} so as to mean a pull when it is positive, and when the stress $\mathbf{P}_N$ works in the same sense with $\mathbf{q}$, energy is transferred against the motion, to the matter which is pulled.

The convergence of the energy-flux, or the divergence of $q\mathbf{Q}_q$, is therefore the activity per unit volume. Thus

$$\begin{aligned}\operatorname{div}(\mathbf{Q}_1 q_1 + \mathbf{Q}_2 q_2 + \mathbf{Q}_3 q_3) \\ &= \mathbf{q}(\mathbf{i}\operatorname{div}\mathbf{Q}_1 + \mathbf{j}\operatorname{div}\mathbf{Q}_2 + \mathbf{k}\operatorname{div}\mathbf{Q}_3) + (\mathbf{Q}_1\nabla q_1 + \mathbf{Q}_2\nabla q_2 + \mathbf{Q}_3\nabla q_3) \qquad (47) \\ &= \mathbf{q}(\nabla_1\mathbf{P}_1 + \nabla_2\mathbf{P}_2 + \nabla_3\mathbf{P}_3) + \mathbf{P}_1\nabla_1\mathbf{q} + \mathbf{P}_2\nabla_2\mathbf{q} + \mathbf{P}_3\nabla_3\mathbf{q}, \qquad (48)\end{aligned}$$

where the first form (47) is generally most useful. Or

$$\operatorname{div}\Sigma\,\mathbf{Q}q = \mathbf{Fq} + \Sigma\,\mathbf{Q}\nabla q\,; \qquad (49)$$

where the first term on the right is the translational activity, and the rest is the sum of the distortional and rotational activities. To separate the latter introduce the strain-velocity vectors (analogous to $\mathbf{P}_1$, $\mathbf{P}_2$, $\mathbf{P}_3$)

$$\mathbf{p}_1 = \tfrac{1}{2}(\nabla q_1 + \nabla_1\mathbf{q}), \qquad \mathbf{p}_2 = \tfrac{1}{2}(\nabla q_2 + \nabla_2\mathbf{q}), \qquad \mathbf{p}_3 = \tfrac{1}{2}(\nabla q_3 + \nabla_3\mathbf{q})\,; \qquad (50)$$

and generally

$$\mathbf{p}_N = \tfrac{1}{2}(\nabla.\mathbf{qN} + \mathbf{N}\nabla.\mathbf{q}). \qquad (51)$$

Using these we obtain

$$\begin{aligned}\Sigma\,\mathbf{Q}\nabla\mathbf{q} &= \mathbf{Q}_1\mathbf{p}_1 + \mathbf{Q}_2\mathbf{p}_2 + \mathbf{Q}_3\mathbf{p}_3 + \mathbf{Q}_1\frac{\nabla q_1 - \nabla_1\mathbf{q}}{2} + \mathbf{Q}_2\frac{\nabla q_2 - \nabla_2\mathbf{q}}{2} + \mathbf{Q}_3\frac{\nabla q_3 - \nabla_3\mathbf{q}}{2} \\ &= \Sigma\,\mathbf{Qp} + \tfrac{1}{2}\mathbf{Q}_1\mathrm{V}\mathbf{i}\operatorname{curl}\mathbf{q} + \tfrac{1}{2}\mathbf{Q}_2\mathrm{V}\mathbf{j}\operatorname{curl}\mathbf{q} + \tfrac{1}{2}\mathbf{Q}_3\mathrm{V}\mathbf{k}\operatorname{curl}\mathbf{q} \\ &= \Sigma\,\mathbf{Qp} + \epsilon\operatorname{curl}\mathbf{q}. \qquad (52)\end{aligned}$$

Thus $\Sigma\,\mathbf{Qp}$ is the distortional activity and $\epsilon\operatorname{curl}\mathbf{q}$ the rotational activity. But since the distortion and the rotation are quite independent, we may put $\Sigma\,\mathbf{Pp}$ for the distortional activity; or else use the self-conjugate stress, and write it $\tfrac{1}{2}\Sigma(\mathbf{P}+\mathbf{Q})\mathbf{p}$.

§ 12. In an ordinary "elastic solid," when isotropic, there is elastic resistance to compression and to distortion. We may also imaginably have elastic resistance to translation and to rotation; nor is there, so far as the mathematics is concerned, any reason for excluding dissipative resistance to translation, distortion, and rotation; and kinetic energy may be associated with all three as well, instead of with the translation alone, as in the ordinary elastic solid.

Considering only three elastic moduli, we have the old k and n of Thomson and Tait (resistance to compression and rigidity), and a new coefficient, say n_1, such that

$$\epsilon = n_1\operatorname{curl}\mathbf{D}, \qquad (53)$$

if $\mathbf{D}$ be the displacement and 2ϵ the torque, as before.

The stress on the i-plane (any plane) is

$$\mathbf{P}_1 = n(\nabla D_1 + \nabla_1 \mathbf{D}) + \mathbf{i}(k - \tfrac{2}{3}n)\,\mathrm{div}\,\mathbf{D} + n_1 \mathrm{V}\,\mathrm{curl}\,\mathbf{D}.\mathbf{i}$$

$$= (n + n_1)\nabla_1 \mathbf{D} + (n - n_1)\nabla D_1 + (k - \tfrac{2}{3}n)\mathbf{i}\,\mathrm{div}\,\mathbf{D}\,; \quad \text{..........(54)}$$

and its conjugate is

$$\mathbf{Q}_1 = n(\nabla D_1 + \nabla_1 \mathbf{D}) + \mathbf{i}(k - \tfrac{2}{3}n)\,\mathrm{div}\,\mathbf{D} - n_1(\nabla_1 \mathbf{D} - \nabla D_1)$$

$$= (n - n_1)\nabla_1 \mathbf{D} + (n + n_1)\nabla D_1 + \mathbf{i}(k - \tfrac{2}{3}n)\,\mathrm{div}\,\mathbf{D}\,; \quad \text{..........(55)}$$

from which

$$F_1 = \mathrm{div}\,\mathbf{Q}_1 = (n - n_1 + k - \tfrac{2}{3}n)\nabla_1\,\mathrm{div}\,\mathbf{D} + (n + n_1)\nabla^2 D_1 \quad \text{......(56)}$$

is the **i**-component of the translational force; the complete force **F** is therefore

$$\mathbf{F} = (n + n_1)\nabla^2 \mathbf{D} + (k + \tfrac{1}{3}n - n_1)\nabla\,\mathrm{div}\,\mathbf{D}\,; \quad \text{..............(57)}$$

or, in another form, if $P = -k\,\mathrm{div}\,\mathbf{D}$,

P being the isotropic pressure,

$$\mathbf{F} = -\nabla P + n(\nabla^2 \mathbf{D} + \tfrac{1}{3}\nabla\,\mathrm{div}\,\mathbf{D}) - n_1\,\mathrm{curl}^2\,\mathbf{D}, \quad \text{..........(58)}$$

remembering (23) and (53).

We see that in (57) the term involving div **D** may vanish in a compressible solid by the relation $n_1 = k + \tfrac{1}{3}n$; this makes

$$n + n_1 = k + \tfrac{4}{3}n, \qquad n_1 - n = k - \tfrac{2}{3}n, \quad \text{..............(59)}$$

which are the moduli, longitudinal and lateral, of a simple longitudinal strain; that is, multiplied by the extension, they give the longitudinal traction, and the lateral traction required to prevent lateral contraction.

The activity per unit volume, other than translational, is

$$\begin{aligned}\Sigma\,\mathbf{Q}\nabla q &= (n - n_1)(\nabla_1\mathbf{D}.\nabla q_1 + \nabla_2\mathbf{D}.\nabla q_2 + \nabla_3\mathbf{D}.\nabla q_3)\\ &\quad + (n + n_1)(\nabla D_1.\nabla q_1 + \nabla D_2.\nabla q_2 + \nabla D_3.\nabla q_3)\\ &\quad + (k - \tfrac{2}{3}n)\,\mathrm{div}\,\mathbf{D}\,\mathrm{div}\,\mathbf{q}\\ &= n(\nabla_1\mathbf{D}.\nabla q_1 + \nabla_2\mathbf{D}.\nabla q_2 + \nabla_3\mathbf{D}.\nabla q_3 + \nabla D_1\nabla q_1 + \nabla D_2\nabla q_2 + \nabla D_3\nabla q_3)\\ &\quad + (k - \tfrac{2}{3}n)\,\mathrm{div}\,\mathbf{D}\,\mathrm{div}\,\mathbf{q} + n_1\,\mathrm{curl}\,\mathbf{D}\,\mathrm{curl}\,\mathbf{q}\,; \quad \text{........................(60)}\end{aligned}$$

or, which is the same,

$$\begin{aligned}\Sigma\,\mathbf{Q}\nabla q = \frac{d}{dt}\Big[&\tfrac{1}{2}k(\mathrm{div}\,\mathbf{D})^2 + \tfrac{1}{2}n_1(\mathrm{curl}\,\mathbf{D})^2 - \tfrac{1}{3}n(\mathrm{div}\,\mathbf{D})^2\\ &+ \tfrac{1}{2}n\Big((\nabla D_1)^2 + (\nabla D_2)^2 + (\nabla D_3)^2 + \nabla D_1.\nabla_1\mathbf{D} + \nabla D_2.\nabla_2\mathbf{D} + \nabla D_3.\nabla_3\mathbf{D}\Big)\Big], \quad (61)\end{aligned}$$

where the quantity in square brackets is the potential energy of an *infinitesimal* distortion and rotation. The italicized reservation appears to be necessary, as we shall see from the equation of activity later, that the convection of the potential energy destroys the completeness of the statement

$$\Sigma\,\mathbf{Q}\nabla q = \dot{U},$$

if U be the potential energy.

In an elastic solid of the ordinary kind, with $n_1 = 0$, we have

$$\left.\begin{aligned} \mathbf{P}_N &= n(2 \operatorname{curl} \mathrm{V}\mathbf{D}\mathbf{N} + \mathrm{V}\mathbf{N} \operatorname{curl} \mathbf{D}),\\ \mathbf{F} &= -n \operatorname{curl}^2 \mathbf{D}. \end{aligned}\right\} \qquad (62)$$

In the case of a medium in which n is zero but n_1 finite (Sir W. Thomson's rotational ether),

$$\left.\begin{aligned} \mathbf{P}_N &= n_1 \mathrm{V} \operatorname{curl} \mathbf{D}.\mathbf{N},\\ \mathbf{F} &= -n_1 \operatorname{curl}^2 \mathbf{D}. \end{aligned}\right\} \qquad (63)$$

Thirdly, if we have both $k = -\frac{4}{3}n$ and $n = n_1$, then

$$\left.\begin{aligned} \mathbf{P}_N &= 2n \operatorname{curl} \mathrm{V}\mathbf{D}\mathbf{N},\\ \mathbf{F} &= -2n \operatorname{curl}^2 \mathbf{D}, \end{aligned}\right\} \qquad (\epsilon = n \operatorname{curl} \mathbf{D}), \qquad (64)$$

i.e., the sums of the previous two stresses and forces.

§ 13. As already observed, the vector flux of energy due to the stress is

$$-\Sigma\, \mathbf{Q}q = -\mathbf{Q}_q q = -(\mathbf{Q}_1 q_1 + \mathbf{Q}_2 q_2 + \mathbf{Q}_3 q_3). \qquad (65)$$

Besides this, there is the flux of energy

$$\mathbf{q}(U + T)$$

by convection, where U is potential and T kinetic energy. Therefore,

$$\mathbf{W} = \mathbf{q}(U + T) - \Sigma\, \mathbf{Q}q \qquad (66)$$

represents the complete energy-flux, so far as the stress and motion are concerned. Its convergence increases the potential energy, the kinetic energy, or is dissipated. But if there be an impressed translational force $\mathbf{f}$, its activity is $\mathbf{fq}$. This supply of energy is independent of the convergence of $\mathbf{W}$. Hence

$$\mathbf{fq} = Q + \dot{U} + \dot{T} + \operatorname{div}\,[\mathbf{q}(U + T) - \Sigma\, \mathbf{Q}q] \qquad (67)$$

is the equation of activity.

But this splits into two parts at least. For (67) is the same as

$$(\mathbf{f} + \mathbf{F})\mathbf{q} + \Sigma\, \mathbf{Q}\nabla q = Q + \dot{U} + \dot{T} + \operatorname{div} \mathbf{q}(U + T), \qquad (68)$$

and the translational portion may be removed altogether. That is,

$$(\mathbf{f} + \mathbf{F})\mathbf{q} = Q_0 + \dot{U}_0 + \dot{T}_0 + \operatorname{div} \mathbf{q}(U_0 + T_0), \qquad (69)$$

if the quantities with the zero suffix are only translationally involved. For example, if

$$\mathbf{f} + \mathbf{F} = \rho \frac{\partial \mathbf{q}}{\partial t}, \qquad (70)$$

as in fluid motion, without frictional or elastic forces associated with the translation, then

$$(\mathbf{f} + \mathbf{F})\mathbf{q} = \rho \mathbf{q} \frac{\partial \mathbf{q}}{\partial t} = \dot{T} + \operatorname{div} \mathbf{q}T, \qquad (71)$$

if $T = \frac{1}{2}\rho q^2$, the kinetic energy per unit volume. The complete form (69) comes in by the addition of elastic and frictional resisting forces. So, deducting (69) from (68), there is left

$$\Sigma\, \mathbf{Q}\nabla q = Q_1 + \dot{U}_1 + \dot{T}_1 + \operatorname{div} \mathbf{q}(U_1 + T_1), \qquad (72)$$

where the quantities with suffix unity are connected with the distortion and the rotation, and there may plainly be two sets of dissipative terms, and of energy (stored) terms. Thus the relation

$$\epsilon=\left(n_1+n_2\frac{d}{dt}+n_3\frac{d^2}{dt^2}\right)\text{curl }\mathbf{D} \qquad (73)$$

will bring in dissipation and kinetic energy, as well as the former potential energy of rotation associated with n_1.

That there can be dissipative terms associated with the distortion is also clear enough, remembering Stokes's theory of a viscous fluid. Thus, for simplicity, do away with the rotating stress, by putting $\epsilon=0$, making $\mathbf{P}_N$ and $\mathbf{Q}_N$ identical. Then take the stress on the i-plane to be given by

$$\mathbf{P}_1=\left(n+\mu\frac{d}{dt}+\nu\frac{d^2}{dt^2}\right)(\nabla D_1+\nabla_1\mathbf{D})-\mathbf{i}\left\{P+\tfrac{2}{3}\left(n+\mu\frac{d}{dt}+\nu\frac{d^2}{dt^2}\right)\text{div }\mathbf{D}\right\}, \qquad (74)$$

and similarly for any other plane; where $P=-k\,\text{div }\mathbf{D}$.

When $\mu=0$, $\nu=0$, we have the elastic solid with rigidity and compressibility. When $n=0$, $\nu=0$, we have the viscous fluid of Stokes. When $\nu=0$ only, we have a viscous elastic solid, the viscous resistance being purely distortional, and proportional to the speed of distortion. But with n, μ, ν, all finite, we still further associate kinetic energy with the potential energy and dissipation introduced by n and μ.

We have

$$\Sigma\,\mathbf{P}\nabla q=Q_2+\dot{U}_2+T_2$$

for infinitesimal strains, omitting the effect of convection of energy; where

$$T_2=\tfrac{1}{2}\nu\left[-\tfrac{2}{3}(\text{div }\mathbf{q})^2+\nabla q_1(\nabla q_1+\nabla_1\mathbf{q})+\nabla q_2(\nabla q_2+\nabla_2\mathbf{q})+\nabla q_3(\nabla q_3+\nabla_3\mathbf{q})\right], \qquad (75)$$

$$Q_2=\mu\left[-\tfrac{2}{3}(\text{div }\mathbf{q})^2+\nabla q_1(\nabla q_1+\nabla_1\mathbf{q})+\nabla q_2(\nabla q_2+\nabla_2\mathbf{q})+\nabla q_3(\nabla q_3+\nabla_3\mathbf{q})\right], \qquad (76)$$

$$U_2=\tfrac{1}{2}n\left[\left(\frac{k}{n}-\tfrac{2}{3}\right)(\text{div }\mathbf{D})^2+\nabla D_1(\nabla D_1+\nabla_1\mathbf{D})+\nabla D_2(\nabla D_2+\nabla_2\mathbf{D})+\nabla D_3(\nabla D_3+\nabla_3\mathbf{D})\right] \quad (77)$$

Observe that T_2 and Q_2 only differ in the exchange of μ to $\frac{1}{2}\nu$; but U_2, the potential energy, is not the same function of n and $\mathbf{D}$ that T_2 is of ν and $\mathbf{q}$. But if we take $k=0$, we produce similarity. An elastic solid having no resistance to compression is also one of Sir W. Thomson's ethers.

When $n=0$, $\mu=0$, $\nu=0$, we come down to the frictionless fluid, in which

$$\mathbf{f}-\nabla P=\rho\frac{\partial\mathbf{q}}{\partial t}, \qquad (78)$$

and

$$\Sigma\,\mathbf{P}\nabla q=-P\,\text{div }\mathbf{q}, \qquad (79)$$

with the equation of activity

$$\mathbf{fq}=\dot{U}+\dot{T}+\text{div }(U+T+P)\mathbf{q}, \qquad (80)$$

the only parts of which are not always easy to interpret are the $P\mathbf{q}$ term, and the proper measure of U. By analogy, and conformably with more general cases, we should take

$$P = -k \operatorname{div} \mathbf{D}, \qquad \text{and} \qquad U = \tfrac{1}{2}k(\operatorname{div} \mathbf{D})^2,$$

reckoning the expansion or compression from some mean condition.

The Electromagnetic Equations in a Moving Medium.

§ 14. The study of the forms of the equation of activity in purely mechanical cases, and of their interpretation, is useful, because in the electromagnetic problem of a moving medium we have still greater generality, and difficulty of safe and sure interpretation. To bring it as near to abstract dynamics as possible, all we need say regarding the two fluxes, electric displacement **D** and magnetic induction **B**, is that they are linear functions of the electric force **E** and magnetic force **H**, say

$$\mathbf{B} = \mu\mathbf{H}, \qquad \mathbf{D} = c\mathbf{E}, \qquad \text{.......................(81)}$$

where c and μ are linear operators of the symmetrical kind, and that associated with them are the stored energies U and T, electric and magnetic respectively (per unit volume), given by

$$U = \tfrac{1}{2}\mathbf{ED}, \qquad T = \tfrac{1}{2}\mathbf{HB}. \qquad \text{.......................(82)}$$

In isotropic media c is the permittivity, μ the inductivity. It is unnecessary to say more regarding the well-known variability of μ and hysteresis than that a magnet is here an ideal magnet of constant inductivity.

As there may be impressed forces, **E** is divisible into the force of the field and an impressed part; for distinctness, then, the complete **E** may be called the "force of the flux" **D**. Similarly as regards **H** and **B**.

There is also waste of energy (in conductors, namely) at the rates

$$Q_1 = \mathbf{EC}, \qquad Q_2 = \mathbf{HK}, \qquad \text{......................(83)}$$

where the fluxes **C** and **K** are also linear functions of **E** and **H** respectively; thus

$$\mathbf{C} = k\mathbf{E}, \qquad \mathbf{K} = g\mathbf{H}, \qquad \text{......................(84)}$$

where, when the force is parallel to the flux, and k is scalar, it is the electric conductivity. Its magnetic analogue is g, the magnetic conductivity. That is, a magnetic conductor is a (fictitious) body which cannot support magnetic force without continuously dissipating energy.

Electrification is the divergence of the displacement, and its analogue, magnetification, is the divergence of the induction; thus

$$\rho = \operatorname{div} \mathbf{D}, \qquad \sigma = \operatorname{div} \mathbf{B}, \qquad \text{....................(85)}$$

are their volume-densities. The quantity σ is probably quite fictitious, like **K**.

According to Maxwell's doctrine, the true electric current is always circuital, and is the sum of the conduction-current and the current of displacement, which is the time-rate of increase of the displacement.

But, to preserve circuitality, we must add the convection-current when electrification is moving, so that the true current becomes

$$\mathbf{J} = \mathbf{C} + \dot{\mathbf{D}} + \mathbf{q}\rho, \quad \text{..........................(86)}$$

where $\mathbf{q}$ is the velocity of the electrification ρ. Similarly

$$\mathbf{G} = \mathbf{K} + \dot{\mathbf{B}} + \mathbf{q}\sigma \quad \text{..........................(87)}$$

should be the corresponding magnetic current.

§ 15. Maxwell's equation of electric current in terms of magnetic force in a medium at rest, say,

$$\text{curl}\, \mathbf{H}_1 = \mathbf{C} + \mathbf{D},$$

where $\mathbf{H}_1$ is the force of the field, should be made, using $\mathbf{H}$ instead,

$$\text{curl}\, (\mathbf{H} - \mathbf{h}_0) = \mathbf{C} + \dot{\mathbf{D}} + \mathbf{q}\rho,$$

and here $\mathbf{h}_0$ will be the intrinsic force of magnetization, such that $\mu\mathbf{h}_0$ is the intensity of intrinsic magnetization. But I have shown that when there is motion, another impressed term is required, viz., the motional magnetic force

$$\mathbf{h} = \text{V}\mathbf{D}\mathbf{q}, \quad \text{..............................(88)}$$

making the first circuital law become

$$\text{curl}\, (\mathbf{H} - \mathbf{h}_0 - \mathbf{h}) = \mathbf{J} = \mathbf{C} + \dot{\mathbf{D}} + \mathbf{q}\rho. \quad \text{................(89)}$$

Maxwell's other connection to form the equations of propagation is made through his vector-potential **A** and scalar potential Ψ. Finding this method not practically workable, and also not sufficiently general, I have introduced instead a companion equation to (89) in the form

$$-\text{curl}\, (\mathbf{E} - \mathbf{e}_0 - \mathbf{e}) = \mathbf{G} = \mathbf{K} + \dot{\mathbf{B}} + \mathbf{q}\sigma, \quad \text{..............(90)}$$

where $\mathbf{e}_0$ expresses intrinsic force, and $\mathbf{e}$ is the motional electric force given by

$$\mathbf{e} = \text{V}\mathbf{q}\mathbf{B}, \quad \text{..............................(91)}$$

which is one of the terms in Maxwell's equation of electromotive force. As for $\mathbf{e}_0$, it includes not merely the force of intrinsic electrization, the analogue of intrinsic magnetization, but also the sources of energy, voltaic force, thermoelectric force, etc.

(89) and (90) are thus the working equations, with (88) and (91) in case the medium moves; along with the linear relations before mentioned, and the definitions of energy and waste of energy per unit volume. The fictitious **K** and σ are useful in symmetrizing the equations, if for no other purpose.

Another way of writing the two equations of curl is by removing the **e** and **h** terms to the right side. Let

$$\left.\begin{aligned} \text{curl}\, \mathbf{h} &= \mathbf{j}, \qquad & \mathbf{J} + \mathbf{j} &= \mathbf{J}_0, \\ -\text{curl}\, \mathbf{e} &= \mathbf{g}, \qquad & \mathbf{G} + \mathbf{g} &= \mathbf{G}_0. \end{aligned}\right\} \quad \text{.....................(92)}$$

Then (89) and (90) may be written

$$\left.\begin{aligned} \text{curl}\, (\mathbf{H} - \mathbf{h}_0) &= \mathbf{J}_0 = \mathbf{C} + \dot{\mathbf{D}} + \mathbf{q}\rho + \mathbf{j}, \\ -\text{curl}\, (\mathbf{E} - \mathbf{e}_0) &= \mathbf{G}_0 = \mathbf{K} + \dot{\mathbf{B}} + \mathbf{q}\sigma + \mathbf{g}. \end{aligned}\right\} \quad \text{.................(93)}$$

So far as circuitality of the current goes, the change is needless, and still further complicates the make-up of the true current, supposed now to be $\mathbf{J}_0$. On the other hand, it is a simplification on the left side, deriving the current from the force of the flux or of the field more simply.

A question to be settled is whether $\mathbf{J}$ or $\mathbf{J}_0$ should be the true current. There seems only one crucial test, viz., to find whether $\mathbf{e}_0\mathbf{J}$ or $\mathbf{e}_0\mathbf{J}_0$ is the rate of supply of energy to the electromagnetic system by an intrinsic force $\mathbf{e}_0$. This requires, however, a full and rigorous examination of all the fluxes of energy concerned.

The Electromagnetic Flux of Energy in a stationary Medium.

§ 16. First let the medium be at rest, giving us the equations

$$\text{curl}\,(\mathbf{H}-\mathbf{h}_0)=\mathbf{J}=\mathbf{C}+\dot{\mathbf{D}}, \qquad (94)$$

$$-\text{curl}\,(\mathbf{E}-\mathbf{e}_0)=\mathbf{G}=\mathbf{K}+\dot{\mathbf{B}}. \qquad (95)$$

Multiply (94) by $(\mathbf{E}-\mathbf{e}_0)$, and (95) by $(\mathbf{H}-\mathbf{h}_0)$, and add the results. Thus,

$$(\mathbf{E}-\mathbf{e}_0)\mathbf{J}+(\mathbf{H}-\mathbf{h}_0)\mathbf{G}=(\mathbf{E}-\mathbf{e}_0)\,\text{curl}\,(\mathbf{H}-\mathbf{h}_0)-(\mathbf{H}-\mathbf{h}_0)\,\text{curl}\,(\mathbf{E}-\mathbf{e}_0),$$

which, by the formula (25), becomes

$$\mathbf{e}_0\mathbf{J}+\mathbf{h}_0\mathbf{G}=\mathbf{EJ}+\mathbf{HG}+\text{div}\,\text{V}(\mathbf{E}-\mathbf{e}_0)(\mathbf{H}-\mathbf{h}_0)\,;$$

or, by the use of (82), (83),

$$\mathbf{e}_0\mathbf{J}+\mathbf{h}_0\mathbf{G}=Q+\dot{U}+\dot{T}+\text{div}\,\mathbf{W}, \qquad (96)$$

where the new vector $\mathbf{W}$ is given by

$$\mathbf{W}=\text{V}(\mathbf{E}-\mathbf{e}_0)(\mathbf{H}-\mathbf{h}_0). \qquad (97)$$

The form of (96) is quite explicit, and the interpretation sufficiently clear. The left side indicates the rate of supply of energy from intrinsic sources. Thus, $(Q+\dot{U}+\dot{T})$ shows the rate of waste and of storage of energy in the unit volume. The remainder, therefore, indicates the rate at which energy is passed out from the unit volume; and the flux $\mathbf{W}$ represents the flux of energy necessitated by the postulated localization of energy and its waste, when $\mathbf{E}$ and $\mathbf{H}$ are connected in the manner shown by (94) and (95).

There might also be an independent circuital flux of energy, but, being useless, it would be superfluous to bring it in.

The very important formula (97) was first discovered and interpreted by Professor Poynting, and independently discovered and interpreted a little later by myself in an extended form. It will be observed that in my mode of proof above there is no limitation as to homogeneity or isotropy as regards the permittivity, inductivity, and conductivity. But c and μ should be symmetrical. On the other hand, k and g do not require this limitation in deducing (97).*

* The method of treating Maxwell's electromagnetic scheme employed in the text (first introduced in "Electromagnetic Induction and its Propagation," *The Electrician*, January 3, 1885, and later) may, perhaps, be appropriately termed the

It is important to recognize that this flux of energy is not dependent upon the translational motion of the medium, for it is assumed explicitly to be at rest. The vector **W** cannot, therefore, be a flux of the kind $\mathbf{Q}_q q$ before discussed, unless possibly it be merely a rotating stress that is concerned.

The only dynamical analogy with which I am acquainted which seems at all satisfactory is that furnished by Sir W. Thomson's theory of a rotational ether. Take the case of $\mathbf{e}_0=0$, $\mathbf{h}_0=0$, $k=0$, $g=0$, and c and μ constants, that is, pure ether uncontaminated by ordinary matter. Then

$$\text{curl}\,\mathbf{H}=c\dot{\mathbf{E}}, \qquad (98)$$

$$-\text{curl}\,\mathbf{E}=\mu\dot{\mathbf{H}}. \qquad (99)$$

Now, let **H** be velocity, μ density; then, by (99), $-\text{curl}\,\mathbf{E}$ is the translational force due to the stress, which is, therefore, a rotating stress; thus,

$$\mathbf{P}_N=\mathrm{V}\mathbf{E}\mathbf{N}, \qquad \mathbf{Q}_N=\mathrm{V}\mathbf{N}\mathbf{E}; \qquad (100)$$

and $2\mathbf{E}$ is the torque. The coefficient c represents the compliancy or reciprocal of the quasi-rigidity. The kinetic energy $\frac{1}{2}\mu\mathbf{H}^2$ represents the magnetic energy, and the potential energy of the rotation represents the electric energy; whilst the flux of energy is $\mathrm{V}\mathbf{E}\mathbf{H}$. For the activity of the torque is

$$2\mathbf{E}.\frac{\text{curl}\,\mathbf{H}}{2}=\mathbf{E}\,\text{curl}\,\mathbf{H},$$

and the translational activity is

$$-\mathbf{H}\,\text{curl}\,\mathbf{E}.$$

Their sum is $\qquad -\text{div}\,\mathrm{V}\mathbf{E}\mathbf{H},$

making $\mathrm{V}\mathbf{E}\mathbf{H}$ the flux of energy.*

All attempts to construct an elastic-solid analogy with a distortional stress fail to give satisfactory results, because the energy is wrongly localized, and the flux of energy incorrect. Bearing this in mind, the above analogy is at first sight very enticing. But when we come to

Duplex method, since its characteristics are the exhibition of the electric, magnetic, and electromagnetic relations in a duplex form, symmetrical with respect to the electric and magnetic sides. But it is not merely a method of exhibiting the relations in a manner suitable to the subject, bringing to light useful relations which were formerly hidden from view by the intervention of the vector-potential and its parasites, but constitutes a method of working as well. There are considerable difficulties in the way of the practical employment of Maxwell's equations of propagation, even as they stand in his treatise. These difficulties are greatly magnified when we proceed to more general cases, involving heterogeneity and eolotropy and motion of the medium supporting the fluxes. The duplex method supplies what is wanted. Potentials do not appear, at least initially. They are regarded strictly as auxiliary functions which do not represent any physical state of the medium. In special problems they may be of great service for calculating purposes; but in general investigations their avoidance simplifies matters greatly. The state of the field is settled by **E** and **H**, and these are the primary objects of attention in the duplex system.

* This form of application of the rotating ether I gave in *The Electrician*, January 23, 1891, p. 360.

remember that the d/dt in (98) and (99) should be $\partial/\partial t$, and find extraordinary difficulty in extending the analogy to include the conduction current, and also remember that the electromagnetic stress has to be accounted for (in other words, the known mechanical forces), the perfection of the analogy, as far as it goes, becomes disheartening. It would further seem, from the explicit assumption that $\mathbf{q}=0$ in obtaining $\mathbf{W}$ above, that no analogy of this kind can be sufficiently comprehensive to form the basis of a physical theory. We must go altogether beyond the elastic solid with the additional property of rotational elasticity. I should mention, to avoid misconception, that Sir W. Thomson does not push the analogy even so far as is done above, or give to μ and c the same interpretation. The particular meaning here given to μ is that assumed by Professor Lodge in his "Modern Views of Electricity," on the ordinary elastic-solid theory, however. I have found it very convenient from its making the curl of the electric force be a Newtonian force (per unit volume). When impressed electric force $\mathbf{e}_0$ produces disturbances, their real source is, as I have shown, not the seat of $\mathbf{e}_0$, but of curl $\mathbf{e}_0$. So we may with facility translate problems in electromagnetic waves into elastic-solid problems by taking the electromagnetic source to represent the mechanical source of motion, impressed Newtonian force.

Examination of the Flux of Energy in a Moving Medium, and Establishment of the Measure of "True" Current.

§ 17. Now pass to the more general case of a moving medium with the equations

$$\text{curl}\,\mathbf{H}_1 = \quad \text{curl}\,(\mathbf{H}-\mathbf{h}_0-\mathbf{h}) = \mathbf{J} = \mathbf{C}+\dot{\mathbf{D}}+\mathbf{q}\rho, \quad \text{.........(101)}$$

$$-\text{curl}\,\mathbf{E}_1 = -\text{curl}\,(\mathbf{E}-\mathbf{e}_0-\mathbf{e}) = \mathbf{G} = \mathbf{K}+\dot{\mathbf{B}}+\mathbf{q}\sigma, \quad \text{.........(102)}$$

where $\mathbf{E}_1$ is, for brevity, what the force $\mathbf{E}$ of the flux becomes after deducting the intrinsic and motional forces; and similarly for $\mathbf{H}_1$.

From these, in the same way as before, we deduce

$$(\mathbf{e}_0+\mathbf{e})\mathbf{J}+(\mathbf{h}_0+\mathbf{h})\mathbf{G} = \mathbf{EJ}+\mathbf{HG}+\text{div}\,\mathbf{V}\mathbf{E}_1\mathbf{H}_1\,; \quad \text{.........(103)}$$

and it would seem at first sight to be the same case again, but with impressed forces $(\mathbf{e}+\mathbf{e}_0)$ and $(\mathbf{h}+\mathbf{h}_0)$ instead of $\mathbf{e}_0$ and $\mathbf{h}_0$, whilst the Poynting flux requires us to reckon only $\mathbf{E}_1$ and $\mathbf{H}_1$ as the effective electric and magnetic forces concerned in it.*

*It will be observed that the constant 4π, which usually appears in the electrical equations, is absent from the above investigations. This demands a few words of explanation. The units employed in the text are rational units, founded upon the principle of continuity in space of vector functions, and the corresponding appropriate measure of discontinuity, viz., by the amount of divergence. In popular language, the *unit* pole sends out *one* line of force, in the rational system, instead of 4π lines, as in the irrational system. The effect of the rationalization is to introduce 4π into the formulæ of central forces and potentials, and to abolish the swarm of 4π's that appears in the practical formulæ of the practice of theory on Faraday-Maxwell lines, which receives its fullest and most appropriate expression in the rational method. The rational system was explained by me in *The Electrician* in 1882, and applied to the general theory of potentials

But we must develop $(Q+\dot{U}+\dot{T})$ plainly first. We have, by (86), (87), used in (103),

$$\mathbf{e}_0\mathbf{J}+\mathbf{h}_0\mathbf{G}=\mathbf{E}(\mathbf{C}+\dot{\mathbf{D}}+\mathrm{q}\rho)+\mathbf{H}(\mathbf{K}+\dot{\mathbf{B}}+\mathrm{q}\sigma)-(\mathbf{eJ}+\mathbf{hG})+\mathrm{div}\,\mathrm{V}\mathbf{E}_1\mathbf{H}_1. \quad (104)$$

Now here we have

$$\left.\begin{aligned}\dot{U}&=\frac{d}{dt}\tfrac{1}{2}\mathbf{ED}=\tfrac{1}{2}\mathbf{E}\dot{\mathbf{D}}+\tfrac{1}{2}\mathbf{D}\dot{\mathbf{E}}=\mathbf{E}\dot{\mathbf{D}}+\tfrac{1}{2}(\mathbf{D}\dot{\mathbf{E}}-\mathbf{E}\dot{\mathbf{D}})\\&=\mathbf{E}\dot{\mathbf{D}}-\tfrac{1}{2}\mathbf{E}\dot{c}\mathbf{E}=\mathbf{E}\dot{\mathbf{D}}-\dot{U}_c.\end{aligned}\right\}\quad\text{...........}(105)$$

Comparison of the third with the second form of (105) defines the generalized meaning of $\dot{c}$ when c is not a mere scalar. Or thus,

$$\begin{aligned}\dot{U}_c&=\tfrac{1}{2}\mathbf{E}\dot{c}\mathbf{E}=\tfrac{1}{2}\frac{d}{dt}(\mathbf{ED})_c\\&=\tfrac{1}{2}\dot{c}_{11}E_1^2+\tfrac{1}{2}\dot{c}_{22}E_2^2+\tfrac{1}{2}\dot{c}_{33}E_3^2+\dot{c}_{12}E_1E_2+\dot{c}_{23}E_2E_3+\dot{c}_{31}E_1E_3,\quad\text{......}(106)\end{aligned}$$

representing the time-variation of U due to variation in the c's only.

Similarly $$\dot{T}=\mathbf{H}\dot{\mathbf{B}}-\tfrac{1}{2}\mathbf{H}\dot{\mu}\mathbf{H}=\mathbf{H}\dot{\mathbf{B}}-\dot{T}_\mu,\quad\text{..................}(107)$$

with the equivalent meaning for $\dot{\mu}$ generalized.

Using these in (104), we have the result

$$\begin{aligned}\mathbf{e}_0\mathbf{J}+\mathbf{h}_0\mathbf{G}=(Q+\dot{U}+\dot{T})+\mathrm{q}(\mathbf{E}\rho+\mathbf{H}\sigma)+(\tfrac{1}{2}\mathbf{E}\dot{c}\mathbf{E}+\tfrac{1}{2}\mathbf{H}\dot{\mu}\mathbf{H})\\-(\mathbf{eJ}+\mathbf{hG})+\mathrm{div}\,\mathrm{V}\mathbf{E}_1\mathbf{H}_1.\quad(108)\end{aligned}$$

Here we have, besides $(Q+\dot{U}+\dot{T})$, terms indicating the activity of a

and connected functions in 1883. (Reprint, vol. 1, p. 199, and later, especially p. 262.) I then returned to irrational formulæ because I did not think, then, that a reform of the units was practicable, partly on account of the labours of the B.A. Committee on Electrical Units, and partly on account of the ignorance of, and indifference to, theoretical matters which prevailed at that time. But the circumstances have greatly changed, and I do think a change is now practicable. There has been great advance in the knowledge of the meaning of Maxwell's theory, and a diffusion of this knowledge, not merely amongst scientific men, but amongst a large body of practicians called into existence by the extension of the practical applications of electricity. Electricity is becoming, not only a master science, but also a very practical one. It is fitting, therefore, that learned traditions should not be allowed to control matters too greatly, and that the units should be rationalized. To make a beginning, I am employing rational units throughout in my work on "Electromagnetic Theory," commenced in *The Electrician* in January, 1891, and continued as fast as circumstances will permit; to be republished in book form. In Section XVII. (October 16, 1891, p. 655) will be found stated more fully the nature of the change proposed, and the reasons for it. I point out, in conclusion, that as regards theoretical treatises and investigations, there is no difficulty in the way, since the connection of the rational and irrational units may be explained separately; and I express the belief that when the merits of the rational system are fully recognised, there will arise a demand for the rationalization of the practical units. We are, in the opinion of men qualified to judge, within a measurable distance of adopting the metric system in England. Surely the smaller reform I advocate should precede this. To put the matter plainly, the present system of units contains an absurdity running all through it of the same nature as would exist in the metric system of common units were we to define the unit area to be the area of a circle of unit diameter. The absurdity is only different in being less obvious in the electrical case. It would not matter much if it were not that electricity is a practical science.

translational force. Thus, $\mathbf{E}\rho$ is the force on electrification ρ, and $\mathbf{Eq}\rho$ its activity. Again,

$$\frac{\partial c}{\partial t} = \dot{c} + \mathbf{q}\nabla.c\,;$$

so that we have

$$\left.\begin{aligned} \dot{c} &= \frac{\partial c}{\partial t} - \mathbf{q}\nabla.c, \\ \dot{\mu} &= \frac{\partial \mu}{\partial t} - \mathbf{q}\nabla.\mu, \end{aligned}\right\} \qquad (109)$$

and, similarly, (second line above)

the generalized meaning of which is indicated by

$$-\frac{\partial U_c}{\partial t} + \tfrac{1}{2}\mathbf{E}\dot{c}\mathbf{E} = -\tfrac{1}{2}\mathbf{E}(\mathbf{q}\nabla.c)\mathbf{E} = -\mathbf{q}\nabla U_c\,; \qquad (110)$$

where, in terms of scalar products involving $\mathbf{E}$ and $\mathbf{D}$,

$$-\mathbf{q}\nabla U_c = -\tfrac{1}{2}(\mathbf{E}.\mathbf{q}\nabla.\mathbf{D} - \mathbf{D}.\mathbf{q}\nabla.\mathbf{E}). \qquad (111)$$

This is also the activity of a translational force. Similarly,

$$-\frac{\partial T_\mu}{\partial t} + \tfrac{1}{2}\mathbf{H}\dot{\mu}\mathbf{H} = -\mathbf{q}\nabla T_\mu \qquad (112)$$

is the activity of a translational force. Then again,

$$-(\mathbf{eJ} + \mathbf{hG}) = -\mathbf{JVqB} - \mathbf{GVDq} = \mathbf{q}(\mathbf{VJB} + \mathbf{VDG}) \qquad (113)$$

expresses a translational activity. Using them all in (108), it becomes

$$\mathbf{e}_0\mathbf{J} + \mathbf{h}_0\mathbf{G} = (Q + \dot{U} + \dot{T}) + \mathbf{q}(\mathbf{E}\rho + \mathbf{H}\sigma - \nabla U_c - \nabla T_\mu + \mathbf{VJB} + \mathbf{VDG}) + \operatorname{div} \mathbf{VE}_1\mathbf{H}_1 + \frac{\partial}{\partial t}(U_c + T_\mu). \qquad (114)$$

It is clear that we should make the factor of $\mathbf{q}$ be the complete translational force. But that has to be found; and it is equally clear that, although we appear to have exhausted all the terms at disposal, the factor of $\mathbf{q}$ in (114) is not the complete force, because there is no term by which the force on intrinsically magnetized or electrized matter could be exhibited. These involve $\mathbf{e}_0$ and $\mathbf{h}_0$. But as we have

$$\mathbf{q}(\mathbf{Vj}_0\mathbf{B} + \mathbf{VDg}_0) = -(\mathbf{ej}_0 + \mathbf{hg}_0), \qquad (115)$$

a possible way of bringing them in is to add the left member and subtract the right member of (115) from the right member of (114); bringing the translational force to $\mathbf{f}$, say, where

$$\mathbf{f} = \mathbf{E}\rho + \mathbf{H}\sigma - \nabla U_c - \nabla T_\mu + \mathbf{V}(\mathbf{J} + \mathbf{j}_0)\mathbf{B} + \mathbf{VD}(\mathbf{G} + \mathbf{g}_0). \qquad (116)$$

But there is still the right number of (115) to be accounted for. We have

$$-\operatorname{div}(\mathbf{Veh}_0 + \mathbf{Ve}_0\mathbf{h}) = \mathbf{ej}_0 + \mathbf{hg}_0 + \mathbf{e}_0\mathbf{j} + \mathbf{h}_0\mathbf{g}, \qquad (117)$$

and, by using this in (114), through (115), (116), (117), we bring it to

$$\mathbf{e}_0\mathbf{J} + \mathbf{h}_0\mathbf{G} = (Q + \dot{U} + \dot{T}) + \mathbf{fq} - (\mathbf{e}_0\mathbf{j} + \mathbf{h}_0\mathbf{g}) + \operatorname{div}(\mathbf{VE}_1\mathbf{H}_1 - \mathbf{Veh}_0 - \mathbf{Ve}_0\mathbf{h}) + \frac{\partial}{\partial t}(U_c + T_\mu)\,; \qquad (118)$$

or, transferring the $\mathbf{e}_0$, $\mathbf{h}_0$ terms from the right to the left side,

$$\mathbf{e}_0\mathbf{J}_0+\mathbf{h}_0\mathbf{G}_0=Q+\dot{U}+\dot{T}+\mathbf{fq}+\text{div}\,(\mathrm{V}\mathbf{E}_1\mathbf{H}_1-\mathrm{V}\mathbf{eh}_0-\mathrm{V}\mathbf{e}_0\mathbf{h})+\frac{\partial}{\partial t}(U_c+T_\mu). \quad (119)$$

Here we see that we have *a* correct form of activity equation, though it may not be *the* correct form. Another form, equally probable, is to be obtained by bringing in $\mathrm{V}\mathbf{eh}$; thus

$$\text{div}\,\mathrm{V}\mathbf{eh}=\mathbf{h}\,\text{curl}\,\mathbf{e}-\mathbf{e}\,\text{curl}\,\mathbf{h}=-(\mathbf{ej}+\mathbf{hg})=\mathbf{q}(\mathrm{V}\mathbf{jB}+\mathrm{V}\mathbf{Dg}), \quad (120)$$

which converts (119) to

$$\mathbf{e}_0\mathbf{J}_0+\mathbf{h}_0\mathbf{G}_0=Q+\dot{U}+\dot{T}+\mathbf{Fq}+\text{div}(\mathrm{V}\mathbf{E}_1\mathbf{H}_1-\mathrm{V}\mathbf{eh}-\mathrm{V}\mathbf{eh}_0-\mathrm{V}\mathbf{e}_0\mathbf{h})+\frac{\partial}{\partial t}(U_c+T_\mu), \quad (121)$$

where $\mathbf{F}$ is the translational force

$$\mathbf{F}=\mathbf{E}\rho+\mathbf{H}\sigma-\nabla U_c-\nabla T_\mu+\mathrm{V}\,\text{curl}\,\mathbf{H}.\mathbf{B}+\mathrm{V}\,\text{curl}\,\mathbf{E}.\mathbf{D}, \quad \ldots\ldots(122)$$

which is perfectly symmetrical as regards $\mathbf{E}$ and $\mathbf{H}$, and in the vector products utilizes the fluxes and their complete forces, whereas former forms did this only partially. Observe, too, that we have only been able to bring the activity equation to a correct form (either (119) or (122)) by making $\mathbf{e}_0\mathbf{J}_0$ be the activity of intrinsic force $\mathbf{e}_0$, which requires that $\mathbf{J}_0$ should be the true electric current, according to the energy criterion, not $\mathbf{J}$.

§18. Now, to test (119) and (121), we must interpret the flux in (121), or say

$$\mathbf{Y}=\mathrm{V}\mathbf{E}_1\mathbf{H}_1-\mathrm{V}\mathbf{eh}-\mathrm{V}\mathbf{eh}_0-\mathrm{V}\mathbf{e}_0\mathbf{h}, \quad \ldots\ldots\ldots\ldots\ldots(123)$$

which has replaced the Poynting flux $\mathrm{V}\mathbf{E}_1\mathbf{H}_1$ when $\mathbf{q}=0$, along with the other changes. Since $\mathbf{Y}$ reduces to $\mathrm{V}\mathbf{E}_1\mathbf{H}_1$ when $\mathbf{q}=0$, there must still be a Poynting flux when $\mathbf{q}$ is finite, though we do not know its precise form of expression. There is also the stress flux of energy and the flux of energy by convection, making a total flux

$$\mathbf{X}=\mathbf{W}+\mathbf{q}(U+T)-\Sigma\,\mathbf{Q}q+\mathbf{q}(U_0+T_0), \quad \ldots\ldots\ldots\ldots(124)$$

where $\mathbf{W}$ is the Poynting flux, and $-\Sigma\,\mathbf{Q}q$ that of the stress, whilst $\mathbf{q}(U_0+T_0)$ means convection of energy connected with the translational force. We should therefore have

$$\mathbf{e}_0\mathbf{J}_0+\mathbf{h}_0\mathbf{G}_0=(Q+\dot{U}+\dot{T})+(Q_0+\dot{U}_0+\dot{T}_0)+\text{div}\,\mathbf{X} \quad \ldots\ldots(125)$$

to express the continuity of energy. More explicitly

$$\mathbf{e}_0\mathbf{J}_0+\mathbf{h}_0\mathbf{G}_0=Q+\dot{U}+\dot{T}+\text{div}\,[\mathbf{W}+\mathbf{q}(U+T)]$$
$$+Q_0+\dot{U}_0+\dot{T}_0+\text{div}\,[-\Sigma\,\mathbf{Q}q+\mathbf{q}(U_0+T_0)]. \quad \ldots\ldots\ldots(126)$$

But here we may simplify by using the result (69) (with, however, $\mathbf{f}$ put $=0$), making (126) become

$$\mathbf{e}_0\mathbf{J}_0+\mathbf{h}_0\mathbf{G}_0=(Q+\dot{U}+\dot{T})+\mathbf{Fq}+\mathbf{Sa}+\text{div}[\mathbf{W}+\mathbf{q}(U+T)-\Sigma\mathbf{Q}q], \quad (127)$$

where $\mathbf{S}$ is the torque, and $\mathbf{a}$ the spin.

Comparing this with (121), we see that we require

$$\mathbf{W}+\mathbf{q}(U+T)-\Sigma\,\mathbf{Q}q=\mathrm{V}\mathbf{E}_1\mathbf{H}_1-\mathrm{V}\mathbf{eh}-\mathrm{V}\mathbf{e}_0\mathbf{h}-\mathrm{V}\mathbf{eh}_0, \quad \ldots\ldots(128)$$

with a similar equation when (119) is used instead; and we have now to separate the right member into two parts, one for the Poynting flux, the other for the stress flux, in such a way that the force due to the stress is the force **F** in (121), (122), or the force **f** in (119), (116); or similarly in other cases. It is unnecessary to give the failures; the only one that stands the test is (121), which satisfies it completely.

I argued that

$$\mathbf{W} = \mathrm{V}(\mathbf{E} - \mathbf{e}_0)(\mathbf{H} - \mathbf{h}_0) \quad \text{......................(129)}$$

was the probable form of the Poynting flux in the case of a moving medium, not $\mathrm{V}\mathbf{E}_1\mathbf{H}_1$, because when a medium is endowed with a *uniform* translational motion, the transmission of disturbances through it takes place just as if it were at rest. With this expression (129) for **W**, we have, identically,

$$\mathrm{V}\mathbf{E}_1\mathbf{H}_1 - \mathrm{V}\mathbf{eh} - \mathrm{V}\mathbf{e}_0\mathbf{h} - \mathrm{V}\mathbf{eh}_0 = \mathbf{W} - \mathrm{V}\mathbf{eH} - \mathrm{V}\mathbf{Eh}. \quad \text{.......(130)}$$

Therefore, by (128) and (130), we get

$$\Sigma\,\mathbf{Q}q = \mathrm{V}\mathbf{eH} + \mathrm{V}\mathbf{Eh} + \mathbf{q}(U+T), \quad \text{................(131)}$$

to represent the negative of the stress flux of energy, so that, finally, the fully significant equation of activity is

$$\mathbf{e}_0\mathbf{J}_0 + \mathbf{h}_0\mathbf{G}_0 = Q + \dot{U} + \dot{T} + \mathbf{Fq} + \mathbf{Sa} + \mathrm{div}[\mathrm{V}(\mathbf{E} - \mathbf{e}_0)(\mathbf{H} - \mathbf{h}_0) + \mathbf{q}(U+T)] - \mathrm{div}[\mathrm{V}\mathbf{eH} + \mathrm{V}\mathbf{Eh} + \mathbf{q}(U+T)]. \quad (132)$$

This is, of course, an identity, subject to the electromagnetic equations we started from, and is only one of the multitude of forms which may be given to it, many being far simpler. But the particular importance of this form arises from its being the only form apparently possible which shall exhibit the principle of continuity of energy without outstanding terms, and without loss of generality; and this is only possible by taking $\mathbf{J}_0$ as the proper flux for $\mathbf{e}_0$ to work upon.*

* In the original an erroneous estimate of the value of $(\partial/\partial t)(U_c + T_\mu)$ was used in some of the above equations. This is corrected. The following contains full details of the calculation. We require the value of $(\partial/\partial t)U_c$, or of $\frac{1}{2}\mathbf{E}(\partial c/\partial t)\mathbf{E}$, where $\partial c/\partial t$ is the linear operator whose components are the time-variations (for the same matter) of those of c. The calculation is very lengthy in terms of these six components. But vectorially it is not difficult. In (27), (28) we have

$$\left.\begin{aligned}\mathbf{D} = c\mathbf{E} &= \mathbf{i}.\mathbf{c}_1\mathbf{E} + \mathbf{j}.\mathbf{c}_2\mathbf{E} + \mathbf{k}.\mathbf{c}_3\mathbf{E}\\ &= (\mathbf{i}.\mathbf{c}_1 + \mathbf{j}.\mathbf{c}_2 + \mathbf{k}.\mathbf{c}_3)\mathbf{E},\end{aligned}\right\} \quad \text{........................(132a)}$$

if the vectors $\mathbf{c}_1$, $\mathbf{c}_2$, $\mathbf{c}_3$ are given by

$$\mathbf{c}_1 = \mathbf{i}c_{11} + \mathbf{j}c_{12} + \mathbf{k}c_{13}, \qquad \mathbf{c}_2 = \mathbf{i}c_{21} + \mathbf{j}c_{22} + \mathbf{k}c_{23}, \qquad \mathbf{c}_3 = \mathbf{i}c_{31} + \mathbf{j}c_{32} + \mathbf{k}c_{33}.$$

We, therefore, have

$$\mathbf{E}\frac{\partial c}{\partial t}\mathbf{E} = \mathbf{E}\left(\frac{\partial \mathbf{i}}{\partial t}.\mathbf{c}_1 + \frac{\partial \mathbf{j}}{\partial t}.\mathbf{c}_2 + \frac{\partial \mathbf{k}}{\partial t}.\mathbf{c}_3\right)\mathbf{E} + \mathbf{E}\left(\mathbf{i}.\frac{\partial \mathbf{c}_1}{\partial t} + \mathbf{j}.\frac{\partial \mathbf{c}_2}{\partial t} + \mathbf{k}.\frac{\partial \mathbf{c}_3}{\partial t}\right)\mathbf{E}. \quad (132b)$$

The part played by the dots is to clearly separate the scalar products.

Now suppose that the eolotropic property symbolized by c is intrinsically unchanged by the shift of the matter. The mere translation does not, therefore, affect it, nor does the distortion; but the rotation does. For if we turn round an eolotropic portion of matter, keeping **E** unchanged, the value of U is altered by the rotation of the principal axes of c along with the matter, so that a torque is required.

In equation (132*a*), then, to produce (132*b*), we keep **E** constant, and let the six

Derivation of the Electric and Magnetic Stresses and Forces from the Flux of Energy.

§ 19. It will be observed that the convection of energy disappears by occurring twice oppositely signed; but as it comes necessarily into the expression for the stress flux of energy, I have preserved the cancelling terms in (132). A comparison of the stress flux with the Poynting flux is interesting. Both are of the same form, viz., vector products of the electric and magnetic forces with convection terms; but whereas in the latter the forces in the vector-product are those of the field (*i.e.*, only intrinsic forces deducted from **E** and **H**), in the former we have the motional forces **e** and **h** combined with the complete **E** and **H** of the fluxes. Thus the stress depends entirely on the fluxes, however they be produced, in this respect resembling the electric and magnetic energies.

To exhibit the stress, we have (131), or

$$\mathbf{Q}_1 q_1 + \mathbf{Q}_2 q_2 + \mathbf{Q}_3 q_3 = \mathrm{V}\mathbf{eH} + \mathrm{V}\mathbf{Eh} + \mathbf{q}(U+T). \qquad \text{..........(133)}$$

In this use the expressions for **e** and **h**, giving

$$\begin{aligned}\Sigma\, \mathbf{Q}q &= \mathrm{V}\mathbf{H}\mathrm{V}\mathbf{Bq} + \mathrm{V}\mathbf{E}\mathrm{V}\mathbf{Dq} + \mathbf{q}(U+T)\\ &= \mathbf{B}.\mathbf{Hq} - \mathbf{q}.\mathbf{HB} + \mathbf{D}.\mathbf{Eq} - \mathbf{q}.\mathbf{ED} + \mathbf{q}(U+T)\\ &= (\mathbf{B}.\mathbf{Hq} - \mathbf{q}T) + (\mathbf{D}.\mathbf{Eq} - \mathbf{q}U)\,; \qquad \text{........................(134)}\end{aligned}$$

where observe the singularity that $\mathbf{q}(U+T)$ has changed its sign. The first set belongs to the magnetic, the second to the electric stress, since we see that the complete stress is thus divisible.

vectors **i**, **j**, **k**, $\mathbf{c}_1$, $\mathbf{c}_2$, $\mathbf{c}_3$ rotate as a rigid body with the spin $\mathbf{a} = \frac{1}{2}\operatorname{curl}\mathbf{q}$. But when a vector magnitude **i** is turned round in this way, its rate of time-change $\partial\mathbf{i}/\partial t$ is $\mathrm{V}\mathbf{ai}$. Thus, for $\partial/\partial t$, we may put $\mathrm{V}\mathbf{a}$ throughout. Therefore, by (132*b*),

$$\mathbf{E}\frac{\partial c}{\partial t}\mathbf{E} = \mathbf{E}\Big(\mathrm{V}\mathbf{ai}.\mathbf{c}_1 + \mathrm{V}\mathbf{aj}.\mathbf{c}_2 + \mathrm{V}\mathbf{ak}.\mathbf{c}_3\Big)\mathbf{E} + \mathbf{E}\Big(\mathbf{i}.\mathrm{V}\mathbf{ac}_1 + \mathbf{j}.\mathrm{V}\mathbf{ac}_2 + \mathbf{k}.\mathrm{V}\mathbf{ac}_3\Big)\mathbf{E}. \quad (132c)$$

In this use the parallelepipedal transformation (12), and it becomes

$$\begin{aligned}\mathbf{E}\frac{\partial c}{\partial t}\mathbf{E} &= \mathrm{V}\mathbf{Ea}\Big(\mathbf{i}.\mathbf{c}_1 + \mathbf{j}.\mathbf{c}_2 + \mathbf{k}.\mathbf{c}_3\Big)\mathbf{E} + \mathbf{E}\Big(\mathbf{i}.\mathbf{c}_1 + \mathbf{j}.\mathbf{c}_2 + \mathbf{k}.\mathbf{c}_3\Big)\mathrm{V}\mathbf{Ea}\\ &= (\mathrm{V}\mathbf{Ea})c\mathbf{E} + \mathbf{E}c(\mathrm{V}\mathbf{Ea}) = (\mathbf{D} + \mathbf{D}')\mathrm{V}\mathbf{Ea}, \qquad \text{...........................(132}d\text{)}\end{aligned}$$

by (132*a*), if $\mathbf{D}'$ is conjugate to $\mathbf{D}$; that is, $\mathbf{D}' = c'\mathbf{E} = \mathbf{E}c$. So, when $c = c'$, as in the electrical case, we have

$$\left.\begin{aligned}\frac{\partial U_c}{\partial t} &= \tfrac{1}{2}\mathbf{E}\frac{\partial c}{\partial t}\mathbf{E} = \mathbf{D}\mathrm{V}\mathbf{Ea} = \mathbf{a}\mathrm{V}\mathbf{DE},\\ \text{and similarly}\qquad \frac{\partial T_\mu}{\partial t} &= \tfrac{1}{2}\mathbf{H}\frac{\partial \mu}{\partial t}\mathbf{H} = \mathbf{B}\mathrm{V}\mathbf{Ha} = \mathbf{a}\mathrm{V}\mathbf{BH}.\end{aligned}\right\} \qquad \text{........................(132}e\text{)}$$

Now the torque arising from the stress is (see (139))

$$\mathbf{S} = \mathrm{V}\mathbf{DE} + \mathrm{V}\mathbf{BH},$$

so we have

$$\frac{\partial}{\partial t}(U_c + T_\mu) = \mathbf{Sa} = \text{torque} \times \text{spin}. \qquad \text{......................(132}f\text{)}$$

The variation allowed to **i**, **j**, **k** may seem to conflict with their constancy (as reference vectors) in general. But they merely vary for a temporary purpose,

The divergence of $\Sigma \mathbf{Q}q$ being the activity of the stress-variation per unit volume, its N-component is the activity of the stress per unit surface, that is,

$$(\mathbf{NB.Hq} - \mathbf{Nq}.T) + (\mathbf{ND.Eq} - \mathbf{Nq}.U)$$
$$= \mathbf{q}(\mathbf{H.BN} + \mathbf{E.DN} - \mathbf{N}U - \mathbf{N}T) = \mathbf{P}_N\mathbf{q}. \quad \text{.........(135)}$$

The stress itself is therefore

$$\mathbf{P}_N = (\mathbf{E.DN} - \mathbf{N}U) + (\mathbf{H.BN} - \mathbf{N}T), \quad \text{..............(136)}$$

divided into electric and magnetic portions. This is with restriction to symmetrical μ and c, and with persistence of their forms as a particle moves, but is otherwise unrestricted.

Neither stress is of the symmetrical or irrotational type in case of eolotropy, and there appears to be no getting an irrotational stress save by arbitrary assumptions which destroy the validity of the stress as a correct deduction from the electromagnetic equations. But, in case of isotropy, with consequent directional identity of **E** and **D**, and of **H** and **B**, we see, by taking **N** in turns parallel to, or perpendicular to **E** in the electric case, and to **H** in the magnetic case, that the electric stress consists of a tension U parallel to **E** combined with an equal lateral pressure, whilst the magnetic stress consists of a tension T parallel to **H** combined with an equal lateral pressure. They are, in fact, Maxwell's stresses in an isotropic medium homogeneous as regards μ and c. The difference from Maxwell arises when μ and c are variable (including abrupt changes from one value to another of μ and c), and

being fixed in the matter instead of in space. But we may, perhaps better, discard **i**, **j**, **k** altogether, and use any independent vectors, **l**, **m**, **n** instead, making

$$\mathbf{D} = (\mathbf{l}.\mathbf{c}_1 + \mathbf{m}.\mathbf{c}_2 + \mathbf{n}.\mathbf{c}_3)\mathbf{E}, \quad \text{..............................(132}g\text{)}$$

wherein the **c**'s are properly chosen to suit the new axes. The calculation then proceeds as before, half the value of $\partial U_e/\partial t$ arising from the variation of **l**, **m**, **n**, and the other half from the **c**'s, provided c is irrotational.

Or we may choose the three principal axes of c in the body, when **l**, **m**, **n** will coincide with, and therefore move with them.

Lastly, we may proceed thus:—

$$\mathbf{E}\frac{\partial c}{\partial t}\mathbf{E} = \mathbf{E}\frac{\partial \mathbf{D}}{\partial t} - \mathbf{D}\frac{\partial \mathbf{E}}{\partial t} = \mathbf{E}\mathrm{V}\mathbf{a}\mathbf{D} - \mathbf{D}\mathrm{V}\mathbf{a}\mathbf{E} = 2\mathbf{a}\mathrm{V}\mathbf{D}\mathbf{E}. \quad \text{......................(132}h\text{)}$$

That is, replace $\partial/\partial t$ by Va when the operands are **E** and **D**. This is the correct result, but it is not easy to justify the process directly and plainly; although the clue is given by observing that what we do is to take a difference, from which the time-variation of **E** disappears.

If it is **D** that is kept constant, the result is 2aV**ED**, the negative of the above.

It is also worth noticing that if we split up **E** into $\mathbf{E}_1 + \mathbf{E}_2$ we shall have

$$\left.\begin{aligned}\mathbf{E}_1\frac{\partial c}{\partial t}\mathbf{E}_2 &= \mathbf{a}\Big[\mathrm{V}(\mathbf{E}_1 c)\mathbf{E}_2 - \mathrm{V}\mathbf{E}_1(c\mathbf{E}_2)\Big],\\ \mathbf{E}_2\frac{\partial c}{\partial t}\mathbf{E}_1 &= \mathbf{a}\Big[\mathrm{V}(\mathbf{E}_2 c)\mathbf{E}_1 - \mathrm{V}\mathbf{E}_2(c\mathbf{E}_1)\Big].\end{aligned}\right\} \quad \text{........................(132}i\text{)}$$

These are only equal when $c = c'$, or $\mathbf{E}c = c\mathbf{E}$; so that, in the expansion of the torque,

$$\mathrm{V}\mathbf{DE} = \mathrm{V}\mathbf{D}_1\mathbf{E}_1 + \mathrm{V}\mathbf{D}_2\mathbf{E}_2 + \mathrm{V}\mathbf{D}_2\mathbf{E}_1 + \mathrm{V}\mathbf{D}_1\mathbf{E}_2,$$

the cross-torques are not $\mathrm{V}\mathbf{D}_2\mathbf{E}_1$ and $\mathrm{V}\mathbf{D}_1\mathbf{E}_2$, which are unequal, but are each equal to half the sum of these vector-products.

when there is intrinsic magnetization, Maxwell's stresses and forces being then different.

The stress on the plane whose normal is **VEH**, is

$$\frac{\mathbf{E}.\mathbf{D}\mathrm{V}\mathbf{EH}+\mathbf{H}.\mathbf{B}\mathrm{V}\mathbf{EH}-(U+T)\mathrm{V}\mathbf{EH}}{\mathrm{V}_0\mathbf{EH}}=\frac{\mathbf{E}.\mathbf{H}\mathrm{V}\mathbf{DE}+\mathbf{H}.\mathbf{E}\mathrm{V}\mathbf{HB}-(U+T)\mathrm{V}\mathbf{EH}}{\mathrm{V}_0\mathbf{EH}}, \quad (137)$$

reducing simply to a pressure $(U+T)$, in lines parallel to **VEH**, in case of isotropy.

§ 20. To find the force **F**, we have

$$\begin{aligned}\mathbf{FN}&=\mathrm{div}\,\mathbf{Q}_N=\mathrm{div}\,(\mathbf{D}.\mathbf{EN}-\mathbf{N}U+\mathbf{B}.\mathbf{HN}-\mathbf{N}T)\\&=\mathbf{EN}.\rho+\mathbf{D}\nabla.\mathbf{EN}-\tfrac{1}{2}\mathbf{E}.\mathbf{N}\nabla.\mathbf{D}-\tfrac{1}{2}\mathbf{D}.\mathbf{N}\nabla.\mathbf{E}+\text{etc.}\\&=\mathbf{EN}.\rho+\mathbf{D}(\nabla.\mathbf{EN}-\mathbf{N}\nabla.\mathbf{E})+\tfrac{1}{2}(\mathbf{D}.\mathbf{N}\nabla.\mathbf{E}-\mathbf{E}.\mathbf{N}\nabla.\mathbf{D})+\text{etc.}\\&=\mathbf{N}[\mathbf{E}\rho+\mathrm{V}\,\mathrm{curl}\,\mathbf{E}.\mathbf{D}-\nabla U_e+\text{etc.}], \quad \ldots\ldots(138)\end{aligned}$$

where the unwritten terms are the similar magnetic terms. This being the **N**-component of **F**, the force itself is given by (122), as is necessary.

It is $\mathrm{V}\,\mathrm{curl}\,\mathbf{h}_0.\mathbf{B}$ that expresses the translational force on intrinsically magnetized matter, and this harmonizes with the fact that the flux **B** due to any impressed force $\mathbf{h}_0$ depends solely upon $\mathrm{curl}\,\mathbf{h}_0$.

Also, it is $-\nabla T_\mu$ that explains the forcive on elastically magnetized matter, *e.g.*, Faraday's motion of matter to or away from the places of greatest intensity of the field, independent of its direction.

If **S** be the torque, it is given by

$$\mathrm{V}\mathbf{SN}=\mathbf{P}_N-\mathbf{Q}_N=\mathbf{E}.\mathbf{DN}-\mathbf{D}.\mathbf{EN}+\text{etc.}=\mathrm{V}\mathbf{N}(\mathrm{V}\mathbf{ED}+\mathrm{V}\mathbf{HB});$$

therefore $$\mathbf{S}=\mathrm{V}\mathbf{DE}+\mathrm{V}\mathbf{BH}. \quad \ldots\ldots(139)$$

But the matter is put more plainly by considering the convergence of the stress flux of energy and dividing it into translational and other parts. Thus

$$\mathrm{div}\,\Sigma\,\mathbf{Q}q=\mathbf{Fq}+(\mathbf{E}.\mathbf{D}\nabla.\mathbf{q}-U\,\mathrm{div}\,\mathbf{q})+(\mathbf{H}.\mathbf{B}\nabla.\mathbf{q}-T\,\mathrm{div}\,\mathbf{q}), \quad \ldots(140)$$

where the terms following **Fq** express the sum of the distortional and rotational activities.

Shorter Way of going from the Circuital Equations to the Flux of Energy, Stresses, and Forces.

§ 21. I have given the investigation in §§ 17 to 19 in the form in which it occurred to me before I knew the precise nature of the results, being uncertain as regards the true measure of current, the proper form of the Poynting flux, and how it worked in harmony with the stress flux of energy. But knowing the results, a short demonstration may be easily drawn up, though the former course is the most instructive. Thus, start now from

$$\left.\begin{aligned}\mathrm{curl}\,(\mathbf{H}-\mathbf{h}_0)&=\mathbf{J}_0,\\-\mathrm{curl}\,(\mathbf{E}-\mathbf{e}_0)&=\mathbf{G}_0,\end{aligned}\right\} \quad \ldots\ldots(141)$$

on the assumptional understanding that $\mathbf{J}_0$ and $\mathbf{G}_0$ are the currents which make $\mathbf{e}_0\mathbf{J}_0$ and $\mathbf{h}_0\mathbf{G}_0$ the activities of $\mathbf{e}_0$ and $\mathbf{h}_0$ the intrinsic forces. Then

$$\mathbf{e}_0\mathbf{J}_0+\mathbf{h}_0\mathbf{G}_0=\mathbf{E}\mathbf{J}_0+\mathbf{H}\mathbf{G}_0+\operatorname{div}\mathbf{W}, \qquad (142)$$

where

$$\mathbf{W}=\mathrm{V}(\mathbf{E}-\mathbf{e}_0)(\mathbf{H}-\mathbf{h}_0); \qquad (143)$$

and we now assume this to be the proper form of the Poynting flux. Now develop $\mathbf{E}\mathbf{J}_0$ and $\mathbf{H}\mathbf{G}_0$ thus:—

$$\begin{aligned}\mathbf{E}\mathbf{J}_0+\mathbf{H}\mathbf{G}_0 &= \mathbf{E}(\mathbf{C}+\dot{\mathbf{D}}+\mathbf{q}\rho+\operatorname{curl}\mathbf{h})+\mathbf{H}(\mathbf{K}+\dot{\mathbf{B}}+\mathbf{q}\sigma-\operatorname{curl}\mathbf{e}), \quad \text{by (93)};\\ &= Q_1+\dot{U}+\dot{U}_c+\mathbf{E}\mathbf{q}\rho+\mathbf{E}\operatorname{curl}\mathrm{V}\mathbf{D}\mathbf{q}\\ &\quad +Q_2+\dot{T}+\dot{T}_\mu+\mathbf{H}\mathbf{q}\sigma+\mathbf{H}\operatorname{curl}\mathrm{V}\mathbf{B}\mathbf{q}, \quad \text{by (88) and (91)};\\ &= Q_1+\dot{U}+\dot{U}_c+\mathbf{E}\mathbf{q}\rho+\mathbf{E}(\mathbf{D}\operatorname{div}\mathbf{q}+\mathbf{q}\nabla.\mathbf{D}-\mathbf{q}\operatorname{div}\mathbf{D}-\mathbf{D}\nabla.\mathbf{q})\\ &\quad +Q_2+\dot{T}+\dot{T}_\mu+\mathbf{H}\mathbf{q}\sigma+\mathbf{H}(\mathbf{B}\operatorname{div}\mathbf{q}+\mathbf{q}\nabla.\mathbf{B}-\mathbf{q}\operatorname{div}\mathbf{B}-\mathbf{B}\nabla.\mathbf{q}), \text{by (26)}\\ &= Q_1+\dot{U}+\dot{U}_c+2U\operatorname{div}\mathbf{q}+\mathbf{E}.\mathbf{q}\nabla.\mathbf{D}-\mathbf{E}.\mathbf{D}\nabla.\mathbf{q}\\ &\quad +\text{magnetic terms},\\ &= (Q_1+\dot{U}+\operatorname{div}\mathbf{q}U)+(U\operatorname{div}\mathbf{q}-\mathbf{E}.\mathbf{D}\nabla.\mathbf{q})+(\dot{U}_c-\mathbf{q}\nabla.U+\mathbf{E}.\mathbf{q}\nabla.\mathbf{D})\\ &\quad +\text{magnetic terms}. \qquad (144)\end{aligned}$$

Now here

$$\mathbf{q}\nabla.U=\tfrac{1}{2}\mathbf{E}.\mathbf{q}\nabla.\mathbf{D}+\tfrac{1}{2}\mathbf{D}.\mathbf{q}\nabla.\mathbf{E},$$

so that the terms in the third pair of brackets in (144) represent

$$\dot{U}_c+\mathbf{q}\nabla.U_c=\frac{\partial U_c}{\partial t}=\tfrac{1}{2}\mathbf{E}\frac{\partial c}{\partial t}\mathbf{E},$$

with the generalized meaning before explained. So finally

$$\begin{aligned}\mathbf{E}\mathbf{J}_0+\mathbf{H}\mathbf{G}_0 &= Q+\dot{U}+\dot{T}+\operatorname{div}\mathbf{q}(U+T)+\frac{\partial}{\partial t}(U_c+T_\mu)\\ &\quad +(U\operatorname{div}\mathbf{q}-\mathbf{E}.\mathbf{D}\nabla.\mathbf{q})+(T\operatorname{div}\mathbf{q}-\mathbf{H}.\mathbf{B}\nabla.\mathbf{q}), \quad \ldots(145)\end{aligned}$$

which brings (142) to

$$\begin{aligned}\mathbf{e}_0\mathbf{J}_0+\mathbf{h}_0\mathbf{G}_0 &= Q+\dot{U}+\dot{T}+\operatorname{div}\{\mathbf{W}+\mathbf{q}(U+T)\}\\ &\quad +\frac{\partial}{\partial t}(U_c+T_\mu)+(U\operatorname{div}\mathbf{q}-\mathbf{E}.\mathbf{D}\nabla.\mathbf{q})+(T\operatorname{div}\mathbf{q}-\mathbf{H}.\mathbf{B}\nabla.\mathbf{q}), \quad (146)\end{aligned}$$

which has to be interpreted in accordance with the principle of continuity of energy.

Use the form (127), first, however, eliminating $\mathbf{F}\mathbf{q}$ by means of

$$\operatorname{div}\Sigma\,\mathbf{Q}q=\mathbf{F}\mathbf{q}+\Sigma\,\mathbf{Q}\nabla q,$$

which brings (127) to

$$\mathbf{e}_0\mathbf{J}_0+\mathbf{h}_0\mathbf{G}_0=Q+\dot{U}+\dot{T}+\operatorname{div}\{\mathbf{W}+\mathbf{q}(U+T)\}-\Sigma\,\mathbf{Q}\nabla q+\mathbf{S}\mathbf{a}; \quad (147)$$

and now, by comparison of (147) with (146) we see that

$$-\mathbf{S}\mathbf{a}+\Sigma\,\mathbf{Q}\nabla q=(\mathbf{E}.\mathbf{D}\nabla.\mathbf{q}-U\operatorname{div}\mathbf{q})-\frac{\partial U_c}{\partial t}+(\mathbf{H}.\mathbf{B}\nabla.\mathbf{q}-T\operatorname{div}\mathbf{q})-\frac{\partial T_\mu}{\partial t}; \quad (148)$$

from which, when μ and c do not change intrinsically, we conclude that

$$\left.\begin{aligned}\mathbf{Q}_N &= \mathbf{B}.\mathbf{H}\mathbf{N}-\mathbf{N}T+\mathbf{D}.\mathbf{E}\mathbf{N}-\mathbf{N}U,\\ \mathbf{P}_N &= \mathbf{H}.\mathbf{B}\mathbf{N}-\mathbf{N}T+\mathbf{E}.\mathbf{D}\mathbf{N}-\mathbf{N}U,\end{aligned}\right\} \qquad (149)$$

as before. In this method we lose sight altogether of the translational force which formed so prominent an object in the former method as a guide.

Some Remarks on Hertz's Investigation relating to the Stresses.

§ 22. Variations of c and μ in the same portion of matter may occur in different ways, and altogether independently of the strain-variations. Equation (146) shows how their influence affects the energy transformations; but if we consider only such changes as depend on the strain, *i.e.*, the small changes of value which μ and c undergo as the strain changes, we may express them by thirty-six new coefficients each (there being six distortion elements, and six elements in μ, and six in c), and so reduce the expressions for $\partial U_c/\partial t$ and $\partial T_\mu/\partial t$ in (148) to the form suitable for exhibiting the corresponding change in $\mathbf{Q}_N$ and in the stress function $\mathbf{P}_N$. As is usual in such cases of secondary corrections, the magnitude of the resulting formula is out of all proportion to the importance of the correction-terms in relation to the primary formula to which they are added.

Professor H. Hertz* has considered this question, and also refers to von Helmholtz's previous investigation relating to a fluid. The c and μ can then only depend on the density, or on the compression, so that a single coefficient takes the place of the thirty-six. But I cannot quite follow Hertz's stress investigation. First, I would remark that in developing the expression for the distortional (*plus* rotational) activity, he assumes that all the coefficients of the spin vanish identically; this is done in order to make the stress be of the irrotational type. But it may easily be seen that the assumption is inadmissible by examining its consequence, for which we need only take the case of c and μ intrinsically constant. By (139) we see that it makes $\mathbf{S}=0$, and therefore (since the electric and magnetic stresses are separable), $\mathbf{VHB}=0$, and $\mathbf{VED}=0$; that is, it produces directional identity of the force $\mathbf{E}$ and the flux $\mathbf{D}$, and of the force $\mathbf{H}$ and the flux $\mathbf{B}$. This means isotropy, and, therefore, breaks down the investigation so far as the eolotropic application, with six μ and six c coefficients, goes. Abolish the assumption made, and the stress will become that used by me above.

Another point deserving of close attention in Hertz's investigation, relates to the principle to be followed in deducing the stress from the electromagnetic equations. Translating into my notation it would appear to amount to this, the *à priori* assumption that the quantity

$$\frac{1}{v}\frac{\partial}{\partial t}\left(Tv\right), \quad \text{.............................(150)}$$

where v indicates the volume of a moving unit element undergoing distortion, may be taken to represent the distortional (*plus* rotational) activity of the magnetic stress. Similarly as regards the electric stress.

Expanding (150) we obtain

$$\frac{\partial T}{\partial t}+\frac{T}{v}\frac{\partial v}{\partial t}=\mathbf{H}\frac{\partial \mathbf{B}}{\partial t}+T\operatorname{div}\mathrm{q}-\frac{\partial T_\mu}{\partial t}. \quad \text{..............(151)}$$

* *Wiedemann's Annalen*, v. 41, p. 369.

Now the second circuital law (90) may be written

$$-\operatorname{curl}(\mathbf{E}-\mathbf{e}_0)=\mathbf{K}+\frac{\partial\mathbf{B}}{\partial t}+(\mathbf{B}\operatorname{div}\mathbf{q}-\mathbf{B}\nabla.\mathbf{q}). \quad\text{..........(152)}$$

Here ignore $\mathbf{e}_0$, $\mathbf{K}$, *and ignore the curl of the electric force*, and we obtain, by using (152) in (151),

$$\mathbf{H}.\mathbf{B}\nabla.\mathbf{q}-\mathbf{HB}\operatorname{div}\mathbf{q}+T\operatorname{div}\mathbf{q}-\frac{\partial T_\mu}{\partial t}=\mathbf{H}.\mathbf{B}\nabla.\mathbf{q}-T\operatorname{div}\mathbf{q}-\frac{\partial T_\mu}{\partial t}, \quad (153)$$

which represents the distortional activity (my form, not equating to zero the coefficients of curl $\mathbf{q}$ in its development). We *can*, therefore, derive the magnetic stress in the manner indicated, that is, from (150), with the special meaning of $\partial\mathbf{B}/\partial t$ later stated, and the ignorations or nullifications.

In a similar manner, from the first circuital law (89), which may be written

$$\operatorname{curl}(\mathbf{H}-\mathbf{h}_0)=\mathbf{C}+\frac{\partial\mathbf{D}}{\partial t}+(\mathbf{D}\operatorname{div}\mathbf{q}-\mathbf{D}\nabla.\mathbf{q}), \quad\text{...........(154)}$$

we can, by ignoring the conduction-current and the curl of the magnetic force, obtain

$$\frac{1}{v}\frac{\partial}{\partial t}(vU)=\mathbf{E}.\mathbf{D}\nabla.\mathbf{q}-U\operatorname{div}\mathbf{q}-\frac{\partial U_e}{\partial t}, \quad\text{.............(155)}$$

which represents the distortional activity of the electric stress.

The difficulty here seems to me to make it evident *à priori* that (150), with the special reckoning of $\partial\mathbf{B}/\partial t$, *should* represent the distortional activity (*plus* rotational understood); this interesting property should, perhaps, rather be derived from the magnetic stress when obtained by a safe method. The same remark applies to the electric stress. Also, in (150) to (155) we overlook the Poynting flux. I am not sure how far this is intentional on Professor Hertz's part, but its neglect does not seem to give a sufficiently comprehensive view of the subject.

The complete expansion of the magnetic distortional activity is, in fact,

$$\mathbf{H}.\mathbf{B}\nabla.\mathbf{q}-T\operatorname{div}\mathbf{q}-\frac{\partial T_\mu}{\partial t}=Q_2+\dot{T}+\operatorname{div}\mathbf{q}T-\mathbf{HG}_0; \quad\text{......(156)}$$

and similarly, that of the electric stress is

$$\mathbf{E}.\mathbf{D}\nabla.\mathbf{q}-U\operatorname{div}\mathbf{q}-\frac{\partial U_e}{\partial t}=Q_1+\dot{U}+\operatorname{div}\mathbf{q}U-\mathbf{EJ}_0. \quad\text{......(157)}$$

It is the last term of (156) and the last term of (157), together, which bring in the Poynting flux. Thus, adding these equations,

$$\Sigma\,\mathbf{Q}\nabla q-\frac{\partial}{\partial t}(U_e+T_\mu)=Q+\dot{U}+\dot{T}+\operatorname{div}\mathbf{q}(U+T)-(\mathbf{EJ}_0+\mathbf{HG}_0), \quad (158)$$

where

$$(\mathbf{EJ}_0+\mathbf{HG}_0)=(\mathbf{e}_0\mathbf{J}_0+\mathbf{h}_0\mathbf{G}_0)-\operatorname{div}.\mathbf{W}; \quad\text{.............(159)}$$

and so we come round to the equation of activity again, in the form (146), by using (159) in (158).

Modified Form of Stress-Vector, and Application to the Surface separating two Regions.

§ 23. The electromagnetic stress, $\mathbf{P}_N$ of (149) and (136) may be put into another interesting form. We may write it

$$\mathbf{P}_N = \tfrac{1}{2}(\mathbf{E}.\mathbf{ND} + \mathrm{V}.\mathrm{V}\mathbf{NE}.\mathbf{D}) + \tfrac{1}{2}(\mathbf{H}.\mathbf{NB} + \mathrm{V}.\mathrm{V}\mathbf{NH}.\mathbf{B}). \quad \text{.....(160)}$$

Now, **ND** is the surface equivalent of div **D** and **NB** of div **B**; whilst V**NE** and V**NH** are the surface equivalents of curl **E** and curl **H**. We may, therefore, write

$$\mathbf{P}_N = \tfrac{1}{2}(\mathbf{E}\rho' + \mathrm{V}\mathbf{DG}') + \tfrac{1}{2}(\mathbf{H}\sigma' + \mathrm{V}\mathbf{J}'\mathbf{B}), \quad \text{............(161)}$$

and this is the force, reckoned as a pull, on unit area of the surface whose normal is **N**. Here the accented letters are the surface equivalents of the same quantities unaccented, which have reference to unit volume.

Comparing with (122) we see that the type is preserved, except as regards the terms in **F** due to variation of c and μ in space. That is, the stress is represented in (161) as the translational force, due to **E** and **H**, on the fictitious electrification, magnetification, electric current, and magnetic current produced by imagining **E** and **H** to terminate at the surface across which $\mathbf{P}_N$ is the stress.

The coefficient $\frac{1}{2}$ which occurs in (161) is understandable by supposing the fictitious quantities ("matter" and "current") to be distributed uniformly within a very thin layer, so that the forces **E** and **H** which act upon them do not then terminate quite abruptly, but fall off gradually through the layer from their full values on one side to zero on the other. The mean values of **E** and **H** through the layer, that is, $\frac{1}{2}$**E** and $\frac{1}{2}$**H**, are thus the effective electric and magnetic forces on the layer as a whole, per unit volume-density of matter or current; or $\frac{1}{2}$**E** and $\frac{1}{2}$**H** per unit surface-density when the layer is indefinitely reduced in thickness.

Considering the electric field only, the quantities concerned are electrification and magnetic current. In the magnetic field only they are magnetification and electric current. Imagine the medium divided into two regions A and B, of which A is internal, B external, and let **N** be the unit normal from the surface into the external region. The mechanical action between the two regions is fully represented by the stress $\mathbf{P}_N$ over their interface, and the forcive of B upon A is fully represented by the **E** and **H** in B acting upon the fictitious matter and current produced on the boundary of B, on the assumption that **E** and **H** terminate there. If the normal and $\mathbf{P}_N$ be drawn the other way,

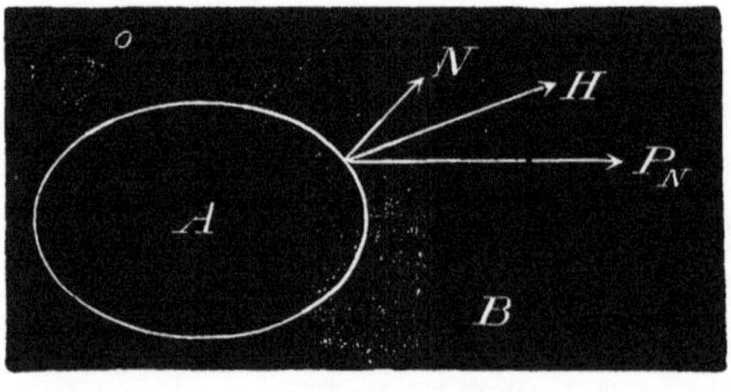

thus negativing them both, as well as the fictitious matter and current on the interface, then it is the forcive of A on B that is represented by the action of **E** and **H** in A on the new interfacial matter and current. That is, the **E** and **H** in the region A may be done away with altogether, because their abolition will immediately introduce the fictitious matter and current equivalent, so far as B is concerned, to the influence of the region A. Similarly **E** and **H** in B may be abolished without altering them in A. And, generally, any portion of the medium may be taken by itself and regarded as being subjected to an equilibrating system of forces, when treated as a rigid body.

§ 24. When c and μ do not vary in space, we do away with the forces $-\frac{1}{2}E^2\nabla c$ and $-\frac{1}{2}H^2\nabla\mu$, and make the form of the surface and volume translational forces agree. We may then regard every element of ρ or of σ as a source sending out from itself displacement and induction isotropically, and every element of **J** or **G** as causing induction or displacement according to Ampère's rule for electric current and its analogue for magnetic current. Thus

$$\mathbf{E} = \sum \frac{\rho/c + \mathrm{V}\mathbf{r}_1\mathbf{G}}{4\pi r^2}, \quad \text{.........................(162)}$$

$$\mathbf{H} = \sum \frac{\sigma/\mu + \mathrm{V}\mathbf{J}\mathbf{r}_1}{4\pi r^2}, \quad \text{.........................(163)}$$

where $\mathbf{r}_1$ is a unit vector drawn from the infinitesimal unit volume in the summation to the point at distance r where **E** or **H** is reckoned. Or, introducing potentials,

$$\mathbf{E} = -\nabla \sum \frac{\rho/c}{4\pi r} - \operatorname{curl} \sum \frac{\mathbf{G}}{4\pi r}, \quad \text{..................(164)}$$

$$\mathbf{H} = -\nabla \sum \frac{\sigma/\mu}{4\pi r} + \operatorname{curl} \sum \frac{\mathbf{J}}{4\pi r}. \quad \text{..................(165)}$$

These apply to the whole medium, or to any portion of the same, with, in the latter case, the surface matter and current included, there being no **E** or **H** outside the region, whilst within it **E** and **H** are the same as due to the matter and current in the whole region ("matter," ρ and σ; "current," **J** and **G**). But there is no known general method of finding the potentials when c and μ vary.

We may also divide **E** and **H** into two parts each, say $\mathbf{E}_1$ and $\mathbf{H}_1$ due to matter and current in the region A, and $\mathbf{E}_2$, $\mathbf{H}_2$ due to matter and current in the region B surrounding it, determinable in the isotropic homogeneous case by the above formulæ. Then we may ignore $\mathbf{E}_1$ and $\mathbf{H}_1$ in estimating the forcive on the matter and current in the region A; thus,

$$\Sigma\,(\mathbf{H}_2\sigma_1 + \mathrm{V}\mathbf{J}_1\mathbf{B}_2) + \Sigma\,(\mathbf{E}_2\rho_1 + \mathrm{V}\mathbf{D}_2\mathbf{G}_1), \quad \text{..............(166)}$$

where $\sigma_1 = \operatorname{div}\mathbf{B}_1 = \operatorname{div}\mathbf{B}$, and $\mathbf{J}_1 = \operatorname{curl}\mathbf{H}_1 = \operatorname{curl}\mathbf{H}$ in region A, is the resultant force on the region A, and

$$\Sigma\,(\mathbf{H}_1\sigma_2 + \mathrm{V}\mathbf{J}_2\mathbf{B}_1) + \Sigma\,(\mathbf{E}_1\rho_2 + \mathrm{V}\mathbf{D}_1\mathbf{G}_2) \quad \text{...............(167)}$$

is the resultant force on the region B; the resultant force on A due to its own **E** and **H** being zero, and similarly for B. These resultant forces are equal and opposite, and so are the equivalent surface-integrals

$$\Sigma(\mathbf{H}_2\sigma_1' + \mathrm{V}\mathbf{J}_1'\mathbf{B}_2) + \Sigma(\mathbf{E}_2\rho_1' + \mathrm{V}\mathbf{D}_2\mathbf{G}_1'), \quad \ldots\ldots\ldots\ldots\ldots(168)$$

and

$$\Sigma(\mathbf{H}_1\sigma_2' + \mathrm{V}\mathbf{J}_2'\mathbf{B}_1) + \Sigma(\mathbf{E}_1\rho_2' + \mathrm{V}\mathbf{D}_1\mathbf{G}_2'), \quad \ldots\ldots\ldots\ldots\ldots(169)$$

taken over the interface. The quantity summed is that part of the stress-vector, $\mathbf{P}_N$, which depends upon products of the **H** of one region and the **B** of the other, etc. Thus, for the magnetic stress only,

$$\begin{aligned}\mathbf{H}.\mathbf{BN} - \mathbf{N}.\tfrac{1}{2}\mathbf{HB} = (\mathbf{H}_1.\mathbf{B}_1\mathbf{N} - \mathbf{N}.\tfrac{1}{2}\mathbf{H}_1\mathbf{B}_1) + (\mathbf{H}_1.\mathbf{B}_2\mathbf{N} - \mathbf{N}.\tfrac{1}{2}\mathbf{H}_1\mathbf{B}_2)\\ + (\mathbf{H}_2.\mathbf{B}_2\mathbf{N} - \mathbf{N}.\tfrac{1}{2}\mathbf{H}_2\mathbf{B}_2) + (\mathbf{H}_2.\mathbf{B}_1\mathbf{N} - \mathbf{N}.\tfrac{1}{2}\mathbf{H}_2\mathbf{B}_1), \quad (170)\end{aligned}$$

and it is the terms in the second and fourth brackets (which, be it observed, are not equal) which together make up the magnetic part of (168) and (169) or their negatives, according to the direction taken for the normal; that is, since $\mathbf{H}_1\mathbf{B}_2 = \mathbf{H}_2\mathbf{B}_1$,

$$\begin{aligned}\Sigma\mathbf{P}_N &= \Sigma(\mathbf{H}_1.\mathbf{B}_2\mathbf{N} + \mathbf{H}_2.\mathbf{B}_1\mathbf{N} - \mathbf{N}.\mathbf{H}_1\mathbf{B}_2) = \Sigma(\mathbf{H}.\mathbf{BN} - \mathbf{N}.\tfrac{1}{2}\mathbf{HB})\\ &= \Sigma(\mathbf{H}_1\sigma_2' + \mathrm{V}\mathbf{J}_2'\mathbf{B}_1) = \Sigma(\mathbf{H}_2\sigma_1' + \mathrm{V}\mathbf{J}_1'\mathbf{B}_2) = \Sigma(\mathbf{H}\sigma' + \mathrm{V}\mathbf{J}'\mathbf{B})\\ &= \Sigma\mathbf{F} = \Sigma(\mathbf{H}_1\sigma_2 + \mathrm{V}\mathbf{J}_2\mathbf{B}_1) = \Sigma(\mathbf{H}_2\sigma_1 + \mathrm{V}\mathbf{J}_1\mathbf{B}_2) = \Sigma(\mathbf{H}\sigma + \mathrm{V}\mathbf{JB}), \quad (171)\end{aligned}$$

where the first six expressions are interfacial summations, and the four last summations throughout one or the other region, the last summation applying to either region. No special reckoning of the sign to be prefixed has been made. The notation is such that $\mathbf{H} = \mathbf{H}_1 + \mathbf{H}_2$, $\sigma = \sigma_1 + \sigma_2$, etc., etc.

The comparison of the two aspects of electromagnetic theory is exceedingly curious; namely, the precise mathematical equivalence of "explanation" by means of instantaneous action at a distance between the different elements of matter and current, each according to its kind, and by propagation through a medium in time at a finite velocity. But the day has gone by for any serious consideration of the former view other than as a mathematical curiosity.

Quaternionic Form of Stress-Vector.

§ 25. We may also notice the Quaternion form for the stress-function, which is so vital a part of the mathematics of forces varying as the inverse square of the distance, and of potential theory. Isotropy being understood, the electric stress may be written

$$\mathbf{P}_N = \tfrac{1}{2}c[\mathbf{EN}^{-1}\mathbf{E}], \quad \ldots\ldots\ldots\ldots\ldots\ldots\ldots\ldots(172)$$

where the quantity in the square brackets is to be understood quaternionically. It is, however, a pure vector. Or,

$$\left[\frac{\mathbf{P}_N}{\mathbf{E}}\right] = \frac{c}{2}\left[\frac{\mathbf{E}}{\mathbf{N}}\right], \quad \ldots\ldots\ldots\ldots\ldots\ldots\ldots\ldots(173)$$

that is, not counting the factor $\frac{1}{2}c$, the quaternion $\left[\frac{\mathbf{P}_N}{\mathbf{E}}\right]$ is the same as the quaternion $\left[\frac{\mathbf{E}}{\mathbf{N}}\right]$; or the same operation which turns **N** to **E** also turns **E** to $\mathbf{P}_N$. Thus **N**, **E**, and $\mathbf{P}_N$ are in the same plane, and the angle between **N** and **E** equals that between **E** and $\mathbf{P}_N$; and **E** and $\mathbf{P}_N$ are on the same side of **N** when **E** makes an acute angle with **N**. Also, the tensor of $\mathbf{P}_N$ is U, so that its normal and tangential components are

$$U\cos 2\theta \quad \text{and} \quad U\sin 2\theta, \quad \text{if} \quad \theta = \widehat{\mathbf{NE}}.$$

Otherwise,

$$\mathbf{P}_N = -\tfrac{1}{2}c[\mathbf{ENE}], \qquad \text{(174)}$$

since the quaternionic reciprocal of a vector has the reverse direction. The corresponding volume translational force is

$$\mathbf{F} = -c\mathrm{V}[\mathbf{E}\nabla\mathbf{E}], \qquad \text{(175)}$$

which is also to be understood quaternionically, and expanded, and separated into parts to become physically significant. I only use the square brackets in this paragraph to emphasize the difference in notation. It rarely occurs that any advantage is gained by the use of the quaternion, in saying which I merely repeat what Professor Willard Gibbs has been lately telling us; and I further believe the disadvantages usually far outweigh the advantages. Nevertheless, apart from practical application, and looking at it from the purely quaternionic point of view, I ought to also add that the invention of quaternions must be regarded as a most remarkable feat of human ingenuity. Vector analysis, without quaternions, could have been found by any mathematician by carefully examining the mechanics of the Cartesian mathematics; but to find out quaternions required a genius.

Remarks on the Translational Force in Free Ether.

§ 26. The little vector **Veh**, which has an important influence in the activity equation, where **e** and **h** are the motional forces

$$\mathbf{e} = \mathrm{V}\mathbf{qB}, \qquad \mathbf{h} = \mathrm{V}\mathbf{Dq},$$

has an interesting form, viz., by expansion,

$$\mathrm{V}\mathbf{eh} = \mathbf{q}\,.\,\mathbf{q}\mathrm{V}\mathbf{DB} = \frac{\mathbf{q}}{v^2}\,.\,\mathbf{q}\mathrm{V}\mathbf{EH}, \qquad \text{(176)}$$

if v be the speed of propagation of disturbances. We also have, in connection therewith, the equivalence

$$\mathbf{eD} = \mathbf{hB}, \qquad \text{(177)}$$

always.

The translational force in a non-conducting dielectric, free from electrification and intrinsic force, is

$$\mathbf{F} = \mathrm{V}\mathbf{JB} + \mathrm{V}\mathbf{DG} + \mathrm{V}\mathbf{jB} + \mathrm{V}\mathbf{Dg},$$

or, approximately [vol. II., p. 509],

$$= \mathrm{V}\dot{\mathbf{D}}\mathbf{B} + \mathrm{V}\mathbf{D}\dot{\mathbf{B}} = \frac{d}{dt}\mathrm{V}\mathbf{DB} = \frac{1}{v^2}\frac{d}{dt}\mathrm{V}\mathbf{EH} = \frac{\dot{\mathbf{W}}}{v^2}. \qquad \text{(178)}$$

The vector **VDB**, or the flux of energy divided by the square of the speed of propagation, is, therefore, the momentum (translational, not magnetic, which is quite a different thing), provided the force **F** *is* the complete force from all causes acting, and we neglect the small terms **VjB** and **VDg**.

But have we any right to safely write

$$\mathbf{F} = m\frac{\partial \mathbf{q}}{\partial t}, \qquad (179)$$

where m is the density of the ether? To do so is to assume that **F** is the only force acting, and, therefore, equivalent to the time-variation of the momentum of a moving particle.*

Now, if we say that there is a certain forcive upon a conductor supporting electric current; or, equivalently, that there is a certain distribution of stress, the magnetic stress, acting upon the same, we do not at all mean that the accelerations of momentum of the different parts are represented by the translational force, the "electromagnetic force." It is, on the other hand, a dynamical problem in which the electromagnetic force plays the part of an impressed force, and similarly as regards the magnetic stress; the actual forces and stresses being only determinable from a knowledge of the mechanical conditions of the conductor, as its density, elastic constants, and the way it is constrained. Now, if there is any dynamical meaning at all in the electromagnetic equations, we must treat the ether in precisely the same way. But we do not know, and have not formularized, the equations of motion of the ether, but only the way it propagates disturbance through itself, with due allowance made for the effect thereon of given motions, and with formularization of the reaction between the electromagnetic effects and the motion. Thus the theory of the stresses and forces in the ether and its motions is an unsolved problem, only a portion of it being known so far, *i.e.*, assuming that the Maxwellian equations do express the known part.

When we assume the ether to be motionless, there is a partial similarity to the theory of the propagation of vibrations of infinitely small range in elastic bodies, when the effect thereon of the actual translation of the matter is neglected.

But in ordinary electromagnetic phenomena, it does not seem that the ignoration of **q** can make any sensible difference, because the speed of propagation of disturbances through the ether is so enormous, that if the ether were stirred about round a magnet, for example, there would be an almost instantaneous adjustment of the magnetic induction to what it would be were the ether at rest.

Static Consideration of the Stresses.—Indeterminateness.

§ 27. In the following the stresses are considered from the static point of view, principally to examine the results produced by changing the form of the stress-function. Either the electric or the magnetic stress alone may be taken in hand. Start then, from a knowledge that the

* Professor J. J. Thomson has endeavoured to make practical use of the idea, *Phil. Mag.*, March, 1891. See also my article, *The Electrician*, January 15, 1886 [vol. I., pp. 547-8].

force on a magnetic pole of strength m is $\mathbf{R}m$, where $\mathbf{R}$ is the polar force of any distribution of intrinsic magnetization in a medium, the whole of which has unit inductivity, so that

$$\operatorname{div} \mathbf{R} = m = \operatorname{conv} \mathbf{h}_0 \qquad (180)$$

measures the density of the fictitious "magnetic" matter; $\mathbf{h}_0$ being the intrinsic force, or, since here $\mu = 1$, the intensity of magnetization. The induction is $\mathbf{B} = \mathbf{h} + \mathbf{R}$. This rudimentary theory locates the force on a magnet at its poles, superficial or internal, by

$$\mathbf{F} = \mathbf{R} \operatorname{div} \mathbf{R}. \qquad (181)$$

The $\mathbf{N}$-component of $\mathbf{F}$ is

$$\mathbf{FN} = \mathbf{RN}.\operatorname{div} \mathbf{R} = \operatorname{div}\{\mathbf{R}.\mathbf{RN} - \mathbf{N}.\tfrac{1}{2}\mathbf{R}^2\}, \qquad (182)$$

because $\operatorname{curl} \mathbf{R} = 0$. Therefore

$$\mathbf{P}_N = \mathbf{R}.\mathbf{RN} - \mathbf{N}.\tfrac{1}{2}\mathbf{R}^2 \qquad (183)$$

is the appropriate stress of irrotational type. Now, however uncertain we may be about the stress in the interior of a magnet, there can be no question as to the possible validity of this stress in the air outside our magnet, for we know that the force $\mathbf{R}$ is then a polar force, and that is all that is wanted, m and $\mathbf{h}$ being merely auxiliaries, derived from $\mathbf{R}$.

Now consider a region A, containing magnets of this kind, enclosed in B, the rest of space, also containing magnets. The mutual force between the two regions is expressed by $\Sigma \mathbf{P}_N$ over the interface, which we may exchange for $\Sigma \mathbf{R}m$ through either region A or B, still on the assumption that $\mathbf{R}$ remains polar.

But if we remove this restriction upon the nature of $\mathbf{R}$, and allow it to be arbitrary, say in region B or in any portion thereof, we find

$$\mathbf{NF} = \operatorname{div} \mathbf{P}_N = \mathbf{RN} \operatorname{div} \mathbf{R} + \mathbf{N}\mathrm{V}(\operatorname{curl} \mathbf{R}).\mathbf{R};$$

or

$$\mathbf{F} = \mathbf{R}m + \mathrm{V}\mathbf{JR},$$

if $\mathbf{J} = \operatorname{curl} \mathbf{R}$. This gives us, from a knowledge of the external magnetic field of polar magnets only, the mechanical force exerted by a magnet on a region containing $\mathbf{J}$, whatever that may be, provided it be measureable as above; and without any experimental knowledge of electric currents, we could now predict their mechanical effects in every respect by the principle of the equality of action and reaction, not merely as regards the mutual influence of a magnet and a closed current, but as regards the mutual influence of the closed currents themselves; the magnetic force of a closed current, for instance, being the force on unit of m, is equivalently the force exerted by m on the closed current, which, by the above, we know. Also, we see that according to this magnetic notion of electric current, it is necessarily circuital.

At the same time, it is to be remarked that our real knowledge must cease at the boundary of the region containing electric current, a metallic conductor for instance; the surface over which $\mathbf{P}_N$ is reckoned, on one side of which is the magnet, on the other side electric current, can only be pushed up as far as the conductor. The stress $\mathbf{P}_N$ may therefore cease altogether on reaching the conductor, where it forms a distribution of surface force fully representing the action of the magnet on the conductor. Similarly, we need not continue the stress into the interior of the magnet. Then, so far as the resultant force on the magnet as a whole, in translating or rotating it, and, similarly, so far as the action on the conductor is concerned, the simple stress $\mathbf{P}_N$ of constant tensor $\frac{1}{2}\mathbf{R}^2$, varying from a tension parallel to $\mathbf{R}$ to an equal pressure laterally, acting in the medium between the magnet and conductor, accounts, by its terminal pulls or pushes, for the mechanical forces on them. The lateral pressure is especially prominent in the case of conductors, whilst the tension goes more or less out of sight, as the immediate cause of motion. Thus, when parallel currents appear to attract one another, the conductors are really pushed together by the lateral pressure on each conductor being greater on the side remote from the other than on the near side: whilst if the currents are oppositely directed, the pressure on the near sides is greater than on the remote sides, and they appear to repel one another.

The effect of continuing the stress into the interior of a conductor of unit inductivity, according to the same law, instead of stopping it on its boundary, is to distribute the translational force bodily, according to the formula $\Sigma \mathrm{V}\mathbf{JR}$, instead of superficially, according to $\Sigma\,\mathbf{P}_N$. In either case, of course, the conductor must be strained by the magnetic stress, with the consequent production of a mechanical stress. But the strain (and associated stress) will be different in the two cases, the applied forces being differently localized. The effect of the stress on a straight portion of a wire supporting current, due to its own field only, is to compress it laterally, and to lengthen it. Besides this, there will be resultant force on it arising from the different pressures on its opposite sides due to the proximity of the return conductor or rest of the circuit, tending to move it so as to increase the induction through the circuit per unit current, that is, the inductance of the circuit.

§ 28. If, now, we bring an elastically magnetizable body into a magnetic field, it modifies the field by its presence, causing more or less induction to go through it than passed previously in the air it replaces, according as its inductivity exceeds or is less than that of the air. The forcive on it, considered as a rigid body, is completely accounted for by the simple stress $\mathbf{P}_N$ in the air outside it, reckoned according to the changed field, and supposed to terminate on the surface of the disturbing body. This is true whether the body be isotropic or heterotropic in its inductivity; nor need the induction be a linear function of the magnetic force. It is also true when the body is intrinsically magnetized; or is the seat of electric current. In short, since the external stress depends upon the magnetic force outside the body, when we take the external field as we may find it, that is, as modified by any known or unknown causes within the body, the corresponding stress, terminated upon its boundary, fully represents the forcive on the body, as a whole, due to magnetic causes. This follows from the equality of action and reaction; the force on the body due to a unit pole is the opposite of that of the body on the pole.

If we wish to continue the stress into the interior of the body, surrounded on all sides by the unmagnetized medium of unit inductivity, as we must do if we wish to arrive ultimately at the mutual actions of its different parts, and how they are modified by variations of inductivity, by intrinsic magnetization, and by electric current in the body, we may, so far as the resultant force and torque on it are concerned, do it in any way we please, provided we do not interfere with the stress outside. For the internal stress, of any type, will have no resultant force or torque on the body, and there is merely left the real external stress.

Practically, however, we should be guided by the known relations of magnetic force, induction, magnetization, and current, and not go to work in a fanciful manner; furthermore, we should always choose the stress in such a way that if, in its expression, we take the inductivity to be unity, and the intrinsic magnetization zero, it must reduce to the simple Maxwellian stress in air (assumed to represent ether here). But as we do not know definitely the forcive arising from the magnetic stress in the interior of a magnet, there are several formulæ that suggest themselves as possible.

Special Kinds of Stress Formulæ statically suggested.

§ 29. Thus, first we have the stress (183); let this be quite general, then

$$(1)\begin{cases}\mathbf{P}_N = \mathbf{R}.\mathbf{RN} - \mathbf{N}.\tfrac{1}{2}\mathbf{R}^2, & \text{.....................(184)}\\ \mathbf{F} = \mathbf{R}\,\mathrm{div}\,\mathbf{R} + \mathrm{V}\mathbf{JR}. & \text{.....................(185)}\end{cases}$$

Here $\mathbf{R}$ is the magnetic force of the field, not of the flux $\mathbf{B}$. If $\mu=1$, $\mathrm{div}\,\mathbf{R}$ is the density of magnetic matter—the convergence of the intrinsic magnetization—but not otherwise. In general, it is the density of the matter of the magnetic potential, calculated on the assumption $\mu=1$. The force on a magnet is located in this system at its poles, whether the magnetization be intrinsic or induced. The second term in (185) represents the force on matter bearing electric current ($\mathbf{J}=\mathrm{curl}\,\mathbf{R}$), but has to be supplemented by the first term, unless $\mathrm{div}\,\mathbf{R}=0$ at the place.

§ 30. Next, let the stress be μ times as great for the same magnetic force, but be still of the same simple type, μ being the inductivity, which is unity outside the body, but having any positive value, which may be variable, within it. Then we shall have

$$(2)\begin{cases}\mathbf{P}_N = \mathbf{R}.\mathbf{N}\mu\mathbf{R} - \mathbf{N}.\tfrac{1}{2}\mathbf{R}\mu\mathbf{R}, & \text{.................(186)}\\ \mathbf{F} = \mathbf{R}m + \mathrm{V}\mathbf{J}\mu\mathbf{R} - \tfrac{1}{2}\mathbf{R}^2\nabla\mu, & \text{.................(187)}\end{cases}$$

where $m=\mathrm{conv}\,\mu\mathbf{h}_0=\mathrm{div}\,\mu\mathbf{R}$ is the density of magnetic matter, $\mu\mathbf{h}_0$ being the intensity of intrinsic magnetization.

The electromagnetic force is made μ times as great for the same magnetic force; the force on an intrinsic magnet is at its poles; and there is, in addition, a force wherever μ varies, including the intrinsic magnet, and not forgetting that a sudden change in μ, as at the boundary of a magnet, has to count. This force, the third term in

(187), explains the force on inductively magnetized matter. It is in the direction of most rapid decrease of μ.

§ 31. Thirdly, let the stress be of the same simple type, but taking **H** instead of **R**, **H** being the force of the flux $\mathbf{B}=\mu\mathbf{H}=\mu(\mathbf{R}+\mathbf{h}_0)$, where $\mathbf{h}_0$ is as before. We now have

$$(3)\begin{cases}\mathbf{P}_N=\mathbf{H}.\mathbf{NB}-\mathbf{N}.\tfrac{1}{2}\mathbf{HB}, & \text{.................(188)}\\ \mathbf{F}\ =\mathrm{V}\mathbf{JB}+\mathrm{V}\mathbf{j}_0\mathbf{B}-\tfrac{1}{2}\mathbf{H}^2\nabla\mu, & \text{.................(189)}\end{cases}$$

where $\mathbf{j}_0=\text{curl}\,\mathbf{h}_0$ is the distribution of fictitious electric current which produces the same induction as the intrinsic magnetization $\mu\mathbf{h}_0$, and **J** is, as before, the real current.

It is now *quasi*-electromagnetic force that acts on an intrinsic magnet, with, however, the force due to $\nabla\mu$, since a magnet has usually large μ compared with air.

The above three stresses are all of the simple type (equal tension and perpendicular pressure), and are irrotational, unless μ be the eolotropic operator. No change is, in the latter case, needed in (186), (188), whilst in the force formulæ (187), (189), the only change needed is to give the generalized meaning to $\nabla\mu$. Thus, in (189), instead of $\mathbf{H}^2\nabla\mu$, use $2\nabla T_\mu$, or $\nabla_\mu(\mathbf{H}\mu\mathbf{H})$, or

$$\mathbf{i}\left(\mathbf{H}\frac{d\mu}{dx}\mathbf{H}\right)+\mathbf{j}\left(\mathbf{H}\frac{d\mu}{dy}\mathbf{H}\right)+\mathbf{k}\left(\mathbf{H}\frac{d\mu}{dz}\mathbf{H}\right),$$

or
$$(\nabla_B-\nabla_H)\mathbf{HB},$$

or
$$\mathbf{i}(\mathbf{H}\nabla_1\mathbf{B}-\mathbf{B}\nabla_1\mathbf{H})+\mathbf{j}(\mathbf{H}\nabla_2\mathbf{B}-\mathbf{B}\nabla_2\mathbf{H})+\mathbf{k}(\mathbf{H}\nabla_3\mathbf{B}-\mathbf{B}\nabla_3\mathbf{H}),$$

showing the **i**, **j**, **k** components.

Similarly in the other cases occurring later.

The following stresses are not of the simple type, though all consist of a tension parallel to **R** or **H** combined with an isotropic pressure.

§ 32. Alter the stress so as to locate the force on an intrinsic magnet bodily upon its magnetized elements. Add $\mathbf{R}.\mu\mathbf{h}_0\mathbf{N}$ to the stress (186), and therefore $\mu\mathbf{h}_0.\mathbf{RN}$ to its conjugate; then the divergence of the latter must be added to the **N**-component of the force (187). Thus we get, if $\mathbf{I}=\mu\mathbf{h}_0$,

$$(4)\begin{cases}\mathbf{P}_N=\mathbf{R}.\mathbf{BN}-\mathbf{N}.\tfrac{1}{2}\mathbf{R}\mu\mathbf{R}, & \text{.................(190)}\\ \mathbf{F}\ =\mathbf{I}\nabla.\mathbf{R}+\mathrm{V}\mathbf{J}\mu\mathbf{R}-\tfrac{1}{2}\mathbf{R}^2\nabla\mu. & \text{.................(191)}\end{cases}$$

But here the sum of the first two terms in **F** may be put in a different form. Thus,

$$\mathbf{I}\nabla.\mathbf{R}=I_1\nabla_1\mathbf{R}+I_2\nabla_2\mathbf{R}+I_3\nabla_3\mathbf{R}=\mathbf{i}.\mathbf{I}\nabla R_1+\mathbf{j}.\mathbf{I}\nabla R_2+\mathbf{k}.\mathbf{I}\nabla R_3.$$

Also
$$\mathbf{I}\nabla R_1=\mathbf{I}\nabla_1\mathbf{R}+\mathbf{I}(\nabla R_1-\nabla_1\mathbf{R})=\mathbf{I}\nabla_1\mathbf{R}+\mathbf{i}\mathrm{V}\mathbf{JI}.$$

These bring (191) to

$$\mathbf{F}=(\mathbf{i}.\mathbf{I}\nabla_1\mathbf{R}+\mathbf{j}.\mathbf{I}\nabla_2\mathbf{R}+\mathbf{k}.\mathbf{I}\nabla_3\mathbf{R})+\mathrm{V}\mathbf{JB}-\tfrac{1}{2}\mathbf{R}^2\nabla\mu, \quad \text{.........(192)}$$

where the first component (the bracketted part) is Maxwell's force on intrinsic magnetization, and the second his electromagnetic force. The third, as before, is required where μ varies.

§ 33. To the stress (190) add $-\mathbf{N}.\frac{1}{2}\mathbf{RI}$, without altering the conjugate stress, making

$$(5)\begin{cases}\mathbf{P}_N=\mathbf{R}.\mathbf{BN}-\mathbf{N}.\frac{1}{2}\mathbf{RB}, & \text{(193)}\\ \mathbf{F}\ =\mathbf{VJB}-\frac{1}{2}\{\mathbf{i}(\mathbf{R}\nabla_1\mathbf{B}-\mathbf{B}\nabla_1\mathbf{R})+\mathbf{j}(\mathbf{R}\nabla_2\mathbf{B}-\mathbf{B}\nabla_2\mathbf{R})+\mathbf{k}(\mathbf{R}\nabla_3\mathbf{B}-\mathbf{B}\nabla_3\mathbf{R})\} & \\ \quad =\mathbf{VJB}-(\nabla_B-\nabla_R)\frac{1}{2}\mathbf{RB}. & \text{(194)}\end{cases}$$

This we need not discuss, as it is merely a transition to the next form.

§ 34. To the stress (193) add $\mathbf{h}_0.\mathbf{NB}$; we then get

$$(6)\begin{cases}\mathbf{P}_N=\mathbf{H}.\mathbf{NB}-\mathbf{N}.\frac{1}{2}\mathbf{RB}, & \text{(195)}\\ \mathbf{F}\ =\mathbf{VJB}+\{\mathbf{i}.\mathbf{B}\nabla h_1+\mathbf{j}.\mathbf{B}\nabla h_2+\mathbf{k}.\mathbf{B}\nabla h_3\} & \\ \qquad -\frac{1}{2}\{\mathbf{i}.(\mathbf{R}\nabla_1\mathbf{B}-\mathbf{B}\nabla_1\mathbf{R})+\mathbf{j}.(\mathbf{R}\nabla_2\mathbf{B}-\mathbf{B}\nabla_2\mathbf{R})+\mathbf{k}.(\mathbf{R}\nabla_3\mathbf{B}-\mathbf{B}\nabla_3\mathbf{R})\} & \\ \quad =\mathbf{VJB}+\mathbf{B}\nabla.\mathbf{h}_0-(\nabla_B-\nabla_R)\frac{1}{2}\mathbf{RB}, & \text{(196)}\end{cases}$$

where h_1, h_2, h_3 are the components of $\mathbf{h}_0$.

Now if to this last stress (195) we add $-\mathbf{N}.\frac{1}{2}\mathbf{h}_0\mathbf{B}$, we shall come back to the third stress, (188), of the simple type.

Perhaps the most instructive order in which to take the six stresses is (1), (2), (4), (5), (6), and (3); merely adding on to the force, in passing from one stress to the next, the new part which the alteration in the stress necessitates.

To the above we should add Maxwell's general stress, which is

$$(7)\begin{cases}\mathbf{P}_N=\mathbf{R}.\mathbf{NB}-\mathbf{N}.\frac{1}{2}\mathbf{R}^2, & \text{(197)}\\ \mathbf{F}\ =\mathbf{VJB}+\{\mathbf{i}.\mathbf{I}\nabla_1\mathbf{R}+\mathbf{j}.\mathbf{I}\nabla_2\mathbf{R}+\mathbf{k}.\mathbf{I}\nabla_3\mathbf{R}\} & \\ \qquad\qquad +\{\mathbf{i}.\mathbf{M}\nabla_1\mathbf{R}+\mathbf{j}.\mathbf{M}\nabla_2\mathbf{R}+\mathbf{k}.\mathbf{M}\nabla_3\mathbf{R}\}, & \text{(198)}\\ \quad =\mathbf{VJB}+\nabla_R[\mathbf{R}(\mathbf{I}+\mathbf{M})], & \end{cases}$$

where $\mathbf{M}=(\mu-1)\mathbf{R}=$ intensity of induced magnetization. There is a good deal to be said against this stress; some of which later.

Remarks on Maxwell's General Stress.

§ 35. All the above force-formulæ refer to the unit volume; whenever, therefore, a discontinuity in the stress occurs at a surface, the corresponding expression per unit surface is needed; *i.e.*, in making a special application, for it is wasted labour else. It might be thought that as Maxwell gives the force (198), and in his treatise usually gives surface-expressions separately, so none is required in the case of his force-system (198). But this formula will give entirely erroneous results if carried out literally. It forms no exception to the rule that all the expressions require surface-additions.

Maxwell's general stress has the apparent advantage of simplicity. It merely requires an alteration in the tension parallel to $\mathbf{R}$, from $\mathbf{R}^2$ to $\mathbf{RB}$, whilst the lateral pressure remains $\frac{1}{2}\mathbf{R}^2$, when we pass from unmagnetized to magnetized matter. The force to which it gives rise is also apparently simple, being merely the sum of two forces, one the electromagnetic, $\mathbf{VJB}$, the other a force on magnetized matter whose $\mathbf{i}$-component is $(\mathbf{I}+\mathbf{M})\nabla_1\mathbf{R}$, both per unit volume, the latter being accompanied (in case of eolotropy) by a torque. Now **I** is the intrinsic and **M** the induced magnetization, so the force is made irrespective of the proportion in which the magnetization exists as intrinsic or induced. In fact, Maxwell's "magnetization" is the sum of the two without reservation or distinction. But to unite them is against the whole behaviour of induced and intrinsic magnetization in the electromagnetic scheme of Maxwell, as I interpret it. Intrinsic magnetization (using Sir W. Thomson's term) should be regarded as impressed ($\mathbf{I} = \mu\mathbf{h}_0$, where $\mathbf{h}_0$ is the equivalent impressed magnetic force); on the other hand, "induced" magnetization depends on the force of the field $\{\mathbf{M} = (\mu - 1)\mathbf{R}\}$. Intrinsic magnetization keeps up a field of force. Induced magnetization is kept up by the field. In the circuital law **I** and **M** therefore behave differently. There may be absolutely no difference whatever between the magnetization of a molecule of iron in the two cases of being in a permanent or a temporary magnet. That, however, is not in question. We have no concern with molecules in a theory which ignores molecules, and whose element of volume must be large enough to contain so many molecules as to swamp the characteristics of individuals. It is the resultant magnetization of the whole assembly that is in question, and there is a great difference between its nature according as it disappears on removal of an external cause, or is intrinsic. The complete amalgamation of the two in Maxwell's formula must certainly, I think, be regarded as a false step.

We may also argue thus against the probability of the formula. If we have a system of electric current in an unmagnetizable ($\mu = 1$) medium, and then change μ everywhere in the same ratio, we do not change the magnetic force at all, the induction is made μ times as great, and the magnetic energy μ times as great, and is similarly distributed. The mechanical forces are, therefore, μ times as great, and are similarly distributed. That is, the translational force in the $\mu = 1$ medium, or $\mathbf{VJR}$, becomes $\mathbf{VJ}\mu\mathbf{R}$ in the second case in which the inductivity is μ, without other change. But there is no force brought in on magnetized matter *per se*.

Similarly, if in the $\mu = 1$ medium we have intrinsic magnetization **I**, and then alter μ in any ratio everywhere alike, keeping **I** unchanged, it is now the induction that remains unaltered, the magnetic force becoming μ^{-1} times, and the energy μ^{-1} times the former values, without alteration in distribution (referring to permanent states, of course). Again, therefore, we see that there is no translational force brought in on magnetized matter merely because it is magnetized.

Whatever formula, therefore, we should select for the stress-function, it would certainly not be Maxwell's, for cumulative reasons. When, some six years ago, I had occasion to examine the subject of the stresses, I was unable to arrive at any very definite results, except outside of magnets or conductors. It was a perfectly indeterminate problem to find the magnetic stress inside a body from the existence of a known, or highly probable, stress outside it. All one could do was to examine the consequences of assuming certain stresses, and to reject those which did not work well. After going into considerable detail, the only two which seemed possible were the second and third above (those of equations (186) and (188) above). As regards the seventh (Maxwell's stress, equation (198) above), the apparent simplicity produced by the union of intrinsic and induced magnetization, turned out, when examined into its consequences, to lead to great complication and unnaturalness. This will be illustrated in the following example, a simple case in which we can readily and fully calculate all details by different methods, so as to be quite sure of the results we ought to obtain.

A worked-out Example to exhibit the Forcives contained in Different Stresses.

§ 36. Given a fluid medium of inductivity μ_1, in which is an intrinsic magnet of the same inductivity. Calculate the attraction between the magnet and a large solid mass of different inductivity μ_2. Here it is only needful to calculate the force on a single pole, so let the magnet be infinitely thin and long, with one pole of strength m at distance a from the medium μ_2, which may have an infinitely extended plane boundary. By placing a fictitious pole of suitable strength at the optical image in the second medium of the real pole in the first, we may readily obtain the solution.

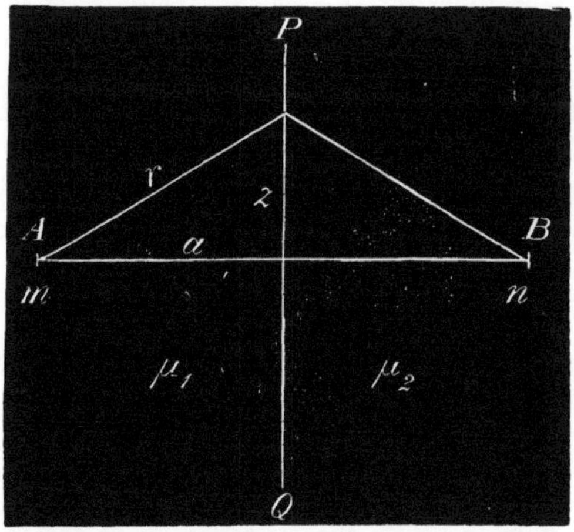

Let PQ be the interface, and the real pole be at A, and its image at B. We have first to calculate the distribution of $\mathbf{R}$, magnetic force, in both media due to the pole m, as disturbed by the change of inductivity. We have $\operatorname{div} \mu_1 \mathbf{R}_1 = m$ in the first medium, and $\operatorname{div} \mu_2 \mathbf{R}_2 = 0$ in the second, therefore $\mathbf{R}$ has divergence only on the interface. Let σ be the surface-density of the fictitious interfacial matter to correspond; its force goes symmetrically both ways; the continuity of the normal induction therefore gives, at distance r from A, the condition

$$\mu_1\left(\frac{ma}{4\pi\mu_1 r^3} - \tfrac{1}{2}\sigma\right) = \mu_2\left(\frac{ma}{4\pi\mu_1 r^3} + \tfrac{1}{2}\sigma\right), \quad \text{.............(199)}$$

because $m/4\pi\mu_1 r^2$ is the tensor of the magnetic force due to m in the μ_1 medium when of infinite extent. Therefore

$$\sigma = \frac{\mu_1 - \mu_2}{\mu_1 + \mu_2} \cdot \frac{ma}{2\pi\mu_1 r^3}. \qquad \text{.......................(200)}$$

The magnetic potential Ω, such that $\mathbf{R} = -\nabla\Omega$ is the polar force in either region, is therefore the potential of m/μ_1 at A and of σ over the interface.

But if we put matter n at the image B, of amount

$$n = \frac{\mu_1 - \mu_2}{\mu_1 + \mu_2} \cdot \frac{m}{\mu_1}, \qquad \text{.........................(201)}$$

the normal component of $\mathbf{R}_1$ on the μ_1 side due to n and the pole m will be

$$\frac{ma}{4\pi\mu_1 r^3} - \frac{na}{4\pi r^3} = \frac{ma}{4\pi\mu_1 r^3} - \tfrac{1}{2}\sigma, \qquad \text{..................(202)}$$

the same value as before; the force $\mathbf{R}_1$ on the μ_1 side is, therefore, the same as that due to matter m/μ_1 at A and matter n at B; whilst on the μ_2 side the force $\mathbf{R}_2$ is that due to matter m/μ_1 at A and matter n also at A, that is, to matter $\dfrac{2m}{\mu_1 + \mu_2}$ at A. Thus, in the μ_2 medium the force $\mathbf{R}_2$ is radial from A as if there were no change of inductivity, though altered in intensity.

The repulsion between the pole m and the solid mass is not the repulsion between the matters m/μ_1 and n of the potential, but is

$= m \times$ magnetic force at A due to matter n at B,

$= n \times$ magnetic force at B due to matter m at A,

$$= \frac{mn}{4\pi(2a)^2} = \frac{\mu_1 - \mu_2}{\mu_1 + \mu_2} \cdot \frac{m^2}{4\pi\mu_1(2a)^2}, \qquad \text{..................(203)}$$

becoming an attraction when $\mu_2 > \mu_1$, making n negative. When $\mu_2 = 0$, the repulsion is

$$\frac{m^2}{4\pi\mu_1(2a)^2};$$

when $\mu_2 = \infty$, it is turned into an attraction of equal amount.

Similarly, if we consider the attraction to be the resultant force between m and the interfacial matter σ, we shall get the same result by

$$\Sigma \frac{\sigma ma}{4\pi r^3}, \qquad \text{.............................(204)}$$

the quantity summed (over the interface) being $\sigma \times$ normal component of magnetic force due to matter m in a medium of unit inductivity, or the normal component of induction due to m in its own medium. For this is

$$\int \frac{ma}{4\pi r^3} \cdot \frac{\mu_1 - \mu_2}{\mu_1 + \mu_2} \cdot \frac{ma}{2\pi\mu_1 r^3} 2\pi r\, dr = \frac{m^2 a^2}{4\pi\mu_1} \cdot \frac{\mu_1 - \mu_2}{\mu_1 + \mu_2} \int \frac{dr}{r^5} = (203) \text{ again.}$$

Another way is to calculate the variation of energy made by displacing either the pole m or the μ_2 mass. The potential energy is expressed by

$$\tfrac{1}{2}(P+p)m = \tfrac{1}{2}Pm + \tfrac{1}{2}\Sigma P\sigma\mu, \qquad \text{.................(205)}$$

where $P = m/4\pi\mu_1 r$ and $p = \Sigma\, \sigma/4\pi r$, the potentials of matter m/μ_1 and σ, where r is the distance from m or from σ to the point where P and p are reckoned.

The value of the second part in (205), depending upon σ, comes to

$$\tfrac{1}{2}\frac{\mu_1-\mu_2}{\mu_1+\mu_2}\cdot\frac{m^2}{4\pi\mu_1 . 2a}, \qquad \text{........................(206)}$$

and its rate of decrease with respect to a expresses the repulsion between the pole and the μ_2 region. This gives (203) again.

A fourth way is by means of the *quasi*-electromagnetic force on fictitious interfacial electric current, instead of matter, the current being circular about the axis of symmetry AB. The formula for the attraction is

$$\Sigma\, \mathrm{V\,curl\,\mathbf{B}}\,.\,\mathbf{R}_0, \qquad \text{..........................(207)}$$

if $\mathbf{R}_0$ be the radial magnetic force from m in its own medium, tensor $m/4\pi\mu_1 r^2$. Here the curl of $\mathbf{B}$ is represented by the interfacial discontinuity in the tangential induction, or

$$\frac{2zm}{4\pi r^3}\cdot\frac{\mu_1-\mu_2}{\mu_1+\mu_2}.$$

Also the tangential component of $\mathbf{R}_0$ is $mz/4\pi\mu_1 r^3$. Therefore the repulsion is

$$\int\frac{2mz}{4\pi r^3}\cdot\frac{\mu_1-\mu_2}{\mu_1+\mu_2}\,\frac{mz}{4\pi\mu_1 r^3}2\pi r\,dr = \frac{m^2}{4\pi\mu_1}\cdot\frac{\mu_1-\mu_2}{\mu_1+\mu_2}\int_a^\infty\frac{r^2-a^2}{r^5}dr$$

$$= \frac{m^2}{4\pi\mu_1}\cdot\frac{\mu_1-\mu_2}{\mu_1+\mu_2}\cdot\frac{1}{4a^2}, \qquad \text{.........(208)}$$

as before, equation (203). This method (207) is analogous to (204).

§ 37. There are several other ways of representing the attraction, employing fictitious matter and current; but now let us change the method, and observe how the attraction between the magnetic pole and the iron mass is accounted for by a stress-distribution, and its space-variation. The best stress is the third, equation (188), § 31. Applying this, we have simply a tension of magnitude $\tfrac{1}{2}\mu_1 R_1^2 = T_1$ in the first medium and $\tfrac{1}{2}\mu_2 R_2^2 = T_2$ in the second, parallel to $\mathbf{R}_1$ and $\mathbf{R}_2$ respectively, each combined with an equal lateral pressure, so that the tensor of the stress-vector is constant.

But, so far as the attraction is concerned, we may ignore the stress in the second medium altogether, and consider it as the $\Sigma\,\mathbf{P}_N$ of the stress-vector in the first medium over the surface of the second medium. The tangential component summed has zero resultant; the attraction is therefore the sum of the normal components, or $\Sigma\, T_1 \cos 2\theta_1$, where θ_1 is the angle between $\mathbf{R}_1$ and the normal. This is the same as

$\Sigma \frac{1}{2}\mu_1(R_N^2 - R_T^2)$, if R_N and R_T are the normal and tangential components of $\mathbf{R}_1$; or

$$\int_a^\infty 2\pi r\,dr\,\tfrac{1}{2}\mu_1\left[\left(\frac{ma}{4\pi\mu_1 r^3}\cdot\frac{2\mu_2}{\mu_1+\mu_2}\right)^2 - \left(\frac{mz}{4\pi\mu_1 r^3}\cdot\frac{2\mu_1}{(\mu_1+\mu_2)}\right)^2\right]; \quad (209)$$

which on evaluation gives the required result (203).

But this method does not give the true distribution of translational force due to the stresses. In the first medium there is no translational force, except on the magnet. Nor is there any translational force in the second μ_2 medium. But at the interface, where μ changes, there is the force $-\frac{1}{2}R^2\nabla\mu$ per unit volume, and this is represented by the stress-difference at the interface. It is easily seen that the tangential stress-difference is zero, because

$$T\sin 2\theta = \mu R_N R_T, \quad \text{.........................(210)}$$

and both the normal induction and the tangential magnetic force are continuous. The real force is, therefore, the difference of the normal components of the stress-vectors, and is, therefore, normal to the interface. This we could conclude from the expression $-\frac{1}{2}R^2\nabla\mu$. But since the resultant of the interfacial stress in the second medium is zero, we need not reckon it, so far as the attraction of the pole is concerned. The normal traction on the interface, due to both stresses, is of amount

$$\frac{m^2}{8\pi^2 r^6}\cdot\frac{\mu_2-\mu_1}{(\mu_1+\mu_2)^2}\left(r^2+a^2\frac{\mu_2-\mu_1}{\mu_1}\right) \quad \text{................(211)}$$

per unit area. Summed up, it gives (203) again.

That (211) properly represents the force $-\frac{1}{2}R^2\nabla\mu$ when μ is discontinuous, we may also verify by supposing μ to vary continuously in a very thin layer, and then proceed to the limit.

The change from an attraction to a repulsion as μ_2 changes from being greater to being less than μ_1, depends upon the relative importance of the tensions parallel to the magnetic force and the lateral pressures operative at different parts of the interface. In the extreme case of $\mu_2 = 0$, we have $\mathbf{R}_1$ tangential, with, therefore, a pressure everywhere. For the other extreme, $\mathbf{R}_1$ is normal, and there is a pull on the second medium everywhere. When μ_2 is finite there is a certain circular area on the interface within which the translational force due to the stress in the medium containing the pole m is towards that medium, whilst outside it the force is the other way. But when both stresses are allowed for, we see that when $\mu_2 > \mu_1$ the pull is towards the first medium in all parts of the interface, and that this becomes a push in all parts when $\mu_1 > \mu_2$.

A Definite Stress only obtainable by Kinetic Consideration of the Circuital Equations and Storage and Flux of Energy.

§ 38. We see that the stress considered in the last paragraph gives a rationally intelligible interpretation of the attraction or repulsion. The same may be said of other stresses than that chosen. But the use of

Maxwell's stress, or any stress leading to a force on inductively magnetized matter as this stress does, leads us into great difficulties. By (198) we see that there is first a bodily force on the whole of the μ_2 medium, because it is magnetized, unless $\mu_2 = 1$. When summed up, the resultant does not give the required attraction. For, secondly, the μ_1 medium is also magnetized, unless $\mu_1 = 1$, and there is a bodily force throughout the whole of it. When this is summed up (not counting the force on the magnet), its resultant added on to the former resultant still does not make up the attraction (*i.e.*, equivalently, the force on the magnet). For, thirdly, the stress is discontinuous at the interface (though not in the same manner as in the last paragraph). The resultant of this stress-discontinuity, added on to the former resultants, makes up the required attraction. It is unnecessary to give the details relating to so improbable a system of force.

Our preference must naturally be for a more simple system, such as the previously considered stress. But there is really no decisive settlement possible from the theoretical statical standpoint, and nothing short of actual experimental determination of the strains produced and their exhaustive analysis would be sufficient to determine the proper stress-function. But when the subject is attacked from the dynamical standpoint, the indeterminateness disappears. From the two circuital laws of variable states of electric and magnetic force in a moving medium, combined with certain distributions of stored energy, we are led to just one stress-vector, viz. (136). It is, in the magnetic case, the same as (188): that is, it reduces to the latter when the medium is kept at rest, so that $\mathbf{J}_0$ and $\mathbf{G}_0$ become $\mathbf{J}$ and $\mathbf{G}$.

It is of the simple type in case of isotropy (constant tensor), but is a rotational stress in general, as indeed are all the statically probable stresses that suggest themselves. The translational force due to it being divisible conveniently into (*a*), the electromagnetic force on electric current, (*b*), the ditto on the fictitious electric current taking the place of intrinsic magnetization, (*c*), force depending upon space-variation of μ; we see that the really striking part is (*b*). Of all the various ways of representing the forcive on an intrinsic magnet it is the most extreme. The magnetic "matter" does not enter into it, nor does the distribution of magnetization; it is where the intrinsic force $\mathbf{h}_0$ has curl that the translational force operates, usually on the sides of a magnet. From actual experiments with bar-magnets, needles, etc., one would naturally prefer to regard the polar regions as the seat of translational force. But the equivalent forcive $\Sigma \mathbf{j}_0\mathbf{B}$ has one striking recommendation (apart from the dynamical method of deducing it), viz., that the induction of an intrinsic magnet is determined by curl $\mathbf{h}_0$, not by $\mathbf{h}_0$ itself; and this, I have shown, is true when $\mathbf{h}_0$ is imagined to vary, the whole varying states of the fluxes $\mathbf{B}$, $\mathbf{D}$, $\mathbf{C}$ due to impressed force being determined by the curls of $\mathbf{e}_0$ and $\mathbf{h}_0$, which are the sources of the disturbances (though not of the energy).

The rotational peculiarity in eolotropic substances does not seem to be a very formidable objection. Are they not solid?

As regards the assumed constancy of μ, a more complete theory must, to be correct, reduce to one assuming constancy of μ, because, as Lord Rayleigh* has shown, the assumed law has a limited range of validity, and is therefore justifiable as a preparation for more complete views. Theoretical requirements are not identical with those of the practical engineer.

But, for quite other reasons, the dynamically determined stress might be entirely wrong. Electric and magnetic "force" and their energies are facts. But it is the total of the energies in concrete cases that should be regarded as the facts, rather than their distribution; for example, that, as Sir W. Thomson proved, the "mechanical value" of a simple closed current C is $\frac{1}{2}LC^2$, where L is the inductance of the circuit (coefficient of electromagnetic capacity), rather than that its distribution in space is given by $\frac{1}{2}\mathbf{HB}$ per unit volume. Other distributions may give the same total amount of energy. For example, the energy of distortion of an elastic solid may be expressed in terms of the square of the rotation and the square of the expansion, if its boundary be held at rest; but this does not correctly localize the energy. If, then, we choose some other distribution of the energy for the same displacement and induction, we should find quite a different flux of energy. But I have not succeeded in making any other arrangement than Maxwell's work practically, or without an immediate introduction of great obscurities. Perhaps the least certain part of Maxwell's scheme, as modified by myself, is the estimation of magnetic energy as $\frac{1}{2}\mathbf{HB}$ in intrinsic magnets, as well as outside them, that is, by $\frac{1}{2}\mathbf{B}\mu^{-1}\mathbf{B}$, however $\mathbf{B}$ may be caused. Yet, only in this way are thoroughly consistent results apparently obtainable when the electromagnetic field is considered comprehensively and dynamically.

APPENDIX.

Received June 27, 1891.

Extension of the Kinetic Method of arriving at the Stresses to cases of Non-linear Connection between the Electric and Magnetic Forces and the Fluxes. Preservation of Type of the Flux of Energy Formula.

§ 39. It may be worth while to give the results to which we are led regarding the stress and flux of energy when the restriction of simple proportionality between "forces" and "fluxes," electric and magnetic respectively, is removed. The course to be followed, to obtain an interpretable form of the equation of activity, is sufficiently clear in the light of the experience gained in the case of proportionality.

First assume that the two circuital laws (89) and (90), or the two in (93), hold good generally, without any initially stated relation between the electric force $\mathbf{E}$ and its associated fluxes $\mathbf{C}$ and $\mathbf{D}$, or between the

* *Phil. Mag.*, January, 1887.

magnetic force **H** and its associated fluxes **K** and **B**. When written in the form most convenient for the present application, these laws are

$$\text{curl}\,(\mathbf{H}-\mathbf{h}_0)=\mathbf{J}_0=\mathbf{C}+\frac{\partial\mathbf{D}}{\partial t}+(\mathbf{D}\,\text{div}\,\mathbf{q}-\mathbf{D}\nabla.\mathbf{q}),\quad\ldots\ldots\ldots(212)$$

$$-\,\text{curl}\,(\mathbf{E}-\mathbf{e}_0)=\mathbf{G}_0=\mathbf{K}+\frac{\partial\mathbf{B}}{\partial t}+(\mathbf{B}\,\text{div}\,\mathbf{q}-\mathbf{B}\nabla.\mathbf{q}).\quad\ldots\ldots\ldots(213)$$

Now derive the equation of activity in the manner previously followed, and arrange it in the particular form

$$\mathbf{e}_0\mathbf{J}_0+\mathbf{h}_0\mathbf{G}_0+\text{conv}\,\mathrm{V}(\mathbf{E}-\mathbf{e}_0)(\mathbf{H}-\mathbf{h}_0)=(\mathbf{EC}+\mathbf{HK})+\left(\mathbf{E}\frac{\partial\mathbf{D}}{\partial t}+\mathbf{H}\frac{\partial\mathbf{B}}{\partial t}\right)$$
$$+(\mathbf{E}.\mathbf{D}\nabla.\mathbf{q}-\mathbf{ED}\,\text{div}\,\mathbf{q})+(\mathbf{H}.\mathbf{B}\nabla.\mathbf{q}-\mathbf{HB}\,\text{div}\,\mathbf{q}),\quad\ldots\ldots(214)$$

which will best facilitate interpretation.

Although independent of the relation between **E** and **D**, etc., of course the dimensions must be suitably chosen so that this equation may really represent activity per unit volume in every term.

Now, guided by the previous investigation, we can assume that $(\mathbf{e}_0\mathbf{J}_0+\mathbf{h}_0\mathbf{G}_0)$ represents the rate of supply of energy from intrinsic sources, and also that $\mathrm{V}(\mathbf{E}-\mathbf{e}_0)(\mathbf{H}-\mathbf{h}_0)$, which is a flux of energy independent of **q**, is the correct form in general. Also, if there be no other intrinsic sources of energy than $\mathbf{e}_0$, $\mathbf{h}_0$, and no other fluxes of energy besides that just mentioned except the convective flux and that due to the stress, the equation of activity should be representable by

$$(\mathbf{e}_0\mathbf{J}_0+\mathbf{h}_0\mathbf{G}_0)+\text{conv}\,[\mathrm{V}(\mathbf{E}-\mathbf{e}_0)(\mathbf{H}-\mathbf{h}_0)+\mathbf{q}(U+T)]$$
$$=(Q+\dot{U}+\dot{T})+\mathbf{Fq}+\text{conv}\,\mathbf{Q}_q\mathbf{q}$$
$$=(Q+\dot{U}+\dot{T})+\Sigma\,\mathbf{Q}\nabla q,\quad\ldots\ldots\ldots\ldots\ldots\ldots\ldots\ldots\ldots\ldots\ldots\ldots(215)$$

where **Q** is the conjugate of the stress-vector, **F** the translational force, and Q, U, and T the rate of waste and the stored energies, whatever they may be.

Comparing with the preceding equation (214), we see that we require

$$\Sigma\,\mathbf{Q}\nabla q=(Q-\mathbf{EC}-\mathbf{HK})+\left(\frac{\partial U}{\partial t}-\mathbf{E}\frac{\partial\mathbf{D}}{\partial t}\right)+\left(\frac{\partial T}{\partial t}-\mathbf{H}\frac{\partial\mathbf{B}}{\partial t}\right)$$
$$+[\mathbf{E}.\mathbf{D}\nabla.\mathbf{q}-(\mathbf{ED}-U)\,\text{div}\,\mathbf{q}]+[\mathbf{H}.\mathbf{B}\nabla.\mathbf{q}-(\mathbf{HB}-T)\,\text{div}\,\mathbf{q}].\quad(216)$$

Now assume that there is no waste of energy except by conduction; then

$$Q=\mathbf{EC}+\mathbf{HK}.\quad\ldots\ldots\ldots\ldots\ldots\ldots\ldots\ldots(217a)$$

Also assume that
$$\frac{\partial U}{\partial t}=\mathbf{E}\frac{\partial\mathbf{D}}{\partial t},\qquad\frac{\partial T}{\partial t}=\mathbf{H}\frac{\partial\mathbf{B}}{\partial t}.\quad\ldots\ldots\ldots\ldots\ldots(217b)$$

These imply that the relation between **E** and **D** is, for the same particle of matter, an invariable one, and that the stored electric energy is

$$U=\int_0^D\mathbf{E}\,d\mathbf{D},\quad\ldots\ldots\ldots\ldots\ldots\ldots\ldots\ldots\ldots(218)$$

where **E** is a function of **D**. Similarly,

$$T=\int_0^B\mathbf{H}\,d\mathbf{B}\quad\ldots\ldots\ldots\ldots\ldots\ldots\ldots\ldots\ldots(219)$$

expresses the stored magnetic energy, and **H** must be a definite function of **B**.

On these assumptions, (216) reduces to

$$\Sigma \mathbf{Q}\nabla q = [\mathbf{E}.\mathbf{D}\nabla.\mathbf{q} - (\mathbf{ED} - U)\,\mathrm{div}\,\mathbf{q}] + [\mathbf{H}.\mathbf{B}\nabla.\mathbf{q} - (\mathbf{HB} - T)\,\mathrm{div}\,\mathbf{q}], \quad (220)$$

from which the stress-vector follows, namely,

$$\mathbf{P}_N = [\mathbf{E}.\mathbf{DN} - \mathbf{N}(\mathbf{ED} - U)] + [\mathbf{H}.\mathbf{BN} - \mathbf{N}(\mathbf{HB} - T)]. \quad \text{......(221)}$$

Or, $$\mathbf{P}_N = (\mathrm{V}\mathbf{D}\mathrm{V}\mathbf{EN} + \mathbf{N}U) + (\mathrm{V}\mathbf{B}\mathrm{V}\mathbf{HN} + \mathbf{N}T). \quad \text{..................(222)}$$

Thus, in case of isotropy, the stress is a tension U parallel to **E** combined with a lateral pressure $(\mathbf{ED} - U)$; and a tension T parallel to **H** combined with a lateral pressure $(\mathbf{HB} - T)$.

The corresponding translational force is

$$\mathbf{F} = \mathbf{E}\,\mathrm{div}\,\mathbf{D} + \mathbf{D}\nabla.\mathbf{E} - \nabla(\mathbf{ED} - U) + \mathbf{H}\,\mathrm{div}\,\mathbf{B} + \mathbf{B}\nabla.\mathbf{H} - \nabla(\mathbf{HB} - T), \quad (223)$$

which it is unnecessary to put in terms of the currents.

Exchange **E** and **D**, and **H** and **B**, in (221) or (222) to obtain the conjugate vector $\mathbf{Q}_N$; from which we obtain the flux of energy due to the stress,

$$-q\mathbf{Q}_q = \mathbf{D}.\mathbf{Eq} - \mathbf{q}(\mathbf{ED} - U) + \mathbf{B}.\mathbf{Hq} - \mathbf{q}(\mathbf{HB} - T)$$
$$= \mathrm{V}\mathbf{E}\mathrm{V}\mathbf{Dq} + \mathrm{V}\mathbf{H}\mathrm{V}\mathbf{Bq} + \mathbf{q}(U + T), \quad \text{..................(224)}$$

or $$-q\mathbf{Q}_q = \mathrm{V}\mathbf{eH} + \mathrm{V}\mathbf{Eh} + \mathbf{q}(U + T), \quad \text{...........................(225)}$$

where **e** and **h** are the motional electric and magnetic forces, of the same form as before, (88) and (91); so that the complete form of the equation of activity, showing the fluxes of energy and their convergence, is

$$\mathbf{e}_0\mathbf{J}_0 + \mathbf{h}_0\mathbf{G}_0 + \mathrm{conv}\,[\mathrm{V}(\mathbf{E} - \mathbf{e}_0)(\mathbf{H} - \mathbf{h}_0) + \mathbf{q}(U + T)]$$
$$- \mathrm{conv}\,[\mathrm{V}\mathbf{eH} + \mathrm{V}\mathbf{Eh} + \mathbf{q}(U + T)] = \mathbf{Fq} + (Q + \dot{U} + \dot{T}), \quad (226)$$

where **F** has the above meaning.

There is thus a remarkable preservation of form as compared with the corresponding formulæ when there is proportionality between force and flux. For we produce harmony by means of a Poynting flux of identical expression, and a flux due to the stress which is also of identical expression, although U and T now have a more general meaning, of course.*

* As the investigation in this Appendix has some pretensions to generality, we should try to settle the amount it is fairly entitled to. No objection is likely to be raised to the use of the circuital equations (212), (213), with the restriction of strict proportionality between **E** and **H** and the fluxes **D** and **B**, or **C** and **K** entirely removed; nor to the estimation of $\mathbf{J}_0$ and $\mathbf{G}_0$ as the "true" currents; nor to the use of the same form of flux of electromagnetic energy when the medium is stationary. For these things are obviously suggested by the preceding investigations, and their justification is in their being found to continue to work, which is the case. But the use in the text of language appropriate to linear functions, which arose from the notation, etc., being the same as before, is unjustifiable. We may, however, remove this misuse of language, and make the equation (226), showing the flux of energy, rest entirely upon the two circuital equations. In fact, if we substitute in (226) the relations (217*a*), (217*b*), it becomes merely a particular way of writing (214).

It is, therefore, to (217*a*), (217*b*) that we should look for limitations. As regards

Example of the above, and Remarks on Intrinsic Magnetization when there is Hysteresis.

§ 40. In the stress-vector itself (for either the electric or the magnetic stress) the relative magnitude of the tension and the lateral pressure varies unless the curve connecting the force and the induction be a straight line. Thus, if the curve be of the type shown in the first figure, the shaded area will represent the stored energy and the tension, and the remainder of the rectangle will represent the lateral pressure. They are equal when **H** is small; later on the pressure preponderates, and more and more so the bigger **H** becomes.

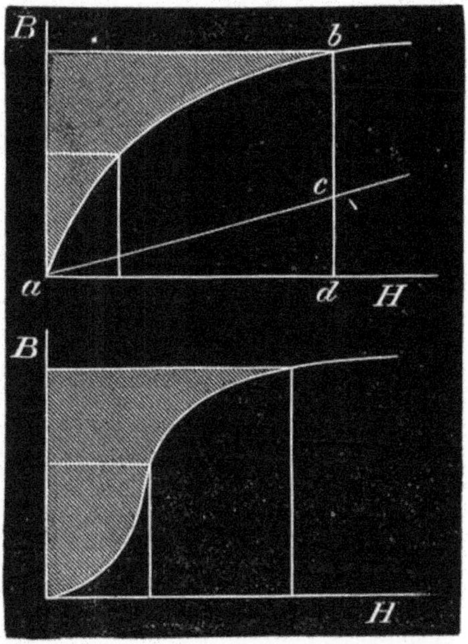

But if the curve be of the type shown in the second figure, then, after initial equality, the tension preponderates; though, later on, when **H** is very big, the pressure preponderates.

To obtain an idea of the effect, take the concrete example of an infinitely long rod, uniformly axially inductized by a steady current in an overlapping solenoid, and consider the forcive on the rod. Here both **H** and **B** are axial or longitudinal; and so, by equation (223), the translational force would be a normal force on the surface of the rod, acting outwards, of amount

$$(\mathbf{HB} - T) - \tfrac{1}{2}\mathbf{H}_0\mathbf{B}_0$$

per unit area; this being the excess of the lateral pressure in the rod over $\frac{1}{2}H_0B_0$, the lateral pressure just outside it.

In case of proportionality of force to flux, the first pressure is $\frac{1}{2}HB$, and, if there is no intrinsic magnetization, H and H_0 are equal. The

(217*a*), there does not seem to be any limitation necessary. That is, there is no kind of relation imposed between **E** and **C**, and **H** and **K**. This seems to arise merely from Q meaning energy wasted for good, and having no further entry into the system. But as regards (217*b*), the case is different. For it seems necessary, in order to exclude terms corresponding to $\mathbf{E}(\partial c/\partial t)\mathbf{E}$ and $\mathbf{H}(\partial \mu/\partial t)\mathbf{H}$ in the linear theory, when there is rotation, that **E** and **D** should be parallel, and likewise **H** and **B**. At any rate, if such terms be allowed, some modification may be required in the subsequent reckoning of the mechanical force. In other respects, it is merely implied by (217*b*) that **E** and **D** are definitely connected, likewise **H** and **B**, so that there is no waste of energy other than that expressed by Q.

outward force is therefore positive for paramagnetic, and negative for diamagnetic substances, and the result would be lateral expansion or contraction, since the infinite length would prevent elongation.

But if the curve in the rod be of the type of the first figure, and the straight line *ac* be the air-curve to correspond, it is the area *abc* that now represents the outward force per unit area when the magnetic force has the value *ad*. If the straight line can cross the curve *ab*, we see that by sufficiently increasing H we can make the external air-pressure preponderate, so that the rod, after initially expanding, would end by contracting.

If the rod be a ring of large diameter compared with its thickness, the forcive would be approximately the same, viz., an outward surface-force equal to the difference of the lateral pressures in the rod and air. The result would then be elongation, with final retraction when the external pressure came to exceed the internal.

Bidwell found a phenomenon of this kind in iron, but it does not seem possible that the above supposititious case is capable of explaining it, though of course the true explanation may be in some respects of a similar nature. But the circumstances are not the same as those supposed. The assumption of a definite connection between H and B, and elastic storage of the energy T, is very inadequate to represent the facts of magnetization of iron, save within a small range.

Magneticians usually plot the curve connecting $H-h_0$ and B, not between H and B, or which would be the same, between $H-h_0$ and $B-B_0$, where B_0 is the intrinsic magnetization. Now when an iron ring is subjected to a given gaussage (or magnetomotive force), going through a sequence of values, there is no definite curve connecting $H-h_0$ and B, on account of the intrinsic magnetization. But, with proper allowance for h_0, it might be that the resulting curve connecting H and B in a given specimen would be approximately definite, at any rate, far more so than that connecting $H-h_0$ and B. Granting perfect definiteness, however, there is still insufficient information to make a theory. The energy put into iron is not wholly stored; that is, in increasing the coil-current we increase B_0 as well as B, and in doing so dissipate energy; but although we know, by Ewing's experiments, the amount of waste in cyclical changes, it is not so clear what the rate of waste is at a given moment. There is also the further peculiarity that the energy of the intrinsic magnetization at a given moment, though apparently locked up, and really locked up temporarily, however loosely it may be secured, is not wholly irrecoverable, but comes into play again when H is reversed. Now it may be that the energy of the intrinsic magnetization plays, in relation to the stress, an entirely different part from that of the elastic magnetization. It is easy to make up formulæ to express special phenomena, but very difficult to make a comprehensive theory.

But in any case, apart from the obscurities connected with iron, it is desirable to be apologetic in making any application of Maxwell's stresses or similar ones to practice when the actual strains produced are in question, bearing in mind the difficulty of interpreting and harmonizing with Maxwell's theory the results of Kerr, Quincke, and others.

LIII. THE POSITION OF 4π IN ELECTROMAGNETIC UNITS.

[*Nature*, July 28, 1892, p. 292.]

THERE is, I believe, a growing body of opinion that the present system of electric and magnetic units is inconvenient in practice, by reason of the occurrence of 4π as a factor in the specification of quantities which have no obvious relation with circles or spheres.

It is felt that the number of lines from a pole should be m rather than the present $4\pi m$, that "ampere turns" is better than $4\pi nC$, that the electromotive intensity outside a charged body might be σ instead of $4\pi\sigma$, and similar changes of that sort; see, for instance, Mr. Williams's recent paper to the Physical Society.

Mr. Heaviside, in his articles in *The Electrician* and elsewhere, has strongly emphasized the importance of the change and the simplification that can thereby be made.

In theoretical investigations there seems some probability that the simplified formulæ may come to be adopted—

$$\mu \text{ being written instead of } 4\pi\mu, \text{ and } k \text{ instead of } \frac{4\pi}{K};$$

but the question is whether it is or is not too late to incorporate the practical outcome of such a change into the units employed by electrical engineers.

For myself I am impressed with the extreme difficulty of now making any change in the ohm, the volt, etc., even though it be only a numerical change; but in order to find out what practical proposal the supporters of the redistribution of 4π had in their mind, I wrote to Mr. Heaviside to inquire. His reply I enclose; and would merely say further that in all probability the general question of units will come up at Edinburgh for discussion.

OLIVER J. LODGE.

My dear Lodge,—I am glad to hear that the question of rational electrical units will be noticed at Edinburgh—if not thoroughly discussed. It is, in my opinion, a very important question, which must, sooner or later, come to a head and lead to a thoroughgoing reform. Electricity is becoming not only a master science, but also a very practical science. Its units should therefore be settled upon a sound and philosophical basis. I do not refer to practical details, which may be varied from time to time (Acts of Parliament notwithstanding), but to the fundamental principles concerned.

If we were to define the unit area to be the area of a circle of unit diameter, or the unit volume to be the volume of a sphere of unit diameter, we could, on such a basis, construct a consistent system of units. But the area of a rectangle or the volume of a parallelepiped would involve the quantity π, and various derived formulæ would possess the same peculiarity. No one would deny that such a system was an absurdly irrational one.

I maintain that the system of electrical units in present use is founded upon a similar irrationality, which pervades it from top to bottom. How this has happened, and how to cure the evil, I have considered in my papers—first in 1882-83, when, however, I thought it was hopeless to expect a thorough reform; and again in 1891, when I have, in my "Electromagnetic Theory," adopted rational units from the beginning, pointing out their connection with the common irrational units separately, after giving a general outline of electrical theory in terms of the rational.

Now, presuming provisionally that the first and second stages to Salvation (the Awakening and Repentance) have been safely passed through, which is, however, not at all certain at the present time, the question arises, How proceed to the third stage, Reformation? Theoretically this is quite easy, as it merely means working with rational formulæ instead of irrational; and theoretical papers and treatises may, with great advantage, be done in rational formulæ at once, and irrespective of the reform of the practical units. But taking a far-sighted view of the matter, it is, I think, very desirable that the practical units themselves should be rationalized as speedily as may be. This must involve some temporary inconvenience, the prospect of which, unfortunately, is an encouragement to shirk a duty; as is, likewise, the common feeling of respect for the labours of our predecessors. But the duty we owe to our followers, to lighten their labours permanently, should be paramount. This is the main reason why I attach so much importance to the matter; it is not merely one of abstract scientific interest, but of practical and enduring significance; for the evils of the present system will, if it continue, go on multiplying with every advance in the science and its applications.

Apart from the size of the units of length, mass, and time, and of the dimensions of the electrical quantities, we have the following relations between the rational and irrational units of voltage V, electric current C, resistance R, inductance L, permittance S, electric charge Q, electric force E, magnetic force H, induction B. Let x^2 stand for 4π, and let the suffixes $_r$ and $_i$ mean rational and irrational (or ordinary). Also let the presence of square brackets signify that the "absolute" unit is referred to. Then we have—

$$x=\frac{[E_r]}{[E_i]}=\frac{[V_r]}{[V_i]}=\frac{[H_r]}{[H_i]}=\frac{[B_r]}{[B_i]}=\frac{[C_i]}{[C_r]}=\frac{[Q_i]}{[Q_r]};$$

$$x^2=\frac{[R_r]}{[R_i]}=\frac{[L_r]}{[L_i]}=\frac{[S_i]}{[S_r]}.$$

The next question is, what multiples of these units we should take to make the practical units. In accordance with your request I give my ideas on the subject, premising, however, that I think there is no finality in things of this sort.

First, if we let the rational practical units be the same multiples of the "absolute" rational units as the present practical units are of *their* absolute progenitors, then we would have (if we adopt the centimetre, gramme, and second, and the convention that $\mu = 1$ in ether)

$$[R_r] \times 10^9 = \text{new ohm} = x^2 \text{ times old.}$$
$$[L_r] \times 10^9 = \text{new mac} = x^2 \quad ,,$$
$$[S_r] \times 10^{-9} = \text{new farad} = x^{-2} \quad ,,$$
$$[C_r] \times 10^{-1} = \text{new amp} = x^{-1} \quad ,,$$
$$[V_r] \times 10^8 = \text{new volt} = x \quad ,,$$
$$10^7 \text{ ergs} = \text{new joule} = \text{old joule.}$$
$$10^7 \text{ ergs per sec.} = \text{new watt} = \text{old watt.}$$

I do not, however, think it at all desirable that the new units should follow on the same rules as the old, and consider that the following system is preferable :—

$$[R_r] \times 10^8 = \text{new ohm} = \frac{x^2}{10} \times \text{old ohm.}$$
$$[L_r] \times 10^8 = \text{new mac} = \frac{x^2}{10} \times \text{old mac.}$$
$$[S_r'] \times 10^{-8} = \text{new farad} = \frac{10}{x^2} \times \text{old farad.}$$
$$[C_r] \times 1 = \text{new amp} = \frac{10}{x} \times \text{old amp.}$$
$$[V_r] \times 10^8 = \text{new volt} = x \times \text{old volt.}$$
$$10^8 \text{ ergs} = \text{new joule} = 10 \times \text{old joule.}$$
$$10^8 \text{ ergs per sec.} = \text{new watt} = 10 \times \text{old watt.}$$

It will be observed that this set of practical units makes the ohm, mac, amp, volt, and the unit of elastance, or reciprocal of permittance, all larger than the old ones, but not greatly larger, the multiplier varying roughly from $1\frac{1}{4}$ to $3\frac{1}{2}$.

What, however, I attach particular importance to is the use of one power of 10 only, viz., 10^8, in passing from the absolute to the practical units; instead of, as in the common system, no less than four powers, 10^1, 10^7, 10^8, and 10^9 I regard this peculiarity of the common system as a needless and (in my experience) very vexatious complication. In the 10^8 system I have described, this is done away with, and still the practical electrical units keep pace fairly with the old ones. The multiplication of the old joule and watt by 10 is, of course, a necessary accompaniment. I do not see any objection to the change. Though not important, it seems rather an improvement. (But transformations of units are so treacherous, that I should wish the whole of the above to be narrowly scrutinized.)

It is suggested to make 10^9 the multiplier throughout, and the results are :—

$$[R_r] \times 10^9 = \text{new ohm} = x^2 \times \text{old ohm.}$$
$$[L_r] \times 10^9 = \text{new mac} = x^2 \times \text{old mac.}$$
$$[S_r] \times 10^{-9} = \text{new farad} = x^{-2} \times \text{old farad.}$$
$$[C_r] \times 1 = \text{new amp} = \frac{10}{x} \times \text{old amp.}$$
$$[V_r] \times 10^9 = \text{new volt} = 10x \times \text{old volt.}$$
$$10^9 \text{ ergs} = \text{new joule} = 10^2 \times \text{old joule.}$$
$$10^9 \text{ ergs p. sec.} = \text{new watt} = 10^2 \times \text{old watt.}$$

But I think this system makes the ohm inconveniently big, and has some other objections. But I do not want to dogmatize in these matters of detail. Two things I would emphasize :—First, rationalize the units. Next, employ a single multiplier, as, for example, 10^8.

OLIVER HEAVISIDE.

PAIGNTON DEVON, July 18, 1892.

CORRECTIONS. VOL. II.

p. 69, equation (51*b*), change sign of last term from − to +, as in (73), p. 192.

p. 69, equation (52*b*), change sign of last term from + to −, as in (72), p. 192, and for)[read $)^2$[, to agree with (72), p. 192.

p. 316, equation (40*g*), the lower limit should be z_0.

p. 355, last line, *for* 361 *read* 301.

p. 387, seventh line, *for* 153 *read* 393.

p. 400, second line, *for* fraction to *read* fraction of to.

INDEX.

END OF VOL. II.

ROBERT MACLEHOSE, PRINTER TO THE UNIVERSITY, GLASGOW.

www.ingramcontent.com/pod-product-compliance
Ingram Content Group UK Ltd.
Pitfield, Milton Keynes, MK11 3LW, UK
UKHW040700180125
453697UK00010B/316

9 781108 028578